Transport Processes in Pharmaceutical Systems

DRUGS AND THE PHARMACEUTICAL SCIENCES

Executive Editor

James Swarbrick

AAI, Inc.
Wilmington, North Carolina

Advisory Board

DRUGS AND THE PHARMACEUTICAL SCIENCES

A Series of Textbooks and Monographs

1. Pharmacokinetics, *Milo Gibaldi and Donald Perrier*
2. Good Manufacturing Practices for Pharmaceuticals: A Plan for Total Quality Control, *Sidney H. Willig, Murray M. Tuckerman, and William S. Hitchings IV*
3. Microencapsulation, *edited by J. R. Nixon*
4. Drug Metabolism: Chemical and Biochemical Aspects, *Bernard Testa and Peter Jenner*
5. New Drugs: Discovery and Development, *edited by Alan A. Rubin*
6. Sustained and Controlled Release Drug Delivery Systems, *edited by Joseph R. Robinson*
7. Modern Pharmaceutics, *edited by Gilbert S. Banker and Christopher T. Rhodes*
8. Prescription Drugs in Short Supply: Case Histories, *Michael A. Schwartz*
9. Activated Charcoal: Antidotal and Other Medical Uses, *David O. Cooney*
10. Concepts in Drug Metabolism (in two parts), *edited by Peter Jenner and Bernard Testa*
11. Pharmaceutical Analysis: Modern Methods (in two parts), *edited by James W. Munson*
12. Techniques of Solubilization of Drugs, *edited by Samuel H. Yalkowsky*
13. Orphan Drugs, *edited by Fred E. Karch*
14. Novel Drug Delivery Systems: Fundamentals, Developmental Concepts, Biomedical Assessments, *Yie W. Chien*
15. Pharmacokinetics: Second Edition, Revised and Expanded, *Milo Gibaldi and Donald Perrier*
16. Good Manufacturing Practices for Pharmaceuticals: A Plan for Total Quality Control, Second Edition, Revised and Expanded, *Sidney H. Willig, Murray M. Tuckerman, and William S. Hitchings IV*
17. Formulation of Veterinary Dosage Forms, *edited by Jack Blodinger*
18. Dermatological Formulations: Percutaneous Absorption, *Brian W. Barry*
19. The Clinical Research Process in the Pharmaceutical Industry, *edited by Gary M. Matoren*
20. Microencapsulation and Related Drug Processes, *Patrick B. Deasy*
21. Drugs and Nutrients: The Interactive Effects, *edited by Daphne A. Roe and T. Colin Campbell*
22. Biotechnology of Industrial Antibiotics, *Erick J. Vandamme*
23. Pharmaceutical Process Validation, *edited by Bernard T. Loftus and Robert A. Nash*

ADDITIONAL VOLUMES IN PREPARATION

Transport Processes
in Pharmaceutical
Systems

edited by
Gordon L. Amidon
The University of Michigan
Ann Arbor, Michigan

Ping I. Lee
Schering-Plough Corporation
Kenilworth, New Jersey

Elizabeth M. Topp
University of Kansas
Lawrence, Kansas

CRC Press
Taylor & Francis Group
Boca Raton London New York

CRC Press is an imprint of the
Taylor & Francis Group, an **informa** business

CRC Press
Taylor & Francis Group
6000 Broken Sound Parkway NW, Suite 300
Boca Raton, FL 33487-2742

First issued in paperback 2019

© 2009 by Taylor & Francis Group, LLC
CRC Press is an imprint of Taylor & Francis Group, an Informa business

No claim to original U.S. Government works

ISBN-13: 978-0-8247-6610-8 (hbk)
ISBN-13: 978-0-367-39922-1 (pbk)

Visit the Taylor & Francis Web site at
http://www.taylorandfrancis.com

and the CRC Press Web site at
http://www.crcpress.com

Preface

A fundamental understanding of transport phenomena has become increasingly important to pharmaceutical scientists during the past 20 years. Applications range from drug and nutrient transport across cell membranes, drug dissolution and absorption across biological membranes, whole body kinetics, and drug release from polymer reservoirs and matrices to heat and mass transport associated with freeze drying and hygroscopicity. Two factors often obscure a basic understanding in this field. First is the diversity of applications, with their differing language and objectives. The second is the often blurred distinction between the transport processes and the thermodynamics and kinetics of the system under consideration. A full development of any system requires that both the thermodynamic and kinetic reaction factors, as well as the transport processes, be fully understood. It is the aim of this volume to present a unified approach such that the common basic principles involved in many applications are clearly presented.

Our purpose is to discuss those areas of transport phenomena that have direct relevance and application to the pharmaceutical sciences. The book can be divided into roughly four sections, with Part I, Chapters 1–5, presenting basic transport processes, including drug dissolution; Part II, Chapters 6–10, presenting various aspects of biological transport; Part III, Chapters 11–15, presenting transport in polymers and drug delivery systems; and Part IV, Chapters 16 and 17, presenting heat and mass transfer in freeze drying and hygroscopicity. This illustrates the breadth of coverage in this monograph.

In our courses, we cover some aspects of all topics in this book. This is appropriate for students who become involved in drug discovery and research and development. The chapters are comprehensive in covering the diversity of applications of each subject. Each section of this book could be a monograph on its own. However, with the material presented in these chapters a scientist would be able to enter the entire literature and understand current research papers.

The level of mathematics and physical chemistry required for understanding each chapter varies greatly. Chapters 3, 4, 6–9, 11, 13, and 14 require only basic algebra and physical chemistry at an undergraduate level. However, we have attempted to minimize the actual mathematical detail. The solution to mathematical problems in this book and arising from research problems, in general, is greatly aided by the availability of sophisticated mathematical software that can handle any of the problems presented in this book on a personal computer. Students who have taken the time to learn one or more of these software packages have found them useful for handling problems in this book, as well as for much more complex problems involving numerical solutions to ordinary and partial differential equations. We encourage the use of this software but do not require it in our graduate courses.

We want to thank the authors for their contributions to this monograph. In addition, we would like to thank the following graduate students in the Department of Pharmaceutical Chemistry at the University of Kansas: Melissa Beck-Westemeyer, Jutima Boonleang, Sirirat Choosakoonkriang, Joshua Cooper, Anita Freed, Andrew Gawron, Victor Guarino, Karen Hamilton, Bradley Hanson, Susan Hovorka, Erin Hugger, Lisa Kueltzo, Brian Lobo, Antonie Rice, Sun Wei, Waree Tiyaboonnai, Christopher Wiethoff, Anne Wolka and Jerry Yang. These students helped to edit several of the chapters of this book, and their concern with clarity and attention to detail are much appreciated. Finally, in addition to individuals who have contributed chapters to this monograph, we particularly want to thank Ms. Iris Templin, whose management, follow-up, and editing made possible the successful completion of this monograph.

Gordon L. Amidon
Ping I. Lee
Elizabeth M. Topp

Contents

Contributors

Anthony Adson, Ph.D.* Pharmacia & Upjohn, Inc., Kalamazoo, Michigan, and Department of Pharmaceutical Chemistry, School of Pharmacy, University of Kansas, Lawrence, Kansas

Gordon L. Amidon, Ph.D. Professor, College of Pharmacy, University of Michigan, Ann Arbor, Michigan

Mandana Asgharnejad, Ph.D. Research Fellow, Merck Research Laboratories, West Point, Pennsylvania

Kenneth L. Audus, Ph.D. Professor, Department of Pharmaceutical Chemistry, School of Pharmacy, University of Kansas, Lawrence, Kansas

You Han Bae, Ph.D.† College of Pharmacy, University of Utah, Salt Lake City, Utah

Craig L. Barsuhn, Ph.D. Pharmacia & Upjohn, Inc., Kalamazoo, Michigan

Ronald T. Borchardt, Ph.D. Solon F. Summerfield Professor, Department of Pharmaceutical Chemistry, School of Pharmacy, University of Kansas, Lawrence, Kansas

Philip S. Burton, Ph.D. Senior Scientist, Pharmacia & Upjohn, Inc., Kalamazoo, Michigan

John R. Cardinal, Ph.D.‡ Vice President, Oakmont Pharmaceuticals, North Wales, Pennsylvania

John R. Crison, Ph.D. PORT Systems LLC, Ann Arbor, Michigan

* *Current affiliation*: Alpharm USPD, Baltimore, Maryland.
† *Current affiliation*: Kwangju Institute of Science and Technology, Kwangju, Korea.
‡ *Current affiliation*: Applied Analytical Industries, Inc., Wilmington, North Carolina.

Sarma Duddu, Ph.D.   Inhale Therapeutics, Palo Alto, California

David Fleisher, Ph.D.   Associate Professor, College of Pharmacy, University of Michigan, Ann Arbor, Michigan

James M. Gallo, Ph.D.   Department of Pharmacology, Fox Chase Cancer Center, Philadelphia, Pennsylvania

Larry Gatlin, Ph.D.   Director, Pharmaceutical Science and Technology, Biogen, Inc., Cambridge, Massachusetts

Stevin H. Gehrke, Ph.D.   Professor and Head, Tom H. Barnett University Faculty Chair, Department of Chemical Engineering, Kansas State University, Manhattan, Kansas

David J. W. Grant, D.Sc.   Professor, College of Pharmacy, University of Minnesota, Minneapolis, Minnesota

Norman F. H. Ho, Ph.D.*   Pharmacia & Upjohn, Inc., Kalamazoo, Michigan

Sung Wan Kim, Ph.D.   College of Pharmacy, University of Utah, Salt Lake City, Utah

Uday B. Kompella, Ph.D.†   University of Southern California, Los Angeles, California

Jim H. Kou, Ph.D.   Principal Scientist, Schering-Plough Research Institute, Kenilworth, New Jersey

Vincent H. L. Lee, Ph.D.   Professor and Chairman, Department of Pharmaceutical Sciences, and Associate Dean for Research and Graduate Affairs, University of Southern California, Los Angeles, California

John W. Mauger, Ph.D.   Professor and Dean, College of Pharmacy, University of Utah, Salt Lake City, Utah

Daniel P. McNamara, Ph.D.   Senior Principal Scientist, Boehringher Ingelheim, Ridgefield, Connecticut

Manesh J. Nerurkar, Ph.D.   Research Fellow, Merck & Company, West Point, Pennsylvania

Michael J. Pikal, Ph.D.   Professor, School of Pharmacy, University of Connecticut, Storrs, Connecticut

---

* *Current affiliation*: University of Utah, Salt Lake City, Utah.
† *Current affiliation*: University of Nebraska Medical Center, Omaha, Nebraska.

Thomas J. Raub, Ph.D. Senior Research Scientist, Pharmacia & Upjohn, Inc., Kalamazoo, Michigan

J. Howard Rytting, Ph.D. Professor, Department of Pharmaceutical Chemistry, University of Kansas, Lawrence, Kansas

Elizabeth M. Topp, Ph.D. Associate Professor, Department of Pharmaceutical Chemistry, University of Kansas, Lawrence, Kansas

Michael L. Vieira, M.S. PORT Systems LLC, Ann Arbor, Michigan

James Wright, Ph.D. Alkermes, Inc., Cambridge, Massachusetts

Lawrence X. Yu, Ph.D.* Research Investigator, Glaxo Wellcome Research and Development, Research Triangle Park, North Carolina

* *Current affiliation*: Food and Drug Administration, Rockville, Maryland.

Transport Processes in Pharmaceutical Systems

1
Principles of Mass Transfer

Elizabeth M. Topp
The University of Kansas, Lawrence, Kansas

I. INTRODUCTION TO MASS TRANSFER

Mass transfer is the movement of the molecules of a fluid in space in response to applied driving forces. This chapter describes the fundamental principles on which mass transfer theory is based. The chapter begins by describing the thermodynamic basis for mass transfer, together with an approach to posing and solving mass transfer problems (Sections I.A and I.B). Diffusion and convection are then defined, and the theories and models used to describe diffusive (Section III) and convective (Section II) mass transfer are presented. Finally, the effects of multiple driving forces on mass transfer are discussed (Section IV). This discussion includes information on combined convection and diffusion, on the influences of electrical potential gradients, and on combined heat and mass transfer. The chapter is intended to be an overview of fundamental principles that will be assumed in subsequent chapters. To that end, references to the later chapters are provided whenever possible, a glossary of common terms in mass transfer is included (Section VI), and references to recommended texts are provided (Section VII).

A. Thermodynamic Basis for Mass Transfer

Mass transfer is a kinetic process, occurring in systems that are not at equilibrium. To understand mass transfer from a thermodynamic perspective, consider the isolated system shown in Figure 1. The system is bounded by an impermeable insulating wall which prevents the transfer of matter, heat, or mechanical energy between the system and the external environment. The system is subdivided into

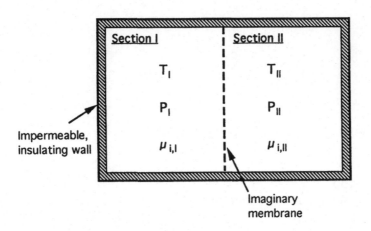

Figure 1 Isolated system consisting of two sections separated by an imaginary permeable membrane. At equilibrium, the temperatures (T), pressures (P), and chemical potentials of each species i (μ_i) are equal in the two sections.

two sections of equal volume, labeled I and II. The sections are separated by an imaginary membrane, indicated by the dashed line, which is completely permissive to the transfer of mass, heat, and mechanical energy. Thermodynamics assures us that the following conditions are both necessary and sufficient for section I to be in equilibrium with section II: (1) their temperatures are equal $(T_I = T_{II})$, (2) their pressures are equal $(P_I = P_{II})$, and (3) the chemical potentials of the species present are equal $(\mu_{i,I} = \mu_{i,II}$ for all species $i)$.

If this isolated system is unperturbed, it will remain at this thermodynamic equilibrium indefinitely. Consider now a particular hypothetical perturbation of the system away from this equilibrium condition, in which the pressure in section I is increased (so that $P_I > P_{II}$) while the temperatures and all chemical potentials remain equal in the two sections. Thermodynamics asserts that as a result of this perturbation the system will change in an attempt to establish a new equilibrium condition. This can happen in a number of ways, through changes in any or all of the variables shown in Figure 1. However, one of the simplest changes that can be imagined is the re-equilibration of the two pressures P_I and P_{II}, with no change in the values of the temperatures or chemical potentials. If the imaginary membrane is fully permeable to the fluids in the system, the re-equilibration of the pressures will occur through the flow of fluid from section I to section II, which will continue until the values of P_I and P_{II} are again equal. This flow of fluid in response to a spatial gradient in pressure is called *convection*.

Similarly, the system can be perturbed from equilibrium by altering the equality of chemical potentials. Suppose that the chemical potential of one of the species, A, is now increased in section I, so that $\mu_{A,I} > \mu_{A,II}$. Recall that the chemical potential of A is related to its concentration, the ideality of the solution, and the temperature, so that this perturbation may occur, for example, by increasing the concentration of A in section I. Again, the system will change in order to establish a new equilibrium, and again, one of the simplest changes that can be imagined is the re-equilibration of the chemical potentials of species A, with no change in the other variables. If the imaginary membrane is permeable to species A, the re-equilibration will occur through the movement of A from section I to section II, which will continue until the chemical potentials are again equal. In this case, the movement occurs as a result of the random, thermally induced Brownian motion of the molecules of A, since there is no gradient in pressure to cause bulk flow of the fluid. An analogous process will occur in a box containing black and white marbles, initially segregated by a vertical divider, when the divider is removed and the box is shaken vigorously. This movement of mass in response to a spatial gradient in chemical potential, and as the result of the random thermal motion of molecules, is called *diffusion*.

B. The Systems Approach to Mass Transfer Problems

The theories that have been developed to describe mass transfer arise from the law of conservation of mass, which states that mass can be neither created nor destroyed. According to this law, the total mass in a particular region in space can increase only by the addition of mass from the surroundings and can decrease only by the loss of mass back to them. Processes such as radioisotope decay and nuclear fission are exceptions to this law, since they involve the interconversion of matter and energy. In the absence of nuclear decay, however, the law of conservation of mass holds and is broadly applicable to mass transfer problems.

In order to apply this law, the "region in space" must be defined carefully and specifically for the problem of interest. In the discussion that follows, this carefully defined region in space will be called the *system*. While the definition of the system is subject to a degree of investigator discretion, a judicious choice often can simplify a seemingly complex problem. Generally speaking, the system should be selected so that it is a single-phase, homogeneous region with well-defined boundaries and (if possible) spatially invariant physical properties. For example, in describing mass transfer in a chromatographic column, the system might be defined as the fluid in the column, excluding the solid packing material and the column walls.

When the system has been defined, an equation can be written to state the law of conservation of mass. Such a "total mass balance equation" can be given in verbal form as

$$\begin{array}{llll}
\text{Rate of change} & \text{Rate of mass} & \text{Rate of mass} & \\
\text{of total mass in} & = \text{gain from} & - \text{loss to} & \text{(1)} \\
\text{the system} & \text{surroundings} & \text{surroundings} &
\end{array}$$

Note that the equation includes rates of gain and loss rather than total amounts. As a result, the mathematical form will be a differential equation rather than an algebraic one. The differential form is preferred in almost all mass transfer problems, because variations in the rates with position and time can be incorporated accurately. Each term in the equation will take on a specific functional form depending on the parameters and mass transfer characteristics of the problem of interest.

In addition to the total mass balance, equations can be written to describe changes in each of the individual chemical species, or ''components,'' that are present. As with the total mass, the mass of a component can be altered by exchange with the surroundings. However, it can also be affected by chemical reactions occurring within the system, converting one component to another. The total mass of the system is not affected by such interconversions, since the mass of reactants consumed is exactly equal to the mass of products formed. In verbal form, the ''component mass balance'' for a particular component A in the system is

$$\begin{array}{llll}
\text{Rate of change} & \text{Rate of gain} & \text{Rate of loss} & \text{Rate of gain} \\
\text{of mass A} & = \text{of A from} & - \text{of A to} & \pm \text{ or loss of A} \quad \text{(2)} \\
\text{in the system} & \text{surroundings} & \text{surroundings} & \text{by reaction}
\end{array}$$

Note that in the component mass balance the kinetic rate laws relating reaction rate to species concentrations become important and must be specified. As with the total mass balance, the specific form of each term will vary from one mass transfer problem to the next. A complete description of the behavior of a system with n components includes a total mass balance and $n - 1$ component mass balances, since the total mass balance is the sum of the individual component mass balances. The solution of this set of equations provides relationships between the dependent variables (usually masses or concentrations) and the independent variables (usually time and/or spatial position) in the particular problem. Further manipulation of the results may also be necessary, since the natural dependent variable in the problem is not always of the greatest interest. For example, in describing drug diffusion in polymer membranes, the concentration of the drug within the membrane is the natural dependent variable, while the cumulative mass transported across the membrane is often of greater interest and can be derived from the concentration.

In many problems, the system ultimately will reach a time-invariant state in which the total mass and/or the masses of the components are no longer changing. This condition is called the *steady state*. In mass transfer problems, the mathematical condition for the steady state is that all derivatives with respect to time are equal to zero. This is a fairly restrictive definition of the steady state. Oscillatory steady states, in which the dependent variable changes in a periodic manner with constant frequency and amplitude, are allowed in fields such as clinical pharmacokinetics (e.g., oral dosing to steady state) and electrical engineering. The simpler ''zero time derivatives'' definition is sufficient for most mass transfer problems, however. Steady-state solutions to mass transfer problems can be obtained by solving the set of differential equations when time derivatives are set equal to zero. This simplifies the mathematics considerably, in many cases reducing ordinary differential equations (ODEs) to algebraic ones, and partial differential equations to ODEs. While these steady-state solutions contain no information about the dynamics of the system, they often provide useful information about its long-term, steady behavior.

Solving mass transfer problems can be of great practical value to pharmaceutical scientists. At the simplest level, the solutions can describe experimental data in terms of fundamental material properties. In many cases, the numerical values of the material properties can then be calculated. For example, for the membrane transport problem described above, the solution of a mass transfer problem allows the investigator to relate the experimentally measured permeability of the membrane to the diffusion and partition coefficients, fundamental material properties of the system. On a more basic level, solving mass transfer problems enables the investigator to develop predictive models of system behavior in the absence of data. These can then be used as a theoretical guide to the design of experiments, identifying important parameters and their probable relationships.

The remaining sections of this chapter provide examples of mass transfer models, presented using the systems approach described above. In many cases, the models are of such importance that they are regarded as theories in their own right. These basic models are also the foundation for the more specific applications in the subsequent chapters of this book.

II. CONVECTION

Convection is mass transfer that is driven by a spatial gradient in pressure. This section presents two simple models for convective mass transfer: the stirred tank model (Section II.A) and the plug flow model (Section II.B). In these models, the pressure gradient appears implicitly as a spatially invariant fluid velocity or volumetric flow rate. However, in more complex problems, it is sometimes necessary to develop an explicit relationship between fluid velocity and pressure gradients. Section II.C describes the methods that are used to develop these relationships.

A. Stirred Tank Models

One of the simplest models for convective mass transfer is the stirred tank model, also called the continuously stirred tank reactor (CSTR) or the mixing tank. The model is shown schematically in Figure 2. As shown in the figure, a fluid stream enters a filled vessel that is stirred with an impeller, then exits the vessel through an outlet port. The stirred tank represents an idealization of mixing behavior in convective systems, in which incoming fluid streams are instantly and completely mixed with the system contents. To illustrate this, consider the case in which the inlet stream contains a water-miscible blue dye and the tank is initially filled with pure water. At time zero, the inlet valve is opened, allowing the dye to enter the

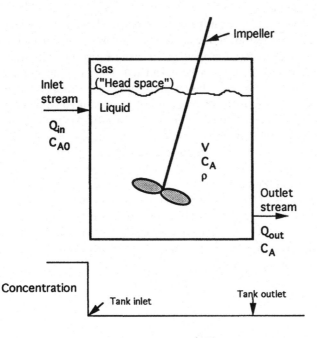

Concentration Profile

Figure 2 The stirred tank, a simple model for convective mass transfer. The liquid in the tank is characterized by its volume (V), density (ρ), and the concentrations of the components (C_A). Liquid enters through the inlet stream at a flow rate Q_{in} and concentration C_{A0}. Liquid exits through the outlet stream at volumetric flow rate Q_{out} and concentration identical to that in the tank (C_A). The concentration profile below the tank shows the step change in concentration encountered as the inlet stream is mixed with tank contents of lower concentration.

tank. As the first drop of dye enters the tank, it is mixed instantly and completely with the water, imparting a uniform light blue color. Because the distribution of dye in the tank is uniform, the fluid entering the exit stream at this instant also is of the same light blue color. The continued addition of dye to the tank through the flowing inlet stream will gradually darken the color of the fluid in the tank, until both the tank and the exit stream are the same color as the inlet stream.

This simple example illustrates two important features of stirred tanks: (1) the concentration of dissolved species is uniform throughout the tank, and (2) the concentration of these species in the exit stream is identical to their concentration in the tank. Note that a consequence of the well-stirred behavior of this model is that there is a step change in solute concentration from the inlet to the tank, as shown in the concentration profile in Figure 2. Such idealized behavior cannot be achieved in real stirred vessels; even the most enthusiastically stirred will not display this step change, but rather a smoother transition from inlet to tank concentration. It should also be noted that stirred tank models can be used when chemical reactions occur within the tank, as might occur in a flow-through reaction vessel, although these do not occur in the simple dye dilution example.

Mass balance equations can be developed to describe mass transfer in the stirred tank. The system is defined as the fluid within the tank, excluding any headspace above the fluid and the solid tank itself. The total mass balance on this sytem, given above in verbal form as Eq. (1), takes the following form when fluid density (ρ) is assumed to be constant:

$$\frac{d(\rho V)}{dt} = \rho(Q_{in} - Q_{out}) \tag{3}$$

where V is the volume of fluid in the tank, t is time, and Q_{in} and Q_{out} are the volumetric flow rates of the inlet and exit streams. When the fluid volume, V, is constant, the derivative on the left-hand side is equal to zero, and the total mass balance states simply that the inlet and exit flow rates are equal ($Q_{in} = Q_{out} = Q$). Similarly, a component mass balance can be written for each chemical species that enters or exits the tank. For example, for a species A that is present in the inlet stream at a constant concentration C_{A0}, the component mass balance is

$$\frac{d(C_A V)}{dt} = Q(C_{A0} - C_A) - R \tag{4}$$

where C_A is the uniform concentration of dye in the tank and R is the rate of loss or gain of A due to chemical reaction. In this equation, note that fluid flowing from the tank has a concentration C_A, a consequence of the well-stirred model. Solution of this equation requires the specification of the kinetics of the reaction as well as an initial condition of the form "At $t = 0$, $C_A = \beta$," where β is a constant. When the reaction is either zero- or first-order in C_A, the differential equation is linear and the solution contains the exponential terms common to linear systems. For example, when A is consumed by an irreversible first-order reaction with rate constant k and the initial concentation of A in the tank is zero, the solution is

$$C_A = \frac{C_{A0}}{1 + \alpha} \left\{ 1 - \exp\left[-\left(\frac{Q}{V} + k \right)t \right] \right\} \tag{5}$$

where $\alpha = kV/Q$. The rate constant for the system is the constant multiplier $Q/V + k$ appearing in the exponential term. Rates of change of concentration in the stirred tank thus depend on flow rate and tank volume as well as on the rate of reaction.

Stirred tank models have been widely used in pharmaceutical research. They form the basis of the compartmental models of traditional and physiological pharmacokinetics and have also been used to describe drug bioconversion in the liver [1,2], drug absorption from the gastrointestinal tract [3], and the production of recombinant proteins in continuous flow fermenters [4]. In this book, a more detailed development of stirred tank models can be found in Chapter 3, in which pharmacokinetic models are discussed by Dr. James Gallo. The conceptual and mathematical simplicity of stirred tank models ensures their continued use in pharmacokinetics and in other systems of pharmaceutical interest in which spatially uniform concentrations exist or can be assumed.

B. Plug Flow Models

A second simple model for convective flow is the plug flow model, an idealization of flow in closed cylindrical vessels such as pipes and tubes. In plug flow, fluid is imagined to move down the tube in discrete "plugs" of fluid, as shown in Figure 3. Within each plug, fluid mixing is instantaneous and complete, as if each plug were a tiny stirred tank moving down the tube. No exchange of fluid is allowed between adjacent plugs of fluid, however. Plug flow thus is characterized by complete mixing in the radial direction (r in Fig. 3), with no mixing in the axial direction (z). Chemical reaction may occur in the plug flow model, increasing or decreasing the concentrations of various dissolved solutes as the plugs move down the tube.

As with the stirred tank model, mass balance equations can be developed to describe mass transfer in plug flow. In this case, it is convenient to define the system as a differential cylindrical section of the tube, with length Δz and volume $\pi D^2 \, \Delta z/4$, where D is the tube diameter. This system is fixed in space and may

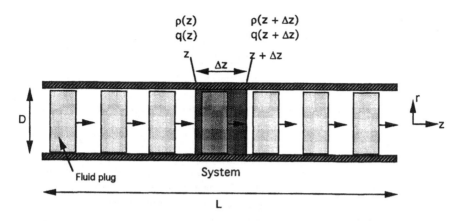

Figure 3 The plug flow model for convective mass transfer. The drawing shows a tube of diameter D and length L. Discrete fluid plugs (shaded rectangles) move down the tube; fluid within each plug is completely mixed, while there is no mixing between adjacent plugs. The system is a cylindrical section of the tube between $z + z$ and Δz and is fixed in space. Fluid enters the system at z with density $\rho(z)$ and volumetric flow rate $q(z)$; fluid exits the system at $z + \Delta z$ with density $\rho(z + \Delta z)$ and volumetric flow rate $q(z + \Delta z)$.

be conceptualized as a ''window'' in the tube through which the moving fluid plugs are observed. A total mass balance for this system is

$$\frac{d(\rho \pi D^2 \, \Delta z/4)}{dt} = \rho(z)q(z) - \rho(z + \Delta z)q(z + \Delta z) \tag{6}$$

where $q(z)$ is the volumetric flow rate of fluid entering the system at position z, $q(z + \Delta z)$ is the volumetric flow rate of fluid exiting the system at position $z + \Delta z$, and ρ is the fluid density, which may also vary with position. Further manipulation of this equation and the definition of fluid velocity (v) as the volumetric flow rate divided by the cross-sectional area of the tube ($\pi D^2/4$) yields the relationship

$$\frac{\partial \rho}{\partial t} + v \frac{\partial \rho}{\partial z} = -\rho \frac{\partial v}{\partial z} \tag{7}$$

where ∂ indicates the partial derivative with respect to position (z) or time (t). This equation demonstrates that if fluid density does not change with position in the tube or with time (i.e., $\partial \rho/\partial t = \partial \rho/\partial z = 0$), the fluid velocity in the tube must be constant along its length ($\partial v/\partial z = 0$). The development of the equation also highlights the need to consider both position and time as independent variables in this model.

When fluid velocity is constant, a component mass balance for a chemical species A in plug flow can be written as

$$\frac{\partial C_A}{\partial t} = -v \frac{\partial C_A}{\partial z} \pm r_A \tag{8}$$

where C_A is the concentration of A as a function of position (z) and time (t), and r_A is the rate of loss or gain of A by chemical reaction. Complete solution of this equation requires the specification of: (1) an initial condition of the form "At $t = 0$, $C_A = \beta$," where β is a constant or a function that is independent of t; (2) a boundary condition of the form "At $z = \omega$, $C_A = \gamma$," where ω is a constant and γ is either a constant or a function independent of z; and (3) the functional form of the dependence of the rate of reaction on concentration. Often in plug flow systems, the behavior of the system at steady state ($\partial C_A / \partial t = 0$) is of interest, in part because analytical solutions of the component mass balance can be obtained easily. For example, when A is degraded by an irreversible first-order reaction and the concentration of A at the tube inlet ($z = 0$) is C_{A0}, the steady-state solution to the component mass balance given above is

$$C_A = C_{A0} \exp[-(k/v)z] \tag{9}$$

where k is the rate constant for the reaction. Interestingly, this equation has the same functional form as that for the first-order degradation of A in a closed vessel, but with the rate constant modified by the fluid velocity (v) and with the independent variable for spatial position in the tube (z) replacing time (t).

An understanding of plug flow is important for pharmaceutical scientists involved in drug analysis, because a close approximation to this idealized flow is desired in chromatographic and capillary electrophoresis columns. For example, when dispersion is minimal, the flow of an injected sample in an HPLC column will approximate plug flow. Plug flow models have also been used to describe drug metabolism in the liver (the "parallel tube" model) [1,2] and drug absorption from the gastrointestinal tract [3].

C. Complex Flows

While the simple stirred tank and plug flow models are adequate to describe convective transport in many cases, a more complete description of fluid flow is sometimes needed. For example, an accurate description of tablet dissolution in a stirred vessel may require information about the changing fluid velocity near the tablet surface. Neither the stirred tank nor the plug flow models can address these velocity changes, since both assume that velocity is independent of position and time. In such cases, a more detailed description of fluid flow can be developed using the Navier–Stokes equations, which describe the effects of pressure, viscosity, and applied forces on fluid velocity in three dimensions. In a rectangular coordinate geometry, the Navier–Stokes equations are

$$\rho\left(\frac{\partial u}{\partial t} + u\frac{\partial u}{\partial x} + v\frac{\partial u}{\partial y} + w\frac{\partial u}{\partial z}\right) = -\frac{\partial P}{\partial x} + F_x$$
$$+ \mu\left(\frac{\partial^2 u}{\partial x^2} + \frac{\partial^2 u}{\partial y^2} + \frac{\partial^2 u}{\partial z^2}\right) \tag{10}$$

$$\rho\left(\frac{\partial v}{\partial t} + u\frac{\partial v}{\partial x} + v\frac{\partial v}{\partial y} + w\frac{\partial v}{\partial z}\right) = -\frac{\partial P}{\partial y} + F_y$$
$$+ \mu\left(\frac{\partial^2 v}{\partial x^2} + \frac{\partial^2 v}{\partial y^2} + \frac{\partial^2 v}{\partial z^2}\right) \tag{11}$$

$$\rho\left(\frac{\partial w}{\partial t} + u\frac{\partial w}{\partial x} + v\frac{\partial w}{\partial y} + w\frac{\partial w}{\partial z}\right) = -\frac{\partial P}{\partial z} + F_z$$
$$+ \mu\left(\frac{\partial^2 w}{\partial x^2} + \frac{\partial^2 w}{\partial y^2} + \frac{\partial^2 w}{\partial z^2}\right) \tag{12}$$

Here, u, v, and w are the components of the velocity vector in the x, y, and z directions, respectively. Note that velocity is treated as a vector quantity, so that the vector sum $u\mathbf{i} + v\mathbf{j} + w\mathbf{k}$ (where $\mathbf{i}$, $\mathbf{j}$, and $\mathbf{k}$ are the unit vectors in the x, y, and z directions) represents both the direction and magnitude of the fluid velocity at a particular position and time. The symbol P represents fluid pressure, μ is the fluid viscosity, ρ is the fluid density, and the F parameters are the components of a "body force" acting on the fluid in the x, y, and z directions. (A *body force* is a force that acts on the fluid as a result of its mass rather than its surface area; gravity is the most common body force.)

The Navier–Stokes equations have a complex form due to the necessity of treating many of the terms as vector quantities. To understand these equations, however, one need only recognize that they are not mass balances but an elaboration of Newton's second law of motion for a flowing fluid. Recall that Newton's second law states that the vector sum of all the forces acting on an object ($\Sigma\mathbf{F}$) will be equal to the product of the object's mass (m) and its acceleration (a), or $\Sigma\mathbf{F} = m\mathbf{a}$. Now consider the first of the three Navier–Stokes equations listed above, Eq. (10). The "object" in this case is a differential fluid element, that is, a small cube of fluid with volume $dx\,dy\,dz$ and mass $\rho(dx\,dy\,dz)$. The left-hand side of the equation is essentially the product of mass and acceleration for this fluid element ($m\mathbf{a}$), while the right-hand side represents the sum of the forces acting on it ($\sum\mathbf{F}$). These include forces resulting from fluid pressure (P), retarding forces due to fluid viscosity (μ), and body forces such as gravity (F_x).

In pharmaceutics, the Navier–Stokes equations have been used to predict fluid velocity profiles in studies of tablet dissolution. In addition, concepts derived from the Navier–Stokes equations often are employed in discussing various systems in which convection occurs. Three of the most important of these concepts are laminar flow, turbulent flow, and boundary layer flow. In laminar flow, the viscous terms (i.e., those containing μ) in the Navier–Stokes equations are large relative to those describing fluid momentum (those containing fluid velocity and density). Laminar flow occurs at slow flow rates or in viscous fluids. In contrast, turbulent flow occurs when fluid momentum dominates over fluid viscosity, as might occur at high flow rates or in relatively inviscid fluids. In turbulent flow, the viscous terms of the Navier–Stokes equations are negligible. Finally, boundary layer flow occurs in fluids moving over stationary solid surfaces. In boundary layer flow, fluid velocity increases from zero at the solid surface to the mean velocity in the bulk fluid at some distance above the solid. The region in which this transition occurs is called the *boundary layer*. An understanding of these flow patterns is important in chromatography, in describing drug delivery from moving fluids such as the gastrointestinal tract or bloodstream, and, as noted above, in tablet dissolution.

III. DIFFUSION

Diffusion is the movement of mass due to a spatial gradient in chemical potential and as a result of the random thermal motion of molecules. While the thermodynamic basis for diffusion is best apprehended in terms of chemical potential, the theories describing the rate of diffusion are based instead on a simpler and more experimentally accessible variable, concentration. The most fundamental of these theories of diffusion are Fick's laws. Fick's first law of diffusion states that in the presence of a concentration gradient, the observed rate of mass transfer is proportional to the spatial gradient in concentration. In one dimension (x), the mathematical form of Fick's first law is

$$J = -D\frac{dC}{dx} \tag{13}$$

where J is the rate of mass transfer per unit area, called the *flux*; D is the diffusion coefficient; and dC/dx is the gradient in concentration (C) in the x direction. Fick's first law asserts that steep gradients in concentration will result in rapid diffusion, while no diffusion will occur when the concentration gradient is completely flat (i.e., $dC/dx = 0$). In Fick's first law, the diffusion coefficient appears as a proportionality constant relating the flux to the concentration gradient. Its value is a fundamental material property of the system and is dependent on the solute, the temperature, and the medium through which diffusion occurs. By developing Fick's first law from the more fundamental assertion that flux is proportional to the gradient in chemical potential, it is possible to show that the "constant" diffusion coefficient also is influenced by solution nonideality, expressed as the activity coefficient (γ), and to a lesser extent by solute concentration (C).

Fick's second law of diffusion can be derived from Fick's first law by using a mass balance approach. Consider the differential fluid element shown in Figure 4. This "differential fluid element" is simply a small cube of liquid or gas, with volume $\Delta x\, \Delta y\, \Delta z$, and will be defined as the "system" for the mass balance. Assume now that component A enters the cube at position x by diffusion and exits the cube at $x + \Delta x$ by the same mechanism. For the moment, assume that no diffusion occurs in the y or z directions and that the faces of the cube that are perpendicular to the y and z axes thus are impermeable to the diffusion of A. Under these conditions, the component mass balance for A in this system is

$$\frac{d(S\, \Delta x\, C_A)}{dt} = S[J(x) - J(x + \Delta x)] \tag{14}$$

where S is the surface area of the x-directed faces of the cube ($S = \Delta y\, \Delta z$).

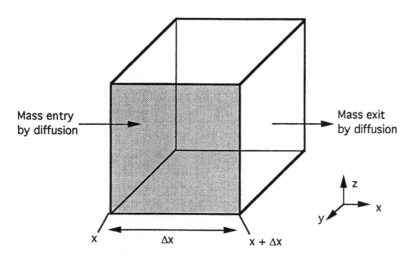

Mass entry by diffusion

Mass exit by diffusion

x Δx $x + \Delta x$

Figure 4   Differential fluid element (system) used for the development of Fick's second law. Diffusion occurs only in the x direction, as shown. The front face of the cube is shaded for contrast.

When S is constant, dividing both sides of the equation by Δx and taking the limit as $\Delta x \to 0$ gives

$$\frac{\partial C_A}{\partial t} = -\frac{\partial J}{\partial x} \tag{15}$$

where the partial derivative (∂) replaces the ordinary derivative (d) since two independent variables $(x$ and $t)$ appear in the equation. Substituting the relationship for flux that appears in Fick's first law gives

$$\frac{\partial C_A}{\partial t} = D\frac{\partial^2 C_A}{\partial x^2} \tag{16}$$

This is Fick's second law of diffusion, the equation that forms the basis for most mathematical models of diffusion processes. The simple form of the equation shown above is applicable only to diffusion in one dimension (x) in systems of rectangular geometry. The mathematical form of the equation becomes more complex when diffusion is allowed to occur in more than one dimension or when the relationship is expressed in cylindrical or spherical coordinate geometries. Since the "simple" form shown above is itself a second-order partial differential equation, the threat of added complexity is an unpleasant proposition at best.

However, as with the plug flow model, considerable physical insight can be gained by examining the steady-state solution $(\partial C_A/\partial t = 0)$ of Eq. (16). At steady state, the relationship becomes $\partial^2 C_A/\partial x^2 = 0$, which has the solution $C_A = mx + B$, where m and B are constants of integration that can be determined from boundary conditions posed for the particular problem of interest. Note that the result implies that in a rectangular coordinate geometry with diffusion in one dimension, the concentration is a linear function of position and the flux [Eq. (13)] is constant (i.e., independent of position). This linear concentration profile and spatially invariant mass flux are characteristic of steady-state diffusion in a rectangular geometry in the absence of chemical reaction. The cumulative mass (M_t) that has been transported through the system is also a linear function of time and takes the form $M_t = JSt$. When the diffusing species is consumed or produced by a chemical reaction occurring in the system, the concentration profile becomes concave or convex, respectively, and the diffusional flux varies with position. Interestingly, the steady-state concentration profiles also show different functional dependencies on the spatial variable in other coordinate geometries: In a cylindrical system with diffusion in the radial direction, the concentration is linearly related to the natural logarithm of the radial distance from the center $(\ln r)$, while in spherical coordinates it is linearly related to the inverse of the radial distance $(1/r)$.

In the pharmaceutical sciences, diffusion processes are central to drug delivery. They are particularly important in the release of drugs from controlled-release devices, in the dissolution of solid tablets and particles, and in the transport of drugs across biological barriers within the body. Many of the remaining chapters of this book deal specifically with diffusion. Theories of diffusion and their analytical solutions are developed more fully in Chapter 2. Methods that are used to measure diffusion coefficients and the application of diffusion theory to models of drug dissolution is described in this book. Part II of the book discusses biological transport processes, which in many cases rely on passive diffusion or on diffusion occurring in parallel with active transport mechanisms. Part III describes transport in polymer systems, such as those used in controlled release; the importance of diffusion in these polymer systems is evident, particularly in Chapter 12. In fact, diffusion is so crucial to the pharmaceutical sciences that "mass transfer" is practically synonymous with "diffusion" in many conversations. The reason for its importance may be that diffusion is usually slower than convection and so is more likely to limit the overall rate of transport in complex systems.

IV. MASS TRANSPORT WITH MULTIPLE DRIVING FORCES

The theories presented above describe convection or diffusion processes occurring independently. In many systems of pharmaceutical interest, however, both diffusion and convection can occur and are important in determining the overall rates of mass transfer. Similarly, externally applied driving forces other than pressure and concentration gradients may influence mass transfer rates. Gradients in electrical potential are important in iontophoresis, for example, while temperature gradients may be important in lyophilization. This section describes approaches to mass transfer when multiple driving forces are present, first addressing combined convection and diffusion (Section IV.A), then presenting methods of analyzing mass transfer when other driving forces are involved (Section IV.B).

A. Combined Convection and Diffusion

Two major approaches to describing combined convection and diffusion have been used by pharmaceutical scientists. These are the convective diffusion approach and the nonequilibrium thermodynamics approach, described in Sections IV.A.1 and IV.A.2, respectively.

1. Convective Diffusion Approach

Fick's first and second laws of diffusion can be modified to include terms describing fluid convection. As with the discussion of Fick's laws, the equations presented below are for transport in one dimension (x) in a rectangular coordinate geometry. For a fluid flowing in the x direction with velocity v, the flux (J) of a dissolved species A is

$$J = -D_A \frac{dC_A}{dx} + vC_A \tag{17}$$

where D_A is the diffusivity of A and C_A is its concentration. This equation is analogous to Fick's first law of diffusion, with the added term vC_A describing the convective movement of the solute. Using this description of flux, an equation comparable to Fick's second law can be developed:

$$\frac{\partial C_A}{\partial t} = D_A \frac{\partial^2 C_A}{\partial x^2} - \frac{\partial(vC_A)}{\partial x} \tag{18}$$

Note that when the fluid velocity (v) is constant, the description of convection given by the second term on the right-hand side of this equation is identical to that of the plug flow model [Eq. (8)]. In more complex systems, a spatially varying fluid velocity may by incorporated by using the Navier–Stokes equations [Eqs. (10)–(12)] to describe velocity profiles.

   The convective diffusion equations presented above have been used to model tablet dissolution in flowing fluids and the penetration of targeted macromolecular drugs into solid tumors [5]. In comparison with the nonequilibrium thermodynamics approach described below, the convective diffusion equations have the advantage of theoretical rigor. However, their mathematical complexity dictates a numerical solution in all but the simplest cases.

2. Nonequilibrium Thermodynamics Approach

Nonequilibrium thermodynamics provides a second approach to combined convection and diffusion problems. The Kedem–Katchalsky equations, originally developed to describe combined convection and diffusion in membranes, form the basis of this approach [6,7]:

$$J_v = L_p \, \Delta P - \sigma L_p \, \Delta \Pi \tag{19}$$

$$J_s = c_s J_v (1 - \sigma) + \omega \, \Delta \Pi \tag{20}$$

Here J_v is the volumetric flow rate of fluid per unit surface area (the volume flux), and J_s is the mass flux for a dissolved solute of interest. The driving forces for mass transfer are expressed in terms of the pressure gradient (ΔP) and the osmotic pressure gradient ($\Delta \Pi$). The osmotic pressure (Π) is related to the concentration of dissolved solutes (c); for dilute ideal solutions, this relationship is given by the van't Hoff law as $\Pi = RTc$, where R is the ideal gas constant and T is temperature. The term c_s is the arithmetic average of the solute concentrations on either side of the membrane. The constants L_p, σ, and ω describe the mass transfer characteristics of the membrane. L_p is the hydraulic conductivity and is inversely related to the resistance to fluid flow provided by the membrane. σ is the reflection coefficient, a measure of the membrane's permissivity to solute transport. σ values of 1.0 correspond to an ideal semipermeable membrane that is completely impermeable to the solute, while values of zero indicate free and unrestricted solute transport. ω is the solute permeability and is related to the diffusive permeability measured in studies of pure diffusion.

Note that in the Kedem–Katchalsky equations, both pressure gradients (ΔP) and concentration gradients ($\Delta \Pi$) appear in the equations for solute and volume flux. In pure convection, volume flux depends only on gradients in pressure; here, gradients in concentration also can drive the bulk flow of fluid through osmosis, as expressed in the term $\sigma L_p \, \Delta \Pi$. Similarly, in pure diffusion, the transport of solute depends only on gradients in concentration, while the Kedem–Katchalsky equations allow for the convective movement of solute in the flowing fluid through the term $c_s J_v (1 - \sigma)$. Also note that the relationships between the transport rates (J_v, J_s) and the driving forces (ΔP, $\Delta \Pi$) are linear in these equations. This contrasts sharply with the nonlinearities and mathematical complexities of the convective diffusion approach described earlier. One way to view the Kedem–Katchalsky equations is as a linearization of more complex and theoretically complete models. As such, they are "phenomenological" equations, valid for a particular experimental system, and should not be viewed as models whose parameters are fundamental material properties of that system.

In the pharmaceutical sciences, the nonequilibrium thermodynamics approach has been particularly important in the design of osmotic drug delivery devices, as discussed in Chapter 11. It has also been used to describe the convective transport of a binding antibody in an in vitro model of a solid tumor [8]. As our appreciation of the roles of convection and osmosis in drug delivery increases, the nonequilibrium thermodynamics approach may find wider appeal.

B. Other Driving Forces

Mass transfer may be influenced by gradients in variables other than concentration and pressure. In the pharmaceutical sciences, gradients in electrical potential and in temperature are two important examples of these other driving forces. Section IV.B.1 describes the effect of electrical potential gradients on the transport of ions, and Section IV.B.2 discusses mass transport in the presence of temperature gradients, known as combined heat and mass transfer.

1. Gradients in Electrical Potential

Mass transfer involving electrolytes may be influenced by gradients in electrical potential as well as by gradients in concentration or pressure. For example, in electrodiffusion, both concentration gradients and electrical potential gradients can drive the movement of an ion in space. In describing electrodiffusion, the chemical potential gradient of species i (μ_i) must be replaced with the more general electrochemical potential ($\tilde{\mu}_i$), defined as

$$\tilde{\mu}_i = \mu_i + z_i F \psi \tag{21}$$

where z_i is the net charge on species i, F is Faraday's constant, and ψ is the electrostatic potential. Using this definition of electrochemical potential, an equation analogous to Fick's first law of diffusion can be developed; it is called the electrodiffusion equation or the Nernst–Planck equation. In one dimension, the Nernst–Planck equation takes the form

$$J_i = -D_i \left(\frac{dC_i}{dx} - C_i z_i \frac{F}{RT} \frac{d\psi}{dx} \right) \tag{22}$$

where J_i is the flux of species i, D_i is the diffusion coefficient for species i, R is the ideal gas constant, T is absolute temperature, and $d\psi/dx$ is the gradient in electrostatic potential in the x direction. Note that when the species i is uncharged ($z_i = 0$) or when the electrostatic potential gradient is zero ($d\psi/dx = 0$), the Nernst–Planck equation reduces to Fick's first law of diffusion [Eq. (13)]. Convective forces can be included in the Nernst–Planck equation by adding a term of the form vC_i, where v is the fluid velocity, to produce an equation analogous to the convective diffusion equation described above [Eq. (17)]. Mass transfer problems involving electrolytes are further complicated by the requirement of local electroneutrality, which states that the sum of positive and negative charges must be zero at all positions in the system.

In the pharmaceutical sciences, electrical potential gradients are important in the iontophoretic delivery of drugs through the skin [9] and in drug analysis by capillary electrophoresis [10]. Given that many drugs are charged, it is reasonable to expect that our understanding and exploitation of electrical potential gradients in drug delivery and drug analysis will increase in the future.

2. Combined Heat and Mass Transfer

Problems involving combined heat and mass transfer provide a final example of multiple driving forces but differ in important ways from the previous examples. In those examples, all the gradients involved (concentration, pressure, electrical potential) were directly responsible for driving mass transfer. In combined heat and mass transfer, however, both heat and mass are being transported. The transfer of heat may drive mass transfer indirectly, as in the loss of a volatile solvent from a beaker when it is heated by evaporation. Thus, problems in heat and mass transfer have not only multiple driving forces but also multiple quantities (heat, mass) being transported.

Recall that in the isolated system shown in Figure 1, the temperatures in the two sections were equal at equilibrium. Heat transfer occurs in response to a spatial gradient in temperature and can occur through the mechanisms of thermal convection, thermal conduction, or radiation. Thermal convection is analogous to the convection of mass discussed above and occurs as a result of bulk movement of a fluid. Thermal conduction is analogous to diffusion and occurs due to energy transfer on the molecular scale. Radiation has no counterpart in mass transfer and involves the transfer of energy through space in the absence of molecular motion, as in the radiant transfer of heat from the sun.

The analogies between heat and mass transfer are reflected in the equations used to describe them. Thermal conduction is described by Fourier's law, which in one dimension is

$$h = -k_T \frac{dT}{dx} \tag{23}$$

where h is the heat flux, T is temperature, and k_T is the thermal conductivity. This equation is identical to Fick's law of diffusion [Eq. (13)], with heat flux replacing mass flux, temperature gradient replacing concentration gradient, and thermal conductivity replacing diffusivity. (Historically, the similarities in the two equations are not coincidental, since Fick patterned his laws of diffusion after Fourier's laws of thermal conduction [11].) Similarly, thermal convection can be described for simple systems by stirred tank and plug flow models. In these models, the total energy (internal + potential + kinetic) in the system is treated as the conserved quantity, rather than total mass. When pressure and volume changes can be neglected, the resulting "energy balance" equations describe the change in system enthalpy due to flowing fluid streams of differing temperature and heat capacity and also incorporate the generation or consumption of thermal energy by chemical reaction.

In pharmaceutical systems, both heat and mass transfer are involved whenever a phase change occurs. Lyophilization (freeze-drying) depends on the solid–vapor phase transition of water induced by the addition of thermal energy to a frozen sample in a controlled manner. Lyophilization is described in detail in Chapter 16. Similarly, the adsorption of water vapor by pharmaceutical solids liberates the heat of condensation, as discussed in Chapter 17.

V. SUMMARY

This chapter has presented an overview of mass transfer principles of relevance to the pharmaceutical scientist. Mass transfer is defined as the movement of mass in space, occurring by the two major mechanisms of convection and diffusion. Convection is mass transfer occurring in response to a spatial gradient in pressure. In the simplest cases, convective mass transfer can be described by stirred tank or plug flow models. These models assume constant fluid velocities or flow rates; when constant velocity cannot be assumed, the Navier–Stokes equations can be used to describe more complex velocity profiles. Diffusion is mass transfer occurring in response to a spatial gradient in concentration and as the result of random thermal motion of molecules. Fick's laws of diffusion are the basis for solving mass transfer problems involving diffusion. In complex systems, mass transfer may be driven by multiple driving forces; combined convective diffusion and electrodiffusion are examples of relevance to the pharmaceutical scientist.

An overview of this kind is, of necessity, limited in detail. Readers interested in a more thorough development of mass transfer principles are encouraged to consult the references listed at the end of the chapter. In particular, Cussler's excellent textbook on diffusion is an accessible introduction to the subject geared toward the physical scientist [11]. Those with a more biological orientation may prefer Friedman's text on biological mass transfer [12], which is also exceptional. A classic reference in the field is Crank's *Mathematics of Diffusion* [13], which contains solutions to many important diffusion problems.

VI. GLOSSARY OF TERMS

Diffusion   Mass transfer driven by a gradient in concentration.

Component   A chemical species present in the system.

Component mass balance   An equation relating the rate of change of the mass of a component in the system to the rates of inflow, outflow, and chemical reaction. See Eq. (2).

Convection   Mass transfer driven by a gradient in pressure.

Fick's laws   Equations that describe the relationship between gradients in concentration and the rate of diffusion. See Eqs. (13) and (16).

Flux   Rate of mass transfer per unit surface area.

Fourier's law for thermal conduction   An equation describing the relationship between the rate of heat flux and the temperature gradient. See Eq. (23).

Kedem–Katchalsky equations   Phenomenological equations for combined convection and diffusion, derived from nonequilibrium thermodynamics. See Eqs. (19) and (20).

Laminar flow   Fluid flow pattern at low flow rate and/or high viscosity.

Navier–Stokes equations   A series of differential equations derived from Newton's second law of motion ($\Sigma \mathbf{F} = m\mathbf{a}$) that describe the relationship between fluid velocity and applied forces in a moving fluid. See Eqs. (10)–(12).

Osmotic pressure A driving force for convective and diffusive mass transfer that is related to solute concentration.

Permeability A rate of mass transfer, usually expressed per unit surface area. For Fickian diffusion in a membrane, the permeability is proportional to the diffusion coefficient and inversely proportional to the membrane thickness.

Plug flow A simple convective flow pattern in pipes and tubes that is characterized by a fluid velocity independent of radial position, complete mixing in the radial direction, and no mixing in the axial direction. Also called the "parallel tube model" or "tubular flow." See Eqs. (7) and (8) and Figure 3.

Steady state A time-invariant condition for a differential equation in which all time derivatives are zero.

Stirred tank model A simple convective flow pattern in tanks, characterized by complete and instantaneous mixing in all directions. Also called the "continuously stirred tank reaction" or the "mixing tank model." See Eqs. (3) and (4) and Figure 2.

System A well-defined region in space for which mass balance equations are written.

Total mass balance An equation relating the rate of change of the total mass in the system to the rates of inflow and outflow.

Turbulent flow Fluid flow pattern at high flow rate and/or low viscosity.

REFERENCES

1. KS Pang, M Rowland. Hepatic clearance of drugs. I. Theoretical considerations of a "well-stirred" model and a "parallel tube" model. Influence of hepatic blood flow, plasma and blood cell binding, and the hepatocellular enzymatic activity on hepatic drug clearance. J Pharmacokin Biopharm 5/6:625–653, 1977.

2. KS Pang, M Rowland. Hepatic clearance of drugs. II. Experimental evidence for acceptance of the "well-stirred" model over the "parallel tube" model using lidocaine in the perfused rat liver in situ preparation. J Pharmacokin Biopharm 5/6:655–699, 1977.

3. PJ Sinko, GD Leesman, GL Amidon. Predicting fraction dose absorbed in humans using a macroscopic mass balance approach. Pharm Res 8(8):979–988, 1991.

4. PM Hayter, EMA Curling, ML Gould, AJ Baines, N Jenkins, I Salmon, PG Strange, AT Bull. The effect of the dilution rate on CHO cell physiology and recombinant interferon gamma production in glucose limited chemostat culture. Biotechnol Bioeng 42:1077–1085, 1993.

5. LT Baxter, RK Jain. Transport of fluid and macromolecules in tumors. I. Role of interstitial pressure and convection. Microvasc Res 37:77–104, 1989.

6. O Kedem, A Katchalsky. A physical interpretation of the phenomenological coefficients of membrane permeability. J Gen Physiol 45:143–179, 1961.

7. TG Kaufmann, EF Leonard. Studies of intramembrane transport: A phenomenological approach. AIChE J 14:110–117, 1968.
8. BS DeSilva, TL Hendrickson, EM Topp. Development of a cell culture system to study antibody convection in tumors. J Pharm Sci 86:858–864, 1997.
9. JE Riviere, MC Heit. Electrically-assisted drug delivery (review). Pharm Res 14(6): 687–697, 1997.
10. LA Holland, NP Chetwyn, MD Perkins, SM Lunte. Capillary electrophoresis in pharmaceutical analysis. Pharm Res 14(4):372–387, 1997.
11. EL Cussler. Diffusion: Mass Transfer in Fluid Systems. 2nd ed. New York: Cambridge University Press, 1996.
12. MH Friedman. Principles and Models of Biological Transport. New York: Springer-Verlag, 1986.
13. J Crank. The Mathematics of Diffusion. 2nd ed. New York: Oxford University Press, 1975.

2

Analytical Solutions to Mass Transfer

Lawrence X. Yu*
Glaxo Wellcome Research and Development, Research Triangle Park, North Carolina

Gordon L. Amidon
The University of Michigan, Ann Arbor, Michigan

I. INTRODUCTION

Mass transfer phenomena exist everywhere in nature and are important in the pharmaceutical sciences. We may think of drug synthesis; preformulation studies; dosage form design and manufacture; and drug absorption, distribution, metabolism, and excretion. Mass transfer plays a significant role in each. Mass transfer is referred to as the movement of molecules caused not only by diffusion but also by convection [1].

Diffusion is the process by which solute molecules are transported from one part of a system to another as a result of random molecular motion [2]. It can be observed with the naked eye when a drop of dye is carefully and slowly placed at the bottom of a beaker filled with water. At first the colored part is separated from the clear by a sharp, well-defined boundary. Later the upper part turns colored, and the color becomes fainter toward the top while the lower part becomes correspondingly less intensely colored. After sufficient time, the whole solution has a uniform color. There is evidently, therefore, a net transfer of dye molecules from the lower part to the upper part of the beaker. The dye molecules have diffused into the water. This diffusion process is primarily due to random molecular motion.

---

* Current affiliation: Food and Drug Administration, Rockville, Maryland.

In the example above, if the drop of dye is not carefully placed at the bottom of the beaker, the water is disturbed and convection currents are set up. Consequently, it can be observed that the dye is transported to the upper part at a much faster rate. It generally takes less time for the solution to achieve a uniform color because of convection caused by external forces.

This chapter provides analytical solutions to mass transfer problems in situations commonly encountered in the pharmaceutical sciences. It deals with diffusion, convection, and generalized mass balance equations that are presented in typical coordinate systems to permit a wide range of problems to be formulated and solved. Typical pharmaceutical problems such as membrane diffusion, drug particle dissolution, and intrinsic dissolution evaluation by rotating disks are used as examples to illustrate the uses of mass transfer equations.

II. DIFFUSION EQUATIONS

A. Fick's First Law

Fick first recognized the analogy among diffusion, heat conduction, and electrical conduction and described diffusion on a quantitative basis by adopting the mathematical equations of Fourier's law for heat conduction or Ohm's law for electrical conduction [1]. Fick's first law relates flux of a solute to its concentration gradient, employing a constant of proportionality called a diffusion coefficient or diffusivity:

$$J = Aj = -AD\frac{\partial c}{\partial z} \tag{1}$$

where J is the total flux, A is the area through which diffusion occurs, j is the flux per unit area, c is concentration, z is distance, and D is the diffusion coefficient. If c is given in mg/cm^3, A is in cm^2, z is in cm, t is in sec, and j is given in mg/(cm^2 · sec), D then has units of cm^2/sec. In some cases, such as diffusion in dilute (ideal) solutions, D can be reasonably considered as constant, while in others, such as diffusion in concentrated solutions, D may significantly depend on concentration (Chapter 4 of this book).

Equation (1) is the one-dimensional form of Fick's first law in Cartesian coordinates. In cylindrical and spherical coordinates, the form of Fick's first law for radial diffusion is

$$j = -D\frac{\partial c}{\partial r} \tag{2}$$

The negative sign in Fick's first law suggests that diffusion occurs in the opposite direction to increasing concentration. In other words, diffusion occurs in the direction of decreasing concentration of diffusing molecules.

B. Fick's Second Law

Fick's first law is a concise mathematical statement; however, it is not directly applicable to solutions of most pharmaceutical problems. Fick's second law presents a more general and useful equation in resolving most diffusion problems. Fick's second law can be derived from Fick's first law.

    Consider an element of volume of a rectangular box of dimensions $dx\ dy\ dz$, as shown in Figure 1. It is assumed that diffusion occurs only in the z direction. At the center (x, y, z) of the box, the concentration is c and the flux is j_z. The rate at which diffusing molecules enter the face at $z - dz/2$ is

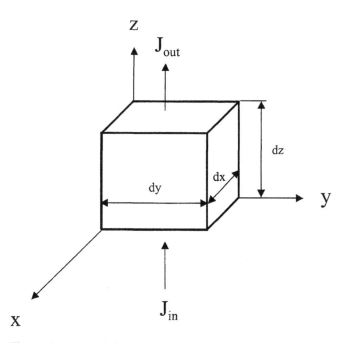

Figure 1   Diffusion through a rectangular box of dimensions $dx\ dy\ dz$. It is assumed that diffusion occurs only in the z direction. At the center (x, y, z) of the box the concentration is c and the flux is j_z. J_{in} is the flux into the box at $z - dz/2$, and J_{out} is the flux out of the box at $z + dz/2$.

$$J_{in} = dx\,dy\left(j_z - \frac{dz}{2}\frac{\partial j_z}{\partial z}\right) \tag{3}$$

and the rate at which diffusing molecules leave the face at $z + dz/2$ is

$$J_{out} = dx\,dy\left(j_z + \frac{dz}{2}\frac{\partial j_z}{\partial z}\right) \tag{4}$$

The net contribution to the rate of accumulation of diffusing molecules in the element from these two faces is thus equal to

$$J_{in} - J_{out} = -dx\,dy\,dz\,\frac{\partial j_z}{\partial z} \tag{5}$$

The rate of accumulation of diffusing molecules in the element is

$$dx\,dy\,dz\,\frac{\partial c}{\partial t} \tag{6}$$

If neither convection nor chemical reaction occurs within the element, the rate of accumulation of diffusing molecules is then equal to the net contribution by diffusion:

$$dx\,dy\,dz\,\frac{\partial c}{\partial t} = -dx\,dy\,dz\,\frac{\partial j_z}{\partial z} \tag{7}$$

or

$$\frac{\partial c}{\partial t} = -\frac{\partial j_z}{\partial z} \tag{8}$$

Substitution of Fick's first law into Eq. (8) gives

$$\frac{\partial c}{\partial t} = D\frac{\partial^2 c}{\partial z^2} \tag{9}$$

Equation (9) is Fick's second law of diffusion, derived on the assumption that D is constant. Fick's second law essentially states that the rate of change in concentration in a volume within the diffusional field is proportional to the rate of change in the spatial concentration gradient at that point in the field, the proportionality constant being the diffusion coefficient.

III. CONVECTION AND GENERALIZED MASS BALANCE EQUATIONS

In many diffusion problems of practical importance in the pharmaceutical sciences, such as intrinsic dissolution studies and drug release from solid dosage forms, the medium under consideration is not at rest. In addition to concentration changes due to diffusion, there are concentration changes by convection. External forces, such as pressure gradients and temperature differences, can cause convective flows. Although convection can also be caused by diffusion itself, our discussion is limited to convection caused by external forces, since convection produced by diffusion is negligible (less than 10%) for most pharmaceutical problems.

We restrict our derivation to the one-dimensional case. The flux caused by diffusion alone can be described by Fick's first law:

$$j_z = -D \frac{\partial c}{\partial z} \tag{10}$$

If there is an external force acting in the same direction on solute molecules, the velocity of these molecules is v_z and the resulting flux is cv_z. Therefore, the total flux, n_z, due to both diffusion and convection is

$$n_z = j_z + cv_z \tag{11}$$

With this equation, we can now discuss a generalized mass balance equation. We still use Figure 1 to show the derivation. Based on Eq. (5), the net contribution by diffusion and convection now becomes

$$(n_z)_{in} - (n_z)_{out} = -dx \, dy \, dz \frac{\partial n_z}{\partial z} \tag{12}$$

The rate of accumulation of solute molecules in the element is still

$$dx \, dy \, dz \frac{\partial c}{\partial t} \tag{13}$$

In addition, reaction is considered. The rate of production of solute molecules by homogeneous reaction is

$$dx \, dy \, dz \, R \tag{14}$$

where R is the rate per unit volume of a homogeneous chemical reaction producing solute. From mass balance (the rate of accumulation = net contribution by diffusion and convection + rate of production), we have

$$dx\,dy\,dz\,\frac{\partial c}{\partial t} = -dx\,dy\,dz\,\frac{\partial n_z}{\partial z} + dx\,dy\,dz\,R \tag{15}$$

or

$$\frac{\partial c}{\partial t} = -\frac{\partial n_z}{\partial z} + R \tag{16}$$

Substitution of Eq. (11) into Eq. (16) yields

$$\frac{\partial c}{\partial t} = -\left(\frac{\partial j_z}{\partial z} + \frac{\partial(cv_z)}{\partial z}\right) + R \tag{17}$$

If the velocity and the diffusion coefficient are constant, combining Eq. (17) with Fick's first law, Eq. (1), gives

$$\frac{\partial c}{\partial t} + v_z\frac{\partial c}{\partial z} = D\frac{\partial^2 c}{\partial z^2} + R \tag{18}$$

This is a generalized mass balance equation in one dimension. If diffusion and convection occur in other directions, the generalized mass balance equation becomes

$$\frac{\partial c}{\partial t} + v_x\frac{\partial c}{\partial x} + v_y\frac{\partial c}{\partial y} + v_z\frac{\partial c}{\partial z} = D\left(\frac{\partial^2 c}{\partial x^2} + \frac{\partial^2 c}{\partial y^2} + \frac{\partial^2 c}{\partial z^2}\right) + R \tag{19}$$

In vectorial notation, the generalized equation can be written as

$$\frac{\partial c}{\partial t} + v \cdot \nabla c = D\,\nabla^2 c + R \tag{20}$$

where

$$\nabla = \frac{\partial}{\partial x}i + \frac{\partial}{\partial y}j + \frac{\partial}{\partial z}k \tag{21}$$

and

$$\nabla^2 = \frac{\partial^2}{\partial x^2} + \frac{\partial^2}{\partial y^2} + \frac{\partial^2}{\partial z^2} \tag{22}$$

A generalized mass balance equation in other coordinate systems is sometimes useful. In the cylindrical system, Eq. (19) becomes

$$\frac{\partial c}{\partial t} + v_r\frac{\partial c}{\partial r} + v_\theta\frac{1}{r}\frac{\partial c}{\partial \theta} + v_z\frac{\partial c}{\partial z} = D\left[\frac{1}{r}\frac{\partial}{\partial r}\left(r\frac{\partial c}{\partial r}\right) + \frac{1}{r^2}\frac{\partial^2 c}{\partial \theta^2} + \frac{\partial^2 c}{\partial z^2}\right] + R \quad (23)$$

and in the spherical coordinate system, Eq. (19) becomes

$$\frac{\partial c}{\partial t} + v_r\frac{\partial c}{\partial r} + v_\theta\frac{1}{r}\frac{\partial c}{\partial \theta} + v_\theta\frac{1}{r\sin\theta}\frac{\partial c}{\partial \phi}$$

$$= D\left[\frac{1}{r^2}\frac{\partial}{\partial r}\left(r^2\frac{\partial c}{\partial r}\right) + \frac{1}{r^2\sin\theta}\frac{\partial}{\partial \theta}\left(\sin\theta\frac{\partial c}{\partial \theta}\right) + \frac{1}{r^2\sin^2\theta}\left(\frac{\partial^2 c}{\partial \phi^2}\right)\right] + R \quad (24)$$

IV. MEMBRANE DIFFUSION

In the previous section, we detailed diffusion equations and generalized mass balance equations. We now turn to their practical uses in the pharmaceutical sciences. Mass transport problems can be classified as "steady" or "unsteady." In steady mass transport there is no change of concentration with time [3], characterized mathematically by

$$\frac{\partial c}{\partial t} = 0 \qquad (25)$$

In contrast, unsteady mass transport means that there exists a concentration change with time, depicted mathematically by

$$\frac{\partial c}{\partial t} \neq 0 \qquad (26)$$

Membrane transport represents a major application of mass transport theory in the pharmaceutical sciences [4]. Since convection is not generally involved, we will use Fick's first and second laws to find flux and concentration across membranes in this section. We begin with the discussion of steady diffusion across a thin film and a membrane with or without aqueous diffusion resistance, followed by steady diffusion across the skin, and conclude this section with unsteady membrane diffusion and membrane diffusion with reaction.

A. Steady Diffusion Across a Thin Film

Figure 2 illustrates steady diffusion across a thin film of thickness h. The solutions on both sides of the film are dilute, so the diffusion coefficient can be considered constant. The solute molecules diffuse from the well-mixed higher concentration c_1 to the well-mixed lower concentration c_2. The concentrations on both sides of the film are kept constant. After sufficient time, a steady state is reached in which

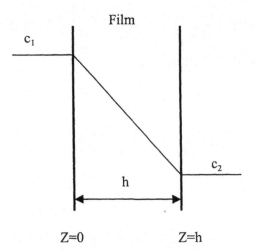

Figure 2 Diffusion across a thin film. The solute molecules diffuse from the well-mixed higher concentration c_1 to the well-mixed lower concentration c_2. The concentrations on both sides of the film are kept constant. At steady state, the concentrations remain constant at all points in the film. The concentration profile inside the film is linear, and the flux is constant.

the concentrations remain constant at all points of the film, as shown in Figure 2. The one-dimensional form of Fick's second law is applicable:

$$\frac{\partial c}{\partial t} = D\frac{\partial^2 c}{\partial z^2} \tag{27}$$

At steady state,

$$D\frac{\partial^2 c}{\partial z^2} = 0 \tag{28}$$

Integrating this equation twice yields

$$c = A + Bz \tag{29}$$

where A and B are integration constants. For the problem considered here, we have two boundary conditions

$$z = 0, \quad c = c_1 \tag{30}$$
$$z = h, \quad c = c_2 \tag{31}$$

Substitution of these two boundary conditions into the general solution yields A and B, resulting in the following concentration profile across the film:

$$c = c_1 + (c_2 - c_1)\frac{z}{h} \tag{32}$$

This equation shows that the concentration changes linearly from c_1 to c_2 through the film. The flux is the same across all sections of the film and is given by

$$j = \frac{D}{h}(c_1 - c_2) \tag{33}$$

The term h/D is often called the diffusional resistance, denoted by R. The flux equation, therefore, can be written as

$$j = \frac{c_1 - c_2}{R} \tag{34}$$

Steady diffusion across a thin film is mathematically straightforward but physically subtle. Dissolution film theory, suggested initially by Nernst and Brunner, is essentially based on steady diffusion across a thin film.

B. Steady Diffusion Across a Membrane

Figure 3 shows a steady diffusion across a membrane. As in the previous case, the membrane separates two well-mixed dilute solutions, and the diffusion coefficient D_m is assumed constant. However, unlike the film, the membrane has different physicochemical characteristics than the solvent. As a result, the diffusing solute molecules may preferentially partition into the membrane or the solvent.

As before, applying Fick's second law to diffusion across a membrane, we have

$$D_m \frac{\partial^2 c_m}{\partial z^2} = 0 \tag{35}$$

where subscript m is referred to the membrane throughout this chapter. The boundary conditions are the membrane concentrations at $z = 0$ and $z = h_m$, which are unknown. However, the membrane concentrations at the surfaces can be related to the solution concentrations by the partition coefficient provided that equilibrium exists at the interfaces of the membrane and the solutions. The boundary conditions, therefore, can be written as

$$z = 0, \qquad c_m = Kc_1 \tag{36}$$
$$z = h_m, \qquad c_m = Kc_2 \tag{37}$$

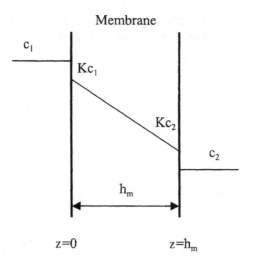

Membrane

Figure 3 Diffusion across a membrane. The solute molecules diffuse from the well-mixed higher concentration c_1 to the well-mixed lower concentration c_2. Equilibrium is assumed at the interfaces of membrane and solutions. The concentrations on both sides of the membrane are kept constant. At steady state, the concentrations c_m remain constant at all points in the membrane. The concentration profile inside the membrane is linear, and the flux is constant.

where K is the partition coefficient. Solving Eq. (35) with the boundary conditions of Eqs. (36) and (37) results in the concentration profile

$$c_m = Kc_1 + K(c_2 - c_1)\frac{z}{h_m} \tag{38}$$

and the flux

$$j = \frac{KD_m}{h_m}(c_1 - c_2) \tag{39}$$

The term KD_m/h_m is often referred to as the permeability coefficient [1]. The concentrations c_1 and c_2 are assumed to be independent of time in Eqs. (36) and (37). This may experimentally imply that the volume of the membrane is negligible compared to that of solutions on either side of the membrane. Practically, neither these concentrations nor the concentration gradient is exactly constant. In diffusion or Caco-2 cell experiments, for example, a solution of higher concentration is initially placed in the donor compartment and a solution of lower concentration is placed in the receptor compartment. Even after sufficient time, the concentration will still fall in the donor compartment and rise in the receptor compartment. Therefore, we are not actually dealing with a steady state in the strict sense of the word, but rather a quasi-steady state. The premise of a steady state is approximately satisfied. Such an approximation will not cause a serious error, however.

In general, it is easier to measure the permeability coefficient than the partition coefficient, diffusion coefficient, and membrane thickness. Once permeability is measured, it is difficult to separate the contributions of the individual variables. We often see permeability coefficients being reported in the literature instead of the more fundamental partition coefficient and diffusion coefficient being reported separately.

C. Steady Diffusion Across a Membrane with Aqueous Diffusion Layers

In the example above, the solutions are assumed to be well stirred and mixed; the aqueous resistance is negligible, and the membrane is the only transport barrier. However, in any real case, the solutions on both sides of the membrane become less and less stirred as they approach the surface of the membrane. The aqueous diffusion resistance, therefore, very often needs to be considered. For example, for very highly permeable drugs, the resistance to absorption from the gastrointestinal tract is mainly aqueous diffusion. In the section, we give a general solution to steady diffusion across a membrane with aqueous diffusion resistance [5].

A stagnant diffusion layer is often assumed to approximate the effect of aqueous transport resistance. Figure 4 shows a membrane with a diffusion layer on each side. The bulk solutions are assumed to be well mixed and therefore of uniform concentrations c_{b1} and c_{b2}. Adjacent to the membrane is a stagnant diffusion layer in which a concentration gradient of the solute may exist between the well-mixed bulk solution and the membrane surface; the concentrations change from c_{b1} to c_{s1} for solution 1 and from c_{b2} to c_{s2} for solution 2. The membrane surface concentrations are c_{m1} and c_{m2}. The membrane has thickness h_m, and the aqueous diffusion layers have thickness h_1 and h_2.

Applying the concentration profile of Eq. (32) obtained from a thin film to both aqueous diffusion layers at steady state, we have

$$c_1 = c_{b1} + (c_{s1} - c_{b1}) \frac{z}{h_1} \tag{40}$$

$$c_2 = c_{s2} + (c_{b2} - c_{s2}) \frac{z - h_1 - h_m}{h_2} \tag{41}$$

Similarly, applying the concentration profile of Eq. (38) to membrane diffusion gives

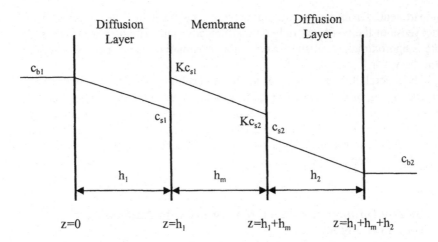

Figure 4 Diffusion across a membrane with aqueous diffusional layers. c_{b1} and c_{b2} are the concentrations of bulk solutions 1 and 2, respectively. The thicknesses of the aqueous diffusion layers are h_1 and h_2. The membrane has a thickness of h_m. Equilibrium is assumed at the interfaces of the membrane and the aqueous diffusion layers. At steady state, the concentrations remain constant at all points in the membrane and in the aqueous diffusion layers. The concentration profiles inside the membrane and aqueous diffusion layers are linear, and the flux is constant.

$$c_m = Kc_{s1} + K(c_{s2} - c_{s1})\frac{z - h_1}{h_m} \tag{42}$$

In Eqs. (40)–(42), c_{b1} and c_{b2} are experimentally measurable and the aqueous diffusion layer thickness can be estimated theoretically. Therefore, the only unknowns are the solute concentrations at the interfaces, c_{s1} and c_{s2}. Their estimation is shown below.

The steady flux across each aqueous diffusion layer is

$$j_1 = \frac{D}{h_1}(c_{b1} - c_{s1}) \tag{43}$$

$$j_2 = \frac{D}{h_2}(c_{s2} - c_{b2}) \tag{44}$$

The steady flux across the membrane is

$$j_m = \frac{KD_m}{h_m}(c_{s1} - c_{s2}) \tag{45}$$

At steady state, the flux of solute across each diffusion layer or membrane must be the same; therefore,

$$j_1 = j_m = j_2 \tag{46}$$

Solving Eqs. (43)–(46), the solute concentration at each interface is given by

$$c_{s1} = \frac{(Dh_m + KD_m h_2)c_{b1} + KD_m h_1 c_{b2}}{Dh_m + KD_m(h_1 + h_2)} \tag{47}$$

and

$$c_{s2} = \frac{KD_m h_2 c_{b1} + (Dh_m + KD_m h_1)c_{b2}}{Dh_m + KD_m(h_1 + h_2)} \tag{48}$$

Substituting Eqs. (47) and (48) into Eq. (45) yields

$$j_1 = j_m = j_2 = \frac{KDD_m(c_{b1} - c_{b2})}{Dh_m + KD_m(h_1 + h_2)} \tag{49}$$

Very often, we may see in the literature that flux is expressed as

$$j = P_{\text{eff}}(c_{b1} - c_{b2}) = \frac{c_{b1} - c_{b2}}{R_{\text{tot}}} \tag{50}$$

where P_{eff} is called the effective permeability coefficient and R_{tot} is the total resistance. The product of P_{eff} and R_{tot} is equal to 1.

Let

$$P_1 = \frac{1}{R_1} = \frac{D}{h_1}, \qquad P_m = \frac{1}{R_m} = \frac{KD_m}{h_m}, \qquad P_2 = \frac{1}{R_2} = \frac{D}{h_2} \tag{51}$$

Comparing Eqs. (49) and (50) yields

$$R_{\text{tot}} = R_1 + R_m + R_2 \tag{52}$$

or

$$\frac{1}{P_{\text{eff}}} = \frac{1}{P_1} + \frac{1}{P_m} + \frac{1}{P_2} \tag{53}$$

Equation (52) shows an important property of diffusion across barriers in series: The diffusional resistance is additive. Therefore, in the case of diffusion involving n barriers in series, the total resistance can be calculated as

$$R_{\text{tot}} = R_1 + R_2 + \cdots + R_n \tag{54}$$

The corresponding flux can still be calculated by using Eq. (50).

It is interesting to examine two limits of Eq. (49). First, when the membrane's diffusional resistance is much greater than the total aqueous resistance (that is, $R_m > R_1 + R_2$ by a factor of at least 10), the rate-determining step is the diffusion across the membrane: $R_{tot} = R_m = h_m/KD_m$. Consequently, Eq. (49) is reduced to

$$j = \frac{c_{b1} - c_{b2}}{R_{tot}} = \frac{c_{b1} - c_{b2}}{R_m} = \frac{KD_m}{h_m}(c_{b1} - c_{b2}) \tag{55}$$

which is identical to Eq. (39), where only membrane diffusion is considered. Second, when the aqueous diffusion layer resistance is much greater than the membrane's diffusional resistance (that is, $R_1 + R_2 > R_m$ by a factor of at least 10), the rate-determining step becomes the diffusion across the aqueous diffusion layers: $R_{tot} = R_1 + R_2 = (h_1 + h_2)/D$. Equation (49) is then reduced to

$$j = \frac{D}{h_1 + h_2}(c_{b1} - c_{b2}) \tag{56}$$

which is similar to Eq. (33), where diffusion across a thin film is considered.

D. Steady Diffusion in Parallel

The section discusses diffusion across a number of diffusion barriers in parallel. Diffusion across the skin represents one of the best examples to illustrate steady diffusion involving two or more independent diffusional pathways in parallel [6].

When a drug is applied topically to the skin, it will partition into either the stratum corneum or the sebum-filled ducts of the pilosebaceous glands. Two principal diffusion routes are then employed for drug absorption; one is the transepidermal route involving diffusion across the stratum corneum, and the other is the transfollicular route involving diffusion through the follicular pore. The solute molecules from either route finally reach the viable epidermal and dermal points of entry. A concentration gradient is established across the skin up to the outer reaches of the skin's microcirculation, where the solute is swept away by the capillary blood flow and distributed throughout the body. A critical factor in transdermal delivery is the rate at which the solute molecules reach the systemic circulation.

First, consider the transepidermal (TE) route. The solute molecules diffuse across the stratum corneum and the viable tissues located above the capillary bed. Considering the stratum corneum and the viable tissues as two diffusion barriers in series, the total resistance is given by

$$R_{TE} = R_{m1-TE} + R_{m2-TE} = \frac{h_{m1-TE}}{D_{m1-TE}K_{m1-TE}} + \frac{h_{m2-TE}}{D_{m2-TE}} \tag{57}$$

where the subscripts $m1$ and $m2$ refer to the stratum corneum and the viable tissues, respectively. The flux by the transepidermal route is

$$j_{TE} = \frac{f_{TE}}{R_{TE}}(c_{b1} - c_{b2}) = f_{TE}P_{TE}(c_{b1} - c_{b2}) \tag{58}$$

where f_{TE} is the fractional area of this route, which is virtually equal to 1.0 since this route constitutes the bulk of the area available for diffusion. Similarly, for the transfollicular (TF) route, the drug molecules diffuse through sebaceous pores and viable tissues. The total transfollicular resistance is given by

$$R_{TF} = R_{m1-TF} + R_{m2-TF} = \frac{h_{m1-TF}}{D_{m1-TF}K_{m1-TF}} + \frac{h_{m2-TF}}{D_{m2-TF}} \tag{59}$$

The corresponding flux by the transfollicular route is given by

$$j_{TF} = \frac{f_{TF}}{R_{TF}}(c_{b1} - c_{b2}) = f_{TF}P_{TF}(c_{b1} - c_{b2}) \tag{60}$$

Thus, the overall flux by both transepidermal and transfollicular routes is

$$j = j_{TE} + j_{TF} = (f_{TE}P_{TE} + f_{TF}P_{TF})(c_{b1} - c_{b2}) \tag{61}$$

In terms of the effective permeability or the total resistance, we have

$$P_{eff} = f_{TE}P_{TE} + f_{TF}P_{TF} \tag{62}$$

or

$$\frac{1}{R_{tot}} = \frac{f_{TE}}{R_{TE}} + \frac{f_{TF}}{R_{TF}} \tag{63}$$

Equation (63) shows that, unlike diffusion in series, the total diffusional resistance for diffusion in parallel is no longer additive. For diffusion in series, the total diffusional resistance is higher than any individual resistance, whereas for diffusion in parallel, the total diffusional resistance is between the minimum and maximum individual resistances.

E. Unsteady Diffusion into a Semi-Infinite Membrane

We have discussed steady diffusion across a membrane with or without aqueous diffusion resistance. If the membrane is extremely thick or if solute diffusion in the membrane is extremely slow, the membrane may behave as if it is almost impermeable. In this case, diffusion on one side will not affect the other. As a result, the impermeable membrane acts like a semi-infinite slab. We want to determine the flux and the concentration profile when one side of the membrane is subject to a concentration increase [1].

Figure 5 shows the diffusion of a solute into such an impermeable membrane. The membrane initially contains no solute. At time zero, the concentration of the solute at $z = 0$ is suddenly increased to c_1 and maintained at this level. Equilibrium is assumed at the interface of the solution and the membrane. Therefore, the corresponding membrane concentration at $z = 0$ is Kc_1. Since the membrane is impermeable, the concentration on the other side will not be affected by the change at $z = 0$ and will still be free of solute. This abrupt increase produces a time-dependent concentration profile as the solute penetrates into the membrane. If the solution is assumed to be dilute, Fick's second law Eq. (9) is applicable:

$$\frac{\partial c_m}{\partial t} = D_m \frac{\partial^2 c_m}{\partial z^2} \tag{64}$$

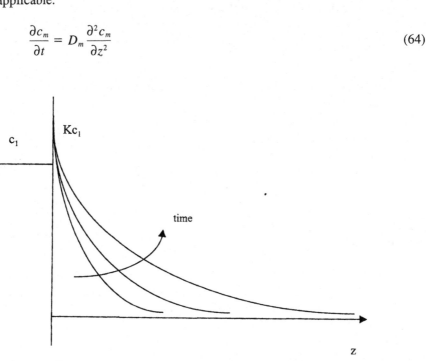

Figure 5 Diffusion into a semi-infinite membrane. The membrane initially contains no solute. At time zero, the concentration of the solution at $z = 0$ is suddenly increased to and maintained at c_1. This abrupt increase produces time-dependent concentration profiles as the solute penetrates into the membrane.

The initial condition is

$$t = 0, \qquad z \geq 0, \qquad c_m = 0 \tag{65}$$

The boundary conditions are

$$t > 0, \qquad z = 0, \qquad c_m = Kc_1 \tag{66}$$

$$t > 0, \qquad z = \infty, \qquad c_m = 0 \tag{67}$$

Let

$$\zeta = \frac{z}{\sqrt{4D_m t}} \tag{68}$$

Equation (64) then becomes

$$\frac{d^2 c_m}{d\zeta^2} + 2\zeta \frac{dc_m}{d\zeta} = 0 \tag{69}$$

and the initial and boundary conditions, Eqs. (65)–(67), become

$$\zeta = 0, \qquad c_m = Kc_1 \tag{70}$$

$$\zeta = \infty, \qquad c_m = 0 \tag{71}$$

where Eq. (70) is transformed from Eq. (66) and Eq. (71) is transformed from either Eq. (65) or Eq. (67). Integrating Eq. (69) twice gives

$$c_m = A \int_0^\zeta e^{-u^2} \, du + B \tag{72}$$

where A and B are integration constants. Applying the boundary conditions Eqs. (70) and (71) to Eq. (72) yields

$$A = \frac{-Kc_1}{\displaystyle\int_0^\infty e^{-\zeta^2} \, d\zeta} = -\frac{2Kc_1}{\sqrt{\pi}}, \qquad B = Kc_1 \tag{73}$$

Substitution of A and B into Eq. (72) gives

$$c_m = Kc_1 \, \mathrm{erfc}(\zeta) = Kc_1[1 - \mathrm{erf}(\zeta)] \tag{74}$$

where $\mathrm{erf}(\zeta)$ is the error function, which by definition is

$$\mathrm{erf}(\zeta) = \frac{2}{\sqrt{\pi}} \int_0^\zeta e^{-u^2} \, du \tag{75}$$

The flux is

$$j = -D_m \frac{\partial c_m}{\partial z} = \sqrt{\frac{D_m}{\pi t}} K c_1 \exp\left[-\frac{z^2}{4D_m t}\right] \tag{76}$$

Note that, unlike the steady-state case, the flux in the semi-infinite membrane decreases with increasing time and distance. The flux at the interface is

$$j|_{z=0} = \sqrt{\frac{D_m}{\pi t}} K c_1 \tag{77}$$

It may be appropriate here to introduce film theory. As mentioned in reference to the steady diffusion across a thin film, we often hypothesize a film called an "unstirred layer" to account for the aqueous diffusion resistance to mass transfer. Film theory is valuable not only because of its simplicity but also because of its practical utility. However, the thickness of the film is often difficult to determine. In the following, we try to answer the question, What does the thickness of the film represent?

If we assume that the thickness of the aqueous film is δ, the diffusional flux across the film is

$$j|_{z=0} = \frac{D_m}{\delta} c_1 \tag{78}$$

Coupling Eq. (77) with Eq. (78) gives

$$\delta = \sqrt{\pi D_m t} \tag{79}$$

Note that $K = 1$ and D_m becomes D in aqueous media. Equation (79) suggests that under steady-state conditions the thickness of the film is a function of time. The tangent equation of the concentration curve at $z = 0$ is

$$c = c_1 + \frac{\partial c}{\partial z}\bigg|_{z=0} z \tag{80}$$

or

$$c = c_1 - \frac{c_1}{\sqrt{\pi D t}} z \tag{81}$$

The z-axis intercept is

$$z = \sqrt{\pi D t} \tag{82}$$

Therefore, the film thickness is the z-axis intercept of the tangent curve of the concentration profile. The calculated thickness of the aqueous diffusion layer is

Table 1 Calculated Thickness
of the Aqueous Diffusion
Layer Based on Eq. (82),
Assuming a Diffusion
Coefficient of 5×10^{-6} cm^2/sec

Time (sec)	Film thickness (μm)
0.1	13
1	40
10	125
30	217
60	307
120	434

given in Table 1, assuming that the diffusion coefficient is 5×10^{-6} cm^2/sec. Table 1 shows that the film thickness increases significantly with time. This occurs partly due to our assumption of impermeability.

F. Unsteady Diffusion Across a Membrane

In this section we want to discuss unsteady diffusion across a permeable membrane. In other words, we are interested in how concentration and flux change before reaching the steady state discussed in Section IV.B. The membrane is initially free of solute. At time zero, the concentrations on both sides of the membrane are increased, to c_1 and c_2. Equilibrium between the solution and the membrane interface is assumed; therefore, the corresponding concentrations on the membrane surfaces are Kc_1 and Kc_2. Fick's second law is still applicable:

$$\frac{\partial c_m}{\partial t} = D_m \frac{\partial^2 c_m}{\partial z^2} \tag{83}$$

with the following initial and boundary conditions

$$t = 0, \quad 0 \leq z \leq h, \quad c_m = 0 \tag{84}$$

$$t > 0, \quad z = 0, \quad c_m = Kc_1 \tag{85}$$

$$t > 0, \quad z = h_m, \quad c_m = Kc_2 \tag{86}$$

This problem can be solved by the method of separation of variables. The details can be found in Jost [3]. The final solution is

$$c_m = Kc_1 + K(c_2 - c_1)\frac{z}{h_m}$$

$$+ \frac{2}{\pi}\sum_{n=1}^{\infty}\frac{Kc_2(-1)^n - Kc_1}{n}\sin\frac{n\pi z}{h_m}\exp\left[-\frac{D_m n^2 \pi^2 t}{h_m^2}\right] \tag{87}$$

The flux across the interface $z = h$ is given by Fick's first law,

$$j = -D_m\frac{\partial c_m}{\partial z}\bigg|_{z=h} = \frac{KD_m(c_1 - c_2)}{h_m}$$

$$+ \frac{2D_m}{h_m}\sum_{n=1}^{\infty}[Kc_1(-1)^n - Kc_2]\exp\left[-\frac{D_m n^2 \pi^2 t}{h_m^2}\right] \tag{88}$$

The concentration profiles are shown in Figure 6. As time approaches infinity, the term involving the exponential vanishes and the diffusion process approaches the steady state; Eqs. (87) and (88) are then reduced to the steady concentration profile, Eq. (38), and flux, Eq. (39).

By integrating the flux with respect to time, we obtain the cumulative mass of diffusing solute molecules per unit area:

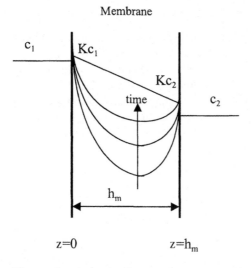

Figure 6 Unsteady diffusion across a membrane. The membrane is initially free of solute. At time zero, the concentrations on the two sides of the membrane are increased to and maintained at c_1 and c_2. The solute penetrates into the membrane from both sides, resulting in time-dependent concentration profiles within the membrane.

$$M = \frac{KD_m(c_1 - c_2)t}{h_m}$$

$$+ \frac{2h_m}{\pi^2} \sum_{n=1}^{\infty} \frac{Kc_1(-1)^n - Kc_2}{n^2} \left\{ 1 - \exp\left[-\frac{D_m n^2 \pi^2 t}{h_m^2} \right] \right\} \tag{89}$$

If $c_1 \gg c_2$, Eq. (89) is then reduced to

$$M = \frac{KD_m c_1 t}{h_m} - \frac{h_m Kc_1}{6} - \frac{2h_m Kc_1}{\pi^2} \sum_{n=1}^{\infty} \frac{(-1)^n}{n^2} \exp\left[-\frac{D_m n^2 \pi^2 t}{h_m^2} \right] \tag{90}$$

When time approaches infinity, the cumulative mass approaches a linear function of time:

$$M = \frac{KD_m c_1}{h_m}\left(t - \frac{h_m^2}{6D_m} \right) \tag{91}$$

Equation (91) represents the situation where the permeation has attained a steady state so that the amount penetrating per unit time is constant.

In addition, the straight line expressed by Eq. (91) has an intercept on the time axis of

$$t_L = \frac{h_m^2}{6D_m} \tag{92}$$

This intercept is referred to as the lag time [4], and it provides a means for estimating the diffusion coefficient provided the membrane thickness is known and with the assumption that other time-dependent processes, such as membrane hydration, do not occur during the lag phase.

G. Effect of Reaction on Membrane Diffusion

A biologically important factor affecting drug absorption is drug metabolism or reaction coincident with diffusion across a membrane. The reaction often produces inactive or less potent products than the parent drug. It is conceivable that the reaction will also reduce the drug flux into the systemic circulation. We are interested in the effect of reaction on membrane diffusion.

For simplicity, only an irreversible first-order reaction is considered. We begin with the generalized mass balance equation in rectangular coordinates with mass transfer in one dimension:

$$\frac{\partial c_m}{\partial t} + v_z \frac{\partial c_m}{\partial z} = D_m \frac{\partial^2 c_m}{\partial z^2} - k_1 c_m \tag{93}$$

The negative sign before the reaction term $k_1 c_m$ indicates that the drug is consumed rather than being produced as in the derivation of the generalized mass balance equation.

We deal with membrane diffusion; no convection is involved ($v_z = 0$). We also assume that the system is at steady state ($\partial c_m/\partial t = 0$). The generalized mass balance equation, therefore, can be reduced to

$$D_m \frac{\partial^2 c_m}{\partial z^2} - k_1 c_m = 0 \tag{94}$$

with the boundary conditions

$$z = 0, \qquad c_m = Kc_1 \tag{95}$$
$$z = h_m, \qquad c_m = 0 \tag{96}$$

A general solution of Eq. (94) is

$$c_m = A \cosh[\sqrt{k_1/D_m}\, z] + B \sinh[\sqrt{k_1/D_m}\, z] \tag{97}$$

Using the boundary conditions to determine the integration constants A and B yields

$$\frac{c_m}{Kc_1} = \frac{\sinh[\sqrt{k_1/D_m}\, (h_m - z)]}{\sinh[\sqrt{k_1/D_m}\, h_m]} \tag{98}$$

The flux is then

$$j = -D_m \frac{dc_m}{dz} = Kc_1 \sqrt{D_m k_1} \frac{\cosh[\sqrt{k_1/D_m}\, (h_m - z)]}{\sinh[\sqrt{k_1/D_m}\, h_m]} \tag{99}$$

In order to understand the effect of reaction on membrane diffusion, we use the membrane without reaction as a reference. The corresponding flux [Eq. (39), where c_2 is equal to zero] is

$$j_0 = \frac{KD_m c_1}{h_m} \tag{100}$$

Therefore, we have the ratio of steady flux with reaction to steady flux without reaction across a membrane:

$$\frac{j}{j_0} = h_m \sqrt{k_1/D_m} \frac{\cosh[\sqrt{k_1/D_m}\, (h_m - z)]}{\sinh[\sqrt{k_1/D_m}\, h_m]} \tag{101}$$

At $z = 0$, the flux ratio becomes

$$\frac{j|_{z=0}}{j_0} = h_m \sqrt{k_1/D_m} \coth[\sqrt{k_1/D_m}\, h_m] \tag{102}$$

and at $z = h_m$, the flux ratio is

$$\frac{j|_{z=h_m}}{j_0} = \frac{h_m \sqrt{k_1/D_m}}{\sinh[\sqrt{k_1/D_m}\, h_m]} \tag{103}$$

These two equations represent the effect of reaction on membrane diffusion.

Two limits are instructive. First, when the reaction is very slow and k_1 is close to zero or $h_m \sqrt{k_1/D_m} \ll 1$, Eqs. (102) and (103) become

$$\frac{j|_{z=0}}{j_0} = 1 \tag{104}$$

and

$$\frac{j|_{z=h_m}}{j_0} = 1 \tag{105}$$

These results suggest that the effect of reaction is insignificant when the reaction is relatively slow. Second, when the reaction is fast, k_1 is large or or $h_m \sqrt{k_1/D_m} \gg 1$. As a result,

$$\frac{j|_{z=0}}{j_0} = h_m \sqrt{k_1/D_m} \gg 1 \tag{106}$$

and

$$\frac{j|_{z=h_m}}{j_0} = 0 \tag{107}$$

Therefore, when the reaction is fast, the rate of solute entry into the membrane is significantly increased. But the solute flux leaving the membrane (e.g., entering the systemic circulation in the case of drug absorption) is nearly zero, since most of the solute is consumed in the reaction (metabolism).

V. DIFFUSION IN A CYLINDER

Membrane diffusion represents over 80% of the diffusion problems of pharmaceutical interest. Membrane diffusion may be called "diffusion in a plane sheet," since it involves one-dimensional diffusion in a medium bounded by two parallel planes. In this and the next section, we discuss diffusion in a cylinder and a sphere. Diffusion in a cylinder is less common in the pharmaceutical sciences; however, we discuss it briefly for the sake of completeness.

Consider a long circular cylinder in which a solute diffuses radially. The concentration is a function of radial position (r) and time (t). In the case of constant diffusion coefficient, the diffusion equation is

$$\frac{\partial c}{\partial t} = \frac{D}{r} \frac{\partial}{\partial r} \left(r \frac{\partial c}{\partial r} \right) \tag{108}$$

If the cylinder is hollow with inner and outer radii r_1 and r_2, at steady state we have

$$\frac{D}{r} \frac{\partial}{\partial r} \left(r \frac{\partial c}{\partial r} \right) = 0, \qquad r_1 < r < r_2 \tag{109}$$

The general solution is

$$c = A + B \ln r \tag{110}$$

where A and B are integration constants. If the surface at $r = r_1$ has a constant concentration c_1 and the surface at $r = r_2$ has a constant concentration c_2, the boundary conditions are then

$$r = r_1, \qquad c = c_1 \tag{111}$$
$$r = r_2, \qquad c = c_2 \tag{112}$$

Using the boundary conditions to determine the constants A and B, we have

$$c = \frac{c_1 \ln(r_2/r) + c_2 \ln(r/r_1)}{\ln(r_2/r_1)} \tag{113}$$

The flux can still be obtained from Fick's first law:

$$j = -D \frac{\partial c}{\partial r} = \frac{D(c_1 - c_2)}{r \ln(r_2/r_1)} \tag{114}$$

Note that, unlike steady flux in a membrane, steady flux in a hollow cylinder is a function of radius. However, the total flux per unit length is constant:

$$J = 2\pi r j = \frac{2\pi D(c_1 - c_2)}{\ln(r_2/r_1)} \tag{115}$$

VI. DISSOLUTION OF A SPHERE

Diffusion in a sphere may be more common than that in a cylinder in the pharmaceutical sciences. The example we may think of is the dissolution of a spherical particle. Since convection is normally involved in solute particle dissolution in reality, the dissolution rate estimated by considering only diffusion often underestimates experimental values. Nevertheless, we use it as an example to illustrate the solution of the differential equations describing diffusion in the spherical coordinate system [1].

A sphere is assumed to be a poorly soluble solute particle and therefore to have a constant radius r_0. However, the solid solute quickly dissolves, so the concentration on the surface of the sphere is equal to its solubility. Also, we assume we have a large volume of dissolution medium so that the bulk concentration is very low compared to the solubility (sink condition). The diffusion equation for a constant diffusion coefficient in a spherical coordinate system is

$$\frac{\partial c}{\partial t} = D\left(\frac{\partial^2 c}{\partial r^2} + \frac{2}{r}\frac{\partial c}{\partial r}\right) \tag{116}$$

At steady state, we have

$$D\left(\frac{\partial^2 c}{\partial r^2} + \frac{2}{r}\frac{\partial c}{\partial r}\right) = 0 \tag{117}$$

with boundary conditions

$$r = r_0, \quad c = c_s \tag{118}$$
$$r = r_0, \quad c = 0 \tag{119}$$

where c_s is the solubility of solute. A general solution of Eq. (117) is

$$c = B + \frac{A}{r} \tag{120}$$

Using the boundary conditions to determine the integration constants A and B, we have the concentration profile

$$c = \frac{r_0}{r}c_s \tag{121}$$

The steady dissolution flux is

$$j|_{r=r_0} = -D\frac{\partial c}{\partial r}\bigg|_{r=r_0} = \frac{D}{r_0}c_s \tag{122}$$

The total amount of solute dissolved from time 0 to time t is

$$M = \int_0^t \frac{AD}{r_0}c_s\,dt = \frac{AD}{r_0}c_s t \tag{123}$$

where A is the surface area of a spherical particle and equal to $4\pi r^2$. Equation (123) shows a linear relationship between the amount dissolved and the solubility. It must be pointed out, however, that the size of the spherical solute particle is assumed to remain constant.

Now we want to determine the time to reach the steady state. To answer this question, we return to the mass balance equation, Eq. (116), subject to the following initial and boundary conditions:

$$t = 0, \qquad r > r_0, \qquad c = 0 \tag{124}$$

$$t > 0, \qquad r = r_0, \qquad c = c_s \tag{125}$$

$$t > 0, \qquad r = \infty, \qquad c = 0 \tag{126}$$

Let

$$u = cr \tag{127}$$

Equation (116) then becomes

$$\frac{\partial u}{\partial t} = D\frac{\partial^2 u}{\partial r^2} \tag{128}$$

with the modified initial and boundary conditions

$$t = 0, \qquad r > r_0, \qquad u = 0 \tag{129}$$

$$t > 0, \qquad r = r_0, \qquad u = r_0 c_s \tag{130}$$

$$t > 0, \qquad r = \infty, \qquad u = 0 \tag{131}$$

Equation (128) with the initial and boundary conditions of Eqs. (129)–(131) is very similar to the model equations for diffusion in a semi-infinite membrane [Eqs. (64)–(67)]. The concentration profile is

$$c = \frac{r_0 c_s}{r}\operatorname{erfc}\frac{r - r_0}{\sqrt{4Dt}} = \frac{r_0 c_s}{r}\left(1 - \operatorname{erf}\frac{r - r_0}{\sqrt{4Dt}}\right) \tag{132}$$

The flux is then

$$j = \frac{Dc_s}{r_0}\left(1 + \frac{r_0}{\sqrt{\pi Dt}}\right) \tag{133}$$

A comparison of Eq. (133) with Eq. (123) suggests that the term $r_0/\sqrt{\pi Dt}$ is due to the effect of the unsteady state. It can be used to estimate the time to achieve steady state. For example, if $r_0 = 0.01$ cm and $D = 5 \times 10^{-6}$ cm²/sec, then it will take about 43 min for the term $r_0/\sqrt{\pi Dt}$ to become equal to 0.05, significantly less than 1.

VII. INTRINSIC DISSOLUTION

The intrinsic dissolution rate is the rate of mass transfer from the solid phase to the liquid phase. Information on the intrinsic dissolution rate is important in early drug product development. It has been suggested that drugs with intrinsic dissolution rates of less than 0.1 mg/(min · cm²) will have dissolution rate–limited absorption, while drugs with intrinsic dissolution rates greater than 0.1 mg/ (min · cm²) are unlikely to have dissolution rate problems.

The intrinsic dissolution rate is usually evaluated by using a rotated disk method (Fig. 7). The pure powdered solute is compressed in a die under high pressure, in the absence of any excipients. The resulting nondisintegrating disk is then transferred to a dissolution cell which has sufficient volume to maintain sink conditions. The die is rotated at a certain speed, and the rate of drug dissolution is then measured.

We solve this problem by using a generalized mass balance equation in cylindrical coordinates, centered on the disk:

$$\frac{\partial c}{\partial t} + v_r \frac{\partial c}{\partial r} + \frac{v_\theta}{r}\frac{\partial c}{\partial \theta} + v_z \frac{\partial c}{\partial z}$$
$$= \frac{D}{r}\left[\frac{\partial}{\partial r}\left(r\frac{\partial c}{\partial r}\right) + \frac{\partial}{\partial \theta}\left(\frac{1}{r}\frac{\partial c}{\partial \theta}\right) + \frac{\partial}{\partial z}\left(r\frac{\partial c}{\partial z}\right)\right] + R \tag{134}$$

At steady state, the concentration does not change with time. It is recognized that the concentration profile is angularly symmetrical; the concentration, therefore, does not change with θ. We also assume that the disk is infinitely wide so that the concentration is a function of z only. This assumption can be justified since the diameter of the solute loaded is usually much smaller than that of the disk. Also, no production of drug occurs during dissolution testing. Consequently, Eq. (134) can be reduced to

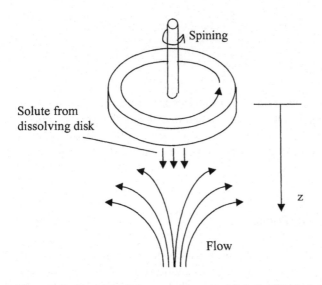

Figure 7 Rotating disk to evaluate the intrinsic dissolution rate of compounds. The amount of drug dissolving per unit area is the same everywhere on the disk surface. This simplification makes the disk a powerful experimental tool in drug discovery and development.

$$v_z \frac{dc}{dz} = D \frac{d^2c}{dz^2} \tag{135}$$

subject to the boundary conditions

$$z = 0, \quad c = c_s \tag{136}$$

$$z = \infty, \quad c = 0 \tag{137}$$

where c_s is the saturated concentration or solubility of the solute. Note that v_z is not a constant; it is a function of z. Therefore, we have to find v_z before we can integrate Eq. (135) to have the concentration profile. The finding of v_z is a problem in fluid mechanics detailed by Levich [7]. The expression for v_z is

$$v_z = -\frac{0.51 \, \omega^{3/2} z^2}{v^{1/2}} \tag{138}$$

where ω is the angular velocity of the rotating disk and v is the kinematic viscosity of the dissolution medium. Combining Eqs. (135) and (138), we have

$$-\frac{0.51 \, \omega^{3/2} z^2}{D v^{1/2}} \frac{dc}{dz} = \frac{d^2c}{dz^2} \tag{139}$$

Integrating Eq. (139) once gives

$$\frac{dc}{dz} = A \exp\left[-\frac{\omega^{3/2} z^3}{5.88\, D v^{1/2}}\right] \tag{140}$$

and a second integration results in

$$c = A \int_0^z \exp\left[-\frac{\omega^{3/2} z^3}{5.88\, D v^{1/2}}\right] dz + B \tag{141}$$

where A and B are integration constants. Subsitution of the boundary conditions, Eqs. (136) and (137), gives

$$\frac{c}{c_s} = 1 - \frac{\displaystyle\int_0^z \exp\left[-\frac{\omega^{3/2} z^3}{5.88\, D v^{1/2}}\right] dz}{\displaystyle\int_0^\infty \exp\left[-\frac{\omega^{3/2} z^3}{5.88\, D v^{1/2}}\right] dz} \tag{142}$$

where

$$\int_0^\infty \exp\left[-\frac{\omega^{3/2} z^3}{5.88\, D v^{1/2}}\right] dz = \frac{1.81\, D^{1/3} v^{1/6}}{\omega^{1/2}}$$

$$\times \int_0^\infty \exp\left[-\left(\frac{\omega^{1/2} z}{1.81\, D^{1/3} v^{1/6}}\right)^3\right] d\left(\frac{\omega^{1/2} z}{1.81\, D^{1/3} v^{1/6}}\right)$$

$$= \frac{1.61\, D^{1/3} v^{1/6}}{\omega^{1/2}} \tag{143}$$

Therefore, Eq. (142) becomes

$$\frac{c}{c_s} = 1 - \frac{0.62\, \omega^{1/2}}{D^{1/3} v^{1/6}} \int_0^z \exp\left[-\frac{\omega^{3/2} z^3}{5.88\, D v^{1/2}}\right] dz \tag{144}$$

Equation (144) is the concentration profile as a function of the distance from the disk surface. The diffusion flux (intrinsic dissolution rate) is

$$j = -D \frac{\partial c}{\partial z}\bigg|_{z=0} = 0.62 \left(\frac{D^{2/3} \omega^{1/2}}{v^{1/6}}\right) c_s \tag{145}$$

Therefore, the final equation is relatively simple despite the fact that the derivation is complex. Equation (145) shows that the intrinsic dissolution rate depends on the diffusion coefficient and solubility of the drug, disk rotational speed, and the viscosity of the dissolution medium. The amount of drug dissolving per unit area is the same everywhere on the disk's surface. This makes the disk a powerful experimental tool in drug discovery and development.

VIII. SUMMARY

This chapter has introduced Fick's law for dilute solutions and has shown how this law can be combined with mass balances to calculate concentrations and fluxes. The mass balances are made on a small rectangular box. When the box becomes very small, the mass balances become the differential equations used to solve various pharmaceutical transport problems. Thus, this chapter discussed many mathematical equations. Although these mathematical equations are tedious, they are essential to each specific application of mass transfer in pharmaceutical systems to be discussed in the rest of this book.

Membrane diffusion illustrates the uses of Fick's first and second laws. We discussed steady diffusion across a film, a membrane with and without aqueous diffusion layers, and the skin. We also discussed the unsteady diffusion across a membrane with and without reaction. The solutions to these diffusion problems should be useful in practical situations encountered in pharmaceutical sciences, such as the development of membrane-based controlled-release dosage forms, selection of packaging materials, and experimental evaluation of absorption potential of new compounds. Diffusion in a cylinder and dissolution of a sphere show the solutions of the differential equations describing diffusion in cylindrical and spherical systems. Convection was discussed in the section on intrinsic dissolution. Thus, this chapter covered fundamental mass transfer equations and their applications in practical situations.

REFERENCES

1. EL Cussler. Diffusion. New York: Cambridge University Press, 1984, pp 18, 24, 32, 33, 42–44, 55, 81, 82.
2. J Crank. The Mathematics of Diffusion. 2nd ed. Oxford, UK: Clarendon Press, 1975.
3. W Jost. Diffusion in Solids, Liquids, Gases. New York: Academic Press, 1960, pp 8, 42–45.
4. GL Flynn, SH Yalkowsky, TJ Roseman. Mass transfer phenomena and models: Theoretical concepts. J Pharm Sci 63:479–510 (1974).

5. GE Amidon, WI Higuchi, NSF Ho. Theoretical and experimental studies of transport of micelle-solubilized solutes. J Pharm Sci 71:77–84 (1982).
6. GL Flynn. Topical drug absorption and topical pharmaceutical systems. In: GB Banker, CT Rhodes, eds. Modern Pharmaceutics. New York: Marcel Dekker, 1990, pp 263–326.
7. VG Levich. Physicochemical Hydrodynamics. Englewood Cliffs, NJ: Prentice-Hall, 1962, pp 39–72.

3

Pharmacokinetics: Model Structure and Transport Systems

James M. Gallo
Fox Chase Cancer Center, Philadelphia, Pennsylvania

I. PROPERTIES OF PHARMACOKINETIC MODELS

Drug transport in pharmacokinetics is based on material exchanges between and irreversible losses from compartments. The compartment serves as a homogeneous anatomic and physiological region, representing one or more organs, with time and spatial dependencies in drug concentrations possible. Normally, drug concentrations within each compartment are assumed to be homogeneous, and only time dependencies in concentration are sought. The mathematical description and inherent assumptions of drug transport within each compartment determine the type of pharmacokinetic model and analysis.

All compartmental representations of in vivo drug transport are a simplification of the real system. In reality, drug transport in any organ involves transport across different types of cell membranes under different physiological conditions of blood flow, protein and receptor concentrations, and pH. The compartment model lumps or averages potentially diverse regions into one or more homogeneous units that are uniform with respect to drug disposition. The extent of lumping and representation of model parameters, in effect, determines the type of drug transport model.

A. Compartmental Models

Pharmacokinetics emerged as a discipline in the 1960s with its foundation in compartmental modeling, although earlier origins of pharmacokinetics can be traced [1]. Mammillary compartmental models provided the framework for pharmacokinetics, with catenary compartmental models serving as an ancillary approach. The mammillary compartmental model requires all peripheral compartments to be reversibly linked to a central compartment, whereas a catenary compartmental model may have compartments arranged in series (see Fig. 1) [2]. Other than blood being identified in the central compartment, the compartments represent one or more tissues that have similar blood flow and affinity for the drug. The organs represented by any one compartment behave kinetically the same with parallel drug concentration–time profiles. There is no need or purpose to identify which organs comprise a particular compartment. Thus, the compartments form a system of black boxes representing all or a portion of the body with material (drug) transferred between and eliminated from the compartments. Compartmental models may represent both linear and nonlinear transport processes, although the former case is the most common. Michaelis–Menten elimination is the typical nonlinear process observed.

The compartmental modeling approach, after receiving considerable attention in the 1970s, has receded in popularity due to the development of other data analysis techniques, most notably noncompartmental and system analysis techniques [3–5]. Noncompartmental methods, in offering considerable flexibility in data analysis, have supplanted compartmental modeling as a tool to initially assess a drug's pharmacokinetic properties. Although compartmental modeling

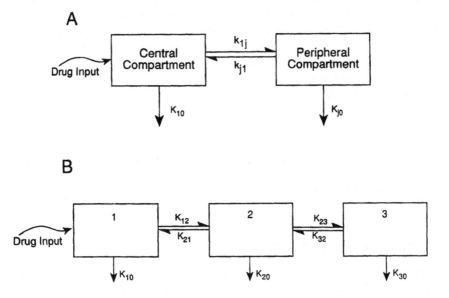

Figure 1 Mammillary (A) and catenary (B) compartmental models.

may not be used as the first step in characterizing a drug's pharmacokinetics, it is the basis of population modeling approaches that are becoming increasingly important and integral to the drug development process. Each technique (i.e., compartmental modeling, noncompartmental and system analysis) has associated advantages and disadvantages, although only compartmental models are based on differential mass balance equations and readily identified as a means to model drug transport.

B. Physiologically Based Pharmacokinetic Models

The origins of physiologically based pharmacokinetic (PB-PK) models may be traced to Teorell [6,7], Bellman et al. [8], Bischoff and Brown [9], and Dedrick and Bischoff [10]. The combined investigations of these individuals established the idea of characterizing drug distribution to specific organs in terms of relevant anatomic and physiological variables. In contrast to classical compartmental modeling, tissue compartments represent specific anatomic volumes connected by the blood circulation. Drug uptake into organs is a function of both thermodynamic (i.e., tissue-to-blood partition coefficients, drug protein binding) and membrane transport (i.e., mass transfer coefficient) properties, rather than simple first-order transfer rate constants like those used in compartmental models. The PB-PK model parameters are cast into differential mass balance equations that are solved numerically to provide predicted drug concentrations for each organ and compartment.

Individual organs usually have either a blood flow–limited or membrane-limited model structure. A blood flow–limited tissue compartment consists of a single compartment and assumes that blood flow, as opposed to membrane transport, is the rate-limiting step in drug uptake into tissue. Thus, the tissue is a well-mixed homogeneous medium identified by a single tissue concentration. A membrane-limited organ structure normally comprises two homogeneous subcompartments, either an extracellular–intracellular configuration or a vascular–extravascular configuration. In either case, drug transport is rate-limited by a membrane rather than by blood flow to the tissue. A vascular–extravascular arrangement is useful for representing blood-brain barrier transport, whereas extracellular–intracellular arrangements are typical for tissues other than brain. In an extracellular–intracellular organ representation, vascular and interstitial spaces form the extracellular compartment, and thus it is assumed that drug is instantaneously equilibrated between the vascular and interstitial spaces. Vascular–extravascular arrangements lump intracellular and interstitial spaces into a single homogeneous compartment assuming there is no concentration gradient between these milieus. Three subcompartment organ structures, depicting vascular, interstitial, and intracellular compartments, have also been presented in the PB-PK model format [11]. The level of organ complexity will depend on the number of discrete tissue spaces that can be identified along with the associated drug transport parameters. Division of an intracellular compartment into subcompartments representing different cell types, one possibly serving as a receptor binding site for the drug, is possible in the PB-PK modeling approach. Alternatively, the intracellular compartment could describe free and receptor-bound drug concentrations. Figure 2 illustrates some potential organ representations in PB-PK models.

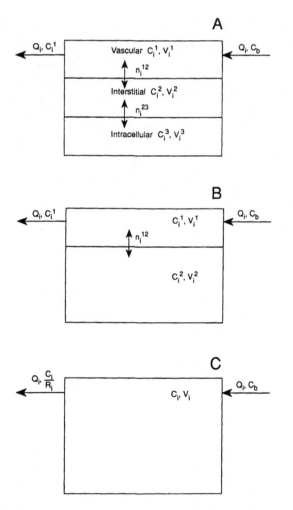

Figure 2 Individual organ representations for a three-subcompartment (A), two-subcompartment (B), or typical membrane-linked and blood flow–limited (C) physiologically based pharmacokinetic model. See text for definition of symbols.

Representation of the whole body by individual organs connected anatomically through the blood circulation is referred to as a "global" PB-PK model. A "hybrid" PB-PK model focuses on a single organ or small group of organs. The organ models are analogous to those found in a global model, yet the blood concentration of drug entering the compartment is described by an exponential equation or forcing function. The forcing function is attained by fitting a classical compartmental model or polyexponential equation to observed plasma drug concentration–time data, justifying the descriptor "hybrid." Advantages of hybrid models are that fewer model parameters have to be estimated and comprehensive tissue drug concentration–time data that are somewhat requisite in the global approach are unnecessary, yet the organ(s) of interest maintain a physiological representation.

Objectives for development of a PB-PK model can be varied. Most often the endpoint of model development is subjective agreement between observed and predicted concentrations, although more rigorous model validation methods are available (see Section III.B). Acceptance of the model implicitly characterizes drug dispositions as being linear or nonlinear with respect to clearance, membrane transport, and protein binding. Although this endpoint is valuable, such models also provide a springboard to enable PB-PK models to reach their full potential as a predictive tool. Model predictions can be conducted under numerous conditions with a driving force to answer "what if" questions (e.g., What are target tissue drug concentrations if blood flow or protein binding is altered?). Some applications of PB-PK models are outlined below.

The advent of targeted drug delivery systems promotes the use of PB-PK models through their ability to characterize drug disposition at the tissue or cell level. Since tissue uptake and cellular interactions of the drug delivery system will be crucial to successful drug targeting, a PB-PK model capable of characterizing carrier and drug transport can furnish feedback on the design of the delivery system and can be used as a quantitative tool to evaluate the delivery system.

Physiologically based pharmacokinetic models provide a format to analyze relationships between model parameters and physicochemical properties for a series of drug analogues. Quantitative structure–pharmacokinetic relationships based on PB-PK model parameters have been pursued [12,13] and may ultimately prove useful in the drug development process. In this venue, such relationships, through predictions of tissue distribution, could expedite drug design and discovery.

A hallmark of PB-PK models is the ability to scale up animal-based models to humans, thus allowing tissue drug concentrations to be predicted in the absence of data that are difficult or impossible to collect. Initial efforts to apply interspecies extrapolations to anticancer drugs have been greatly extended to chemical risk assessment based on PB-PK models [14]. Empirical allometric equations based on animal body weight have been the mainstay to scale organ weights and blood flow as well as macro pharmacokinetic variables such as clearance and volume of distribution [15–17]. Few have endeavored to study scale-up relationships for drug transport and thermodynamic parameters. Without these relationships, prediction of human tissue drug concentrations will be unsupported. The lack of substantive investigations on the scale-up of drug-specific parameters may be due to the preception, possibly unfounded, that large quantities of data are required to develop scale-up relationships. Utilization of in vitro systems from different species to measure metabolic rates, partition coefficients, and membrane transport parameters could lead to scale-up relationships and decrease the number of animal experiments. Establishment of optimized in vivo study protocols across different animal species may also limit the number of animal studies necessary to develop allometric relationships. Currently, PB-PK models should be viewed as a powerful means to represent drug transport based on mass transport principles, yet further demonstration of applicability to predicting human tissue pharmacokinetics would be a certain enhancement to their use.

C. Distributed Parameter Models

Distributed parameter models allow spatial dependencies in drug distribution to be examined as well as time dependencies [18,19]. Accordingly, the models express diffusive and convective drug transport in terms of partial differential equations. A motivating force in developing a distributed parameter model is nonuniform blood flow and variable drug diffusivities, situations that occur in solid tumors. Distributed parameter models, in contrast to PB-PK models, are more mathematically explicit in representing drug and fluid transport parameters and utilize diffusion coefficients and blood velocity profiles. Distributed parameter models are normally based on drug distribution in a single organ with a time-dependent input function, possibly polyexponential, based on blood concentrations of drug. A distributed parameter model allows one to assess factors influencing regional drug distribution and to analyze how modulation of the factors change drug concentrations. Regional differences in drug concentrations, such as those attained in a necrotic core of a tumor versus that at the tumor periphery, would serve as a basis to pursue development of a distributed parameter model.

II. MATHEMATICAL TREATMENT OF PHARMACOKINETIC MODELS

A. Compartmental Models

Extensive literature is available on general mathematical treatments of compartmental models [2]. The compartmental system based on a set of differential equations may be solved by Laplace transform or integral calculus techniques. By far the most pertinent compartmental model in pharmacokinetics is the linear, time-invariant mammillary model that is described by linear first-order differential equations. Figure 1 illustrates such a model, and the corresponding differential equations are

$$\frac{dX_1}{dt} = \sum_{j=2}^{n} k_{j1}X_j - \sum_{j=2}^{n} k_{1j}X_1 - k_{10}X_1 + U_1(t) \tag{1}$$

and

$$\frac{dX_j}{dt} = k_{1j}X_1 - (k_{j1} + k_{j0})X_j \tag{2}$$

where

X_1, X_j = amount of drug in compartment 1 and j, respectively
k_{1j}, k_{j1} = first-order transfer rate constants from compartment 1 to j and from compartment j to 1, respectively
k_{10}, k_{j0} = first-order exit rate constants from compartment 1 and j, respectively
$U_1(t)$ = input function into compartment 1
n = number of compartments

The first-order transfer and exit rate constants can be replaced by nonlinear terms dependent on the amount or concentration of drug in a particular compartment. For instance, saturable metabolism of drug in compartment 1 (the central compartment) would result in the Michaelis–Menten equation

$$k_{10} = \frac{V_m}{k_m + X_1} \tag{3}$$

where V_m equals the maximum metabolic rate and k_m, the Michaelis constant, equals the amount of drug at one-half V_m. The input function, $U_1(t)$, in Eq. (1) characterizes the rate of drug input into the central compartment and can describe both vascular and extravascular administration regimens. Solution of the system of differential equations describing the model provides an analytical solution for the amount of drug in any compartment. Division of the solution to Eq. (1) by the volume of compartment 1 results in a function for the blood or plasma drug concentration, a measurement routinely available.

It has been shown that linear mammillary compartment models can readily be represented by products of input and disposition functions in the Laplace domain [20]. Solutions for the drug concentration or amount in any compartment are obtained by taking the inverse of the Laplace function. This approach avoids the use of differential equations and their potentially tedious solution. The Laplace transform of the amount of drug in compartment 1, $a_{s,1}$, and analogous to Figure 1 and Eq. (1) is

$$a_{s,1} = \text{in}_s \, d_{s,1} \tag{4}$$

where in_s = Laplace transform of input, the function describing rate of drug input into the blood, and $d_{s,1}$ = Laplace transform of the disposition function for compartment 1.

The $d_{s,1}$ function characterizes the dynamics of drug transport once the drug enters the blood circulation and is given by

$$d_{s,1} = \frac{\displaystyle\prod_{i=2}^{n}(s + E_i)}{\displaystyle\prod_{i=1}^{n}(s + E_i) - \sum_{j=2}^{n}\left[k_{1j}k_{j1}\prod_{\substack{m=2;\\m\neq j}}^{n}(s + E_m)\right]} \tag{5}$$

where

$\qquad \prod$ = continued product where any term is defined as equal to 1 when the index takes a forbidden value, i.e., $i = 1$ in the numerator or $m = j$ in the denominator

$\qquad \sum$ = summation where any term is defined as equal to zero when the index j takes a forbidden value, i.e., $j = 1$

$\quad k_{1j}, k_{j1}$ = first-order intercompartmental transfer rate constants

$\quad E_i, E_m$ = sum of exit rate constants from compartments i or m

$\qquad n$ = number of driving force compartments in the disposition model, i.e., compartments having exit rate constants

The disposition function can be expressed more conveniently as

$$d_{s,1} = \frac{\displaystyle\prod_{i=2}^{n}(s + E_i)}{\displaystyle\prod_{i=1}^{n}(s + \lambda_i)} \tag{6}$$

where λ_i = disposition rate constant for the ith phase.

Disposition rate constants are functions of the intercompartmental transfer rate constants and exit rate constants and can be expressed as such by equating the denominators of Eqs. (5) and (6). Common input functions, in_s, are as follows.

For intravascular bolus administration:

$in_s = Dose$

For zero-order infusions:

$$in_s = \frac{k_0(1 - e^{-Ts})}{s} \tag{7a}$$

For first-order absorption:

$$in_s = \frac{k_a F \text{ Dose}}{s + k_a} \tag{7b}$$

where

k_0 = infusion rate, amount/time
T = duration of zero-order infusion started at time 0
k_a = first-order absorption rate constant
F = absolute bioavailability
s = Laplace operator

Multiplication of in_s [Eq. (7)] and $d_{s,1}$ [Eq. (6)] gives $a_{s,1}$ in the Laplace domain. Solution of $a_{s,1}$, yielding A_1, the amount of drug in compartment 1, is obtained by taking the inverse of the Laplace transform by use of Heaviside's expansion theorems [21]. Expressing $a_{s,1}$ as the quotient of two polynomials, $a_{s,1} = p(s)/q(s)$, if $q(s)$ of greater degree than $p(s)$ and contains the unrepeated linear factor $(s + \lambda_i)$, the inverse Laplace transform, L^{-1}, is

$$L^{-1}\left[\frac{p(s)}{q(s)}\right] = \sum_{i=1}^{n} \frac{p(\lambda_i)}{q_i(\lambda_i)} e^{-\lambda_i t} \tag{8}$$

where the λ_i's are the roots of the polynomial $q(s)$. $q_i(\lambda_i)$ is the value of the denominator when λ_i is substituted for all s terms except for the term originally containing λ_i. It can be seen that the time domain function obtained in Eq. (8) for the amount of drug in compartment 1 can be written in terms of the plasma concentration, C, by dividing by the volume of compartment 1, V_1. The resulting polyexponential equation is conveniently expressed as

$$C = \sum_{i=1}^{n} A_i e^{-\lambda_i t} \tag{9}$$

Equation (9) is most often associated with intravenous bolus administration, although proper definition of A_i allows this equation to apply to extravascular and intravenous infusion administrations.

The input–disposition function approach can also be used to solve for the amount of drug in any peripheral compartment, A_{pj}. The Laplace transform of the amount of drug in a peripheral compartment, $a_{s,j}$, is

$$a_{s,j} = \text{in}_{s,j}\, a_{s,1}$$

where

$$\text{in}_{s,j} = \frac{k_{1j}}{s + E_j} \tag{10}$$

(E_j = sum of exit rate constants from compartment j) and

$$a_{s,1} = \text{in}_s\, d_{s,1}$$

Taking the inverse Laplace transform will result in the polyexponential equation

$$A_{pj} = \sum_{i=1}^{n} \frac{k_{1j}}{E_j - \lambda_i} A_i e^{-\lambda_i t} \tag{11}$$

B. Physiologically Based Pharmacokinetic Models

Figure 2A illustrates a noneliminating tissue compartment, i, divided into three anatomically relevant subcompartments, each homogeneous with respect to drug concentration. The corresponding differential mass balance equations are

$$V_i^1 \frac{dC_i^1}{dt} = Q_i(C_b - C_i^1) - n_i^{12} \tag{12}$$

$$V_i^2 \frac{dC_i^2}{dt} = n_i^{12} - n_i^{23} \tag{13}$$

$$V_i^3 \frac{dC_i^3}{dt} = n_i^{23} \tag{14}$$

where

C_i^1 = concentration (amount/volume) in compartment 1 of tissue i
C_i^2 = concentration (amount/volume) in compartment 2 of tissue i
C_i^3 = concentration (amount/volume) in compartment 3 of tissue i
C_b = blood concentration entering tissue i
n_i^{12} = net mass flux from compartment 1 to 2, amount/time
n_i^{23} = net mass flux from compartment 2 to 3, amount/time
V_i^1 = volume of compartment 1 in tissue i
V_i^2 = volume of compartment 2 in tissue i
V_i^3 = volume of compartment 3 in tissue i

Compartment 1 represents the vascular space of the tissue and can be identified as either a blood, plasma, or serum space. Compartment 2 is the interstitial space, and compartment 3 the intracellular tissue volume. Assuming first-order or linear membrane transport for the drug, the flux terms can be expanded as

$$n_i^{12} = h_i^{12}(C_i^1 - C_i^2/R_i^{12}) \tag{15}$$

$$n_i^{23} = h_i^{23}(C_i^2/R_i^{23} - C_i^3/R_i^{23}) \tag{16}$$

where

h_i^{12} = mass transfer coefficient (volume/time) characterizing drug transport from compartment 1 to compartment 2

h_i^{23} = mass transfer coefficient (volume/time) characterizing drug transport from compartment 2 to compartment 3

$R_i^{12} = f_u^1/f_u^2$

$R_i^{23} = f_u^3/f_u^2$

f_u = the unbound fraction of drug in compartment 1, 2, or 3

The mass transfer coefficients may also be expressed in units of time^{-1} by multiplying by the appropriate compartmental volume term. Irreversible drug elimination from the tissue requires the addition of an expression to the differential equation that represents the subcompartment in which elimination occurs. For instance, hepatic drug elimination would be described by a linear or nonlinear expression added to the intracellular liver compartment mass balance equation since this compartment represents the hepatocytes. Formal elimination terms are given below for the simplified tissue models.

The three-compartment tissue model is ordinarily simplified by lumping all three subcompartments, lumping subcompartments 1 and 2, or lumping subcompartments 2 and 3. These simplifications result in the blood flow–limited (i.e., lumping all three subcompartments) and the membrane-limited (i.e., lumping any two subcompartments) tissue models. Differential mass balance equations for a noneliminating membrane-limited compartment are

$$V_i^1 \frac{dC_i^1}{dt} = Q_i(C_b - C_i^1) - n_i^{12} \tag{17}$$

$$V_i^2 \frac{dC_i^2}{dt} = n_i^{12} \tag{18}$$

where all terms have the same basic definitions as before. Combining the vascular and interstitial subcompartments defines subcompartment 1 as the extracellular compartment and subcompartment 2 as the intracellular compartment. Lumping interstitial and intracellular subcompartments, as might be done for a brain compartment, defines compartment 1 as the vascular space and compartment 2 as the extravascular compartment. Under conditions of linear membrane transport,

$$n_i^{12} = h_i^{12}\left(C_i^1 - \frac{C_i^2}{R_i}\right) \tag{19}$$

Nonlinear or saturable membrane transport is represented as

$$n_i^{12} = \frac{a_i C_i^1}{b_i + C_i^1} - \frac{a_i(C_i^2/R_i)}{b_i + C_i^2/R_i} \tag{20}$$

where a_i = maximum transport rate, amount/time and b_i = drug concentration at $a_i/2$, amount/volume.

Membrane transport of drugs by both linear and nonlinear mechanisms is represented by addition of Eqs. (19) and (20).

Drug elimination from a membrane-limited tissue compartment requires subtraction of the rate of elimination, q_i, from the appropriate mass balance equation, typically from subcompartment 2.

For linear or first-order drug elimination,

$$q_i = k_i V_i^2 C_i^2 \tag{21}$$

and for nonlinear elimination,

$$q_i = \frac{V_m(C_i^2/R_i)}{k_m + C_i^2/R_i} \tag{22}$$

where

k_i = first-order elimination rate constant, time^{-1}
V_m = maximum elimination rate, amount/time
k_m = Michaelis constant, equal to the concentration at $V_m/2$

Lumping compartments 1, 2, and 3 into a single homogeneous tissue compartment implies the blood flow–limited model. The tissue mass balance equation for a noneliminating organ is

$$V_i \frac{dC_i}{dt} = Q_i\left(C_b - \frac{C_i}{R_i}\right) \tag{23}$$

where $R_i = C_i/C_0$, with C_0 being the effluent venous blood drug concentration. Drug elimination terms are now

$$q_i = k_i V_i C_i \tag{24}$$

$$q_i = \frac{V_m(C_i/R_i)}{k_m + C_i/R_i} \tag{25}$$

for linear and nonlinear elimination, respectively. Equation (24) is frequently represented as

$$q_i = \frac{CL_i^I C_i}{R_i} \tag{26}$$

where CL_i^I = total intrinsic organ clearance. Equation (26) is based on the following relationships, where CL_i equals organ clearance:

$$CL_i = k_i V_i \tag{27}$$

$$CL_i^I = k_i V_i R_i \tag{28}$$

Thus,

$$q_i = k_i V_i R_i C_0 \tag{29}$$

$$q_i = CL_i^I C_0 \tag{30}$$

$$q_i = \frac{CL_i^I C_i}{R_i} \tag{31}$$

A complete or global tissue distribution model consists of individual tissue compartments connected by the blood circulation. In any global model, individual tissues may be blood flow–limited, membrane-limited, or more complicated structures. The venous and arterial blood circulations can be connected in a number of ways depending on whether separate venous and arterial blood compartments are used or whether right and left heart compartments are separated. The two most common methods are illustrated in Figure 3 for blood flow–limited tissue compartments. The associated mass balance equations for Figure 3A are

$$V_l \frac{dC_l}{dt} = \left[\sum_{i=1}^{n} \frac{Q_i C_i}{R_i} \right] - \frac{Q_l C_l}{R_l} \tag{32}$$

$$V_b \frac{dC_b}{dt} = \frac{Q_l C_l}{R_l} - Q_b C_b \tag{33}$$

where l and b subscripts refer to lung and arterial blood compartments, and n equals the number of compartments providing venous return. For Figure 3B the equations are

$$V_v \frac{dC_v}{dt} = \left[\sum_{i=1}^{n} \frac{Q_i C_i}{R_i} \right] - Q_v C_v \tag{34}$$

$$V_l \frac{dC_l}{dt} = Q_v \left(C_v - \frac{C_l}{R_l} \right) \tag{35}$$

$$V_b \frac{dC_b}{dt} = Q_v \frac{C_l}{R_l} - Q_b C_b \tag{36}$$

A

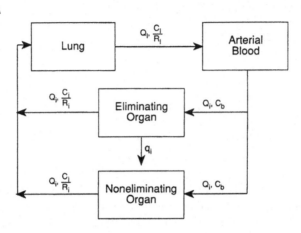

B

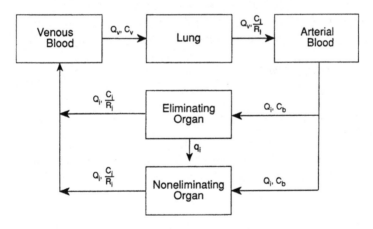

Figure 3 Possible blood circulation connections in a physiologically based pharmacokinetic model. (A) Venous return incorporated into lung mass balance equation; (B) separate venous blood compartment. See text for definition of symbols.

where subscript v indicates the venous blood compartment. Whatever representation is used for the venous–arterial system, balance of blood flow has to be maintained; thus, above,

$$\sum_{i=1}^{n} Q_i = Q_l \quad \text{and} \quad \sum_{i=1}^{n} Q_i = Q_v = Q_b \tag{37}$$

for Figures 3A and 3B, respectively.

Membrane transport of drugs by passive diffusion, although a function of free or unbound drug concentrations, is most often expressed in terms of total drug concentration, as presented in Eqs. (19) and (20). Casting these equations in terms of total drug concentrations requires reparameterization of the flux expressions using unbound drug concentrations [22]. For example, Eq. (19) is arrived at by

$$n_i^{12f} = h_i^f(C_i^{1f} - C_i^{2f}) \tag{38}$$
$$C_i^{1f} = C_i^1/R_i^1 \tag{39}$$
$$C_i^{2f} = C_i^2/R_i^2 \tag{40}$$

where superscript f refers to the free drug, and $R_i^1 = 1/f_u^1$ and $R_i^2 = 1/f_u^2$. Substituting Eqs. (39) and (40) into (38) gives

$$n_i^{12f} = h_i^f(C_i^1/R_i^1 - C_i^2/R_i^2) \tag{41}$$

Setting $h_i = h_i^f/R_i^1$ and $R_i = R_i^2/R_i^1$ permits Eq. (41) to be written as

$$n_i^{12} = h_i(C_i^1 - C_i^2/R_i^2) \tag{42}$$

which is equivalent to Eq. (19).

Although drug fluxes and tissue concentrations are, for practical purposes, presented in terms of total drug concentrations, division into unbound and protein-bound concentrations is possible. This approach may be beneficial for a drug that undergoes nonlinear protein binding over the concentration range of interest or when pharmacodynamic issues are of concern. Problems encountered in representing free and bound drug mass balance equations have been discussed and center on the use of an effective protein fraction capable of binding drug in tissues [23]. Based on Figure 4, for a membrane-limited noneliminating tissue the following equations can be written:

$$V_i^1 f_i^{1f} \frac{dC_i^{1f}}{dt} + V_i^1 f_i^{1b} \frac{dC_i^{1b}}{dt} = Q_i[f_i^{1f}C_b^f + f_i^{1b}C_b^b - f_i^{1f}C_i^{1f} - f_i^{1b}C_i^{1b}]$$
$$- h_i^f(C_i^{1f} - C_i^{2f}) \tag{43}$$

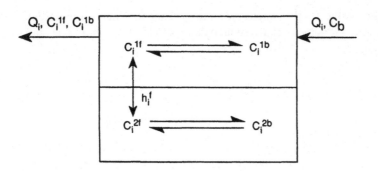

Figure 4 Representation of protein-bound and unbound drug in a noneliminating membrane-limited organ structure. See text for definition of symbols.

$$V_i^2 f_i^{2f} \frac{dC_i^{2f}}{dt} + V_i^2 f_i^{2b} \frac{dC_i^{2b}}{dt} = h_i^f (C_i^{1f} - C_i^{2f}) \tag{44}$$

where

f_i^{1f}, f_i^{2f} = water fraction in subcompartments 1 and 2 in tissue i, dimensionless

f_i^{1b}, f_i^{2b} = effective drug protein-binding fraction in subcompartments 1 and 2 in tissue i, dimensionless

C_i^{1f}, C_i^{2f} = free or unbound drug concentration in subcompartments 1 and 2 in tissue i

C_i^{1b}, C_i^{2b} = bound drug concentration in subcompartments 1 and 2 in tissue i

C_b^f, C_b^b = free and bound drug concentrations in blood

h_i^f = mass transfer coefficient for free drug, volume/time

Simplification to a blood flow–limited tissue compartment can be made by removing the flux term and redefining water and effective protein-binding fractions for the whole tissue. Drug elimination from the tissue requires the addition of an appropriate q_i term. Solution of a system of equations as expressed in Eqs. (43) and (44) requires a function relating free drug to bound drug concentrations in each subcompartment [24]. Assuming that the drug–protein binding can be characterized by a Langmuir binding equilibrium, the concentration of bound drug is

$$C_i^{xb} = \sum_{j=1}^{n} \frac{B_i K_i C_i^{xf}}{1 + K_i C_i^{xf}} \tag{45}$$

where

B_j = maximum binding capacity for jth class of binding sites, moles
per unit volume

K_j = equilibrium association constant for jth class of binding sites,
moles per unit volume

C_i^{xf} = free drug concentration in compartment x of tissue i, moles per
unit volume

C_i^{xb} = bound drug concentration in compartment x of tissue i, moles per
unit volume

N = number of independent classes of binding sites

Combined use of Eqs. (43)–(45) allows free drug concentrations to be predicted for each subcompartment. This approach to modeling free drug concentrations would make use of protein binding parameters (i.e., B_i, K_i) obtained from in vitro experiments.

Input functions [i.e., $I(t)$], describing the rate at which the administered dose enters a compartment, may have various forms depending on the administration schedule. The input function $I(t)$ is added to the appropriate mass balance equation and can describe any drug administration pattern. First-order absorption may be expressed as

$$I(t) = (F \times \text{Dose} \times k_a)e^{-k_a t} \tag{46}$$

where F = fraction of the administered dose entering compartment and k_a = first-order absorption rate constant.

Constant rate or zero-order infusions are

$$I(t) = k_0 u(t)$$

$$u(t) = \begin{cases} 1 & \text{for } t \leq T \\ 0 & \text{for } t > T \end{cases} \tag{47}$$

where k_0 = zero-order infusion rate constant, amount/time, and T = duration of infusion.

Bolus administration is usually characterized by a normalized injection function that effectively inputs the dose in a bell-shaped curve manner. Since bolus administration is not instantaneous, this approach is considered more realistic. $I(t)$ for bolus administration is

$$I(t) = Mg(t) \tag{48}$$

where

M = dose

$g(t)$ = normalized injection function [= $30 \lambda(\lambda t)^2(1 - \lambda t)^2$]

λ = reciprocal of the injection duration

System analysis techniques have been used to generate input functions for PB-PK models. Oral administration of carbon tetrachloride in different vehicles was successfully described by absorption input functions obtained by deconvolution and disposition decomposition methods [25,26].

Comparisons of predicted and observed drug concentrations are ordinarily based on total drug concentrations in the tissue, since most analytical methods make use of total tissue homogenates. Differentation of cellular and extracellular drug concentrations is not possible with the destructive tissue sample homogenization procedures routinely employed. Measurements of subcompartmental drug concentrations will become more realistic as noninvasive quantitation methods become refined and accessible. A volume-averaged total tissue drug concentration is calculated from predicted subcompartmental drug concentrations to allow comparisons to observed total tissue drug concentrations. For a three-subcompartment organ structure as depicted in Figure 2, the predicted total tissue concentration, C_t, is

$$C_t = \frac{V_i^1 C_i^1 + V_i^2 C_i^2 + V_i^3 C_i^3}{V_t} \tag{49}$$

where V_t = total tissue volume.

Total drug concentrations in a two-subcompartment structure are

$$C_t = \frac{V_i^1 C_i^1 + V_i^2 C_i^2}{V_t} \tag{50}$$

Tissue volumes may be expressed in mass or volume units. When volume units are used, a tissue density of 1 is normally assumed. Predicted concentrations from a blood flow–limited compartment may be compared directly to observed values.

Representation of a hybrid model structure employs features of both classical compartmental models and PB-PK models. The target organ or organs in a hybrid model can assume any PB-PK structure, such as the one depicted in Figure 2. The differential mass balance for the target organ or organ systems would be analogous to those presented above corresponding to Figure 2. The key difference in the hybrid approach is that the incoming blood drug concentration is characterized by an empirical function. This function is readily characterized by a polyexponential equation [see Eq. (9)] as is done for the classical compartmental approach. The function describing the blood concentration, C_b, as a function of time is substituted into the target organ mass balance equations for C_b, permitting the target organ equations to be solved analytically or numerically. The function for C_b is referred to as a forcing function since predicted target organ drug concentrations will be highly dependent on the nature of the forcing function.

Model development and parameter estimation methods for PB-PK models are presented in Section III.B.

C. Distributed Parameter Models

The development of distributed parameter models has been primarily confined to investigations of drug transport within solid tumors where inhomogeneities in blood flow and vascular permeability are well documented [18,19]. Collection of drug concentration data as a function of space and time is a rigorous exercise that limits pursuit of these types of models. However, as noninvasive measurements of regional drug concentrations improve, these models may be examined more avidly. The mass conservation equation in each compartment can be expressed, in general, as

$$\frac{\partial C}{\partial t} = \nabla J + S \tag{51}$$

where C = drug concentration, J = flux vector, and S = source/sink term.

The flux vector accounts for mass transport by both convection (i.e., blood flow, interstitial fluid flow) and conduction (i.e., molecular diffusion), whereas S describes membrane transport between adjacent compartments and irreversible elimination processes. For the three-subcompartment organ model presented in Figure 2, with concentration both space- and time-dependent, the conservation equations are

$$\frac{\partial C^1}{\partial t} = -\nabla J^1 - \frac{n^{12}}{V^1} - \frac{q^1}{V^1} \tag{52}$$

$$\frac{\partial C^2}{\partial t} = -\nabla J^2 + \frac{n^{12}}{V^2} - \frac{n^{23}}{V^2} - \frac{q^2}{V^2} \tag{53}$$

$$\frac{\partial C^3}{\partial t} = -\nabla J^3 - \frac{n^{23}}{V^3} - \frac{q^3}{V^3} \tag{54}$$

where flux, (n^{12}, n^{23}), volume (V^1, V^2, V^3), and elimination terms (q_i) may be defined as above [Eqs. (12)–(14), (24), (25)]. The flux term, J, is governed by Fick's law according to

$$J = UC - D \nabla C \tag{55}$$

where U = bulk velocity and D = diffusivity of drug.

Expansion of J for each subcompartment will depend on the coordinate system (i.e., Cartesian, rectangular, or spherical), dimensionality of transport (i.e., transport in one or more directions), and the need to include both convection and diffusion terms. For example, under the assumption of drug transport in the axial direction for a cylindrical coordinate system, and thus ignoring radial diffusion and convection, J in the vascular subcompartment is

$$J = -V \frac{\partial C^1}{\partial z} + D \frac{\partial^2 C^1}{\partial z^2} \tag{56}$$

where V = average axial velocity.

Analysis of drug transport in a solid tumor compartment could be represented in spherical coordinates as well as cylindrical [19] as depicted in Eq. (56). In this case, and assuming that drug diffusion occurs only in the radial direction, Eq. (53) can be written as

$$\frac{\partial C^3}{\partial t} = \frac{D}{r^2} \frac{\partial}{\partial r} \left[r^2 \frac{\partial C^3}{\partial r} \right] + \frac{n^{23}}{V^3} \tag{57}$$

where r = distance from the tumor center.

Solution of the conservation equations requires boundary conditions that are dictated by the specifics of the experiment and assumptions of the model.

III. MODEL DEVELOPMENT

A. Compartmental Models

Model development is intimately linked to correctly assigning model parameters to avoid problems of identifiability and model misspecification [27–29]. A full understanding of the objectives of the modeling exercise, combined with carefully planned study protocols, will limit errors in model identification. Compartmental models, as much as any other modeling technique, have been associated with overzealous interpretation of the model and parameters.

Studies interested in the determination of macro pharmacokinetic parameters, such as total body clearance or the apparent volume of distribution, can be readily calculated from polyexponential equations such as Eq. (9) without assignment of a specific model structure. Parameters (i.e., A_i, λ_i) associated with such an equation are initially estimated by the method of residuals followed by nonlinear least squares regression analyses [30].

Assignment of a specific model structure to a real data set should, among other things, take into account the identifiability problem. Simply stated, the identifiability problem occurs when more than one model structure and associated parameter sets are able to describe the actual data, typically drug plasma concentration–time data. In other words, how unique is the parameter set, and what physical interpretation does it have? Tests for model identifiability include transfer function analysis and result in the model and parameters being classified as unidentifiable, locally identifiable, or globally identifiable [2]. Rigorous transfer function analysis requires the determination of the transfer function matrix. In practical terms, this can be accomplished by equating the denominators of Eqs. (5) and (6) and determining the number of independent equations relating λ's to k's. If the number of equations is equal to the number of rate constants (i.e., k's), then the model is identifiable. When only one possible or unique solution is available, then the model is globally identifiable, whereas a finite set of solutions indicates that the model is locally identifiable. In most pharmacokinetic investigations, models are locally identifiable. An illustrative case is a two-compartmental model fitted to observe plasma concentration–time data. The same model could have central only, peripheral only, or both central and peripheral drug elimination. A simple criterion, $R = 2(n - 1) + 1$, is used to determine the maximum number of solvable rate constants, R, for a linear mammillary compartment model with n driving force compartments [20]. This criterion could be used to distinguish between an unidentifiable and identifiable model structure but not between a local or globally identifiable model. Study protocols that incorporate different routes of administration, different doses, and if appropriate, drug metabolite measurements will greatly facilitate model identification [31]. Compartmental models specifying individual organ compartments, such as a brain compartment, should be based on both plasma and individual organ drug concentrations.

Determination of model parameters for a compartmental model proceeds in a manner similar to that for fitting empirical equations. Preliminary parameter estimates are obtained by a curve-stripping method, such as the method of residuals, and are then used as initial parameter estimates in a nonlinear regression technique. The data are usually weighed in some manner to account for differences in variances of the measured concentrations. Through a series of interactions, and using convergence criteria, final parameter estimates are obtained. A variety of optimization algorithms for regression analysis are available through different computer programs [32–34]. All regression analyses provide a variety of statistical properties of the parameters and model that are used to evaluate the quality of the fit. In cases where more than one model (i.e., two vs. three compartments) adequately describe the data, a statistical test, such as the F-test or Akaike criteria, can be instituted to determine which model best describes the data [35,36]. These tests are regularly based on the weighted sum of squared residuals.

Model-fitting procedures are usually based on analytical solutions of the model; however, model parameters may be estimated by fitting the differential equations describing the model. Since the numerical solution of the differential equations introduces another source of error, fitting of differential equations is usually limited to cases where nonlinearities are present.

B. Physiologically-Based Pharmacokinetic Models

No doubt one of the most important and most controversial issues facing PB-PK modeling is parameter estimation. The often large number and diverse types of parameters certainly have contributed to the problem of parameter estimation. Another source of concern, and in part a philosophical one, is the question of whether parameter estimates should be based on in vitro or in vivo data. It would seem from a historical and idealistic perspective that it is preferable to estimate certain model parameters (i.e., partition coefficients, protein binding parameters, mass transfer coefficients) from in vitro studies. Combined with literature estimates of organ blood flows and volumes, model predictions can be made in lieu of extensive animal experimentation. One would be left with the task of validating or verifying the model predictions. Nonetheless, the use of in vivo data for parameter estimation appears to predominate in model development strategy. Measured tissue drug concentration–time data serve two purposes: parameter estimation and model validation. The question still remains as to how much in vivo data is required. Determination of parameters from single-dose or even steady-state experiments does not ensure that the model will adequately predict drug tissue concentrations at different doses, particularly if nonlinearities exist.

At this point, no universal parameter estimation protocol has been adopted, although standard methods have been applied to individual models. The great flexibility that PB-PK models offer in terms of characterizing all types of drug transport makes routine parameter estimation methods unnecessary and unattractive. As with any modeling endeavor, the modeler should have a clear idea of which parameter estimation procedures will be employed so that experiments can be designed properly.

1. Organ Blood Flows and Volumes

Numerous compilations of organ blood flows and volumes in different animals are available [37–40]. Organ blood flows may be measured by radioactive test substances or microspheres or from the organ clearance of highly extracted compounds [41–43]. For example, renal plasma flow can be equated to the clearance of *para*-aminohippurate. Total tissue volumes are simply determined by weight, assuming a density of 1, and fractionated into subcompartment volumes depending on the model structure [9]. There are tables available to estimate blood and water volumes of organs to permit corrections of blood contamination and to assist in determining effective protein binding fractions [37,44,45]. There have been enough independent investigations of organ blood flow and volumes to develop allometric relationships based on animal body weight [46]. Most PB-PK modeling investigators choose to use literature estimates for organ blood flows, in particular, and at times for tissue volumes. Total tissue volumes can be readily obtained from experiments in which drug concentrations are measured. One should realize that estimates for these parameters, particularly organ blood flows, vary greatly and should not be overlooked as sources of error in the model.

2. Partition Coefficients

Estimation methods for tissue-to-blood partition coefficients (i.e., R_i) have been the most prolific, no doubt due to the need for this parameter in most organ models. Both in vitro and in vivo parameter estimation techniques are available.

Lin et al. [47], on the basis of in vitro equilibrium dialysis binding studies of a diluted tissue homogenate, derived an equation for blood flow–limited compartmental partition coefficients. Assuming Langmuir-type drug–protein binding and equality of unbound plasma and tissue drug concentrations, the partition coefficient, R_i, is

$$R_i = \frac{1 + \alpha_i}{\gamma(1 + \alpha_p)} \tag{58}$$

where $\alpha_{i,p}$ = bound to free drug concentration ratio in tissue i or plasma and γ = whole blood to plasma drug concentration ratio.

The authors found that in their model for ethoxybenzamide, estimates of R obtained by use of Eq. (58) and in vivo techniques were comparable.

Chen and Gross [48] derived equations to calculate partition coefficients for blood flow–limited compartments from either constant rate infusion (i.e., steady-state conditions) or intravenous bolus regimens. For a noneliminating organ under steady-state conditions,

$$R_i = \frac{C_i^{ss}}{C_p^{ss}} \tag{59}$$

and following an intravenous bolus dose,

$$R_i = \frac{Q_i C_i^0}{Q_i C_p^0 + \alpha V_i C_i^0} \tag{60}$$

where

C_i^{ss} = steady-state drug concentration in tissue i
C_p^{ss} = steady-state drug concentration in plasma
C_i^0 = terminal phase y-axis intercept from tissue drug concentration–time plot
C_p^0 = initial plasma drug concentration
α = terminal phase slope from tissue drug concentration–time plot

Gallo et al. [49] developed the area method for calculation of partition coefficients for both blood flow–limited and membrane-limited compartments.

Following intravenous bolus administration, the partition coefficients for a non-eliminating blood flow–limited compartment is

$$R_i = \frac{\mathrm{AUC}_i}{\mathrm{AUC}_p} \tag{61}$$

where AUC_i = total area under the tissue drug concentration–time curve and AUC_p = total area under the drug concentration–time curve.

Under the same conditions for a membrane-limited compartment, R_i is

$$R_i = \frac{V_i \, \mathrm{AUC}_i}{V_i^2 \, \mathrm{AUC}_p} - \frac{V_i^1}{V_i^2} \tag{62}$$

where V_i = total tissue volume and V_i^1, V_i^2 = subcompartment 1 or 2 volumes for tissue i.

Both the Chen and Gross [48] and the Gallo et al. [49] methods have been applied to eliminating compartments. Both derivation methods are based on the specific mass balance equations for the given model structure. Monte Carlo investigations have demonstrated that both methods provide reasonably accurate and precise estimates of partition coefficients from concentration–time data sets containing error, data one is likely to encounter from in vivo studies.

3. Mass Transfer Coefficients

Mass transfer coefficients (i.e., h_i) may be obtained from in vitro experiments by utilizing various membrane permeability methodologies [50]. These include rapid mixing and separation, cell volume, and spectroscopic techniques. A general approach is to add drug to a cell suspension, rapidly mix and separate the cells, and measure drug concentration in the suspending medium after different mixing times. This experiment is designed to measure the rate of influx into the cell, and if linear bidirectional membrane transport is assumed, the rate of drug efflux would be equal to the rate of influx. Specific efflux rates could be obtained by performing the study with drug-loaded cells. Different starting drug concentrations should be used to examine the possibility of nonlinear membrane transport.

Shah et al. [51] demonstrated the use of a donor–receptor compartment apparatus separated by a cell monolayer to estimate membrane transport parameters. Permeability coefficients, P, were calculated as

$$P = \frac{KV}{AC_0} \tag{63}$$

where

  K = flux rate of solute, amount/volume
  V = volume of donor chamber

A = cross-sectional area of cell surface

C_0 = initial solute concentration in the donor chamber

The flux rate is obtained from the slope of the receptor chamber solute concentration versus time plot. The mass transfer coefficient, h_i, is equal to PA, where A is the surface area of the membrane separating the two subcompartments.

One should realize that the intracellular compartment as depicted in Figure 2 represents multiple cell types, whereas in vitro studies normally utilize a single cell type pertinent to characterizing specific attributes of drug transport in that cell system. The method of Shah et al. [51] would be of great benefit to investigating blood-brain barrier transport, consistent with a vascular–extravascular subcompartment brain model.

In vivo experiments can also be utilized to estimate mass transfer coefficients. Under the condition of steady-state blood concentrations of drug, a linear mass transfer coefficient can be estimated from [10]

$$\ln\left[1 - \frac{C_i^2}{R_i C_b}\right] = \exp\left[\frac{-h_i}{V_i V_i^2}t\right] \tag{64}$$

where all terms are defined as above for a two-subcompartment organ model. The mass transfer coefficient is based on measured C_b and C_i values with h_i obtained from the slope of Eq. (64).

A moment method to estimate linear mass transfer coefficients for noneliminating organs was derived by Gallo et al. [52]. h_i is estimated as

$$h_i = \left[\frac{V_i \, \text{AUMC}_i - (V_i^1 + R_i V_i^2) \, \text{AUMC}_p}{(V_i^2 R_i)^2 \, \text{AUC}_p} - \frac{(V_i^1)^2 + 2V_i^1 V_i^2 R_i + (V_i^2 R_i)^2}{Q(V_i^2 R_i)^2}\right]^{-1} \tag{65}$$

where the new terms are defined as

AUMC_i = area under the first-moment curve for tissue i

AUMC_p = area under the first-moment curve for plasma

AUC_p = area under the plasma concentration–time curve

The method compared favorably to a forcing function technique for estimating h [49].

4. Clearances

Organ clearances are usually based on in vivo rather than in vitro studies. Drug metabolic parameters, V_m and K_m, can be estimated from in vitro hepatic enzyme and hepatocyte preparations [53]. Knowledge of total clearance and fractional elimination by each pathway, based on in vivo data, readily allows calculation of organ and intrinsic drug clearances. In general,

$$CL_i = f_i CL \tag{66}$$

where

CL_i = organ clearance
f_i = fraction of dose eliminated by organ i
CL = total systemic clearance

Blood drug concentration and amount of drug in urine versus time data are usually sufficient to obtain parameters for Eq. (65). No doubt for many drugs this information is available in the literature, and independent studies may be unnecessary. Intrinsic organ clearance, CL_i', as presented in Eq. (26), is obtained from

$$CL_i' = \frac{CL_i\, Q_i}{Q_i - CL_i} \tag{67}$$

5. Protein Binding

Protein binding parameters B_i and K_i, as depicted in Eq. (44), are obtained from in vitro experiments utilizing filtration, centrifugation, and dynamic dialysis or equilibrium dialysis methods. These techniques have been reviewed elsewhere [54,55].

6. Miscellaneous

Forcing function methods may be used to estimate any parameters associated with the system mass balance equations [52,56–58]. The procedures may be applied in the context of either a hybrid or global PB-PK model. The method requires nonlinear regression analysis based on tissue and blood drug concentration–time data. The blood or plasma concentration–time data would serve to obtain a forcing function, usually a polyexponential equation, which is then used as an input function in the mass balance equation for the tissue. The model describing the individual tissue's drug disposition is then fit to the observed tissue concentration–time data with one or more parameters to be estimated. The forcing function procedure would be particularly useful for estimating parameters in which literature estimates are unavailable. Application of the method to a global PB-PK model would require sequentially fitting individual organ models to obtain parameter estimates and then reconstructing the global model. For a hybrid PB-PK model, the model-fitting procedures result in the final model.

A modification of the forcing function approach makes use of linear systems analysis for individual tissue compartments [59]. Parametric or nonparametric functions are fitted to observed blood drug concentration–time data and are then combined with tissue drug concentration–time measurements deconvolved to obtain a disposition function for drug kinetics in the tissue. From the tissue disposition function, certain parameters for blood flow–limited organ models can be obtained.

A new idea has recently been presented that makes use of Monte Carlo simulations [60,61]. By defining a range of parameter values, the parameter space can be examined in a random fashion to obtain the best model and associated parameter set to characterize the experimental data. This method avoids difficulties in achieving convergence through an optimization algorithm, which could be a formidable problem for a complex model. Each set of simulated concentration–time data can be evaluated by a goodness-of-fit criterion to determine the models that predict most accurately.

7. Distinguishing Between Blood Flow–Limited and Membrane-Limited Organ Models

The use of a membrane-limited two- or three-subcompartment organ model is indicated if tissue drug concentrations do not decline in parallel with drug concentrations in plasma or blood. A formal criterion has been given by Dedrick and Bischoff [10] such that if

$$\frac{Q_i}{h_i} < \left(1 + \frac{V_i^1}{R_i V_i^2}\right)^2 \tag{68}$$

then the organ can be represented as blood flow–limited. If this condition is not true, then the organ is best represented as being membrane-limited. Equation (68) can be approximated as $Q_i/h_i \ll 1$, since $V_i^1/R_i V_i^2$ is much less than 1. Thus, determining whether a blood flow–limited or membrane-limited compartment is compatible with the data requires estimates for R_i and h_i. Often it is assumed that a blood flow–limited representation is correct in the absence of a formal analysis.

8. Model Validation

Two issues present themselves when the question of PB-PK model validation is raised. The first issue is the accuracy with which the model predicts actual drug concentrations. The actual concentration–time data have most likely been used to estimate certain total parameters. Quantitative assessment, via goodness-of-fit tests, should be done to assess the accuracy of the model predictions. Too often, model acceptance is based on subjective evaluation of graphical comparisons of observed and predicted concentration values.

Goodness-of-fit tests may be a simple calculation of the sum of squared residuals for each organ in the model [26] or calculation of a log likelihood function [60]. In the former case,

$$\text{SSR} = \sum_{i=1}^{N} (C_i^o - C_i^p)^2 \tag{69}$$

and in the latter case,

$$\text{LL} = \sum_{i=1}^{N} \frac{n_i}{2} \ln\left[1 + \frac{(C_i^o - C_i^p)^2}{S_i^2} \right] \tag{70}$$

where

- SSR = sum of squared residuals
- N = number of observations
- C_i^o = mean of the observed drug concentration
- C_i^p = predicted drug concentration
- n_i = number of experimental repetitions
- S_i^2 = variance of the observed concentrations at each data point
- LL = log likelihood function

Lower values of SSR and LL indicate closer agreement between observed and predicted concentrations for that organ.

The second issue that arises on model validation is the accuracy of model predictions in the absence of experimental data. The question is somewhat meaningless and lacks a definitive answer, since experimental data are unavailable to compare to predicted results. Nonetheless, the predictive power of PB-PK models facilitates obtaining prediction of drug concentrations under conditions different than those used to generate the experimental data. For instance, it may be useful to simulate drug concentrations for different dosing regimens or possibly following alterations in physiological parameters. No single assessment method would indicate the validity of such simulations, yet examination of the model assumptions and available data could provide partial support for the simulations and their inferences. Were studies conducted to differentiate between linear and nonlinear kinetic processes? What criteria were used to establish a blood flow–limited or membrane-limited organ structure? Were enough animal data collected to develop allometric relationships for model parameters to allow predictions of human drug tissue concentrations? Naturally, support for the model is gained by favorable agreement with observed data, even if only at a few selected time points.

C. Distributed Parameter Models

Development of a distributed parameter model will rely on data obtained in vivo. Time and spatial dependencies of drug concentration in a target organ are used as the basis to estimate parameters by nonlinear regression analysis. Distribution parameter models, in the context of examining drug distribution in vivo are complex and specific, limiting generalizations regarding parameter estimation methods.

IV. SUMMARY

Pharmacokinetic analysis of drug disposition in vivo offers a variety of techniques that vary from simple (i.e., noncompartmental analysis) to complex (i.e., distributed parameter models). It behooves the experimentalist to delineate the pharmacokinetic information that is to be collected from each study and the intended method of analysis to ensure that the study design is robust and that the stated objectives can be achieved. All of the compartmental modeling approaches require analytical or numerical solutions to a system of differential equations. Analytical solutions readily incorporate statistical optimization functions helpful in model selection. Numerical solutions indicative of nonlinear systems or organ-based models, such as physiological pharmacokinetic models, do not rely on a single dogma in model development but rather on greater input from the investigator. Nonetheless, these models should still utilize and take advantage of statistical tools to develop and validate the models. Regardless of the modeling strategy, a plethora of computer programs are available and provide an optimum format for model development. The combination of modeling methods and computer packages provides ample tools for drug disposition to be addressed at any level of sophistication.

REFERENCES

1. JG Wagner. History of pharmacokinetics. Pharmacol Ther 12:537–562, 1981.
2. K Godfrey. Compartmental Models and Their Application. London: Academic Press, 1983.
3. EA Nuesch. Noncompartmental approaches in pharmacokinetics using moments. Drug Metab Rev 15:103–131, 1984.
4. JJ DiStefano III. Noncompartmental vs. compartmental analysis: Some bases for choice. Am J Physiol 243:R1–R6, 1982.
5. P Veng-Pedersen. Linear and nonlinear systems approach in pharmacokinetics: How much do they have to offer? I. General considerations. J Pharmacokin Biopharm 16:413–472, 1988.
6. T Teorell. Kinetics of distribution of substances administered to the body. I. The extravascular modes of administration. Arch Int Pharmacodyn Ther 57:226–255, 1937.
7. T Teorell. Kinetics of distribution of substances administered to the body. II. The

intravascular modes of administration. Arch Int Pharmacodyn Ther 57:226–240, 1937.

8. R Bellman, JA Jacquez, R Kalaba. Some mathematical aspects of chemotherapy. I. One organ models. Bull Math Biophys 22:181–198, 1960.

9. KB Bischoff, RG Brown. Drug distribution in mammals. Chem Eng Prog Symp 62: 33–45, 1966.

10. RL Dedrick, KB Bischoff. Pharmacokinetics in applications of the artificial kidney. Chem Eng Prog Symp Ser 64:32–44, 1968.

11. JM Gallo, CT Hung, PK Gupta, DG Perrier. Physiological pharmacokinetic model of adriamycin delivered via magnetic albumin microspheres in the rat. J Pharmacokin Biopharm 17:305–326, 1989.

12. JC Dearden. Molecular structure and drug transport. In: CA Ramsden, ed. Comprehensive Medicinal Chemistry, Vol. 4, Quantitative Drug Design. New York: Pergamon Press, 1990, pp 375–411.

13. GE Blakey, IA Nestorov, PA Arundel, LJ Aarons, M Rowland. Quantitative structure–pharmacokinetics relationships. I. Development of a whole body physiologically based model to characterize changes in pharmacokinetics across a homologous series of barbiturates in the rat. J Pharmacokin Biopharm 25:277–312, 1997.

14. KH Watanabe, FY Bois. Interspecies extrapolations of physiological pharmacokinetic parameter distributions. Risk Anal 16:741–754, 1996.

15. H Boxenbaum, R Ronfeld. Interspecies pharmacokinetic scaling and the Dedrick plots. Am J Physiol 245:R768–R774, 1983.

16. Y Igari, Y Sugiyama, Y Sawada, T Iga, M Hanano. Prediction of diazepam disposition in the rat and man by a physiologically based pharmacokinetic model. J Pharmacokin Biopharm 11:577–593, 1983.

17. BA Patel, FD Boudinot, RF Schinazi, JM Gallo, CK Chu. Comparative pharmacokinetics and interspecies scaling of 3-azido-3-deoxythymidine (AZT) in several mammalian species. J Pharmacobiol Dynam 13:206–211, 1990.

18. RK Jain, LE Gerlowski. Extravascular transport in normal and tumor tissues. CRC Crit Rev Oncol Hematol 5:115–170, 1985.

19. RK Jain, J Wei. Dynamics of drug transport in solid tumors: Distributed parameter models. J Bioeng 1:313–330, 1977.

20. LZ Benet. General treatment of linear mammillary models with elimination from any compartment as used in pharmacokinetics. J Pharm Sci 6:536–541, 1972.

21. CR Wylie. Advanced Engineering Mathematics. 4th ed. New York: McGraw-Hill, 1975, pp 289–294.

22. JM Weissbrod. A comprehensive approach to whole body pharmacokinetics in mammalian systems: Applications to methotrexate, streptozotocin and zinc. DSc Eng dissertation, Columbia University, New York, 1979, pp 110–112.

23. D Shen, M Gibaldi. Critical evaluation of use of effective protein fractions in developing pharmacokinetic models for drug distribution. J Pharm Sci 63:1698–1702, 1974.

24. KB Bischoff. Some fundamental considerations of the applications of pharmacokinetics to cancer chemotherapy. Cancer Chemother Rep 59:777–793, 1975.

25. WR Gillespie, LL Cheung, HJ Kim, JV Bruckner, JM Gallo. Application of system analysis to toxicology: Characterization of carbon tetrachloride oral absorption ki-

netics. In: TR Gererity, CJ Henry, eds. Principles of Route-to-Route Extrapolation for Risk Assessment. Amsterdam, Netherlands: Elsevier Science, 1990, pp 285–295.

26. JM Gallo, LL Cheung, HJ Kim, JV Bruckner, WR Gillespie. A physiological and system analysis hybrid pharmacokinetic model to characterize carbon tetrachloride blood concentrations following administration in different oral vehicles. J Pharmacokin Biopharm 21:551–574, 1993.

27. K Zierler. A critique of compartmental models. Ann Rev Biophys Bioeng 10:531–562, 1981.

28. G Cobelli, JJ DiStefano III. Parameter and structural identifiability concepts and ambiguities: A critical review and analysis. Am J Physiol 239:R7–R24, 1980.

29. DM Gibon, ME Taylor, WA Colburn. Curve fitting and unique parameter identification. J Pharm Sci 76:658–659, 1987.

30. M Gibaldi, DG Perrier. Pharmacokinetics. 2nd ed. New York: Marcel Dekker, 1982, pp 433–444.

31. KR Godfrey, RP Jones, RF Brown. Identifiable pharmacokinetic models: The role of extra inputs and measurements. J Pharmacokin Biopharm 8:633–648, 1980.

32. CM Metzler, GL Elfring, AJ McEwen. A User's Manual for NONLIN and Associated Programs. Kalamazoo, MI: The Upjohn Co., 1974.

33. DZ D'Argenio, A Schumitzky. ADAPT II User's Guide. Biomedical Simulations Resource. Los Angeles, CA: University of Southern California, 1990.

34. M Berman, WF Beltz, PC Greif, R Chabay, RC Boston. CONSAM User's Guide. Bethesda, MD: US Department of Health and Human Services, 1983.

35. HG Boxenbaum, S Riegelman, RM Elashoff. Statistical estimation in pharmacokinetics. J Pharmacokin Biopharm 2:123–148, 1974.

36. K Yamaoka, T Nakagawa, T Uno. Application of Akaike's information criteria (AIC) in the evaluation of linear pharmacokinetic model. J Pharmacokin Biopharm 6:165–175, 1978.

37. PL Altman, DS Dittmer, eds. Dittmer Biology Data Book, Vols 1 and 3. 2nd ed. Bethesda, MD: Federation of American Societies for Experimental Biology, 1972 and 1974.

38. LE Gerlowski, RK Jain. Physiologically based pharmacokinetic modeling: Principles and applications. J Pharm Sci 72:1103–1127, 1983.

39. WS Spector, ed. Handbook of Biological Data. Philadelphia, PA: WB Saunders, 1956.

40. RP Brown, MD Delp, SL Lindstedt, LR Rhomberg, RP Beliles. Physiological parameter values for physiologically based pharmacokinetic models. Toxicol Ind Health 13:407–484, 1997.

41. L Jansky, JS Hart. Cardiac output and organ blood flow in warm and cold-acclimated rats exposed to cold. Can J Physiol Pharmacol 46:653–659, 1968.

42. NA Lassen, W Perl. Tracer Kinetic Methods in Medical Physiology. New York: Raven Press, 1979.

43. MD Delp, RO Manning, JV Bruckner, RB Armstrong. Distribution of cardiac output during diurnal changes of activity in rats. Am J Physiol 261:H1–H7, 1991.

44. PL Altman, DS Dittmer, eds. Respiration and Circulation. Bethesda, MD: Federation of American Societies for Experimental Biology, 1971.

45. CN Chen, JD Andrade. Pharmacokinetic model for simultaneous determination of drug levels in organs and tissues. J Pharm Sci 65:717–724, 1976.
46. AD Arms, CC Travis. Reference Physiological Parameters in Pharmacokinetic Modeling. Washington, DC: US Environmental Protection Agency, February 1988.
47. JH Lin, Y Sugiyama, S Awazu, M Hanano. In vitro and in vivo evaluation of the tissue-to-blood partition coefficients for physiological pharmacokinetic models. J Pharmacokin Biopharm 10:637–647, 1982.
48. HSG Chen, JF Gross. Estimation of tissue-to-plasma partition coefficients used in physiological pharmacokinetic models. J Pharmacokin Biopharm 7:117–125, 1979.
49. JM Gallo, FC Lam, DG Perrier. Area method for the estimation of partition coefficients used in physiological pharmacokinetic models. J Pharmacokin Biopharm 15: 271–280, 1987.
50. WD Stein. Transport and Diffusion Across Cell Membranes. Orlando, FL: Academic Press, 1986, pp 52–68.
51. MV Shah, KL Audus, RT Borchardt. The application of bovine brain microvessel endothelial-cell monolayers grown onto polycarbonate membranes in vitro to estimate the potential permeability of solutes through the blood-brain barrier. Pharm Res 6:624–627, 1989.
52. JM Gallo, FC Lam, DG Perrier. Moment method for the estimation of mass transfer coefficients for physiological pharmacokinetic models. Biopharm Drug Dispos 12: 127–137, 1991.
53. R Brown, H Seifried. Extrapolation of in vivo metabolic rate constants from in vitro pharmacokinetic data. Washington, DC: US Environmental Protection Agency, 1988.
54. CF Chignell. Protein binding. In: ER Garrett, JL Hirtz, eds. Drug Fate and Metabolism. New York: Marcel Dekker, 1977, pp 187–228.
55. MC Meyer, DE Guttman. Dynamic dialysis as a method for studying protein binding. II: Evaluation of the method with a number of binding systems. J Pharm Sci 59: 39–48, 1970.
56. JM Weissbrod, RK Jain, FM Sirotnak. Pharmacokinetics of methotrexate in leukemia cells: Effect of dose and mode of injection. J Pharmacokin Biopharm 6:487–503, 1978.
57. FG King, RL Dedrick. Physiological model for the pharmacokinetics of 2′-deoxycoformycin in normal and leukemic mice. J Pharmacokin Biopharm 9:519–534, 1981.
58. JM Gallo, EE Hassan, DG Groothius. Targeting anticancer drugs to the brain. II. Physiological pharmacokinetic model of oxantrazole following intraarterial administration to rat glioma-2 (RG-2) bearing rats. J Pharmacokin Biopharm 21:575–592, 1993.
59. D Verotta, LB Sheiner, WF Ebling, DR Stanski. A semiparametric approach to physiological flow models. J Pharmacokin Biopharm 17:463–491, 1989.
60. FY Bois, TJ Woodruff, RC Spear. Comparison of three physiologically based pharmacokinetic models of benzene disposition. Toxicol Appl Pharmacol 110:79–88, 1991.
61. RS Thomas, WE Lytle, TJ Keefe, AA Constan, RSH Yang. Incorporating Monte Carlo simulation into physiologically based pharmacokinetic models using advanced continuous simulation language (ACSL): A computational method. Fundam Appl Toxicol 31:19–28, 1996.

4

Experimental Methods to Evaluate Diffusion Coefficients and Investigate Transport Processes of Pharmaceutical Interest

John W. Mauger
University of Utah, Salt Lake City, Utah

I. INTRODUCTION

Experimental methods which yield precise and accurate data are essential in studying diffusion-based systems of pharmaceutical interest. Typically the investigator identifies a mechanism and associated mass transport model to be studied and then constructs an experiment which is consistent with the hypothesis being tested. When mass transport models are explicitly involved, experimental conditions must be physically consistent with the initial and boundary conditions specified for the model. Model testing also involves recognition of the assumptions and constraints and their effect on experimental conditions. Experimental conditions in turn affect the maintenance of sink conditions, constant surface area for mass transport, and constant and known hydrodynamic conditions.

Experimental measurements must be interpreted in connection with all external influences, including fluid flow, bulk viscosity, and pH of the system. Consideration must be given to whether the mass transport is occurring in one or more dimensions and whether mass transport is affected by pressure gradients and/or osmotic pressure gradients.

The specific methods discussed in this chapter are representative of those which have found use in the study of mass transport in systems of pharmaceutical interest. Many of these methods represent updated versions of methods taken from the biological, chemical, and engineering sciences and adapted to the study of pharmaceutically relevant systems. These methods have been useful in determining diffusion coefficients, in studying fundamental research questions related to physicochemical and biological aspects of diffusion theory, in evaluating mechanisms related to release rates for drug delivery systems and dosage forms, and in connection with transport through synthetic and biological membranes.

II. THEORETICAL CONSIDERATIONS RELATED TO EXPERIMENTAL CONDITIONS

One important differentiating feature between methods, related to both theoretical considerations and experimental conditions, is whether non-steady-state or steady-state conditions prevail. For the steady-state case in one dimension,

$$\frac{\partial^2 c}{\partial x^2} = 0; \qquad \frac{\partial c}{\partial x} = a \tag{1}$$

where c is the concentration, x is the spatial variable, and a is a constant. Thus, the concentration gradient is constant in this case and invariant with position.

For the non-steady-state case, the governing equation for diffusion in one dimension is

$$\frac{\partial c}{\partial t} = D \frac{\partial^2 c}{\partial x^2} \tag{2}$$

and a simple integral analysis leads to

$$x \propto \sqrt{Dt} \tag{3}$$

Thus, a known relationship between x and t emerges which is relevant to the considerations for the anticipated experiment.

Differentiating between these cases is critical to the selection or development of an experimental method which corresponds to the case of interest. Incorrectly interpreting the data from a transport study by assuming one case when another prevails will lead to erroneous conclusions. Other theoretical aspects which affect the experimental outcome are discussed in connection with each method.

III. DISCUSSION OF METHODS

A. Liquid/Liquid Interface

1. Miscible Phases

Capillary methods are useful to determine diffusion coefficients. The method depends on the formation of a sharp boundary between the diffusing species in the glass capillary and a receptor solvent in which the capillary or tube is placed. After the capillary has been filled with drug solution, it is submerged in a receptor fluid held at constant temperature to initiate the experiment. Based on a boundary condition for the solution of the governing equation for this method, the concentration at the end of the capillary or tube must be maintained at zero. This condition is met by mildly stirring a volume of receptor fluid which is large enough to ensure sink conditions. At the termination of the experiment, the capillary is removed from the receptor sink and its contents are mixed. Contents from the tubes are then analyzed, and the average concentration, C_{ave}, is determined.

Under certain constraints the following equation can be used to calculate the diffusion coefficient.

$$D = \frac{\pi L^2}{4t} \left(1 - \frac{C_{ave}}{C_0} \right)^2 \tag{4}$$

Since C_{ave} can be experimentally determined and C_0, L (tube length), and t are known, the diffusion coefficient, D, can be directly determined from this equation.

Error analysis of Eq. (4) shows that the value for D is sensitive to small changes in C_{ave} because the term in parentheses is squared. Therefore, it is of critical important to minimize variation related to assay or any other experimental source.

Tubes are much more sensitive to convection effects than capillaries, but capillaries contain much smaller amounts of solution for analysis. Transport by convection when tubes are used can be accounted for by experimental evaluation. A disadvantage of both methods is the amount of time required for the experiment, which may be hundreds of hours. The investigator must address the possibility of adsorption of the diffusing solute onto the glass surfaces. In addition, the dimensions of the glass capillary must be known with considerable accuracy.

Farng and Nelson [1] applied the capillary method to the determination of the diffusion coefficient of salicylic acid in the presence of polyelectrolytes. The reported variability in terms of the coefficient of variation ranged from 0.89% to 8.3%. Stout et al. [2] showed the tube method to be useful for determining diffusion coefficients of water-insoluble pharmaceuticals such as sulfonamides and steroids. The coefficient of variation associated with the diffusion coefficient for sulfisoxazole is 5.5%.

Capillary zone electrophoresis, an up-to-date high resolution separation method useful for proteins and peptides, has been shown to be a useful method for determining electrophoretic mobilities and diffusion coefficients of proteins [3]. Diffusion coefficients can be measured from peak widths of analyte bands. The validity of the method was demonstrated by measuring the diffusion coefficients for dansylated amino acids and myoglobin.

The Polson method [4] is a closed system version of the tube or capillary method which involves the formation of a free boundary for diffusion by bringing a drug solution into contact with a receptor solvent. The unique aspect of this method is the design of the apparatus, which consists of two Plexiglas or stainless steel circular sections containing six holes of known diameter and length drilled around the rim of each section. The upper and lower sections are joined by a central bolt. Each hole in the upper and lower sections comprises one diffusion cell when the two sections are mated and the holes aligned.

The device is prepared for a transport experiment by aligning the holes of the upper and lower sections and then filling the cells in the lower section with diffusing solution of concentration C_0. Then the upper unit is turned on the central bolt so that the holes in the upper unit are no longer aligned with holes in the bottom section, which permits filling the cells in the upper unit with receptor solvent. After allowing the unit to reach thermal equilibrium, the experiment is initiated by carefully turning the upper unit on the central bolt so that the holes in the upper and lower section are aligned to form a liquid/liquid boundary between the donor cell and the receptor column. At the termination of the experiment, the upper unit is again carefully turned so that the holes in the upper unit are unaligned with those in the lower unit, and the solution in the upper cell is removed for analysis.

The diffusion coefficient, D, can be calculated via

$$D = \frac{C(t)^2}{C_0^2}\left(\frac{\pi}{t}\right)h^2 \tag{5}$$

where $C(t)$ is the concentration of drug diffusing through cross-sectional area A in time t and h is the receptor column length. An advantage of this method is that the volume available for analysis is large relative to that of the capillary or tube method. Limitations include the length of time required for the experiment, difficulty in achieving a tight seal between the two halves of the device, and formation of a sharp boundary for diffusion at the initiation of the experiment.

Using the data set reported by Polson [4], the coefficient of variation for the diffusion coefficient for a virus solution is 32%. This value is for a set of experiments in which the time for the experiment was changed. It is expected that this value would be improved if the experimental conditions were kept constant. This method was originally proposed for the determination of diffusion coefficients for large molecular weight biologically active substances, such as proteins, but should also be applicable to small molecular weight drugs and compounds.

The method of hydrodynamic stability involves bringing two solutions of different concentrations into contact in a capillary tube, forming a stable and measurable concentration gradient. The diffusion coefficient is a function of the resulting gradient, solution viscosity and density, and some capillary dimensions [5].

The diffusion coefficient can be experimentally determined via this method using the equations

$$D = \frac{\alpha g R^4 C_0}{67.9 \mu L_c} \tag{6}$$

and

$$\alpha = \frac{\rho_s - \rho_0}{C_0} \tag{7}$$

The terms in Eq. (6) include the gravitational constant, g, the tube radius, R, the fluid viscosity, μ, the solute concentration in the donor phase, C_0, and the penetration depth, L_c. The density difference between the solution and solvent ($\rho_s - \rho_0$) is critical to the calculation of α. Thus, this method is dependent upon accurate measurement of density values and close temperature control, particularly when C_0 represents a dilute solution. This method has been shown to be sensitive to different diffusion coefficients for various ionic species of citrate and phosphate [5]. The variability of this method in terms of the coefficient of variation ranged from 19% for glycine to 2.9% for ortho-aminobenzoic acid.

2. Immiscible Phases

The study of transport between immiscible phases requires a device which (1) yields contacting surfaces with known contact area, (2) has a stationary interface, and (3) has known hydrodynamics. Stowe and Shaewitz [6] developed and evaluated a cell which meets these criteria. Each liquid is stirred with a rotating disk. As noted by these investigators, if the fluids are rotated in opposite directions by the rotating disks, then the torque on the interface cancels. The criterion for operation of the stirred cell with a stationary fluid interface is

$$\frac{\Omega_1}{\Omega_2} = \frac{(\rho_2 \mu_2)^{1/3}}{(\rho_1 \mu_1)^{1/3}} \tag{8}$$

where the ratio of torques (Ω_1/Ω_2) is inversely proportional to the density times the viscosity for each fluid. The authors state that this cell can be operated under realistic experimental conditions within 5% of theoretical behavior.

Byron and coworkers [7,8] developed and evaluated a transport cell which proved to be useful in predicting interfacial transfer kinetics between aqueous and organic boundary layers. In this system, stirring is generated by using paddles in each phase. These investigators demonstrated that successful prediction of the transfer kinetics of any homologue in a series was possible in all cases from

knowledge of the partition coefficient and transfer kinetics of the parent compound, the partition coefficient of the homologue, and some easily determined
system parameters.

A simple rocking device was tested for routine determination of distribution
coefficients [9]. Sample cells were constructed for two-phase [9] and three-phase
[10] systems. The investigators claim that the rocking action causes the shape
of each phase to vary slowly and constantly and that the precision associated
with the distribution coefficient is similar to that for shake-out methods. The
three-phase cell was tested as an in vitro model to simulate factors involved in
the absorption process. Rates of drug transfer and equilibrium drug distribution
were evaluated under conditions in which one aqueous phase was maintained at
pH 7.4 and the other phase was maintained at another pH.

B. Liquid/Membrane/Liquid Interface

1. Miscible Liquid Phases

Several investigators have used the two-chamber diffusion cell configuration.
This experimental method has been found useful to determine diffusion coefficients [11] and to study drug transport from drug delivery devices [12].

Transport in these side-by-side membrane cells is described by the equation

$$V\frac{dC_r}{dt} = \frac{D_r}{h}A(C_d - C_r) \tag{9}$$

where V is the volume of the receptor fluid, C_r is the concentration of the drug
in the receptor fluid, D_r is the diffusivity of the drug in the receptor fluid, h is
the diffusion layer thickness, A is the effective area for mass transfer, and C_d is
the concentration of the drug in the donor fluid. As noted by this equation, the
concentrations in the donor and receptor cells are important determinants of the
transport rate. Thus, the investigator must carefully control initial concentration
conditions in both cells, determine whether sink conditions prevail, and note
whether drug concentration in the receptor cell is kept constant during the experiment.

Rigorous calibration is a requirement for the use of the side-by-side membrane diffusion cell for its intended purpose. The diffusion layer thickness, h, is
dependent on hydrodynamic conditions, the system geometry, the spatial configuration of the stirrer apparatus relative to the plane of diffusion, the viscosity of
the medium, and temperature. Failure to understand the effects of these factors
on the mass transport rate confounds the interpretation of the data resulting from
the mass transport experiments.

When using this device to determine diffusion coefficients, Goldberg and

Higuchi [11] determined a cell constant, L(cell), using compounds with known diffusion coefficients according to the equation

$$L(\text{cell}) = \frac{G(\text{known})}{D(\text{known})(C_2 - C_1)} \tag{10}$$

where G is the rate of transport and C_2 and C_1 are the concentrations of diffusants in the donor and receptor cell, respectively. L(cell) was then measured using potassium chloride and sucrose as compounds with known diffusion coefficients. Cell constants were determined under conditions of varying diffusant concentrations and at 25 and 30°C. The coefficient of variation for the cell constant was less than 1% for one set of experiments and slightly greater than 1% for another set.

Another set of experiments used benzoic acid to determine the diffusion coefficient. In this case, the following equation was used.

$$D(\text{unknown}) = \frac{G(\text{unknown})}{L(\text{cell})(C_2 - C_1)} \tag{11}$$

Values for G(unknown) were experimentally determined by using the previously calibrated cells, and these data were used to calculate values for D(unknown) using the cell constants. The overall average value of D(unknown) was 1.11×10^{-5}, which compares well with a reported value of 1.1×10^{-5}. The coefficient of variation associated with the diffusion coefficient was 2.7% for one cell and 1.7% for a second cell. This calibration procedure thus provided information about the accuracy and precision of the method as well as the effect of temperature and concentration on the determination of the diffusion coefficient.

The side-by-side diffusion cell has also been calibrated for drug delivery mass transport studies using polymeric membranes [12]. The mass transport coefficient, D/h, was evaluated with diffusion data for benzoic acid in aqueous solutions of polyethylene glycol 400 at 37°C. By varying the polyethylene glycol 400 content incrementally from 0 to 40%, the kinematic viscosity of the diffusion medium, saturation solubility for benzoic acid, and diffusivity of benzoic acid could be varied. The resulting mass transport coefficients, D/h, were correlated with the Sherwood number (Sh), Reynolds number (Re), and Schmidt number (Sc) according to the relationships

$$\text{Sh} = \text{const} \times \text{Re}^m \, \text{Sc}^n \tag{12a}$$

$$\text{Sh} = \frac{Dd}{hDr}; \qquad \text{Re} = \frac{Nd^2\rho}{\mu}; \qquad \text{Sc} = \frac{\mu}{\rho Dr} \tag{12b}$$

The stirring rate is given by N in the Reynolds number expression. A characteristic length of the system is given by d in the Reynolds number expression and was defined as the length of the agitator.

Using the density, viscosity, and diffusivity data along with experimentally determined values for D/h, values for the Sherwood, Reynolds and Schmidt numbers were calculated. In turn, these values were used to determine the constant and the exponents in Eq. (12a). These values were found to be 0.154, 0.61, and 0.33, respectively. Thus, the device was quantitatively calibrated and modeled so that values of the Sherwood number could be calculated for varying conditions of stirring rate, viscosity, and drug concentration. By using the correlation equation and a correction factor, the intrinsic permeation rate of testosterone through silicone membranes was extracted from the experimental data under various non-ideal mixing conditions.

Thus, the side-by-side device, when properly calibrated, is a versatile and useful method to determine diffusion coefficients, to evaluate mass transport mechanisms, and to evaluate up-to-date drug delivery systems.

Aguiar and Weiner [13] developed a modified device in which the receptor cell allows for continuous flow of receptor solvent, which maintains sink conditions. Flynn and Smith [14] designed and evaluated a unique device having a high ratio of diffusional area to diffusional volume and precisely controlled stirring.

Garrett and Chemburkar [15] described a membrane device which provides for steady-state conditions by continuously circulating fresh donor solution into the donor cell.

Pirt [16] described a diffusion capsule for the addition of a solute at a constant rate to a liquid medium. The diffusion capsule consists of a small cylindrical container that can be completely filled with solution and then sealed with a small semipermeable membrane at one end. The device was designed for obtaining the low substrate feed rates required for multiple cultures of microbes.

Srinivasan et al. [17] have described a four-electrode potentiostat system which is suited to maintaining a constant voltage drop across a membrane in a two-chamber diffusion cell. This system was evaluated in connection with transdermal iontophoretic drug delivery of polypeptides.

2. Immiscible Liquid Phases

Koizumi and Higuchi [18] evaluated the mass transport of a solute from a water-in-oil emulsion to an aqueous phase through a membrane. Under conditions where the diffusion coefficient is expected to depend on concentration, the cumulative amount transported, Q, is predicted to follow the relationship

$$Q = 2a \sqrt{t} \tag{13}$$

where a is a constant.

An experimental test demonstrated the validity of the square root relationship. Experimental conditions affecting data resulting from this device include stirring rate, temperature control, and sink conditions.

C. Liquid/Semisolid Interface (Free Boundary)

1. Miscible Phases

Water-miscible semisolids, such as some gels, in direct contact with an aqueous donor or receptor fluid represent a free boundary system, with transport occurring across a liquid/gel interface.

A diffusion cell was used to measure the diffusion of inorganic salts, sugars, amino acids, and proteins into agar gel from a donor solution [19]. The donor solution and diffusion cell containing the gel are first brought to thermal equilibrium. Then, any portion of gel extending beyond the end of the syringe is cut off and the syringe is dipped into the solvent just before initiating the experiment is initiated. The experiment is begun by placing the syringe in contact with the donor solution so that the gel just touches the surface. Care must be taken that air bubbles do not form on the surface of the gel. The donor solution is stirred throughout the experiment. At the end of the diffusion experiment, the gel can be extruded from the syringe and sectioned at known length intervals.

When gel sections are analyzed for solute concentration, a value C' is obtained, and when the solution bathing the gel is analyzed, a value C_0 is obtained. The ratio C'/C_0 is plotted versus gel section distance, x, on arithmetic probability paper, and the slope of this plot is used to determine the diffusion coefficient, D', in the gel as

$$D' = \frac{1}{(\text{slope})^2 (2t)} \tag{14}$$

where t is time.

This method can, in principle, be used to determine the transport characteristics of drugs dissolved or suspended in pharmaceutical gels. A potential problem with this method is the possibility for the gel to dissolve into the aqueous phase, with resulting disruption of the free boundary. These changing experimental conditions would lead to lack of precision for the experimental results and difficulty in interpreting them.

2. Immiscible Phases

Water-immiscible semisolids, such as oleaginous bases, in direct contact with an aqueous receptor or donor fluid represent a free boundary transport system, with transport occurring across the semisolid/liquid interface.

A device was designed for use in studying the release of corticosteroids suspended in an oleaginous ointment base [20]. The apparatus consists of a Teflon dish which floats on the surface of the stirred receptor fluid. After the system has been brought to thermal equilibrium, a known amount of ointment is evenly spread on the bottom surface of the Teflon dish. At particular time intervals, known volumes of the receptor fluid are removed for analysis and replaced with an equal volume of fresh receptor fluid.

Following a lag time, the cumulative amount of drug released per unit area, Q, was found to follow the relationship

$$Q = k \sqrt{t} \qquad (15)$$

where k is a constant and t is time.

In principle, this method would also be useful for transport experiments in which the drug is dissolved rather than suspended in the semisolid. Since the stirring rate can be varied, this system provides for studying stirring rate dependence. An additional advantage is that the transport results are not complicated by the use of a membrane to separate the donor and receptor phases.

3. Multiphase Systems

An apparatus has been described to study the transport of an organic solute into hexadecane–gelatin–water matrices [21]. The experiment consisted of preparing a hexane-in-water emulsion stabilized with gelatin. The uptake experiments were carried out by adding an aqueous solution of the organic solute to the beaker containing the prepared matrix. Stirring was initiated and temperature maintained. Samples were removed at predetermined time intervals for analysis.

This method provides for one-dimensional diffusion and should be useful for studying mass transport to or from a variety of multiphase systems. The method provides for studying stirring rate dependence and the mass transport mechanisms related to the system under study.

D. Liquid/Membrane/Semisolid Interface

Various diffusion cell configurations have been used to investigate drug transport from ointments, creams, and gels through a membrane to an aqueous receptor phase.

A device designed to determine the transport of an epidermal growth factor from a Pluronic gel, a Carbopol gel, and a vanishing cream base was shown capable of demonstrating formulation-dependent release kinetics [22]. This device consists of a release cell which has a membrane to minimize release due to mechanical breakdown and is placed in a stirred and temperature-controlled receptor fluid.

In a similar diffusional device, the ointment containing the drug sample was placed on a membrane over a stirred and thermostated receptor fluid [23]. Potential advantages of this device are that the unit construction fixes the distance between the stirrer and the membrane and the stirrer symmetry relative to the membrane is less susceptible to variation. This device was used to evaluate the mass transport of salicylic acid suspended in petrolatum. The resulting mass transport data were found to be dependent upon the type of membrane used, the initial concentration of salicylic acid, and the crystallinity and viscosity of the petrolatum.

A flow-through cell was used to evaluate the release of drugs dissolved in gels [24]. This device is unique in that its design permits the study of mass transport of drug from thin films of the semisolid matrix to a receptor fluid. The apparatus consists of a stainless steel membrane support screen, a membrane, a circular template, and a glass cover. The thickness of the template (75–1600 μm) determines the thickness of the application. Once the film is applied and is assembled, the cell fresh solvent is continuously pumped through the cell and collected at preset intervals in an automated fraction collector. Important to the study of mass transport of drug from a semisolid is the finding that this device sorts out partitioning, film thickness, and membrane thickness effects.

E. Liquid/Solid Interface

1. Mass Transfer from Constant Surface Area Solids

Solid–liquid mass transfer studies are of particular importance in understanding the physical, chemical, and hydrodynamic determinants of drug release from solid pharmaceuticals. When considering an appropriate method, attention must be given to hydrodynamic conditions, geometric and spatial considerations related to stirring source and dissolving surface, constancy of surface area of the dissolving surface, and physical or chemical factors which will affect the mass transport process. When considering these criteria, the rotating disk method is favored because momentum transport is known through fluid mechanics principles, and the associated mass transport is predictable via convective diffusion theory. In this case the flux, J, is given by

$$J = 0.62DC_s\omega^{1/2}v^{-1/6} \tag{16}$$

Thus, the flux is a function of the diffusion coefficient, drug solubility, C_s, angular velocity of the disk, ω, and the kinematic viscosity of the medium, v.

The power of this method is that mechanisms associated with transport processes can be identified, since the underlying hydrodynamics are known a priori and do not confound the data. For example, Singh et al. [25] demonstrated the value of this method in evaluating the predominant transport mechanism for benzocaine–polysorbate 80 systems. Mooney et al. [26] successfully used the rotating disk method to determine release rates of carboxylic acids as a function of pH. These release rates were then interpreted using a model for mass transport with chemical reaction. Johnson and Amidon [27] studied the effect of enzymatic reaction on dissolution rate using the rotating disk configuration. The coefficient of variation for the flux associated with N-benzoyl tyrosine ethyl ester with no enzyme present is 10.7% at 100 rpm and 6.7% at 300 rpm. With modification of the kinematic viscosity term, Hansford and Litt [28] applied the method to a study of solid–liquid mass transfer of benzoic acid into non-Newtonian polymeric solutions. Thus, the power and versatility of the rotating disk method have been amply demonstrated. It should be noted that the rotating disk also allows for the direct determination of the drug diffusion coefficient when J is determined experimentally and drug solubility is known.

To facilitate the use of the rotating disk configuration, Wood et al. [29] designed a unit consisting of a tablet mold and a stirring shaft. The tablet mold is used to form a disk from powdered material using a hydraulic press. Once formed, the disk remains in the tablet mold, and the tablet mold is then attached to a stirring shaft. After the shaft is fitted into a stirring motor, the unit is lowered into the temperature-controlled liquid medium to initiate the experiment. Important considerations for successful results depend on vertical alignment of the stirring shaft, precise control of stirring speed during the experiment, formation of disks having a high degree of physical integrity, and wettability of the tablet surface. It is prudent to examine the disk surface before and after the experiment to determine whether wear patterns are evident which would lead to data variability or unacceptable results.

An alternative to the rotating disk method in a quiescent fluid is a stationary disk placed in a rotating fluid. This method, like the rotating disk, is based on fluid mechanics principles and has been studied using benzoic acid dissolving into water [30]. Khoury et al. [31] applied the stationary disk method to the study of the mass transport of steroids into dilute polymer solutions. Since this method assumes that the rotating fluid near the disk obeys solid body rotation, the stirring device and the distance of the stirrer from the disk become important considerations when it is used. A similar device was developed by Braun and Parrott [32], who used stationary spherical tablets in a stirred liquid to study the effect of various parameters on the mass transport of benzoic acid.

Shah and Nelson [33] introduced a convective mass transport device in which fluid is introduced through one portal and creates shear over the dissolving surface as it travels in laminar flow to the exit portal. They demonstrated that this device produces expected fluid flow characteristics and yields mass transfer data for pharmaceutical solids which conform to convective diffusion equations.

2. Mass Transfer from Finely Divided Drug Particles

Mass transport from suspensions of pharmaceutical solids represents a challenging and important area for methodological development.

Two general methods have evolved. One uses a stirred tank system, while the other depends on the measurement of particle size or surface area change via particle counting or image analysis techniques.

The stirred tank method must be able to (1) maintain the particles in homogeneous suspension, (2) provide a homogeneous hydrodynamic environment throughout the liquid bulk phase, and (3) permit rapid and efficient sampling of the bulk to accurately assess the amount of drug released. To meet the last criterion, a nonclogging filter which allows for a high turnover of filtered fluid is required.

Shah et al. [34] described a rotating assembly in which a cylindrical filter acts as both the filter and the stirring device. The surface area of the filter is large and tends to be nonclogging. Large amounts of liquid can be rapidly filtered, assayed and returned to the bulk phase using this device. The configuration of the overall system is a rotating cylindrical filter in a stationary cylindrical tank. The similarity in shape is important because stable and predictable fluid flow patterns are generated by this configuration.

Mauger et al. [35] used the rotating filter assembly to assess the mass transport kinetics of particle populations of a steroid and demonstrated the applicability of a proposed diffusion model used to interpret the data.

An alternative to the stirred tank system is a column-type device which provides for constant fluid flow through a powder bed. The mass transport process was shown to be primarily determined by the length and cross-sectional area of the cylinder and the fluid flow rate [36].

The second general method, involving the evaluation of particle size or surface area change of the dissolving drug particle, gives a direct physical measure of the mass transport process as a function of time. Sunada et al. [37] used a LUZEX image analyzer to measure surface area changes of particles of paraben derivatives during the mass transport process. They demonstrated that simulated release profiles estimated from the surface area data coincided well with the measured dissolution profiles over almost the entire dissolution process.

F. Liquid/Cultured Cell Interface

Transport studies of small molecular weight drugs or therapeutic macromolecules using cultured epithelial and endothelial cells are relevant to devising strategies for optimal delivery of these species to targeted biological sites. The devices currently validated for drug transport studies involving cultured cells were reviewed by Audus et al. [38]. These devices have the following features in common: (1) liquid donor compartment, (2) cell culture on a microporous membrane treated with an appropriate matrix material, and 3) liquid acceptor compartment. An important difference in these methods is the presence or absence of stirring. The cell insert devices do not provide for stirring, with the result that an added resistance of the unstirred water layer (UWL) to transport may be problematic for transcellular transport of lipophilic molecules. In contrast, both side-by-side methods provide for stirring. Hidalgo et al. [39] compared an insert device with the gas-lift cell and demonstrated that the gas-lift cell yields permeability coefficients for testosterone and for mannitol which are stirring-dependent. As noted by these researchers, "the need to reduce the UWL will probably result in the development of new diffusion devices."

IV. ESTIMATING DIFFUSION COEFFICIENTS AND MASS TRANSPORT QUANTITIES

A. Diffusion Coefficients

It is useful to be able to estimate diffusion coefficients either to supplement mass transport data or to compare with experimentally determined values. A theoretically based method to estimate the diffusion coefficient includes upper and lower bounds for small molecules and large diffusants, respectively [40]. The equation

$$D = \frac{RT}{f} \tag{17}$$

is based on the relationship of the diffusion coefficient to the frictional coefficient, f, of the hydrodynamic particle as it diffuses through the liquid. As expressed by this equation, the diffusion coefficient is directly proportional to the thermal motion and inversely proportional to the frictional coefficient. The frictional coefficient can be related to the hydrodynamic radius and the solvent viscosity. When the particle is spherical, the following equations result.

When the hydrodynamic particle is small relative to the solvent molecule,

$$D = \frac{kT}{4\pi\eta} \left(\frac{(4\pi N)^{1/3}}{(3v)^{1/3}} \right) \tag{18}$$

where k is the Boltzmann constant (1.38×10^{-16} erg/K per molecule), T is absolute temperature, N is Avogadro's number (6.02×10^{23}), η is solvent viscosity (0.0089 poise for water at 25°C), and γ is the molar volume of the diffusant (mL/mol).

When the hydrodynamic particle is large relative to the solvent molecule, the appropriate equation is

$$D = \frac{kT}{6\pi\eta} \left(\frac{(4\pi N)^{1/3}}{(3v)^{1/3}} \right) \tag{19}$$

Equations (18) and (19) show that the diffusion coefficient is inversely proportional to both solvent viscosity and the molar volume of the hydrodynamic particle to the one-third power. Therefore, the molar volume scale is compressed, so that relatively large changes in molar volume are required to significantly change the diffusion coefficient.

Using the data for water at 25°C, the equations for estimating the diffusion coefficient become

$$D = \frac{4.95 \times 10^{-5}}{V^{1/3}} \quad \text{and} \quad D = \frac{3.30 \times 10^{-5}}{V^{1/3}} \tag{20}$$

The diffusion coefficient can be estimated if the molar volume of the diffusant is known or can be estimated. Flynn et al. [40] provided a useful table of partial molal volume data of some atoms and groups. Thus, one can estimate the molar volume from the chemical structure of the compound. Stout et al. [2] calculated a molar volume for hydrocortisone of 279 mL/mol using these data. The resulting estimates for D are 5.1×10^{-6} to 7.6×10^{-6} cm^2/sec. An experimentally determined value for hydrocortisone is 4.2×10^{-6} cm^2/sec [2]. If it is assumed that the molar volume is approximately the molecular weight divided by the density, then the molar volume can be calculated if density values are available. For many small molecular weight pharmaceuticals, the true density ranges from 1.2 to 1.5 g/mL. Using benzoic acid (MW = 122) as an example, the molar volume is estimated to be 94 mL/mol for a density of 1.3. The resulting range of diffusion coefficients is 7.3×10^{-6}–1.09×10^{-5}, which compares well with the experimentally determined value of 1.08×10^{-5} [32].

Although these examples demonstrate the feasibility of using calculated values as estimates, several constraints and assumptions must be kept in mind. First, the diffusant molecules are assumed to be in the dilute range where Henry's law applies. Thus, the diffusant molecules are presumed to be in the unassociated form. Furthermore, it is assumed that other materials, such as surfactants, are not present. Self-association or interaction with other molecules will tend to lower the diffusion coefficient. There may be differences in the diffusion coefficient for molecules in the neutral or charged state, which these equations do not account for. Finally, these equations only relate diffusion to the bulk viscosity. Therefore, they do not apply to polymer solutions where microenvironmental viscosity plays a role in diffusion.

Other empirical relationships have been proposed for estimating the diffusion coefficient. Higuchi et al. [41] plotted the logarithm of the molecular weight versus the diffusion coefficient in water for a number of organic compounds selected randomly from International Critical Tables. The result was a linear relationship with a negative slope close to $1/2$. According to Higuchi et al., this suggests a relationship of the form

$$Dm^{1/2} = \text{constant} \tag{21}$$

On the basis of this relationship, the ratio of the diffusion coefficient of molecule X to the diffusion coefficient of a particular molecule, A, is given by

$$\frac{D_X}{D_A} = \frac{M_A^{1/2}}{M_X^{1/2}} \tag{22}$$

Higuchi applied this method to 2-naphthol as molecule A and used the resulting ratio for several compounds and complexes to treat mass transport data for complexation using a particular model. In general, ad hoc methods may be useful, particularly when dealing with a homologous series or a series of related compounds.

As noted by Liu et al. [42], there is a resurgence of interest in measuring diffusion coefficients using NMR spectroscopy, particularly in complex biological fluids which contain a number of molecules having a broad range of molecular weights and concentrations. These authors demonstrated the use of two-dimensional diffusion-edited total correlation NRM spectroscopy to measure diffusion coefficients in human blood samples.

B. Mass Transport Quantities

The flux, amount per unit time per unit surface area, is a valuable quantity to be able to estimate a priori. The rotating disk model enables this calculation

$$J = 0.62 D^{2/3} C_s \omega^{1/2} \nu^{-1/6} \tag{23}$$

By inspection, the flux is directly proportional to the solubility to the first power and directly proportional to the diffusion coefficient to the two-thirds power. If, for example, the proposed study involves mass transport measurements for series of compounds in which the solubility and diffusion coefficient change incrementally, then the flux is expected to follow this relationship when the viscosity and stirring rate are held constant. This model allows the investigator to simulate the flux under a variety of conditions, which may be useful in planning experiments or in estimating the impact of complexation, self-association, and other physicochemical phenomena on mass transport.

The Higuchi–Hiestand model [43] permits the a priori estimation of the mass transport coefficient for the dissolution of finely divided drug particles. The model relates the particle radius, $a(t)$, with time according to

$$a(t) = \left(a_0^2 - \frac{2DC_s t}{\rho} \right)^{1/2} \tag{24}$$

Given the solubility, the diffusion coefficient, and the density of the solid, the lifetime of a particle of initial size a_0 can be estimated. Furthermore, this equation can be integrated over particle size to simulate the effect of a multisized population of fine particles on mass transport [44].

In comparison to the two models just discussed, a commonly used mass transport model is

$$J = \frac{DAC_s}{h} \tag{25}$$

where h is the diffusion layer thickness. An unfortunate aspect of this model is that h is dependent upon experimental conditions (hydrodynamics, viscosity, geometry) and cannot be calculated a priori. The investigator would have to calibrate the intended method using a compound having known solubility and diffusion coefficient values in order to calculate values of h under relevant experimental conditions. This would be accomplished by measuring the flux for several stirring rates and then using these data to calculate values for h. Assuming that the calculated values for h do not vary greatly for the test compounds, these values would be used along with known diffusion coefficient and solubility data to estimate the flux.

V.   ERROR ANALYSIS AND METHOD VALIDATION

The ability of any experimental method to produce accurate and reproducible results and provide the sensitivity needed to discern differences between transport mechanisms depends on minimizing variability intrinsic to the method. However, formal error analysis is rarely undertaken, even for commonly used methods. Fawcett and Caton [45] performed an error analysis of the capillary method for determining diffusion coefficients more than 25 years after the method was introduced. The value of the analysis is that it reveals which factors contribute the greatest variability to the dependent variable of interest. In the case of transport studies, the dependent variable of primary interest is diffusant concentration, $C(t)$, where

$$C(t) = f(A, t, h) \tag{26}$$

where A is the area for diffusion and h is the diffusional path length. The goal is to evaluate the variability of $C(t)$ as a function of some or all of these factors. Thus,

$$dC(t) = \frac{\partial C(t)}{\partial A} dA + \frac{\partial C(t)}{\partial h} dh + \frac{\partial C(t)}{\partial t} dt \tag{27}$$

An analysis of the Polson cell method [4] is instructive. The concentration of the diffusant, $C(t)$, in the receptor column of length h is measured at time t, and then the diffusion coefficient is calculated using this value. Thus, the quality of the diffusion coefficient value is dependent on the accuracy and reproducibility of the concentration data. The relationship between the diffusion coefficient and $C(t)$ is

$$D = \frac{(C(t))^2}{C_0^2} h^2 \left(\frac{\pi}{t}\right) \qquad (28)$$

Since D depends on $C(t)$ to a power of 2, small changes in $C(t)$ will cause significant variability in D. Thus, the analysis should focus on strategies to minimize this variability. Rearranging Eq. (28) gives

$$\frac{C(t)}{C_0} = \frac{(Dt)^{1/2}}{\pi^{1/2}} \left(\frac{1}{h}\right) \qquad (29)$$

which shows that $C(t)$ increases in direct proportion to the square root of time.

The relationship of interest is the rate of change of $C(t)$, which is given by

$$\frac{\partial C(t)}{\partial t} = 0.5 \frac{D^{1/2}}{(\pi t)^{1/2}} \left(\frac{C_0}{h}\right) \qquad (30)$$

It is evident from this expression that the variation of $C(t)$ will decrease in proportion to the inverse of the square root of time. Since $C(t)$ changes rapidly during the early part of the sampling period, there is an expected increase in variability in $C(t)$ which will, in turn, affect the variability of calculated values for D. Therefore, this simple but useful error analysis shows that the expected variation in calculated values for the diffusion coefficient will be diminished if the concentrations are measured later during the relevant sampling period.

In addition to performing an error analysis, each experimental method should be validated for its application to pharmaceutical systems using diffusants having known physicochemical or biological properties before being used to experimentally evaluate new drug species. For example, diffusants with different molar volumes might be selected so as to test the sensitivity of the transport method to changes in the hydrodynamic radius. Using diffusants having known diffusion coefficients permits an evaluation of the accuracy and precision of the method. Finally, an evaluation of the pertinent operating variables (stirring rate, effect of membrane thickness, assessment of lag time) should be included as part of the validation process. The diffusional methods evaluated by Stout et al. [2] and Flynn and Smith [14] are examples of the application of these validation principles.

Table 1 Summary of Experimental Methods for Evaluating Diffusion Coefficients and Investigating Mass Transport Processes of Pharmaceutical Interest

Method	Application	Ref.
1. Capillaries and tubes (free boundary method)	Diffusion coefficient determination	1,2
2. Polson cell (free boundary method)	Diffusion coefficient determination	3
3. Hydrodynamic stability (free boundary method)	Diffusion coefficient determination	5
4. Liquid/liquid stirred cell (free boundary method)	Mass transport between immiscible phases	6–8
5. Rocking device with two- and three-phase cells (free boundary method)	Nonemulsifying method for determining partition coefficient	9,10
6. Side-by-side diffusion cell (membrane method)	Diffusion coefficient determination; mass transport studies	10–14
7. Diffusion cell for heterogeneous systems (membrane method)	Mass transport studies from emulsions	18
8. Diffusion cell for gels (free boundary method)	Mass transport studies from water-miscible gels	19
9. Method for oleaginous ointment bases (free boundary method)	Mass transport from water-immiscible ointment bases	20
10. Method for semisolids (membrane method)	Mass transport from water-miscible gels or water-immiscible ointment bases	22,23
11. Thin-film transport cell (membrane method)	Drug transport from thin films of topical dosage forms	24
12. Rotating disk method (free boundary method)	Diffusion coefficient determination from solids; mass transport studies	25–29
13. Convective diffusion cell (free boundary method)	Mass transport studies from constant surface area solids	33
14. Rotating filter assembly (free boundary method)	Mass transport studies from finely divided drug powders	34,35
15. Column-type cell (free boundary method)	Mass transport studies from finely divided drug powders	36
16. Cell insert (membrane method)	Mass transport studies using cultured cells (nonstirred system)	38
17. Side-by-side diffusion cell (membrane method)	Mass transport studies using cultured cells (stirred system)	38
18. NMR spectroscopy	Diffusion measurements of complex samples of macromolecules; also provides information on small molecule–macromolecule interactions	42

VI. SUMMARY

Table 1 summarizes several of the experimental methods discussed in this chapter. A need exists for new or revised methods for transport experimentation, particularly for therapeutic proteins or peptides in polymeric systems. An important criterion for the new or revised methods includes in situ sampling using micro techniques which simultaneously sample, separate, and analyze the sample. For example, capillary zone electrophoresis provides a micro technique with high separation resolution and the potential to measure the mobilities and diffusion coefficients of the diffusant in the presence of a polymer. Combining the separation and analytical components adds considerable power and versatility to the method. In addition, up-to-date separation instrumentation is computer-driven, so that methods development is optimized, data are acquired according to a predetermined program, and data analysis is facilitated.

REFERENCES

1. K Farng, K Nelson. Effects of polyelectrolytes on drug transport I: Diffusion. J Pharm Sci 62:1435, 1973.
2. P Stout, N Khoury, J Mauger, S Howard. Evaluation of a tube method for determining diffusion coefficients for sparingly soluble drugs. J Pharm Sci 75:65, 1986.
3. Y Walbroehl, J Jorgenson. Capillary zone electrophoresis for determination of electrophoretic mobilities and diffusion coefficients. J Microcolumn Separ 1:41, 1989.
4. A Polson. New method for measuring diffusion constants of biologically active substances. Nature 154:823, 1944.
5. M Southard, L Dias, K Himmelstein, V Stella. Experimental determinations of diffusion coefficients in dilute aqueous solution using the method of hydrodynamic stability. Pharm Res 8:1489–1491, 1991.
6. L Stowe, J Shaewitz. Hydrodynamics and mass transfer characteristics of a liquid/liquid stirred cell. Chem Eng Commun 11:17, 1981.
7. P Byron, M Rathbone. Prediction of interfacial transfer kinetics. I. Relative importance of diffusional resistance in aqueous and organic boundary layers in two-phase transfer cell. Int J Pharm 21:107, 1984.
8. P Bryon, R Notari, E Tomlinson. Calculation of partition coefficients of an unstable compound using kinetic methods. J Pharm Sci 69:527, 1980.
9. D Reese, G Irwin, L Dittert, C Chong, J Swintosky. Nonemulsifying method for measuring distribution coefficients. J Pharm Sci 53:591, 1964.
10. J Doluisio, J Swintosky. Drug partitioning II. In vitro model for drug absorption. J Pharm Sci 53:597, 1964.
11. A Goldberg, W Higuchi. Improved method for diffusion coefficient determinations employing the silver membrane filter. J Pharm Sci 57:1583, 1968.
12. K Tojo, Y Sun, M Ghannam, Y Chien. Characterization of a membrane permeation system for controlled drug delivery studies. AIChE J 31:741, 1985.

13. A Aguiar, M Weiner. Percutaneous absorption studies of chloramphenicol solutions. J Pharm Sci 58:210, 1969.

14. G Flynn, E Smith. Membrane diffusion I: Design and testing of a new multifeatured diffusion cell. J Pharm Sci 11:1713, 1971.

15. E Garrett, P Chemburkar. Evaluation, control and prediction of drug diffusion through polymeric membranes I. Methods and reproducibility of steady-state diffusion studies. J Pharm Sci 57:944, 1968.

16. S Pirt. The diffusion capsule, a novel device for the addition of a solute at a constant rate to a liquid medium. Biochem J 121:293, 1971.

17. V Srinivasan, W Higuchi, S Sims, A Ghanem, C Behl. Transdermal iontophoretic drug delivery: Mechanistic analysis and application to polypeptide delivery. J Pharm Sci 78:370, 1989.

18. T Koizumi, W Higuchi. Analysis of data on drug release from emulsions II. Pyridine release from water-in-oil emulsions as a function of pH. J Pharm Sci 57:87, 1968.

19. E Schantz, M Lauffer. Diffusion measurements in agar gel. Biochemistry 1:658, 1962.

20. Z Chowhan, R Pritchard. Release of corticoids from oleaginous ointment bases containing drug in suspension. J Pharm Sci 64:754, 1975.

21. A Ghanem, W Higuchi, A Simonelli. Interfacial barriers in interphase transport III: Transport of cholesterol and other organic solutes into hexadecane–gelatin–water matrices. J Pharm Sci 59:659, 1970.

22. M DiBase, C Rhodes. Investigations of epidermal growth factor in semisolid formulations. Pharm Acta Helv 66:165, 1991.

23. M Kneczke, L Landersjo, P Lundgren, C Fuhrer. In vitro release of salicylic acid from two different qualities of white petrolatum. Acta Pharm Suec 23:193, 1986.

24. W Addicks, G Flynn, N Weiner, C Chiang. Drug transport from thin applications of topical dosage forms: Development of methodology. Pharm Res 6:377, 1988.

25. P Singh, S Desai, D Flanagan, A Simonelli, W Higuchi. Mechanistic study of the influence of micelle solubilization and hydrodynamic factors on the dissolution rate of solid drugs. J Pharm Sci 57:959, 1968.

26. K Mooney, M Mintun, K Himmelstein, V Stella. Solution kinetics of carboxylic acids I. Effect of pH under unbuffered conditions. J Pharm Sci 70:13, 1981.

27. D Johnson, G Amidon. The effect of enzymatic reaction on dissolution rate: Theoretical analysis and experimental test. J Pharm Sci 75:195, 1986.

28. G Hansford, M Litt. Mass transport from a rotating disk into power-law liquids. Chem Eng Sci 23:849, 1968.

29. J Wood, J Syarto, H Letterman. Improved holder for intrinsic dissolution rate studies. J Pharm Sci 54:1068, 1965.

30. K Smith, C Colton. Mass transfer to a rotating fluid, Part II. Transport from the base of an agitated cylindrical tank. AIChE J 18:958, 1972.

31. N Khoury, J Mauger, S Howard. Dissolution rate studies from a stationary disk/ rotating fluid system. Pharm Res 5:495, 1988.

32. R Braun, E Parrot. Effect of various parameters upon diffusion-controlled dissolution of benzoic acid. J Pharm Sci 61:592, 1972.

33. A Shah, K Nelson. Evaluation of a convective diffusion drug dissolution rate model. J Pharm Sci 64:1518, 1975.

34. A Shah, C Peot, J Ochs. Design and evaluation of a rotating filter–stationary basket in vitro dissolution test apparatus I: Fixed fluid volume system. J Pharm Sci 64:671, 1973.

35. J Mauger, S Howard, K Amin. Dissolution profiles for finely divided drug suspensions. J Pharm Sci 70:190, 1983.

36. F Langenbucher. In vitro dissolution kinetics: Description and evaluation of a column-type method. J Pharm Sci 58:1265, 1969.

37. H Sunada, A Yamamoto, A Otsuka, Y Yonezawa. Changes of surface area in the dissolution process of crystalline substances. Chem Pharm Bull 36:2557, 1989.

38. K Audus, R Bartel, I Hidalgo, R Borchardt. The use of cultured epithelial and endothelial cells for drug transport and metabolism studies. Pharm Res 7:435, 1990.

39. I Hidalgo, K Hillgreen, G Grass, R Borchardt. Characterization of the unstirred water layer in Caco-2 cell monolayers using a novel diffusion apparatus. Pharm Res 8:222, 1991.

40. G Flynn, S Yalkowsky, T Roseman. Mass transport phenomena and models: Theoretical concepts. J Pharm Sci 63:479, 1974.

41. T Higuchi, S Dayal, I Pitman. Effects of solute–solvent complexation reactions on dissolution kinetics: Testing of a model by using a concentration jump technique. J Pharm Sci 61:695, 1972.

42. M Lui, JK Nicholson, JA Parkinson, JC Lindon. Measurement of bimolecular diffusion coefficients in blood plasma using two-dimensional ^{1}H-^{1}H diffusion-edited total-correlation NMR spectroscopy. Anal Chem 69:1504–1509, 1997.

43. W Higuchi, E Hiestand. Dissolution rates of finely divided drug powders. J Pharm Sci 52:67, 1963.

44. J Mauger, S Howard. Model systems for dissolution of finely divided (multisized) drug powders. J Pharm Sci 65:1042, 1976.

45. N Fawcett, R Caton. Analysis of errors in the capillary method for determining diffusion coefficients. Anal Chem 48:228, 1976.

5

Dissolution of Pharmaceuticals in Simple and Complex Systems

Daniel P. McNamara
Boehringer Ingelheim Pharmaceuticals, Ridgefield, Connecticut

Michael L. Vieira and John R. Crison
PORT Systems, LLC, Ann Arbor, Michigan

I. INTRODUCTION

Pharmaceuticists (a term coined by an emeritus senior pharmaceutics professor) have always been interested in understanding factors that influence the dissolution rate and mass transfer of drugs from the solid state into solution. This may seem an obvious comment given the current trend of suspecting most low solubility new chemical entities of exhibiting dissolution rate–limited absorption. But the fact is that dissolution phenomena have been an active area of pharmaceutical research for decades. This research evolved from a practical premise that if one could understand how factors such as drug solubility, fluid stirring rate, exposed solid surface area, and added excipient ingredients influence dissolution, then perhaps clever formulators could apply this information to control and modify the dissolution of drugs and ultimately influence drug absorption and therapy. In this chapter several areas of dissolution research are critically reviewed, including: (1) the dissolution of acidic and basic drugs in unbuffered and buffered solutions, (2) the dissolution of multicomponent noninteracting and multicomponent interacting solids, (3) dissolution in micellar and emulsion systems, and (4) particle dissolution.

In the first sections of the chapter, we begin with discussions of simpler systems, wherein typically only pure solid drug dissolves and then reacts with species in the dissolution media. Then we proceed to the more complex, but practically relevant, situation of multicomponent solid mixtures (i.e., drugs plus excipients) dissolving and reacting. Since the simpler dissolution systems have been studied by more investigators for longer periods of time, the literature is extensive and thorough. Fortunately, the arduous task of winnowing through the archives of these dissolution papers yields some simple but elegant expressions. These hard-won insights, gained through the toil of generations of graduate students, are applicable to the dissolution of acidic and basic drugs in unbuffered and buffered media. The dissolution of multicomponent solid mixtures containing multiple excipients and drugs has received much less scrutiny and study, despite the practical relevance of this topic to dosage forms which always contain multiple components. Consequently, the multicomponent dissolution literature is more anecdotal than fundamental. Close examination of some more recent papers in the pharmaceutical literature discloses a promising trend toward increasing numbers of researchers who are venturing into the multicomponent dissolution breach.

II. DISSOLUTION OF ACIDIC AND BASIC DRUGS IN UNBUFFERED MEDIA

In 1955, Parrott, Wurster, and T. Higuchi published a paper entitled "Investigation of Drug Release from Solids I. Some Factors Influencing the Dissolution Rate" [1]. This paper was important because it focused attention on physicochemical properties such as intrinsic solubility of the drug, which determine mass transfer from the solid crystalline state, while deemphasizing the role of tablet disintegration, which had so dominated the drug release research of that day. This paper also introduced, albeit with vague explanation, the phenomenon that dissolution of a weak acid (benzoic acid) could be increased by addition of base (urea) to the dissolution media.

By 1958, W. Higuchi, Parrot, Wurster, and T. Higuchi (HPWH) had sufficient experimental evidence and enough theoretical insight to publish their classic paper on the influence of bases on the dissolution of acidic solids [2]. In the derivation of the HPWH dissolution and reaction model, the authors adopted a Nernst–Brunner stagnant film hypothesis, which employed the concept that a stagnant film of fluid adhered to the surface of the dissolving solid. They also assumed that (1) the thickness of the film adhering to the solid surface (δ) was independent of diffusivities at constant stirring rate, viscosity, and geometry of the system; (2) diffusion coefficients were concentration-independent; and (3) chemical reaction of the dissolved acid was rapid compared to diffusion. The dissolution model developed was for a weak solid acid (HA) dissolving into aqueous media containing basic species (B). The acid, HA, once dissolved, could react according to the scheme

$$HA + B \rightleftharpoons A^- + HB^+ \tag{1}$$

where HB^+ is the conjugate acid of the base, B, and A^- is the conjugate base of the dissolving solid weak acid, HA. The reaction in Eq. (1) can be characterized by an equilibrium constant (K) as

$$K = \frac{k_f}{k_r} = \frac{C_{A^-} C_{HB^+}}{C_{HA} C_B} \tag{2}$$

where k_f and k_r represent the forward and reverse second-order micro rate constants associated with the reaction in Eq. (1) and C_i is the concentration of species i. Figure 1 is a schematic representation of the HPWH model.

The steady-state continuity equations which describe mass balance over a fluid volume element for the species in the stagnant film which are subject to uniaxial diffusion and reaction in the z direction are

$$D_{HA} \frac{d^2 C_{HA}}{dz^2} - \kappa_f C_{HA} C_B + \kappa_r C_{A^-} C_{HB^+} = 0 \tag{3}$$

$$D_{A^-} \frac{d^2 C_{A^-}}{dz^2} + \kappa_f C_{HA} C_B - \kappa_r C_{A^-} C_{HB^+} = 0 \tag{4}$$

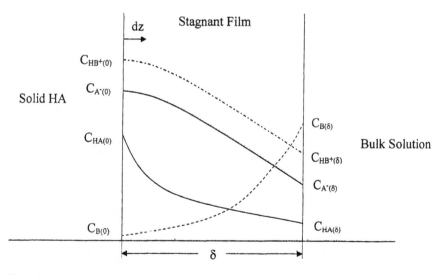

Figure 1 Stagnant film model for dissolution of acidic drug (HA) in solution with base (B). (From Ref. 2.)

$$D_B \frac{d^2 C_B}{dz^2} - \kappa_f C_{HA} C_B + \kappa_r C_{A^-} C_{HB^+} = 0 \tag{5}$$

$$D_{HB^+} \frac{d^2 C_{HB^+}}{dz^2} + \kappa_f C_{HA} C_B - \kappa_r C_{A^-} C_{HB^+} = 0 \tag{6}$$

where D_i represents the diffusion coefficient of species i. The reader will note that Eqs. (3)–(6) are a set of simple second-order differential equations. Olander [3] pointed out that a clever technique that can be used to simplify this set of equations is to add equations so the reaction terms will cancel. Equation (7) shows how Eqs. (3) and (4) can be combined as a material component balance for the acid species HA.

$$D_{HA} \frac{d^2 C_{HA}}{dz^2} + D_{A^-} \frac{d^2 C_{A^-}}{dz^2} = 0 \tag{7}$$

This combined equation represents a differential total material balance of a component, whether present as HA or the reaction product A$^-$, within the reacting phase. The reader is referred to Olander's original paper for a more complete rationale for generating these differential component material balances for systems of reacting species near equilibrium. By using Olander's technique, the system of four differential equations with reaction terms can be simplified significantly to two differential equations with no reaction terms.

The boundary conditions for this early dissolution model included saturated solubility for HA at the solid surface ($C_{HA(0)}$) with sink conditions for both HA and A$^-$ at the outer boundary of a stagnant film ($C_{HA(\delta)} = C_{A(\delta)} = 0$). Since diffusion is the sole mechanism for mass transfer considered and the process occurs within a hypothesized stagnant film, these types of models are colloquially referred to as "film" models. Applying the simplifying assumption that the base concentration at the solid surface is negligible relative to the base concentration in the bulk solution ($C_{B(\delta)} \gg C_{B(0)}$), it is possible to derive a simplified scaled expression for the relative flux (N/N_0) from HPWH's original expressions:

$$N/N_0 = 1 + D_B C_{B(\delta)} / D_{HA} C_{HA(0)} \tag{8}$$

where N_0 is the flux of HA without reaction and N is the flux of HA with reaction.

When the relative flux is 1, reaction of the basic species with the acid does not contribute to the dissolution of HA since the concentration of B is too low relative to the saturated solubility of HA to impact the dissolution rate ($C_{B(\delta)} \ll C_{HA(0)}$). As the concentration of the basic species in bulk increases ($C_{B(\delta)} \approx C_{HA(0)}$), the relative flux will approximately double and the reaction of B with HA will begin to dominate the dissolution process. The utility and elegance of Eq. (8) is that one can estimate the increase in dissolution of an acid due to reaction with a base by simply knowing the base concentration in the dissolution media and the intrinsic solubility of the acid. It should be noted that this same result was derived originally by Hatta (see Ref. 4) in the chemical engineering literature.

Table 1 lists experimental results from HPWH's original paper [2] in which the dissolution of benzoic acid tablets in aqueous sodium hydroxide solutions was measured gravimetrically. Results from other dissolution experiments in acetate, phosphate, carbonate, and tetraborate buffers, where agreement between theory and experiment were comparable to those listed in Table 1, established this paper and the theoretical HPWH model as the premier reference for dissolution with reaction in pharmaceutics in the 1960s and throughout the 1970s.

More recently, in 1981 Mooney, Mintun, Himmelstein, and Stella (MMHS) [5] published a significantly improved film model for solid dissolution and reaction. By the 1980s, analytical techniques used to detect the amount of drug dissolved had advanced to the point where discrepancies with the earlier HPWH model could not be ignored. MMHS noticed specifically that when they measured the dissolution of a weak acids into unbuffered water at pH's near the pK_a of the acid (pH 4–5), the flux increased by a factor of 2 compared to the flux at pH 2. There was no way to explain this result using the simplified HPWH-derived relative flux expression [Eq. (8), with $C_{B(\delta)} \ll C_{HA(0)}$]. MMHS developed their own film model using the same assumptions as the HWPH model, but they insightfully modified the complexity of the reaction scheme. In the new MMHS model, the acidic drug, HA, could react with the base according to Eq. (1) but the acid could also ionize:

$$HA \rightleftharpoons H^+ + A^- \tag{9}$$

Table 1 Relative Flux (N/N_0) for Benzoic Acid Tablets in Sodium Hydroxide Solutions at 25°C

pH	Relative flux (N/N_0)	
	Theoretical[a]	Experimental
12.1	1.96	1.65
12.5	3.39	3.29
12.8	4.78	5.91
13.1	9.56	10.75
13.35	18.0	17.8
13.4	19.1	20.6

[a] $D_B = 2.4 \times 10^{-5}$ cm²/sec, calculated by fitting data to the theoretical expression.
Source: Ref. 2.

Additionally, the base, B, could react with a proton:

$$B + H^+ \rightleftharpoons HB^+ \tag{10}$$

These subtle additions to the reaction scheme, coupled with boundary conditions similar to those used by HWPH, led to the following more complex expression describing the relative flux of HA:

$$\frac{N}{N_0} = 1 + \frac{D_H(C_{H^+(0)} - C_{H^+(\delta)})}{D_{HA}C_{HA(0)}} + \frac{D_B(C_{B(\delta)} - C_{B(0)})}{D_{HA}C_{HA(0)}} \tag{11}$$

There are obvious similarities in the derived relative flux expressions for the MMHS model and the precursor HWPH model. The second term on the right-hand side of Eq. (11) accounts for the increase in dissolution observed near the pK_a of the acid. Addition of the ionization reaction [Eq. (9)] provides the added flexibility and accountability so that dissolution at low pH's can be accurately predicted. The agreement between theoretical predictions and experimental results for three carboxylic acids with intrinsic solubilities ranging from 10^{-2} to 10^{-6} M over a pH range of 2–12 was good (see Fig. 2). Computationally, the MMHS model was also quite reasonable to use; the roots for a quadratic expres-

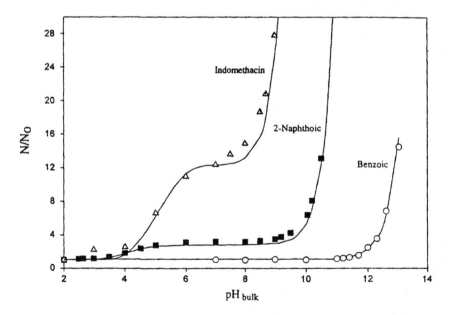

Figure 2   Relative flux (N/N_0) versus pH_{bulk} for several carboxylic acids at 25°C. N_0 is the respective flux at pH 2. (From Ref. 5.)

sion in $C_{H^+(0)}$ were required. The model qualitatively showed that the less soluble acids could not buffer themselves against hydroxide ions that would stream in, through the boundary layer, from the bulk and react with the acid near the solid surface. Hence, an acid like indomethacin, with an intrinsic solubility of 10^{-6} M, had an increase in relative flux of some 10–20-fold once the hydroxide ion concentration in the bulk was comparable to the intrinsic acid solubility in the region of pH 8.

The derivation and experimental verification of the MMHS model represented a significant accomplishment and a natural plateau for film models. To be sure, there are general criticisms of film models and more specific criticisms of the MMHS model [6]. However, overall the MMHS model should be recognized as a robust but simply applicable model which serves to demonstrate how factors such as intrinsic solubility of the acid drug, ionization and pK_a of the drug, and concentration of the reactive base all contribute to increasing the dissolution rate and mass transfer.

As an alternative to film models, McNamara and Amidon [6] included convection, or mass transfer via fluid flow, into the general solid dissolution and reaction modeling scheme. The idea was to recognize that diffusion was not the only process by which mass could be transferred from the solid surface through the boundary layer [7]. McNamara and Amidon constructed a set of steady-state convective diffusion continuity equations such as

$$D_i \frac{d^2 C_i}{dz^2} - v_z \frac{dC_i}{dz} = 0 \tag{12}$$

where v_z is the fluid velocity in the axial z direction. The second term in Eq. (12) represents the convective contribution to mass transfer, while the first term is the familiar diffusive term. Note that there are no reaction terms in the continuity equations. The well-controlled hydrodynamics of the rotating disk experimental setup afforded McNamara and Amidon a known fluid velocity profile in the boundary layer. They were able to show that dissolution and reaction could be modeled as a process driven by diffusion plus convection where convection accounted for 40% of the relative mass transfer capacity through the fluid boundary layer. The drawback to including convection explicitly in the model was that the chemical reactions of the dissolved species had to be described as heterogeneous reactions. These heterogeneous reactions were confined to reaction surfaces or "planes" within the boundary layer. Cussler's book [8] contains an excellent discussion of heterogeneous reactions. The satisfaction of including convection in the physical understanding of the dissolution process and abandoning the stagnant film concept was somewhat tempered by the artificial treatment of the chemical reactions and restricting them to planes. The convective diffusion approach used by McNamara and Amidon fit the experimental data well for acidic drugs.

The convective diffusion approach for weak basic drugs dissolving in acidic media was previously derived by Litt and Serad [9] and, as shown by McNamara and Amidon, could be applied to dissolution of the free base of papaverine.

Other researchers used flow between two parallel plates as the experimental and theoretical system to incorporate diffusion plus convection into their dissolution modeling and avoid film model approximations [10]. Though they did not consider adding reactions to their model, these workers did show that convection was an important phenomenon to consider in the mass transfer process associated with solid dissolution. In fact, the dissolution rate was found to correlate with flow as

$$\text{Dissolution rate} \propto Q^{1/3} \tag{13}$$

where Q is the volumetric flow rate of the fluid over a solid compact.

A paper by Ozturk, Palsson, and Dressman (OPD), reporting a refinement of the MMSH model, did create some controversy. OPD developed a film model with reaction in spherical coordinates and applied quasi-steady-state assumptions to the boundary conditions at the solid surface [11]. They theorized that the flux of all species at the solid surface must be zero, except for HA, or the other species (A^-, H^+, OH^-) would penetrate the solid surface. A debate by correspondence in the Letters to the Editor columns of *Pharmaceutical Research* ensued [12,13]. The reader is invited to evaluate which author's arguments are more convincing. What is difficult to evaluate is whether the OPD model produces dissolution results which are different from those which would be predicted using the MMSH model cast in comparable spherical geometry. Simply, these authors never graphically demonstrate how their model predictions compare to the MMSH model. Algebraically, the solutions to both models appear comparable.

In 1992, Southard et al. [14] revisited the dissolution and reaction problem and employed numerical methods to solve the complete problem of dissolution, diffusion, convection, and reaction of an acidic solid from the rotating disk. They should be recognized for their determination, but the results of their efforts were less than completely successful, though they did solve the complete problem, explicitly including effects due to diffusion, convection, and reaction, neither employing films nor confining reactions to planes. However, the model they developed was limited in that it required the use of microscopic reaction rates which were orders of magnitude smaller than known values for proton transfer reactions as described by Eqs. (1), (9), and (10). Attempts to use more realistic reaction rates resulted in numerical instability of the model and lack of convergence. The reaction rate limitations forced the authors to concede that the concentration profiles which were calculated with the model represented ''nonequilibrium'' concentrations. Other serious limitations of the numerical model included the inability to estimate dissolution effects over the entire pH range, especially with slightly soluble acids (results for indomethacin could not be estimated below pH 6). Given the shortcomings of this complete numerical model, it is debatable if the model represents a real improvement in the prediction and understanding of dissolution and reaction or whether it is more of a frustrating exercise in numerical analysis and computing. Today, the opportunity is still available, for those inclined to numerical analysis, to derive a robust numerical dissolution model capable of including contributions due to diffusion, convection, and reaction which can encompass the entire pH scale with the use of true micro rate constants.

III. DISSOLUTION OF ACIDIC AND BASIC DRUGS INTO BUFFERED SOLUTION

Although Section II of this chapter dealt with dissolution into unbuffered water, it has long been recognized that drug dissolution is affected by buffer type and concentration. Higuchi et al. [2] (HPWH) did apply their original 1958 film model to the case of benzoic acid dissolving in acetate, phosphate, carbonate, and tetraborate buffers. However, their approach for modeling HA dissolution into a buffered, as opposed to an unbuffered, solution was simply to treat the basic species B concentration as the acetate, phosphate, carbonate, or tetraborate conjugate base concentration and to ignore the hydroxide ion bulk concentration ($C_{OH^-(\delta)} \approx 0$). This approach was reasonable as long as the basic species in the buffer was orders of magnitude more concentrated than hydroxide ion ($C_{B(\delta)} \gg C_{OH^-(\delta)}$).

Again, Mooney et al. [15] really defined the effects of buffers on the dissolution of acidic drugs with their 1981 manuscript on buffer dissolution. These authors considered the dissolution of acidic drugs into acetate, phosphate, and citrate buffers using the same film model approach that served them so well in the unbuffered case. Unlike HPWH's earlier attempt, MMHS specifically included all of the reactions that the basic buffer species might have with HA, and they also included all of the buffer ionizations in the reaction schemes. It is no small trick to keep track of each of the buffer species in all of the reactions. The result was a film model with some additional algebraic complexity over the unbuffered case, but again, MMHS were able to show that the data were in good agreement with the results. The computational price one has to pay to use the MMHS model in buffered solutions is that a quartic equation in $C_{H^+(0)}$ must be solved. While this involves some computational effort, the resulting model estimates are quite good. The practical consequence of applying this dissolution model is that, in order to understand whether reaction of the acid is controlling dissolution, it is not enough to specify the pH and the buffer solution used during dissolution studies. To truly understand the controlling factors one must know the solution pH, the buffer employed, and the buffer concentration, in addition

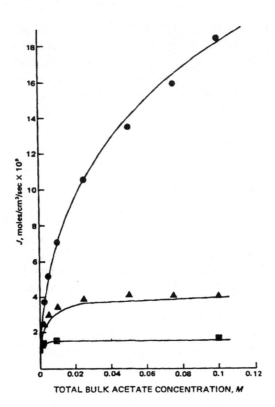

Figure 3 Plot of dissolution rate of 2-naphthoic acid from solid disk at 450 rpm into 25°C buffer media of varying acetate concentration and pH_{bulk}. (■) pH_{bulk} 4.5; (▼) pH_{bulk} 5.0; (●) pH_{bulk} 6.0. (From Ref. 15.)

to the intrinsic solubility and pK_a of the drug. Figure 3 shows the predictions and experimental results from the MMHS model for 2-naphthoic acid dissolution as a function of acetate buffer concentration at bulk pH 4.5, 5.0, and 6.0. As demonstrated by MMHS, and shown graphically in Figure 3, all of these factors are key to determining the rate of drug dissolution. In an auxiliary publication, Aunins et al. [16] again applied their improved film model and showed that the model was applicable to acids dissolving in polyionizable buffers.

An interesting paper that attempted to relate dissolution of a poorly soluble acidic drug (naproxen) to simulated gastrointestinal flow in the presence of buffers was published by Chakrabarti and Southard [17]. In addition to showing that buffer type (citrate, phosphate, or acetate) had a significant impact on naproxen dissolution, these authors unexpectedly found that elevating a solid tablet into the flow channel of the flow-between-two-plates apparatus resulted in a substantial increase in dissolution. This increase in dissolution was greater than what would be predicted on the basis of increased surface area exposure from the sides of the elevated tablet. Chakrabarti and Southard postulated that the presence of the tablet in the flow stream created complex hydrodynamic nonlaminar flow patterns that increased convective removal of the drug from the boundary layer and hence increased dissolution.

IV. DISSOLUTION OF MULTICOMPONENT SOLIDS

Since dosage forms contain more than just active drug, it is of practical interest to understand how the various components from a multicomponent solid influence their own dissolution and release. Nelson [18] was one of the first pharmaceuticists to ponder this question and perform the initial dissolution studies. Unfortunately, Nelson initially considered the dissolution of interacting solids (benzoic acid + trisodium phosphate), which is a more complicated and more complex situation than simple multicomponent dissolution of noninteracting solids. Nelson did show that for his benzoic acid and trisodium phosphate pellets, there was a maximum increase in benzoic acid dissolution in water at a mole fraction ratio of 2:1 (benzoic acid:trisodium phosphate) and that the benzoic acid dissolution rate associated with the maximum rate was some 40 times greater than that of benzoic acid alone.

W. Higuchi, Mir, and Desai (HMD) followed up on Nelson's initial work and proposed a general multicomponent dissolution model for noninteracting and interacting solids which is shown in Figure 4 [19]. The model is a simple one and allows for three general cases. Case I covers the situation where there is an excess of component A in the solid compact (B is more soluble than A), so that a porous surface layer of pure A forms as the compact containing both components dissolves. The surface layer of A has a characteristic depth or length depicted as L_A in Figure 4. Case II represents the analogous situation in which the mass fraction of B is high relative to its rate of dissolution (A is more soluble), and a porous layer of pure B forms on the surface. Finally, Case III represents a special critical concentration, with the mass fractions of A and B in the solid exactly equal to the ratio of the relative aqueous solubilities and diffusion coefficients of the two components. Case III is special in that neither solid dissolves faster than the other and the solid surface simply recedes uniformly as both components dissolve from the compact.

There are several assumptions inherent in the HMD multicomponent derivation that should be understood. In the model derivation, the authors made a steady-state approximation, which means that the solution is applicable for Cases I and II only after the surface films have formed. The model is not applicable during the initial early dissolution phase. HMD also point out that the time re-

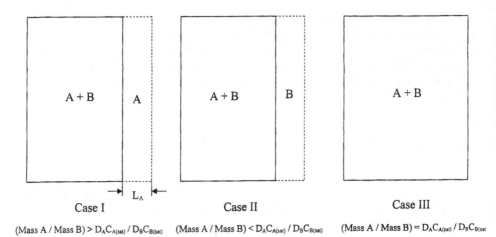

Figure 4   Dissolution model for multicomponent mixtures. (From Ref. 19.)

quired to get to steady state, where the model can be applied, is primarily dependent on the ratio of the solubilities of the two components. If the components have comparable solubilities (within less than one order of magnitude), the time to reach steady state is short, on the order of minutes. Compacts made from components that have tremendously different solubilities may never reach steady-state dissolution. The other conceptual assumption inherent in the model is that the porous surface films do not change in any substantive way, other than becoming larger, during dissolution. That is, for Case I, the growing surface layer of A does not become more or less porous to component B as dissolution proceeds. Figure 5 is from HMD's original work showing the results for solid binary compacts of salicylic acid and benzoic acid. The critical mixture corresponding to Case III dissolution is at roughly 35% salicylic acid.

Several papers have been published showing good agreement between experimental data and theoretical predictions using the HMD model [20,21]. The model has also been further revised to cover dissolution from three-component mixtures [22,23]. Healy and Corrigan [24] showed experimentally that if the condition of comparable solubilities of the two components is disregarded, dissolution of the compact varies with time in a non-steady-state manner. The HMD model has not been applied successfully to predict dissolution of interactive mixtures like acidic drugs combined with basic excipients, but given the solubility differences that exist for a two-component system like benzoic acid and trisodium phosphate, the HMD model could be expected to fail.

In an effort to modernize the multicomponent dissolution HMD dissolution model, Neervannan, Dias, Southard, and Stella (NDSS) adopted the flow-

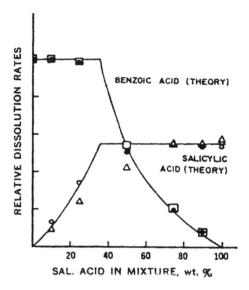

Figure 5 Dissolution rates of salicylic acid–benzoic acid mixtures. (□) Benzoic acid from melt; (○) salicylic acid from melt; (△) salicylic acid from mechanical mix; (●) Benzoic acid from mechanical mix. (From Ref. 19.)

between-two-plates experimental design to model the dissolution of noninteracting two-component solids [25]. NDSS wanted to take advantage of the controlled hydrodynamics and convective profile of this system and to computationally incorporate a numerical solution. Figure 6a schematically shows the NDSS model for a two-component noninteracting solid. Since the authors used solids with similar solubilities and particle sizes, the concept of the model is that particles of A or B randomly associate on the solid surface in a manner that is consistent with the relative mass amounts of each component in the solid blend. The effective length of the A-containing slab L_A is calculated by multiplying the average particle size of A times the calculated number of A particles that would randomly associate. For instance, the authors calculate that at a 25:75 weight ratio (A:B), on average, 1.3 particles of A and 3.96 particles of B would align within the slab. This would translate to an average length of the A-containing slab of 1.3 times the average particle size of A ($L_A = 1.3 \times 112\ \mu m$). NDSS used a numerical solution to make dissolution rate predictions with the model. The theoretical results for the numerical routine were in good agreement with the experimental results and a better fit than those derived from the simple HMD model. The authors did notice that increasing the particle size of both components tended to decrease dissolution rates. They explained this effect as being characteristic of

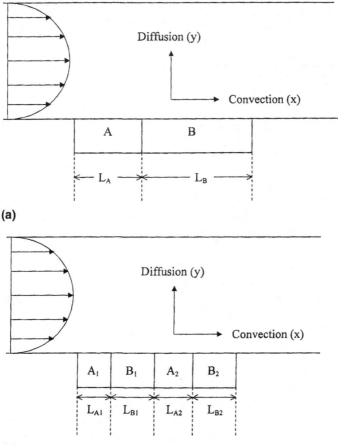

Figure 6 (a) Two-component noninteracting slab dissolution model. (b) Refined two-component noninteracting slab dissolution model. (From Ref. 26.)

a "carryover" phenomenon. Carryover occurs when the fluid near the surface becomes saturated with a particular component and some amount of this material is carried over into the boundary layer of the next particle down the slab. Carryover may be more easily conceptualized by viewing Figure 6b. Some of the mass of A which dissolves in the A_1 segment of the solid will be convectively transported down the slab, or carried over, into the dissolving A_2 segment. From a modeling standpoint, carryover represents a violation of the idealized boundary conditions used by NDSS. In testing the model, the authors did use model drugs whose solubilities were all similar so as to avoid significant pitting or erosion of a single component. In a follow-up publication, Neervannan et al. [26] suggest that the dissolution rate for each component of the two-component co-compressate is complicated by carryover and may not be related simply to exposed surface area.

The theoretical multicomponent dissolution models proposed to date, with their restrictions and inherent assumptions, are of limited practical usefulness to the formulator in the lab faced with the task of making tablets. In particular, the models are applicable only to situations where the drug and excipient have similar solubilities, yet the common experience of most formulators is that more soluble excipients give faster dissolution. Studies by Westerberg et al. [27,28] suggest that for noninteracting solid mixtures with a slightly soluble drug, the way to increase dissolution is to use soluble carriers or excipients at a weight ratio that ensures that the surface of the excipient is accessible and not covered by the drug.

For the original problem of dissolution for interacting solids (acidic or basic drugs with acidic or basic excipients) proposed by Nelson, even less concrete information is available from the pharmaceutical literature. A recent paper by Chakrabarti and Southard [29] experimentally examined the dissolution of an acidic drug (naproxen) from co-compressates with basic buffer-type excipients (calcium phosphate, calcium carbonate, and calcium citrate). As in their previous publications, the authors employed the flow-between-two-plates dissolution apparatus. Although they did not propose a theoretical model to explain how incorporation of basic excipients can accelerate the dissolution of an acidic drug, their experimental work was well enough designed that they could comment on effects due to particle size and the relative ratio of excipient to drug. Chakrabarti and Southard's general recommendations to increase dissolution of an acidic drug include (1) choice of a basic buffer excipient that is less soluble than the active drug, (2) use of roughly a 50:50 % w/w (drug:excipient) ratio, and (3) selection of drug and excipient particles which both have smallest particle size available.

Unfortunately for the formulator, our general understanding of multicomponent dissolution is rudimentary. At present, there are no pharmaceutical recipes for multicomponent formulations with well-designed and well differentiated dissolution profiles. Formulators, like chefs, are left to create their own secret formulas. As described previously, the published multicomponent dissolution models are limited in their scope and application. We do know that excipients which control pH of the solid/liquid interface can have tremendous impact on the dissolution rate of ionizable drugs. Javaid and Cadwallader [30] showed that aspirin-containing tablets formulated with buffering agents which produce carbon dioxide (sodium bicarbonate, magnesium carbonate, and calcium carbonate) dissolved faster by orders of magnitude than simple lactose- and starch-containing tablets. Doherty and York [31] physically measured the pH at the solid/liquid surface, using a pH microelectrode for solid dispersions containing an acidic drug plus phosphate and citrate excipients, and demonstrated that excipients affect the recorded pH.

The published research to date for multicomponent dissolution does offer some hints to the formulator for very specific situations (acidic drug with basic excipients), but there is scant and disconnected information concerning the use and effects of wetting and fluid wicking excipients. While there has been progress, a rugged quantitative understanding of the effects of factors such as excipient solubility, wetting, and particle size on dissolution is still to be developed.

V. DISSOLUTION OF POORLY SOLUBLE COMPOUNDS IN SURFACTANT SOLUTIONS

As the number of NMEs (new molecular entities) with poor aqueous solubility increases, so does the interest in developing dissolution media for evaluating the performance of dosage forms containing these drugs. In the past, this interest primarily focused on testing the product for batch-to-batch variability and reducing the dissolution time or the volume of dissolution media required. More recently, however, the industry began a search for dissolution media that not only reflects the dosage form's in vitro performance regarding manufacturing variability but also, if possible, reflects the in vivo performance. Consequently, numerous surfactants, both synthetic and natural, are being incorporated into dissolution media to enhance the dissolution rate and mimic the in vivo dissolution (32–47). For example, several dissolution media have been developed that use lecithin and bile salts in ratios similar to that of the GI tract [46,47]. The question of whether to use natural or synthetic surfactant systems has been debated, since the lecithin/bile salt systems are significantly more expensive than the synthetic systems (5–10 times greater depending on the purity and type of materials used) [48]. A possible alternative may include two dissolution tests, one for routine analysis that uses synthetic surfactants and one that uses natural surfactants for in vitro–in vivo correlations. As with the above theory pertaining to acid–base dissolution, this section focuses on the most commonly used models for describing the dissolution of poorly soluble drugs.

The basic mechanism for surfactants to enhance solubility and dissolution is the ability of surface-active molecules to aggregate and form micelles [35]. While the mathematical models used to describe surfactant-enhanced dissolution may differ, they all incorporate micellar transport. The basic assumption underlying micelle-facilitated transport is that no enhanced dissolution takes place below the critical micelle concentration of the surfactant solution. This assumption is debatable, since surfactant molecules below the critical micelle concentration may improve the wetting of solids by reducing the surface energy.

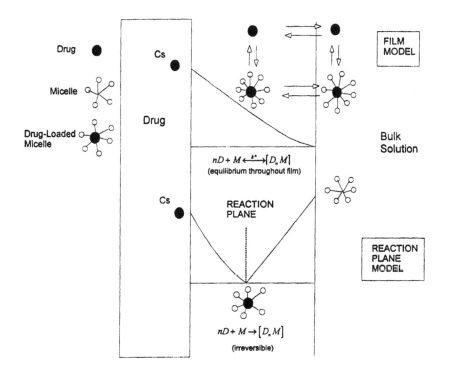

Figure 7 Stagnant film model (top) and reaction plane model (bottom) for micelle-facilitated dissolution.

Mathematical approaches used to describe micelle-facilitated dissolution include film equilibrium and reaction plane models. The film equilibrium model assumes simultaneous diffusive transport of the drug and micelle in equilibrium within a common stagnant film at the surface of the solid as shown in Figure 7. The reaction plane approach has also been applied to micelle-facilitated dissolution and has the advantage of including a convective component in the transport analysis. While both models adequately predict micelle-facilitated dissolution, the scientific community perceives the film equilibrium model to be more mathematically tractable, so this model has found greater use.

A. Film Equilibrium Model

The kinetic scheme used to describe micelle-facilitated dissolution is

$$n\mathrm{S} + \mathrm{M} \underset{k}{\overset{k^*}{\rightleftharpoons}} \mathrm{S}_n\mathrm{M} \tag{14}$$

where

$$k* = \frac{[S_n M]}{[nS][M]} \tag{15}$$

Here, S is the free solute, M is the micelle, n is the number of solute molecules per micelle, SM is the solute–micelle complex, and $k*$ is the equilibrium coefficient [39].

As with acid–base reactions, the equation of continuity used to describe the mass transport of a solid in a fluid is

$$\frac{\partial c_i}{\partial t} = D_i \nabla^2 c_i - v \nabla c_i + R_i \tag{16}$$

where D_i, c_i, and R_i are the diffusion coefficient, concentration, and reaction terms for species i, v is the convective velocity of the fluid, and t is time [49]. Simplification of Eq. (16) to the steady-state, one-dimensional form gives

$$0 = D_i \frac{d^2 c_i}{dx^2} - v_z \frac{dc_i}{dx} + R_i \tag{17}$$

Solving Eq. (16) with the boundary conditions of saturation solubility at the solid surface, $C_{s(0)}$, sink conditions in the bulk solution, and assuming no convection or reaction contributions, yields

$$N = D_i C_{s(0)}/\delta \tag{18}$$

where N is the flux and δ is the stagnant film layer thickness. Higuchi used this analysis to define the diffusion coefficient of the solute–micelle complex as an *effective* diffusion coefficient as follows.

$$\frac{D_{eff} C_{total}}{\delta_{eff}} = \frac{D_s C_s}{\delta_s} + \frac{D_m C_m}{\delta_m} \tag{19}$$

where $D_s C_s/\delta_s$ is the flux of the free solute and $D_m C_m/\delta_m$ is the flux of the micelle. Assuming that $\delta_{eff} = \delta_s = \delta_m$, then,

$$D_{eff} = \frac{D_s C_s + D_m C_m}{C_{total}} \tag{20}$$

Finally, inserting Eq. (15) and rearranging Eq. (20) yields

$$D_{eff} = \frac{D_s + k* D_m C_m}{1 + k* C_m} \tag{21}$$

where

$$C_{total} = C_s + C_m \tag{22}$$

Gibaldi et al. [45] postulated that convective forces may be present in the GI tract during in vivo dissolution. This study took advantage of the well-defined hydrodynamics of the rotating disk, incorporating the solutions for the velocity profile and transport equations of Cochran [50] and Levich [51] to obtain

$$N = 0.62 D_i^{2/3} v^{-1/6} \omega^{1/2} C_{s(0)} \tag{23}$$

where v is the kinematic viscosity, ω is the rotational speed of the disk, and $C_{s(0)}$ is the saturation solubility of the solid at the disk surface. Gibaldi then defined an effective boundary layer thickness as

$$\frac{D_{eff} C_{total}}{\delta_{eff}} = \frac{D_s c_{s(0)}}{\delta_s} + \frac{D_{sm} c_{sm(0)}}{\delta_{sm}} \tag{24}$$

and dividing by the concentration of the solute, c_s, to obtain

$$D_{eff} = \frac{D_s + k^* c_{m(b)} D_{sm}}{1 + k^* c_{m(b)}} \tag{25}$$

where D_i and c_i are the diffusivity and concentration for species i and δ_i is the diffusion boundary layer thickness for the solute (s) and the micelle (m) or the effective boundary layer thickness of the solute and micelle (sm).

A method used to describe the enhanced dissolution rate following micelle-facilitated dissolution is to compare the dissolution of the drug in the surfactant solution to that of the dissolution rate in water; this is often termed the "reaction factor" method. The reaction factor, ϕ_{FM}, which is the total flux of the micelle-solubilized solute plus the free solute divided by the flux of the free solute, is given by

$$\phi_{FM} = \frac{N_{total}}{N_{s(0)}} = \frac{0.62 D_{eff}^{2/3} \omega^{1/2} v^{-1/6} c_{total}}{0.62 D_s^{2/3} \omega^{1/2} v^{-1/6} c_s} = \frac{D_{eff}^{2/3} c_{total}}{D_s^{2/3} c_s} \tag{26}$$

$$\phi_{FM} = \frac{D_{eff}^{2/3}}{D_s^{2/3}} (1 + k^* c_{m(b)}) \tag{27}$$

The limitation of using such a model is the assumption that the diffusional boundary layer, as defined by the effective diffusivity, is the same for both the solute and the micelle [45]. This is a good approximation when the diffusivities of all species are similar. However, if the micelle is much larger than the free solute, then the difference between the diffusional boundary layer of the two species, as defined by Eq. (24), is significant since δ_i is directly proportional to the diffusion coefficient. If known, the thickness of the diffusional boundary layer for each species can be included directly in the definition of the effective diffusivity. This approach is similar to the reaction plane model which has been used to describe acid–base reactions.

Crison et al. [52] presented an alternative derivation of Eq. (29) that included individualized transport of the solute and micelle while still maintaining the basic assumption of equilibrium. This was accomplished by rewriting Eq. (24) to include the magnitude of the individual diffusional boundary layers for free drug and drug-loaded micelle according to Eq. (10), as follows:

$$\frac{D_{eff} C_{total}}{\delta_{L,eff}} = \frac{D_s C_{s(0)}}{\delta_{L,s}} + \frac{D_{sm} C_{sm(0)}}{\delta_{L,sm}}$$ (28)

From Eq. (28), the diffusional boundary layer for the rotating disk, Eq. (29), can be substituted to give

$$D_{eff} = \frac{D_s C_s + D_m C_m}{C_{total}}$$ (29)

$$D_{eff}^{2/3} C_{total} = D_{s(0)}^{2/3} C_{s(0)} + D_{sm}^{2/3} C_{sm}$$ (30)

$$D_{eff}^{2/3} = \frac{D_s^{2/3} + k^* c_{m(b)} D_{sm}^{2/3}}{1 + k^* c_{m(b)}}$$ (31)

Furthermore, the reaction factor now becomes

$$\phi_{FM} = 1 + \frac{D_{sm}^{2/3}}{D_s^{2/3}} k^* c_{m(b)}$$ (32)

where the slope of Eq. (32) is now a function of the ratio of the diffusivity of the drug-loaded micelle to the diffusivity of the free drug. This model was used to evaluate the effect of surfactant purity on the dissolution rate enhancement. The authors showed that lauryl alcohol impurities reduced the dissolution rate enhancement by 10–15% for a range of surfactant concentrations, demonstrating that impurities in the surfactant can lead to inconsistent results in dissolution tests [53].

B. Reaction Plane Model

The reaction plane model with heterogeneous reactions was discussed at length for acid–base reactions in the previous section. The same modeling technique, of confining the reactions to planes, can be applied to micelle-facilitated dissolution. As with the acid–base model, one starts with a one-dimensional steady-state equation for mass transfer that includes diffusion, convection, and reaction. This equation is then applied to the individual species i, i.e., the solute, s, the micelle, m, and the drug-loaded micelle, sm, to yield

$$0 = D_s \frac{d^2 c_s}{dx^2} - v_z \frac{dc_s}{dx} + R_s$$ (33)

$$0 = D_{sm}\frac{d^2c_{sm}}{dx^2} - v_x\frac{dc_{sm}}{dx} + R_{sm} \tag{34}$$

Litt and Serad [9] made the acid–base form of the above equations dimensionless to facilitate the solution, resulting in the equation

$$0 = \frac{d^2C_i^*}{d\eta^2} - Sc_i H\frac{dC_i^*}{d\eta} + \frac{Da_i}{Re} \tag{35}$$

where C_i^*, Sc_i, H, Re, Da_i, and η are dimensionless variables for the concentration, diffusivity, fluid velocity, kinematic viscosity, reaction, and distance, respectively. Incorporating the reaction terms as boundary conditions at planes gives

$$\frac{d^2C_s^*}{d\eta^2} - Sc_s H\frac{dC_s^*}{d\eta} = 0, \qquad 0 \le \eta \le \eta_R \tag{36}$$

$$\frac{d^2C_{sm}^*}{d\eta^2} - Sc_{sm} H\frac{dC_{sm}^*}{d\eta} = 0, \qquad \eta_R \le \eta \le \infty \tag{37}$$

where η_R is the dimensionless position of the reaction plane. Following the assumptions made by Litt and Serad, the reaction factor for this model is

$$\phi_{RPM} = \frac{\theta_{s(\infty)}}{\theta_{sm(\infty)}}\left(1 + \frac{Sc_s ac_{m(\infty)}}{Sc_m bc_{s(0)}}\right) = \frac{D_s^{1/3}}{D_{sm}^{1/3}}\left(1 + \frac{D_{sm} ac_{m(\infty)}}{D_s bc_{s(0)}}\right) \tag{38}$$

$$\phi_{RPM} = \frac{D_s^{1/3}}{D_{sm}^{1/3}} + \frac{D_{sm}^{2/3} ac_{m(\infty)}}{D_s^{2/3} bc_{s(0)}} \tag{39}$$

where a and b are the stoichiometric coefficients for the solute and micelle, respectively, $c_{m(\infty)}$ is the concentration of the micelle in the bulk solution, $c_{s(0)}$ is the concentration of the solute at the surface of the solid, and D_{sm} and D_s are the diffusion coefficients of the solute and drug-loaded micelle, respectively. The reaction factor, Eq. (39), is similar to that for the film model in that both are functions of the micelle concentration.

VI. DISSOLUTION OF POORLY SOLUBLE DRUGS IN EMULSION SYSTEMS

As with micelle-facilitated dissolution, emulsion-facilitated dissolution has gained renewed interest due to its application to water-insoluble drug delivery and enhanced absorption. Over the years, emulsion systems have been developed and used to either model the in vivo dissolution process or mimic the intestinal surfactant system to enhance drug delivery of poorly soluble compounds [54–66]. Emulsions have also been used as vehicles for drug delivery, e.g., to protect

enzyme-sensitive drugs, however, this section addresses solubilization and dissolution effects.

Dissolution of water-insoluble drugs into emulsions is similar to micelle-
facilitated dissolution in that there is a partitioning of the drug into the emulsion
droplet based on the affinity of the drug for the internal phase. Since emulsion
particles are significantly larger than micelles, being approximately 100–1000-
fold greater in diameter, the loading capacity can be much greater [65]. However,
while the size of the emulsion droplet is much greater, the transport of emulsion
particles through liquid media is often dominated by convective forces rather
than diffusion. In addition to the large particles, most emulsion systems contain
a micellar or liposomal component, depending on the materials used to make the
emulsion [65]. The inclusion of this species greatly complicates modeling drug

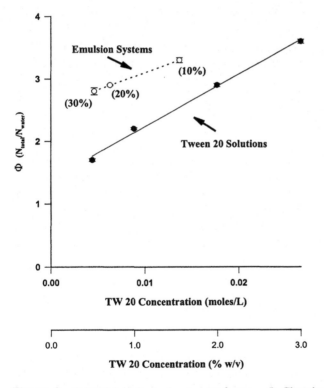

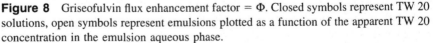

Figure 8   Griseofulvin flux enhancement factor = Φ. Closed symbols represent TW 20
solutions, open symbols represent emulsions plotted as a function of the apparent TW 20
concentration in the emulsion aqueous phase.

transport into emulsions, since it is difficult to separate the contributions of the individual components to the total dissolution process experimentally.

Goldberg and coworkers [56,57] described an interfacial resistance at the surface of the emulsion droplet for solute molecule transport into the internal phase. In this model the interfacial resistance was defined as the reciprocal of the effective diffusivity and the diffusional boundary layer at the surface of the droplet. More recently, Dressman and coworkers [46,47] compared the effect of bile salt concentration and lecithin–bile salt mixtures on the dissolution of several poorly soluble drug salts. From this work a correlation was developed between the logarithm of the octanol:water partition coefficient and the solubility of several poorly soluble steroids in 15 mM sodium taurocholate. Also important was the finding that no significant increase in solubility was observed at "fasted" concentrations of the bile salt. These results may be useful in developing in vitro–in vivo correlations under fed and fasted conditions. Additional work by Dressman showed that a 4:1 ratio of bile salt to lecithin could be used as a dissolution medium for water insoluble drugs. Routine use of such media for determining batch-to-batch variability during manufacturing may be costly; however, the ben-

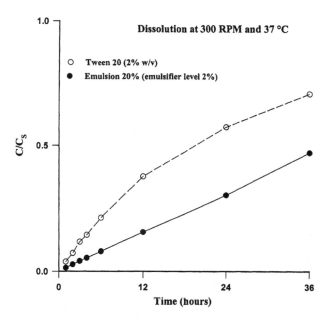

Figure 9 The fraction of griseofulvin saturation concentration achieved versus time determined using the rotating disk. C = dissolution concentration, C_s = saturation concentration in the dissolution medium at 37°C.

efit may outweigh the cost when establishing in vitro–in vivo correlations for scale-up and postapproval changes (SUPAC).

Recent efforts have focused on the mechanism by which emulsion systems contribute to the dissolution of poorly water-soluble drugs. Vieira and coworkers [67] are studying the solubilization and dissolution of griseofulvin, into emulsion formulations formulated with the nonionic surfactant Tween 20. Griseofulvin solubility was markedly improved in the emulsion systems as compared to the surfactant solutions (i.e., 3 to 9 times more drug was solubilized by the different emulsion systems at 37°C). However, the drug dissolution rate enhancement in the emulsions, as evaluated using the rotating disk system, improved only slightly (Fig. 8). The comparative dissolution of griseofulvin in a 2% w/v surfactant solution versus a 20% emulsion system emulsified using 2% surfactant showed distinct time dependence for the emulsion facilitated dissolution case in that the surfactant solution exhibited a higher rate of dissolution than the emulsion system (Fig. 9). The conclusion from their studies was that the emulsified oil droplets have a smaller effect on the initial rate of dissolution of a poorly water-soluble drug than the surfactant solutions. However the emulsion systems can have a significant effect on the extent of dissolution since these systems solubilize a considerably higher level of drug.

VII. PARTICLE DISSOLUTION

Up to now, dissolution of solids under various conditions, i.e., acid–base, micellar, and emulsion systems, has been derived based on a flat plate or slab configuration. However, in practice most pharmaceutical preparations are particulate in nature, either as solid drug particles or as spherical modified release matrices, and the effect of the particle size on the dissolution rate is important [69–79]. Abbreviated derivations of more complex mathematical expressions have been used successfully to describe particle dissolution providing the environment, i.e., dissolution media, fluid velocity profile, temperature gradient, etc., is simple as well. However, more recently dissolution testing has evolved from a quality control tool to a method for establishing bioequivalence between two drug products. In order to achieve high levels of differentiation, dissolution tests may include changes in pH, surfactants, enzymes, and use of oil/water emulsions. With the many different combinations possible, it is evident that mathematical models will need to be able to approximate the different layers of complexities associated with the new dosage forms and dissolution requirements. Therefore, the following sections provide a basic understanding of the important variables that control the dissolution of spherical particles. The effect of the shape of the particle and surface irregularities on dissolution are not discussed here in detail, and the reader is encouraged to review the literature on this topic.

A. Boundary Layer Thickness

Of all the different variables to consider when modeling particle dissolution, defining the thickness of the diffusional boundary layer is generally the most complex. The boundary layer at the surface of the particle is a function of the viscosity of the dissolution media, the fluid velocity, e.g., mixing speed, agitation rate, GI motility, etc., and density and size of the particle, as well as the diffusivity, solubility, surface area, and temperature. While the diffusivity, solubility, temperature, density, and fluid viscosity can be maintained as experimental constants, the size, surface area, and fluid velocity profile (at the surface) change as the dissolution process progresses. Hence, the magnitude of the boundary layer may also change throughout the dissolution process, and a single expression may not be appropriate for the entire dissolution process.

Microscopically, the boundary layer results from reduced fluid flow at the solid surface due to frictional forces between the fluid and the particle. The thickness of the boundary layer at the surface of a solid that is axially symmetrical is defined as the height for which the deviation of the peripheral velocity is equal to 2% of the free stream [80]. Calculation of the velocity distribution as a function of the distance from the sphere surface requires some knowledge of the hydrodynamics surrounding the particles, e.g., laminar vs. turbulent flow [49]. In the case of in vitro testing, controlling the fluid velocity or mixing speed during dissolution has been part of the validation process for quite some time. Some dissolution apparatuses are designed so that the fluid velocity profile at the liquid/solid interface is well characterized [50,51]. However, these devices are limited to flat surfaces and are not applicable to small particles in a stirred vessel.

Two regimes exist regarding the boundary layer thickness for describing particle dissolution. In one the boundary layer thickness is smaller than the particle radius; in the other the boundary layer thickness is approximately equal to the particle radius. In the simplest case, the boundary layer for small particles starts out very small, and the thickness of the boundary layer can be approximated by the particle radius throughout the entire dissolution process. On the other hand, for large particles, the boundary layer may be smaller than the particle radius initially, but as dissolution proceeds the boundary layer dimension may become comparable to the particle radius. A convenient method for determining which category particles fall into is to compare the viscous forces with the buoyancy forces, i.e., calculate the Grashof number (Gr) as follows [8]:

$$Gr = \frac{D_p^3 g \, \Delta\rho/\rho}{v^2} \tag{40}$$

where D_p is the particle diameter, g is the acceleration due to gravity, $\Delta\rho/\rho$ is the fractional density change between the solid and fluid, and v is the kinematic viscosity. For Gr > 1, the thickness of the boundary will be smaller than the radius, and for Gr $\leq$ 1, the thickness is approximately equal to the radius. As a point of reference, assuming a solid density range of 1.2–1.4 g/cm^3, a kinematic viscosity of 0.007 poise for water at 37°C, gravity of 980 cm/s^2, and water density at 37°C of 0.993 gm/cm^3, then the corresponding particle diameter range for Gr = 1 is 50–63 μm. Therefore, under "average" conditions the cutoff point for a particle changing from large to small with respect to the boundary layer thickness is approximately 50–60 μm.

B. Mathematical Models for Particle Dissolution: Monodisperse Populations

The simplest case for modeling particle dissolution is to assume that the particles are monodisperse. Under these conditions, only one initial radius is required in the derivation of the model. Further simplification is possible if the assumption is made that mass transport from a sphere can be approximated by a flat surface or a slab, as was the case for the derivation for the Hixson–Crowell cube root law [70]. Using the Nernstian expression for uniaxial flux from a slab (ignoring radial geometry or mass balance), one can derive the expression

$$\frac{dm}{dt} = -SD\frac{C_{s(0)} - C_{b(\delta)}}{\delta} \tag{41}$$

where m stands for the mass dissolved, D is the diffusion coefficient of the dissolving species, C represents concentration, S represents surface area of exposed solid, and t is time.

In Eq. (41), the concentration gradient is expressed as a difference between the surface concentration, $C_{s(0)}$, and the bulk concentration, $C_{b(\delta)}$, divided by the diffusion layer thickness, δ. If sink dissolution conditions are assumed ($C_b \approx 0$) and the solid has a uniform density ($\rho = m/V$), then Eq. (42) can be derived.

$$\rho dV = -\frac{D}{\delta} SC_{s(0)} dt \tag{42}$$

where V is the volume of the dissolving sphere. Since the differential volume change for a dissolving sphere can be related to differential radius change ($dV = 4\pi r^2 dr$), Eq. (42) can be rewritten as

$$dr = -\frac{DC_{s(0)}}{\delta\rho} dt \tag{43}$$

Equation (43) can be integrated and evaluated at the obvious initial conditions (time = 0, radius = r_0) and final conditions (time = t, radius = r) as

$$\int_{r_0}^{r} dr = -\int_0^t \frac{DC_{s(0)}}{\delta\rho} dt \tag{44}$$

The result is

$$r = r_0 - \frac{DC_{s(0)}}{\delta\rho} t \tag{45}$$

To finish the derivation, one needs to recognize that the initial mass of solid (m_0) is related to the total number of initial particles (n_0) as

$$m_0 = n_0 \rho \frac{4}{3} \pi r_0^3 \tag{46}$$

After substituting Eq. (46) into Eq. (45), it can be shown that Eq. (45) can be rearranged to give

$$m^{1/3} - m_0^{1/3} = -\left(n_0 \rho \frac{4}{3} \pi\right)^{1/3} \frac{DC_s}{\delta\rho} t \tag{47}$$

Finally, the first term on the right-hand side of Eq. (47) can be rewritten as

$$\left(n_0 \rho \frac{4}{3} \pi\right)^{1/3} = \frac{m_0^{1/3}}{r_0} \tag{48}$$

and Eq. (47) can be written as

$$1 - \left(\frac{m}{m_0}\right)^{1/3} = \frac{DC_s}{r_0 \delta\rho} t \tag{49}$$

Since m is the mass of solid remaining at time t, the quantity m/m_0 is the fraction undissolved at time t. The time to total dissolution ($m/m_0 = 0$) of all the particles is easily derived. Equation (49) is the classic cube root law still presented in most pharmaceutics textbooks. The reader should note that the cube root law derivation begins with misapplication of the expression for flux from a slab (Cartesian coordinates) to describe flux from a sphere. The error that results is insignificant as long as $r_0 \gg \delta$.

Alternatively, a simple model describing particle dissolution can be derived using spherical coordinates as follows:

$$\frac{\partial c}{\partial t} = \frac{D}{r^2} \frac{\partial}{\partial r}\left(r^2 \frac{\partial c}{\partial r}\right) \tag{50}$$

where r is the radial distance from the origin at the center of the particle, D is the diffusion coefficient, c is the concentration of the solute, and t is time (Ref.

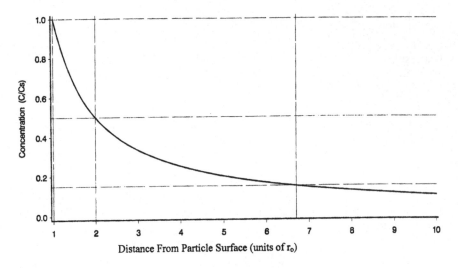

Figure 10   Concentration profile of a spherical particle.

68, Fig. 10). Under steady-state conditions and assuming the boundary conditions (note that no assumptions about the stagnant film thickness have been made)

At $r = r_0$, $c = c_{s(r_0)}$ and   at $r = \infty$, $c = 0$

where r_0 is the particle radius, Eq. (50) becomes

$$0 = \frac{d}{dr} r^2 \frac{dc}{dr} \tag{51}$$

Following integration, assuming sink conditions with no convection, Eq. (51) yields the flux, N, as

$$N = \frac{D}{r_0} c_{s(r_0)} \qquad \text{at } r = r_0 \tag{52}$$

Equation (52) applies to dissolution under sink conditions and no convection. Successful modeling of particle dissolution requires a good understanding of the environment and circumstances for dissolution. Equation (52) describes a relatively simple system (Fig. 11).

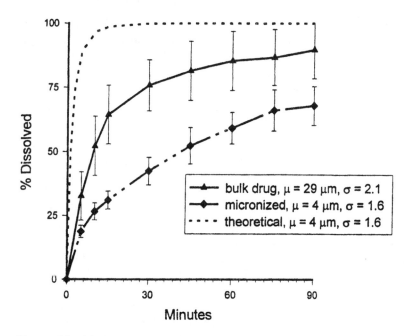

Figure 11   Dissolution of two particle size distribution fractions of a water-insoluble drug (solubility in water at 37°C < 10 μ/mL). Equation (41) was used to calculate the values shown by the dashed line, which represents the estimated dissolution of the drug based on the particle size distribution and solubility determined experimentally.

C. Mathematical Models for Particle Dissolution: Polydisperse Populations

Several mathematical models are available for predicting the dissolution of particles of mixed size. Some are more complex than others and require lengthy calculations. The size of polydisperse drug particles can be represented with a distribution function. During the milling of solids, the distribution of particle sizes most often results in a log-normal distribution.* A log-normal distribution is positively skewed such that there can exist a significant tail on the distribution, hence a number of large particles. The basic equation commonly used to describe the particle distribution is the log-normal function,

$$f(D_p) = \frac{1}{\ln(\sigma_g \sqrt{2\pi}\, D_p)} \exp \frac{-[\ln(D_p/D_{pg})]^2}{2(\ln \sigma_g)^2} \tag{53}$$

---

* Solids that are crushed generally result in a Rosin–Rammler distribution, $f(D_p) = (n/D_p)$ $(D_p/D_p)_{n-1} \exp[(D_p/D_p)^n]$, where n is a polydispersity index [73].

where D_p and D_{pg} are the diameter and geometric mean diameter of the particles and σ_g is the geometric standard deviation of the distribution [73]. The percent of mass dissolved for a given time based on Eq. (53) at time $t = T$ is

$$\text{Percent mass} = 1 - \frac{\displaystyle\int_0^\infty D_p^3 f(D_{p_{t=T}})\,dD_p}{\displaystyle\int_0^\infty D_p^3 f(D_{p_{t=0}0})\,dD_p} \tag{54}$$

There has been much discussion regarding the significance of the tail of the log-normal distribution [73]. Truncating the distribution simplifies the math and allows for easier handling of the calculations. Indeed, in some cases, truncation has little effect on the estimating of total dissolution of the powder sample, as was shown by Brooke [74]. In any case, a powder sample with a small mean diameter and large size distribution may have a deceptively low dissolution rate owing to a large number of big particles at the high end of the distribution which disproportionately contain the bulk of the powder mass.

Leblanc and Fogler developed a population balance model for the dissolution of polydisperse solids that included both reaction controlled and diffusion-controlled dissolution . This model allows for the handling of continuous particle size distributions. The following population balance was used to develop this model.

$$R(D_p)F(D_p, t)|_{D_p} - R(D_p)F(D_p, t)|_{D_p+\Delta D_p} = \frac{\partial}{\partial t}[F(D_p, t)\Delta D_p] \tag{55}$$

where $R(D_p)$ is the particle growth rate and the population balance on a differential in D_p to $D_p + \Delta D_p$. The utility of this model is that it applies to particle dissolution and crystal growth as well.

Johnson and Swindell [77] developed a method for evaluating the complete particle distribution and its effect on dissolution. This method divided the distribution into discrete, noncontinuous partitions, from which Johnson and Swindell determined the dissolution of each partition under sink conditions. The dissolution results from each partition value were then summed to give the total dissolution. Oh et al. [82] and Crison and Amidon [83] performed similar calculations using an expression for non-sink conditions based on a macroscopic mass balance model for predicting oral absorption. The dissolution results from this approach could then be tied to the mass balance of the solution phase to predict oral absorption.

Many studies have been conducted to look at the effect of particle size on dissolution and the bioavailability of insoluble drugs [69–84]. One study looked at the effect of surface area on the bioavailability of griseofulvin, which has an aqueous solubility of 0.014 mg/mL and an average dose of 500 mg. The authors showed that an approximately sixfold increase in the surface area led to a 2.5-

fold increase in bioavailability [84]. From these results, particle size reduction for poorly soluble drugs seems to be the method of choice for improving absorption. However, one of the problems often encountered in the micronization process is agglomeration and aggregation due to drug hydrophobicity and the increased surface energy, resulting in a decrease of effective surface area (see Ref. 45, Fig. 11).

$$\Delta G = S\, \gamma_{SL} \tag{56}$$

where ΔG is the free energy of the system, S is the surface area of the solid, and γ_{SL} is the surface tension between the liquid and solid phases.

With griseofulvin, since the dissolution rate is directly proportional to the surface area, one would expect a greater than 2.5-fold increase in bioavailability from a sixfold increase in surface area, assuming that absorption is dissolution rate limited. The less than expected result may have been due to particle aggregation in vivo. Furthermore, the hydrophobic surface of the drug particle or particle aggregates tends to absorb air and become difficult to wet or to disperse by aqueous medium. Therefore, simple micronization of insoluble drugs does not always guarantee improved dissolution or absorption.

VIII. CONCLUSIONS

This chapter has examined some of the many topics that have occupied those who have been active in dissolution research over the years. As one way to summarize this review, it may be worthwhile to condense the foregoing material into a loose set of dissolution truisms which may comfort, or continue to confound, those of us who ponder these phenomena and research results.

1. The accelerated dissolution of soluble pure acidic or basic drugs, through the addition of acidic or basic components to the dissolution media, becomes significant when the concentration of added components approaches the intrinsic solubility of the drug.
2. Even for experimentally simple systems, a quantitative understanding of the dissolution of ionizable drugs is possible only if drug solubility, drug ionization constant, buffer concentration, buffer species, and buffer pH are known.
3. Dissolution film models, though conceptually simple and instructive, are inherently unsatisfying in their application to complex dissolution phenomena or systems.
4. Multicomponent dissolution effects have been historically underappreciated, but these phenomena are finally getting some scrutiny. Multicomponent effects are likely to influence in vivo–in vitro correlations.
5. Particle dissolution models, which presume spherical monodisperse particle size, inadequately depict the complexity of typical pharmaceutical powders.

6. It is possible to derive a simple particle dissolution model where diffusional film thickness is not explicitly required. However, the boundary layer concentration profile derived from this model will extend for distances which cover an order of magnitude of the initial particle radius.

7. Surfactants added to dissolution media of insoluble drugs may produce dissolution results which are more indicative of the in vivo fed state than of the fasted state.

8. Variability in particle aggregation during dissolution, the consequence of dissimilarities between in vivo and in vitro dissolution conditions and media, may confound in vivo and in vitro correlations.

Symbols

Symbol	Description	Units (L[=]length, M[=]mass, T[=]time)
δ	stagnant film thickness	L
K	equilibrium constant	unitless
k^*	equilibrium coefficient for micelle complex	$(M/L^3)^{-1}$
C_i	concentration of species i	M/L^3
$C_{j(k)}$	concentration of species j at position k	M/L^3
$C_{j(0)}$	concentration of species j at solid surface	M/L^3
$C_{j(\delta)}$	concentration of species j at end of stagnant film	M/L^3
$C_{s(0)}$	saturated solubility of dissolving species at solid surface	M/L^3
k_f	forward micro-rate constant	L^3/MT
k_r	reverse micro-rate constant	L^3/MT
D_i	diffusion coefficient of i	L^2/T
m	mass	M
M	molar concentration	Moles/liter
N	flux of HA with reaction	M/L^2T
N_o	flux of HA without reaction	M/L^2T
N/N_o	relative flux	unitless
n	number of particles	unitless
r	radial axis	L
S	surface area	L^2
V	volume	L^3
v_z	fluid velocity in z axis	L/T
Q	volumetric flow rate	L^3/T
ϕ	reaction factor	unitless
ν	kinematic viscosity	L^2/T
ρ	density	M/L^3
ω	radial velocity	L/T

Dimensionless Numbers

Symbol	Name	Description	Definition
C^*	Concentration	C/C_0	C = concentration with reaction
			C_0 = concentration without reaction
Sc	Schmidt number	ν/D	ν = kinematic viscosity
			D = diffusion coefficient
H	Velocity	$w/(\nu\omega)^{1/2}$	w = axial velocity
			ν = kinematic viscosity
			ω = radial velocity
η	Distance	$z(\omega/\nu)^{1/2}$	z = axial distance
			ν = kinematic viscosity
			ω = radial velocity
Da	Second Damkohler number	$\kappa l^2/D$	κ = first-order reaction rate constant
			l = characteristic length
			D = diffusion coefficient
Gr	Grashof number	$(D_p^3 g\, \Delta\rho/\rho)/\nu^2$	D_p = particle diameter
			g = acceleration due to gravity
			$\Delta\rho$ = solid–fluid density difference
			ρ = fluid density
Re	Reynolds number	$\omega l^2/\nu$	ω = radial velocity
			ν = kinematic viscosity
			l = characteristic length

REFERENCES

1. EL Parrot, DE Wurster, T Higuchi. Investigation of drug release from solids I. Some factors influencing the dissolution rate. J Am Pharm Assoc 44:269–273, 1955.
2. W Higuchi, EL Parrot, DE Wurster, T Higuchi. Investigation of drug release from solids II. Theoretical and experimental study of influences of bases and buffers on rates of dissolution of acidic solids. J Am Pharm Assoc 47:376–383, 1958.
3. DR Olander. Simultaneous mass transfer and equilibrium chemical reaction. AIChE J 6:233–239, 1960.
4. G Astarita. Mass Transfer with Chemical Reaction. Amsterdam: Elsevier, 1967, p 21.
5. KG Mooney, MA Mintun, KJ Himmelstein, VJ Stella. Dissolution kinetics of carboxylic acids I: Effect of pH under unbuffered conditions. J Pharm Sci 70:13–22, 1981.
6. DP McNamara, GL Amidon. Dissolution of acidic and basic compounds from the rotating disk: Influence of convective diffusion and reaction. J Pharm Sci 75:858–868, 1986.

7.   H Grijseels, DJA Crommelin, CJ De Blaey. Hydrodynamic approach to dissolution rate. Pharm Weekbl 3:129–144, 1981.
8.   EL Cussler. Diffusion Mass Transfer in Fluid Systems. 2nd ed. New York: Cambridge Univ Press, 1997, pp 371–389.
9.   M Litt, G Serad. Chemical reactions on a rotating disk. Chem Eng Sci 19:867–884, 1964.
10. AC Shah, KG Nelson. Evaluation of a convective diffusion drug dissolution rate model. J Pharm Sci 64:1518–1520, 1975.
11. SS Ozturk, BO Palsson, JB Dressman. Dissolution of ionizable drugs in buffered and unbuffered solutions. Pharm Res 5:272–282, 1988.
12. KJ Himmelstein. Letter to the editor. Pharm Res 6:436–437, 1989.
13. S Ozturk, B Palsson, JB Dressman. Letter to the editor. Pharm Res 6:438–439, 1989.
14. MZ Southard, DW Green, VJ Stella, KJ Himmelstein. Dissolution of ionizable drugs into unbuffered solution: A comprehensive model for mass transport and reaction in the rotating disk geometry. Pharm Res 9:58–69, 1992.
15. KG Mooney, MA Mintun, KJ Himmelstein, VJ Stella. Dissolution kinetics of carboxylic acids II: Effect of buffers. J Pharm Sci 70:22–32, 1981.
16. JG Aunins, MZ Southard, RA Meyers, KJ Himmelstein, VJ Stella. Dissolution of carboxylic acids III: The effect of polyionizable buffers. J Pharm Sci 74:1305–1316, 1985.
17. S Chakrabarti, MZ Southard. Control of poorly soluble drug dissolution in conditions simulating the gastrointestinal tract flow. 1. Effect of tablet geometry in buffered medium. J Pharm Sci 85:313–319, 1996.
18. E Nelson. Dissolution rate of mixtures of weak acids and tribasic sodium phosphate. J Am Pharm Assoc 47:300–302, 1958.
19. WI Higuchi, NA Mir, SJ Desai. Dissolution rate of polyphase mixtures. J Pharm Sci 54:1405–1410, 1965.
20. SA Shah, EL Parrott. Dissolution of two-component solids. J Pharm Sci 65:1784–1790, 1976.
21. GR Carmichael, SA Shah, EL Parrott. General model for dissolution rates of n-component, nondisintegrating spheres. J Pharm Sci 70:1331–1338, 1981.
22. M Simpson, EL Parrott. Dissolution kinetics of a three-component solid I: Ethylparaben, phenacetin, and salicylamide. J Pharm Sci 72:757–764, 1983.
23. EL Parrott, M Simpson, DR Flanagan. Dissolution kinetics of a three-component solid II: Bezoic acid, salicylic acid, and salicylamide. J Pharm Sci 72:765–768, 1983.
24. AM Healy, OI Corrigan. Predicting the dissolution rate of ibuprofen–acidic excipient compressed mixtures in reactive media. Int J Pharm 84:167–173, 1992.
25. S Neervannan, LS Dias, MZ Southard, VJ Stella. A convective-diffusion model for dissolution of two non-interacting drug mixtures from co-compressed slabs under laminar hydrodynamic conditions. Pharm Res 11:1228–1295, 1994.
26. S Neervannan, MZ Southard, VJ Stella. Dependence of dissolution rate on surface area: Is a simple linear relationship valid for co-compressed drug mixtures? Pharm Res 11:1391–1395, 1994.
27. M Westerberg, B Jonsson, C Nystrom. Physicochemical aspects of drug release. IV.

The effect of carrier particle properties on the dissolution rate from ordered mixtures. Int J Pharm 28:23–31, 1986.

28. M Westerberg, C Nystrom. Physicochemical aspects of drug release. XVII. The effect of drug surface area coverage to carrier materials on drug dissolution from ordered mixtures. Int J Pharm 90:1–17, 1993.

29. S Chakrabarti, MZ Southard. Control of poorly soluble drug dissolution in conditions simulating the gastrointestinal tract flow. 2. Cocompression of drugs with buffers. J Pharm Sci 86:465–469, 1997.

30. KA Javaid, DE Cadwallader. Dissolution of aspirin from tablets containing various buffering agents. J Pharm Sci 61:1370–1373, 1972.

31. C Doherty, P York. Microenvironmental pH control of drug dissolution. Int J Pharm 50:223–232, 1989.

32. HW Davenport. Physiology of the Digestive Tract. 5th ed. Chicago: Year Book Medical Publishers, 1982.

33. HW Davenport. A Digest of Digestion. 2nd ed. Chicago: Year Book Medical Publishers, Chap 15.

34. VP Shah, JJ Konecny, RL Everett, B McCullough, AC Noorizadeh, JP Skelly. In vitro dissolution profile of water insoluble drug dosage forms in the presence of surfactants. Pharm Res 6:612–618, 1989.

35. GE Amidon, WI Higuchi, NFH Ho. Theoretical and experimental studies of transport of micelle-solubilized solutes. J Pharm Sci 71:77–84, 1982.

36. C Huang, DF Evans, EL Cussler. Linoleic acid solubilization with a spinning liquid disc. J Colloid Interface Sci 82:499–506, 1981.

37. AF Chan, DF Evans, EL Cussler. Explaining solubilization kinetics. AIChE J 22:1006–1012, 1976.

38. M Gilbaldi, S Feldman, R Wynn, ND Weiner. Dissolution rates in surfactant solutions under stirred and static conditions. J Pharm Sci 57:787–791, 1968.

39. WI Higuchi. Effects of interacting colloids on transport rates. J Pharm Sci 53:532–535, 1964.

40. C Nystrom, J Mazur, MI Barnett, M Glazer. Dissolution rate measurements of sparingly soluble compounds with the Coulter counter model TAII. J Pharm Pharmacol 37:217–221, 1985.

41. PW Taylor Jr, DE Wurster. Dissolution kinetics of certain crystalline forms of prednisolone II: Influence of low concentrations of sodium lauryl sulfate. J Pharm Sci 54:1654–1658, 1965.

42. K Shirahama, M Hayashi, R Matuura. The effects of organic additives on the solubilities and CMC's of potassium alkyl sulfates in water. I. The effects of several hydroxy compounds. Chem Bull Jpn 42:1206–1212, 1969.

43. K Shirahama, M Hayashi, R Matuura. The effects of organic additives on the solubilities and CMC's of potassium alkyl sulfates in water. II. Effects of some nonhydroxy compounds. Chem Bull Jpn 42:2123–2128, 1969.

44. JH Collett, L Koo. Interaction of substituted benzoic acids with Polysorbate 20 micelles. J Pharm Sci 64:1253–1255, 1975.

45. M Gibaldi, S Feldman, ND Weiner. Hydrodynamics and diffusional considerations in assessing the effects of surface active agents on the dissolution rate of drugs. Chem Pharm Bull 18:715–723, 1970.

46. SD Mithani, V Bakatselou, CN TenHoor, JB Dressman. Estimation of the increase in solubility of drugs as a function of bile salt concentration. Pharm Res 13:163–167, 1996.

47. LJ Naylor, V Bakatselou, JB Dressman. Comparison of the mechanism of dissolution of hydrocortisone in simple and mixed micelle systems. Pharm Res 10:865–870, 1993.

48. Sigma Catalog, PO Box 14508, St. Louis, MO 63178, 1998.

49. RB Bird, WE Stewart, EN Lightfoot. Transport Phenomena. New York: Wiley, 1960.

50. WG Cochran. The flow due to a rotating disc. Proc Cambridge Phil Soc 30:365–375, 1934.

51. VG Levich. Physico-Chemical Hydrodynamics. Englewood Cliffs NJ: Prentice-Hall, 1962, pp 39–72.

52. JR Crison, VP Shah, JP Skelly, GL Amidon. Drug dissolution into micellar solutions: Development of a convective diffusion model and comparison to the film equilibrium model with application to surfactant-facilitated dissolution of carbamazepine. J Pharm Sci 85:1005–1011, 1996.

53. JR Crison, ND Weiner, GL Amidon. Dissolution media for in vitro testing of water-insoluble drugs: Effect of surfactant purity on in vitro dissolution of carbamazepine in aqueous solutions of sodium lauryl sulfate. J Pharm Sci 86:384–388, 1997.

54. T Koizumi, WI Higuchi. Analysis of data on drug release from emulsions II. J Pharm Sci 57:87–92, 1968.

55. T Koizumi, WI Higuchi. Analysis of data on drug release from emulsions III. J Pharm Sci 57:93–96, 1968.

56. AH Goldberg, WI Higuchi. Mechanisms of interphase transport II: Theoretical considerations and experimental evaluation of interfacially controlled transport in solubilized systems. J Pharm Sci 58:1341–1352, 1969.

57. AH Goldberg, WI Higuchi, NFH Ho, G Zografi. Mechanism of interphase transport I: Theoretical considerations of diffusion and interfacial barriers in transport of solubilized systems. J Pharm Sci 56(11):1432–1437, 1967.

58. A Brodin, A Nyqvist-Mayer. In vitro release studies on lidocaine aqueous solutions, micellar solutions, and o/w emulsions. Acta Pharm Suec 19:267–284, 1982.

59. K Umezawa, A Karino, M Hayashi, K Tahara, A Kimura, S Awaza. Hepatic uptake of lipid-soluble drugs from fat emulsion. J Pharmacobio-Dynam 14:591–598, 1991.

60. AB Bikhazi, WI Higuchi. Interfacial barrier limited interphase transport of cholesterol in the aqueous polysorbate 80–hexadecane system. J Pharm Sci 59:744–748, 1970.

61. JR Crison, GD Leesman, JP Skelly, VP Shah, GL Amidon. Dissolution of carbamazepine in a soybean oil/water emulsion. AAPS Sixth Annual Meeting, Washington DC, Poster Presentation, November 1991.

62. JH DeSmidt, JCA Offringa, DJA Crommelin. Dissolution kinetics of griseofulvin in sodium dodecylsulphate solutions. J Pharm Sci 76:711–714, 1987.

63. JE Staggers, O Hernell, RJ Stafford, MC Carey. Physical-chemical behavior of dietary and biliary lipids during intestinal digestion and absorption. 1. Phase

behavior and aggregation states of model lipid systems patterned after aqueous duodenal contents of healthy adult human beings. Biochemistry 29:2028–2040, 1990.

64. O Hernell, JE Staggers, MC Carey. Physical-chemical behavior of dietary and biliary lipids during intestinal digestion and absorption. 2. Phase analysis and aggregation states of luminal lipids during duodenal fat digestion in healthy adult human beings. Biochemistry 29:2041–2056, 1990.

65. J du Plessis, LR Tiedt, AF Kotze, CJ van Wyk, C Ackerman. A transmission electron microscope method for determination of droplet size in parenteral fat emulsions using negative staining. Int J Pharm 46, 1988.

66. KA Yoon, DJ Burgess. Effect of nonionic surfactant on transport of model drugs in emulsions. Pharm Res 13:433–439, 1996.

67. JR Crison, ML Vieira, VP Shah, LJ Lesko, GL Amidon. Variability in the dissolution of water insoluble drugs under simulated fed and fasted conditions with application to establishing in vivo–in vitro correlations. AAPS Annual Meeting, Oct 27–31, Seattle WA, 1996.

68. J Crank. The Mathematics of Diffusion. 2nd ed. Oxford UK: Oxford Sci, Clarendon Press, 1986.

69. P Harriott. Mass transfer to particles: Part I. Suspended in agitated tanks. AIChE J 8:93–102, 1962.

70. AW Hixson, JH Crowell. Dependence of reaction velocity upon surface and agitation. I. Theoretical consideration. Ind Eng Chem 23:923–931, 1931.

71. RH Guy, J Hadgraft, IW Kellaway, MJ Taylor. Calculations of drug release from spherical particles. Int J Pharm 11:199–207, 1982.

72. SS Kornblum, JO Hirschorn. Dissolution of poorly water-soluble drugs I. Some physical parameters related to method of micronization and tablet manufacture of a quinazolinone compound. J Pharm Sci 59:606–609, 1970.

73. SE LeBlanc, HS Fogler. Population balance modeling of the dissolution of polydisperse solids: Rate limiting regimes. AIChE J 33:54–63, 1987.

74. D Brooke. Dissolution profile of log-normal powders: Exact expression. J Pharm Sci 65:795–798, 1973.

75. RB Hintz, KC Johnson. The effect of particle size on dissolution rate and oral absorption. Int J Pharm 51:9–17, 1989.

76. WI Higuchi, EN Heistand. Dissolution rates of finely divided drug powders I. Effect of a distribution of particle sizes in a diffusion-controlled process. J Pharm Sci 52:67–71, 1963.

77. KC Johnson, AC Swindell. Guidance in the setting of drug particle size specifications to minimize variability in absorption. Pharm Res 13:1795–1798, 1996.

78. SK Friedlander. Mass and heat transfer to single spheres and cylinders at low Reynolds numbers. AIChE J 3:43–48, 1957.

79. RG Cox, SG Mason. Suspended particles in fluid flow through tubes. Annu Rev Fluid Mech 3:291–315, 1971.

80. H Schlichting. Boundary Layer Theory. 7th ed. New York: McGraw-Hill, 1979, pp 24–44.

81. JB Dressman, D Fleisher. Mixing-tank model for predicting dissolution rate control of oral absorption. J Pharm Sci 75:109–116, 1986.

82. D Oh, RL Curl, GL Amidon. Estimating the fraction dose absorbed from suspension of poorly soluble compounds in humans: A mathematical model. Pharm Res 10: 264–270, 1993.

83. JR Crison, GL Amidon. Expected variation in bioavailability for water insoluble drugs. In: HH Blume, KK Midha, eds. Bio-International 2, Bioavailability, Bioequivalence, and Pharmacokinetic Studies. Stuttgart, Germany: Medpharm, 1995, pp 301–310.

84. RM Atkinson, C Bedford, KJ Child, EG Tomich. Effect of particle size on blood griseofulvin levels in man. Nature 193(4815):588–589, 1962.

6
Biological Transport Phenomena in the Gastrointestinal Tract: Cellular Mechanisms

David Fleisher
The University of Michigan, Ann Arbor, Michigan

I. INTRODUCTION

Since this chapter focuses on transport at the tissue and cellular level, a broad spectrum of biological transport topics is included with a general transport perspective. For the reader seeking more detailed information with regard to a particular cellular transport pathway, driving force, or experimental methodology, numerous references are cited. The thermodynamic and kinetic foundations for cellular transport mechanisms can be found in earlier chapters as well as in the references for this chapter. While the transport foundations will not change subsequent to publication, advances in cellular biology will quickly date this chapter and the chapter citations. Readers are encouraged to update these references utilizing Citation Index.

II. GASTROINTESTINAL EPITHELIA—STRUCTURE AND FUNCTION

An appropriate starting point for any discussion of drug transport in the gastrointestinal (GI) tract at the cellular level requires some introductory remarks on the structure and function of GI tissue. As a class of tissue, epithelia demarcate body entry points (skin, eye, respiratory, urinary, and GI organ systems), predisposing a general barrier function with respect to solute entry and translocation. In addition to their barrier function, the epithelia lining the GI tract serve specialized functions promoting efficient nutrient digestion and absorption, and support other organ systems in body water, electrolyte, and bile salt homeostasis. The dual absorption–protection function of GI epithelia requires solute recognition mechanisms to amplify nutrient transport and to attenuate epithelial entry of potentially noxious solutes. The homeostatic demands on GI tissue dictated by this dual function may pose special transport considerations compared to solute translocation across biologically inert barriers.

Geometrically idealizing the GI tract as a cylindrical tube (Fig. 1), regional differences in function provide both axial (x) and radial (r) heterogeneity to solute and water transport at the tube boundary. In the axial direction, upper and lower GI epithelia present variable surface area and availability of transbarrier solute pathways (Figs. 1 and 2) as a function of both regional differences in anatomy and physiology and differences in solute physicochemical properties. In the radial direction, the solute encounters a diverse set of transport resistances in series. These include a hydrodynamic boundary layer, mucus layer, membrane ionic microclimate layers, glycocalyx (glycoprotein cover, Fig. 3), microvillus mucosal brush-border membrane in parallel with a tight junction barrier, enterocyte cytosol and cellular organelles, basolateral membrane in parallel with the lateral intercellular space, intervillus space, basement membrane, and parallel lamina propria endothelial membranes of the blood capillaries and lymphatic central lacteal. It is often stated that the mucosal membrane plus tight junctions provide the rate-limiting barrier to solute transport. However, the contributions of other resistances and intestinal metabolism and secretion may significantly alter absorption kinetics dependent on solute properties and GI biochemistry [1]. The magnitude of some of these resistances varies with GI region and physiological conditions as a function of oral input and homeostatic response. Additional solute transport variability is predicated by regional and physiological variations in solute removal from GI tissue as provided by villus blood capillary and lymphatic duct density and flow (Figs. 1.4 and 4).

Crypt stem cell differentiation and maturation as well as variable cellular migration rates within individual villi may result in solute transport heterogeneity along the crypt–villus axis (Fig. 3). This factor will be revisited in other chapter sections on absorbing surface area and coupled transport. Stem cells in the crypt region provide continuous renewal of this rapid turnover tissue, differentiating into paneth, goblet, endocrine, and absorptive cells [2]. Paneth cells migrate deeper into the crypt region and produce lysozyme as part of intestinal defense to bacterial translocation across the gut. Goblet cells, which secret mucus, migrate up the villus to serve an additional protective function. A number of enteroendocrine cells secrete signal peptides into the lumen (paracrine) and bloodstream (endocrine) to provide feedback control for efficient nutrient absorption and de-

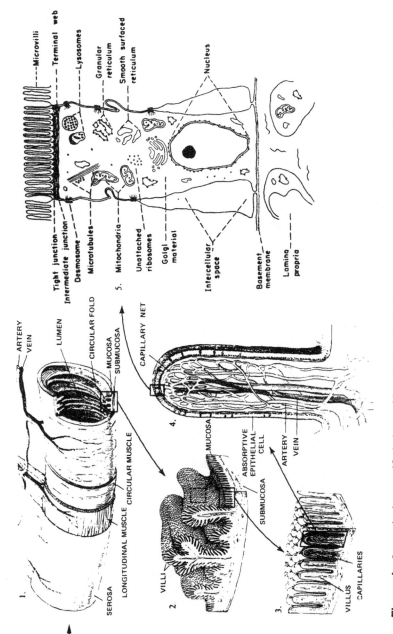

Figure 1 Increasingly magnified views of intestinal epithelia from cylindrical tube to enterocyte. (From Ref. 76.)

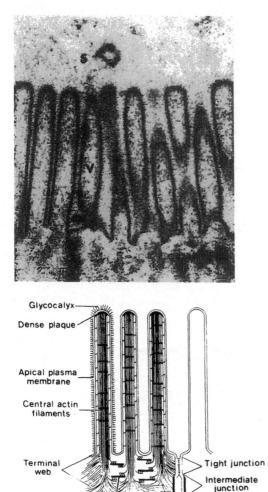

Figure 2 (Left) Microscopic magnification (×46,500) of mucosal surface (S) associated with intestinal microvilli (V). (Right) Schematic of the microvilli illustrating the cytoskeleton elements including junctional complexes. (From Ref. 76.)

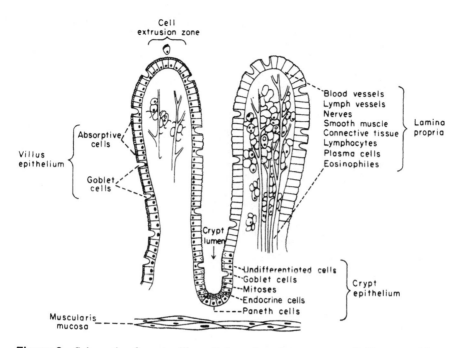

Figure 3 Schematic of crypt–villus cellular axis and components of villus central lacteal. (From Ref. 77.)

fense from toxic solutes. Immunological M cells are grouped in lymphoid tissue as follicle-associated epithelia known as Peyer's patches (Fig. 4). These regions (visibly distinguishable since they lack developed microvilli and have a folded surface) are prominent in human ileum, while they are distributed throughout the intestine in rodents. The M cells signal nearby T and B lymphocytes and contain vesicles involved in pinocytosis. They further support input defense systems through antigen recognition and initiation of IgA mucosal immunity in connection with the lamina propria [3].

The absorptive cells (enterocytes) dominate the cellular population of the villus epithelium and serve the major role in GI solute transport. However, even these cells provide structural and functional heterogeneity in accordance with radial position and axial region. Within the range of regional heterogeneity, the spatial dimensions of individual structures providing resistance to solute transport across this tissue are fairly consistent between GI regions and even across mammalian species at the cellular level. These dimensions are given in Table 1 so that relative contributions to transport can be estimated based on resistance path lengths and solute physical chemistry [4,5].

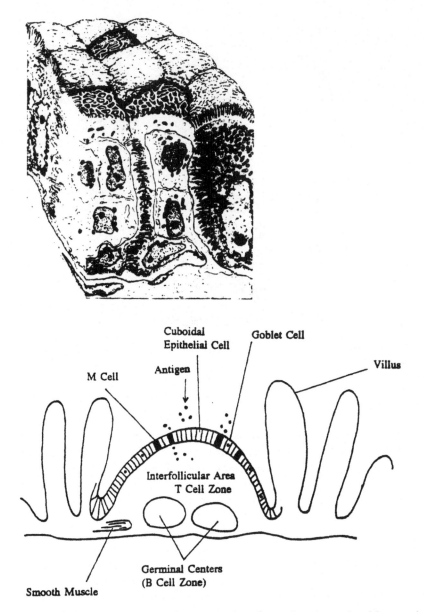

Figure 4 Epithelium overlying a human Peyer's patch and associated immunological and lymphoid cell regions. (From Ref. 77.)

Table 1 Transport Dimensions in Human Intestinal Epithelia

Lumenal radius	1 cm
Mean aqueous boundary layer	
In vivo	≤40 μm
In situ and in vitro systems	100–900 μm
Mucus layer measured in vitro	100–500 μm
Jejunal pH microclimate layer	500–900 μm
Villus height	500–800 μm
Intervillus space	50 μm
Microvillus height	1.4 μm
Microvillus width	80 μm
Glycocalyx	0.1–0.2 μm
Epithelial cell height	30 μm
Epithelial cell width	8 μm
Mucosal bilayer membrane thickness	10–11 nm
Basolateral bilayer membrane thickness	7 nm
Polar headgroup-associated aqueous region	1–2 nm
Hydrophobic lipid tail (hydrophobic core)	3–5 nm
Basement membrane thickness	30 nm
Tight junction depth (villus)	400 nm
Tight junction depth (crypt)	250 nm
Normal lateral intercellular space width	15–30 nm
Under stimulated fluid absorption	2–3 μm
Distance basement membrane to capillaries	0.5 μm

Source: Refs. 4 and 5.

III. TRANSPORT PATHWAYS AND SOLUTE STRUCTURE

A. Precellular Barriers

With respect to solute transport across GI tissue, the monolayer of epithelial cells lining the GI lumen generally dominates solute transport by controlling solute entry. However, for solutes demonstrating high membrane permeability, lumenal aqueous resistance may control transport rate. In the case of the GI tract, aqueous boundary layer resistance (diffusion layer resistance, unstirred water layer resistance) to solute transport is a function of lumenal fluid flow rate as determined by GI motility. The magnitude of this resistance may be further modified by epithelial water absorption or secretion and possibly by villus contractions. A maximal estimate [6] of mean diffusion layer thickness of 40 μm in dog and human intestine in vivo is similar in magnitude to the human epithelial cell thickness (30 μm). In theory, under membrane sink conditions, this resistance is unambiguously defined as the ratio of the mean aqueous diffusion layer thickness (L_{aq})

to solute aqueous diffusivity (D_{aq}) , where L_{aq} is a function of the fluid flow velocity (v) in the GI lumen.

$$R_{aq} = L_{aq}/D_{aq} \tag{1}$$

In practice, estimation of L_{aq} requires information on the rate of solute removal at the membrane since aqueous resistance is calculated from experimental data defining the solute concentration profile across this barrier [7]. Mean L_{aq} values calculated from the product of aqueous diffusivity (at body temperature) and aqueous resistance obtained from human and animal intestinal perfusion experiments in situ are in the range of 100–900 µm, compared to lumenal radii of 0.2 cm (rat) and 1 cm (human). These estimates will necessarily be a function of perfusion flow rate and choice of solute. The lower L_{aq} estimated in vivo is rationalized by better mixing within the lumen in the vicinity of the mucosal membrane [6].

The role of intestinal mucus as a barrier to solute transport is dependent on both solute physical chemistry and physiological factors controlling mucus production and secretion. While specific binding of some drugs to intestinal mucin has been reported [8], a more general role for mucus as a diffusion barrier has been cited based on reduction of this barrier to solute transport by mucolytic agents [9]. The mean thickness of this glycoprotein gel layer is highly variable [10], 100–500 µm (compared to intestinal villus heights of 500–800 µm), and a number of biochemical entities control its viscosity and removal from the lumenal side and its secretion and viscosity from the cellular side. Again, these large values measured following invasive procedures may not reflect normal in vivo mucus layer thickness. While its relative resistance to general solute transport compared to aqueous boundary layer resistance has been the subject of scientific debate [11], this viscous gel will contribute greater resistance to solute transport than an aqueous boundary layer of equivalent mean thickness [12]. In situ and in vitro flow rates and stirring conditions, utilized to reduce the size and resistance of the aqueous boundary layer, may also clear mucus. In vivo, these regions certainly overlap defining a lumped precellular resistance parameter to solute transport. In addition to its protective function, mucus serves to slow the diffusion of secreted ions away from the epithelial membrane, providing maintenance of mucosal ionic microclimates [13]. The influence of such microclimates on drug transport is discussed later in this chapter.

B. Lipid Barriers, Transcellular Partition Path, Surface Area

Theoretical aspects of diffusional resistances are discussed in Part I of this volume. Following these discussions, parallel ''transcellular'' and ''paracellular'' pathways (Fig. 5) specify two parallel resistances to solute transport in GI epithelia. The lipid bilayers composing mucosal and basolateral epithelial cell membranes (Fig. 6) define the primary transcellular diffusional resistance to solute transport across GI epithelia. Distinct lipid (and protein) composition in the exoplasmic mucosal and basolateral membranes is maintained by restrictions on lateral diffusion and mixing of membrane components by the tight junctions. Solute partition and diffusion across membrane lipid and solute translocation via membrane carrier proteins represent parallel transcellular pathways. (The paracellular pathway between epithelial cells and mediated transcellular transport are discussed in subsequent sections.)

In series with a desolvation energy barrier required to disrupt aqueous solute hydrogen bonds [14], the lipid bilayer offers a practically impermeable barrier to hydrophilic solutes. It follows that significant transepithelial transport of water-soluble molecules must be conducted paracellularly or mediated by solute translocation via specific integral membrane proteins (Fig. 6). Transcellular permeability of lipophilic solutes depends on their solubility in GI membrane lipids relative to their aqueous solubility. This lumped parameter, "membrane" permeability,

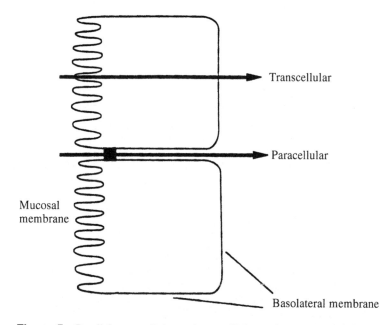

Figure 5 Parallel transcellular and paracellular pathways through intestinal epithelial cell monolayer.

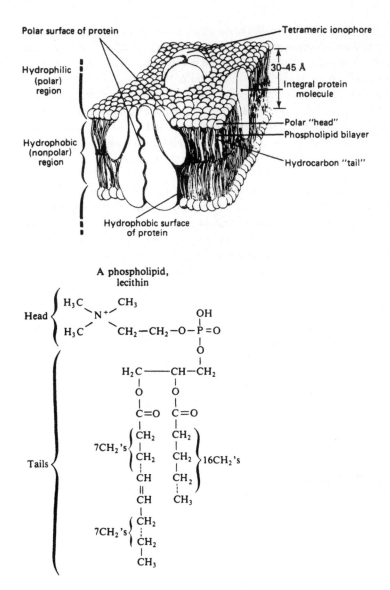

Figure 6 Intestinal cell membrane model with integral membrane proteins embedded in lipid bilayer. The phospholipid bilayer is 30–45 Å thick, and membrane proteins can span up to 100 Å through the bilayer. The structure of a typical phospholipid membrane constituent, lecithin is illustrated. (From Ref. 76.)

is defined as the ratio of the product of membrane diffusivity (D_m) and membrane/water partition coefficient (K_p) to the membrane thickness (L_m):

$$P_m = K_p D_m / L_m \tag{2}$$

In whole tissue or cell monolayer experiments, transcellular "membrane" resistance $(R_m = P_m^{-1})$ lumps mucosal to serosal compartment elements in series with aqueous resistance $(R_{aq} = P_{aq}^{-1})$. The operational definition of L_m depends on the experimental procedure for solute transport measurement (see Section VII), but its magnitude can be considered relatively constant within any given experimental system. Since the K_p range dwarfs the range of D_m, solute differences in partition coefficient dominate solute differences in transcellular membrane transport. The lumped precellular resistance and lumped membrane resistance add in series to define an effective resistance to solute transport:

$$R_{eff} = P_{eff}^{-1} = R_{aq} + R_m \tag{3}$$

Following this analysis, variations in transcellular membrane permeability for a particular solute will depend on the fluid (viscous) state of the lipid membrane. Changes in lipid viscosity (fluidity) modulate solute membrane partition and diffusion processes as well as conformation and subsequent solute translocation by integral membrane proteins. The lumenal monolayer of the intestinal brush border mucosal membrane bilayer is high in sphingolipid content, providing tight lipid packing through significant intermolecular hydrogen bonding forces [15]. The cytosolic leaflet of the mucosal membrane contains a greater percentage of glycerol and inositol phospholipids, permitting cytosolic control of inner bilayer fluidity and transcellular signaling. Changes in cholesterol/phospholipid ratios modulate membrane lipid viscosity (fluidity), which in turn regulates membrane enzyme (e.g., alkaline phosphatase) and carrier protein (e.g., glucose transporter) activities [16]. The total mucosal microvillus membrane thickness of 10–11 nm is broader than most plasma membranes and might be attributed to its high glycosylation and high protein/lipid ratio as well as its distinct lipid composition [17].

Basolateral bilayer membrane thickness (7 nm) is more typical of plasma membrane dimensions. Similar to the mucosal membrane, the hydrophobic core region comprising the lipid hydrocarbon chain tail regions spans 3–5 nm of the total membrane thickness (Fig. 6). Greater fluidity changes are observed in basolateral membranes where membrane protein interactions (i.e., Na-K ATPase) with circulating hormones provide homeostatic control of transepithelial ion and water transport [18]. Intestinal membrane fluidity changes and resultant transport function are further regulated by unsaturated fatty acid content in response to diet [18]; these fatty acids are also more prominent in basolateral membranes than in mucosal membranes. In general, mucosal membranes are less fluid and show smaller fluidity changes than basolateral membranes. Furthermore, lower intestinal (ileal and colonic) membranes are less fluid and show smaller fluidity changes than upper intestinal (duodenal and jejunal) membranes [18]. Such fluidity changes may indirectly influence the transport of both passively and actively absorbed xenobiotics.

Precellular solute ionization dictates membrane permeability dependence on mucosal pH. Therefore, lumenal or cellular events that affect mucosal "microclimate" pH may alter the membrane transport of ionizable solutes. The mucosal microclimate pH is defined by a region in the neighborhood of the mucosal membrane in which pH is lower than in the lumenal fluid. This is the result of proton secretion by the enterocytes, for which outward diffusion is slowed by intestinal mucus. (In fact, mucosal secretion of any ion coupled with mucus-restricted diffusion will provide an ionic microclimate.) Important differences in solute transport between experimental systems may be due to differences in intestinal ions and mucus secretion. It might be anticipated that microclimate pH effects would be less pronounced in epithelial cell culture (devoid of goblet cells) transport studies than in whole intestinal tissue.

Membrane uptake of nonionized solute is favored over that of ionized solute by the membrane/water partition coefficient (K_p). If $K_p = 1$ for a nonionized solute, membrane permeability should mirror the solute ionization curve (i.e., membrane permeability should be half the maximum value when mucosal pH equals solute pK_a). When the K_p is high, membrane uptake of nonionized solute shifts the ionization equilibrium in the mucosal microclimate to replace nonionized solute removed by the membrane. As a result, solute membrane permeability (absorption rate) versus pH curves are shifted toward the right for weak acids and toward the left for weak bases (Fig. 7).

These sigmoidal curve shifts can be described by shifts in the inflection point pH as a function of the aqueous and membrane permeability [19].

$$\text{pH}_{\text{inflection}} = pK_a \pm \log\left(1 + \frac{P_m}{P_{\text{aq}}}\right) \qquad (4)$$

At mucosal pH where drug is 100% ionized and Eq. (4) has accounted for pH partition shifts, nonzero permeability suggests that ionized drug (K_p close to zero) permeates the membrane by a route other than via lipid partition/diffusion. A discussion of such alternative pathways follows in Section III.C.

An important extension of lipid–solute interaction components [20] to membrane partitioning is provided by solute molecular structure. Spacing between polar and nonpolar regions (Fig. 8) within a solute molecule may result in significant distortion of the $K_p D_m$ product across the membrane polar headgroup/lipid core interface [21]. Such interactions may be responsible for deviations from projected transport predictions based on simple partitioning theory translating to deviations from predicted absorption kinetics [1].

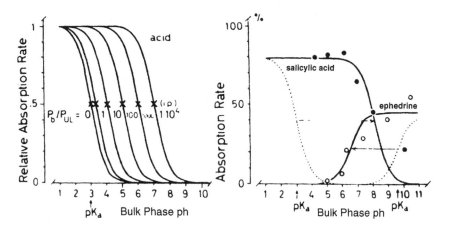

Figure 7 (Left panel) Relative absorption rate for a weak acid ($pK_a = 3$) as a function of mucosal pH for increasing barrier (membrane) permeability (P_b) with fixed unstirred aqueous layer permeability (P_{ul}). $X = pH_{inflection\ point}$ in Eq. (4). (Right panel) Partition coefficient–dependent absorption rates for salicylic acid and the weak base ephedrine. (From Ref. 19.)

The lumped material resistance properties of intestinal membranes as defined by permeability can be described macroscopically by the ratio of mass flux (J) through the membrane to the solute concentration difference across the membrane. Permeability (reciprocal resistance) has units of length per unit time.

$$\frac{J}{\Delta C} = P \quad \left(\frac{[(\text{mass/area}) \times \text{time})]}{\text{mass/volume}} \right) \tag{5}$$

This phenomenological description requires that the process of solute transport be normalized for membrane surface area. As total mucosal surface area is amplified by the villi and microvilli (Fig. 1), calculations based on estimates of cylindrical absorbing surface as generated from intestinal radius would overestimate solute permeability. However, it has been suggested that the relative contribution of intestinal membrane surface area to solute transport depends, in turn, on relative solute membrane permeability [22]. It is projected that for solutes possessing high membrane permeability, villus tip absorption is adequate for complete membrane uptake. Under these conditions, the assumption of cylindrical geometry to estimate absorbing surface area should prove adequate, and, furthermore, intestinal absorption may be controlled by aqueous resistance. High permeability solutes include lipophilic solutes under passive membrane transport and hydrophilic solutes under mediated membrane transport. (Membrane transport mediators ma-

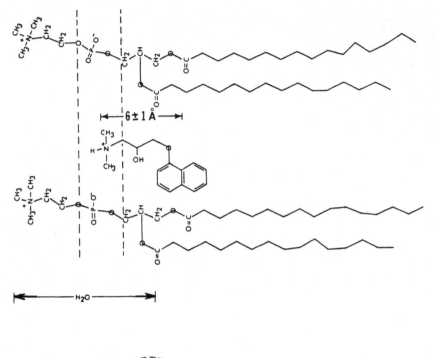

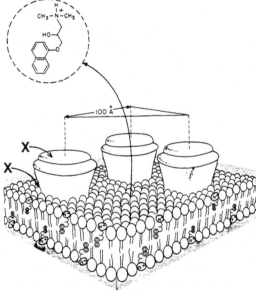

Figure 8 Localization of solute (propranolol) within the lipid bilayer. This solute–membrane interaction has been shown to influence the conformation and activity of a calcium-pump protein (X) embedded in the bilayer. (From Ref. 78.)

ture to maximum activity in the upper villus.) For more poorly membrane permeable solutes, absorption may be projected to depend more strongly on total available intestinal membrane surface area. As a result, a plot of mucosal membrane uptake as a function of solute partition coefficient may appear sigmoidal with strong surface area dependency between the plateau regions. Furthermore, regional differences in uptake of poorly permeable solutes may be more pronounced as the duodenum and jejunum present more mucosal surface than is the case for the ileum and colon.

C. Paracellular Transport—Equivalent Pore and Circuit Theory

The paracellular pathway (in parallel with the transcellular pathway) comprises the tight junction at the apical cellular boundary in series with the lateral intercellular space (Figs. 5 and 9). The tight junctions are thought to be responsive to cytoskeletal elements [16] in the terminal web (Fig. 2). It has been projected, on the basis of membrane surface area, that the paracellular pathway is three orders of magnitude less available to solute transport than the lipophilic transcellular route. As a consequence of this morphological feature, paracellular transport does not play a significant role in the absorption of lipophilic solutes. While this pathway provides an aqueous route for hydrophilic solutes, it is both size- (pore theory) and charge- (circuit theory) selective. GI epithelia are classified as ''leaky'' or ''tight'' based either on transepithelial electrical resistance (TEER) to charge transport or the relative conductance of the paracellular (shunt) pathway to the total tissue conductance [23] as measured in Ussing chambers (see Section VII). Gastric mucosa are classed as tight epithelia, with TEER values of 2000 $\Omega \cdot cm^2$ and an ionic shunt conductance less than 50% of the total conductance. Small intestinal tissues are described as leaky epithelia, with TEER values ranging from 50 to 100 $\Omega \cdot cm^2$ and shunt conductances of 85–95% of the total, with higher conductance in the jejunum than in the ileum. The colonic TEER of 300–400 $\Omega \cdot cm^2$ corresponds to a relative paracellular conductance of 60–80% and is consistent with TEER measurements in colon cancer cell culture (Caco-2) monolayers of 300–400 $\Omega \cdot cm^2$ 12–25 days postconfluence [24]. In addition, crypt epithelia show a higher resistance than villus epithelia, and amphibian intestinal epithelia tend to be tighter than mammalian intestinal epithelia.

With respect to the size and charge selectivity of paracellular pathways, ''equivalent'' pore theory has been utilized to calculate an effective radius based on the membrane transport of uncharged hydrophilic molecules, while ''equivalent'' circuit theory has been used to separate mediated from paracellular membrane transport of small ions. The term ''equivalent'' should be emphasized, as selectivity parameters are obtained from membrane transport data, so phenomenological information is used to quantitate the magnitude of aqueous pathways

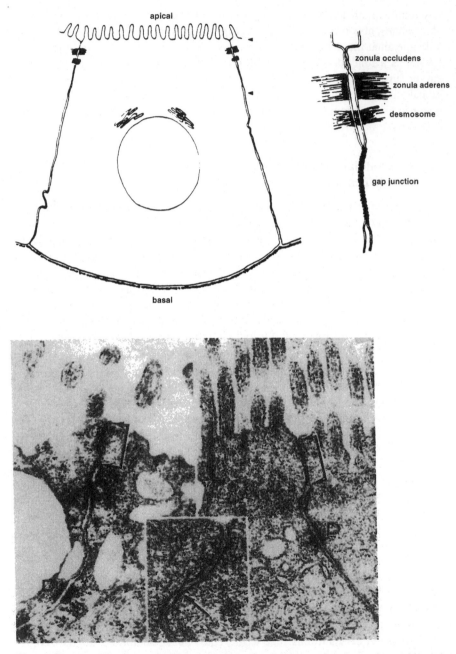

Figure 9   Magnification (×52,700) and schematic of tight junctional complex and lateral intercellular space. (From Ref. 74.)

through the lipid barrier. An "equivalent" pore radius is calculated assuming transport of the solute through homogeneously distributed cylindrical pores of uniform radius and length equal to the membrane thickness. It is further assumed that these pores represent the only possible solute route through the membrane.

Membrane transport data for evaluating pore size selectivity are generated by comparing hydrophilic solute filtration ($P_f = J_f/[(C_m + C_s)/2]$) versus diffusion ($P_d = J_d/\Delta C$) permeabilities. Filtration permeabilities are obtained under conditions generating osmotically, or hydrostatically, driven water transport, where J_f defines hydraulic conductivity or convective flux. (Note that the filtration permeability is defined in accordance with the arithmetic mean of solute mucosal (C_m) and serosal (C_s) concentrations across the barrier. A logarithmic mean concentration is more accurate, but the arithmetic mean serves as a good approximation.) This differs from the diffusive flux, for which the driving force is the concentration gradient or difference (ΔC) across the barrier. The driving force for hydraulic conductivity is the pressure gradient defining the fluid velocity embodied in P_f. In further discussions of solvent drag of a solute through paracellular pathways it is assumed that convective flux dominates solute diffusion in the pore. In addition, it is typical to assume that sink conditions ($C_s = 0$) hold. (In subsequent discussions on paracellular and mediated transport, these assumptions are implicit and $C_m = C$ is utilized to simplify the analysis.) Diffusion permeabilities are determined under iso-osmotic conditions free of hydrostatic pressure gradients.

Utilizing such data, the mucosal surface of mammalian ileum and colon have been found to be heteroporous with wide "neutral" pathways of 65 Å equivalent radius and smaller cation-selective pathways of 8 Å equivalent radius [25,26]. The larger pathway may correspond to the lateral intercellular space, which admits large molecules such as horseradish peroxidase (MW 40,000) from the serosal side up to the occluding tight junction (Figs. 5 and 9). The smaller pathway may be accommodated within the structure of the tight junction [26]. This pathway has been further characterized by estimating the equivalent pore size (R_p) restriction to the transport of nonelectrolyte hydrophilic solutes of varying molecular radius [25].

$$R_p = -R_w + (2R_w^2 + k)^{1/2}$$
$$k = \frac{8\gamma_w D_w RT}{V_w}\left(\frac{P_f}{P_d}\right) \tag{6}$$

where R_w, γ_w, D_w, and V_w represent the molecular radius, viscosity, diffusivity, and partial molar volume of water. Experimentally, P_f is obtained by monitoring concentration changes of an impermeable solute under the influence of osmotically induced water flux. P_d is evaluated from the flux of tritiated water from the mucosal compartment under iso-osmotic conditions. Consistent with transepithelial resistance measurements, the effective radius of this pathway in the jejunum has been calculated to be twice that estimated in the lower intestine [27]. Such calculations do not equate equivalent pore radius with the molecular radius of the largest molecule permeating under filtration conditions [28]. Both steric and viscous hindrance reduce (by as much as fourfold) the permeability of pore pathways to idealized "spherical" molecules (see Table 1). It is standard practice to lump hindrance factors, including solute molecular geometry, into a reflection or exclusion coefficient (δ). Assuming that the pore does not restrict the transport of water (molecular radius 1.5 Å), the pore reflection coefficient for a given solute is defined by

$$\delta = 1 - A_{wf}/A_{sf} \tag{7}$$

where A_{wf} and A_{sf} are the "effective" surface areas for filtration of water and solute, respectively, and the magnitude of δ varies from zero to one [29]. The steric hindrance projected for this equivalent pore size indicates that the colonic pathway would provide close to total exclusion of the polysaccharide inulin (MW 5000, effective hydrodynamic radius 12–15 Å, reflection coefficient close to one) and substantial restriction to a monosaccharide like mannitol (MW 182, effective hydrodynamic radius 4.2 Å). However, evidence for transient in situ expansion of the paracellular pathways by enhancers in the lower intestinal epithelia [25] and by sodium/nutrient cotransport in the upper intestinal epithelia [30] has been reported (see dimensions of lateral intercellular space in Table 1).

The leaky epithelia of the small intestine exploit the accessibility of the paracellular pathway to small inorganic ions like sodium and potassium (effective hydrodynamic radii 1.2–1.8 Å [31], reflection coefficient close to zero) in maintaining electrolyte homeostasis. It has been further hypothesized that this pathway permits serosal to mucosal recirculation of sodium to drive nutrient transport [32]. The general cationic selectivity of this pathway is the result of a lining of negative charge (carboxylate, phosphate, and sulfate) that discriminates against passive anion transport. This negatively charged lining has an effective pK_a in the range of 4–5, providing for reversals in paracellular pathway charge selectivity as a function of local mucosal sodium–proton exchange [23]. An investigation of size and charge selectivity of the paracellular pathway in rat ileum demonstrated a relatively high shunt pathway permeability to small organic cations [33] consistent with ileal microclimate pH.

Ion transport across membranes can be evaluated by using mucosal and serosal electrodes to read transepithelial current (I) and potential difference (Ψ). With these parameters, equivalent circuit analysis can be utilized to account for the relative contributions of transcellular and paracellular pathways. Ionic flux (J) is defined by the Nernst–Planck equation,

$$J_z = -D\frac{dC}{dr} + \frac{zF}{RT}DC\frac{d\Psi}{dr} \tag{8}$$

where z and C represent the valence and concentration of the ionic solute, F is Faraday's constant, RT the temperature dependent kinetic energy, and dr the differential distance across the membrane. Equivalent to Ohm's law, the membrane conductance (G, reciprocal resistance) is defined by

$$G = \frac{I}{\Delta \Psi - E} \qquad (9)$$

where $\Delta \Psi$ is the integrated potential difference across the membrane, based on the transmembrane gradient of C, and E is the Nernst potential, based on the equilibrium charge distribution (initial condition emf from equivalent circuit analysis). Paracellular transport of an ionic solute can be separated from passive transcellular transport by plotting the transepithelial flux versus the exponential applied potential [34].

$$J = J_0 + J_s \, \exp\left[\frac{-zF\Psi}{2RT}\right] \qquad (10)$$

The intercept defines ion flux across the epithelia (J_0) while the slope provides a measure of the magnitude of the shunt pathway (J_s) under short-circuit ($\Psi = 0$) conditions. However, as discussed in the following section, mediated solute transport may also be voltage-dependent.

Additional epithelial aqueous pathways of significantly smaller radius (<3 Å) have also been documented utilizing both equivalent pore and circuit theory [25]. These pathways may correspond to specific channels through lipid membranes as opposed to paracellular pathways. Osmotically activated ion channels [35] and even specific water channels [36] have been characterized in renal epithelia. In intestinal epithelia, mucosal chloride channels have been studied in secreting crypt cells, and basolateral potassium channels in colonic epithelia serve cellular ion and volume homeostatic functions.

D. Membrane Transport Facilitators: Carrier, Channel, and Endosomal Proteins

Supplementing the solute transport selectivity of the membrane lipid bilayer and paracellular pathways, specialized epithelial binding, carrier, and channel proteins mediate sensitive biochemical recognition and screening for solute entry and exit and membrane transport (Figs. 6 and 8). Operation and regulation of these transport mediators is controlled by a number of factors, which modulate their conformation. These factors include coupling to transmembrane ion gradients and cellular biochemical energy gradients as well as interactions with membrane lipids and other biochemical modifiers. Such couplings and interactions permit (1) faster rates of solute membrane transport than are possible by free diffusion through membrane lipid bilayers, (2) transport against transmembrane solute concentration gradient driving forces, (3) transport control through ion and energy gradient dissipation, and (4) transport control through membrane cis–trans solute competition and mediator cooperativity.

The classification of membrane transport mediators as channels (water-filled solute-conducting pores) and carriers (solute-binding membrane translocators) has become somewhat blurred in light of recent findings. Both channels and carriers span the bilayer, and, in fact, some carrier membrane–spanning domains have been shown to form fluid-filled channels. Both channel gating and carrier selectivity operate through mediator conformational changes. A major distinction between the two transport mediators is the rate of solute transport. A single carrier translocates solute across the membrane at a rate of 10^2–10^4 solute molecules per second. Single-channel solute transport rates are typically 10^6–10^9 molecules (ions) per second.

Transmembrane glycoprotein antigens, which initiate solute transport through receptor-mediated endocytotic pathways, represent another class of membrane transporters. While this process may be targeted for drug delivery in the liver and pancreas [37], its role in intestinal transport may be limited [38]. The process of transcytosis, in which solute endocytotic transport occurs across intestinal cells, has been documented in the ileum of developing rats [39] to transport milk proteins. Endocytotic transport of insulin and other growth factors in adult rat ileum has also been reported [40], and ileal transport of intrinsic factor–cobalamin complex is thought to be endocytotically mediated [41].

In addition to faster solute transport rates, the major experimental features of membrane-facilitated transport that distinguish it from membrane diffusion include (1) specificity and selectivity; (2) saturability; (3) inhibition, activation, and cooperativity; (4) transmembrane effects; and (5) greater temperature sensitivity than is characteristic of membrane diffusion [42].

IV. TRANSPORT KINETICS

A. Kinetics of Cellular Diffusion

Diffusion provides an effective basis for net migration of solute molecules over the short distances encountered at cellular and subcellular levels. Since the diffusional flux is linearly related to the solute concentration gradient across a transport barrier [Eq. (5)], a mean diffusion time constant (reciprocal first-order rate constant) can be obtained as the ratio of the mean squared migration distance (L) to the effective diffusivity in the transport region of interest.

$$t = k^{-1} = L^2/D_{eff} \tag{11}$$

At the cellular level, this transport region is composed of aqueous and membrane phases in series, requiring that the diffusivity be modified by the phase distribution constant (an equilibrium constant defined by desolvation and aqueous–membrane phase transfer energy barriers). As will be discussed under solute–solvent coupling, solvent convection may dominate diffusion of small molecules through paracellular pathways. In these special cases, a solute transport time constant can be obtained as the ratio of the effective length (L) of the paracellular pathway to the solvent convective velocity through this pathway.

$$t = L/v = k^{-1} \tag{12}$$

Other kinetic parameters can be specified as a function of system capacity. For example, from information on lumenal volume (V), kinetic uptake by intestinal epithelia can be defined as a clearance (Cl) from the lumenal compartment [43].

$$Cl = Vk \tag{13}$$

If initial solute uptake rate is determined from intestinal tissue incubated in drug solution, uptake must be normalized for intestinal tissue weight. Alternative capacity normalizations are required for vesicular or cellular uptake of solute (see Section VII). Cellular transport parameters can be defined either in terms of kinetic rate–time constants or in terms of concentration normalized flux [Eq. (5)]. Relationships between kinetic and transport descriptions can be made on the basis of information on solute transport distances. Note that division of Eq. (11) or (12) by transport distance defines a transport resistance of reciprocal permeability (conductance).

B. Active and Facilitated Transport

Within the scope of biological membrane components which mediate solute transport beyond free diffusion are those that participate in ''active'' transport. This process is defined by facilitator-mediated solute translocation that is driven by energy derived from cellular metabolism. The driving force for both free diffusion and nonactive facilitated transport is the concentration gradient or thermodynamic solute potential across the membrane; these are classified as ''passive'' transport processes. Active transport processes, in contrast, require transduction of a cellular metabolic energy driving force into transport work. Solute transport coupled directly to metabolic energy is defined as ''primary'' active transport. The sodium–potassium exchanger at the basolateral membrane (Fig. 10) is coupled directly to release of phosphate from ATP as mediated by an ATPase enzyme and represents a primary active transport component. Many active transport systems also convert chemical energy into electrostatic potential energy by contributing to the potential difference across the membrane. Solute transport coupled to

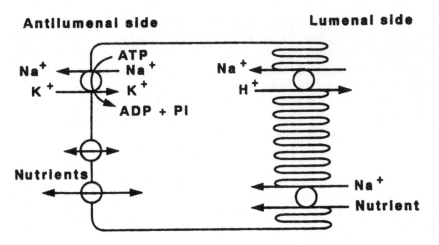

Figure 10   Schematic of cotransporters and countertransporters (or exchangers) in enterocyte mucosal and basolateral membranes.

such a potential generated by cellular metabolism is referred to as a "secondary" active transport process. Glucose and amino acids are translocated across intestinal mucosal membranes coupled to transporter conformational changes driven by transmucosal sodium gradients (Fig. 10). Since the ion gradient required to drive mucosal nutrient transport is generated by the basolateral Na^+-K^+ ATPase-coupled exchanger, sodium/nutrient cotransport is a secondary active transport process. Experimental manipulations, which compromise cellular energy sources, will limit active transport.

C.   Mediated Transport Kinetics: Saturation and Inhibitors

Two distinguishing features of gastrointestinal active and facilitated transport processes are that they are capacity-limited and inhibitable. Passive transcellular solute flux is proportional to mucosal solute concentration (C), where the proportionality constant is the ratio of the product of membrane diffusion coefficient (D_m) and distribution coefficient (K_d) to the length of the transcellular pathway (L_m).

$$\frac{V}{A}\frac{dC}{dT} = J = \frac{D_m K_d}{L_m} C \qquad (14)$$

V represents the volume of the mucosal compartment and A the surface area of the mucosal barrier. Passive paracellular solute flux is also proportional to mucosal solute concentration, where the proportionality constant is the ratio of the aqueous diffusion coefficient to the paracellular path length (L_{aq}). When significant fluid convection dominates aqueous diffusion through the paracellular pathway, the proportionality constant is the volume-averaged fluid velocity (v).

$$J = D_{aq}/L_{aq}\ C \tag{15}$$

$$J = vC \tag{16}$$

Passive transport flux is therefore linearly dependent on mucosal solute concentration, provided the transported solute is readily removed by the villus blood supply (sink conditions).

Carrier-mediated transport is linear with mucosal solute concentration until this concentration exceeds the number of available carriers. At this point the maximal solute flux (J_{max}) is independent of further increases in mucosal solute concentration. In the linear range of solute flux versus mucosal concentration (C), the proportionality constant is the ratio of J_{max} to the solute–carrier affinity constant (K_m). This description of Michaelis–Menten kinetics is directly analogous to time changes in mass per unit volume (velocity of concentration change) found in enzyme kinetics, while here the appropriate description is the time change in solute mass per unit surface area of membrane supporting the carrier.

$$J = \frac{J_{max}\ C}{K_m + C} + P_m C \tag{17}$$

Lack of perfect specificity in carrier–solute recognition provides for the possibility that structurally similar solutes may compete for carrier availability. Analysis of competitive [Eq. (18)] and noncompetitive [Eq. (19)] inhibition as well as cooperativity effects (allosteric modulation by structurally dissimilar solutes) on carrier-mediated solute flux is equivalent to assessment of the velocity of enzyme reactions.

Competitive inhibition:

$$J = \frac{J_{max}\ C}{K_m(1 + I/K_i) + C} + P_m C \tag{18}$$

Noncompetitive inhibition:

$$J = \frac{J_{max}\ C}{K_m + (1 + I/K_i)C} + P_m C \tag{19}$$

where the mucosal inhibitor concentration and the inhibitor–carrier affinity constant are I and K_i, respectively. Kinetic analysis of allosteric modulation of carriers utilizing the Hill equation is likewise analogous to the use of this equation in enzyme kinetics.

D. Exchange Phenomena: Cis and Trans Effects

Both secondary active transport and positive cooperativity effects enhance carrier-mediated solute flux, in contrast to negative cooperativity and inhibition phenomena, which depress this flux. Most secondary active transport in intestinal epithelia is driven by transmembrane ion gradients in which an inorganic cation is cotransported with the solute (usually a nutrient or inorganic anion). Carriers which translocate more than one solute species in the same direction across the membrane are referred to as cotransporters. Carriers which translocate different solutes in opposite directions across the membrane are called countertransporters or exchangers (Figs. 10 and 11).

In order to maintain low intracellular sodium and the resultant high trans-mucosal sodium gradient to drive nutrient transport, a basolateral ATPase pump "exchanges" intracellular sodium for extracellular potassium. Energy-independent mucosal sodium–proton and chloride–bicarbonate exchangers serve in the homeostasis of intracellular pH and the maintenance of mucosal ionic microclimates (Fig. 11). (Maintenance of low microclimate pH (Fig. 12) is known to drive the carrier-mediated transport of a number of di- and tripeptides and peptide-like drugs [44].) While such processes are often referred to as "exchange diffusion," they are usually carrier-mediated. As is the case for competitive inhibition, trans-

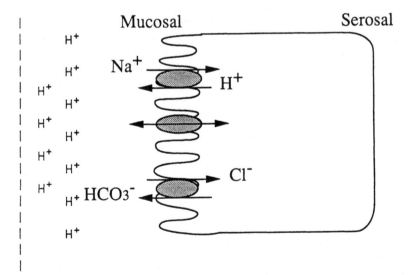

Figure 11 Schematic of mucosal membrane sodium–proton exchanger and chloride–bicarbonate exchanger responsible for pH homeostasis in enterocyte cytosol. Microclimate pH is maintained by mucosal slowing of proton diffusion away from the lumenal membrane.

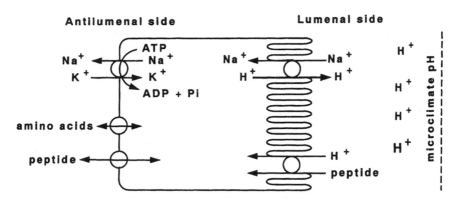

Figure 12 Schematic of generation of mucosal microclimate pH as a transmucosal proton-gradient driving force for di- and tripeptide carrier-mediated translocation across the mucosal membrane into the enterocyte.

porter recognition of more than one solute provides for competitive and cooperative solute transport effects on either side of a membrane-associated exchanger. Such effects operating on the same side of the membrane are termed "cis" effects, while those operating on opposite sides of the membrane are "trans" effects.

Cis stimulation of solute flux occurs by cooperativity or by a cotransported

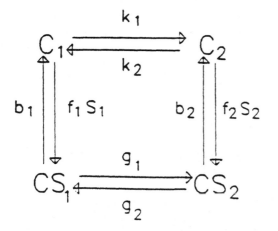

Figure 13 Mediated transport kinetic scheme. C = carrier, S = solute; 1 and 2 represent sides of the membrane; g_i are rate constants for changes in conformation of solute-loaded carrier; k_i are rate constants for conformational changes of unloaded carrier; f_i and b_i are rate constants for formation and separation of carrier–solute complex. (From Ref. 73.)

ion mediating a favorable carrier conformation for solute binding and/or solute translocation. Trans stimulation of solute flux can be generated by an event that promotes favorable carrier conformation for the solute from the opposite side of the membrane. Countertransport is also classified as a trans stimulation effect. Similarly, cis inhibition of solute flux is provided by competitive and noncompetitive reduction of solute binding and translocation, while trans inhibition serves to limit membrane transport through buildup of translocated co-ions or the presence of another molecule which binds the carrier from the opposite side of the membrane. Trans inhibition prevents a return of the carrier to a conformation favorable to solute translocation across the membrane. A simplified model from which kinetic equations can be derived for carrier-mediated transmembrane solute (S) transport is shown in Figure 13. Separation of the solute from the carrier or mediator protein is rapid, and conformational reorientation tends to limit the mediated transport rate.

V. TRANSPORT COUPLING AND DRIVING FORCES

Control of nutrient transport dictates significant coupling between transported components in GI epithelia. This complicates solute transport analysis by requiring a multicomponent description. Flux equations written for each component constitute a nonlinear system in which the coupling nonlinearities are embodied in the coefficients modifying individual transport contributions to flux.

Possible driving forces for solute flux can be enumerated as a linear combination of gradient contributions [Eq. (20)] to solute potential across the membrane barrier (see Part I of this volume). These transbarrier gradients include chemical potential (concentration gradient–driven diffusion), hydrostatic potential (pressure gradient–driven convection), electrical potential (ion gradient–driven cotransport), osmotic potential (osmotic pressure–driven convection), and chemical potential modified by chemical or biochemical reaction.

$$J_i = \sum_{j=1}^{n} L_{ij} X_j, \qquad i = 1, \dots, n \tag{20}$$

The relative contribution of each driving force (X) generated by component j to the flux of solute i (J_i) is expressed by coefficients L_{ij} in this phenomenological description of parallel transport processes.

A. Ion Gradients: Solute–Solute Coupling

Carrier-mediated transport of nutrients in small intestinal epithelia is often promoted by the maintenance of transmucosal ion gradients. A mathematical description of cotransport of an uncharged solute with an inorganic ion requires that the solute chemical potential be coupled to an electrical potential as specified in Eq. (21) where ion [I] and nutrient [N] concentrations in mucosal (m) and cytosolic (c) compartments are given with respect to their transport stoichiometry (n) and valence (z).

$$\frac{[N]_c}{[N]_m} \le \left(\frac{[I]_m}{[I]_c}\right)^{nI/nN} \exp\left[\frac{F}{RT}(\Psi_m - \Psi_c)\left(\frac{nI}{nN}\right)zI\right] \tag{21}$$

$\Psi_m - \Psi_c$ defines the electrical potential gradient across the mucosal membrane where potential inside the enterocyte is negative.

In the small intestine, the transport of elementary nutrients such as glucose and amino acids and anions such as phosphate is directly coupled through sodium-modulated carrier conformational changes [44]. Maintenance of low (10–15 mM) cytosolic sodium levels (by Na^+-K^+ ATPase) is fundamental in providing a high transmucosal gradient driving force for intestinal nutrient absorption in the upper intestine and ion and water homeostasis in the lower intestine. Similarly, Na^+–H^+ exchange is fundamental in maintaining a low mucosal microclimate pH to drive small intestinal transport of di- and tripeptides.

It has been documented that carrier-mediated transmucosal transport of folic acid, dipeptides, and dipeptide-like drugs (cephalosporins and ACE inhibitors) is coupled to transmucosal proton gradients [44]. Solute transport coupled to hydrogen ion potential can alternatively be described as a function of pH. While an extracellular to intracellular sodium gradient is maintained for most cell types through sodium-potassium ATPase pumps, transmucosal proton gradients are more region-specific. Furthermore, since local pH influences solute ionization and subsequent aqueous to membrane partitioning, regions of high mucosal proton potential have been described as a pH microclimate.

In rat jejunum, the highest mucosal proton concentrations (214–224 nM, pH 6.67–6.65) have been measured 10–100 µM down from the villus tip, while in the crypt region mucosal pH is in the range of cytosolic pH [45]. Jejunal absorption of glucose has been associated with decreasing jejunal upper villus microclimate pH to values as low as 5.7–5.8 [46]. Upper villus mucosal pH has been measured at pH 7.44 in rat ileum, documenting that a transmucosal proton gradient does not exist in this region [47]. The region-dependent maintenance of an acid microclimate may play a significant role in the variable absorption of weakly acidic and basic drugs (see Section III.B), for which mucosal ionization couples with nonionized drug membrane partitioning to influence variable membrane uptake. In this regard, pH partition theory describes membrane uptake as a sigmoidal function of mucosal pH, mirroring the fraction of nonionized drug and shifted by the magnitude of the membrane intrinsic drug partition coefficient [20] (Fig. 7).

B. Osmotic Gradients: Solute–Solvent Coupling

Recent work has suggested that sodium/nutrient cotransport generates epithelial water absorption as driven by cellular osmolality gradients propagated by sodium transport [5]. Furthermore, electron microscopy has provided some evidence that this process is coupled to dilation of tight junctions and a broadening of lateral intercellular spaces, promoting solvent drag of small nutrient molecules through the paracellular pathway [49]. Teleological arguments have been made to support this finding on the basis of conservation of cellular energy required to maintain transcellular sodium gradients. That this process would serve to diminish the barrier function by reducing the specificity for transport of hydrophilic solutes argues against the proposed mechanism.

"Net" water flux is typically measured in intestinal perfusion experiments by measuring dilution (net water secretion into the intestinal lumen) or concentration (net water absorption from the intestinal lumen) of a "nonabsorbable" marker (typically a very large hydrophilic molecule). Karino et al. [50] suggested that solvent drag of solutes is more appropriately correlated with unidirectional water flux (mucosal to serosal) in the direction of water absorption (J_{ms}) determined by including tritiated water in the perfusion solution. This analysis requires that back flux (serosal to mucosal) of tritiated water is negligible. The dual water marker technique has generated more realistic solute reflection coefficient (δ) values via Eq. (22) than via data generated using Eq. (23) in which single-marker net water flux (J_w) is utilized.

Dual marker net flux:

$$J_s = P_m C + \delta J_{ms} C \tag{22}$$

Single-marker net flux:

$$J_s = P_m C + \delta J_w C \tag{23}$$

The solute flux (J_s) is the linear sum of diffusive and convective flux across the membrane, where the convective flux is generated by an osmotic gradient and restricted by the magnitude of the reflection coefficient. Further support for using unidirectional water flux comes from information on enterocyte physiology [51], suggesting that water secretion (osmotically driven by chloride secretion) is a function performed by the crypt cells, while water absorption (osmotically driven by sodium absorption) is dominated by villus tip cells (Fig. 14). Net water transport averages these simultaneous processes. As discussed under paracellular transport (Section III.C), osmotic homeostasis in leaky epithelia suggests that reflection coefficients for water and small inorganic ions are zero. At issue for solvent drag of solute through paracellular pathways is the magnitude of the reflection coefficient as a function of solute molecular size and the physiological conditions for which tight-junctional solute sieving might be relaxed [52,53].

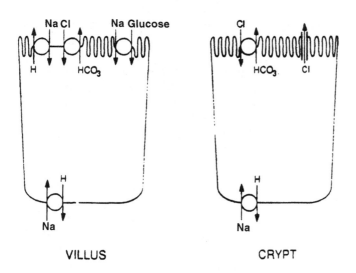

VILLUS   CRYPT

Figure 14 Ion transport pathways responsible for water flux across intestinal epithelia. Sodium absorption in villus tip cells (left) stimulates water absorption, while chloride channel exit in crypt cells (right) stimulates water secretion.

C. Series and Parallel Transport Processes

The coupled processes described by Eqs. (8), (14), (17), and (22) can be added in (20) as parallel solute transport pathways across the membrane. The phenomenological coefficients (L_{ij}) describe the membrane permeability by these pathways [potential-dependent, Eq. (8); via membrane lipid partition and diffusion, Eq. (14); carrier-mediated, Eq. (17); and convectively coupled, Eq. (22)]. These pathways define parallel resistances through the intestinal barrier in series with precellular resistances to solute transport.

D. The Role of Metabolism and Intestinal Secretion: Reaction–Transport Coupling

The coupling of solute transport in the GI lumen with solute lumenal metabolism (homogeneous reaction) and membrane metabolism (heterogeneous reaction) has been discussed by Sinko et al. [54] and is more generally treated in Cussler's text [55]. At the cellular level, solute metabolism can occur at the mucosal membrane, in the enterocyte cytosol, and in the endoplasmic reticulum (or microsomal compartment). For peptide drugs, the extent of hydrolysis by lumenal and membrane-bound peptidases reduces drug availability for intestinal absorption [56]. Preferential hydrolysis (metabolic specificity) has been targeted for reconversion of soluble peptide, amino acid, and phosphate prodrugs [57]. Chemical modification of poorly soluble drugs with these progroups permits higher lumenal concentration gradients to drive transport toward intestinal membranes. Such progroups are cleaved via membrane-bound peptidases and phosphatases to release parent drug in the vicinity of the mucosal membrane. Peptide prodrugs can also be targeted for hydrolysis by specific cytosolic peptidases following transport by the mucosal peptide carrier [58]. These oral drug delivery strategies take advantage of digestion and transport processes geared for intestinal nutrient absorption [59].

Microsomal and cytosolic enzymes, on the other hand, underlie general intestinal clearance mechanisms, which have evolved to aid the liver in preventing the entry of potentially toxic compounds into the systemic circulation. Both phase I and phase II metabolism have been reported to occur in intestinal epithelia [60]. Since these processes serve to reduce further entry by increasing molecular polarity, transporters to export metabolites past enterocyte membranes should exist corresponding to those found in bile canalicular membranes in hepatocytes [61]. The coupling of these enzymes with carrier-mediated export has become a crucial consideration with respect to biological contributions to drug permeability and absorption variability in the intestine. Coupling between cytochrome P-450 3A enzymes in enterocyte endoplasmic reticulum with metabolite and drug export by mucosal P-glycoprotein has been observed to diminish cyclosporin intestinal permeability and contribute to variable absorption [62]. The involvement of other intestinal enzymes, such as flavin monooxygenase and other mucosal exporters, may contribute to the variable intestinal absorption of a number of weakly basic drugs [63]. Since these intestinal clearance mechanisms are often region- or site-dependent, drug–nutrient and drug–drug interactions may provide an additional contribution to drug absorption variability via competition for these coupled elimination mechanisms [64,65].

VI. TRANSPORT ADAPTATION AND REGULATION

Regulation of solute transport in the GI tract is served by a host of homeostatic mechanisms, many of which have been discussed in previous sections. Homeostatic alterations of precellular barriers include changes in mucus secretion, changes in aqueous resistance (through mechanisms controlling GI and villus motility), and changes in ionic microclimates in response to cellular metabolism. At the cellular level, modulation of tight junctional resistance coupled with homeostatic mechanisms generating water absorption or secretion serve to regulate paracellular solute transport. Regulation of transcellular solute transport is achieved through modulations of membrane lipid fluidity influencing permeation and diffusion of lipophilic solutes, as well as through the activity of membrane transporters controlling transcellular transport of specific hydrophilic solutes.

Membrane fluidity is subject to regulation by circulating plasma hormones operating at the basolateral membrane and through cytosolic second messengers. Changes in membrane lipid fluidity control the conformational states available to membrane-bound transporters dictating regulations with respect to carrier affinity for specific solutes. Cis and trans regulation modulate carrier affinity as described by Eq. (17). It is likely that a number of these mechanisms operate via upper intestinal gut peptide endocrine release to homeostatically regulate lower intestinal response [66]. As opposed to these carrier regulation mechanisms which operate over short time frames to regulate carrier affinity, longer term adaptive mechanisms include "up and down regulation" of transporters to control transport capacity (J_{max}). These processes include carrier insertion in the membrane from vesicular storage sites in the cytosol and endocytosis-mediated carrier removal from the mucosal or basolateral membrane. Longer term up and down regulation is mediated by carrier biosynthesis, turnover, and degradation at the level of nuclear, endoplasmic, and lysosomal subcellular compartments [67].

VII. ISOLATED EXPERIMENTAL METHODS OF STUDY AND CORRELATIONS WITH IN VIVO ABSORPTION

While in vivo studies assess absorption rates as process-lumped time constants from blood level versus time data, these rate parameters encompass the kinetics of dosage-form release, GI transit, metabolism, and membrane permeation. The use of isolated tissue and cellular preparations to screen for drug absorption potential and to evaluate absorption rate limits at the tissue and cellular levels has been expanded by the pharmaceutical industry over the past several years. For more detail in this regard, the reader is referred to an article by Stewart et al. [68] for references on these preparations and for additional details on the various experimental techniques outlined below.

In situ perfusion studies assess absorption as lumenal clearance or membrane permeability and provide for isolation of solute transport at the level of the intestinal tissue. Controlled input of drug concentration, perfusion pH, osmolality, composition, and flow rate combined with intestinal region selection allow for separation of aqueous resistance and water transport effects on solute tissue permeation. This system provides for solute sampling from GI lumenal and plasma (mesenteric and systemic) compartments. A sensitive assay can separate metabolic from transport contributions.

In vitro studies permit further isolation of parallel transport processes and can provide a reduction in experimental variability. Rate of absorption assessment can be measured as intestinal uptake or flux across an intestinal barrier at both the tissue and cell monolayer levels. Experimental variability is also reduced by the fact that a large number of tissue samples can be used from the same experimental animal and a large number of samples can be run under the same cell culture conditions. The value of flux chamber techniques is that they provide for both mucosal and serosal drug addition and drug and metabolite sampling measurements. Ussing chambers provide additional electrical parameter information via tissue mucosal (cell monolayer brush border) and tissue serosal (cell monolayer basolateral) surface electrodes with which to evaluate intestinal ion transport. Everted intestinal sacs can also be used to measure solute flux across intestinal tissue.

Studies involving measurements of drug uptake by excised intestinal tissue are classified as tissue accumulation methods. Incubation of everted intestinal rings and sleeves [69] in drug solution has provided an excellent system for distinguishing active from passive transport. The simplicity of this method permits experiments to be carried out over a relatively short time frame required to ensure tissue viability. The use of small rings (20–50 mg wet weight) provides a large number of samples from the same experimental animal to statistically compare treatment differences. It is routine to first determine drug uptake by rings as a function of time. Uptake is normalized by the weight of the tissue ring sample. Drug accumulation by small rings reaches influx–efflux equilibrium fairly rapidly as they represent a small-capacity drug uptake system. Accurate assessment of kinetic uptake parameters as a function of drug incubation concentration requires that the initial rate of uptake (before back flux becomes significant) be utilized. Drug uptake treatment studies should be carried out at a fixed time when the initial drug uptake rate is linear as a function of time. Transport dependence of solute concentration can be performed in situ and in vitro. Again the large number of samples obtainable from a single animal permits a statistically reliable assessment of transport parameters. This is especially valuable for accurately determining active transport parameters, requiring that a number of concentrations be studied close to the K_m.

Solute uptake can also be evaluated in isolated cell suspensions, cell monolayers, and enterocyte membrane vesicles. In these preparations, uptake is normalized by enzyme activity and/or protein concentration. While the isolation of cells in suspension preparations is an experimentally easy procedure, disruption of cell monolayers causes dedifferentiation and mucosal-to-serosal polarity is lost. While cell monolayers from culture have become a popular drug absorption screening tool, differences in drug metabolism and carrier-mediated absorption [70], export, and paracellular transport may be cell-type- and condition-dependent.

Mucosal brush border membrane vesicles and basolateral membrane vesicles can be isolated to study solute uptake across specific enterocyte boundaries. These more isolated vesicle systems allow for investigation of solute transport across a particular membrane barrier and permit separation of membrane transport from cellular metabolism. (Intestinal microsomes can be utilized to focus on cellular metabolism [71].) In addition, the vesicles focus on transmembrane solute translocation free of parallel paracellular pathways [72].

The use of vesicle cell membranes, isolated cells, and cell monolayers and intestinal tissue studies has provided valuable correlations with in situ and in vivo drug absorption in animals as well as correlations with drug absorption in clinical studies. Most prominent among the literature sources establishing correlations between in vitro tissue and cellular systems with drug absorption in humans are the work of Dowty and Dietsch [73], Lennernas et al. [74], and Stewart et al. [75].

While experimental isolation serves to investigate the mechanisms of cellular transport, the in vivo and clinical implications are of "bottom-line" importance. Experimental treatment differences may produce significant variations in solute transport at the cellular and tissue levels of investigation. However, these effects may have minimal impact on blood drug levels following oral dose if intestinal membrane transport or cellular metabolism is not the rate-limiting step. The reader is encouraged to keep these basic transport principles in mind for investigations of solute transport at the intestinal tissue or cellular level. The value of obtaining correlations between isolated in vitro studies and in vivo and clinical absorption parameters cannot be overemphasized.

VIII. SUMMARY

The mucosal membrane of the intestinal epithelial cells is regarded as the primary permeability barrier to drug absorption from oral administration. Transport pathways for drug absorption across this barrier are a function of solute physicochemical properties. The pathways for solute transport across intestinal epithelia include lipid membrane partitioning for transcellular transport of lipophilic solutes and paracellular and carrier-mediated transcellular transport of small hydrophilic solutes. Transcytotic pathways exist for only a very limited number of solutes. Since the function of this organ system includes nutrient absorption and maintenance of water and electrolyte homeostasis, close attention should be paid to pathways mediating these functions in studies of drug absorption. Recent research indicates that the intestine may also act as an organ of both systemic and presystemic elimination including drug metabolism mediated by cytochrome P450 3A4 and drug and metabolite secretion mediated by P-glycoprotein [62]. When drug absorption is less than would be projected on the basis of solute physicochemical properties, drug recognition by these elimination pathways may be involved. Scientists working in the area of oral drug delivery will do well to keep up with these advances in biological transport in the gastrointestinal tract.

REFERENCES

1. K Dackson, JA Stone, KJ Palin, WN Charman. Evaluation of the mass balance assumption with respect to the two-resistance model of intestinal absorption by using in situ single-pass intestinal perfusion of theophylline in rats. J Pharm Sci 81:321–325, 1992.
2. JI Gordon. Intestinal epithelial differentiation. J Cell Biol 108:1187–1194, 1989.
3. C Press, S McClure, T Landsverk. Computer-assisted morphometric analysis of absorptive and follicle-associated epithelia of Peyer's patches. Immunology 72:386–392, 1991.
4. RL Elliott. Intestinal absorption: Analysis of theoretical models. PhD thesis, University of Wisconsin, Madison, 1979.
5. JL Madara, JS Trier. Functional morphology of the mucosa of the small intestine. In: LR Johnson, ed. Physiology of the Gastrointestinal Tract. 2nd ed. New York: Raven, 1994.
6. A Strocchi, MD Levitt. A reappraisal of the magnitude and implications of the intestinal unstirred layer. Gastroenterology 101:843–847, 1991.
7. JH Kou, D Fleisher, GL Amidon. Calculation of the aqueous diffusion layer resistance for absorption in a tube: Application to intestinal membrane permeability determination. Pharm Res 8:298–305, 1991.
8. JJ Niibuchi, Y Aramaki, S Tsuchiya. Binding of antibiotics to rat intestinal mucin. Int J Pharm 30:181–187, 1986.
9. FGJ Poelma, R Breas, JJ Tukker. Intestinal absorption of drugs. IV. The influence of taurocholate and L-cysteine on the barrier function of mucus. Int J Pharm 64:161–169, 1990.
10. MA Desai, CV Nicholas, P Vadgama. Electrochemical determination of the permeability of porcine mucus to model solute compounds. J Pharm Pharmacol 43:124–127, 1990.
11. (a) KW Smithson, DB Millar, LR Jacobs, GM Gray. Intestinal diffusion barrier: Unstirred water layer or membrane surface mucous coat? Science 214:1241–1243, 1981. (b) JA DeSimone. Diffusion barrier in the small intestine. Science 220:221–222, 1983; KW Smithson. Reply to 11b. Science 220:222, 1983.
12. D Winne, W Verheyen. Diffusion coefficient in native mucus gel of rat small intestine. J Pharm Pharmacol 42:517–519, 1989.
13. YF Shiau, P Fernandez, MJ Jackson, S McMonagle. Mechanisms maintaining a low-pH microclimate in the intestine. Am J Physiol 248:G608–G617, 1985.
14. MS Karls, RD Rush, KF Wilkinson, TJ Vidmar, PS Burton, MJ Ruwart. Desolvation energy: A major determinant of absorption, but not clearance, of peptides in rats. Pharm Res 8:1477–1481, 1991.
15. JS Patton. Is the intestinal membrane bilayer freely permeable to lipophilic molecules? In: F Alvarado, CH van Os, eds. Ion Gradient-Coupled Transport. INSERM Symposium No. 26. Amsterdam: Elsevier Science, 1986, pp 33–36.
16. TA Brasitus, R Dahiya, PK Dudeja, BM Bissonnette. Cholesterol modulates alkaline phosphatase activity of rat intestinal microvillus membranes. J Biol Chem 263:8592–8597, 1988.

17. S Chapelle, M Gilles-Baillien. Phospholipids and cholesterol in brush border and basolateral membranes from rat intestinal mucosa. Biochim Biophys Acta 753:269–271, 1983.

18. SM Schwartz, HE Bostwick, MS Medow. Estrogen modulates ileal basolateral membrane lipid dynamics and Na^+-K^+ ATPase activity. Am J Physiol 254:G687–G694, 1988.

19. D Winne. Shift of pH-absorption curves. J Pharmacokin. Biopharm 5:53–94, 1977.

20. EL Cussler. Diffusion: Mass Transfer in Fluid Systems. New York: Cambridge University Press, 1984, pp. 172–193.

21. LG Herbette, DG Rhodes, RP Mason. New approaches to drug design and delivery based on drug–membrane interactions. Drug Des Delivery 7:75–118, 1991.

22. D Winne. The permeability coefficient of the wall of a villous membrane. J Math Biol 6:95–108, 1978.

23. DW Powell. Barrier function of epithelia. Am J Physiol 241:G275–G288, 1981.

24. G Wilson, IF Hassan, CJ Dix, I Williamson, R Shah, M Mackay. Transport and permeability properties of human Caco-2 cells: An in vitro model of the intestinal epithelial cell barrier. J Controlled Release 11:25–40, 1990.

25. M Tomita, M Shiga, M Hayashi, S Awazu. Enhancement of colonic drug absorption by the paracellular permeation route. Pharm Res 5:341–346, 1988.

26. RJ Naftalin, S Tripathi. Passive water flows driven across isolated rabbit ileum by osmotic, hydrostatic and electrical gradients. J Physiol 360:27–50, 1985.

27. CA Loehry, J Kingham, J Baker. Small intestinal permeability in animals and man. Gut 14:683–688, 1973.

28. MH Friedman. Principles and Models of Biological Transport. Berlin: Springer-Verlag, 1986, pp 142–152.

29. AK Soloman. Characterization of biological membranes by equivalent pores. J Gen Physiol 51:335s–364s, 1968.

30. JR Pappenheimer, KZ Reiss. Contribution of solvent drag through intracellular junctions to absorption of nutrients by the small intestine of the rat. J Membrane Biol 100:123–136, 1987.

31. SG Schultz, AK Solomon. Determination of the effective hydrodynamic radii of small molecules by viscometry. J Gen Physiol 44:1189–1199, 1961.

32. ML Halperin, SL Wolman, DR Greenberg. Paracellular recirculation of sodium is essential to support nutrient absorption in the GI tract. Clin Invest Med 9:209–211, 1986.

33. PG Ruifrok, WEM Mol. Paracellular transport of inorganic and organic ions across rat ileum. Biochem Pharmacol 32:637–640, 1983.

34. G Barnett, S Hui, LZ Benet. Effects of theophylline on salicylate transport in isolated rat jejunum. Biochim Biophys Acta 507:517–523, 1978.

35. J Ubl, H Murer, HA Kolb. Ion channels activated by osmotic and mechanical stress in membranes of opossum kidney cells. J Membrane Biol 104:223–232, 1988.

36. MM Meyer, AS Verkman. Evidence for water channels in renal proximal tubule cell membranes. J Membrane Biol 96:107–119, 1987.

37. RJ Fallon, AL Schwartz. Receptor-mediated endocytosis and targeted drug delivery. Hepatology 5:899–901, 1985.

38. EJ Hughson, CR Hopkins. Endocytotic pathways in polarized Caco-2 cells: Identifi-

cation of an endosomal compartment accessible from both apical and basolateral surfaces. J Cell Biol 110:337–348, 1990.

39.   JM Wilson, JA Whitney, MR Neutra. Biogenesis of the apical endosome–lysosome complex during differentiation of absorptive epithelial cells in rat ileum. J Cell Sci 100:133–143, 1991.

40.   M Bendayan, E Ziv, R Ben-Sasson, H Bar-On, M Kidron. Morpho-cytochemical and biochemical evidence for insulin absorption by the rat ileal epithelium. Diabetologia 33:197–204, 1990.

41.   RM Donaldson. In: LR Johnson, ed. Physiology of the GI Tract. 2nd ed. Vol. 2. New York: Raven Press, 1987, pp 967–969.

42.   MH Friedman. Principles and Models of Biological Transport. Berlin: Springer-Verlag, 1986, pp 44–45.

43.   H Yuasa, T Iga, M Hanano, J Watanabe. Relationship between in vivo first-order intestinal absorption rate constant and the membrane permeability clearance. J Pharm Sci 78:922–924, 1989.

44.   H Yuasa, D Fleisher, GL Amidon. Noncompetitive inhibition of cephradine uptake by enalapril in rabbit intestinal brush-border membrane vesicles: An enalapril specific inhibitory binding site on the peptide carrier. J Pharmacol Exp Ther 269:1107–1111, 1994.

45.   R Kinne. Epithelial transport: The interplay between ion gradients and cell polarity. In: F Alvarado, CH van Os, eds. Ion Gradient-Coupled Transport. Inserm Symposium No. 26. New York: Elsevier, 1986, pp 255–264.

46.   H Daniel, B Neugebauer, A Kratz, G Rehner. Localization of acid microclimate along intestinal villi of rat jejunum. Am J Physiol 248:G293–G298, 1985.

47.   H Daniel, G Rehner. Effect of metabolizable sugar on the mucosal surface pH of rat intestine. J Nutr 116:768–777, 1986.

48.   H Daniel, C Fett, A Kratz. Demonstration and modification of intervillus pH profiles in rat small intestine in vitro. Am J Physiol 257:G489–G495, 1989.

49.   JL Madara, JR Pappenheimer. Structural basis for physiological regulation of paracellular pathways in intestinal epithelia. J Membrane Biol 100:149–164, 1987.

50.   A Karino, M Hayashi, T Horie, S Awaza, H Minami, M Hanano. Solvent drag effect in intestinal drug absorption. J Pharmacobiol Dynam 5:410–417, 670–677, 1982.

51.   MJ Welsh, PL Smith, M Fromm, RA Frizzell. Crypts are the site of intestinal fluid and electrolyte secretion. Science 218:1219–1221, 1982.

52.   H Lu, J Thomas, D Fleisher. Influence of D-glucose-induced water absorption on rat jejunal uptake of two passively absorbed drugs. J Pharm Sci 81:21–25, 1992.

53.   K Atisook, JL Madara. An oligopeptide permeates intestinal tight junctions at glucose-elicited dilatations. Gastroenterology 100:719–724, 1991.

54.   PJ Sinko, GD Leesman, GL Amidon. Mass balance approaches for estimating the intestinal absorption and metabolism of peptides and analogues: Theoretical development and applications. Pharm Res 10:271–275, 1993.

55.   EL Cussler. Diffusion: Mass Transfer in Fluid Systems. New York: Cambridge University Press, 1984, pp 325–412.

56.   SS Davis. Overcoming barriers to the oral administration of peptide drugs. J Pharm Sci 11:353–355, 1990.

57. D Fleisher, BH Stewart, GL Amidon. Design of prodrugs for improved GI absorption by intestinal enzyme targeting. Methods Enzymol 112:360–381, 1985.

58. JPF Bai, GL Amidon. Structural specificity of mucosal-cell transport and metabolism of peptide drugs. J Pharm Sci 9:969–978, 1992.

59. D Fleisher, R Bong, BH Stewart. Improved oral drug delivery: Solubility limitations overcome by the use of prodrugs. Adv Drug Deliv Rev 19:115–130, 1996.

60. KF Ilett, LBG Tee, PT Reeves, RF Minchin. Metabolism of drugs and other xenobiotics in the gut lumen and wall. Pharm Ther 46:67–93, 1990.

61. RB Sund, F Lauterbach. 1-Naphthol metabolism and metabolite transport in the small and large intestine. Pharmacol Toxicol 60:262–268, 1987.

62. LZ Benet, CY Wu, MF Hebert, VJ Wacher. Intestinal drug metabolism and antitransport processes: A potential paradigm shift in oral drug delivery. J Controlled Release 39:139–143, 1996.

63. YF Hui, J Kolars, Z Hu, D Fleisher. Intestinal clearance of H_2-antagonists. Biochem Pharmacol 48:229–231, 1994.

64. DJ Edwards, FH Bellevue, PM Woster. Identification of 6′,7′-dihydroxybergamottin, a cytochrome P450 inhibitor, in grapefruit juice. Drug Metab Dispos 24:1287–1290, 1996.

65. LH Pao, SY Zhou, C Cook, T Kararli, C Kirchoff, J Truelove, A Karim, D Fleisher. Reduced systemic availability of an antiarrhythmic drug, bidisomide, with meal coadministration: Relationship with region-dependent intestinal absorption. Pharm Res 15:225–231, 1998.

66. HJ Cooke. Intestinal salt and water transport. In: JH Walsh, GJ Dockray, eds. Gut Peptides. New York: Raven Press, 1994, pp 749–764.

67. WH Karasov, JM Diamond. Adaptation of intestinal nutrient transport. In: LR Johnson, ed. Physiology of the GI Tract. 2nd ed. New York: Raven Press, 1987, Vol 2, pp 1489–1497.

68. BH Stewart, OH Chan, N Jezyk, D Fleisher. Discrimination between drug candidates using models for evaluation of intestinal absorption. Adv Drug Deliv Rev 23:27–45, 1997.

69. WH Karasov, JM Diamond. A simple method for measuring intestinal solute uptake in vitro. J Comp Physiol 152:105–116, 1983.

70. N Jezyk, C Li, BH Stewart, X Wu, D Fleisher. Transport of pregabalin in rat intestine and caco-2 monolayers. Pharm Res 16:519–526, 1999.

71. X Lu, C Li, D Fleisher. Cimetidine sulfoxidation in small intestinal microsomes. Drug Metab & Dispos 26:940–942, 1998.

72. N Piyapolrungroj, C Li, RL Pisoni, D Fleisher. Cimetidine transport in brush-border membrane vesicles from rat small intestine. J Pharmacol & Exp Ther 289:346–353, 1999.

73. ME Dowty, CR Dietsch. Improved prediction of in vivo peroral absorption from in vitro intestinal permeability using an internal standard to control for intra- and inter-rat variability. Pharm Res 14:1792–1797, 1997.

74. H Lennernas, S Nylander, AL Ungell. Jejunal permeability: A comparison between the Ussing chamber technique and the single-pass perfusion in humans. Pharm Res 14:667–671, 1997.

75. BH Stewart, OH Chan, RH Lu, EL Reyner, HL Schmid, HW Hamilton, BA

Steinbaugh, MD Taylor. Comparison of intestinal permeabilities in multiple in vitro and in situ models: Relationship to absorption in humans. Pharm Res 12:693–699, 1995.

76. MH Friedman. Principles and Models of Biological Transport. Berlin: Springer-Verlag, 1986.

77. LR Johnson. Physiology of the gastrointestinal tract. Raven Press, New York, NY 1994.

78. L Herbette, AM Katz, JM Sturtevant. Comparisons of the interaction of propranolol and timolol with model biological membrane systems. Molec Pharmacol 24:259–269, 1983.

7
Improving Oral Drug Transport via Prodrugs

Mandana Asgharnejad
Merck Research Laboratories, West Point, Pennsylvania

I. PRODRUG APPROACH

A. Introduction

A prodrug is a chemical derivative of a parent drug molecule that undergoes transformation within the body to release an active moiety. The activation or conversion to the parent drug should be in a controlled, quantitative fashion, producing by-products that are pharmacologically inert and with no intrinsic toxicity. The chemical derivative that satisfies the prodrug requirements, once released, must be an endogenous substance in its own right and rapidly eliminated from the body [1]. The conversion of prodrug to its active moiety usually takes place during or after absorption or at the specific site of the action in the body, although conversion prior to this may also occur. This transformation of prodrug to the original moiety in the body can be accomplished by different reactions. Most oral prodrugs require an enzymatic catalysis in order to release the parent drug molecule [2]. Active drug species containing hydroxyl, carboxyl, and amine groups have been derivatized to produce prodrugs and optimize delivery. These prodrugs are converted back to the original parent drug by the action of enzymes (e.g., esterase) within the body.

A prodrug approach may be used to obtain maximum systemic concentration in the body by the oral route. The prodrug optimization of drug delivery may overcome several limiting factors such as

Low oral bioavailability due to low membrane permeability (1)
Poor solubility in the gastrointestinal (GI) tract (poor drug aqueous solubility)

Drug-induced damage to GI tissue
Chemical instability in the GI tract
Extensive first-pass metabolism

Several types of chemical derivatives have been studied for designing oral prodrugs. The purpose of the next section is to examine various approaches for obtaining oral prodrug forms which improve the absorption and bioavailability of the parent compound.

B. Ester Derivatives as Prodrugs

Esters have been considered extensively as an important class of oral prodrugs. Enzymes capable of hydrolyzing esters are abundant in organisms. Esterases can be found in the blood, liver, and other organs and tissues. Molecules with a hydroxyl or carboxyl group can be esterified in order to produce hydrophilic or lipophilic derivatives. Since 1970 there have been a number of attempts to improve oral bioavailability by using ester prodrugs. Several examples of the use of ester prodrugs as alternatives for optimizing oral drug delivery are presented below.

L-Dopa (L-3,4-dihydroxy phenylalanine) is the most important drug in the management of parkinsonism. However, the oral delivery of L-dopa is complicated by low water solubility, low lipid solubility, and extensive metabolism in the GI tract [4]. As a result, a relatively small amount of L-dopa is absorbed into the blood intact. L-Dopa undergoes decarboxylation by L-aromatic amino acid decarboxylase, which has high activity in the gastric mucosa. In addition, conjugation of the drug takes place in the gastrointestinal-hepatic system. L-Dopa is also metabolized in the blood, in part due to its low protein binding. In order to improve the therapeutic value of L-dopa, Bodor et al. [4] synthesized various derivatives of L-dopa to protect the carboxyl function, the amino moiety, and/ or the catechol system, the three main sites of metabolism of the molecule. The derivatives included carboxyl esters, phenol esters, amides, peptides, and various combinations of these functional groups. The protective groups used in that study for the reactive sites in L-dopa were acetyl and pivalyl for the catechol, methyl benzyl esters and N-terminal peptides for the carboxyl, and formyl and C-terminal dipeptide for the amino group. Multiprotected prodrugs were also synthesized. The prodrugs were administered orally to dogs in order to determine the net bioavailability of L-dopa. Most of the prodrugs increased the systemic availability of L-dopa, but the shapes of the plasma concentration–time curves of L-dopa remained close to those obtained by administration of L-dopa itself. The metabolic patterns of the prodrugs were remarkably close to that of L-dopa.

Others have also tried to develop L-dopa prodrugs. L-Dopa carboxyl esters were studied by Cooper and colleagues [5,6] and were not long-lasting. Garzon et al. [7] reported higher L-dopa plasma levels after administration of L-dopa-dihexadecacarbonylglycerol conjugate, but the molecular weight was 3.11-fold that of L-dopa, making it impractical for clinical use. Tsuchiya et al. [8] studied a catechol mono-o-pivaloyl ester of L-dopa, L-3-(3-hydroxy-4-pivaloyloxy-phenyl)alanine (NB-355). This new promoiety had a molecular weight only 1.5 times that of L-dopa. In bioavailability studies using male Sprague–Dawley rats and beagles, NB-355 approximately doubled the L-dopa plasma lifetime when it was coadministered with carbidopa (which inhibits 1-dopa decarboxylase). In addition, the prodrug increased L-dopa bioavailability over L-dopa itself following oral administration with carbidopa in rats and dogs. The authors concluded that NB-355 is more protective against the first-pass metabolism of L-dopa, extending L-dopa's plasma concentration and increasing L-dopa's bioavailability in rats and dogs. Unlike L-dopa, which is rapidly and almost entirely absorbed from the jejunum and duodenum by an active transport mechanism, NB-355 is absorbed more slowly from these regions. After hydrolysis and deesterification in the gut, the major site for absorption of L-dopa following oral administration of NB-355 was the ileum.

The ability of NB-355 to stimulate locomotor activity and induce dyskinesia in MPTP-treated squirrel monkeys was studied (MPTP induces parkinsonism) [9]. NB-355 was similar to L-dopa in stimulating locomotor activity. Furthermore, NB-355 induced less severe dyskinesia than was seen with L-dopa. Some other prodrugs of L-dopa include short-chain alkyl esters (methyl, ethyl, isopropyl, butyl, hydroxypropyl, and hydroxybutyl) intended for rectal absorption [10]. These esters of L-dopa have high water solubility (>600 mg/mL). Initial bioavailability studies indicated that all of these esters, with the exception of the hydroxypropyl ester, resulted in significantly greater bioavailability than that obtained with L-dopa itself. However, given the high level of esterase activity in the small intestine, the use of these compounds is limited to rectal administration.

Penicillins are also candidates for prodrug design, but simple aliphatic or aromatic esters of penicillin are not sufficiently labile in vivo for efficient prodrug conversion [2]. The reason for this enzymatic stability is that the carboxyl group in penicillin is highly sterically hindered. This problem was solved by synthesis of the acyloxymethyl ester, a double ester, of benzyl penicillin. The terminal ester in the acyloxymethyl derivative is less hindered and therefore more accessible to enzymatic hydrolysis. The ampicillin esters bacampicillin, pivampicillin, and talampicillin are designed on the basis of this principle to improve oral bioavailability compared to the parent drug. Bacampicillin and pivampicillin are entirely hydrolyzed to ampicillin as they pass through the GI wall [11]. The bioavailability of the pivaloyloxymethyl ester (pivampicillin) and the ethoxycarbonyloxyethyl ester (bacampicillin) were examined in cross over experiments in healthy male volunteer subjects [12]. The bioavailability of bacampicillin (86 ± 11%) was significantly greater than that of ampicillin (62 ± 17%). The highest absorption rate belonged to bacampicillin (0.89 ± 0.39% of dose absorbed per minute), followed by pivampicillin (0.64 ± 0.19) and ampicillin (0.511 ± 0.16). A comparison of the bioavailability of bacampicillin and talampicillin (the phathaloyl ester) indicated that bacampicillin has a significantly higher relative bioavailability than talampicillin [13]. All of these prodrugs of ampicillin produce a uniform and reliable absorption [14]. Furthermore, ampicillin esters are physiologically more favorable than ampicillin. Oral ampicillin causes changes in the intestinal microbial ecosystem that lead to adverse local reactions such as diarrhea. The ampicillin prodrugs have no antibacterial activity until they are transformed into active drugs upon absorption. This, and the fact that, unlike ampicillin, the oral absorption of the prodrugs is independent of dose, imbues them with appreciable therapeutic advantage [15].

Carbamate esters, another group with an ester promoiety, may be promising candidates for phenolic drugs. One of the main problems in the design of the ester prodrugs is the prevention of the first-pass metabolism of the promoiety by esterase present in the intestinal wall, blood, and liver. In order to achieve reduced first-pass metabolism of terbutaline, a bronchodilator, a carbamate prodrug bambuteral [1-(3,5-bis-(N,N-dimethylcarbonyloxy)-phenyl)-2-t-butylaminoethanol hydrochloride] was used [16]. A carbamate was chosen because it is known to be a reversible cholinesterase inhibitor and because carbamates are poor substrates for carboxyesterases present in the GI tract. Therefore, the presystemic stability of bambuteral should be high. The molecular structure of bambuteral, the monocarbamate, and terbutaline are presented in Figure 1. Bambuteral is extremely effective in inhibiting cholinesterase when butyrylthiocholin (BuChE) is used as substrate [$I_{50} = (1.7 ± 0.3) × 10^{-5}$ M, $N = 10$] [16]. The inhibition of BuChE by bambuteral is achieved by slow turnover of the reaction between BuChE and the carbamate. Thus, bambuteral retards its own hydrolysis to terbutaline. Bambuteral is extremely selective in its inhibition of cholinesterase; when acetylthiocholine was used as a substrate, the inhibition was decreased 2400-fold. Earlier reports by Olssen and Sevensson [17] also indicated that the carbamate ester of terbutaline (the bis-N,N-dimethyl carbamate of terbutaline) has high stability. The prodrug inhibited its own hydrolysis by reversibly binding to plasma esterase. The oral administration of the prodrug to dogs produced sustained blood levels of terbutaline.

N-Ethyl carbamate esters of fenoldopam, used for treatment of acute circulatory failure, provide another example of a carbamate ester prodrug. The intrinsically short half-life and extensive first-pass metabolism of fenoldopam limit its oral administration. N-Ethyl carbamate esters of fenoldopam provide elevated plasma fenoldopam levels and increases in the renal blood flow of significantly greater duration than when the parent compound is administered orally [30].

Salicylate esters may be promising prodrug candidates for estrogen. β-Estradiol has been used for treatment of various conditions such as menopausal

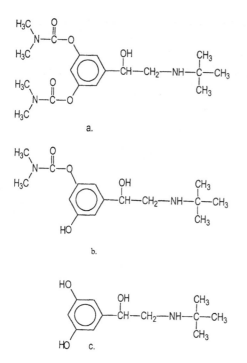

Figure 1 The molecular structure of (a) bambuterol, (b) monocarbamate, and (c) terbutaline. (From Ref. 16.)

symptoms and prevention of the progression of osteoporosis. Considerable first-pass metabolism of estradiol is the main reason for its poor systemic bioavailability by the oral route. The primary routes of first-pass metabolism are oxidation to estrone and conjugation of estradiol and estrone to the sulfate and glucaronides [18]. To reduce conjugative first-pass metabolism and increase the oral bioavailability of β-estradiol, estradiol-3-salicylate, and β-estradiol-3-anthranilate, ester prodrugs were synthesized and their oral bioavailabilities in dogs were evaluated [19]. With these promoieties, the 3-phenolic hydroxy group of β-estradiol (the normally metabolized functional group) was blocked so that first-pass conjugative metabolism could be reduced. The relative bioavailability of estradiol was significantly improved when administered in these prodrug forms. A 17-fold increase was observed with the estradiol-3-salicylate. The β-estradiol-3-anthranilate increased the systemic availability five-fold.

Prodrug approaches have been used to increase the oral bioavailabilities of cephalosporins. Ester prodrugs are formed by reversible esterification of the carboxyl group on these antibiotics [20]. These prodrugs are more orally bioavailable, but the magnitude of such enhancement proved to be small. Bonn and coworkers studied the pharmacokinetics of cefpodoxime in plasma and skin blister fluid following oral dosing of cefpodoxime proxetil. Although the level of cefpodoxime in the blister fluid following oral administration of cefpodoxime proxetil to human volunteers appeared high, the authors themselves concluded that the additional work is needed to validate their experimental model.

Alkoxyalkanoate esters have been used as prodrugs to improve the oral bioavailability of antiviral agents such as (+)-cycladarine (carbocyclic arabinofuranosyl adenine) [41]. (+)-Cycladarine has been shown to be effective against herpes simplex virus in tissue culture at noncytotoxic concentrations. Two prodrugs of (+)-cycladarine, namely, (+)-cycladarine-5′-methoxyacetate (CM) and (+)-cycladarine-5′-ethoxypropionate (CE) (Fig. 2), may be promising candidates

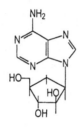

(a)

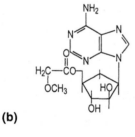

(b)

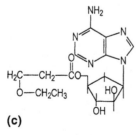

(c)

Figure 2 Chemical structures of (a) (+)-cycladarine, (b) (+)-cycladarine-5′-methoxyacetate, and (c) (+)-cycladarine-5′-ethoxypropionate. (From Ref. 22.)

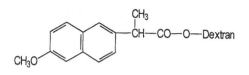

Figure 3 Naproxen-dextran. (From Ref. 23.)

to improve the bioavailability of the antiviral moiety. Following oral administration of these prodrugs to squirrel monkeys, rabbits, and rats, the pharmacokinetic parameters of (+)-cyclaradine were determined. CM resulted in higher serum levels of (+)-cyclaradine in squirrel monkeys, indicating that the methoxyacetate ester may be a good prodrug for (+)-cyclaradine. Furthermore, CE produced a greater area under curve (AUC) of (+)-cyclaradine in squirrel monkeys, rabbits, and rats, suggesting that CE was better absorbed than CM. The improved absorption was probably due to increased lipophilicity of CE. No intact esters were detected in the serum of squirrel monkeys, rabbits, and rats at any time following administration of these prodrugs.

   Macromolecular ester prodrugs also have been used for site-specific delivery of macromolecular drugs. Bioavailability of naproxen was determined [21] after administration of an aqueous solution of a dextran naproxen ester prodrugs in pigs. The dextran prodrugs varied in molecular weight (10,000–500,000) [21] (Fig. 3). In this instance, the dextran backbone protects the attached drug from enzymatic metabolism in the small intestine, so that selective regeneration of the drug takes place in the cecum and the colon, leading to almost 100% bioavailability of the naproxen.

C.   Oral Prodrugs for N—H Acidic Compounds

Acidic compounds with N—H bonds such as amides, carbamates, and hydantoins, may be transformed to *N*-mannich bases to form oral prodrugs [2]. These prodrugs are generally made by reacting an amide, carbamate, or hydantoin with formaldehyde and a primary or secondary aliphatic or aromatic amine (Fig. 4). The *N*-mannich prodrugs tend to have better physicochemical properties than the parent compounds. The derivatives may have increased water solubility, dissolution rate, and/or lipophilicity.

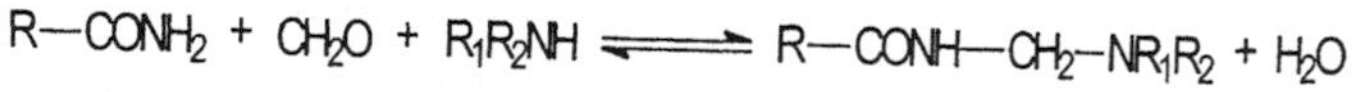

$$R-CONH_2 + CH_2O + R_1R_2NH \rightleftharpoons R-CONH-CH_2-NR_1R_2 + H_2O$$

Figure 4 *N*-Mannich base. (From Ref. 2.)

The concept of *N*-mannich base formation of N—H acidic compounds to yield more soluble oral prodrugs has been utilized in the case of phenytoin (I) [22]. This high melting, weakly acidic, and poorly water soluble anticonvulsant exhibits erratic bioavailability in both parenteral and oral forms [23]. The rapid intravenous injection of sodium phenytoin can be hazardous; intravenous administration must be done slowly and cautiously. Free acid precipitation at the intramuscular injection site leads to prolonged and marginal phenytoin release. To overcome these problems, an *N*-hydroxymethyl derivative of phenytoin, 5,5-diphenhydantoin), was synthesized. Various water-solubilizing esters were added to 3-(hydroxymethyl)-5, 5-diphenyl hydantoin (compound II). The ester derivatives of compound II are cleaved in vivo by esterase to compound II, which in turn breaks down further to compund I and formaldehyde. The ester derivatives of compound II include acetyl (III), *N*,*N*-dimethylglycine, monomethanesulfonate (IV), *N*,*N*-diethyl-β-alanine, mono-2,11-naphthalene sulfate (V), 2-(dimethylamino)ethyl carbonate, monomethanesulfonate (VI), disodium phosphate (VII), 3-(hydroxymethyl)-5,5-diphenylhydantoin sodium succinate (IX), and sodium glutarate (X) (see Fig. 5). Aqueous solubilities of compounds II, IV, V, VI, and VII were 8.5-fold, 3810-fold, 39-fold, 4730-fold, and 4408.5-fold that of compound I, respectively. The rapid precipitation of compounds to phenytoin acid in aqueous solutions limited their usefulness as prodrugs of compound I. All the prodrugs had equivalent activity compared with phenytoin, but all displayed

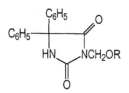

III: R = $COCH_3$

IV: R = $COCH_2NH(CH_3)_2CH_3SO_3^-$

V: R = $CO(CH_2)_2NH(C_2H_5)_2$
        (2-naphthalenesulfonate)

VI: R = $CO_2(CH_2)_2NH(CH_3)_2CH_3SO_3^-$

VII: R = PO_3^{2-}—Na_2^+

VIII: R = SO_3—Na^+

IX: R = $CO(CH_2)_2CO_2^-Na^+$

X: R = $CO(CH_2)_3CO_2^-Na^+$

Figure 5 Phenytoin and water solubilizing functional groups. (From Ref 24.)

significantly higher toxicity (lower TD_{50} values). However, an oral study of IV and VII in mice indicated a lack of early toxicity. Tissue studies showed reasonably facile cleavage of esters IV–VII in vitro in rat, dog, and human. The esters IV, VI, and VII were chosen for further evaluation, and chemical stability studies were performed on these compounds [24]. All the compounds were sufficiently stable for oral dosage, especially in solid dosage forms. Compound VII had the best stability among the three and was chosen as the best candidate for an injectable form of phenytoin based on this. Further studies concentrated on oral and intravenous administration of IV, VI, and VII to beagles to evaluate their pharmacokinetic behavior [25]. Phenytoin displayed nonlinear pharmacokinetics in dogs. All three prodrugs produced higher plasma levels of phenytoin after oral administration than were achieved upon administration of sodium phenytoin. It was concluded that IV would be a better oral prodrug of phenytoin than compound VI, as the latter produced side effects after oral and intravenous dosing. However, compound IV produced significant irritation after subcutaneous and intramuscular administration to rats, limiting its use as a parenteral prodrug [26]. Compound VII produced far superior phenytoin levels after intravenous administration to rats. Eventually, this compound was developed and marketed.

D. Prodrugs for Amines

Dopamine, a vasodilator, has been widely used for treatment of acute circulatory failure. However, since dopamine is rapidly metabolized when administered orally, its use has been limited to intravenous infusion. Murata et al., studied the bioavailability and the pharmacokinetics of orally administered dopamine (DA). The oral administration of DA to dogs resulted in an absolute bioavailability of approximately 3%. To minimize the extensive first-pass metabolism of DA, a dopamine prodrug, N-(N-acetyl-l-methionyl)-o,o-bis(ethoxycarbonyl)dopamine (TA-870), was synthesized [28] (Fig. 6). Since DA is a substrate for both mono-

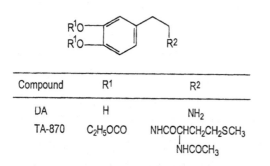

Compound	R1	R2
DA	H	NH_2
TA-870	C_2H_5OCO	$NHCOCHCH_2CH_2SCH_3$
		$NHCOCH_3$

Figure 6 Dopamine (DA) and the dopamine prodrug TA-870. (From Ref. 30.)

amine oxidase (MAO) and catechol O-methyl transferase (COMT), the purpose was to protect the amino group and the catechol system. Oral administration of TA-870 to dogs raised plasma concentrations of DA severalfold over that following DA administration. The absolute bioavailability of TA-870 was 10 times that of orally administered DA. Two protective groups of TA-870 rendered it resistant to first-pass metabolism during the absorption process. The bioactivation of TA-870 to DA took place in the intestine, blood, and liver. Yoshikawa and coworkers [29] studied the disposition of TA-870 in humans and reported that the free DA plasma levels upon oral dosing were two to three times higher than those observed after DA intravenous infusion. In addition, the study indicated that orally administered DA was rapidly metabolized to 3,4-dihydroxyphenylacetic acid by MAO in the small intestine and the liver, while the metabolism of TA-870 was suppressed by the protective groups. After oral dosing of TA-870, the urinary metabolites were nearly identical to the DA metabolites, indicating rapid conversion of TA-870 to DA in the body. In dogs, TA-870 increased renal and mesenteric blood flow, renal vasodilatation and diuresis [29].

5-Aminosalicylic acid (5-ASA) is an amine drug that has been administered in the form of prodrug, sulfasalazine, in the treatment of inflammatory bowel disease. The extensive absorption of 5-ASA after oral dosing prevents its effective concentration in the colon, a place where 5-ASA exerts an anti-inflammatory action. Sulfasalazine is composed of sulfahydrine and 5-ASA joined by an azo bond [2,31,32]. Colonic bacterial azo-reduction splits sulfasalazine into the active moiety 5-ASA and the carrier group sulfapyridine. However, the carrier group sulfapyridine is absorbed after being released in the colon, resulting in side effects in some patients taking the drug. Several other 5-ASA prodrugs based on bioactivation through azo-reductase in the colon have been under clinical investigation [32–36]. The transport group, sulfapyridine, is replaced by 4-aminobenzoyl-β-alanine in balsalazide, by p-aminobenzoate in HB-313, by p-aminohippurate in ipsalazide, by a nonabsorbable sulphanilamideoethylene polymer in poly-ASA, and by another 5-ASA molecule in disodium azodisalicylate (Fig. 7). The transport groups of both ipsalazide and balsalazide as well as 5-ASA were found in the feces following oral administration. Clearly, the same active drug was available for local action. However, further pharmacokinetic, clinical, and therapeutic studies are needed to obtain the necessary information regarding the appropriateness of these approaches. Following oral dosing, azodisalicylate sodium produced low serum concentrations. More than 30% of the daily oral dose was recovered from feces as the prodrug or the active moiety. Azodisalicylate was excreted intact in the urine. In patients with ileostomy, neither prodrug nor the active moiety could be found in the serum or urine, indicating that a GI tract with a normal bacterial environment is required for activity [37]. These prodrugs are promising candidates to achieve site-specific drug delivery through the release of the active substance at the target.

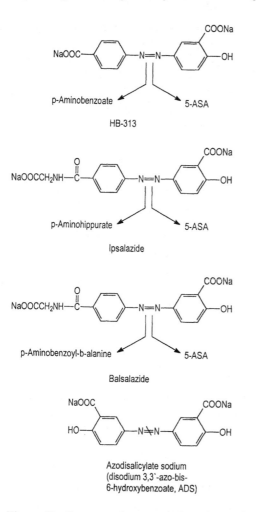

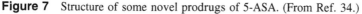

Figure 7 Structure of some novel prodrugs of 5-ASA. (From Ref. 34.)

E. Enzyme Prodrugs Targeting Specific Enzymes

Acyclovir [9-(2-hydroxyethoxymethyl)guanine] (ACV) is clinically useful in the treatment of infections caused by several members of the herpes virus (e.g., herpes simplex, varicella zoster and Epstein-Barr virus [2,38,39]). An enzyme coded for by the virus phosphorylates ACV to a monophosphate intermediate. This species in turn undergoes further phosphorylation to a triphosphate with the aid of normal cell enzymes. Then, ACV triphosphate inhibits the herpes virus DNA polymerase. However, the GI absorption of ACV is inefficient due to its generally poor solubility and high polarity. Its bioavailability is only about 19% $\pm$ 6.5%. Clinical trials have indicated that large doses of ACV are a must if one is to achieve therapeutic effects [40]. To overcome the relatively inefficient absorption of the drug, compounds having minor alterations in position 6 of ACY have been developed. Two such compounds are 2,6-diamino-9-(2-hydroxyethoxymethyl)purine (A134U), with an amino group at position 6, and 2-[(2-amino-9H-purine-9-yl) methoxy]ethanol (desciclovir, DCV), which lacks an oxo group at that position. ACV contains an oxygen group at position 6. The enzyme adenosine deaminase converts A134U to acyclovir by the enzyme xanthine oxidase, which is abundant in the liver and intestine. Oral administration of A134U to human subjects was well tolerated. However, the mean plasma ACV concentration was just marginally higher than if ACY itself had been given. DCV is 18 times more water-soluble than ACV. Furthermore, oral dosing of DCV to human subjects produced mean peak plasma ACV levels 10 times greater than those seen with the same dose of ACV, and the area under the concentration–time curve of ACV was eight-fold greater. Additional studies showed that DCV, at a dose of 250 mg every 8 hr for 10 days, produced excellent ACV levels in plasma and had good clinical tolerance. DCV was converted to ACV rapidly with no sign of renal toxicity or substantial accumulation in the subjects. This oral prodrug provides ACV levels similar to those achieved by parenteral therapy. Harndem et al. [42] took a similar approach with another antiherpes virus agent, acyclonucleoside 9-[4-hydroxy-3-(hydroxymethyl)-but-1-yl]guanine (BRL39123). The 6-deoxy congener of RBL39123 and its O-acyl and other derivatives were synthesized. Some of these had low melting points. Furthermore, several derivatives were absorbed well and were effectively converted to BRL39123 after oral administration in mice.

F. Amino Acid or Peptidyl Derivatives as Prodrugs

Cyanamide is a potent aldehyde dehydrogenase (AIDH) inhibitor used therapeutically as a deterrent agent for alcohol abuse. AIDH is responsible for the metabolism of ethanol to acetaldehyde (AcH) by the liver. Inhibition of AIDH results in elevated blood AcH, leading to an unpleasant physiological reaction. A catalase enzymatically activates cyanamide to an active metabolite (unidentified) which in turn inhibits AIDH. The rapid metabolism of cyanamide to acetyl cyanamide is responsible for the short duration of its pharmacologic activity. To overcome this problem, Kwon and colleagues [43] examined some α-amino acid and peptidyl derivatives of cyanamide that prevent rapid metabolism of the parent drug. N-protected aminoacyl and peptidyl derivatives of cyanamide included N-carbobenzoxyglycyl cyanamide (Z-Gly), hippurylcyanamide (BZ-Gly), N-benzoyl-l-leucyl cyanamide (BZ-L-leu), N-carbobenzoxyglycyl-l-leucyl cyanamide

(Z-Gly-L-leu), *N*-carbobenzoxy-*l*-pyroglutamyl cyanamide (Z-L-pGlu), L-pyro-glutamyl-L-leucyl cyanamide (L-HpGlu-*l*-leu), and L-pyroglutamyl-*l*-phenylalanyl cyanamide (L-HpGLu-L-phe). The prodrugs were administered as single doses, followed by a dose of ethanol. Z-Gly and Z-Gly-L-leu elevated blood AcH more than of 100-fold over controls. BZ-Gly and BZ-L-leu were longer acting and produced significant elevation of ethanol-derived blood AcH. L-HpGlu-L-Leu and Z-L-pGlu were not as active as the others, but L-HpGlu-L-phe was effective over the short term. It is believed that α-aminoacyl and peptidyl groups would be hydrolyzed by nonspecific tissue aminopeptidases to release the cyanamide moiety in vivo. Nitrogen-protected α-aminoacyl and peptidyl derivatives of cyanamide can be considered to be pro-prodrugs, since the cyanamide released must still be further bioactivated to an intermediate metabolite. These prodrugs may be superior to the parent moiety with respect to duration of action and desired pharmacologic response.

L-α-Methyldopa (LMD) is an antihypertensive agent with poor GI absorption. Studies of its oral bioavailability have indicated that its absorption is both concentration- and pH-dependent [45,46]. LMD is an amino acid analogue which is absorbed from the intestine by a concentration-dependent saturable mechanism. Amino acid carriers responsible for the transport of amino acids and their analogues are structurally restrictive. However, the peptide carrier is less restrictive in its structural requirement and more efficient in its transport of substrates [46,47]. The α-methyl group in LMD hinders its binding to the carrier, compromising its absorption. To achieve better absorption of LMD, the possibility of using the peptide carrier was explored [48]. Five dipeptidyl derivatives of LMD, namely, Gly-LMD, Pro-LMD, LMD-Pro, Phe-LMD, and LMD-Phe, were synthesized. The intestinal permeabilities of these derivative dipeptides were studied by an in situ intestinal perfusion method. All of the dipeptides had intestinal wall permeabilities significantly greater than that of LMD. The dipeptide LMD-Phe displayed a wall permeability 20-fold higher than that of LMD and was most resistant to luminal enzymatic attack. Further studies of the LMD-Phe carrier uptake kinetics at low concentration indicated a carrier permeability 20-fold higher than that of LMD and a much lower K_m value. It has been shown that the prolidase (a cytoplasmic dipeptidase) can be used as a prodrug-hydrolyzing enzyme for a dipeptidyl analogue containing proline at its C-terminal end [58]. The potential sites of hydrolysis are the intestine and the erythrocyte, both of which are rich in the prolidase enzyme.

The above strategy was tested by demonstrating the presence of LMD and/or LMD-Phe in the plasma following oral administration of the prodrug LMD-Phe [49,50] (Fig. 8). To compare the extent of absorption of the drug and the prodrug, a rat model ($n = 6$–7) with a chronically isolated intestinal loop was chosen. A single weight equivalent dose of either agent was given inside the jejunal loop (9.5 ± 1.0 cm), and the blood samples were collected from the tail vein of the rat. To determine the plasma level of LMD and LMD-Phe, high

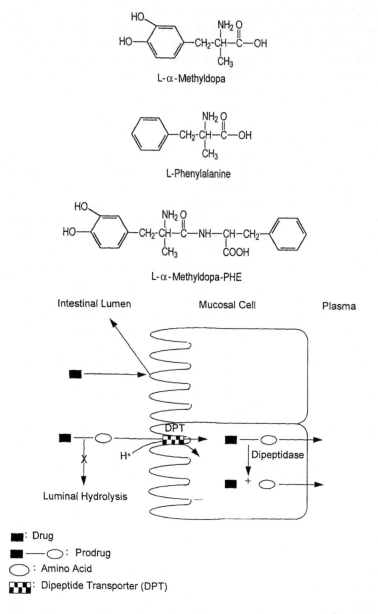

Figure 8   A peptide carrier mediated transport to improve oral absorption.

performance liquid chromatographic assays using electrochemical detection were developed. The mean C_{max} of LMD (235.0 ± 37.4 ng/mL) following the jejunal administration of the prodrug was more than twofold higher than that of the drug (109.0 ± 31.4 ng/mL) (Fig. 9). The prodrug showed a T_{max} value (60 min) four-fold higher than that of LMD (15 min). The results indicated the bioavailability of LMD following the dose of LMD-Phe ($F = 0.94$) was almost three times higher than when an equivalent dose of LMD was administered alone ($F = 0.37$). In addition, the prodrug was not detectable in plasma, indicating the complete bioconversion of the prodrug to the parent drug.

Enalaprilat and SQ27,519 are angiotensin-converting enzyme (ACE) inhibitors with poor oral absorption. Enalapril and fosinopril are dipeptide and amino acid derivatives of enalaprilat and SQ27,519, respectively [51] (Fig. 10). Both prodrugs are converted via deesterification to the active drug by hepatic biotransformation. In situ rat perfusion of enalapril indicated a nonpassive absorption mechanism via the small peptide carrier–mediated transport system. In contrast to the active parent, enalapril renders enalaprilat more peptide-like, with higher apparent affinity for the peptide carrier. The absorption of fosinopril was predominantly passive. Carrier-mediated transport was not demonstrated, but neither was its existence ruled out.

Captopril, like some other ACE inhibitors, is a peptide derivative. The

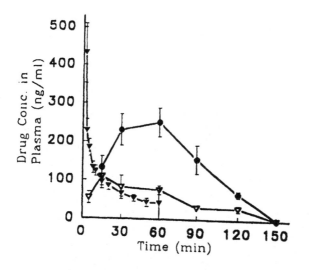

Figure 9 Plasma profile of L-α-methyldopa following intravenous dose of L-α-methyldopa and jejunal dose of L-α-methyldopa-phenylalanine and L-α-methyldopa ($n = 6–7$). (●) L-α-methyldopa following jejunal dose of prodrug; (▽) L-α-methyldopa jejunal dose; (▼) L-α-methyldopa intravenous dose.

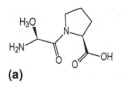

(a)

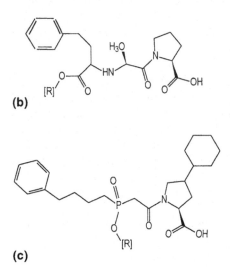

(b)

(c)

Figure 10 The chemical structure of (a) Gly-Pro, (b) enalaprilat ([R] = H) and SQ27,519 ([R] = H) and their respective prodrugs enalapril ([R] = ethyl) and (c) fosino-pril (SQ28,555) ([R] = isobutanediolacetate) (From Ref. 51.)

mechanism of captopril absorption is carrier-mediated via the peptide carrier system with a significant passive component [52]. The rapid metabolism of captopril to captopril disulfide is responsible for its rapid disappearance from the body. The facile metabolism and renal excretion of captopril render it a pharmacokinetically undesirable antihypertensive agent. To improve its bioavailability and half-life, a disulfide dimer conjugate of captopril (CPD) was synthesized [53]. Following the oral and intravenous administration of CPD to rats, captopril appeared rapidly in the systemic circulation, the lung, the kidney, and the liver. Although the plasma levels after administration of the dimer were lower than those following administration of captopril, the disposition of CPD was similar to that obtained following administration of the same dose of captopril. However, the plasma levels of captopril decreased at a slower rate following the oral dosing of CPD. This leveled out the large variation in plasma captopril levels seen when captopril is given orally. The oral bioavailability of CPD was similar to that of captopril.

Table 1 Various Approaches for Obtaining Oral Prodrug Forms

Approach	Drug	Derivative	Ref.
I. Ester derivatives	L-Dopa	Carboxyl ester, e.g., methyl; catechol mono-o-pivaloyl, e.g., pivaloyl ester; short-chain alkyl esters, e.g., methyl, ethyl, isopropyl, hydroxypropyl, and butyl esters	4–6,8–10
	Penicillin	Acyloxymethyl ester, e.g., pivaloyloxymethyl ester, ethoxycarbonyloxyethyl ester, and phatalidyl ester	2,14,15
	Terbutaline	Carbamate ester, e.g., bambuterol, a mono carbamate ester of terbutaline; bis-N,N-dimethyl carbamate	16,17
	Fenoldopam	Carbamate ester, e.g., N-ethyl carbamate ester	18
	β-Estradiol	Salicylate esters	19
	Cephalosporins	Carboxyl esters	21
	(+)-Cyclaradine	Alkoxyalkanoate ester, e.g., methoxyacetate and ethoxy propionate	22
II. N—H acidic compounds	Naproxen	Dextran ester	23
	Phenytoin	N-Mannich base, e.g., N-hydroxymethyl	24,25
	Dopamine	N-Acetyl-L-methionyl	29–31
III. Amines	5-Aminosalicylic acid	Azo bound with sulfahydrine, 4-aminobenzoyl-β-alanine, etc.	23,32,33
IV. Enzyme prodrugs targeting specific enzymes	Acyclovir	2,6-Diamino-9-(2-hydroxyethoxymethyl)purine	39–41
	RBL39123 (antiherpesvirus agent)	6-Deoxy congener of RBL39123, and acyl derivative	42
	Cynamide	N-protected α-amino acyl, and peptidyl derivatives or cynamide, e.g., N-carbobenzoxyglycyl	43
V. Amino acid or peptidyl derivatives as prodrugs	L-α-Methyldopa	Dipeptidyl derivative, e.g., α-methyldopaphenylalanine	46–48
	Enalaprilat	E.g., enalapril and fosinopril	51
	Captopril	Disulfide dimer conjugate of captopril	52,53

In conclusion, CPD represents a prodrug of captopril which seems to lengthen the pharmacokinetic half-life and extend the duration of action of captopril in humans.

Various strategies for obtaining oral prodrug forms are given in Table 1.

II. ENZYME SUBSTRATE SPECIFICITY CONSIDERATIONS IN DESIGN OF PRODRUGS

A. Introduction

Drug derivatization has been an important strategy for oral drugs to optimize their bioavailabilities. In the previous section, the rational design of prodrugs was discussed. In some cases, structural modification of a parent moiety led to reduction or elimination of a delivery problem. In addition, the in vivo conditions which are required to regenerate a parent molecule from its prodrug forms were examined. To design prodrugs more effectively, one must consider the specificity of the activating enzyme for its prodrug substrate.

In order to take advantage of enzyme–substrate specificities in the prodrug approach, a thorough knowledge of the specific enzyme or enzyme system is necessary. The enzyme's distribution and level as well as its kinetics and binding should be explored [54]. The role the enzyme plays in the cellular biochemistry and the reactions it ordinarily catalyzes are of particular importance. Many drugs containing free amino, carboxyl, or hydroxyl groups have been converted to prodrugs with more favorable physicochemical properties for uptake which act in vivo as substrates for hydrolyzing enzymes. As mentioned in the previous section, the most commonly cited prodrug forms are simple organic esters or amides. Such drugs can go under nonenzymatic hydrolysis in regions close to absorption sites or can be targeted for nonspecific enzyme hydrolyses. However, with these approaches the site of reconversion of the prodrug is nonspecific, limiting its use.

The use of nutrient substances as derivatizing groups leads to modification of physicochemical properties of the parent moiety, permitting more specific targeting for enzymes involved in the terminal phases of digestion [55]. These enzymes break down carbohydrates, fats, mineral-containing nutrients, and protein to form sugars, lipids, fatty acids, various electrolyte species, and amino acids. Prodrugs with nutrient-derivatizing species produce nontoxic nutrient by-products upon hydrolysis. For example, a prodrug of α-methyldopa having phenylalanine as a derivatizing group was synthesized to improve α-methyldopa's GI absorption. α-Methyldopa-phenylalanine increased the GI wall permeability 20-fold over that of the parent molecule [49]. Amino acids are wide-ranging in polarity and therefore offer a number of possibilities for modification of the physical properties of drugs. The drug in question can be altered to be more or less polar by choosing an appropriate amino acid, or it may be converted from acidic to basic or vice versa. The diverse possibilities offered by the amino acids in addition to their nontoxicity make them suitable for the prodrug approach. It has been shown that prolidase (a cytoplasmic dipeptide) can be used as a prodrug-hydrolyzing enzyme for a dipeptidyl analogue containing proline at its C-terminal [50]. The potential sites of hydrolysis can be the intestine or can involve the erythrocytes, which are rich in prolidase enzyme.

B. Major Groups of Enzymes

Based on the reactions they catalyze, enzymes are classified into six major categories: hydrolases, isomerases, ligases, lyases, oxidoreductase, and transferases [56,57]. The hydrolases catalyze hydrolysis reactions which involve transfer of functional groups to water. The isomerases are involved in the transfer of groups within molecules to yield isomeric forms. The ligases catalyze formation of C—C, C—S, C—O, and C—N bonds by condensation reactions coupled with ATP cleavage. The lyases catalyze reactions responsible for addition of groups to double bonds or the reverse. The oxidoreductases are involved in reactions leading to the transfer of electrons. Finally, the transferases catalyze group-transfer reactions. The enzymes in the hydrolase class are of particular importance as sites of prodrug reconversion. For example, the hydrolytic action of enzymes in the brush border membrane efficiently breaks down proteins and polypeptides efficiently to dipeptides, tripeptides, and amino acids. Furthermore, many enzymes in the hydrolytic class have been extensively studied, which provides necessary information for prodrug design.

C. The Michaelis–Menten Equation

Enzymes greatly enhance the rates of reactions that would otherwise occur only very slowly [57]. They accomplish this by reducing the activation energy of the chemical reactions. The activation energy is defined as "the amount of energy in calories required to bring all the molecules in one mole of substance at a given temperature to the transition state at the top of the energy barrier" [60]. The rate of any enzymatic reaction is dependent on the concentration of the transition state species. The rate of reaction proceeds slowly at very low concentration of the substrate, but its rate increases with higher substrate concentration.

The first step in an enzymatic reaction is the relatively rapid formation of the enzyme–substrate complex (ES) [57].

$$E + S \underset{k_{-1}}{\overset{k_1}{\rightleftharpoons}} ES \tag{1}$$

The ES complex then breaks down in a second slower reversible reaction to produce a product (P) and the free enzyme (E).

$$ES \underset{k_{-2}}{\overset{k_2}{\rightleftharpoons}} P + E \tag{2}$$

The Michaelis–Menten theory assumes that k_{-2} is sufficiently small that the second step in the process does not affect the equilibrium formation of the ES complex [61]. At steady state the rates of formation and breakdown of ES are equal:

$$k_1([E_t] - [ES])[S] = k_{-1}[ES] + k_2[ES] \tag{3}$$

where k_1, k_{-1}, k_{-2}, and k_2 are the rate constants for the forward and reverse reactions. $[E_t]$ represents the total enzyme concentration (the sum of the free and substrate combined enzyme). $[E_t] - [ES]$ represents the concentration of free enzyme. After a few algebraic manipulations, Eq. (3) is solved for [ES] to give

$$[ES] = \frac{[E_t][S]}{[S] + (k_2 + k_{-1})/k_1} \tag{4}$$

The initial velocity, V_0, is determined by the rate of breakdown of [ES] in Eq. (2):

$$V_0 = k_2[ES] \tag{5}$$

Substituting for [ES] in Eq. (5) from Eq. (4) we get

$$V_0 = \frac{k_2[E_t][S]}{[S] + (k_2 + k_{-1})/k_1} \tag{6}$$

Equation (6) can be further simplified by defining K_M as $(k_2 + k_{-1})/k_1$ and by defining V_{max} as $k_2[E_t]$, to give the Michaelis–Menten equation.

$$V_0 = \frac{V_{max}[S]}{[S] + K_M} \tag{7}$$

This equation is fundamental to all aspects of the kinetics of enzyme action. The Michaelis–Menten constant, K_M, is defined as the concentration of the substrate at which a given enzyme yields one-half of its maximum velocity. V_{max} is the maximum velocity, which is the rate approached at infinitely high substrate concentration. The Michaelis–Menten equation is the rate equation for a one-substrate enzyme-catalyzed reaction. It provides the quantitative calculation of enzyme characteristics and the analysis for a specific substrate under defined conditions of pH and temperature. K_M is a direct measure of the strength of the binding between the enzyme and the substrate. For example, chymotrypsin has a K_M value of 108 mM when glycyltyrosinylglycine is used as its substrate, while the K_M value is 2.5 mM when N-20 benzoyltyrosineamide is used as a substrate

[57, p. 216]. The knowledge of K_M and V_{max} values enables us to calculate the reaction rate of an enzyme at any given concentration of its substrate.

D. Enzymes in General

Many enzymes have absolute specificity for a substrate and will not attack the molecules with common structural features. The enzyme aspartase, found in many plants and bacteria, is such an enzyme [57]. It catalyzes the formation of L-aspartate by reversible addition of ammonia to the double bond of fumaric acid. Aspartase, however, does not take part in the addition of ammonia to any other unsaturated acid requiring specific optical and geometrical characteristics. At the other end of the spectrum are enzymes which do not have specificity for a given substrate and act on many molecules with similar structural characteristics. A good example is the enzyme chymotrypsin, which catalyzes hydrolysis of many different peptides or polypeptides as well as amides and esters.

Enzymes require an optimum pH at which their catalytic activity is maximal. The pH–activity profiles of enzymes indicate the pH at which the catalytic sites are in their necessary state of ionization. The optimal pH of an enzyme may be different from that of its normal environment. The action of enzymes in cells may be regulated by variation in the pH of the surrounding medium.

Different enzymes exhibit different specific activities and turnover numbers. The specific activity is a measure of enzyme purity and is defined as the number of enzyme units per milligram of protein. During the purification of an enzyme, the specific activity increases, and it reaches its maximum when the enzyme is in the pure state. The turnover number of an enzyme is the maximal number of moles of substrate hydrolyzed per mole of enzyme per unit time [63]. For example, carbonic anhydrase, found in red blood cells, is a very active enzyme with a turnover number of 36×10^6/min per enzyme molecule. It catalyzes a very important reaction of reversible hydration of dissolved carbon dioxide in blood to form carbonic acid [57, p. 220].

Enzyme inhibition is an important concept which provides valuable information about the substrate specificity of enzymes, the nature of the functional groups at the active site, and the mechanism of the enzyme action. Most enzymes are inhibited by specific chemical reagents which exert their inhibition in a reversible or irreversible fashion. The irreversible inhibitors combine with or destroy an essential site on the enzyme molecule, hindering its catalytic activity. An example of this is the compound diisopropylfluorophosphate (DFP), which inhibits the enzyme acetylcholinesterase. The DFP combines with the hydroxyl group of serine residue at the active site of acetylcholinesterase to form an inactive derivative, rendering the enzyme functionless. Reversible inhibitors of enzymes are divided into two groups of competitive and noncompetitive inhibitors. A competitive inhibitor competes with the substrate for binding to the enzyme. This competitive inhibition can be reversed by elevating the substrate, concentration. Non-competitive inhibition is also reversible, but not by the substrate because the inhibitor binding site differs from the substrate binding site. The noncompetitive inhibitor alters the conformation of enzyme upon binding, resulting in inactivation of the catalytic site.

E. The Distribution of Enzyme Systems in the GI Tract

The GI tract may be viewed as a cylindrical tube with different distributions of enzymes in the lumen and mucosal cells. The transport of a drug or a prodrug to the sites of absorption along the GI wall is a function of residence time and radial transport time. The residence time is the length of time that the drug spends traveling in a longitudinal direction along the GI tract. The radial transport time is the length of time that the drug spends moving from a point in the tube lumen toward the tube wall.

The distribution, activity, and specificity of enzymes involved in digestion and nutrient transport are to a great extent a function of longitudinal position. In addition, the sharp pH gradient that exists in the longitudinal direction along the GI tract affects enzyme distribution. In the radial direction, the drug or prodrug moves from the lumen through the mucin layer and glycocalyx (the surface coat), then proceeds into the brush border apical membrane of the enterocyte, entering the cytosol and the basal cell membrane. Finally, the drug or prodrug passes through the capillary or lymphatic endothelium. In its transport in the radial direction, the drug or prodrug encounters different enzymes with a specific distribution pattern geared to couple sequential digestion with transport in the process of absorption. Considering the different aqueous and membrane components of the radial region, the site of prodrug reconversion becomes critical in achieving successful drug transport.

The target choice for prodrug reconversion is determined by the drug's physical properties, transport limitations, and enzyme considerations. The radial parameters roughly dictate the rate of absorption, while the axial considerations affect the extent of absorption.

1. The Luminal and Mucosal Cell Enzymes

The prodrug approach has been used to improve oral bioavailability of drugs with poor stability at acidic pH and to decrease gastric irritation of some compounds. Ideally, the prodrug should be stable in the acidic environment of the stomach and undergo hydrolysis to the parent moiety in the intestine. Furthermore, prodrugs may be targeted for pancreatic enzymes in the intestinal lumen. Banerjee, Amidon, and colleagues have demonstrated this strategy [64–68].

Some of the pancreatic enzymes in the lumen include pancreatic amylase, pancreatic lipase, elastase, trypsin, α-chymotrypsin, and carboxypeptidase A. For example, the aspirin derivatives aspirin phenylalanine ethyl ester, aspirin phenyllactic ethyl ester, and aspirin phenylalanine amide have been studied as substrates for carboxypeptidase A [67,68], with the phenylalanine ethyl ester derivative proving to be the best substrate. This study indicated that the carboxypeptidase A may serve as a reconversion site for many drug derivatives.

The prodrugs made to increase drug aqueous solubility may be targeted for enzymes associated with the intestinal brush border membrane such as adsorbed pancreatic enzymes, disaccharidases, aminopeptidases, some carboxypeptidases, and alkaline phosphatase [69–72]. These enzymes are scattered across the brush border membrane in various locations within the membrane. The glycoprotein digestive enzymes penetrate to varying depths into the membrane of microvilli, with the protein component extending to the lipid layer of the membrane and the carbohydrate chains protruding into the lumen. The enzyme disaccharidase, which splits one kind of 12-carbon sugar into two six-carbon fragments, penetrates the first lipid layer of membrane. The enzyme alkaline phosphatase, which hydrolyzes phosphate compounds in food, extends to the depth of the membrane. In contrast, aminopeptidases pass all the way through the membrane, penetrating into the interior of the microvilli, and can remove a single amino acid from one end of a short peptide chain.

Targeting a prodrug for mucosal cell enzymes enhances its transport through the aqueous layer. Upon reconversion at the intestinal wall, the more membrane-permeable parent moiety is released. If the intact prodrug passes through the intestinal wall, the cytosolic enzymes will provide additional reconversion sites. However, the cytosolic enzymes have specificities and activities different from those of brush border enzymes [73]. Intestinal enzyme activities and specificities for prodrug targeting have been the subject of several reviews [54,55,74]. A brief review of the intestinal mucosal peptidases and cytoplasmic peptidases is given below.

F. Intestinal Mucosal Peptidases

The intestinal mucosal peptidases are distributed in the brush border and cytosol of the absorptive cell. There are, however, distinct differences between the brush border and cytosolic peptidases [75]. The tetrapeptidase activity is associated exclusively with the brush border enzyme. Furthermore, brush border peptidases exhibit more activity against tripeptides than dipeptides, whereas the cytosolic enzymes show greater activity against dipeptides. Studies have demonstrated that more than 50% of dipeptidase activity was detected in the cytosol [76] and just 10% in the brush border membrane [77]. The brush border enzymes include

aminopeptidase, aspartaloaminopeptidase, dipeptidylaminopeptidase IV, Y-glu-
tamyltransferase, carboxypeptidase P, ACE, aminopeptidase P, and dipeptidases.
The cytoplasmic peptidases include dipeptidase, aminotripeptidase, prolidase,
prolinase and carnosinase.

1. Brush Border Peptidases

Some peptidases of the brush border membrane are capable of hydrolyzing peptides as large as 10 amino acid residues. These peptidases start the hydrolysis
of the substrate from the N-terminus and proceed down the peptide chain. The
aminooligopeptidases have the highest activity against tetrapeptidases with four
or higher amino acid residues (90–98%) and the lowest activity against dipeptidases (5–12%) [78]. Aminooligopeptidase, the most abundant peptidase in the
brush border membrane, is isolated from the intestine and the renal microvilli
[79] and has two major subgroups. Oligopeptidase II is a major enzyme in brush
border membrane with broad specificity for substrates containing neutral or basic
amino acids residues [75]. It has high affinity for peptides containing an amino
acid with a lipophilic side chain at the N-terminus. Aminooligopeptidase is a
zinc protein, and its activity is inhibited by amino acids with hydrophobic characteristics [80]. Its activity is optimal at pH 7–9 [81,82].

Aminopeptidase A is another brush border membrane enzyme which has
been the subject of various studies [79,81,83–86]. It has been found in the intestinal brush border membrane of humans, rabbits, rats, and pigs and is active against
peptides with acidic amino acids at the amino terminus. Its activity against dipeptides is more limited. Shoaf et al., isolated three rat brush border aminopeptidases
with distinct but somewhat overlapping substrate specificities. These enzymes
had preference for dipeptides containing methionine, arginine, or aspartic acid
and glycine. The optimal pH for activity of aminopeptidase was reported to be
7–8.

Dipeptidyl aminopeptidase IV hydrolyzes substrates with free α-amino
groups. Peptide bonds involving the carboxy group of either Pro or Ala are
cleaved by this enzyme to X-Pro or X-Ala, where X may be any amino acid. It
has been shown that peptides with the X-Pro moiety are hydrolyzed more completely than those with X-Ala [79].

Y-Glutamyltranspeptidase is another brush border membrane enzyme, with
its highest activity at pH 7.8. It is responsible for the transfer of the Y-glutamyl
group to the N-terminus of an amino acid or peptide. Carboxypeptidase P, also
a brush border peptidase, hydrolyzes a polypeptide with a C-terminal amino acid
when proline is in the penultimate position [84,87]. Angiotensin converting
enzyme (ACE) releases dipeptides from the carboxy terminus of peptides,
with the highest activity against peptides with proline at the carboxy terminal
end [88].

2. Cytosol Enzymes

The final phase of protein digestion takes place in the cytoplasm of the intestinal mucosal cells which contain several peptidases. These enzymes, unlike those in the brush border membrane, are capable of hydrolyzing di- and tripeptides with distinct substrate specificities. The enzymes in this region of the intestinal mucosa are designated as dipeptidase, aminotripeptidase, and proline dipeptidase [78,89] and catalyze the hydrolysis of peptides with two to four amino acid residues. Their activities, expressed as percent of total cellular activity, are 80–95% for dipeptides, 30–60% for tripeptides, and up to 10% for tetrapeptides [78]. The activity for higher peptides is negligible. The cytoplasmic enzymes are thus well specialized for completing the hydrolysis of absorbed peptides, complementing the activities of the brush border peptidases. This allows the digestion of di- and tripeptides that escape hydrolysis by the brush border enzymes.

Dipeptidases are abundant in the enterocyte. Glycylleucine and glycylglycine peptidase are two such enzymes that have been isolated and characterized [89]. The enzyme glycylleucine dipeptidase, with a molecular weight of 115,000, was purified from pig and monkey intestine [89,90]. It has a wide specificity and is responsible for most of the cytosol dipeptidase activity, hydrolyzing almost all neutral peptides except Gly-Gly [91]. The inactivation by thiol reagents such as Hg^{2+} and p-hydroxymercuribenzoate indicates that glycylleucine dipeptidase is an SH enzyme [91]. Glycylleucine dipeptidase is inhibited by different SH compounds and EDTA, indicating that it is functionally dependent on a metal ion [89]. The proline was detected as the N-terminal amino acid residue of glycylleucine dipeptidase.

Proline dipeptidase or prolidase is another cytoplasmic enzyme with narrow specificity for dipeptides with the configuration X-Pro or X-Hyp [89]. α-methyldopa-Pro was reported to be a substrate for this enzyme, which hydrolyzes aminoacyl proline peptide bonds. Prolidase is composed of two equal subunits with a total molecular weight of 113,000 and is inactivated by Hg^{2+} and p-hydroxymercuribenzoate [92]. Prolidase may be dependent on zinc ion in order to function properly and is also inhibited by various SH compounds and EDTA. It has been suggested that captopril, with its sulfhydryl group, may inhibit prolidase activity [93]. Prolidase enjoys a wide distribution throughout the various organs and tissues in humans and other animals, concentrating in the brain, erythrocytes, heart, intestinal mucosa, kidney, plasma, and uterine tube.

Aminotripeptidase is also a cytoplasmic peptidase, with its highest activity against Pro-Gly-Gly [94]. In order to be hydrolyzed by aminotripeptidase, the substrate should contain a free α-amino group, an α-carboxy group, and the L configuration for the first two amino acid residues (if neither one is glycine). This tripeptidase has no activity against tripeptides with a charged N-terminal amino acid residue or with proline in the second position.

III. PRODRUGS AND SITE-SPECIFIC DELIVERY

A. Introduction

To achieve optimal drug design one must consider (1) the physical-chemical properties of the drug as they pertain to stability, solubility, and other formulation factors and (2) the pharmacokinetic factors such as the transport form of the drug [95]. Prodrug design should be used to optimize both properties. In addition, the goal in drug design is to treat a defined illness with minimally harmful side effects. The purpose of the following sections is to discuss whether prodrugs can provide a site-specific or targeted drug delivery system.

A drug targeting system is defined as a delivery system that is administered at a point away from the target tissue but finds its way to the site of action, where it releases the drug [96]. The idea that drugs could provide a means of targeting drug delivery was proposed as early as the turn of the century by Ehrlich, when he suggested his "magic bullet" concept. However, the concept has rarely been successful [97].

There have been several hypotheses proposed to achieve targeting via prodrugs [98]. The first hypothesis suggests that if a promoiety is added to a drug agent which exhibits an affinity for a specific tissue, then the prodrug will be selectively transported to that tissue. A number of flaws are associated with this hypothesis. First, the prodrug may not undergo biotransformation to the active moiety in the targeted site. Second, if the drug has an affinity for a cell surface receptor, prodrug internalization and subsequent reconversion may be slow relative to biotransformation in less accessible but more reactive tissues.

The second theory may be called the selective metabolism hypothesis. Based on this hypothesis, to accomplish targeting the derivative that is added to the parent drug should be removed only at a specific site. Thus, the active drug agent can be obtained only in that specific tissue. The difficulty with this theory is that the prodrug must have accessibility to the metabolic site. Prodrug transport to the target organ may be limited due to high polarity [98], for example.

Stella and Himmelstein [95,97,99] examined mathematical models of several hypotheses and concluded that an additional condition is necessary for transport: parent drug site retention. They identified three factors for successful targeting via prodrugs [95]:

1. The ease of transportation of the prodrug to the target site and its rapid uptake
2. The selective reconversion of the prodrug to the active moiety at the target relative to its cleavage at other sites in the body
3. The retention of the active moiety by the target tissue following reconversion

Considering the above factors, a better definition of targeting is the selective delivery of prodrug to its site of action and retention of the active moiety upon reconversion at the target, which leads to a high concentration of the therapeutic moiety, thus minimizing drug burden to the rest of the body.

B. Modeling Prodrugs and a Targeted Drug Delivery System

A hybrid physiological/classical pharmacokinetic model [95,97,98] was used to examine various hypotheses for achieving drug targeting via prodrugs. This model describes a system where a prodrug or a drug is introduced into a body volume (50 L). The drug is then transported to the target organ extracellular fluid (50 L) at a perfusion rate of 10 mL/min, where it can partition into the intracellular volume (50 L) through a clearance term. Finally, the drug is eliminated from the body with a clearance of 250 mL/min. The biological half-life to the specific organ extracellular fluid is approximately 5 min. The clearance was varied from 10 mL/min to 0.1 mL/min to represent a drug that permeates easily from the extracellular fluid into the intracellular fluid and a drug that had access to the intracellular fluid, respectively. In this model, prodrug input (350 mg) is the same as for a drug with biotransformation by the Michaelis–Menten mechanism in the body, extracellular fluid, and intracellular fluid. A Michaelis–Menten constant of 10 μg/mL was chosen arbitrarily for all three sites of metabolism and the V_{max} was varied. This allowed the modeling of the selective metabolism in either the target organ extra- or intracellular spaces.

One of the difficulties with targeted drug delivery is that some drugs do not reach the target organ due to poor permeability characteristics. A significant improvement in drug delivery can be made with a permeable prodrug that undergoes nonselective cleavage to the parent drug at the specific site. An example of this is the delivery of the quaternary aromatic ammonium drugs to the brain as dihydro prodrugs. These prodrugs are uncharged at physiological pH and pass the blood-brain barrier to reach the brain, where they are oxidized to their parent drug [100–101]. Although increasing the polarity of an agent can also be used to improve direct delivery to a target organ, this limits penetration of the prodrug to other sites that require a nonpolar drug.

As mentioned earlier, researchers have focused on the importance of target site accessibility and the target site selective metabolism of the prodrug. However, targeting was not achieved even when both factors were optimized. Stella and Himmelstein [99] showed that parent drug site retention is required to achieve selective delivery. A prodrug that is rapidly transported to the target site with selective rapid metabolism will only slightly improve the drug delivery. This is due to the fact that after enzymatic release, the drug rapidly equilibrates with the rest of the body tissues. If the drug has poor target organ permeability, a significant increase in targeting of the drug is accomplished not only by increasing drug transport to its selective metabolic site but also by drug retention at the site.

Hunt et al. [102] confirmed and extended Stella and Himmelstein's findings. They emphasized that the physicochemical properties of a drug and the properties of both the target and nontarget sites were critical in predicting when a targeted carrier can either improve the therapeutic efficacy or enhance the apparent potency. The assumptions were that targeted carriers can be engineered to release the agent at a reproducible rate and that they will deliver up to 100% of the drug to the given specific site. Two new parameters were introduced, namely therapeutic availability (TA) and drug targeting index (DTI), which allow predictions of the magnitude of the improved therapeutic efficacy.

A model was developed by Hunt et al. [102] to predict the potential benefits of administering a drug as a drug–carrier combination in humans. The model, incorporating several compartments, describes the pharmacokinetic properties of a range of real and hypothetical drugs. One compartment, which is designated the response compartment, represents target sites with effective drug concentration and has effective blood flow. Another compartment, designated the toxicity compartment, contains tissues with nontarget sites where a toxic response is initiated. This compartment also has effective blood flow. The effective concentration of the drug at these sites is identified when the drug is administered as the prodrug. The elimination compartment represents the liver and the kidney with blood flow and drug elimination constant. Finally, another compartment represents the blood and all other tissues. The blood flow to this compartment is the total cardiac output minus the blood flow to the other three compartments. The model identifies the drug concentration in the blood and the release rates (and also input functions) of free drug from the carrier into each of the corresponding compartments. When the release rates from the blood target site and elimination compartments are equal, the drug carrier is considered to be ideal.

Hunt and colleagues [102], in their approach, limited their attention to drugs where either the area measurements or the steady-state levels provided sufficient information for estimating drug effect and toxicity. They defined the DTI as the expected ratio of drug delivered to response and toxicity sites when the drug–carrier combination is used, divided by the corresponding ratio when the drug is administered intravenously. The TA was defined as the ratio of the dose fraction reaching target sites when the dose is administered parenterally as a prodrug to the amount reaching the same sites when an equal dose of the drug is administered intravenously. An increase in therapeutic availability translates to an increase in potency. DTI and TA indicate whether a particular drug is a reasonable drug–carrier candidate. They are also useful in determining when the physiology describing a specific therapeutic situation can be exploited by a drug–carrier combination. The modeling reinforces the fact that the strong interplay between various parameters such as the properties of the target site, the prodrug, and the drug should be recognized in order to achieve optimal site-specific delivery.

In conclusion, site-specific delivery based on the prodrug approach can be successful if a number of factors are taken into consideration. The use of prodrugs to improve targeting will be successful only if the parent moiety is poorly delivered to its site of action. The target site should be examined to specify any unique cell surface properties, membrane transport properties, or metabolic activity properties that may help site-specific delivery. The prodrug may be inactive in in vitro administration but a good in vivo candidate. The drug molecule must have a structural group which permits the synthesis of a derivative. Such a prodrug should have the physical, chemical, and biochemical characteristics that allow the prodrug transport to the target site and subsequent release of the active moiety.

REFERENCES

1. R Kato, RW Eastbrook, MN Cayen. Xenobiotic Metabolism and Disposition. London: Taylor & Francis, 1989, p 109.
2. H Bundguard. ed. Design of Prodrugs. Amsterdam: Elsevier, 1985, pp 1–2.
3. D Fleisher, BH Stewart, GL Amidon. Design of prodrugs for improved gastrointestinal absorption by intestinal enzyme targeting. Methods Enzymol 112: 360–381, 1985.
4. N Bodor, KB Sloan, T Higuchi, K Sasahara. Improved delivery through biological membranes. 4 Prodrugs of L-dopa. J Med Chem 20(11):1435–1445, 1977.
5. DR Cooper, C Marrel, H Van de Waterbeemd, B Testa, P Jenner, C D Marsden. L-Dopa esters as potential prodrugs: Behavioural activity in experimental models of parkinson's disease. J Pharm Pharmacol 39:627–635, 1987.
6. C Marrel, G Boss, B Testa, H Van de Waterbeemd, D Cooper, P Jenner, CD Marsden. L-Dopa esters as potential prodrugs. Part II. Chemical and enzymic hydrolysis. Eur J Med Chem Chim Ther 20:467–470, 1985.
7. A Garzon, A Poupaert, M Claesen, P Dumont. A lymphotropic prodrug of L-dopa: Synthesis, pharmacological properties, and pharmacokinetic behavior of 1,3-dihexadecanoyl-2-[(S)-2-amino-3-(3,4-dihydroxyphenyl) propanoyl] propane-1,2,3-triol. J Med Chem 29:687–691, 1986.
8. Y Tsuchiya, Y Sawasaki, A Hsako, H Takehana, K Tomimoto, M Yano. New potential prodrug to improve the duration of L-dopa: L-3-(3-Hydroxy-4-pivaloyl-oxyphenyl) alanine. J Pharm Sci 78(7):525–529, 1989.
9. SJ Tye, NM Rupniak, T Narase, M Miyaji, SD Iversen. NB-355: A novel prodrug for L-dopa with reduced risk for peak-dose dyskinesias in MPTP-treated squirrel monkeys. Clin Neuropharm 12(5):393–403, 1989.
10. AJ Repta. Short-chain alkyl esters of L-dopa as prodrugs for rectal absorption. Pharm Res 6(6):501–505, 1989.
11. B Lund, JP Kampmann, F Lindahl, MJ Hansen. Pivampicillin and ampicillin in bile, portal and peripheral blood. Clin Pharmacol Ther 19:587–591, 1976.

12. S Ehrnebo, O Nilsson, LO Boreus. Pharmacokinetics of ampicillin and its prodrugs bacampicillin and pivampicillin in man. J Pharm Biopharmacokin 7(5):429–451, 1979.
13. L Magni, J Sjovall, E. Vinnars. Bioavailability of bacampicillin and talampicillin. Antimicrob Agent Chemother 20(6):837–838, 1981.
14. L Nerbist. Triple crossover study on absorption and excretion of ampicillin, talampicillin, and amoxycillin. Antimicrob Agent Chemother 10:173–175, 1976.
15. B Huitfeldt, L Magni, CE Nord, J Sjovall. Effect of beta-lactam prodrugs on human intestinal microflora. J Infect Dis 49:73–84, 1986.
16. T Anders, LA Svensson. Bambuterol, a carbamate ester prodrug of terbutaline, as inhibitor of cholinesterases in human blood. Drug Metabol Dispos 16(5):759–763, 1988.
17. OAT Olsson, LA Svensson. Treatment of chronic ulcerative-colitis with poly-ASA—A new nonabsorbable carrier for release of 5-aminosalicylate in the colon. Pharm Res 19–23, 1984.
18. C Longcope, BG Gorbach, M Woods, J Dwyer, J Warren. The metabolism of estradiol: Oral compared to intravenous administration. J Steroid Biochem 23:1065–1070, 1985.
19. M Hussain, BI Aungst, F Shefter. Prodrugs for improved oral beta-estradiol bioavailability. Pharm Res 5(1):44–47, 1988.
20. AN Saab, LW Dihert, AA Hussain. Isomerization of cephalosporin esters: Implications for the prodrug ester approach to enhancing the oral bioavailabilities of cephalosporins. J Pharm Sci 77(10):906–907, 1988.
21. F Harboe, C Larsen, MJ Johansen, HP Olesen. Macromolecular prodrugs. XV. Colon-targeted delivery—Bioavailability of naproxen from orally administered dextran-naproxen ester prodrugs varying in molecular size in the pig. Pharm Res 6(11):919–923, 1989.
22. SA Varia, S Schuller, KB Sloan, VJ Stella. Phenytoin prodrugs III: Water-soluble prodrugs for oral and/or parenteral use. J Pharm Sci 73(8):1068–1073, 1984.
23. Y Yumiko, RD Roberts, VJ Stella. Low-melting phenytoin prodrugs as alternative oral delivery modes for phenytoin: A model for other high-melting sparingly water-soluble drugs. J Pharm Sci 72(4):400–405, 1983.
24. SA Varia, S Shuller, VJ Stella. Phenytoin prodrugs IV: Hydrolysis of various 3-(hydroxymethyl) phenytoin esters. J Pharm Sci 73(8):1074–1080, 1984.
25. SA Varia, VJ Stella. Phenytoin prodrugs V: In vivo evaluation of some water-soluble phenytoin prodrugs in dogs. J Pharm Sci 73(8):1080–1087, 1984.
26. SA Varia, VJ Stella. Phenytoin prodrugs VI: In vivo evaluation of a phosphate ester prodrug of phenytoin after parenteral administration to rats. J Pharm Sci 73(8):1087–1090, 1984.
27. K Murata, K Noda, K Kohno, M Samejirna. Bioavailability and pharmacokinetics of oral dopamine in dogs. J Pharm Sci 77:565, 1988.
28. K Murata, N Kazuo, K Kohano, MJ Sanejirna. Bioavailability and pharmacokinetics of an oral dopamine prodrug in dogs. J Pharm Sci 78(10):812–814, 1989.
29. I Yamaguchi, S Nishiyama, S Akimato, A Yoshiaki, M Yoshikawa, H Nakajima. A novel orally active dopamine prodrug TA-870. I. Renal and cardiovascular ef-

fects and plasma levels of free dopamine in dogs and rats. J Cardiovasc Pharmacol 13(6):879–886, 1989.

30. D Brooks, PD Depalma, MJ Cyronak, MA Bryant, K Karpinski, B Mico, D Gaitanopoulos, PA Chambers, KF Erhard, J Weinstock. The identification of fenoldopam prodrugs with prolonged renal vasodilator. J Pharmacol Exper Ther 254(3):1084–1089, 1990.

31. W Kwapiszewski, J Kolwas. Synthesis of amino acid derivatives of benzocaine. III. Preparation of N-(N'-N'-dimethylaminoacyl)-benzocaines. Acta Polon Pharm 34:377–382, 1977.

32. U Klotz. Clinical pharmacokinetics of sulfasalazine, its metaboliters and other prodrugs of 5-aminosalicylic acid. Clin Pharmacokin 10:285–302, 1985.

33. RP Chan, DJ Pope, AP Gilbert, PJ Sacra, JH Buron, JE Lennard-Jones. Studies of two novel sulfasalazine analogs, ipsalazide and balsalazide. Digest Dis Sci 28:609–716, 1983.

34. CP Willoughby, JK Aronson, H Agback, NO Bodin, SC Tmelone. Distribution and metabolism in healthy volunteers of disodium azodisalicylate, a potential therapeutic agent for ulcerative colitis. Gut 23:1081–1087, 1982.

35. CP Willoughhy, J Piris, C Tranelone. The effect of topical N-acetyl-5-aminosalicylic acid in ulcerative colitis. 15:715–719, 1980.

36. M Garretro, RH Riddell, CS Winans. Treatment of chronic ulcerative-colitis with poly-ASA—A new nonabsorbable carrier for release of 5-aminosalicylate in the colon. Gastroenterology 84:1162, 1983.

37. H Sandberg-Gretzen, M Ryde, G Jarnerot. Absorption and excretion of azodisal sodium and its metabolites in man after rectal administration of a single 2-g dose. Scan J Gastroenterol 18:571–575, 1983.

38. DB Brigden, P Whitman. The clinical-pharmacology of acyclovir and its prodrugs. Scan J Infect Dis, Suppl 47:33–39, 1985.

39. BG Petty, RI Whitty, S Liao, HC Krasny, LE Rocco, LG Davis, PS Lietman. Pharmacokinetics and tolerance of desciclovir, a prodrug of acyclovir, in healthy human volunteers. Antimicrob Agents Chemother 31:1317–1322, 1987.

40. YT Bryson, M Dillon, G Acuna, S Taylor, JD Cherry, BL Johnson, E Wiesmeier, W Growden, T Greagh-Kirk, R Keeney. Treatment of first episodes of genital herpes simplex virus infection with oral acyclovir. A randomized double-blind controlled trial in normal subjects. N Engl J Med 308: 916–921, 1983.

41. J Schwartz, D Loebenberg, GH Miller, S Symchowicz, J Lim, C Lin. Evaluation of (+) -cyclaradine-5'-esters as prodrugs for (+)-cyclaradine in animals. Antimicrob Agents Chemother 31(7):998–1001, 1987.

42. MR Harnden, RL Jarret, MR Boyed, D Sulton, RA Vene Hodge. Prodrugs of the selective antiherpesvirus agent 9-[4-hydroxy-3-(hydroxymethyl) but 1-yl] guanine (BRL 39123) with improved gastrointestinal absorption properties. J Med Chem 32:1738–1743, 1989.

43. C Kwon, HT Nagasawa, EG DeMastr, FN Shirota. Acyl, N-protected alpha-aminoacyl, and peptidyl derivatives as prodrug forms of the alcohol deterrent agent cyanamide. J Med Chem 29: 1922–1929, 1986.

44. GL Amidon, AE Merfeld, JB Dressman. Concentration and pH dependency of alpha-methyldopa absorption in rat intestine. J Pharm Pharmacol 38:363–368, 1986.

45. AE Merfeld, AR Mlodozeniec, MA Cortese, JB Rhodes, JB Dressman, GL Amidon. The effect of pH and concentration on alpha-methyldopa absorption in man. J Pharm Sci 38:815–822, 1986.
46. SA Adibi, YS Kim. In: LR Johnson, ed. Physiology of the gastrointestinal tract. New York: Raven Press, 1981, p 1073.
47. DI Friedman, GL Amidon. J Controlled Release 13:141–146, 1990.
48. Hu Ming, HI Mosberg, GL Amidon. Use of the peptide carrier system to improve the intestinal absorption of L-alpha-methyldopa: Carrier kinetics, intestinal permeabilities, and in vitro hydrolysis of dipeptidyl derivatives of L-alpha-methyldopa. Pharm Res 6(1):66–69, 1989.
49. M Asgharnejad, GL Amidon. Improved oral delivery via the peptide transport: A dipeptide prodrug of L-αmethyldopa. Pharm Res 9:5–248, 1992.
50. M Asgharnejad. Investigation into intestinal transport and absorption of an amino acid, amino acid analougue and its peptideomimetic prodrug. PhD dissertation, Chapter IV. pp 109–127, The University of Michigan, 1992.
51. DI Friedman, GL Amidon. Passive and carrier-mediated intestinal absorption components of two angiotensin converting enzyme (ACE) inhibitor prodrugs in rats: Enalapril and fosinopril. Pharm Res 6(12):1043–1047, 1989.
52. Hu-Ming. Investigation into drug and drug analogs transported by the peptide carrier system: Intestinal absorption of captopril and of peptidyl derivatives of methyldopa. PhD Thesis, College of Pharmacy, University of Michigan, 1988.
53. OH Drummer, B Jarrott. Captopril disulfide conjugates may act as prodrugs: Disposition of the disulfide dimer of captopril in the rat. Biochem Pharmacol 33(22): 3568–1571, 1984.
54. GL Amidon, RS Pearlman, GD Lessman. In: EB Roche, ed. Design of Biopharmaceutical Properties Through Prodrugs and Analogues. Washington, DC: American Pharmaceutical Association, 1977, pp 281–315.
55. O Fleisher, BH Stewart, GL Amidon. Design of prodrugs for improved gastrointestinal absorption by intestinal enzyme targeting. Methods Enzymol 112:361–381, 1985.
56. M Florkin, EH Stotz, eds. Enzyme Nomenclature. New York: Elsevier, 1972.
57. IH Leninger. Principles of Biochemistry, New York: Worth, 1982, pp 207–247.
58. IH Segel. Enzyme Kinetics. New York: Wiley-Interscience, 1975.
59. M Asgharnejad, GL Amidon. Improved oral delivery via the peptide transporter: A dipeptide prodrug of L-alpha-methyldopa. Pharm Res 9:S-248, 1992.
60. CJ Gray. Enzyme-Catalyzed Reactions. London: Van Nostrand Reinhold, 1972, p 72.
61. EB Roche, ed. The Design of Biopharmaceutical Properties Through Prodrugs and Analogues. Washington, DC: Am Pharm Assoc, 1977, p 283.
62. PK Banerjee, GL Amidon. Physicochemical property modification strategies based on enzyme substrate specificities I: Rationale, synthesis, and pharmaceutical properties of aspirin derivatives. J Pharm Sci 70:1299, 1981.
63. PK Banerjee, GL Amidon. Physicochemical Property modification strategies based on enzyme substrate specificities II: Alpha-chymotrypsin hydrolysis of aspirin derivatives. J Pharm Sci 70:1304, 1981.
64. PK Banerjee, GL Amidon. Physicochemical property modification strategies based

on enzyme substrate specificities III: Carboxypeptidase a hydrolysis of aspirin derivatives. J Pharm Sci 70:1307, 1981.

65. GL Amidon, GD Lessman, RL Elliot. Improving intestinal absorption of water-insoluble compounds: A membrane metabolism strategy. J Pharm Sci 69:1363, 1980.

66. I Dawsen, J Pryse-Davies. Gastroenterology 44(6):754–760, 1963.

67. F Moog. The lining of the small intestine. Sci Am 245:154–159, 1981.

68. N Triadou, J Bataille, J Schmitz. Longitudinal study of the human intestinal brush border membrane proteins. Distribution of the main disaccharidases and peptidases. Gastroenterology 85:1326–1332, 1983.

69. GN MaComb, S Bowers Jr, X Dosen. Alkaline Phosphatase. New York: Plenum, 1979.

70. YS Kim. *Intestinal mucosal hydrolysis of proteins and peptides in peptide transport, and hydrolysis.* A Ciba Foundation Symposium. Exerpta Medica. Amsterdam: Elsevier, 1977, pp 151–169.

71. PK Banerjee, GL Amidon. Design of Prodrugs. New York: Elsevier, pp 93–133.

72. DM Matthews, JW Payne. Transmembrane transport of small peptides. Curr Topics Membrane Transport 14:331–425, 1980.

73. EE Sterchi, JF Woodley. Peptide hydrolases of the human small intestinal mucosa: Distribution of activities between brush border membranes and cytosol. Clin Chim Acta 102:49–56, 1980.

74. YS Kim, SA Adibi. In: CR Johnson, ed. Physiology of the Gastrointestinal Tract. New York: Raven Press, 1981, pp 1073–1094.

75. SA Adibi. In: GL Blackburn, JP Grant, NR Young, J Wright, eds. Amino Acids: Metabolism and Medical Applications, Boston: PSG, 1983, pp 255–263.

76. AJ Kenny, S Maronx. Topology of microvillar membrane hydrolases of kidney and intestine. Physiol Rev 62: 91–128, 1982.

77. EJ Brophy, YS Kim. Effect of amino acids on purified rat intestinal brush-border membrane aminooligopeptidase. Gastroenterology 78:82–87, 1979.

78. N Tobey, W Heizer, R Yeh, T Huang, C Hoffner. Human intestinal brush border peptidases. Gastroentrology 88:913–926, 1985.

79. ST Shresh. Methods Enzymol 1985, pp 113, 471–484.

80. EE Strechi, JF Woodley. Peptide hydrolases of the human small intestinal mucosa: Identification of six distinct enzymes in the brush border membrane. Clin Chim Acta 102:57–65, 1980.

81. S Auricchio, A Stellato, B De Vizia. Development of brush border peptidases in human and rat small intestine. Pediatr Res 15:991–995, 1981.

82. S Auricchio, L Geroco, B de Vizia, V Buonowne. Dipeptidylaminopeptidase and carboxypeptidase activities of the brush border of rabbit small intestine. Gastroenterology 75:1073–1079, 1981.

83. C Shoaf, RM Berko, WD Heizer. Isolation and characterization of four peptide hydrolases from the brush border of rat intestinal mucosa. Biochem Biophys Acta 445:649–719, 1976.

84. M Yoshioka, RH Erickson, YS Kim. Digestion and assimilation of proline-containing peptide by rat intestinal brush border membrane carboxypeptidases. Role

of the combined action of angiotensin-converting enzyme and carboxypeptidase P. J Clin Invest 81:1090–1095, 1988.

85.   M Yoshioka, RH Erikson, JF Woodly, R Gulli, D Guam, YS Kim. Role of rat intestinal brush-border membrane angiotensin-converting enzyme in dietary protein digestion. Am J Physiol 253:G-781–G-786, 1987.

86.   L Josefsson, H Sjostrom, O Noren. Peptide transport and hydrolysis. Amsterdam: Elsevier, 1976, pp 199–207.

87.   O Noren, H Sjostrom, L Josefsson. Studies on a soluble dipeptidase from pig intestinal mucosa. I. Purification and specificity. Biochem Biophys Acta 327:446–456, 1973.

88.   M Das, AN Radhakrishnan. Substrate specificity of a highly active dipeptidase purified from monkey small intestine. Biochem J 128:463–465, 1972.

89.   O Noren, E Dabelsteen, H Sjostrom, L Josefsson. Histological localization of two dipeptidases in the pig small intestine and liver, using immunofluorescence. Gastroenterology 72:87–92, 1977.

90.   O. Noren. Studies on a soluble dipeptidase from pig intestinal mucosa. Enzymatic properties. Acta Chem Scand B28:711–716, 1974.

91.   M Hu. Investigation into drug and drug analogs transported by the peptide carrier system: Intestinal absorption of captopril and of peptidyl derivatives of methyldopa. PhD Thesis, College of Pharmacy, University of Michigan, 1988.

92.   KS Hui, A Lajtha. J Neurochem 30:321–327, 1978.

93.   GF King, NJ Crossley, PW Kuckel. Biomed Biochem Acta 46:302–303, 1987.

94.   C Doumeng, S Maroux. Aminotripeptidase, a cytosol enzyme from rabbit intestinal mucosa. Biochem J 177:801–808, 1979.

95.   VJ Stella, KJ Himmelstein. Directed Drug Delivery, Prodrugs: A Chemical Approach to Targeted Drug Delivery. Clifton NJ: Humana, 1985, pp 247–267.

96.   AT Florence. Rate Control in Drug Therapy, Edinburgh: Churchill Livingston, 1985, p 103.

97.   VJ Stella, KJ Himmelstein. In: Bungaurd, ed. Design of Prodrugs. Amsterdam: Elsevier, 1985, pp 177–198.

98.   VJ Stella. In: R Kato, RW Estabrook, MN Cayen, eds. Xenobiotic Metabolism and Disposition. Philadelphia: Taylor and Francis, 1989, pp 109–116.

99.   VJ Stella, KJ Himmelstein. Prodrugs and site-specific drug delivery. J Med Chem 23:1275–1282, 1980.

100. N Bodor, E Shek, T Higuchi. Delivery of a quaternary pyridinium salt across the blood-brain barrier by its dihydropyridine derivative. Science 190:155–156, 1975.

101. N Bodor, E Shek, T Higuchi. Improved delivery through biological membranes. 1. Synthesis and properties of 1-methyl-1,6-dihydropyridine-2-carbaldoxime, a prodrug of N-methylpyridinium-2-carbaldoxime chloride. J Med Chem 19: 102–107, 1976.

102. AC Hunt, RD MacGregor, RA Siegel. Engineering targeted in vivo drug delivery.1. The physiological and physicochemical principles governing opportunities and limitations. Pharm Res 3(6):333–344, 1986.

8

Quantitative Approaches to Delineate Passive Transport Mechanisms in Cell Culture Monolayers

Norman F. H. Ho,* Thomas J. Raub, Philip S. Burton, and Craig L. Barsuhn
Pharmacia & Upjohn, Inc., Kalamazoo, Michigan

Anthony Adson†
Pharmacia & Upjohn, Inc., Kalamazoo, Michigan, and The University of Kansas, Lawrence, Kansas

Kenneth L. Audus and Ronald T. Borchardt
The University of Kansas, Lawrence, Kansas

I. INTRODUCTION

The intent of this chapter is to establish a comprehensive framework in which the physicochemical properties of permeant molecules, hydrodynamic factors, and mass transport barrier properties of the transcellular and paracellular routes comprising the cell monolayer and the microporous filter support are quantitatively and mechanistically interrelated. We specifically define and quantify the biophysical properties of the paracellular route with the aid of selective hydrophilic permeants that vary in molecular size and charge (neutral, cationic, anionic, and zwitterionic). Further, the quantitative interrelationships of pH, pK_a, partition

* Current affiliation: The University of Utah, Salt Lake City, Utah.
† Current affiliation: Alpharm USPD, Baltimore, Maryland.

coefficient, fractions of nondissociated and charged species, permeability coefficients of the aqueous boundary layer (ABL), filter support, and a cell monolayer with its component transcellular and paracellular routes are defined. The influence of stirring on the permeability coefficients of the ABL and cell monolayer are quantified. The aforementioned interrelationships are developed with data collected from models of permeability using cultured cells.

Methods for quantifying both the transcellular diffusion and concurrent metabolism of drugs and the unusual transcellular diffusion of membrane-interactive molecules coupled with the influence of protein binding are described in detail. To demonstrate the utility of cultured cell monolayers as a tool for basic science investigations, a subsection is devoted to the elucidation of rate-determining steps and factors in the passive diffusion of peptides across biological membranes. The chapter concludes with a discussion on the judicious use of in vitro cell monolayer results to predict in vivo results.

A. General Background on Passive Transport

The passage of molecules across epithelial and endothelial cell barriers occurs by either of two general pathways: paracellularly (between the cells) or transcellularly (through the cells) (Fig. 1). Paracellular flux occurs strictly by passive diffu-

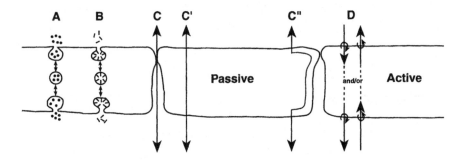

Figure 1 General pathways through which molecules can actively or passively cross a monolayer of cells. (A) Endocytosis of solutes and fusion of the membrane vesicle with the opposite plasma membrane in an active process called transcytosis. (B) Similar to A, but the solute associates with the membrane via specific (e.g., receptor) or nonspecific (e.g., charge) interactions. (C) Passive diffusion between the cells through the paracellular space. (C', C") Passive diffusion (C') through the cell membranes and cytoplasm or (C") via partitioning into and lateral diffusion within the cell membrane. (D) Active or carrier-mediated transport of an otherwise poorly membrane permeable solute into and/or out of a cellular barrier.

sion. Transcellular flux can occur by either passive, facilitated, or active processes (Table 1). While this chapter is restricted to passive processes, it is important to ascertain that the transport mechanism involved in the experiment under consideration is passive before employing the quantitative techniques to be illustrated here. A detailed discussion of nonpassive transport is beyond the scope of this chapter, although the comparative features of the different transport mechanisms are summarized in Table 1. Currently, much research interest is focused on the

Table 1 Transport Mechanisms Common to Cellular Barriers

Mechanism	Characteristics
I. Diffusion	Flux down an electrochemical gradient.
	Energy-independent.
a. Passive	Flux proportional to concentration gradient.
	Rate independent of direction.
b. Facilitated	Carrier-mediated.
	Flux is saturable with increasing concentration.
	Competitive substrates.
	Flux may be asymmetrical.
II. Active carrier-mediated	Flux can be against an electrochemical gradient.
	Energy dependent—directly or indirectly coupled.
	Substrate specificity, competition, saturation.
	Flux is asymmetrical.
III. Endocytosis	Invagination of the plasma membrane forming an internalized membrane vesicle.
	Usually energy-dependent.
	Usually results in solute uptake into and use by a cell; however, transcytosis, in which the vesicle crosses the cell and fuses with the opposite plasma membrane domain, may occur.
	Flux against a gradient.
a. Receptor-mediated (includes potocytosis)	Very substrate-specific.
	High affinity and saturable, but usually low capacity.
	Asymmetrical.
b. Adsorptive	No specific receptor involved; solute nonspecifically adsorbs to cell surface proteins or glycolipids.
	Can be saturable and show competition, often high capacity.
c. Fluid-phase (pinocytosis)	Soluble molecules are internalized with the vesicle volume.
	Nonspecific and nonsaturable.

role of polarized active efflux pathways in the absorption and distribution of drugs in vivo (Lum and Gosland, 1995; Burton et al., 1997; Watkins, 1997). The reader also is referred to Chapter 6 of this book for additional information.

While both paracellular and passive transcellular pathways are available to a solute, the relative contribution of each to the observed transport will depend on the properties of the solute and the membrane in question. Generally, polar membrane-impermeant molecules diffuse through the paracellular route, which is dominated by tight junctions (Section III.A). Exceptions include molecules that are actively transported across one or both membrane domains of a polarized cell (Fig. 2). The tight junction provides a rate-limiting barrier for many ions, small molecules, and macromolecules depending on the shape, size, and charge of the solute and the selectivity and dimensions of the pathway.

Moderately and highly lipophilic molecules passively diffuse across cellular barriers by the transcellular or transmembrane pathways. Passive transcellular diffusion can occur through two different mechanisms. Once a solute has passed through the aqueous boundary or unstirred water layer (Section V) and has partitioned from the aqueous extracellular environment, which includes the net negatively charged glycocalyx, into the cell plasma membrane, it can follow either of two pathways (Fig. 3). The solute can conceivably remain mostly within the hydrocarbon domain of the lipid bilayer on a time average and diffuse laterally around the tight junctions. The tight junction has been shown to restrict the lateral diffusion of lipids within the outer leaflet of the lipid bilayer (van Meer and Simons, 1986; van Meer et al., 1992). Transport of the solute is completed by desorption from the receiver-side membrane or by repartitioning into the aqueous receiver compartment. Examples of this putative pathway are shown in Section VIII.

Alternatively, it has been assumed that the solute can partition out of the donor-side membrane into the cell cytoplasm. Once there, diffusion through the aqueous cytosol and repartitioning into the basolateral cell membrane will result in transport across the cell. However, it should be appreciated that the interior of the cell is a highly organized environment with a number of subcellular organelles with membranes of varying composition. Thus, the solute can sequentially partition its way through the cell depending on its relative affinity for the different intracellular membranes. Further, there may be specific carrier proteins within the cytosol which transfer the solute across the cell or among various subcellular compartments. Although the details of the processes may be different, all of these mechanisms can result in solute crossing through the interior of the cell proper.

In order to probe these mechanistic aspects of solute transport, simplified models of the respective cellular barriers have been developed in recent years. In many cases, the cells that comprise these barriers, or cells that serve to mimic the barrier of interest, can be maintained in culture. The ability of the cells to

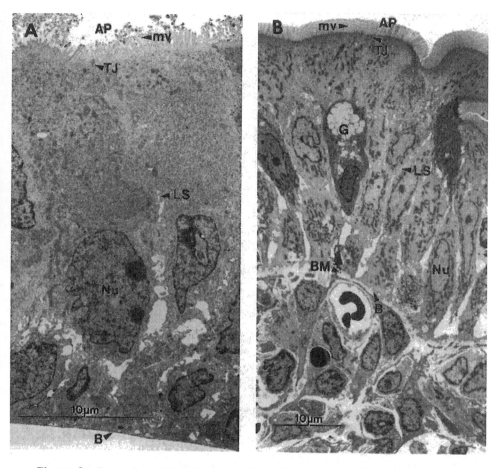

Figure 2 Comparison of intestinal epithelial cells in culture and in situ. (A) Human colon Caco-2 cells grown in culture for 16 days on a semiporous filter. (B) Epithelial layer of rat jejunum. AP, apical or luminal membrane; B, basal or abluminal membrane; BM, basement membrane; G, goblet cell; LS, lateral space; mv, microvilli; Nu, nucleus; TJ, tight junction. Bars equal 10 μm.

form and maintain a confluent monolayer, with characteristics similar to the in vivo barrier being modeled, on some type of semipermeable support is a prerequisite for transport studies. Advantages of using an in vitro model include small sample volumes, low variability between replicates, and the fact that under these conditions many external factors that influence transport can be manipulated eas-

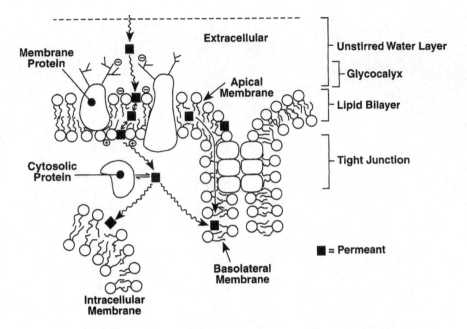

Figure 3 A hydrophobic permeant must negotiate through a complex series of diffusional and thermodynamic barriers as it penetrates into a cell. The lipid and protein compositions and charge distribution of the inner and outer leaflets of the membrane lipid bilayer can play limiting roles, particularly at the tight junction. Depending upon the permeant's characteristics, it may remain within the plasma membrane or enter the cytoplasm, possibly in association with cytosolic proteins, and partition into cytoplasmic membranes.

ily. More important, these cell culture models simplify the transport process by removing additional physical and physiological factors that exist in in vivo and ex vivo experimental models.

Epithelial and endothelial cells of various tissue origins can be grown in culture on filter supports (Cereijido et al., 1987; Shasby and Shasby, 1990; Zweibaum et al., 1991; Miller et al., 1992). Some of the most studied cell lines are listed in Table 2. Recently, these models have been used to (1) screen homologous chemical entities for optimal transmembrane permeability, (2) examine mechanisms of transcellular transport of compounds and especially to understand why certain compound classes are poorly absorbed in vivo, (3) study the effects and mechanisms of additives or pharmacological agents (e.g., adjuvants) that increase compound permeability, and (4) observe metabolism of different drug entities. These model systems are being used within the pharmaceutical industry to aid in the selection of orally active drug candidates (Lee et al., 1997).

Table 2 Most Commonly Used Cell Culture Models for
Measuring Transcellular Flux

Cell name	Species/tissue origin	Cell type
Caco-2	Human colon	Epithelial
HT-29	Human colon	Epithelial
T_{84}	Human colon	Epithelial
MDCK	Canine kidney	Epithelial[a]
LLC-PK$_1$	Pig kidney	Epithelial[a]
BMEC	Bovine brain	Endothelial[b]
BPAEC	Bovine pulmonary artery	Endothelial[b]
BAEC	Bovine aorta	Endothelial[b]
HUVEC	Human umbilical cord	Endothelial[b]

[a] MDCK (Madin–Darby canine kidney) cells are derived from distal tubules, whereas LLC-PK$_1$ are from proximal tubes.
[b] BMEC (brain microvessel endothelial cells) are isolated from capillaries. BPAEC (bovine pulmonary artery endothelial cells), BAEC (bovine aortic endothelial cells), and HUVEC (human umbilical vein endothelial cells) are large vessel endothelia.

As will be discussed in the following section, a variety of experimental designs are available and have been described to conduct such transport studies. While these variations are all available, the choice of a system and design of the experiment will be dictated by the information desired from the study. However, as illustrated in the applications section, if the variables present in the experimental design are taken into proper consideration, it will be possible to extract mechanistic information which is essentially independent of the system used. In this way it should be possible to compare results from one system or laboratory with those of another.

B. Selection of a Transport System and Parameters to Consider

In the selection of an appropriate cell culture system, a number of criteria must be considered (Table 3). These include not only the characteristics of the cell type but also the controllable parameters of the complete transport system such as the permeants, the filter properties, and the assay conditions. In general, most transport experiments employ the experimental design shown schematically in Figure 4 with modifications as discussed below. Typically, the desired cell is seeded onto some sort of semipermeable filter support and allowed to reach confluence. The filter containing the cell monolayer separates the donor and receiver

Table 3 Criteria for the Selection of a Model System

Solute properties	Radioisotope
	Method of detection
	Stability and solubility
	Adsorption
	Intrinsic charge
Stirring	Type of motion
	Speed
	Surface area
	Fluid volumes
	Equality of the hydrodynamics
Filter properties	Thickness
	Pore diameter
	Porosity and pore tortuosity
	Composition
Assay conditions	Sink conditions
	Disappearance, appearance, or both
	Reservoir volume
	Sampling time
	Temperature
Matrix	Composition and thickness
Medium	Composition
	pH
	Protein concentration
Cells	Binding, adsorption, partitioning
	Physical dimensions
	Metabolism
	Monolayer integrity
	Membrane domain characteristics (polarity): surface area; transporters, receptors; lipid composition; charge
	Cell phenotype and culture conditions

compartments of a diffusion chamber. Variations on this basic design involve the type of filter used, the presence or absence of matrix, the mechanisms for agitation of the donor and/or receiver solutions, and the configuration of the diffusion cell. The following sections describe how these factors can impact the experimental design, data analysis, and interpretation. Following this general discussion, selected examples of actual transport data are presented in detail to illustrate how the experimental variables can be dealt with to arrive at meaningful mechanistic information.

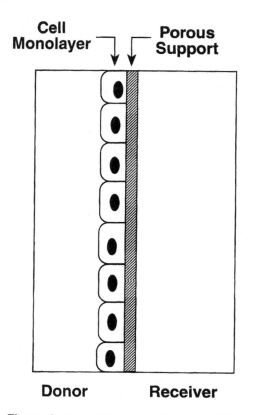

Figure 4 Essential features of a simple diffusion model consisting of a monolayer of cells grown on a porous support which separates fluid-filled donor and receiver compartments.

1. Cell Type

It is shown in Table 3 that there are many parameters to consider with regard to the cells. The general quality of the monolayer and its function as a barrier must be checked regularly to detect changes in the pathways being studied. This is usually accomplished by using a reference permeant, whose permeability is well known, with each experiment. Transepithelial (or transendothelial) electrical resistance (TER) measurements can also provide a reference. It is recommended that the cell cultures be checked regularly for microbial contamination, with mycoplasma being the most common contaminant. Although the permeability of the passive barrier in response to a mycoplasma infection may not be affected, its presence could alter other events such as pharmacological response. As another example, small amounts of exotoxin produced by an undetectable population of bacteria also could be deleterious to monolayer integrity (Hecht et al., 1988; Hirsch and Noske, 1993).

In addition, the physical dimensions of the cells making up the monolayer should be considered. Cell shape can influence the relative contributions of the paracellular and transcellular pathways. For example, junctional density is greater in cells that are narrow or of small diameter than in cells that are wide or spread out on the substrate. The height of the cells can impact the path length traveled by a permeant, as will the morphology of the junctional complex and lateral space (Section III.B.2). It is unknown how the mass of lipid or membrane within a cell influences transcellular flux of a lipophilic permeant.

At a more molecular level, the influences of the composition of the membrane domains, which are characteristic of a polarized cell, on diffusion are not specifically defined. These compositional effects include the differential distribution of molecular charges in the membrane domains and between the leaflets of the membrane lipid bilayer (Fig. 3). The membrane domains often have physical differences in surface area, especially in the surface area that is accessible for participation in transport. For example, the surface area in some cells is increased by the presence of membrane folds such as microvilli (see Figs. 2 and 6). The membrane domains also have differences in metabolic selectivity and capacity as well as in active transport due to the asymmetrical distribution of receptors and transporters.

All of these characteristics can be under the regulation of the cell and influenced by the cell culture conditions. The age of the cell monolayer in culture can have a profound impact on the quality of the barrier. In monolayers with actively dividing cells, resistance increases with time in culture as tight junctions form (see Fig. 15, Section III.C.4). Resistance reaches a plateau, then decreases as cell viability declines (Section III.C.4). Time in culture may also be a factor in the expression of polarity, which is related to tight junction formation as well as the state of differentiation of the cells (e.g., differential gene expression).

In most cases, cultured cells must be passaged over certain time intervals. This procedure involves lifting the cells from a growth substrate and disaggregating them, usually in a cation-poor solution of protease. Often small portions of the total number of cells are then used to start another culture; this involves dilution and can lead to the selection of subpopulations of cells with slightly different characteristics than those of the original parent cell. Since most of the cells in the more commonly used models (Table 2) arose from tumors, some drift in their characteristics can be expected. Therefore, one needs to be aware that identical cell lines from different sources might have different phenotypes and cannot be considered equivalent despite having arisen from the same progenitor cell. Consequently, passage number may also be a critical factor to control in these experiments. The importance of this and the other factors will be dependent upon the cell type under study.

2. Matrix

Most cells in culture generate some form of extracellular matrix, but the amount and composition vary considerably. In general, the matrix assembled by the cells is not an important factor in flux unless it influences the expression of a transporter system. A matrix really affects molecular diffusion only when it is applied to the filter surface for facilitating growth and/or differentiation of the cells. This is true only when the matrix is of significant thickness; thin films will not have a significant effect. For example, rat tail collagen is used to coat filters for attachment and growth of brain microvessel endothelial cells and can form collagen mats of substantial thickness (up to 0.4 μm). The cross-linked collagen fibrils with a fixed net negative charge can act as a molecular sieve with charge selectivity. Under these conditions, the role of the matrix must be accounted for in the model (see Section IV). The presence of a matrix is more representative of the in vivo situation, in which a basement membrane underlies most cellular barriers (Fig. 2).

3. Filter Support

The sole purpose of the filter support and any applied extracellular matrix is simply to provide a surface for cell attachment and thus to provide mechanical support to the monolayer. However, the filter and matrix also can act as serial barriers to solute movement after diffusion through the cell monolayer. The important variables are the chemical composition of the filter, porosity, pore size, and overall thickness. In some cases, pore tortuosity also can be important. It is desired that the filter, with or without an added matrix, provide a favorable surface to which the cells can attach. However, in some cases these properties can also result in an attractive surface for nonspecific adsorption of the transported solute. In these instances, the appearance of the solute in the receiver compartment of the diffusion cell will not be a true reflection of its movement across the monolayer. Such problems must be examined on a case-by-case basis.

To cross the filter, the solute must diffuse through the filter pores. Therefore, pore size, density, and tortuosity must be taken into consideration. Many filter configurations are commercially available and have been employed for these types of studies. In general, the greater the pore size and porosity, the less the potential for the filter to act as a significant diffusion barrier. For small solutes, the filter will probably not present much of a problem. However, as the molecular size of the solute increases and approaches the dimension of the pore, these considerations become more important. In principle, solute diffusion through the filter can be examined in the absence of a cell monolayer. If significant diffusive restriction is found, appropriate corrections to the transport data can be made (see Section IV).

One potential solution to the problem of filter-restricted diffusion of a solute is to use a filter with larger pores. However, there are practical limitations to how porous a filter one can use and still support the monolayer. In some cases, if the pores are too large (e.g., >1–3 μm diameter), the cell can actually migrate through the filter and begin to grow on the basolateral side of the membrane. This results in a poorly defined monolayer and should be avoided whenever possible.

4. Medium or Buffer Composition

The solution or medium bathing the cells must be carefully tailored to the experimental design. Its composition must not only be conducive to maintaining normal cell function, it must also be compatible with the permeants being measured. Ionic strength, osmolarity, buffer type and buffer capacity, pH, and, occasionally, protein content are all important considerations. For example, wide deviations from physiological pH 7.4 for prolonged periods can be detrimental to the cells. However, it is sometimes possible to have differential pH conditions in the donor and receiver solutions. In Caco-2 cell monolayers, a donor solution of pH 4 is tolerated if the receiver solution pH is maintained near physiological pH (Hilgers and Burton, unpublished). This is because most nutrient uptake occurs and many vital transport systems function on the basolateral side of the cell monolayer bathed by the receiver solution. Manipulating the pH is crucial for examining changes in flux of a charged molecule, since pH changes will affect ionization (Section III.B.1). It is also possible that under extreme conditions the solution pH can affect pH at the cell surface; this might affect the partitioning of a molecule into the membrane.

Proteins can be added to the solutions to increase the solubility of a poorly water soluble compound, to decrease nonspecific binding to noncellular or cellular surfaces, or to study the effects of protein binding on permeant flux (Section VIII). Intestinal mucus, albumin, other serum proteins, or serum itself can be used depending upon the intended endpoint. It is also likely that adsorption of proteins to the cell surface, particularly to the region occupied by the paracellular space, alters transjunctional passive diffusion of permeants (Michel, 1988; Schnitzer, 1993). Other additives, such as suspending agents, polyethylene glycols, and cyclodextrin, also may be used after it is established that the additive has no effect on the barrier and solute flux. Keep in mind that any additive may decrease the thermodynamic activity of a compound by decreasing the unbound fraction in solution.

5. Stirring

Depending upon the design of the diffusion cell apparatus, several mechanisms for agitation of the donor and/or receiver solutions are described in the literature. All of them attempt to manipulate the dimensions of the aqueous boundary layer (ABL), which lies adjacent to the cell monolayer. In physical terms, the ABL is a region of unstirred, stagnant fluid through which the solute must diffuse in order to enter or leave the cell monolayer (Section II). Thus, as with the filter support, the ABL can be considered as another diffusion barrier in series with the cell monolayer. The impact of the stagnant layer on transport kinetics will depend upon the dimensions of the ABL and the solute transport rate through the monolayer. In general, the ABL is more significant for rapidly transported solutes, resulting in an underestimation of the actual transcellular flux of such solutes. The effect of stirring is to decrease the thickness of the ABL and thus increase the observed transport rate for such solutes.

Gas bubbling, orbital shakers, and magnetic stirrers have all been used to achieve agitation with different diffusion cell designs. The efficiency of mixing will have to be determined for each system configuration. Further, it is important to ascertain that the cell monolayer is not adversely affected by the process of agitation. Air lift, for example, has been reported to cause significant mechanical shear to certain types of cells depending upon the proximity of the gas stream and the bubbling rate. Again, some types of cells are more susceptible to mechanical stresses than others.

6. Solute Properties

Several examples have already been pointed out in which the properties of the solute itself can impact on the results obtained from a transport experiment. Metabolic instability and propensity for nonspecific adsorption are problems which can frequently be encountered and must be considered any time a new solute is to be studied. In addition to these problems, there are several other solute-related factors which must be considered in the design and interpretation of transport studies.

One of the first decisions to be made when designing an experiment is the method of detection to be used with a particular solute. If radiolabeled material is available, a simple method of analysis is to count the radiolabel appearing in the receiver compartment as a function of time. While convenient, this can be a dangerous practice. Depending upon the type of radioisotope, its position in the molecule, and its specific activity, radiolabeled compounds can be subject to a variety of chemical and solution-catalyzed degradation pathways. If the stock solution contains a significant amount of radioactive impurities or generates them as a result of solution instability, then the possibility for preferential transport of such impurities always exists. Good quality control requires that a specific chemical assay be used to prove that the radioactivity observed is indeed associated with the desired solute. A similar argument can be made for radiolabeled solutes which will be metabolized during the course of the experiment.

The same considerations will apply to other nonspecific methods of detection, such as fluorescence or UV absorbance determinations. Particularly with these methods, it must be appreciated that many of the cells used to form monolayers secrete a variety of products such as lipids and proteins into both the donor and receiver compartments. These substances can result in a variable background in solutions and may interfere with solute quantitation. Even if a chromatographic method is used with fluorescence or UV detection, these products can still interfere with the separation unless specifically accounted for.

Other important factors dictated by the solute are solubility and ionization state. If the compound has very limited solubility either intrinsically or at the experimental pH, it is frequently possible to do a quick calculation to determine if the experiment is even possible. That is, if the donor concentration is very dilute, one can estimate the receiver concentration which would be obtained for a given solute permeability coefficient and determine if it is within the limits of detection of the assay.

7. Assay Conditions

Several factors must be considered in the design and execution of a transport experiment in order to obtain useful quantitative information. In the simplest configuration, a solution of the solute of interest is added to the donor compartment of the diffusion cell (Fig. 4) and incubated for a period of time, with a solute-free buffer in the receiver compartment. At predetermined equivalent intervals, small aliquots are taken for analysis from the receiver and/or donor compartment. Appearance kinetics are determined from the steady-state rate of solute appearance in the receiver compartment or from the rate of disappearance from the donor compartment. Sampling intervals are usually chosen such that only a small amount of solute is transported across the monolayer during that period in order to maintain essentially constant donor concentration. This also minimizes the problem of "back diffusion" of the solute from the receiver compartment into the donor. In this way, sink conditions are maintained during the experiment. Also, equal fluid volumes are used to prevent hydrostatic pressure gradients in either direction. The impact of pressure gradients on solute flux increases as the leakiness of the cell monolayer increases. Depending upon the configuration of the diffusion cell apparatus, buffer may be added to replace the amount taken for sampling to keep the volumes constant. Finally, at the end of the experiment, a mass balance accounting is taken. If the total mass transferred into the receiver along with the mass remaining in the donor is significantly different from the initial mass in the stock solution, this may be evidence for cellular accumulation of the solute. Alternatively, significant adsorption to the diffusion cell apparatus or cellular metabolism of solute may have occurred during the course of the experiment. It is desirable to distinguish among these possibilities.

If adsorption to the apparatus is suspected, a blank experiment can be run with solute in the diffusion apparatus containing the filter but no monolayer. Any disappearance of the solute in this system is evidence for nonspecific adsorption. If cellular accumulation is suspected, techniques can be developed for harvesting and analyzing the cells for quantitation of accumulated solute. Finally, if significant mass still cannot be accounted for, metabolism may be a likely explanation. Positive confirmation of a metabolic pathway requires a specific assay for the suspected metabolite; such an assay may or may not be available. In any event, it is important to do a mass balance accounting for each new solute. A permeability coefficient calculated simply from a donor disappearance analysis or receiver appearance can be misleading if competing processes other than transport are also taking place. Finally, one can miss valuable, interesting information from an experiment by not being rigorous in the analysis. All of these situations are illustrated in the following sections.

Although temperature needs to be controlled, its effects on passive flux are most often subtle. Experiments usually are performed at physiological temperature (37°C) or at room temperature (22–25°C). As the temperature is decreased, the lipid phase transition of a biological membrane is approached. At low temperature, then, molecular interactions with the membrane may be altered greatly. Low temperatures are also known to affect the paracellular route, making it less leaky (González-Mariscal et al., 1984). One needs to be aware of these changes in passive flux, since low temperature (<18°C) is often used to inhibit transport pathways involving energy.

II. CELL MONOLAYER TRANSPORT KINETICS

A. Steady-State Kinetics Under Sink Conditions

The steady-state flux of drug solute across the cell monolayer–filter support system (Fig. 5) is

$$J = AP_e(C_D - C_R) \tag{1}$$

where

$$J = \text{flux, mass/sec}$$
$$A = \text{cross-sectional area, cm}^2$$
$$P_e = \text{effective permeability coefficient, cm/sec}$$
$$C_D, C_R = \text{concentrations in the donor and receiver, respectively}$$

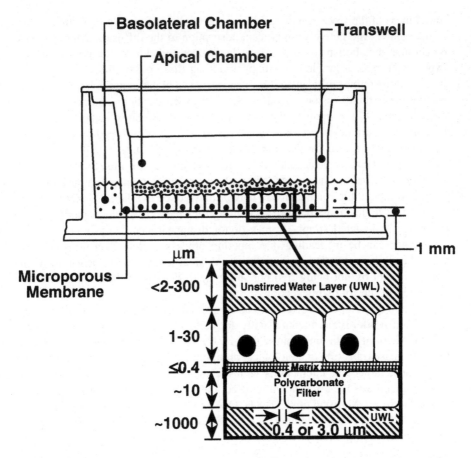

Figure 5 The Costar Transwell system with a cell monolayer grown on a porous poly-carbonate filter that is mounted onto a removable plastic insert forming the apical chamber. Two other systems, (1) the Costar diffusion chamber system, where a filter-grown (Snap-well) cell monolayer is sandwiched between two chambers of equal volume and the bathing solutions are agitated and/or gassed and (2) filter-grown cell monolayers mounted in a two-chamber rotating cylinder device (Imanidis et al., 1996), are not shown.

When sink conditions are imposed, $C_R \approx 0$ and

$$J = AP_e C_D \tag{2}$$

Requiring that the disappearance rate from the donor be equal to the appearance rate in the receiver, it follows that

$$J = -V_D \frac{dC_D}{dt} = V_R \frac{dC_R}{dt} \tag{3}$$

where V_D and V_R are the volumes in the donor and receiver, respectively. The absence of metabolism, avid cytosolic protein binding, significant membrane interaction, and active transport are assumed. The special case of metabolism is treated in Section VII, and that of highly membrane-interactive drugs in Section VIII. Active transport is not dealt with in this chapter; the reader is referred to Ho et al. (1995) for theoretical considerations.

With the initial donor concentration $C_D(0)$, the decrease in donor concentration with time is

$$C_D = C_D(0) \exp\left(-\frac{AP_e t}{V_D}\right) \tag{4}$$

or in terms of mass,

$$M_D = M_D(0) \exp\left(-\frac{AP_e t}{V_D}\right) \tag{5}$$

The concomitant increases in receiver concentration and mass with time are

$$C_R = \frac{V_D C_D(0)}{V_R}\left[1 - \exp\left(-\frac{AP_e t}{V_D}\right)\right] \tag{6}$$

and

$$M_R = M_D(0)\left[1 - \exp\left(-\frac{AP_e t}{V_D}\right)\right] \tag{7}$$

In general, the donor and receiver concentrations are exponential functions of time. It is only within the early time period when no more than 10–15% of $C_D(0)$ has been transported that the kinetics are essentially linear; hence, the truncated Maclaurin's expansion of Eqs. (5) and (7) leads to the linear relationships

$$\frac{M_D}{M_D(0)} = 1 - \frac{AP_e t}{V_D} \tag{8}$$

and

$$\frac{M_R}{M_D(0)} = \frac{AP_e t}{V_D} \tag{9}$$

from which the effective permeability coefficient (P_e) is readily calculated. When transport into the sink is complete, $M_D(\infty) = 0$ and $M_R(\infty) = M_D(0)$.

B. Steady-State Kinetics Under Non-Sink Conditions

Unlike the previous kinetics imposed by the sink condition, steady-state transport kinetics under non-sink conditions will lead to equilibrium partitioning between the aqueous phase of the donor and receiver compartments and the cell monolayer. In contrast to the sink condition wherein $C_R \approx 0$ at any time, under non-sink conditions C_R increases throughout time until equilibrium is attained. As previously stated in Eqs. (1) and (3), the rate of mass disappearing from the donor solution is

$$-V_D \frac{dC_D}{dt} = AP_e(C_D - C_R) \tag{10}$$

and the accompanying rate of appearance into the receiver is

$$V_R \frac{dC_R}{dt} = AP_e(C_D - C_R) \tag{11}$$

The general case is taken in which the volumes of the donor and receiver compartments are not equal $(V_D \neq V_R)$. Mass balance requires that the initial amount in the donor solution be equal to the sum of the mass throughout the closed system with time:

$$M_D(0) = M_D + M_{cell} + M_R \tag{12}$$

and by definition

$$V_D C_D(0) = V_D C_D + V_{cell} C_{cell} + V_R C_R \tag{13}$$

wherein the amount or concentration in the paracellular space is assumed to be negligible. When equilibrium is reached at $t = \infty$, the donor and receiver concentrations are equal, i.e., $C_D(\infty) = C_R(\infty)$. The partition coefficient between the cells and the aqueous solutions is

$$K = \frac{C_{cell}(\infty)}{C_D(\infty) + C_R(\infty)} = \frac{C_{cell}(\infty)}{2C_D(\infty)} = \frac{C_{cell}(\infty)}{2C_R(\infty)} \tag{14}$$

and is readily calculated as

$$K = \frac{V_D C_D(0) - (V_D + V_R)C_D(\infty)}{2V_{cell} C_D(\infty)} \tag{15}$$

At any time t, the partition coefficient is approximately

$$K = \frac{C_{\text{cell}}}{C_D + C_R} \tag{16}$$

Its substitution into Eq. (13) gives

$$C_D = \frac{V_D C_D(0) - (V_R + V_{\text{cell}} K) C_R}{V_D + V_{\text{cell}} K} \tag{17}$$

$$C_R = \frac{V_D C_D(0) - (V_D + V_{\text{cell}} K) C_D}{V_R + V_{\text{cell}} K} \tag{18}$$

The integration of Eqs. (10) and (11), with the use of Eqs. (17) and (18) and initial conditions of $C_D = C_D(0)$ and $C_R = 0$ at $t = 0$, results in expressions for the changes in concentration and mass with time. The kinetic decrease in drug concentrations and mass in the donor solution leading to equilibrium is

$$\frac{C_D}{C_D(0)} = \frac{M_D}{M_D(0)} = \frac{1 + (\beta - 1) \exp(-\alpha\beta t)}{\beta} \tag{19}$$

where

$$\alpha = \frac{AP_e}{V_R + V_{\text{cell}} K} \tag{20a}$$

$$\beta = \frac{V_D + V_R + 2V_{\text{cell}} K}{V_D} \tag{20b}$$

The accompanying kinetic increases of drug concentration and mass in the receiver solution are

$$\frac{C_R}{C_D(0)} = \frac{V_D}{V_R} \left[\frac{1 - \exp(-\alpha'\beta' t)}{\beta'} \right] \tag{21}$$

and

$$\frac{M_R}{M_D(0)} = \frac{1 - \exp(-\alpha'\beta' t)}{\beta'} \tag{22}$$

where

$$\alpha' = \frac{AP_e}{V_D + V_{\text{cell}} K} \tag{23a}$$

$$\beta' = \frac{V_D + V_R + 2V_{\text{cell}} K}{V_R} \tag{23b}$$

Because V_D and V_R were taken to be unequal for the general case, the disappearance rate constant $(\alpha\beta)$ is not equal to the appearance rate constant $(\alpha'\beta')$. However, the rate constants become identical when the volumes of the compartments are also identical.

At equilibrium, the asymptotic expressions in terms of concentration are

$$\frac{C_D(\infty)}{C_D(0)} = \frac{1}{\beta} \tag{24}$$

and

$$\frac{C_R(\infty)}{C_D(0)} = \frac{V_D}{V_R\beta'} \tag{25}$$

where it is seen that $C_D(\infty)/C_D(0)$ and $C_R(\infty)/C_D(0)$ are equal and less than unity.

During an early time period in which approximately less than 5% of the mass is transported, the kinetics are effectively linear. Subsequently, the linear approximations to Eqs. (19) and (22) are

$$\frac{M_D}{M_D(0)} = 1 - \frac{AP_e}{V_D}\left(\frac{V_R + 2V_{cell}K}{V_R + V_{cell}K}\right)t \tag{26}$$

and

$$\frac{M_R}{M_D(0)} = \frac{AP_e t}{V_D + V_{cell}K} \tag{27}$$

It is noteworthy that the partition coefficient is included in the apparent rate constants and may be ignored only when $V_{cell}K$ is much smaller than V_R and V_D. Under these conditions, the above equations reduce to Eqs. (8) and (9) for the sink situation.

C. Permeability Coefficients

The effective permeability coefficient is composed of the permeability coefficients for the various transport barriers in series—the ABLs, the cell monolayer, and the filter support:

$$\frac{1}{P_e} = \frac{1}{P_D} + \frac{1}{P_R} + \frac{1}{P_M} + \frac{1}{P_F} \tag{28}$$

where P_D, P_R = permeability coefficients of the ABL on the donor and receiver sides and P_M, P_F = permeability coefficients of the cell monolayer and the filter support, respectively.

Since the monolayer is composed of transcellular and paracellular pathways in parallel,

$$P_M = P_{cell} + P_{paracell} \tag{29}$$

The permeability coefficients, P_D and P_R, are influenced by hydrodynamics. Depending upon the geometric symmetry or asymmetry of stirring in the donor and receiver chambers, their values may be equal or unequal. To analyze these situations, let us define P_{ABL} as the effective permeability coefficient of the ABLs; therefore, the geometric average of the mass transfer resistance of the ABLs is

$$\frac{1}{P_{ABL}} = \frac{1}{P_D} + \frac{1}{P_R} = \frac{P_D + P_R}{P_D P_R} \tag{30}$$

In the cell monolayer/Transwell system (Fig. 5), agitation is usually accomplished by using some type of orbital mixer. Under these conditions the degree of stirring on the donor side is greater than that in the receiver, where the stirring is significantly dampened. If $P_D > P_R$ by tenfold or more, Eq. (23) reduces to the limiting expression,

$$P_{ABL} \simeq P_R \tag{31}$$

For the gas-lift, side-by-side diffusion cell system (Hidalgo et al., 1991) in which the effects of stirring on the ABLs in the donor and receiver chambers are equivalent,

$$P_{ABL} = P_D/2 \tag{32}$$

Consequently, when the ABLs exert a significant influence on the overall transport kinetics, the symmetrical cell monolayer–filter system gives one the strategic advantage of quantitative control of the hydrodynamics. If the kinetics are controlled by the cell monolayer, then the choice of one transport system design over the other is inconsequential.

It is assumed that the convective flow of water across the ABL, cell monolayer, and filter owing to pressure gradients is negligible and that the cell monolayer is uniformly confluent. When these conditions are not met, Katz and Schaeffer (1991) and Schaeffer et al. (1992) point out that mass transfer resistances of the ABL and filter [as described in Eq. (21)] cannot be used simply without exaggerating the permeability of the cell monolayer, particularly the paracellular route. An additional diffusion cell design was described by Imanidis et al. (1996).

III. PARACELLULAR TRANSPORT KINETICS

A. Morphology of the Paracellular Route

The paracellular route is the pathway of diffusion of a permeant between cells growing in a monolayer. The ability of solutes to move through this space is limited by the presence of the zonula occludens, also called the occluding or tight junction (Fig. 6). The tight junction is a region where the outer leaflets of the lipid bilayer which make up the plasma membrane of neighboring cells are fused. In three dimensions, this region encircles the cell near the apical surface, connecting all adjoining cells much like the plastic rings that hold a six-pack of beverage cans together (Diamond, 1977). The tight junction is comprised of anastomosing "strands" which are actually continuous chains of membrane proteins, the composition of which is largely unknown (Fig. 7). It is hypothesized that this protein chain or "fence" contains "gates" of given dimensions. These gates open and close or appear and disappear, creating a dynamic sieve that regulates molecular diffusion. It is thought that these transient gates, which most likely represent the "pores" described in biophysical models of transjunctional permeability, have a

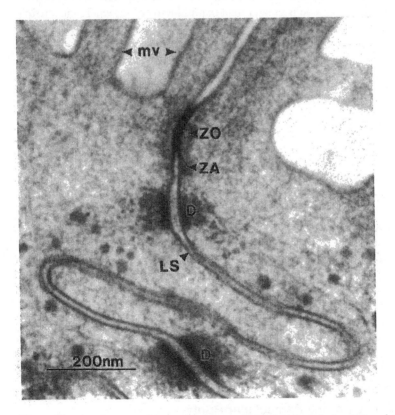

Figure 6 An electron micrograph of intercellular junctions between two human colon Caco-2 cells in culture. D, desmosome; LS, lateral space; mv, microvilli; ZA, zonula adherens; ZO, zonula occludens (i.e., tight junction). Bar equals 200 nm.

A.   **B.**

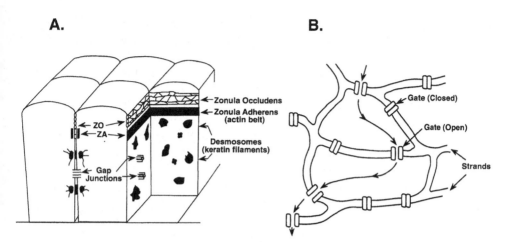

Figure 7 (A) Spatial orientation of intercellular junctions within an epithelial cell monolayer showing how the zonula adherens (tight junction) surrounds each cell. (B) Hypothetical "pores" that comprise the paracellular pathway are represented as dynamic "gates"
within the anastomosing strands of the zonula occludens. [Redrawn from Gumbiner (1987)
and Cereijido (1992) with permission from the publishers.]

fixed net negative charge, thus further conveying an electrical selectivity (Section
III.B.1). The number and complexity of the strands, the number of "gates," and
the fraction that are open at any moment (e.g., porosity) all delineate the path
length which a molecule must travel (Fig. 7).

Some of the proteins that compose tight junctions and their interactions in
the assembly and regulation of tight junctions have only recently been identified
and have been reviewed elsewhere (Lutz and Siahaan, 1997; Denker and Nigam,
1998; Mitic and Anderson, 1998). Only one of these proteins, occludin, is a transmembrane protein that is a functional component of the paracellular pathway.
The number of studies on tight junction regulation or control of the paracellular
pathway has increased with the advent of cell culture models (Section III.C.2).
Pharmacological perturbation of the paracellular pathway has spawned the pursuit
of adjuvants or permeation enhancers which would increase drug delivery
through such formidable barriers as the blood/brain interface.

The tight junction is a component of the junctional complexes which join
cells. Immediately basolateral to the tight junction is the zonula adherens (Figs.
6 and 7). Because the zonula adherens and the gap junctions are focal contact
regions, they do not impact transport by the paracellular pathway. All of these
junctions are specialized regions of the lateral cell membrane which demarcate
the lateral space. In certain types of cells the lateral space is rather narrow and

highly tortuous, whereas in others it is wide, except at junctional regions, and less regular. Passive diffusion through the lateral space will be more influenced in the former configuration (Section III.C.3). Changes in the dimension of the lateral space as a result of changes in cell shape have been shown to influence transjunctional flux (Diamond, 1977; Madara, 1983).

It is best to measure the barrier function of tight junctions using two methods. One is the transjunctional flux of membrane-impermeant molecules that are not actively transported. This approach is not in itself indicative of the formation of tight junctions, since flux would also be reduced significantly if cells separated by relatively large gaps occupied a majority of the surface area or if the cells grew in multiple layers. Therefore, TER is also measured (Section III.C.4). This is the resistance to passage of direct current represented by Ohm's law. In most cases, the pathway of least resistance is through the paracellular route. Because electric current is carried by ions in electrolyte solutions, TER can also be a measure of transcellular ion permeation. To differentiate between these two pathways or relative changes in their importance following a stimulus requires more sophisticated electrophysiological or flux studies (Madara and Hecht, 1989).

B. Biophysical Model Description

1. Molecular Restricted Diffusion Within an Electrostatic Field of Force

Cation and anion flux across cultured cell monolayers by molecular restricted diffusion within an electrostatic field of force across aqueous pores has been described with a model derived by Adson et al. (1994). The ion fluxes per cross-sectional area of the cell monolayer are defined as

$$J^+ = -\varepsilon D^+ F\left(\frac{r}{R}\right)\left(\frac{dC}{dx} + \frac{z^+ e}{kT}\frac{d\psi}{dx}C\right) \tag{33}$$

$$J^- = -\varepsilon D^- F\left(\frac{r}{R}\right)\left(\frac{dC}{dx} + \frac{z^- e}{kT}\frac{d\psi}{dx}C\right) \tag{34}$$

where $x \geq 0$ and

$J^+, J^- = $ flux per unit area of cationic and anionic species, respectively

$\varepsilon = $ porosity or volume fraction of aqueous pores

$F(r/R) = $ molecular sieving or Renkin function, dimensionless

$e = $ unit charge of an ion, 4.8×10^{-10} esu

$kT = $ thermal energy, erg

$z^+, z^- = $ valence of cation and anion, sign included

D^+, D^- = aqueous diffusion coefficients of the cation and anion, respectively, cm^2/sec

dC/dx = concentration gradient of cationic or anionic species

$d\Psi/dx$ = electrical potential gradient, sign included, statvolt/cm

The Renkin function is the dimensionless molecular sieving function for cylindrical channels (Renkin, 1954; Curry, 1984) and is defined as

$$F\left(\frac{r}{R}\right) = \left(1 - \frac{r}{R}\right)^2 \left[1 - 2.104\frac{r}{R} + 2.09\left(\frac{r}{R}\right)^3 - 0.95\left(\frac{r}{R}\right)^5\right] \tag{35}$$

The Renkin function compares the molecular radius (r) with the pore radius (R) where $0 < F(r/R) < 1.0$. With the fixed charges of the pore negative and assuming a linear potential gradient where the sign of the potentials and valences of the ions are accounted for, one gets

$$J^+ = -\varepsilon D F\left(\frac{r}{R}\right)\left(\frac{dC}{dx} - \frac{\kappa}{\delta}C\right) \tag{36}$$

$$J^- = -\varepsilon D F\left(\frac{r}{R}\right)\left(\frac{dC}{dx} + \frac{\kappa}{\delta}C\right) \tag{37}$$

where the dimensionless electrochemical energy function κ across the pore of length δ is

$$\kappa = \frac{ez|\Psi_\delta - \Psi_0|}{kT} = \frac{ez|\Delta\Psi|}{kT} \tag{38}$$

and

$$z = |z^+| = |z^-| \tag{39}$$

$$D = D^+ = D^- \tag{40}$$

The aqueous diffusivities of charged permeants are equivalent to those of uncharged species in a medium of sufficiently high ionic strength. The product $DF(r/R)$ is the effective diffusion coefficient for the pore. It is implicit in κ that adsorption of the cations does not occur, so that the fixed surface charges on the wall of the pore are not neutralized. Adsorption is more likely to occur with multivalent cations than with univalent ones.

The Stokes–Einstein equation gives a good approximation of the molecular radius (r) for spherical or nearly spherical molecules:

$$D = \frac{kT}{6\pi\eta r} \tag{41}$$

Conversely, the diffusion coefficient may be estimated as

$$D = \frac{kT}{6\pi\eta} \left(\frac{4\pi N_{AV}\rho}{3M} \right)^{1/3} \tag{42}$$

where

η = viscosity of the liquid medium, poise
N_{AV} = Avogadro's number, 6.02×10^{23} molecules/mol
ρ = density of the molecule, g/mL
M = molecular weight
r = molecular radius, cm

The differential equations (36) and (37) are solved by the integrating factor method, with the boundary condition $C = C_0$ at $x = 0$.

Consequently, defining the permeability coefficients as

$$J^+ = P^+ C_0 \tag{43}$$
$$J^- = P^- C_0 \tag{44}$$

and requiring $C = 0$ at $x = \delta$ gives the permeability coefficients of the ions as

$$P^+ = \frac{\varepsilon D \, F(r/R)}{\delta} \left(\frac{\kappa}{1 - e^{-\kappa}} \right) \tag{45}$$

$$P^- = \frac{\varepsilon D \, F(r/R)}{\delta} \left(\frac{\kappa}{e^{\kappa} - 1} \right) \tag{46}$$

In the limit of $\kappa = 0$, the permeability coefficient of the neutral molecule is

$$P = \frac{\varepsilon D \, F(r/R)}{\delta} \tag{47}$$

One observes, comparing Eqs. (45) and (46) for permeating ions of comparable size and valence with Eq. (47), that the diffusion of cations across the negatively charged pore is increased by the potential gradient. In contrast, the diffusion of anions is decreased by the electrical forces. In other words, P^+ or P^- is composed of the permeability coefficient of its neutral image upon which the contribution of molecular charge with respect to the negative electric field of the pore is super-imposed.

2. Barriers in Series: Tight Junction and Lateral Space

The paracellular pathway consists of the tight junction (TJ) in series with the tortuous lateral space (LS) (Figs. 6 and 7), i.e., mass transfer resistances in series:

$$\frac{1}{P_{\text{paracell}}} = \frac{1}{P_{\text{TJ}}} + \frac{1}{P_{\text{LS}}} \tag{48}$$

When arranged in the form

$$P_{\text{paracell}} = \frac{P_{\text{TJ}}}{1 + P_{\text{TJ}}/P_{\text{LS}}} \tag{49}$$

it is evident that the tight junction becomes the rate-determining barrier when $P_{\text{LS}} > P_{\text{TJ}}$ by at least tenfold. Since the permeability coefficients are better esti-mated with the use of neutral solutes,

$$P_{\text{TJ}} = \frac{\varepsilon D\, F(r/R)}{\delta} \tag{50}$$

$$P_{\text{LS}} = \frac{\varepsilon' D\, G(r/R')}{\tau \delta'} \tag{51}$$

and

$$\frac{P_{\text{TJ}}}{P_{\text{LS}}} = \left(\frac{\varepsilon}{\delta}\right)_{\text{TJ}} \left(\frac{\tau \delta'}{\varepsilon'}\right)_{\text{LS}} \left(\frac{F(r/R)}{G(r/R')}\right) \tag{52}$$

The paths of the tight junctions and lateral spaces are assumed to be cylin-drical pores and slits, respectively. The Renkin function $F(r/R)$, for the tight junction is defined by Eq. (35), but the function $G(r/R')$ for the lateral space (Curry, 1984) is

$$G\left(\frac{r}{R'}\right) = \left(1 - \frac{r}{R'}\right)\left[1 - 1.004\frac{r}{R'} + 0.418\left(\frac{r}{R'}\right)^3 \right.$$
$$\left. + 0.21\left(\frac{r}{R'}\right)^4 - 0.17\left(\frac{r}{R'}\right)^5\right] \tag{53}$$

where R' = one-half gap width of the slit. The porosity ε' and apparent path length δ' of the lateral space are distinguished from those of the tight junction by the primes, while the tortuosity (τ, unitless) is measured as described in Table 8 (Section III.C.3).

If one desires to assess the relative significance of the tight junction and lateral space for cationic or anionic permeants, charge effects may be superimposed in Eqs. (50) and (51) with Eq. (52) remaining approximately the same.

C. Applications

1. Hydrophilic Extracellular Permeants

The barrier properties of the paracellular route have been assessed using radiolabeled hydrophilic solutes varying in molecular size and charge (Adson, 1992) (Table 4). Kinetics of appearance of these extracellular permeants across the Caco-2 cell monolayer–Transwell system into the receiver maintained at sink conditions are linear (Fig. 8). Effective permeability coefficients P_e, calculated using Eq. (9), are on the order of 10^{-6} cm/sec (Table 5). The kinetics were not affected by stirring, and the calculated permeability coefficient of the collagen-free filter support ($\varepsilon = 0.15$; $R_F = 1.5$ μm; $h_F = 10$ μm) is at least 100-fold larger than P_e, indicating that the transport kinetics are limited by passage of the solutes across the monolayer via the paracellular route.

The permeability coefficients of urea and mannitol decrease with increasing molecular size (Table 5). Similar observations are made for the protonated methylamine and atenolol and between the acetate and hippurate anions. By comparing permeability coefficients among permeants of like charges, e.g., urea and mannitol, or acetate and hippurate, one observes

$$\frac{r_x P_x}{rP} = \frac{F(r_x/R)}{F(r/R)} \tag{54}$$

Table 4 Physicochemical Properties of Paracellular Permeants

Permeant	MW	Density (g/mL)	pK_a	log PC (n-octanol, pH 7.4)	$D \times 10^{-6}$ cm/sec (25°C)[a]	Molecular radius[a] (Å)
Urea	60	1.32	Neutral	−1.09	9.16	2.67
Mannitol	182	1.52	Neutral	<0	6.3	4.10
Acetate	60	1.05	4.76	−0.17	8.48	2.88
Hippurate	179	1.37	3.64	<0	6.44	3.79
D-PheGly	222	1.0	Amphoteric	−2.16	5.40	4.52
D-Phe$_2$Gly	369	1.0	Amphoteric	−1.46	4.56	5.35
D-Phe$_3$Gly	516	1.0	Amphoteric	−0.66	4.08	5.99
Methylamine	31	0.7	10.65	−0.57	9.23	2.65
Atenolol	266	1.0	9.55	−0.2	5.08	4.80

[a] Aqueous diffusion coefficient calculated using Eqs. (41) and 42).
Note: PC = partition coefficient.

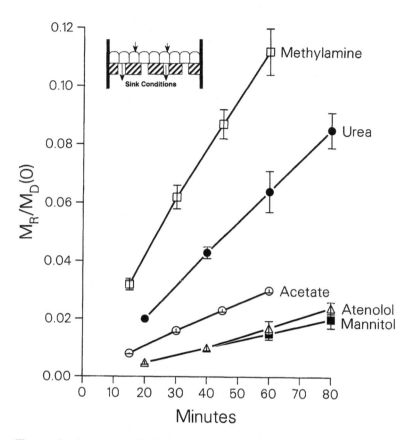

Figure 8 Appearance kinetics of radiolabeled solutes that diffuse across Caco-2 cell monolayers via the paracellular pathway. The Transwell system consisted of a donor and receiver solution at pH 7.4. Stirring by planar rotation up to 100 rpm had no effect. The insert with filter, cell monolayer, and donor were transferred to a new receiver chamber at time intervals to maintain sink conditions.

The permeability coefficients and molecular radii are known. The effective pore radius, R, is the only unknown and is readily calculated by successive approximation. Consequently, unknown parameters (i.e., porosity, tortuosity, path length, electrical factors) cancel, and the effective pore radius is calculated to be 12.0 ± 1.9 Å. Because the Renkin function [see Eq. (35)] is a rapidly decaying polynomial function of molecular radius, the estimation of R is more sensitive to small uncertainties in the calculated molecular radius values than it is to experimental variabilities in the permeability coefficients. The placement of the permeants within the molecular sieving function is shown in Figure 9 for the effective

Table 5 Permeability Coefficients of the Paracellular Pathway and Estimation of the Effective Pore Radius and Molecular Restriction Factor for the Caco-2 Cell Monolayer

Permeant	Molecular radius (Å)	$P_{paracellular} \times 10^{-6}$ cm/sec ($\pm$ SD)	$F(r/R)$	Effective pore radius (Å)
Urea	2.67	4.26 (0.34)	0.299	9.6
Mannitol	4.10	1.17 (0.20)	0.126	
Acetate	2.88	1.66 (0.09)	0.224	10.0
Hippurate	3.79	0.67 (0.002)	0.119	
D-PheGly	4.52	1.06 (0.28)	0.143	12.5
D-Phe$_2$Gly	5.35	0.55 (0.11)	0.088	13.4
D-Phe$_3$Gly	5.99	0.36 (0.07)	0.070	
Methylamine	2.65	6.90 (0.43)	0.424	14.3
Atenolol	4.80	1.50 (0.57)	0.173	

Mean $\pm$ SD = 12.0 $\pm$ 1.9Å.

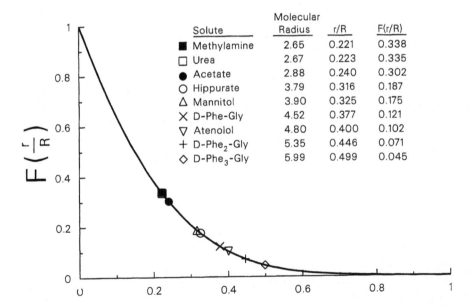

Figure 9 Molecular restriction factor as a function of the ratio of molecular radius to pore radius for paracellular permeants in Caco-2 cell monolayers. Mean pore radius is 1.2 nm.

pore size of 12 Å. Using Eq. (47) and $F(r/12 \text{ Å})$ for urea and mannitol, the average value for ε/δ is found to be 1.22 cm^{-1}. For example, for mannitol,

$$\frac{\varepsilon}{\delta} = \frac{P}{DF(r/R)} = \frac{1.17 \times 10^{-6}}{(6.3 \times 10^{-6})\, F(4.1/12)} = 1.06 \text{ cm}^{-1}$$

and likewise, for urea, $\varepsilon/\delta \approx 1.38$ cm^{-1}.

The plot of permeability coefficient versus molecular radius in Figure 10 shows the interdependence of molecular size and electric charge. The permeability of the solutes decreases with increasing size. The protonated amines permeate the pores faster than neutral solutes of comparable size, and the anions of weak acids permeate the pores at a slower rate. The transport behavior of the ionic permeants is consistent with a net negatively charged paracellular route. These results are phenomenologically identical to those found in the transport kinetics of

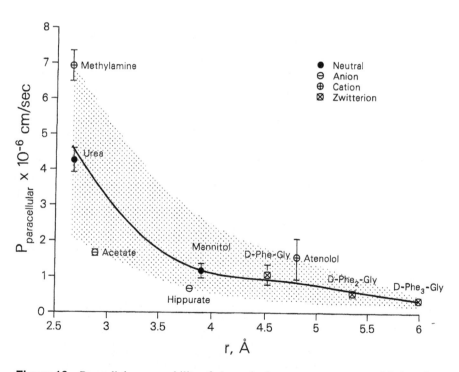

Figure 10 Paracellular permeability of charged solutes as a consequence of their molecular radius illustrates the mechanism of molecular restricted diffusion across negatively charged pores. Solid line depicts the curve drawn through the permeability coefficients of neutral solutes, and the neutral image of positively and negatively charged permeants.

neutral, anionic, and cationic permeants across a water-filled, negatively charged, porous matrix such as the collagen cuticle of the roundworm *Ascaris suum* (Ho et al., 1990).

The quantitative role of electrical factors affecting the transport of charged molecules is obtained by comparing permeability coefficients with the permeability coefficient $P^*_{paracell}$ for molecular size–restricted diffusion independent of the charge on the molecule (i.e., the neutral image). With Eqs. (45) and (46) one obtains

$$P^*_{paracell} = \frac{\varepsilon D \, F(r/R)}{\delta} \tag{55}$$

$$\frac{P^-_{paracell}}{P^*_{paracell}} = \frac{\kappa}{e^\kappa - 1} \tag{56}$$

$$\frac{P^+_{paracell}}{P^*_{paracell}} = \frac{\kappa}{1 - e^{-\kappa}} \tag{57}$$

where $P^*_{paracell}$ is determined by calculation. The value of ε/δ is 1.22 cm^{-1} for Caco-2 cell monolayers grown on Transwell filters, and the values for D and $F(r/R)$ are found in Tables 4 and 5. The dimensionless electrochemical energy parameter, κ, and potential drop across the paracellular barrier, $|\Delta\psi|$, are essentially identical among the cationic and anionic permeants with a single charge (Table 6). The $|\Delta\psi|$ estimate of 17.7 mV is a reasonable value. Note that the

Table 6 Estimations of Electrical and Molecular Size Restriction Factors Influencing the Paracellular Transport of Charged Permeants

Charged permeant	$P^\pm_{paracell}$ ($\times$ 10^{-6} cm/sec)	$P^*_{paracell}$ ($\times$ 10^{-6} cm/sec)	Ratio $P^\pm/P^*$	κ	$|\Delta\Psi|$ (mV)
Anionic					
Acetate	1.66	2.31	0.7186	0.63	16.2
Hippurate	0.67	0.93	0.7204	0.62	15.9
Cationic					
Methylamine	6.90	4.78	1.4435	0.78	20.0
Atenolol	1.50	1.07	1.4019	0.73	18.8
Zwitterionic					
D-PheGly	1.06	0.94	1.1277	—	—
D-Phe$_2$Gly	0.55	0.49	1.1224	—	—
D-Phe$_3$Gly	0.37	0.35	1.0571	—	—

$P^*_{paracell}$ = permeability coefficient accounting for molecular size restricted diffusion independent of the charge on the molecule; calculated from Eq. (47).

negative charge alone on the acetate and hippurate ions is responsible for the 30% reduction in the permeability coefficient expected for molecular restricted diffusion. The positive charge is responsible for an approximately 40% increase in molecular restricted diffusion. Within the uncertainties of the calculations, it is interesting that the permeabilities of the zwitterionic D-PheGly peptides, whose terminal protonated amine and carboxylate groups present discrete positive and negative charges to the pores, are identical to that expected for neutral permeants. This is because molecular size sieving by the pores becomes an increasingly important discriminating factor with large molecules and, in the limit, dominates over the influence of the electric field. Electric field effects play a larger role when the permeating ions are small, e.g., methylamine.

2. Pore Size Changes and Pharmacological Perturbation

As mentioned in Section III.A, little is known about the development of tight junctions and their regulation. However, much evidence exists for the roles of extra- and intracellular signals in controlling tight junction integrity including changes with physiological relevance (Balda, 1992; Bentzel et al., 1992; Contreras et al., 1992; Polak-Charcon, 1992). For example, changes in permeation of the tight junction occur to regulate transcellular diffusion of nutrients, etc. It is this control of tight junction functionality that has gained the attention of pharmaceutical scientists for promoting the penetration of drugs. In fact, it may well be that many so-called drug absorption promoters or enhancers affect tight junctions. There appears to be a direct link between cytoskeletal elements and tight junctions, coupled through phosphorylation–dephosphorylation cascades (Denker and Nigam, 1998; Madara, 1998). The most obvious mediator is calcium; both intracellular and extracellular calcium concentrations are important. The broad activities of many unrelated effectors are attributed to their indirect effect on intracellular calcium. For example, the binding of a molecule to a cell surface receptor may result in a secondary message that ultimately leads to regional changes in calcium levels.

In this section, we show how the pathway of solute diffusion can be studied by perturbing tight junctions and therefore the paracellular route. Treatment of cell monolayers with cytochalasin D, as perturbant of actin microfilaments, results in increased transmonolayer flux of solutes. Morphological studies indicate that cytochalasin D causes condensation of the perijunctional contractile ring that is associated with the tight junction (Madara et al., 1986). This, in turn, alters the structure of the tight junction and the "pores" that make up the paracellular route. Here, we show how one can quantitate these changes in the paracellular pathway.

Figure 11 demonstrates the change in the fluxes of mannitol and sucrose

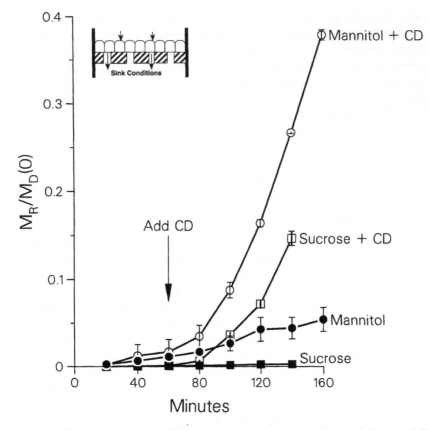

Figure 11   Change in paracellular permeation of sucrose and mannitol across MDCK cell monolayers following addition of 1 μg/mL cytochalasin D.

across the paracellular pathway of the Madin–Darby canine-kidney (MDCK) cell monolayer following the addition of cytochalasin D. There is a 79-fold increase in perturbed monolayer P_{paracell} for sucrose over control and a 12-fold increase for the smaller mannitol molecule (Table 7). From another point of view, the permeability coefficient of mannitol is 23-fold larger than sucrose in the unperturbed monolayer (control) and about 3.5-fold larger in the cytochalasin D–perturbed monolayer.

   To gain insights into the interrelationships of permeability coefficients, let us first determine the pore radius R of the unperturbed monolayer using mannitol (m), sucrose (s), and Eqs. (35) and (50).

Table 7 Permeability Coefficients of the Paracellular Route of Unperturbed and Cytochalasin D–Perturbed MDCK Cell Monolayers at 25°C

Permeant	D (cm²/sec)	Molecular radius (Å)	$P_{paracell}$ ($\times$ 10⁻⁶ cm/sec)		Fold increase
			Control	Perturbed	
Mannitol	6.3×10^{-6}	4.1	2.5	31.0	12.4
Sucrose	4.5×10^{-6}	5.55	0.11	8.7	79.1

$$\frac{F(r_m/R)}{F(r_s/R)} = \frac{r_m P_m}{r_s P_s}$$

$$\frac{F(4.1/R)}{F(5.55/R)} = \frac{4.1}{5.55}\left(\frac{2.5 \times 10^{-6}}{0.11 \times 10^{-6}}\right) = 16.8$$

By successive approximation, $R = 6.1$ Å, and, in turn, $F(4.1/6.1) = 0.0097$ for mannitol and $F(5.55/6.1) = 0.0005$ for sucrose. Next, the pore radius R^* for the perturbed monolayer is calculated using mannitol:

$$\frac{F(r_m/R^*)}{F(r_m/R)} = \frac{P_m^*}{P_m}$$

$$\frac{F(4.1/R^*)}{F(4.1/6.77)} = \frac{31.0 \times 10^{-6}}{2.5 \times 10^{-6}} = 12.4$$

where $R^* = 10.9$ Å. Similarly, with sucrose, $R^* = 11.2$ Å. Between the two close estimates, the average R^* is equal to 11.0 Å. It is assumed in these calculations that ε/δ between the perturbed and nonperturbed monolayers has not changed appreciably.

Since $P_{paracell}$ is proportional to the molecular sieving function $F(r/R)$, the interrelationships of $P_{paracell}$ between mannitol and sucrose, including mannitol–mannitol and sucrose–sucrose, can be put into perspective via a normalized plot of $F(r/R)$ versus r/R for control and perturbed monolayers (Fig. 12). As pointed out before, $F(r/R)$ is a rapid, monotonically decreasing function bounded by 1.0 and zero. One observes that mannitol is less restricted by the pores of the control monolayer than the larger sucrose molecule. However, in the larger pores of the perturbed monolayer, the increase in permeability is less for mannitol than it is

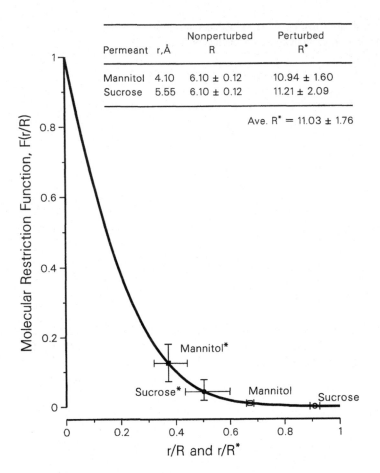

Figure 12 Molecular restriction factor as a function of the ratio of molecular radius to pore radius for mannitol and sucrose flux across MDCK cell monolayers that were untreated or treated with 1 μg/mL cytochalasin D (see Fig. 11).

for sucrose. In summary, there is a loss of molecular size selectivity as the pore dimension increases.

3. Barrier Assessment of the Tight Junction and Lateral Space

To estimate the relative importance of the tight junction and the lateral space composing the paracellular route, let us consider the permeability of mannitol across Caco-2 and MDCK cell monolayers. The results taken from earlier examples are presented below:

		Caco-2	MDCK
Permeability	$P_{paracell}$	1.17×10^{-6} cm/sec	1.17×10^{-6} cm/sec
Pore radius	R	12 Å	6.1 Å
Effective porosity to path length ratio	ε/δ	1.22 cm^{-1}	40.9 cm^{-1}

Interestingly, the permeability coefficients of mannitol in the two cell types are identical, most probably for different reasons, since the physical dimensions of the Caco-2 and MDCK monolayers (Table 8) are markedly different. Compared to the MDCK cell monolayer, the Caco-2 cell monolayer has a taller cell height, a shorter length in tight junctions, longer tortuous path lengths, and smaller slit width in lateral space. One recognizes that

$$P_{paracell} = \frac{1}{\dfrac{1}{P_{TJ}} + \dfrac{1}{P_{LS}}} = 1.17 \times 10^{-6} \text{ cm/sec}$$

for the Caco-2 and MDCK cell monolayers. Although the effective radii of the pores, calculated previously using the cylindrical pore model, are 12 Å for Caco-2 and 6.1 Å for MDCK, these values are functional estimates of the geometrical average of the small pore radius of the tight junction and larger gap width of the lateral space.

Table 8 Physical Dimensions of Caco-2 and MDCK Cell Monolayers

Dimensions[a]	Caco-2	MDCK
Cell, ave. height (μm)	25	7
Ave. width (μm)	6	14
TJ, length (μm)	0.1	0.3
LS, width (Å)	80–160	150–800
Tortuosity[b]	2.0	1.25
Length[c] (μm)	50	8.4
LS/TJ ratio	500	28

[a] TJ, tight junction; LS, lateral space.
[b] Tortuosity is the tortuous length of the lateral space divided by the height of the cell. All physical dimensions are measured by electron microscopy using transverse sections of cell monolayers.
[c] Calculated as (cell height − TJ length) × tortuosity.

Let us attempt to delineate the permeability coefficients of the tight junction and lateral space. Reintroducing Eq. (51), the permeability coefficient of the lateral space is

$$P_{LS} = \frac{\varepsilon' D \, G(r/R')}{\tau \delta'}$$ (58)

Where the terms have been previously described. As a first approximation, we have assumed here that the lateral space is a rectangular slit between cuboidal cells. The porosity, the fractional area of the cell monolayer in the paracellular space, is calculated as

$$\varepsilon' = 1 - \left(\frac{w}{w + 2R'}\right)^2$$ (59)

where w = width of the cuboidal cell (Å) and R' = one-half the slit width (Å).

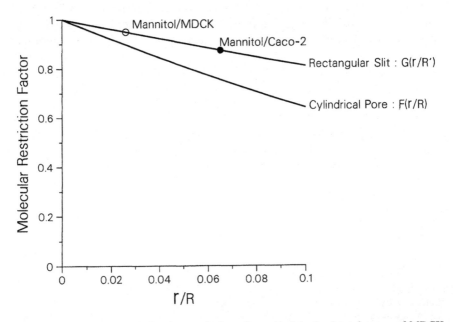

Figure 13 Estimated molecular restriction of mannitol in the lateral space of MDCK and Caco-2 cell monolayers. The lateral space was assumed to be a rectangular slit. Molecular restriction functions for slits and cylindrical pores are put into perspective.

Using the values in Table 8, the porosities are

Caco-2:

$$\varepsilon' = 1 - \left(\frac{6 \times 10^4}{6 \times 10^4 + 140}\right)^2 = 2.5 \times 10^{-3}$$

MDCK:

$$\varepsilon' = 1 - \left(\frac{14 \times 10^4}{14 \times 10^4 + 400}\right)^2 = 2.85 \times 10^{-3}$$

Hence, approximately 0.2% of the monolayer surface is available for paracellular diffusion. The molecular restriction factor (Fig. 13) is $G(4.1/70) \approx 0.886$ for Caco-2 and $G(4.1/200) \approx 0.959$ for MDCK. It follows that for Caco-2,

$$P_{LS} = \frac{(2.5 \times 10^{-3})(6.3 \times 10^{-6}) \times G(4.1/70)}{50 \times 10^{-4}} = 2.79 \times 10^{-6} \text{ cm/sec}$$

and for MDCK,

$$P_{LS} = \frac{(2.85 \times 10^{-3})(6.3 \times 10^{-6}) \times G(4.1/200)}{8.4 \times 10^{-4}} = 20.5 \times 10^{-6} \text{ cm/sec}$$

Upon dividing P_{paracell} by P_{LS}, we estimate that the Caco-2 paracellular transport of mannitol is 40% controlled by the long tortuous lateral spaces and the remaining 60% by the small pore radii of the tight junction. In the MDCK monolayer, diffusion through the relatively short and wide path of the lateral space contributes little to P_{paracell} (5%); the tight junctions dominate the paracellular kinetics to the extent of 95% (Table 9).

Table 9 Theoretical Assessment of Lateral Space and Tight Junction Contributions to the Paracellular Permeability of Mannitol in Caco-2 and MDCK Cell Monolayers[a]

Cell monolayer	Experimental	Predicted		Extent controlled by	
	P_a	P_{L3}	P_{TJ}	LS	TJ
Caco-2	1.17	2.79	2.02	40%	60%
MDCK	1.17	20.5	1.24	5%	95%

[a] Units of permeability coefficients are 10^{-6} cm/sec. LS, lateral space; TJ, tight junction.

4. Correlation of Paracellular Permeability with Transmonolayer Electrical Resistance

In Section III.A, we discussed the use of permeant flux and TER to measure the quality of the barrier formed by a cell monolayer. Madara and Hecht (1989) tell us that the barrier is a circuit of parallel resistors, i.e., tight junctions. The total resistance is the inverse sum of the reciprocals:

$$\frac{1}{R_t} = \frac{1}{R_1} + \frac{1}{R_2} + \frac{1}{R_3} + \cdots + \frac{1}{R_n} \tag{60}$$

The TER measures the transjunctional flux of ions that are much smaller than the tight junction dimensions. Consequently, a small population of leaky tight junctions would dominate permeability. In contrast, permeants that are commonly used for flux studies are close to the dimensions of the tight junction opening, and their diffusion is more a measured average. This relationship is demonstrated well under conditions where tight junctions are perturbed or are undergoing changes. When only a few tight junctions respond or the dimensions

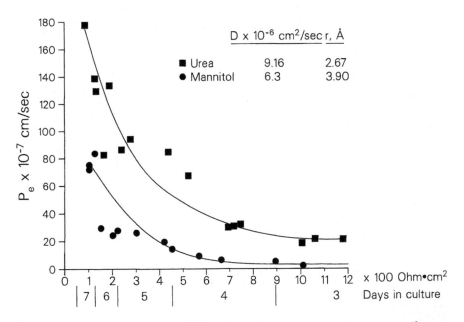

Figure 14   Observed permeability coefficients of urea and mannitol across monolayers of rat alveolar epithelial cells in primary culture in the Transwell system are correlated with transepithelial electrical resistance and days in culture.

change slightly yet uniformly, TER will change dramatically but tracer flux will change only slightly. As the number of affected tight junctions increases or the dimensional changes increase, the effects on TER become less dramatic and the effect on tracer flux becomes more significant.

These changes are illustrated in Figure 14, which shows solute permeability as a function of TER. Here, rat alveolar epithelial cells are plated onto filters at confluence, since they do not divide in culture. As a result, the monolayers begin at their maximal tightness 3 days after plating, and barrier integrity begins to decline after 4–5 days in culture. For other cell culture systems, the cells are plated at subconfluence and are allowed to form tight monolayers with time in culture (Fig. 15). In either case, the relationship between TER and solute perme-

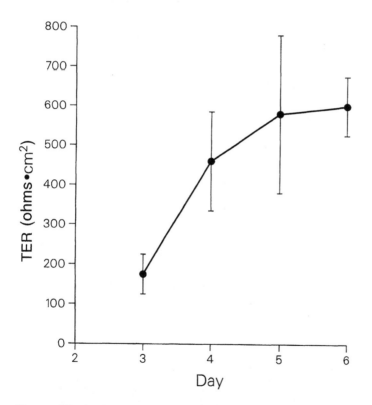

Figure 15 An increase in transepithelial electrical resistance (TER) of MDCK cell monolayers with time in culture reflects the gradual formation of a continuous sheet of epithelia with restrictive tight junctions. [Redrawn from Cho et al. (1989) with permission from the publisher.]

ability remains unchanged. A large reduction in TER valve occurs with little or no change in solute flux, and this is dependent upon the molecular size of the solute. It is during this time that only a subpopulation of tight junctions or ''pores'' are affected. As the affected population increases in size, solute flux increases dramatically with only a slight change in TER.

From these data one can calculate the effective radius of the pores through which solutes diffuse across the junctional strands (Fig. 16). At day 3, the pore radius was ~5.5 Å. This correlates with pore radii of ~10 Å for dog alveolar epithelium (Taylor and Gaar, 1970) and 5 and 8 Å for rabbit and bullfrog gallblad-

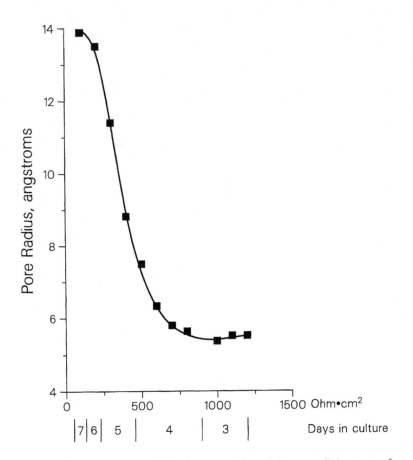

Figure 16   Correlation of effective pore radius of the paracellular route of rat alveolar epithelial cell monolayers with transepithelial electrical resistance and time in culture. Pore radii were calculated from the data shown in Figure 14.

der epithelium, respectively (Diamond, 1977). The relationship depicted in Figure 16 assumes that there is no change in the fractional area of the tight junctions, e.g., no change in cell shape. Moreover, these data imply that the change in pore radius is uniform, whereas in reality the pores are most likely heterogeneous in size. This is a result of the nature of flux, which represents a mean permeability. Therefore, the increased flux of solutes seen in Figure 14 might actually be due to an increase in the number of pores of large radius or the median of frequency of opened tight junctions.

In summary, the relationship between TER and solute permeability shown here and by Madara and Hecht (1989) emphasizes that these two measures of paracellular leakage are related but not directly correlated. The most obvious feature is that permeability as a function of TER is dependent upon the solute characteristics, primarily molecular size but also charge. The degree of correlation becomes worse as the molecular size of the solute increases. Consequently, the interrelationship between TER and solute permeability must be measured for each cell model before a minimum TER value can be selected as a prerequisite for flux studies.

IV. FILTER SUPPORT TRANSPORT KINETICS

A. Permeability Coefficient

The permeability coefficient for the filter is expressed as

$$P_F = \frac{\varepsilon_F D \, F(r/R_F)}{h_F} \tag{61}$$

In the case of the commonly used Transwell system, the microporous polycarbonate filter support is a well-defined membrane with straight cylindrical channels of radius R_F, known porosity ε_F, and thickness h_F. The pore radius, usually ≥ 0.2 μm, is sufficiently large that molecular restricted diffusion is negligible, even for serum albumin molecules with a hydrodynamic radius of 29.8 Å. In effect, $F(r/R_F < 10^{-3}) = 1.0$. Although the walls of the channels may be charged, electrostatic effects on the passage of charged permeants are inconsequential when the pore radius is more than 10^3-fold greater than the geometric mean thickness of the diffuse electrical double layer. This condition is readily achieved at ionic strengths comparable to normal saline for which the double layer thickness is ~10 Å.

When there is no cell-supporting collagen matrix to consider, P_F can be calculated a priori and put into quantitative perspective with other permeability coefficients. Otherwise, P_F needs to be experimentally determined.

B. Applications

Figure 17 depicts the rapid flux of sucrose across the filter into the receiver. Upon treatment of the data with Eq. (7), the effective permeability coefficient (P_e) is found to be 2.29×10^{-4} cm/sec, which takes into account the filter and the ABLs on both sides of the filter; hence,

$$\frac{1}{P_e} = \frac{1}{P_{ABL}} + \frac{1}{P_F} \tag{62}$$

The dimensions of the filter are $\varepsilon = 0.15$, $h = 10$ μm, $R = 0.2$ μm, and $A = 4.71$ cm². The aqueous diffusion coefficient of sucrose is 4.5×10^{-6} cm²/sec at

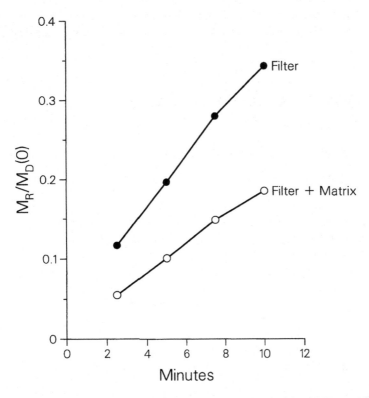

Figure 17 Appearance kinetics of sucrose across untreated Transwell filters (0.4 μm pore size) and those coated with a cross-linked rat tail collagen matrix. In both cases the systems were stirred at 100 rpm.

25°C. With Eq. (61), the permeability coefficient of the filter, P_F, is 6.75×10^{-4} cm/sec, which is several times larger than $P_e = 2.29 \times 10^{-4}$ cm/sec. This means that the mass transfer resistance of the ABLs must be appreciable:

$$\frac{1}{P_{ABL}} = \frac{1}{2.29 \times 10^{-4}} - \frac{1}{6.75 \times 10^{-4}} = 2.89 \times 10^3 \text{ sec/cm}$$

$$P_{ABL} = 3.47 \times 10^{-4} \text{ cm/sec}$$

Upon dividing P_e by P_{ABL}, one finds that transport across the ABLs and filter is about 66% ABL-controlled or 34% filter-controlled.

In Figure 17, the collagen matrix on the filter is shown to decrease the flux of sucrose. The P_e for this situation is 1.12×10^{-4} cm/sec. Accordingly,

$$\frac{1}{P_e} = \frac{1}{P_{ABL}} + \frac{1}{P_F} + \frac{1}{P_{matrix}} \qquad (63)$$

$$\frac{1}{P_{matrix}} = \frac{1}{1.12 \times 10^{-4}} - \frac{1}{3.47 \times 10^{-4}} - \frac{1}{6.75 \times 10^{-4}}$$

Therefore,

$$P_{matrix} = 2.17 \times 10^{-4} \text{ cm/sec}$$

As before, by dividing P_e by P_{ABL}, the transport kinetics is seen to be 32% ABL-controlled, or, upon dividing P_e by P_{matrix}, the overall transport of sucrose is seen to be 51% collagen matrix-controlled by molecular size restricted diffusion.

V. TRANSMONOLAYER AND AQUEOUS BOUNDARY LAYER CONTROLLED KINETICS

A. Permeability Coefficients

In Section III, emphasis was placed on flux kinetics across the cultured monolayer–filter support system where the passage of hydrophilic molecular species differing in molecular size and charge by the paracellular route was transmonolayer-controlled. In this situation, the mass transport barriers of the ABLs on the donor and receiver sides of the Transwell inserts were inconsequential, as evidenced by the lack of stirring effects on the flux kinetics. In this present section, the objective is to give quantitative insights into the permeability of the ABL as a function of hydrodynamic conditions imposed by stirring. The objective is accomplished with selected corticosteroid permeants which have been useful in rat intestinal absorption studies to demonstrate the interplay of membrane and ABL diffusional kinetics (Ho et al., 1977; Komiya et al., 1980).

As described in Section II, the effective permeability coefficient is expressed as

$$\frac{1}{P_e} = \frac{1}{P_{ABL}} + \frac{1}{P_M} + \frac{1}{P_F} \tag{64}$$

where the permeability of ABL on the donor and receiver sides (i.e., P_{ABL}) is

$$\frac{1}{P_{ABL}} = \frac{1}{P_D} + \frac{1}{P_R} \tag{65}$$

and the values of P_D and P_R may be equal or unequal, depending upon the configuration of the donor and receiver compartments (Fig. 4).

In the cell monolayer–Transwell system (Fig. 5) placed on a platform device and stirred at varying orbital speeds, the convective diffusional layer thicknesses on the donor and receiver sides are unequal due to asymmetry in the hydrodynamic conditions. The donor solution is well stirred, but stirring in the receiver is dampened considerably by the confined space nearly filled with solution. In addition, hydrodynamic uniformity is interrupted at various time intervals for sampling of the receiver solution, at which time stirring is stopped and the Transwell insert is transferred to another receiver cup containing fresh buffer solution in order to maintain sink conditions. Consequently, it is important that one maintain consistent sampling intervals throughout the experiment when ABL imposes significant mass transport barriers to the overall flux kinetics. Given this background, let us now attempt to put this complicated hydrodynamic situation into a quantitative framework. By general definition, the permeability coefficient for an ABL, P_{ABL} in cm/sec, is

$$P_{ABL} = \frac{D_{aq}}{h_{ABL}} = \kappa v^x \tag{66}$$

where

D_{aq} = aqueous diffusion coefficient, cm^2/sec
h_{ABL} = effective thickness of the ABL, cm
κ = a constant descriptive of the diffusivity of the permeant, kinematic viscosity, unit conversion, and geometric factors
v = stirring parameter expressed as revolutions per minute (rpm)

The ABL thickness is inversely proportional to v raised to the xth power. Therefore, one arrives at

$$\frac{1}{P_{ABL}} = \frac{1}{\kappa_D v^x} + \frac{1}{\kappa_R v^z} = \frac{1}{K_{ABL} v^n} \tag{67}$$

where $x \neq z$. The last term, $K_{ABL}v^n$, on the right-hand side of Eq. (67) is an operational expression of P_{ABL} wherein the empirical constant K_{ABL} and the nth power are found experimentally. With Eqs. (64) and (67), P_e may be expressed in the form

$$P_e = \frac{K_{ABL}v^n}{1 + (1/P_M + 1/P_F)K_{ABL}v^n} \tag{68}$$

whereupon it is seen that transport across the ABLs may become the rate-determining step, provided P_M and P_F are larger than $K_{ABL}v^n$ by more than tenfold. Another equivalent expression is

$$\frac{1}{P_e} = \frac{1}{P_M} + \frac{1}{P_F} + \frac{1}{K_{ABL}v^n} \tag{69}$$

which permits the determination of $1/K_{ABL}$ and $1/P_M + 1/P_F$ from a linear plot of $1/P_e$ versus $1/v^n$.

In the side-by-side configuration of donor and receiver chambers of equal volume in which stirring is usually achieved by bubbling an O_2–CO_2 mixture at various flow rates, the hydrodynamics are symmetrical, and samples can be readily collected without interruption. Therefore,

$$P_{ABL} = \frac{P_D}{2} = K_{ABL}v^m \tag{70}$$

and

$$\frac{1}{P_e} = \frac{1}{P_M} + \frac{1}{P_F} + \frac{1}{K_{ABL}v^m} \tag{71}$$

Here, the power m is expected to take values between 0.3 and 0.5 (Levich, 1962).

B. Applications

1. Steroids

Factoring the effective permeability coefficient into its component permeability coefficients for the Caco-2 cell monolayer/Transwell system is highlighted in this section (Adson, 1992). Emphasis is placed on the asymmetrical effects of stirring on the donor and receiver ABLs. Table 10 gives the physicochemical properties of hydrocortisone, dexamethasone, testosterone, and progesterone. The intrinsic log partition coefficients (log PC) according to the n-octanol–water scale range from 1.53 to 3.95. Their calculated diffusivities in water at 25°C and their molecular radii are identical, being 5×10^{-6} cm^2/sec and 5 Å, respectively. The flux of hydrocortisone across the Caco-2 cell monolayer system into the receiver

Table 10 Physicochemical Properties of Steroids Used to Assess the Aqueous Boundary Layers[a]

Steroids	MW	Diffusivity (cm^2/sec at 25°C)	Molecular radius (Å)	log PC (n-octanol/H_2O)
Hydrocortisone	362.5	4.88×10^{-6}	5.00	1.53
Dexamethasone	392.5	4.75×10^{-6}	5.14	1.73
Testosterone	288.4	5.25×10^{-6}	4.65	3.31
Progesterone	314.5	4.81×10^{-6}	5.08	3.95

[a] Molecular density $\rho = 1.2$ g/mL.

sink compartment is linear and is affected minimally by stirring (Fig. 18). In contrast, the flux of the more lipophilic testosterone (Fig. 19) is rapid and increases significantly with increasing stirring rate. The profiles for the fraction of the initial donor amount of testosterone appearing in the receiver (sink) versus time are characteristically curvilinear, with the rate tending to decrease as absorption proceeds toward completion or as the steroids in the donor chamber are significantly depleted (Fig. 19). Mass balance, determined from the donor and receiver solutions at various time points for each stirring situation, was nearly 100% for hydrocortisone, dexamethasone, and testosterone. In the case of progesterone (data not shown), the disappearance kinetics were markedly faster than the concurrent appearance kinetics, indicating that there was significant accumulation (>25%) with the cells.

The slow and linear fluxes of hydrocortisone and dexamethasone under various hydrodynamic conditions are quantified in terms of effective permeability coefficients, P_e, by Eq. (9), i.e.,

$$P_e = \frac{V_D}{AM_D(0)} \frac{\Delta M_R}{\Delta t} \qquad (72)$$

Because of the rapid and nonlinear kinetics of testosterone and progesterone transport, P_e is calculated by Eq. (7) in the form

$$\ln\left[1 - \frac{M_R}{M_D(0)}\right] = -\frac{AP_e}{V_D} t \qquad (73)$$

The conditions for Eqs. (72) and (73) are that the disappearance rate must be equal to the appearance rate and there is insignificant binding and metabolism by the cell. These assumptions are satisfied by the steroids with the exception of testosterone and progesterone. Hence, P_e values of testosterone and progesterone based on appearance kinetics will be underestimated unless the enzyme system is saturated.

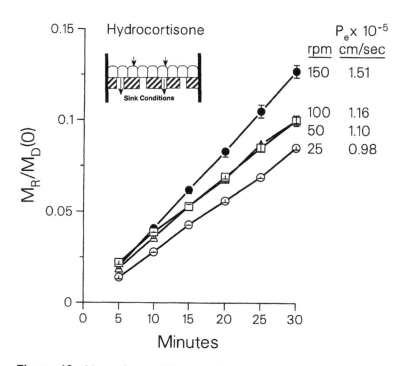

Figure 18 Linear fluxes of hydrocortisone across Caco-2 cell monolayers in the Transwell system into a receiver sink as a function of stirring (rotary platform shaker) rate at 25°C.

The summary of P_e values for the steroids as a function of stirring rates is found in Table 11 and their correlations with log PC (n-octanol–water) in Figure 20. The transport kinetics of the relatively hydrophilic hydrocortisone and dexamethasone are controlled by passive diffusion across the cell monolayer. On the other hand, the P_e values of testosterone and progesterone are highly dependent on stirring rate. The results for testosterone are used to obtain the relationships between the effective permeability coefficients of the ABL on the donor and receiver sides and the stirring rate, using the linear expression (see Eq. (69)]

$$\frac{1}{P_e} = \frac{1}{P_M} + \frac{1}{P_F} + \frac{1}{K_{ABL}v^n} \tag{74}$$

As shown in Figure 21, a double reciprocal plot of P_e versus stirring rate is linear, with

$$\frac{1}{P_e} = 2951 + 2.44 \times 10^5 \left(\frac{1}{v}\right)^{0.8} \tag{75}$$

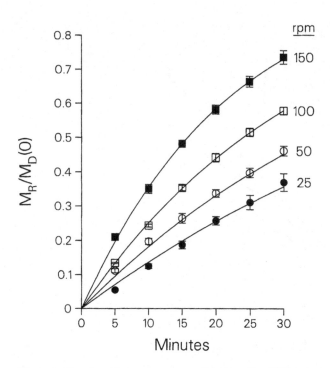

Figure 19 Curvilinear appearance kinetics for [^{14}C]testosterone across Caco-2 cell monolayers in the Transwell system as a function of stirring (rotary platform shaker) rate. Unlabeled testosterone was added at 0.1 mM to saturate metabolism as confirmed by high performance liquid chromatography (Buur and Mørk, 1992).

The stirring dependency was determined to be $(v)^{0.8}$ by statistical best fit. The equation for the line is

$$\frac{1}{K_{ABL}} = 2.44 \times 10^5 \quad \text{or} \quad K_{ABL} = 4.1 \times 10^{-6}$$

and

$$\frac{1}{P_M} + \frac{1}{P_F} = 2951 \text{ sec/cm}$$

It follows that the stirring dependency of the permeability coefficient of the donor and receiver ABLs (i.e., P_{ABL}) is $K_{ABL}(v)^{0.8}$ cm/sec. The calculated values of P_{ABL} and the effective thicknesses of the ABLs, h_{ABL}, are summarized in Table 12. It

Table 11 Delineation of Permeability Coefficients of Steroids in Caco-2 Cell Monolayer System

Steroid	rpm	$P_e{}^a$	P_{ABL}	Cell monolayer		
				P_M	$P_{transcell}$	$P_{paracell}$
Hydrocortisone	25	0.99 (0.02)	5.4	1.30	1.18	0.05
	50	1.14 (0.07)	9.3	1.32	1.28	0.05
	100	1.27 (0.12)	16.3	1.40	1.35	0.05
	150	1.51 (0.07)	22.6	<u>1.65</u>	<u>1.60</u>	0.05
				1.42	1.35	
Dexamethasone	25	0.93 (0.03)	5.4	1.14	1.09	0.05
	50	0.99 (0.03)	9.3	1.12	1.07	0.05
	100	1.12 (0.02)	16.3	1.22	1.17	0.05
	150	1.24 (0.01)	22.6	<u>1.34</u>	<u>1.29</u>	0.05
				1.21	1.16	
Testosterone	25	4.72 (0.24)	5.4	69.9	69.8	0.05
	50	7.07 (0.35)	9.3	52.8	52.8	0.05
	100	10.30 (0.35)	16.3	47.3	47.3	0.05
	150	15.00 (0.86)	22.6	<u>100.1</u>	<u>100.1</u>	0.05
				67.5	67.5	
Progesterone	50	8.75 (0.37)	9.3	—[b]	—	0.05
	100	19.04 (1.32)	16.3	—[b]	—	0.05
	150	24.50 (2.21)	22.6	—[b]	—	0.05

All permeability coefficients have units of 10^{-5} cm/sec; $P_F = 7.5 \times 10^{-4}$ cm/sec.
[a] Values in parentheses are standard deviations, $n = 3$.
[b] Not readily calculated since transport is controlled by aqueous boundary layers and filter.

is interesting that the operational power relationship of the stirring rate is 0.8 when one might expect it to be about 0.5. This is attributed to the asymmetry in the hydrodynamic conditions of the Transwell system (Fig. 5). Karlsson and Artursson (1991) found a first-power relationship of the stirring rate.

Upon taking P_{ABL} and P_e into account, the transmonolayer permeability coefficient (P_M) was quantified and, in turn, delineated into its component permeabilities for the transcellular and paracellular pathways (Table 11). Specifically, $P_{paracell}$ was found as

$$P_{paracell} = \frac{\varepsilon D\ F(r/R)}{\delta}$$
$$= (1.22)\ (5 \times 10^{-6})\ F(5/12)$$
$$= 5.4 \times 10^{-7}\ cm/sec$$

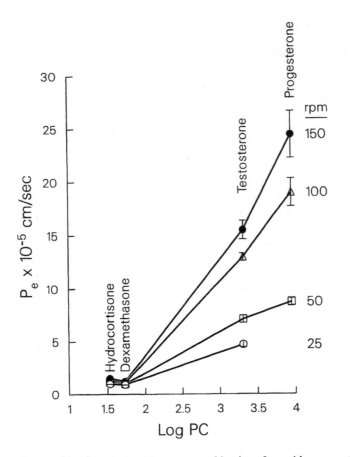

Figure 20 Correlation of appearance kinetics of steroid permeants across Caco-2 cell monolayers in the Transwell system with log partition coefficients (*n*-octanol/water) and stirring dependency.

and P_F for the filter support as

$$P_F = \frac{\varepsilon_F D \, F(r/R_F)}{h_F}$$

$$= \frac{(0.15)(5 \times 10^{-6}) \, F(5/(1.5 \times 10^4))}{10 \times 10^{-4}}$$

$$= 7.5 \times 10^{-4} \text{ cm/sec}$$

As expected, the P_M values for each steroid were independent of stirring rate. They were 1.41 ($\pm$0.18 SD) $\times 10^{-5}$ cm/sec for hydrocortisone, 1.21 ($\pm$0.1) $\times$

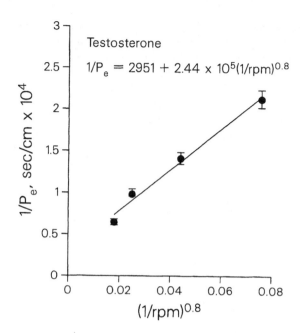

Figure 21 Linearized double reciprocal plot of the effective permeability coefficients and corresponding stirring rates to determine the power dependency of the stirring rate and mass transfer resistances for the aqueous boundary layers and the Caco-2 cell monolayer in the Transwell system.

Table 12 Effective Permeability Coefficients and Thicknesses of the Aqueous Boundary Layer of the Caco-2 Cell Monolayer/ Transwell System as a Function of Stirring by Planar Rotating Shaker[a]

Stirring rate (rpm)	$P_{ABL} = K_{ABL} v^{0.8}$ (cm/sec)	$h_{ABL} = D/K_{ABL} v^{0.8}$	
		cm	μm
25	5.38×10^{-5}	9.29×10^{-2}	929
50	9.37×10^{-5}	5.33×10^{-2}	533
100	16.32×10^{-5}	3.06×10^{-2}	306
150	22.58×10^{-5}	2.21×10^{-2}	221

$K_{ABL} = 4.1 \times 10^{-6}$, $D = 5 \times 10^{-6}$ cm^2/sec in H_2O at 25°C, and v in rpm.
[a] Based on permeation kinetics of testosterone.

10^{-5} cm/sec for dexamethasone, and 67.5 ($\pm$20.6) $\times$ 10^{-5} cm/sec for testosterone. Testosterone (log PC = 3.31), being more lipophilic than hydrocortisone (log PC = 1.53) by about 1.7 log units, was about 50 times more permeable across the monolayer itself. Transport of the steroids across the monolayer occurred principally by the transcellular route (compare P_M with $P_{transcell}$). Depending on the stirring rate, the flux of testosterone was about 75–90% controlled by the ABL, whereas the flux of progesterone was controlled totally by the ABL. These estimates are obtained by dividing P_e by P_{ABL}.

Using the side-by-side diffusion cell system, Hidalgo et al. (1992) quantified the transflux of testosterone in Caco-2 monolayers at 37°C as a function of the flow rate of the O_2/CO_2 gas mixture (Table 13). They concluded that the kinetics were ABL-controlled and proceeded to calculate the ABL thickness on each side of the diffusion chambers using

$$P_e \approx P_{ABL} = \frac{P_D}{2} = \frac{D}{2h_{ABL}} \tag{76}$$

Despite the limited number of flow rate studies, an estimate of the power relationship of the gas flow rate and P_{ABL} could be made using Eq. (70). Thus,

$$\frac{P_{ABL, 40}}{P_{ABL, 15}} = \frac{14.2 \times 10^{-5} \text{ cm/sec}}{10.9 \times 10^{-5} \text{ cm/sec}} = \left(\frac{40 \text{ mL/min}}{15 \text{ mL/min}}\right)^m$$

Therefore,

$$m \approx 0.3$$

Table 13 Influence of Stirring Flow Rate on the Permeability of [^{14}C]Testosterone Across Caco-2 Cell Monolayer/Filter Support in O_2/CO_2 Gas Lift Side-by-Side Diffusion at 37°C

Gas flow rate (mL/min)	P_e[a] ($\times$ 10^{-5} cm/sec)	Aqueous boundary layer thickness[b] (μm)	
		2 h_{ABL}	h_{ABL}
0	4.07 $\pm$ 1.13	1966	983
15	10.90 $\pm$ 1.56	734	367
40	14.18 $\pm$ 1.65	564	282

[a] Assume $P_e \approx P_{ABL}$ and $D = 8 \times 10^{-6}$ cm^2/sec in water at 37°C.
[b] ABL thickness on each side of cell monolayer is h_{ABL}.
Source: Hidalgo et al. (1992).

This indicates that the flow regime is turbulent due to the formation of eddy currents about both surfaces of the cell monolayer–filter support system, which is slightly recessed from the bulk fluid stream.

2. Peptides

In this section, we illustrate the use of cell monolayers to provide insights into peptide transport mechanisms while systematically probing the influence of peptide structure on permeability. For these studies, several series of oligomers were prepared from phenylalanine and glycine which varied in charge, lipophilicity, and chain length in a homologous fashion as shown in Table 14. Further, the use of the D-isomer of phenylalanine stabilized the peptides to hydrolysis by intestinal brush border peptidases. Permeability coefficients were obtained for the transport of the peptides across Caco-2 cells cultured on a Transwell filter support using an unstirred system (Conradi et al., 1991, 1992). The transport results and partition coefficients are summarized in Table 14. Upon comparing the effective and cell monolayer permeability coefficient values, it is clear that the cell monolayer is the principal transport barrier for these peptides. The PheGly oligomers are amphoteric, resulting in increasing charge separation with D-phenylalanyl additions. The peptides penetrate the paracellular route by molecular restricted diffusion and charge interactions (Adson et al., 1994).

Within the neutral series, no obvious relationship between permeability and the n-octanol/water partition coefficient is found, unlike the case with steroids (Section V.B.1) and beta-blockers (Section VI). Similarly, no significant influence of chain length was found. However, there is a strong dependence on the total number of unsubstituted amide nitrogens in the peptide backbone. As the amide bond content increases through the $AcPheNH_2$, $AcPhe_2NH_2$, and $AcPhe_3NH_2$ series, permeability decreases, but, as the amide bonds are methylated in the $AcPhe_3NH_2$ template, permeability increases. These observations support a model in which the principal determinant of transport is the energy required to desolvate the polar amides in the peptide in order for it to enter and diffuse across the cell membrane. Consistent with this idea, a good correlation is found between permeability and the number of potential hydrogen bonds the peptide could make based on an earlier model described by Stein (1967). Therefore, the effect of N-methylation is to remove a hydrogen-bonding donor group from the peptide and effectively reduce the transfer energy from water into the cell membrane. While this simple method of counting the number of potential hydrogen bond forming amides in the peptides is useful, it fails to account for real differences in solute–solvent hydrogen bond strength due to electronic and/or steric influences. One experimental technique for measuring hydrogen bond potential is based on the observation that the difference in solute partitioning into n-octanol as a hydrogen-bonding solvent and another, non-hydrogen-bonding, solvent is directly related to the hydrogen bond potential of the solute:

Table 14 Permeability Coefficients of D-Phenylalanyl Oligomers Across Caco-2 Cell Monolayers and In Vitro Partition Coefficients

Peptide	MW	No. of H bonds	log PC		Δ log PC[a]	Permeability coefficient[b]	
			n-Octanol	Isooctane		Effective	Monolayer[c]
PheGly	222	1	-2.16	—	—	1.05	1.06
Phe$_2$Gly	369	2	-1.46	—	—	0.55	0.56
Phe$_3$Gly	513	3	-0.66	—	—	0.36	0.36
(IV) AcPheNH$_2$	206	5	0.05	-4.92	4.97	7.99	8.85
(V) AcPhe$_2$NH$_2$	353	7	1.19	-5.29	6.48	2.20	2.26
(VI) AcPhe$_3$NH$_2$	500	9	2.30	-5.02	7.32	0.66	0.67
(VII) AcPhe$_2$(N-MePhe)NH$_2$	514	8	2.63	-4.20	6.83	2.78	2.88
(VIII) AcPhe(N-MePhe)$_2$NH$_2$	528	7	2.53	-3.10	5.63	5.68	6.11
(IX) Ac(N-MePhe)$_3$NH$_2$	542	6	2.92	-1.67	4.59	13.80	16.7
(X) Ac(N-MePhe)$_3$NHMe	556	5	3.24	-0.69	3.93	23.80	33.9

[a] Δ log PC = log PC (n-octanol/H_2O) − log PC (isooctane/H_2O).
[b] Permeability coefficients have units of 10^{-6} cm/sec.
[c] Calculated using Eq. (64) where $P_{ABL} = 8 \times 10^{-5}$ cm/sec and P_F is sufficiently large to be ignored.
The roman numerals identify peptides in Figs. 37 and 38.
Source: Conradi et al. (1991, 1992).

$$\Delta \log PC = \log PC_{octanol} - \log PC_{solvent} = \Sigma I_H - b \qquad (77)$$

where I_H is the additive increment to hydrogen bonding by a polar functionality in the solute and b is a constant (Seiler, 1974). Using isooctane as a reference non-hydrogen-bonding solvent, a good correlation is found between Caco-2 cell permeability and $\Delta \log PC$ for these peptides (Fig. 22), support the idea that desolvation of these polar functional groups is a principal determinant of membrane transport (Burton et al., 1992).

These observations can be formulated into the following mechanistic model. In general, the flux of a solute across a cell membrane is determined by the balance of water–solute and membrane–solute forces. For lipophilic solutes, the principal driving force for transfer from water to the membrane will be the

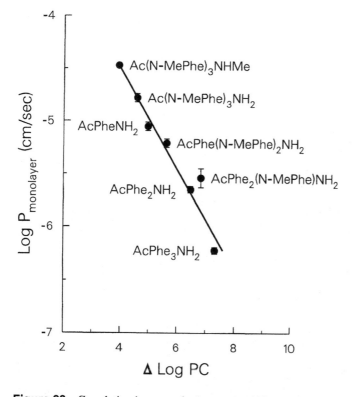

Figure 22 Correlation between the log permeability coefficient for a series of peptides across a Caco-2 cell monolayer in the Transwell system and $\Delta \log PC$, which is defined as log PC(n-octanol/water) $-$ log PC(isooctane/water). [Redrawn from Burton et al. (1992) with permission from the publisher.]

well-known hydrophobic effect. Opposing this tendency will be any specific water–solute interactions which need to be broken and cannot be accommodated in the membrane interior. In the case of a simple nonpolar solute, with minimal solvation, the hydrophobic effect will predominate. However, for solutes containing many polar functionalities, such as a peptide, the attendant desolvation energy can be appreciable.

Considering only the lipid phase as the transport pathway for the peptide, as the solute enters and diffuses across the membrane it will encounter a number of different microenvironments. The first is the aqueous membrane interface (Fig. 23). In this region, the hydrated polar headgroups of the membrane phospholipids separate the aqueous phase from the apolar membrane interior. It has been shown that this region is capable of satisfying up to 70% of the hydrophobic effect

Interface

Apolar Membrane Interior

Figure 23 Representation of a cell membrane according to the fluid mosaic model (Singer, 1974). In this model, the aqueous phospholipid interfacial microdomain separates the water compartment from the apolar membrane interior. [Redrawn from Burton et al. (1992) with permission from the publisher.]

arising from lipophilic residues in a peptide, while allowing the polar groups to remain hydrogen-bonded (Jacobs and White, 1989). Thus, in terms of solvent properties, this region is expected to be n-octanol-like. However, in order to cross the membrane the peptide will have to transfer from the interface to the much more lipophilic membrane interior. This region is expected to have solvent properties more similar to those of isooctane. Since most of the hydrophobic force favoring removal of the peptide from water has already been expended at the interface, this will not be a major factor favoring this transfer. On the other hand, for the transfer to take place, all of the peptide hydrogen bonds with polar groups at the interface will need to be broken. Thus, this step will be more influenced by the desolvation necessary to transfer the hydrogen-bonding groups.

In this model, one can argue that a peptide must have both an affinity for the interface (favorable n-octanol partition coefficient) and small desolvation energy (favorable Δ log PC) in order to efficiently cross a cell membrane. On the other hand, this model also predicts that a peptide with a large n-octanol/water partition coefficient and large desolvation energy, due to a significant number of polar groups, should adsorb and remain at the membrane interface. Both of these predicted events have been observed in the laboratory.

3. Influence of the Perturbed Paracellular Route

The perturbation of monolayers with agents (e.g., disodium ethylenediamine tetraacetate, Ca^{+2}-free medium, sodium citrate, cytochalasin D) to open tight junctions and the effect on the transmonolayer flux of permeants are addressed in this section. It has been observed that permeants taking predominantly the transcellular route are not affected by perturbants of the paracellular route, compared to extracellular or relatively hydrophilic permeants (Artursson and Magnusson, 1990). Let us put these general observations into a quantitative intepretation in the light of the transmonolayer kinetic studies of steroids in this section and of paracellular permeants in Section III. There are three cases to consider: (1) ABL-controlled permeants, (2) monolayer-controlled permeants transported principally by the transcellular route, and (3) monolayer-controlled permeants for which the paracellular route dominates.

To begin, we reintroduce the familiar effective permeability coefficient P_e to put $P_{paracell}$ in context with other permeability coefficients:

$$P_e = \cfrac{1}{\cfrac{1}{K_{ABL}(v)^{0.8}} + \cfrac{1}{P_{cell} + P_{paracell}} + \cfrac{1}{P_F}} \tag{78}$$

and

$$P_{paracell} = \frac{\varepsilon D \, F(r/R)}{\delta} \tag{79}$$

where P_{paracell} attains its maximum velocity when the molecular restriction factor $F(r/R) = 1.0$.

Case 1: Aqueous Boundary Layer–Controlled Permeants

Previously, the rate-determining step in the Caco-2 cell transmonolayer/ Transwell flux of testosterone was shown to be convective diffusion across the ABLs. Using the values for the 50 rpm situation in Table 11, Eqs. (78) and (79) become

$$P_{\text{paracell}} = 1.22 \ (5 \times 10^{-6}) \ F(5/12) = 5 \times 10^{-7} \ \text{cm/sec}$$

$$P_e = \cfrac{1}{\cfrac{1}{9.3 \times 10^{-5}} + \cfrac{1}{(52.8 + 0.05) \times 10^{-5}} + \cfrac{1}{7.5 \times 10^{-4}}}$$

$$= 7.1 \times 10^{-5} \ \text{cm/sec}$$

$$F(5/12) = 0.084$$

If the paracellular route is perturbed to the extent that the pore radius R is changed from 12 Å to 500 Å, then $F(5/500) \approx 1.0$ and $P_{\text{paracell}} = 6.1 \times 10^{-6}$ cm/sec. Although this represents a 12-fold gain from 5×10^{-7} cm/sec, the transcellular route taken by testosterone is yet the dominant pathway of the monolayer. In making the final calculations, we can conclude that the change in the observed P_e will be imperceptible.

Case 2: Monolayer-Controlled Permeants by the Transcellular Route

In this situation, e.g., hydrocortisone, P_e is essentially independent of stirring; therefore,

$$P_e \simeq \cfrac{1}{\cfrac{1}{P_{\text{cell}} + P_{\text{paracell}}} + \cfrac{1}{P_F}} \tag{80}$$

From Table 11 for hydrocortisone at 50 rpm,

$$P_e \simeq \cfrac{1}{\cfrac{1}{(1.28 + 0.05) \times 10^{-5}} + \cfrac{1}{7.5 \times 10^{-4}}} = 1.31 \times 10^{-5} \ \text{cm/sec}$$

With $P_{\text{cell}} = 1.28 \times 10^{-5}$ cm/sec and $P_{\text{paracell}} = 5 \times 10^{-7}$ cm/sec, the transmonolayer permeability is about 96% controlled by the transcellular route. For the hypothetical maximally perturbed situation where $P_{\text{paracell}} = 6.1 \times 10^{-6}$ cm/sec, the extent to which hydrocortisone traverses the cells is reduced to 68%. Upon further calculation, the P_e is estimated to be 1.84×10^{-5} cm/sec. The small change in P_e (1.31×10^{-5} compared to 1.84×10^{-5}) may not be easily observed within the sensitivity of the experimental measurements.

Case 3: Monolayer-Controlled Permeants by the Paracellular Route

In contrast to cases 1 and 2, where the influence of opening tight junctions on P_e was assessed in a hypothetical context, this present case has been demonstrated in MDCK cell transmonolayer kinetic studies employing extracellular permeants, such as mannitol and sucrose, in the presence and absence of cytochalasin D (see Figs. 11 and 12). Here,

$$P_e = \frac{\varepsilon D \, F(r/R)}{\delta} \tag{81}$$

Because P_e is directly proportional to $F(r/R)$, it should be sensitive to changes in pore radius. It is understood that a change in pore radius compared to control does not result in a proportionate change in P_e among permeants (refer to Table 7). Small molecules will be less influenced than larger molecules. Similar results were obtained in Caco-2 cell monolayers using palmitoyl-DL-carnitine as the perturbant (Knipp et al., 1997).

VI. TRANSMONOLAYER FLUX KINETICS OF WEAK ELECTROLYTE PERMEANTS

A. Permeability Coefficients

Figure 24 shows a schematic diagram of the pathways the various species of a permeant (e.g., weak base) take during their passage across the cell monolayer–filter support system. From a general point of view, the effective permeability coefficient may be expressed in the form

$$\frac{1}{P_e} = \underbrace{\frac{1}{K_{ABL}V^n}}_{\substack{\text{aqueous} \\ \text{boundary} \\ \text{layers}}} + \underbrace{\frac{1}{P_M}}_{\substack{\text{monolayer}}} + \underbrace{\frac{1}{P_F}}_{\substack{\text{filter} \\ \text{support}}} \tag{82}$$

where the terms are as defined previously.

The permeability of the cell monolayer consists of parallel transcellular and paracellular pathways. In passive diffusional transport, it is generally taken that uncharged molecules are capable of partitioning into the cell membrane and

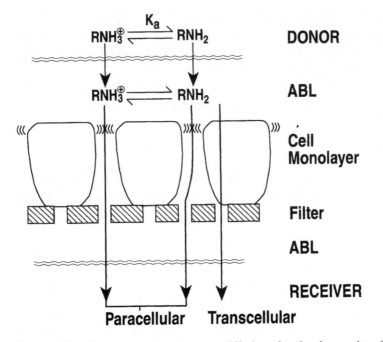

Figure 24 Schematic model of passive diffusion of molecular species of a weak base through the transcellular and paracellular routes of a cell monolayer cultured on a filter support.

diffusing across the cell and that both uncharged and charged species undergo paracellular transport. Thus, the permeability of the monolayer is expressed as

$$P_M = f(P^0_{\text{cell}} + P^0_{\text{paracell}}) + (1 - f)P^{\pm}_{\text{paracell}} \tag{83}$$

where

f = fraction of uncharged species
$1 - f$ = fraction of charged species
P^0_{cell} = permeability of uncharged species for the transcellular route
P^0_{paracell} = permeability of uncharged species for the paracellular route [Eq. (47)]
$P^{\pm}_{\text{paracell}}$ = permeability of charged species (cationic or anionic) for the paracellular route [Eqs. (45) and (46)]

For the monoprotic weak bases,

$$f = \frac{1}{1 + 10^{pK_a - pH}} \tag{84}$$

and for monoprotic weak acids

$$f = \frac{1}{1 + 10^{pH - pK_a}} \tag{85}$$

where pK_a is the negative log of the dissociation constant of the conjugate base or weak acid, as the case may be.

B. Application

The quantitative interplay of pH, pK_a, partition coefficient, the concentrations of drug species, and the permeability coefficients of the ABL, cell monolayer (including the paracellular and transcellular pathways), and filter support is demonstrated with a series of β-blockers (Adson, 1992) (Table 15). These compounds are weak bases with pK_a values ranging between 8.8 and 9.7. They are lipophilic with intrinsic log PC values ranging from 0.27 to 3.30 according to the n-octanol–water scale. Log PC of the compounds becomes significantly smaller at pH 7.4 and 6.5, which are two to three pH units from their respective pK_a's. Diffusivities in water and molecular radii are equivalent. Figure 25 shows the linear appearance kinetics of the β-blockers across the Caco-2 cell monolayer at pH 7.4 and 6.5 at 50 rpm; using these data, P_e values are calculated from Eq. (9).

The transport studies are summarized in Figure 26, in which the P_e values of the compounds are plotted against the apparent log PC values at pH 7.4 and 6.5 as a function of stirring rates. The profile appears to be sigmoidal. The perme-

Table 15 Physicochemical Properties of β-Blocker Compounds[a]

Compound	MW	pK_a	log PC (n-octanol/buffer)			Diffusivity (cm²/sec) at 25°C	Molecular radius (Å)
			Intrinsic	pH 7.4	pH 6.5		
Propranolol	259.3	9.45	3.30	1.25	0.09	5.1×10^{-6}	4.80
Alprenolol	249.3	9.65	3.04	0.79	−0.10	5.2×10^{-6}	4.70
Pindolol	248.3	8.80	1.08	−0.34	−1.22	5.2×10^{-6}	4.70
Atenolol	266.3	9.45	0.27	−1.78	−2.67	5.1×10^{-6}	4.80

[a] PC = partition coefficient; values from Adson et al. (1995). Molecular density $\rho \approx 1.0$ g/mL. Diffusion coefficient and molecular radius calculated by Eqs. (41) and (42).

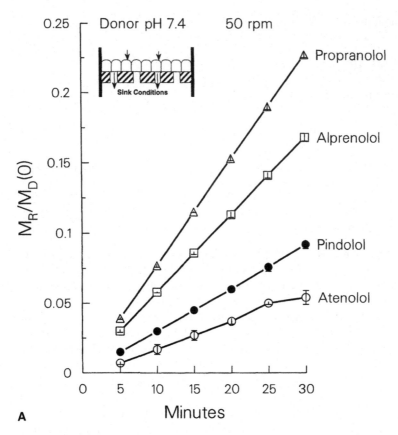

Figure 25 Cumulative fraction of the initial donor concentration of β-blockers that diffused across Caco-2 cell monolayers as a function of donor pH. Transwell systems were used, and stirring was done using a rotary platform shaker. (A), pH 7.4; (B), pH 6.5.

ability of the cell monolayer is governed by the passive diffusion of protonated and nondissociated species as evidenced by the interrelationships of pK_a, lipophilicity, and pH. In the case of pindolol, alprenolol and propranolol, the P_e values correlate with increased partition coefficients. These are more favorable at pH 7.4 than at pH 6.5 because of the larger concentration ratio of nondissociated to protonated species. Existing at nearly 100% in the protonated form at pH 7.4 and pH 6.5, atenolol attained an asymptotic permeability coefficient value of 1×10^{-6} cm/sec for the paracellular route. At pH 6.5, which is two to three pH units below the pK_a of the β-blockers, the P_e versus log PC curve was shifted to the left along the profile for pH 7.4. All the experimental points are seen to fall on a single curve in which P_e approached a minimum plateau value established

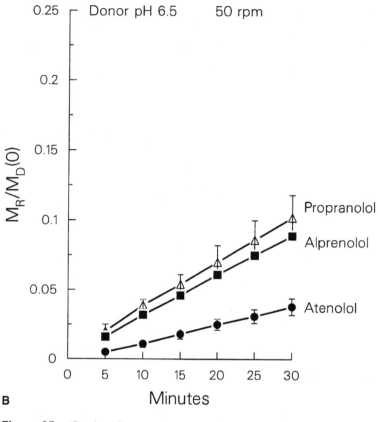

Figure 25 (Continued)

by the protonated atenolol. The effects of stirring (25–100 rpm on a rotating platform shaker) on P_e are negligible for the compounds at negative log PC and tend to be small for the more lipophilic compounds of the series. However, the stirring effects are not statistically significant ($p > 0.05$).

Let us systematically delineate the transport pathways of the nondissociated and protonated species of the β-blockers by applying Eq. (82). The insignificance of the mass transfer resistance of the ABL on the overall transport process, as evidenced by the lack of influence of stirring on P_e, indicates that the passive diffusional kinetics are essentially controlled by the cell monolayer and filter. Therefore, Eq. (82) simplifies to

$$\frac{1}{P_e} = \frac{1}{P_M} + \frac{1}{P_F} \tag{86}$$

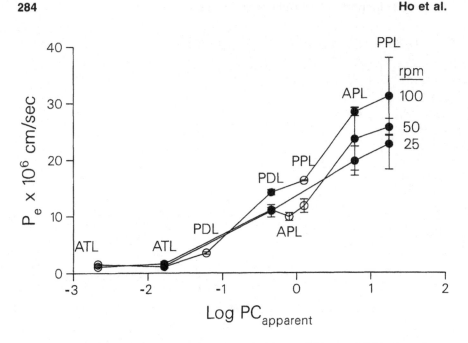

Figure 26 Correlation of the effective permeability coefficients of β-blockers determined at pH 6.5 (○) and 7.4 (●) with stirring rates and log partition coefficients (*n*-
octanol/buffer at pH 6.5 and 7.4). APL, alprenolol; ATL, atenolol; PDL, pindolol; PPL,
propranolol.

The permeability coefficient of the filter support is readily calculated with
Eq. (61) because the aqueous diffusion coefficient ($D = 5.1 \times 10^{-6}$ cm²/sec),
the molecular radius ($r = 4.75$ Å) of the β-blockers, and the porosity ($\varepsilon = 0.15$),
pore radius ($R = 1.5$ μm), and thickness ($h = 10$ μm) of the filter are known.
Consequently, $P_F = 7.6 \times 10^{-4}$ cm/sec. It follows that the permeability coefficient of the cell monolayer, P_M, is estimated as

$$\frac{1}{P_M} = \frac{1}{P_e} - \frac{1}{7.6 \times 10^{-4}} \tag{87}$$

Upon examination of the columns of P_e and P_M values for each drug in
Table 16, one discovers that the influence of P_F is minimal. Therefore, $P_e \approx$
P_M, i.e., permeation of the drugs across the ABL/cell monolayer/filter barriers
is governed by the cell monolayer. The remaining questions are (1) To what
extent is the transmonolayer diffusional process of uncharged and cationic species
gated by the transcellular and paracellular routes? and (2) What are the governing
factors?

Table 16 Permeability Coefficients of β-Blockers for Caco-2 Cell Monolayers[a]

β-Blocker	pH	f	$1-f$	P_e ($\times 10^{-6}$ cm/sec)	P_M ($\times 10^{-6}$ cm/sec)	P_M ($\times 10^{-6}$ cm/sec)		
						$fP^0_{transcell}$	$fP^0_{paracell}$	$(1-f) \times P^+_{paracell}$
Propranolol	7.4	0.009	0.991	26.6	27.6	26.6	0.006	0.99
	6.5	0.001	0.999	14.1	14.4	13.4	0.001	1.00
Alprenolol	7.4	0.005	0.995	24.0	24.8	23.8	0.004	1.00
	6.5	~0.0	~1.00	9.9	10.0	9.0	~0.0	1.00
Pindolol	7.4	0.038	0.962	12.1	12.3	11.3	0.027	0.96
	6.5	0.005	0.995	3.5	3.5	2.5	0.004	1.00
Atenolol	7.4	0.007	0.993	1.3	1.3	0.3	0.005	0.99
	6.5	~0.0	~1.00	1.1	1.1	~0	~0.0	1.00

[a] $P^+_{paracell} = 1 \times 10^{-6}$ cm/sec and $P^0_{paracell} = 0.71 \times 10^{-6}$ cm/sec for all drugs. f = fraction of nonprotonated weak base; P_M = permeability coefficient of the cell monolayer for all molecular species.

Knowing the fraction of nondissociated solutes at pH 7.4 and 6.5, one obtains the permeabilities of the various molecular species across the parallel transcellular and paracellular routes of the cell monolayer. Hence, restating Eq. (83) as

$$P_M = f(P^0_{\text{cell}} + P^0_{\text{paracell}}) + (1 - f)P^+_{\text{paracell}}$$

where P^0_{cell}, P^0_{paracell}, and P^+_{paracell} are the intrinsic permeability coefficients for the solutes for the various routes, as indicated, the extent to which the various routes are utilized is weighted by the fraction of species, determined by the pK_a of the weak base and pH of the solution. Since the β-blockers are equivalent in size and charge, P^+_{paracell} is equal to $\sim 1 \times 10^{-6}$ cm/sec, the value established by atenolol at PH 7.4 and 6.5. Therefore, according to Eq. (45),

$$P^+_{\text{paracell}} = \frac{\varepsilon D\, F(r/R)}{\delta}\left(\frac{\kappa}{1 - e^{-x}}\right) = 1 \times 10^{-6} \text{ cm/sec} \tag{88}$$

for the protonated β-blockers, and

$$P^0_{\text{paracell}} = \frac{\varepsilon D\, F(r/R)}{\delta} \tag{89}$$

for the nondissociated species. From previous calculations (Table 6), the dimensionless energy constant $\kappa/(1 - e^{-\kappa})$ was 1.41 and κ was 0.73. With Eqs. (88) and (89), P^0_{paracell} is 0.71×10^{-6} cm/sec via the molecular size restricted diffusional mechanism. It is generally observed that the fraction of uncharged species, f, is less than 0.01 and the corresponding fraction of charged species, $1 - f$, is greater than 0.99 (Table 16). Upon multiplying the values of f and $1 - f$ with their respective P^0_{paracell} and P^+_{paracell}, one readily obtains the species-weighted permeability coefficients. Taking the case of propranolol at pH 7.4,

$$\begin{aligned} fP^0_{\text{cell}} &= P_M - fP^0_{\text{paracell}} - (1 - f)\,P^+_{\text{paracell}} \\ &= 27.6 \times 10^{-6} - 0.009\,(0.71 \times 10^{-6}) - 0.991\,(1 \times 10^{-6}) \\ &= 26.6 \times 10^{-6} \text{ cm/sec} \end{aligned}$$

In Table 16 and Figure 27, the permeability coefficients of the nondissociated and protonated species of propranolol, alprenolol, pindolol, and atenolol at pH 7.4 and 6.5 are quantitatively delinated for the cell monolayer and its transcellular and paracellular pathways. It is noteworthy that, on the one hand, propranolol traverses the monolayer principally by the transcellular pathway although the fraction of uncharged species is less than 1% of the total drug concentration at donor pH 6.5 and 7.4. This is attributed to two interdependent factors. First, transport by the paracellular route is slow because it represents a small fraction of the accessible area of the monolayer and it discriminates by molecular size and charge. Second, although the fraction of uncharged species available for parti-

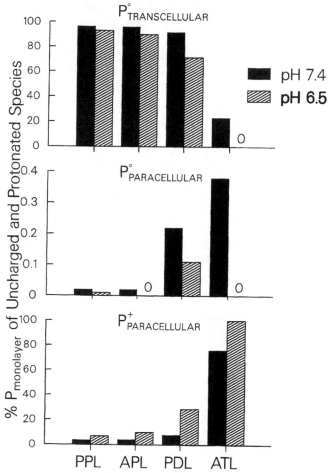

Figure 27 Relative contributions of the transcellular and paracellular routes for diffusion of the neutral and charged molecular species of β-blockers across Caco-2 cell monolayers. APL, alprenolol; ATL, atenolol; PDL, pindolol; PPL, propranolol.

tioning into the apical membrane may be small, the product of this small fraction and the large intrinsic membrane/water partition coefficient of a lipophilic compound is yet reasonably large, particularly given the large surface are for transcellular permeation. Rapid equilibrium between uncharged and charged species is still maintained. On the other hand, atenolol is intrinsically less lipophilic than propranolol; more important, however, atenolol exists totally in the protonated form such that it permeates the monolayer by the paracellular pathway.

VII.   TRANSCELLULAR DIFFUSION AND METABOLISM

A.   Biophysical Model Description

A simple steady-state model is presented for simultaneous passive diffusion and intracellular metabolism of a drug undergoing the irreversible reaction

Drug → metabolite

It should provide the framework to accommodate development of other reaction schemes including metabolism at the brush border surface of the apical membrane.

The model is shown schematically in Figure 28. It consists of the ABL in the donor side in series with the monolayer followed by the sink. The monolayer

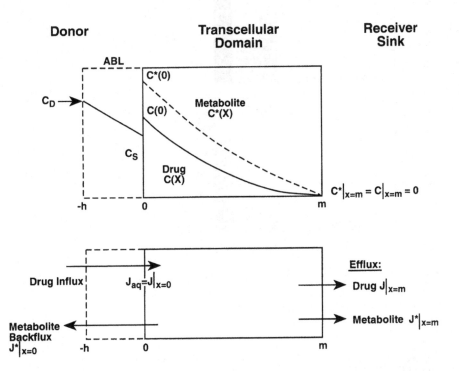

Figure 28   The biophysical model for passive diffusion and concurrent intracellular metabolism of a drug for a simple A-to-B reaction process. Concentration–distance profiles are depicted in the aqueous boundary layer and intracellular domain for the drug and metabolite. The bottom diagram depicts the direction of the fluxes of drug and metabolite viewed from the donor and receiver sides of the cell monolayer. Details of basic assumptions are found in the text.

comprises two parallel pathways: the simultaneous diffusional transport–bioconversion pathway of the cells and the aqueous pore pathway. For general purposes, let the drug be an organic base or acid. It is assumed that (1) quasi-steady-state conditions prevail, (2) metabolism in the donor solution by leaking enzymes, if it occurs, is insignificant, (3) only uncharged species are able to partition into the apical membrane, (4) metabolism occurs homogeneously in the transcellular pathway and, consequently, no metabolism is assumed to take place with surface enzymes of the apical membrane, (5) the paracellular transport of uncharged and charged species occurs without metabolism, and (6) iso-osmotic conditions exist everywhere.

At steady state, changes in the concentrations of drug and metabolite with distance involving diffusional and concurrent bioconversion kinetics are given by

$$D\frac{d^2C}{dx^2} - kC = 0 \qquad (90)$$

and

$$D\frac{d^2C^*}{dx^2} + kC = 0 \qquad (91)$$

for $0 \leq x \leq m$, where

C, C^* = concentrations of drug and metabolite, respectively, as a function of distance x, mol/mL

D = intracellular diffusion coefficient, cm^2/sec, of drug and metabolite taken as equivalent

k = first-order irreversible bioconversion rate constant, sec^{-1}

The boundary conditions are

$C = C(0)$ at $x = 0$
$C = C^* = 0$ at $x = m$

$$D\frac{dC^*}{dx}\bigg|_{x=0} = J^*|_{x=0}$$

where sink conditions on the basolateral side are assumed $J^*|_{x=0}$ = back-flux of metabolite into the donor chamber, mol/(cm$^2 \cdot$ sec). Consequently, the concentration–distance profile is described for the drug by

$$C(x) = \beta \sinh[K(m - x)] \qquad (92)$$

and for the metabolite by

$$C^*(x) = \left[\beta K (\cosh Km) - \frac{J^*|_{x=0}}{D} \right] (m - x) - \beta \sinh[K(m - x)] \quad (93)$$

where the effective surface concentration of drug at the apical membrane side is

$$\beta = \frac{C(0)}{\sinh Km} \ \text{mol/mL} \quad (94)$$

and the diffusion–bioconversion contant is

$$K = \sqrt{k/D} \ \text{cm}^{-1} \quad (95)$$

The efflux of drug into the sink is

$$J|_{x=m} = -D \frac{dC}{dx} \bigg|_{x=m} = \beta DK \quad (96)$$

and the accompanying efflux of metabolite is

$$J^*|_{x=m} = -D \frac{dC^*}{dx} \bigg|_{x=m} = \beta DK[\cosh Km - 1] - J^*|_{x=0} \quad (97)$$

The flux of total drug (nondissociated and charged species) across the ABL, J_{ABL}, is

$$J_{ABL} = P_{ABL}(C_D - C_S) \quad (98)$$

where

$\quad C_D, C_S$ = total drug concentrations in bulk donor solution and at cell
$\qquad\qquad$ surface on the ABL side, mol/mL
$\quad P_{ABL}$ = permeability coefficient of the ABL, cm/sec

The continuity of mass transfer across the water/cell membrane interface requires that

$$J_{ABL} = J|_{x=0} = -D \frac{dC}{dx} \bigg|_{x=0} \quad (99)$$

Hence,

$$P_{ABL}[C_D - C_S] = \frac{DKC(0)}{\tanh Km} \quad (100)$$

The total concentration C_S is related to the nondissociated drug concentration $C(0)$ at the apical membrane side by the effective partition coefficient, or

$$C(0) = K_P X_S C_S \tag{101}$$

where K_P = intrinsic membrane/water partition coefficient for nondissociated species and X_S = fraction of nondissociated drug species at membrane surfaces.

Combining Eqs. (98), (100), and (101), the flux across the ABL, or the disappearance rate per unit area, is

$$J_{ABL} = -\frac{V_D}{A} \frac{dC_D}{dt} = \left[\frac{1}{\dfrac{1}{P_{ABL}} + \dfrac{\tanh Km}{K_P X_S DK}} \right] C_D \tag{102}$$

B. Interrelation of Fluxes and Extent of Metabolism

Let us now interrelate the fluxes in a variety of meaningful ways. All fluxes have units of moles per square centimeter per second. The conservation of mass is satisfied by the requirement that the disappearance rate of drug from the solution is equal to the sum of fluxes of drug and metabolite emerging from the cell:

$$J_{ABL} = J|_{x=m} + J^*|_{x=m} + J^*|_{x=0} \tag{103}$$

The efflux of drug, $J|_{x=m}$, is related to the influx J_{ABL} by

$$J|_{x=m} = \frac{J_{ABL}}{\cosh Km} \tag{104}$$

where the efflux is smaller than the influx of drug by $1/(\cosh Km)$ due to metabolism. Since $J|_{x=m}/J_{ABL}$ is the fraction of drug not metabolized,

$$\text{Fraction of drug metabolized} = 1 - \frac{J|_{x=m}}{J_{ABL}} = \frac{J^*|_{x=m} + J^*|_{x=0}}{J_{ABL}} \tag{105}$$

$$= \frac{\cosh Km - 1}{\cosh Km}$$

Furthermore, upon comparing the flux of the metabolite with efflux of drug, one gets

$$\frac{J^*|_{x=m} + J^*|_{x=0}}{J|_{x=m}} = \cosh Km - 1 \tag{106}$$

The mean intracellular concentration of drug $\langle C \rangle$ is

$$\langle C \rangle = \frac{1}{m} \int_0^m C(x)\, dx = \beta\left(\frac{\cosh\, Km - 1}{Km}\right) \tag{107}$$

and, with Eq. (96),

$$\langle C \rangle = \left(\frac{\cosh\, Km - 1}{DK^2 m}\right) J|_{x=m} \tag{108}$$

The efflux of drug, $J|_{x=m}$, is related to the disappearance rate in the donor by Eq. (104). Consequently, $\langle C \rangle$ is a function of time and approaches zero upon depletion of drug in the donor with time.

The time available for metabolism to occur during diffusion of drug across the length of the cells is approximated by the mean residence time $\langle t \rangle$:

$$\langle t \rangle = \frac{m^2}{D} \tag{109}$$

$\langle t \rangle$ also may be defined as the diffusional time during which molecules enter and leave a planar slab of reaction space. Since the magnitude of Km governs the extent of metabolism and the efflux kinetics of the intact drug and metabolite(s), the relationship of Km with respect to the first-order reaction rate constant and the mean residence time would be useful. Using Eqs. (95) and (109), it follows that $Km = (k\langle t \rangle)^{1/2}$. When $k = 0$ for the case of no metabolism, $Km = 0$ and $\cosh\, Km = 1.0$; consequently, the extent of metabolism is zero. In the other extreme, when Km is fairly large, e.g., 5.0, because of a large rate constant and relatively short residence time or a small rate constant and relatively long residence time, the extent of metabolism is approximately 98.6%.

C. Donor Disappearance and Receiver Appearance Kinetics

The integration of Eq. (102) with the initial condition, $C_D = C_D$ at $t = 0$, gives the change in drug concentration in the donor solution with time:

$$C_D = C_D(0) \exp\left(-\frac{AP_e t}{V_D}\right) \tag{110}$$

where the effective permeability coefficient P_e in the most general form is

$$P_e = \frac{1}{\dfrac{1}{P_{ABL}} + \dfrac{1}{K_P X_S DK/(\tanh\, Km) + P_{paracell}} + \dfrac{1}{P_F}} \tag{111}$$

Here, P_e takes into account the aqueous boundary layers (P_{ABL}), transcellular diffusion with metabolism, the paracellular pathway ($P_{paracell}$), and the filter support (P_F). It is assumed that drug diffusing through the paracellular route escapes metabolism and contributes insignificantly to the appearance of intact drug in the receiver.

Using the relationships of the disappearance rate from the donor and efflux into the receiver (sink) given in Eqs. (96), (102), and (104) and the initial condition of $C_R = 0$ at $t = 0$, the time-dependent change in drug concentration in the receiver is

$$\frac{C_R}{C_D(0)} = \frac{V_D}{V_R}\left(\frac{1}{\cosh Km}\right)\left[1 - \exp\left(-\frac{AP_e t}{V_D}\right)\right] \tag{112}$$

For the linear case (i.e., at small t),

$$\frac{C_R}{C_D(0)} = \frac{A}{V_R}\left(\frac{P_e}{\cosh Km}\right)t \tag{113}$$

With Eqs. (104) and (106), the rate of appearance of metabolite into the receiver sink is

$$\frac{V_R}{A}\frac{dC_R^*}{dt} = \left(\frac{\cosh Km - 1}{\cosh Km}\right)P_e C_D(0)\exp\left(-\frac{AP_e t}{V_D}\right) - J^*|_{x=0} \tag{114}$$

whereupon, with the initial condition $C_R^* = 0$ at $t = 0$, the metabolite concentration with time is

$$\frac{C_R^*}{C_D(0)} = \frac{V_D}{V_R}\left(\frac{\cosh Km - 1}{\cosh Km}\right)\left[1 - \exp\left(-\frac{AP_e t}{V_D}\right)\right]$$
$$- \frac{A}{V_R C_D(0)}\int_0^t J^*|_{x=0}\,dt \tag{115}$$

For the linear case,

$$\frac{C_R^*}{C_D(0)} = \frac{A}{V_R}\left(\frac{\cosh Km - 1}{\cosh Km}\right)P_e t - \frac{A}{V_R C_D(0)}\int_0^t J^*|_{x=0}\,dt \tag{116}$$

In the absence of metabolism, i.e., $K = 0$, Eq. (111) reduces to

$$P_e = \cfrac{1}{\cfrac{1}{P_{ABL}} + \cfrac{1}{K_p DX_S/m + P_{paracell}} + \cfrac{1}{P_F}} \tag{117}$$

and

$$J|_{x=m} = J_{ABL} \tag{118}$$

$$J^*|_{x=m} = J^*|_{x=0} = 0 \tag{119}$$

Experimentally, this could be accomplished with the use of permeable metabolic inhibitors, a stereoisomeric form of the drug (Ho et al., 1977), or excess drug concentrations to saturate the enzyme system.

D. Applications

In this section we demonstrate the utility of the Caco-2 cell monolayer to provide a systematic and quantitative assessment of passive diffusion coupled with intra-

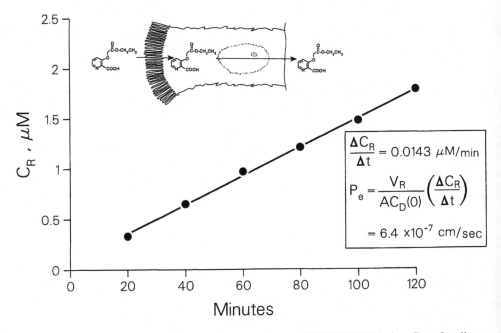

Figure 29 Intrinsic permeability of the monoester of PNU-82,899 using Caco-2 cell monolayers. A Transwell system was used with receiver sink conditions at 25°C. The initial donor concentration was 199 µM, and donor and receiver solutions were at pH 7.4.

cellular metabolism of a prodrug designed to improve the intestinal absorption of the drug. Figure 29 shows the slow (apical-to-basolateral) flux of the monoester drug compared to the vastly improved total flux (Fig. 30) of the diester prodrug PNU-82,899 across the monolayer–filter system. Permeation of the diester is accompanied by extensive metabolism to the monoester, so that the flux of the monoester contributes about 92% of the total flux as viewed from the receiver side. No monoester was detected in the donor solution, and no diacid species was found in either the donor or receiver. Taken together, these observations indicate that metabolism occurs intracellularly by esterases to the monoester, and the monoester is not a substrate for further hydrolysis.

To quantify the metabolic kinetics, the example in Figure 29 is used. The data are

Efflux rate of diester: $\dfrac{\Delta C_R}{\Delta t} = 0.034$ μM/min

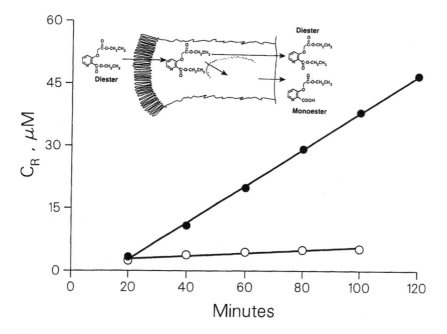

Figure 30 Appearance kinetics of PNU-82,899 and the monoester metabolite into the receiver sink. Diester diffused into the Caco-2 cell, resulting in rapid bioconversion of the diester to the monoester. Membrane surface metabolism was not detected. The initial donor concentration was 145 μM, and donor and receiver solutions were at pH 7.4. ($\bullet$) Monoester. $\Delta C_R^* / \Delta t = 0.44$ μM/min; efflux $P_e^* = 2.68 \times 10^{-5}$ cm/sec. ($\bigcirc$) Diester. $\Delta C_R^* / \Delta t = 0.034$ μM/min; efflux $P_e^* = 0.21 \times 10^{-5}$ cm/sec.

Efflux rate of monoester: $\dfrac{\Delta C_R^*}{\Delta t} = 0.44$ μM/min

Back flux rate of monoester: $J^*|_{x=0} = 0$

Intracellular diffusivity: $D = 1 \times 10^{-6}$ cm^2/sec

Initial donor concentration: $C_D(0) = 145.4$ μM

Cell thickness: $m = 25$ μm

Cross-sectional area: $A = 4.71$ cm^2

Solution volumes: $V_D = 1.5$ mL and $V_R = 2.5$ mL

According to Eq. (106),

$$\cosh Km = \frac{J^*|_{x=m}}{J|_{x=m}} + 1 = \frac{0.44}{0.034} + 1 = 13.95$$

Therefore,

$Km = 3.33$

Since $Km = m\sqrt{k/D}$ by Eq. (95), the bioconversion rate constant is

$$k = \frac{(3.33)^2 D}{m^2} = \frac{(3.33)^2 (1 \times 10^{-6})}{(25 \times 10^{-4})^2} = 1.77 \text{ sec}^{-1}$$

For the prodrug entering and leaving the cell, bioconversion takes place during the time period estimated by the mean residence time [Eq. (109)]:

$$\langle t \rangle = \frac{m^2}{D} = \frac{(25 \times 10^{-4})^2}{1 \times 10^{-6}} = 6.25 \text{ sec}$$

Table 17 Diffusion and Metabolism of PNU-82899E (Diester Prodrug) Across Caco-2 Cell Monolayer

Expt	Intracellular diffusion/ bioconversion		Esterase hydrolysis k (sec^{-1})	Fraction metabolized	Fraction not metabolized
	cosh Km	Km			
1	13.95	3.33	1.77	0.928	0.072
2	18.7	3.62	2.10	0.947	0.053
3	25.0	3.91	2.45	0.960	0.040

Table 18 Delineation of Permeability
Coefficients of PNU-82,899E (Diester Prodrug)[a]

Expt	P_e	P_M	$\dfrac{K_P X_S DK}{\tanh Km}$	$P_{paracell}$
1	2.89	4.52	4.41	0.11
2	2.84	4.40	4.29	0.11
3	3.05	4.93	4.82	0.11

[a] Units of all permeability coefficients are 10^{-5} cm/sec.
Microporous filter support $P_F = 7.5 \times 10^{-4}$ cm/sec;
$P_{ABL} = 9 \times 10^{-5}$ cm/sec; and P_M = permeability coefficient of the cell monolayer.

It follows that

$$\% \text{ Prodrug metabolized} = \frac{\cosh Km - 1}{\cosh Km} \times 100$$

$$= \frac{12.95}{13.95} \times 100 = 92.8\%$$

All bioconversion parameters are found in Table 17.

The effective permeability coefficient of the diester, as viewed from the donor side, can be found by treating the data with Eq. (113); hence,

$$P_e = \frac{V_R}{A} \frac{\Delta C_R}{\Delta t} \frac{\cosh Km}{C_D(0)} = \frac{2.5\,(0.034)\,(13.95)}{4.71\,(145.4)\,(60)} = 2.89 \times 10^{-5} \text{ cm/sec}$$

The permeability coefficient of the diffusional–bioconversion pathway can be delineated with the aid of Eq. (111) once the permeability coefficients of the ABL, filter support, and paracellular routes are known (Table 18). It is seen that 98% of the diester molecules passing through the cell monolayer take the intracellular route.

VIII.   TRANSCELLULAR DIFFUSION OF HIGHLY MEMBRANE INTERACTIVE PERMEANTS

A. Introduction

Membrane-interactive compounds are lipophilic molecules which have high affinity for lipid membranes and consequently possess long membrane residence times. Examples include the lipophilic neutral molecules (cholesterol, lecithin, pesticides, oleyl alcohol, tocopherols, etc.) and large organic cations (chlorpromazine and the antioxidants Tirilazad mesylate and PNU-78,517). Propranolol and progesterone also tend to demonstrate relatively weaker membrane-interactive properties. The transport kinetics are characterized by rapid uptake by epithelial (or endothelial) cells and slow membrane efflux kinetics (Raub et al., 1993; Sawada et al., 1994). This section describes in some systematic detail the trans-monolayer diffusional kinetics of a membrane-interactive permeant, rate-determining factors in the uptake and efflux transport phenomena, and accompanying biophysical models (Fig. 31). For instructive purposes, the model permeant is the lipid membrane antioxidant PNU-78,517, a trolox derivative of a macamine with a pK_a of 6.5 (Raub et al., 1993).

The flux of ^{3}H-labeled PNU-78,517 across MDCK cell monolayers shows the characteristic disparity between the kinetics of disappearance from the donor solution and appearance in the receiver sink (Fig. 32). Drug uptake is rapid and exponential with time and approaches a quasi-equilibrium state; in contrast, the concomitant efflux of drug into the receiver is slow and linear. While maintaining a 3% bovine serum albumin (BSA) concentration in the donor and varying the BSA concentration between 0.5 and 5% in the receiver, the results show that the

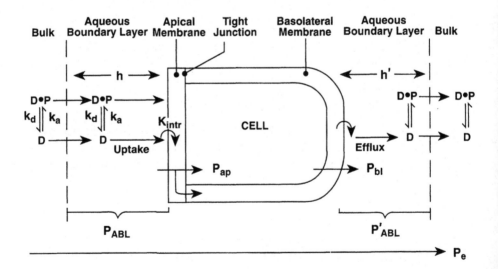

Figure 31 Scheme for the protein-binding, diffusional, and partitioning processes and barriers that are encountered by a highly lipophilic and membrane-interactive drug (D) as it permeates through a cell within a continuous monolayer. h and h', thicknesses of the aqueous boundary layers. k_d and k_a, dissociation and association binding constants, respectively. P, protein molecule. Permeability coefficients: Effective, P_e; aqueous boundary layer, P_{ABL} and P'_{ABL}; apical membrane, P_{ap}; basolateral membrane, P_{bl}.

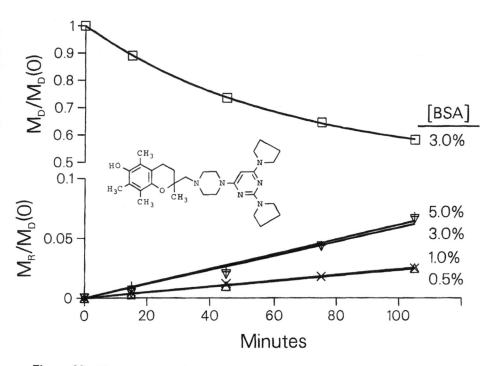

Figure 32 Disappearance and appearance kinetics of transcellular flux of the lipophilic antioxidant PNU-78,517 (pK_a ~ 6.5) across MDCK cell monolayers in Transwell systems at 37°C. Donor solutions contained 3% bovine serum albumin (BSA), and receiver solutions contained 0.5–5% BSA at pH 7.4. [Redrawn from Raub et al. (1993) with permission from the publisher.]

increasing appearance rates with BSA concentration had no obvious influence on the disappearance rate. Autoradiographic findings not only indicated that most of the antioxidant was associated with the plasma membrane, but also that it equilibrated rapidly within the basolateral membrane following uptake by the apical membrane. These results provide evidence that PNU-78,517 readily partitions deep enough into the lipid bilayer to negotiate around the tight junction (Section III.A). The possibility that the antioxidant equilibrated with an intracellular membrane pool is not ruled out.

It is noteworthy that the kinetics viewed from the donor and receiver sides are seemingly independent of each other. As discussed in the next section, this prompts the employment of experimental strategies to understand the events of drug uptake and efflux mechanistically and independently through drug uptake studies with cell monolayers cultured on flat plastic dishes, drug efflux from the apical membrane from the cell monolayer on a plastic dish, and also drug efflux from the basolateral membrane from the cell monolayer cultured on a filter support. Cell monolayers cultured in this manner form functionally unique apical and basolateral membranes.

B. Influx Kinetics and the Apical Membrane Domain

In Figure 33, both the first-order uptake kinetics and the equilibrium plateau levels decrease with increased BSA concentration, indicating that protein binding plays an influential role in the mass transfer process. To put biophysical meaning and quantitation into these observations, the following model is presented.

The total flux of free and BSA-bound drug across the ABL over the monolayer followed by partitioning of free drug into the apical membrane and rapid distribution throughout the cell monolayer (see Fig. 31) is described by

$$ - V_D \frac{dC_D}{dt} = AP_e\left(C_D - \frac{C_{\text{cell}}}{K_e}\right) \tag{120} $$

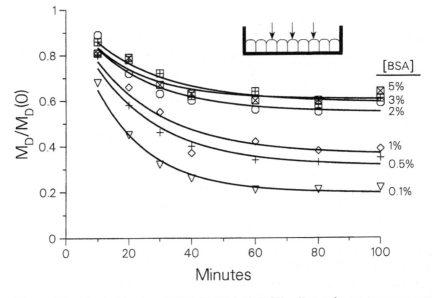

Figure 33 Uptake kinetics of PNU-78,517 by MDCK cell monolayers grown on a solid plastic surface as a function of bovine serum concentration in the bathing solution. [Redrawn from Raub et al. (1993) with permission from the publisher.]

where C_D is the total concentration of drug species in the donor solution of volume V_D, C_{cell} is the drug concentration in the cell monolayer of cross-sectional area A, P_e is the effective permeability coefficient, and K_e is the effective cell monolayer/buffer partition coefficient. The total amount in the system is equal to the sum of amounts in the donor solution and monolayer.

$$V_D C_D(0) = V_D C_D + V_{\text{cell}} C_{\text{cell}} \tag{121}$$

where $C_D(0)$ is the initial donor concentration and V_{cell} is the volume of the monolayer (the product of cell height and cross-sectional area of the monolayer). Subsequently, Eq. (120) becomes

$$\frac{dC_D}{dt} = \alpha[\beta C_D(0) - (1 + \beta)C_D] \tag{122}$$

where the constants are

$$\alpha = \frac{AP_e}{V_D} \text{ min}^{-1} \tag{123}$$

$$\beta = \frac{V_D}{V_{\text{cell}} K_e} \quad \text{(dimensionless)} \tag{124}$$

With the initial condition, $C_D = C_D(0)$ at $t = 0$, we get the fraction of drug remaining in the donor solution with time,

$$\frac{C_D}{C_D(0)} = \frac{\beta + \exp[-\alpha(1 + \beta)t]}{1 + \beta} \tag{125}$$

As time approaches infinity, equilibrium is attained;

$$\frac{C_D(\infty)}{C_D(0)} = \frac{\beta}{1 + \beta} \tag{126}$$

and the effective partition coefficient is

$$K_e = \frac{C_{\text{cell}}(\infty)}{C_D(\infty)} = \frac{V_D}{V_{\text{cell}}} \left[\frac{C_D(0) - C_D(\infty)}{C_D(\infty)} \right] \tag{127}$$

Hence, K_e may be determined experimentally by assaying the concentrations in the donor solution. It is noted that the relative donor concentration in Eqs. (125) and (126) may be replaced by the relative mass in the donor solution.

The values of α and β in Eq. (125) are obtained by nonlinear curve-fitting techniques from which the uptake P_e and K_e values as a function of BSA concentration are calculated. Figure 34 illustrates the fit between and K_e [BSA] (molar concentration of BSA with a molecular weight of ~65 kDa) according to the linear binding expression

$$K_e = \frac{K_{\text{intr}}}{1 + k_a[\text{BSA}]} \tag{128}$$

where K_{intr} is the intrinsic partition coefficient between the cell monolayer and the free drug; k_a is the association constant of the BSA-bound drug (found to be $1.4 \times 10^4 \text{ M}^{-1}$) (see Fig. 31). It follows that the molar ratio of free to total drug is given by $1/(1 + k_a[\text{BSA}])$. The exact nature of the BSA–drug complex is not known; therefore, Eq. (128) is an operational expression to describe the binding phenomenon. The concentration ratio of cell-associated drug to free drug in the external solution (K_{intr}) is 2050 (or log $K_{\text{intr}} = 3.31$). Since the characteristics of PNU-78,517 support a predominantly membrane distribution, K_{intr} value is 1.5×10^5 (or log $K_{\text{intr}} = 5.2$) when it is calculated in terms of total cellular phospholipid content.

The uptake P_e is seen in Figure 34 to decrease monotonically with increasing BSA concentration or decreasing free drug concentration. Because K_e and P_e have relatively large values, one might reason that the rate-limiting step in the kinetic uptake process is convective diffusion of both free and BSA-bound drug across the ABL and not the ensuing step of interphase partitioning of free drug species at the apical membrane with subsequent diffusion into the cell (see Fig. 31). Therefore, the mathematical description (Ho et al., 1983) for P_e is

$$P_e = P^*_{\text{ABL}} + \frac{P_{\text{ABL}} - P^*_{\text{ABL}}}{1 + k_a[\text{BSA}]} \tag{129}$$

where P_{ABL} and P^*_{ABL} are the permeability coefficients of the free drug and BSA–drug complex, respectively, in cm/min. As shown in Figure 35, there is a linear relationship for the experimental data when P_e is plotted against $1/(1 + k_a[\text{BSA}])$, i.e., the fraction of free drug concentration. Extrapolating the line to zero [BSA] gives $P_{\text{ABL}} = 13.8 \times 10^{-3}$ cm/min for the free drug species; extrapolating to infinite [BSA] gives a P^*_{ABL} of 3.6×10^{-3} cm/min for the bound species. It follows that the effective thickness of the ABL is calculated as

$$h = \frac{D}{P_{\text{ABL}}} \quad \text{and} \quad h^* = \frac{D^*}{P^*_{\text{ABL}}} \tag{130}$$

where D is the aqueous diffusivity of the free drug (3.9×10^{-6} cm²/sec at 25°C by the Stokes–Einstein equation) and D^* is the aqueous diffusivity of the bound drug complex [6.9×10^{-7} cm²/sec for BSA (Tanford, 1963)]. The thicknesses h and h^*, whose values are expected to be identical, are 175 and 150 μm, respec-

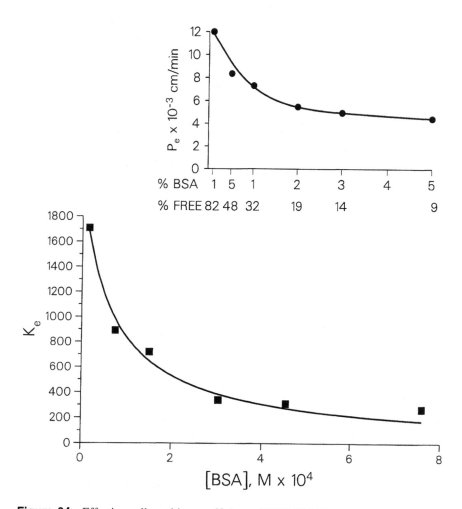

Figure 34 Effective cell partition coefficient of PNU-78,517 as a function of bovine serum albumin concentration. The inset shows the relationship between the effective permeability coefficient (P_e) of appearance and BSA concentration or the fraction of free drug. These data were obtained from the uptake data shown in Figure 33. [Redrawn from Raub et al. (1993) with permission from the publisher.]

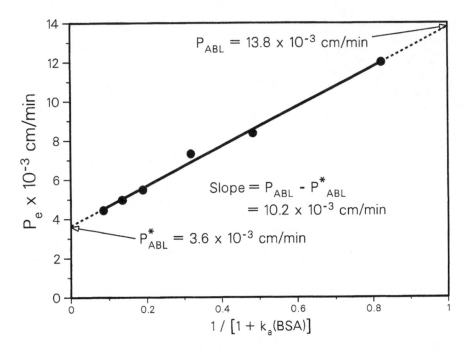

Figure 35 Relationship between the uptake permeability coefficient and the free drug concentration. P_{ABL} is the permeability of the free drug across the aqueous boundary layer, and P^*_{ABL} is the same for the drug–albumin complex. [Redrawn from Raub et al. (1993) with permission from the publisher.]

tively. In conclusion, the uptake kinetics of PNU-78,517 in BSA solutions by the cell monolayer is ABL-controlled.

C. Efflux Kinetics from the Apical Membrane Domain

Given the low permeability of the antioxidant across MDCK cell monolayers and its large membrane partition coefficient, efflux kinetic studies using drug-loaded cell monolayers cultured on plastic dishes could yield useful information when coupled with the following biophysical model. The steady-state flux of drug from the cell monolayer is equal to the appearance rate in the receiver solution:

$$J_{ap} = AP_{ap}(C_{cell} - C_R) = V_R \frac{dC_R}{dt} \tag{131}$$

where J_{ap} is the flux of total drug (free and BSA-bound), mass/min; P_{ap} is the membrane permeability coefficient and is a function of BSA concentration; C_{cell}

is the cell concentration of the drug; C_R is the total concentration in the receiver solution of volume V_R; and A is the cross-sectional area of the monolayer. Most likely, the kinetics are apical membrane controlled; therefore, the permeability of the ABL over the monolayer surface is of negligible significance. The total amount of drug in the system at any time is

$$V_{cell} C_{cell}(0) = V_{cell} C_{cell} + V_{cell} C^*_{cell} + V_R C_R \tag{132}$$

whereupon the substitution of C_{cell} into Eq. (131) gives

$$V_R \frac{dC_R}{dt} = AP_{ap} \left[C_{cell}(0) - C^*_{cell} - \left(1 + \frac{V_R}{V_{cell}} \right) C_R \right] \tag{133}$$

where $C_{cell}(0)$ = initial drug concentration of the cell monolayer and C^*_{cell} = drug concentration in a cell-associated, slow kinetic pool taken to be constant. With the initial condition $C_R = C_R(0)$ at $t = 0$, one gets the change in total concentration in the receiver with time:

$$\frac{C_R}{C_{cell}(0)} = \frac{C_R(0)}{C_{cell}(0)} \exp\left(-\frac{\alpha A P_{ap} t}{V_R} \right) \tag{134}$$

$$+ \frac{1}{\alpha} \left(1 - \frac{C^*_{cell}}{C_{cell}(0)} \right) \left[1 - \exp\left(-\frac{\alpha A P_{ap} t}{V_R} \right) \right]$$

where

$$\alpha = 1 + V_R/V_{cell} \simeq V_R/V_{cell} \tag{135}$$

Generally, the volume of the receiver solution is much larger than the volume of the cell monolayer. In terms of mass, Eq. (134) is multiplied by V_R/V_{cell}; hence,

$$\frac{M_R}{M_{cell}(0)} = \frac{M_R(0)}{M_{cell}(0)} \exp\left(-\frac{\alpha A P_{ap} t}{V_R} \right)$$

$$+ \left(1 - \frac{M^*_{cell}}{M_{cell}(0)} \right) \left[1 - \exp\left(-\frac{\alpha A P_{ap} t}{V_R} \right) \right] \tag{136}$$

The kinetic expressions in Eqs. (134) and (136) take into account the initial burst of drug into the receiver and the possibility of cell-associated drug within a slow kinetic pool which is determined at $t = \infty$.

Figure 36 shows that efflux of PNU-78,517 from the apical membrane is facilitated by BSA. Its permeability coefficients increase with BSA concentration; the magnitudes of the values (10^{-6}–10^{-5} cm/min) and trends indicate that the desorption kinetics are membrane-controlled and that BSA acts as a drug acceptor at the membrane interface via protein binding. The initial mass fraction readily

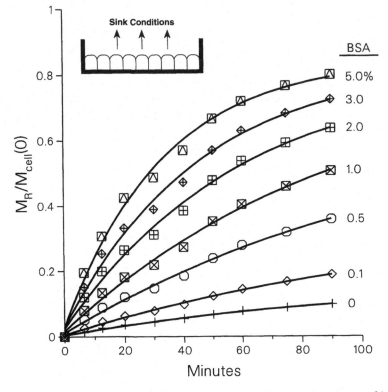

Figure 36 Efflux kinetics of PNU-78,517 from the apical membrane of MDCK cells in monolayer culture on a solid plastic surface as a function of bovine serum albumin concentration. [Redrawn from Raub et al. (1993) with permission from the publisher.]

%BSA	$P_e \times 10^{-6}$ cm/min	$M^*_{cell}/M_{cell}(0)$	$M_R(0)/M_{cell}(0)$	$1 - M^*_{cell}/M_{cell}(0)$
1.0	6.82	0.18	0.01	0.82
2.0	10.8	0.19	0.02	0.81
3.0	14.1	0.16	0.02	0.84
5.0	19.3	0.15	0.02	0.85

desorbed from the membrane surface, $M_R(0)/M_{cell}(0)$, is small (~2%). The mass fraction of drug in some cell-associated, deep kinetic pool, $M^*_{cell}/M_{cell}(0)$, is about 15%. Table 19 shows that apical and basolateral membrane coefficients are quantitatively equivalent in their dependency on BSA concentration. The desorption kinetics are membrane-controlled.

D. Biophysical Model for Transcellular Flux Kinetics

The aforementioned studies employing various kinetic boundary conditions (uptake by the monolayers on a plastic substrate, efflux from the apical (AP) membrane of plastic-grown cells pre-equilibrated with drug, and efflux from the basolateral (BL) membrane of filter-grown cells pre-equilibrated with drug) permit the mechanistic and quantitative dissection of the kinetic process of apical-to-basolateral translocation of membrane-interative compounds in the presence and absence of BSA. The mathematical model for transmonolayer kinetics follows.

The rate of change in total concentration of drug species in the donor is equal to the net rate of uptake by the cells leading to quasi-equilibrium and efflux from the BL membrane and across the filter support into the receiver (sink) (see Fig. 31); thus,

$$-V_D \frac{dC_D}{dt} = AP_e\left(C_D - \frac{C_{cell}}{K_e}\right) \tag{137}$$

Table 19 Comparison Between Permeability Coefficients of the Apical and Basolateral Membranes from Independent Efflux Kinetics from Drug-Loaded MDCK Cell Monolayers

% BSA in receiver solution	Efflux P_{ap} ($\times 10^{-6}$ cm/min)[a]	Efflux P_{bl} ($\times 10^{-6}$ cm/min)[b]
0	0.7	0.6
0.1	5.2	—
0.5	5.4	—
1.0	6.8	10.7
2.0	10.8	—
3.0	14.1	19.6
5.0	19.3	21.0

[a] See Figure 36.
[b] Kinetic data not shown.
Source: Raub et al. (1993).

The concurrent rate of appearance in the receiver is

$$V_R \frac{dC_R}{at} = AP'_{bl} C_{cell} \tag{138}$$

From the mass balance considerations, the cell monolayer concentration is

$$C_{cell} = \frac{V_D}{V_{cell}} \left[C_D(0) - C_D - \frac{V_R}{V_D} C_R \right] \tag{139}$$

whereupon its substitution into Eqs. (137) and (138) gives

$$\frac{dC_D}{dt} = -\alpha C_D + \alpha\beta \left[C_D(0) - C_D - \frac{V_R}{V_D} C_R \right] \tag{140}$$

$$\frac{dC_R}{dt} = \gamma \frac{V_D}{V_R} \left[C_D(0) - C_D - \frac{V_R}{V_D} C_R \right] \tag{141}$$

where α and β are as previously defined. The efflux rate constant γ is

$$\gamma = \frac{AP'_{bl}}{V_{cell}} \text{min}^{-1} \tag{142}$$

where the effective basolateral membrane permeability coefficient, P'_{bl}, takes into account the permeability coefficients of the basolateral membrane itself (P_{bl}), filter (P_F), and aqueous boundary layer (P_{ABL}) and the fractional area (ε) of the basolateral membrane that is readily accessible to acceptor molecules as defined by the porosity of the filter:

$$P'_{bl} = \frac{1}{\dfrac{1}{\varepsilon P_{bl}} + \dfrac{1}{P_F} + \dfrac{1}{P_{ABL}}} \tag{143}$$

The initial conditions are $C_D = C_D(0)$ at $t = 0$ and $C_R = 0$ at $t = 0$. Efforts to obtain analytical solutions are tedious and unnecessary. By applying the change in concentrations (or mass) in the donor and receiver solutions with time to the Laplace transforms of Eqs. (140) and (141), the inverse of the simultaneous transformed equations can be numerically calculated with appropriate software for best estimates of α, β, and γ. It is implicit here that P_e, P_{ap}, P_{bl}, and K_e are functions of protein binding. Upon application of the transmonolayer flux model to the PNU-78,517 data in Figure 32, the effective permeability coefficients from the disappearance and appearance kinetics points of view are in good quantitative agreement with the permeability coefficients determined from independent studies involving uptake kinetics by MDCK cell monolayers cultured on a flat dish

Table 20 Comparison of Transcellular (AP to BL)
Permeability Coefficients with Permeability Coefficients from
Independent Uptake and Efflux Kinetic Studies

BSA concn	AP to BL flux P_e (cm/min)	AP uptake[a] P_e (cm/min)	BL efflux[b] P_{bl} (cm/min)
Donor			
3%	3.2×10^{-3}	5.0×10^{-3}	NA
Receiver			
1%	2.2×10^{-6}	NA[c]	1.6×10^{-6}
3%	4.1×10^{-6}	NA	2.9×10^{-6}
5%	4.6×10^{-6}	NA	3.2×10^{-6}

[a] Uptake of PNU-78,517 in 3% BSA donor solution by MDCK cell monolayer cultured on a plastic dish.
[b] Efflux of PNU-78,517, initially loaded in MDCK cell monolayer cultured
on Transwell filter, into receiver sink solution varying in BSA concentration.
[c] NA = not applicable.
Source: Raub et al. (1993).

and efflux kinetics from drug-loaded cells and across the filter support into the
receiver sink (Table 20).

In conclusion, Eqs. (140) and (141) are applicable to quantify not only the
transmonolayer kinetics of highly membrane interactive permeants, but also the
kinetics of less membrane interactive compounds. Notably, the examples emphasize the importance of simultaneously measuring the disappearance of a compound from the donor solution and its appearance in the receiver and demonstrate
how interactions with proteins on either side of the cellular barrier influence permeability.

IX. USE OF IN VITRO MONOLAYER STUDIES TO PREDICT RESULTS IN VIVO

The identification and characterization of cell culture systems (e.g., Caco-2-cells)
that mimic in vivo biological barriers (e.g., intestinal mucosa) have afforded pharmaceutical scientists the opportunity to rapidly and efficiently assess the permeability of drugs through these barriers in vitro. The results generated from these
types of in vitro studies are generally expressed as effective permeability coefficients (P_e). If P_e is properly corrected to account for the barrier effects of the
filter (P_F) and the aqueous boundary layer (P_{ABL}) as previously described in Section II.C, the results provide the permeability coefficient for the cell monolayer

(P_M). Many scientists have tried to relate these P_e or P_M values for solutes to parameters which are determined from in vivo studies (e.g., oral bioavailability). In vitro–in vivo correlations should be performed judiciously with full appreciation of the nature of the parameters being measured.

As an example, consider the issue of correlating Caco-2 cell permeability data generated in vitro with oral bioavailability data generated in vivo. First, it is important to define oral bioavailability. The simplest definition of oral bioavailability is that it is the fraction of an oral dose reaching the systemic circulation as unchanged drug. Thus, the oral bioavailability of a drug is dependent not only on its permeability across the intestinal mucosa but also on the extent of metabolism which the drug undergoes in the intestinal mucosa. In addition, the oral bioavailability of a drug is dependent upon its propensity to undergo first-pass metabolism and/or clearance by the liver, and on formulation, physicochemical and intestinal physiological factors (Ho et al., 1983). In contrast, the permeability coefficient that is determined from a Caco-2 cell transport experiment measures *only one factor* that determines oral bioavailability. If the rate and/or the pathway of drug metabolism in Caco-2 cells differs from that observed in the intact intestinal mucosa, or if the drug undergoes first-pass metabolism and/or clearance by the liver, then one would not expect a correlation between permeability coefficient values and oral bioavailability. If, however, the drug is neither metabolized by the intestinal mucosa nor metabolized and/or cleared by the liver nor prevented from absorption by limited solubility in the GI tract (dissolution-limited), then permeability through the intestinal mucosa becomes the predominant factor in determining oral bioavailability. Under these circumstances, a correlation between permeability coefficient values and oral bioavailability might be expected.

Instead of using the oral bioavailability of a drug, one can attempt to correlate P_M values with permeability coefficients generated from in situ perfused intestinal preparations. Here, one eliminates the complexities of liver metabolism, clearance, and formulation variables. Recently, this type of in vitro–in situ correlation has been conducted using the model peptides (described previously in Section V.B.2). The permeabilities of these model peptides were determined using a perfused rat intestinal preparation which involved cannulation of the mesenteric vein (Kim et al., 1993). With this preparation, it was possible to measure both the disappearance of the peptides from the intestinal perfusate and the appearance of the peptides in the mesenteric vein. Thus, clearance values (CL_{app}) could be calculated for each peptide. Knowing the effective surface area of the perfused rat ileum, the CL_{app} values could be converted to permeability coefficients (P). When the permeability coefficients of the model peptides were plotted as a function of the lipophilicity of the peptides, as measured by partition coefficients in octanol–water, a poor correlation ($r^2 = 0.02$) was observed. A better correlation was observed between the permeabilities of these peptides and the number of potential hydrogen bonds the peptide can make with water ($r^2 = 0.56$,

Fig. 37), suggesting that desolvation of the polar bonds in the molecule is a major determinant of permeability. Consistent with this, good correlations were found between the permeabilities of these peptides and their partition coefficients between heptane–ethylene glycol ($r^2 = 0.87$) or the differences in partition coefficients between n-octanol–buffer and isooctane–buffer ($r^2 = 0.82$); both these buffer systems provide experimental estimates of hydrogen-bonding potential. These results are qualitatively identical with those described earlier for the permeability of these peptides across Caco-2 cell monolayers.

Having determined the permeability of these model peptides in both the in vitro Caco-2 cell monolayer system and the in situ perfused rat model, we can compare the extent of correlation between these two models. As shown in Figure 38, a good *qualitative* correlation exists between the two permeability coefficients ($r^2 = 0.82$), although the magnitudes of the values differ. The in vitro model gives permeability coefficients which are consistently larger than those obtained in the rat. These results indicate that the data generated from Caco-2 cell monolayers would not have estimated the "absolute" permeability values observed

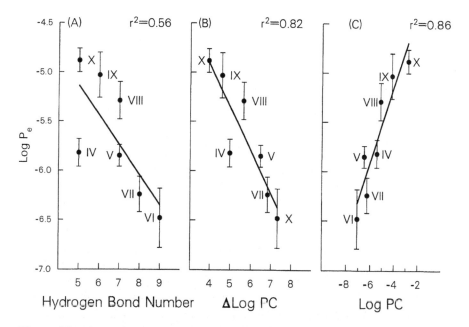

Figure 37   Correlations of log monolayer permeability coefficients for a series of related peptides and various physicochemical indices (see Table 14). In plot (C) the log PC values refer to the immiscible n-heptane/ethylene glycol partitioning system. [Redrawn from Kim et al. (1993) with permission from the publisher.]

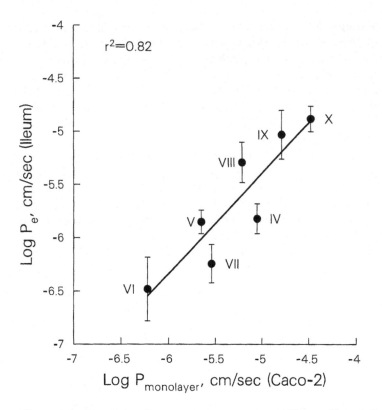

Figure 38 Correlations between appearance permeability coefficients for a related series of peptides measured in mesenteric blood draining perfused rat ileal segments and
Caco-2 cell monolayers in the Transwell system. See Table 14 for identification of the
peptides. The P_e for the rat ileum was not corrected for the aqueous boundary layer and
blood flow effects. [Redrawn from Kim et al. (1993) with permission from the publisher.]

in the perfused rat intestinal model, but the Caco-2 values would have accurately
predicted in the rank order of the model peptide permeabilities in situ.

This type of information about a homologous series of drug candidates,
when considered in light of the propensity of these compounds to undergo first-
pass metabolism and/or liver clearance, allows pharmaceutical scientists to make
more intelligent decisions about which compounds to move into animal studies.
In addition, when an in vitro–in vivo correlation can be demonstrated for a series
of compounds, the results of Caco-2 experiments can be used as a guide by medicinal chemists to make structural modifications to optimize oral bioavailability.

ACKNOWLEDGMENTS

We are indebted to the laboratory personnel who have generated much of the data used to illustrate the concepts. They are Al Hilgers, Bob Conradi, and Geri Sawada of P&U and D. C. Kim, Ph.D., formerly of the University of Kansas. Our special appreciation goes to Sharon Kramer (P&U) for her diligent technical assistance in the preparation of the manuscript.

REFERENCES

Adson A. (1992). Quantitative approaches to delineate passive transport mechanisms in cell culture monolayers. MSc Thesis, University of Kansas, Lawrence, KS.

Adson A, TJ Raub, PS Burton, CL Barsuhn, AR Hilgers, KL Audus, NFH Ho. (1994). Quantitative approaches to delineate paracellular diffusion in cultured epithelial cell monolayers. J Pharm Sci 83:1529–1536.

Adson A, PS Burton, TJ Raub, CL Barsuhn, KL Audus, NFH Ho. (1995). Passive diffusion of weak organic electrolytes across Caco-2 cell monolayers: Uncoupling the contributions of hydrodynamic, transcellular and paracellular barriers. J Pharm Sci 84: 1197–1204.

Artursson P, C Magnusson. (1990). Epithelial transport of drugs in cell culture. II. Effect of extracellular calcium concentration on the paracellular transport of drugs of different lipophilicities across monolayers of intestinal epithelial (Caco-2) cells. J Pharm Sci 79:595–600.

Balda MS. (1992). Intracellular signals in the assembly and sealing of tight junctions. In: M Cereijido, ed. Tight Junctions. Boca Raton, FL: CRC Press, pp 121–137.

Bentzel CJ, CE Palant, M Fromm. (1992). Physiological and pathological factors affecting the tight junction. In: M Cereijido, ed. Tight Junctions. Boca Raton, FL: CRC Press, pp 151–173.

Burton PS, RA Conradi, AR Hilgers, NFH Ho, LL Maggiora. (1992). The relationship between peptide structure and transport across epithelial cell monolayers. J Controlled Release 19:87–98.

Burton PS, JT Goodwin, RA Conradi, NFH Ho, AR Hilgers. (1997). In vitro permeability of peptidomimetic drugs—The role of polarized efflux pathways as additional barriers to absorption. Adv Drug Delivery Rev 23:143–156.

Buur A, N Mørk. (1992). Metabolism of testosterone during in vitro transport across Caco-2 cell monolayers: Evidence for beta-hydroxysteroid dehydrogenase activity in differentiated Caco-2 cells. Pharm Res 9:1290–1294.

Cereijido M. (1992). Introduction: Evolution of ideas on tight junctions. In: M Cereijido. ed. Tight Junctions. Boca Raton, FL: CRC Press, pp 1–22.

Cereijido M, CA Robbins, DD Sabatini. (1987). Polarized epithelial membranes produced in vitro. In: JF Hoffman, ed. Membrane Transport Processes. New York: Raven Press, pp 443–456.

Cho MJ, DP Thompson, CT Cramer, T Vidmar, JF Scieszka. (1989). The Madin–Darby canine kidney (MDCK) epithelial cell monolayer as a model cellular transport barrier. Pharm Res 6:71–77.

Conradi RA, AR Hilgers, NFH HO, PS Burton. (1991). The influence of peptide structure on transport across Caco-2 cells. Pharm Res 8:1453–1460.

Conradi RA, AR Hilgers, NFH Ho, PS Burton. (1992). The influence of peptide structure on transport across Caco-2 cells. II. Peptide bond modification which results in improved permeability. Pharm Res 9:435–439.

Contreras RG, A Ponce, JJ Bolivar. (1992). Calcium and tight junctions. In: M Cereijido, ed. Tight Junctions, Boca Raton, FL: CRC Press, pp 139–149.

Curry FE. (1984). Mechanics and thermodynamics of transcapillary exchange. In: EM Renkin, CC Michel, eds. Handbook of Physiology, Section 2, The Cardiovascular System. Vol. IV, Microcirculation, Part 2. Bethesda, MD: Am Physiol Society, pp 309–374.

Denker BM, SK Nigam. (1998). Molecular structure and assembly of the tight junction. Am J Physiol 274 (Renal Physiol. 43):F1—F9.

Diamond JM. (1977). The epithelial junction: Bridge, gate, and fence. Physiologist 20: 10–18.

González-Mariscal L, BC de Ramirez, M Cereijido. (1984). Effect of temperature on the occluding junctions of monolayers of epithelioid cells (MDCK). J Membrane Biol 79:175–184.

Gumbiner B. (1987). Structure, biochemistry, and assembly of epitheliac tight junctions. Am J Physiol (Cell Physiol) 253:C749–C758.

Hecht G, C Pothoulakis, JT LaMont, JL Madara. (1988). *Clostridium difficile* toxin A perturbs cytoskeletal structure and tight junction permeability of cultured human intestinal epithelial monolayers. J Clin Invest 82:1516–1524.

Hidalgo IJ, KM Hillgren, GM Grass, RT Borchardt. (1991). Characterization of the unstirred water layer in Caco-2 cell monolayers using a novel diffusion apparatus. Pharm Res 8:222–227.

Hidalgo IJ, KM Hillgren, GM Grass, RT Borchardt. (1992). A new side-by-side diffusion cell for studying transport across epithelial cell monolayers. In Vitro Cell Dev Biol 28A:578–580.

Hirsch M, W Noske. (1993). The tight junction: Structure and function. Micron 24:325–352.

Ho NFH, JY Park, W Morozowich, WI Higuchi. (1977). Physical model approach to the design of drugs with improved intestinal absorption. In: EB Roche, ed. Design of Biopharmaceutical Properties Through Prodrugs and Analogs. Washington, DC: APhA/APS, pp 136–227.

Ho NFH, JY Park, PF Ni, WI Higuchi. (1983). Advancing quantitative and mechanistic approaches in interfacing gastrointestinal drug absorption studies in animals and man. In: WG Crouthamel, A Sarapu, eds. Animal Models for Oral Drug Delivery in Man: In Situ and In Vivo Approaches. Washington, DC: APh/APS, pp 27–106.

Ho NFH, DP Thompson, TG Geary, TJ Raub, CL Barsuhn. (1990). Biophysical transport properties of the cuticle of *Ascaris suum*. Mol Biochem Parasitol 41:153–166.

Ho NFH, PS Burton, RA Conradi, CL Barsuhn. (1995). A biophysical model of passive and polarized active transport processes in Caco-2 cells: Approaches to uncoupling apical and basolateral membrane events in the intact cell. J Pharm Sci 84:21–27.

Imanidis G, C Waldner, C Mettler, H Leuenberger. (1996). An improved diffusion cell design for determining drug transport parameters across cultured cell monolayers. J Pharm Sci 85:1196–1203.

Jacobs RE, SH White. (1989). The nature of hydrophobic binding of small peptides at the bilayer interface: Implications for the insertion of transbilayer helices. Biochemistry 28:3421–3437.

Karlsson J, P Artursson. (1991). A method for the determination of cellular permeability coefficients and aqueous boundary layer thickness in monolayers of intestinal epithelial (Caco-2) cells grown in permeable filter chambers. Int J Pharm 7:55–64.

Katz MA, RC Schaeffer Jr. (1991). Convection of macromolecules is the dominant mode of transport across horizontal 0.4 and 0.3 μm filters in diffusion chambers: Significance for biologic monolayer permeability assessment. Microvasc Res 41:149–163.

Kim DC, PS Burton, RT Borchardt. (1993). A correlation between the permeability characteristics of a series of peptides using an in vitro cell culture model (Caco-2) and those using an in situ perfused rat ileum model of the intestinal mucosa. Pharm Res 10:1710–1714.

Knipp GT, NFH Ho, CL Barsuhn, RT Borchardt. (1997). Paracellular diffusion in Caco-2 cell monolayers: Effect of perturbation on the transport of hydrophilic compounds that vary in charge and size. J Pharm Sci 86:1105–1110.

Komiya I, JY Park, NFH Ho, WI Higuchi. (1980). Quantitative mechanistic studies in simultaneous fluid flow and intestinal absorption using steroids as model solutes. Int J Pharm 4:249–262.

Lee CP, RLA Devrueh, PL Smith. (1997). Selection of development candidates based on in vitro permeability measurements. Adv Drug Delivery Rev 23:47–62.

Levich VG. (1962). Physicochemical Hydrodynamics. Englewood Cliffs, NJ: Prentice-Hall, Chapters 2 and 3.

Lum BL, MP Gosland. (1995). MDR expression in normal tissues. Pharmacologic implications for the clinical use of p-glycoprotein inhibitors. Hermatol Oncol Clin North Am 9:319–336.

Lutz KL, T Siahaan. (1997). Molecular structure of the apical junction complex and its contribution to the paracellular barrier. J Pharm Sci 86:977–984.

Madara JL. (1983). Increases in guinea pig small intestinal transepithelial resistance induced by osomotic loads are accompanied by rapid alterations in absorptive-cell tight-junction structure. J Cell Biol 97:125–136.

Madara JL. (1998). Regulation of the movement of solutes across tight junctions. Annu Rev Physiol 60:143–159.

Madara JL, G Hecht. (1989). Tight (occluding) junctions in cultured (and native) epithelial cells. In: KS Matlin, JD Valentich, eds. Functional Epithelial Cells in Culture. New York: Alan Liss, pp 131–163.

Madara JL, D Barenberg, S Carlsson. (1986). Effects of cytochalasin D on occluding junctions of intestinal absorptive cells: Further evidence that the cytoskeleton may influence paracellular permeability and junctional charge selectivity. J Cell Biol 102:2125–2136.

Michel CC. (1988). Capillary permeability and how it may change. J Physiol 404:1–29.

Miller DW, KL Audus, RT Borchardt. (1992). Cultured bovine brain microvessel endothelial cells: A model of the blood brain barrier. J Tiss Cult Meth 14:217–224.

Mitic LL, JM Anderson. (1998). Molecular architecture of tight junctions. Annu Rev Physiol 60:121–142.

Polak-Charcon S. (1992). Proteases and the tight junction. In: M Cereijido, ed. Tight Junctions. Boca Raton, FL: CRC Press, pp 257–277.

Raub TJ, CL Barsuhn, LR Williams, DE Decker, GA Sawada, NFH Ho. (1993). Use of a biophysical-kinetic model to understand the roles of protein binding and membrane partitioning on passive diffusion of highly lipophilic molecules across cellular barriers. J Drug Targeting 1:269–286.

Renkin EM. (1954). Filtration, diffusion and molecular sieving through porous cellulose membranes. J Gen Physiol 38:225–238.

Sawada GA, NFH Ho, LR Williams, CL Barsuhn, TJ Raub. (1994). Transcellular permeability of chlorpromazine demonstrating the roles of protein binding and membrane partitioning. Pharm Res 11:665–673.

Schaeffer JR, RC Gong, MS Bitrick Jr. (1992). Restricted diffusion of macromolecules by endothelial cell monolayers and small pore filters. Am J Physiol 263:L227–L236.

Schnitzer JE. (1993). Update on the cellular and molecular basis of capillary permeability. Trends Cardiovasc Med 3:124–130.

Seiler P. (1974). Interconversions of lipophilicites from hydrocarbon/water systems in the octanol/water systems. Eur J Med Chem 9:473–479.

Shasby DM, SS Shasby. (1990). Endothelial cells grown on filter membranes. In: HM Piper, ed. Cell Culture Techniques in Heart and Vessel Research. Stuttgart: Springer-Verlag, pp 212–219.

Singer SJ. (1974). The molecular organization of membranes. Annu Rev Biochem 43:805–834.

Stein WD. (1967). The Movement of Molecules Across Cell Membranes. New York: Academic Press, pp 65–125.

Tanford C. (1963). Physical Chemistry of Macromolecules. New York: Wiley, p 361.

Taylor AE, KA Gaar Jr. (1970). Estimation of equivalent pore radii of pulmonary capillary and alveolar membranes. Am J Physiol 218:1133–1140.

van Meer G, K Simons. (1986). The function of tight junctions in maintaining differences in lipid composition between the apical and the basolateral cell surface domains of MDCK cells. EMBO J 5:1455–1464.

van Meer G, W van't Hoff, I van Genderen. (1992). Tight junctions and polarity of lipids. In: M Cereijido, ed. Tight Junctions. Boca Raton, FL: CRC Press, pp 187–201.

Watkins PB. (1997). The barrier function of CYP3A4 and P-glycoprotein in the small bowel. Adv Drug Delivery Rev 27:161–170.

Zweibaum A, M Laburthe, E Grasset, D Louvard. (1991). Use of cultured cell lines in studies of intestinal cell differentiation and function. In: M Field, CA Frizzell, eds. Handbook of Physiology, Section 6, The Gastrointestinal System, Vol. IV, Intestinal Absorption and Secretion. Bethesda, MD: Am Physiol Society, pp 223–255.

9

Barriers to Drug Transport in Ocular Epithelia

Uday B. Kompella* and Vincent H. L. Lee
University of Southern California, Los Angeles, California

I. INTRODUCTION

Topically applied drugs can reach their target tissues in the eye via the corneal and/or the conjunctival/scleral pathways, as shown in Scheme 1 [1]. Although the corneal pathway of absorption was once thought to be the exclusive pathway, the work of Doane et al. [2] and Ahmed and Patton [1,3] clearly established that drugs can also gain access to the anterior segment tissues following uptake by the conjunctiva and subsequent diffusion across the sclera. This so-called non-corneal route of penetration was shown to contribute as much as 65% and 80% to the uptake of p-aminoclonidine (log PC = -0.96) and inulin (log PC = -2.98), respectively, by the iris-ciliary body following topical dosing [3,4].

The ocular bioavailability of topically applied ophthalmic drugs is poor, e.g., 0.5–2% for pilocarpine [5], 1.6% for clonidine [6], and 7–10% for flurbiprofen [7]. There are two main reasons: (1) the short residence time of drug in the eye due to nasolacrimal drainage and (2) permeability and metabolic barriers imposed by the ocular epithelia. The major research thrust to improve ocular drug bioavailability during the last decade focused on optimizing the residence time of the applied dose through the use of viscous vehicles [8–10], bioadhesive polymers [11–15], inserts [16,17], in situ gels [18–20], collagen shields [21–24], and ophthalmic rods [25,26]. Modulation of the epithelial barriers of the eye to enhance ocular drug absorption has received less attention. Recent efforts in

---

* Current affiliation: University of Nebraska Medical Center, Omaha, Nebraska

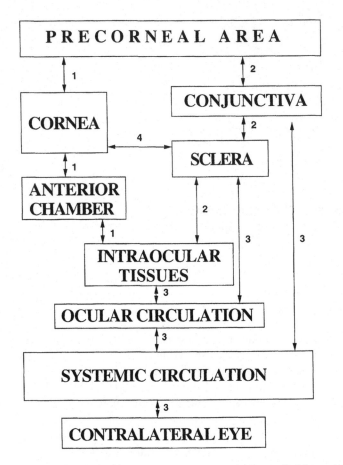

Scheme 1   Ocular penetration routes for topically applied drugs. 1, Transcorneal pathway; 2, noncorneal pathway; 3, systemic return pathway; 4, lateral diffusion. (From Ref. 1.)

this regard have been focused on the corneal pathway of transport through the use of (1) prodrugs [10,27–30], ion-pairing agents [31,32], and penetration enhancers [33–37] to overcome the permeability barrier and (2) enzyme inhibitors [38] to overcome metabolic barriers [39–45]. The exciting discovery that epithelial barrier properties are dynamic and are subject to reversible modulation under physiological conditions [46–50] will undoubtedly open new vistas for improving the ocular bioavailability of topically administered drugs through either the corneal or non-corneal pathway.

The focus of this chapter is on the penetration barrier. The metabolic barrier due to esterases [39–41], aminopeptidases [38], and ketone reductase [43,51] has been reviewed elsewhere [52]. The present chapter reviews and describes the following aspects pertinent to corneal and conjunctival drug transport: (1) morphology and electrophysiology, (2) location and permselectivity of permeability barriers, and (3) conventional and novel approaches to enhance drug transport across the cornea and the conjunctiva. Where appropriate, relevant information obtained from other mucosal epithelia is extrapolated to the cornea and conjunctiva.

II. MORPHOLOGY AND ELECTROPHYSIOLOGY OF THE CORNEA AND THE CONJUNCTIVA

A. Morphology

1. Cornea

The main function of the cornea, 500–600 μm thick, is to provide a refractive interface to the eye within the confines of mechanical rigidity and chemical impermeability. The cornea is composed of six layers [53,54]: (1) the outermost epithelium, forming an impermeable barrier; (2) the basement membrane, forming a boundary between epithelium and stroma; (3) the acellular Bowman's layer, a modified superficial layer of stroma; (4) the stroma, constituting 90% of the corneal thickness and playing a key role in maintaining corneal transparency and hydration; (5) Descemet's membrane, formed from the endothelial cells, which is the basement membrane for the corneal endothelium; and (6) the corneal endothelium, which confers corneal deturgescence through its Na^+ pump activity.

The corneal epithelium is a stratified squamous epithelium like the epidermis of the skin but is nonkeratinized like other mucosal epithelia such as the intestinal and airway epithelia. Although the corneal epithelium, five to seven cells thick, represents less than 10% of the entire corneal thickness, it provides as much as 99% resistance to the diffusion of small electrolytes such as Na^+ and Cl^- [55–57].

Three principal cell types exist in the corneal epithelium, namely, superficial cells, wing cells, and basal cells. A single layer of cuboidal basal cells is the sole site of cell division in the corneal epithelium. By necessity, these cells have more prominent mitochondria and Golgi apparatuses and hence high levels of metabolic and synthetic activities. Immediately above the basal cells is a zone comprising two to three layers of wing cells that are in an intermediate state of differentiation.

The superficial cells are irregular arrays of polygonal cells with a diameter of 40–60 μm and a thickness of 2–6 μm each. These cells, the most differentiated cells of the epithelium, possess microvilli in their apical surfaces, which are covered with a glycocalyx. It is, however, controversial whether mucus exists on their surface [58,59]. As cell division occurs in the basal cells of the cornea, the daughter cells move toward the surface while becoming more differentiated. As the daughter cells migrate toward the outermost layer, the superficial cells are sloughed off from the corneal surface, resulting in a 7 day turnover of the entire epithelium. Thoft and Friend [60] hypothesized that the proliferation rate of basal cells and the centripetal movement of the daughter cells toward the surface should equal the sloughing rate of the superficial cells. Zonula occludens, or tight junctions contributing to the tightness of the corneal epithelium, exist only between the superficial cells. By contrast, only focal tight junctions (macula occludens), that are not entirely occlusive, exist between endothelial cells [61].

2. Conjunctiva

The conjunctiva is a thin, transparent mucous membrane lining the inside of the eyelids and is continuous with the cornea. The conjunctiva occupies a ninefold larger surface area than the cornea in rabbits and a 17 times larger surface area in humans [62]. In addition to being a protective barrier, the conjunctiva secretes mucus believed to be essential for tear film stability [63–65]. The portion of the conjunctiva covering the posterior surface of the lids is referred to as the palpebral conjunctiva, while that covering the anterior surface of the sclera is referred to as the bulbar conjunctiva. The palpebral conjunctiva is firmly attached to the lids, whereas the bulbar conjunctiva is loosely attached to the underlying sclera and is folded several times. The conjunctiva can be divided into three layers [66]: (1) an outer epithelium, forming a permeability barrier; (2) the substantia propria, containing structural and cellular elements, nerves, lymphatics, and blood vessels; and (3) the submucosa, providing a loose attachment to the underlying sclera.

The conjunctival epithelium is a stratified epithelium, squamous in nature at the lids and columnar toward the cornea. The thickness of the conjunctiva varies from region to region, being 10–15 layers thick toward the cornea and five to six layers thick at the lids. As in the cornea, superficial, wing, and basal cells are the three principal epithelial cell types in the conjunctiva. A greater number of wing cell layers in the conjunctiva compared with the corneal epithelium accounts for its greater thickness in the corneal periphery. Interspersed among the superficial cells are round or oval mucus-secreting goblet cells, while the wing and basal cells may contain pigment. Apart from epithelial cells, the conjunctiva contains Langerhans cells, melanocytes, and wandering inflammatory cells [66]. The conjunctival epithelium possesses dense microvilli covered with a glycocalyx and a mucus layer. The surface cells of the conjunctiva are connected by tight junctions which render the epithelium a relatively impermeable barrier [67]. The epithelial cells are further attached to one another by means of desmosomes and to the basal lamina through hemidesmosomes. In the event of conjunctival epithelial injury, these cells will slide over to cover the defect followed by mitosis [68]. The same cells will slide over to repair peripheral corneal epithelial defects, followed by morphological transformation or transdifferentiation to normal corneal epithelial cells after 4–5 weeks [69–71].

Thus, the conjunctiva differs from the cornea in vasculature, mucus secretion, and ability to transdifferentiate. While the conjunctiva is highly vascularized and capable of mucus secretion and transdifferentiation, the cornea is avascular, depends on the conjunctiva for its surface mucus, and is incapable of transdifferentiation. The conjunctiva is, however, similar to the cornea in that the conjunctival epithelium is a nonkeratinized, stratified epithelium with microvilli covered with glycocalyx and mucus [66,72]. Also, the junctional complexes including tight junctions, gap junctions, desmosomes, and hemidesmosomes are present in both ocular epithelia.

B. Electrophysiology

Similar to the intestinal, colonic, and airway epithelia, the epithelial cells of the cornea and the conjunctiva adhere to the neighboring cells through cell adhesion molecules (CAMs) and the apically localized tight junctions [67,73,74]. These intercellular junctions may help maintain the asymmetrical distribution of (1) several proteins including receptors for hormones such as insulin, epidermal growth factor, and transferrin and (2) ion transporters such as Na^+/K^+-ATPase, Cl^- channels, and Na^+-amino acid cotransporters in order to maintain the vectorial transport of hormones and ions, respectively. In addition, the intercellular junctions render the corneal epithelium highly resistant to the passage of small electrolytes such as Na^+ and Cl^- [56]. Establishment of cell–cell contacts is a sign of healing in corneal wounds [75,76]. Despite the importance of the intercellular junctions in wound healing and vectorial transport, little is known about the constitution of the intercellular junctions of the cornea and conjunctiva. While uvomorulin has been identified as a protein capable of establishing cell–cell contacts in a Ca^{2+}-dependent manner in renal epithelia, it has not been characterized in either the cornea or conjunctiva [77–79]. While ZO-1 and cingulin have been identified as the structural units on the cytoplasmic side of the tight junctions in renal and intestinal epithelia [80–82], only ZO-1 has been identified thus far in the cornea [83]. To date, no such information exists in the conjunctiva. Once the structure of the elements constituting intercellular junctions is known, assembly and regulation of these junctional units can be evaluated, which, in turn, may facilitate selective opening of the junctions to enhance drug transport or reassembly of the junctions during wound healing.

The electrophysiology of epithelial tissues can be evaluated at three levels: transepithelial, transmembrane (apical and basolateral), or single-channel level. Transepithelial measurements are usually made in Ussing-type lucite chambers equipped with a voltage-clamp unit and two pairs of calomel and Ag/AgCl electrodes to measure voltage and current, respectively, across the epithelium [84]. In order to obtain correct values of electrical parameters, the electrodes should be connected to the bathing fluids through salt bridges. The ideal salt bridge should have a small electrode-bathing fluid junction independent of the bathing solution composition. KCl (3 M) gelled with 3–4% agar in a fine polyethylene tubing satisfies these requirements [85]. Using this setup, it is possible to obtain simultaneous estimates of transepithelial potential difference (PD), electrical resistance (R_t), and tracer permeability. The estimates of PD and R_t can be made within a few seconds using this method. Furthermore, clamping the PD of the epithelium at 0 mV (i.e., under short-circuit conditions) allows the measurement of short-circuit current (I_{sc}), an index of active ion transport measured as the ratio of PD to R_t.

Maurice [55,75], Klyce [76,86,87], and Klyce and Wong [88] established much of the current understanding of the active ion transport properties of the rabbit cornea using the above technique. The same principle in transepithelial measurements can be applied to the measurement of the electrical parameters of the apical membrane, the basolateral membrane, or a single channel. In order to measure the transmembrane permeability of ions, microelectrodes drawn from fiber-filled borosilicate glass tubing of 1 mm outer diameter and 0.58 mm inner diameter can be used to impale the epithelium mounted as a flat sheet [89]. One microelectrode is placed in the bathing solution while another is placed in the cell cytoplasm for the transmembrane measurements. A micromanipulator should be used to position the microelectrodes, and high-impedance preamplifiers or electrometers should be used for potential measurements. Microelectrodes were used by Klyce [76] to delineate the contribution of intercellular junctions and apical and basolateral membranes of corneal superficial cells to the overall resistance of the cornea, which is discussed subsequently. Ion-selective electrodes can be used in conjunction with microelectrodes to measure both cell potential and ion activities within cells [90,91]. Apical and basolateral membrane permeabilities can also be measured by using membrane vesicles prepared from them. Ion flux into or out of these vesicles can be measured by using atomic absorption spectroscopy, radioactive tracers, or potential-sensitive cyanine dyes [92,93].

The conductance of single ion channels, rate of opening, and regulation can be studied directly using the patch-clamp technique [94,95]. This technique utilizes a salt solution–containing pipet that is heat-polished and a current-to-voltage amplifier to provide the properties and intrinsic regulation of individual ion channels (Fig. 1A). If suction is applied after pressing the micropipet tip on

---

Figure 1 Schematic diagrams illustrating the patch-clamp technique. (A) Overall setup for isolating single ionic channels in an intact patch of cell membrane. P = patch pipet; R = reference microelectrode; I = intracellular microelectrode; V_p = applied patch potential; E_m = membrane potential; $V_m = E_m - V_p$ = potential across the patch; A = patch-clamp amplifier. (From Ref. 90.) (B) Five different recording configurations, and procedures used to establish them. (i) Cell attached or intact patch; (ii) open cell attached patch; (iii) whole cell recording; (iv) excised outside-out patch; (v) excised inside-out patch. Key: i = inside of the cell; o = outside of the cell. (Adapted from Ref. 283.)

A

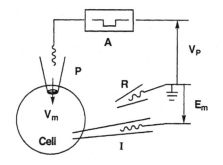

B

(i) <u>Cell Attached</u>

(ii) <u>Open Cell Attached</u>

Saponin

Pull

Suction

(iii) <u>Whole Cell Recording</u>

Air Exposure

Pull

(iv) <u>Excised Outside-Out Patch</u>

(v) <u>Excised Inside-Out Patch</u>

the surface of the epithelial cell, a patch of membrane will be blown into the pipet, forming a high resistance seal (on the order of gigaohms) with the glass. Provided that the resistance of the pipet is very small compared to that of the seal, virtually all of the current flowing through the single channels will be measured using the patch amplifier. Measurements can be made in intact, whole cell, or excised (outside-out or inside-out) patch configurations (Fig. 1B).

Different configurations can be obtained by controlling the extent of suction and pulling of the patch pipet. As shown in Fig. 1B, after the patch pipet is gently placed on the cell surface, application of slight suction will invaginate a patch of membrane into the pipet without dissociating the patch from the cell membrane. This configuration is known as a cell attached or intact patch [Fig. 1B(i)]. In this case, only the composition of the solution in the patch pipet can be varied. There is no control over the intracellular composition. Therefore, liquid junction potential corrections should be invoked in calculating the actual membrane potential difference, as the bathing fluids are not identical in their composition. An intracellular electrode must be inserted into the cell to measure the membrane potential. With high suction in the intact patch configuration, the patch will be detached from the pipet, allowing communication between the inside of the pipet and the inside of the whole cell. This configuration, known as the whole cell configuration [Fig. 1B(iii)], can be used to measure the current across the whole cell membrane. If the pipet is pulled away in this configuration, excised membrane fragments will reseal to form an excised outside-out membrane patch [Fig. 1B(iv)]. If the pipet is pulled away in the intact patch configuration, a closed membrane vesicle as shown in Fig. 1B(v) will be formed in the pipet tip. The outer surface of this vesicle can be disrupted by briefly passing the pipet tip through the air/water interface, thereby forming an excised inside-out patch [Fig. 1B(v)]. With the exception of the intact patch, these configurations allow the maintenance of identical bathing fluids on either side of the patch. In the intact cell attached patch configuration, detergents such as saponin can be used to permeabilize the cell membrane outside the patch membrane, in order to facilitate equilibration of the cell interior and exterior [Fig. 1B(ii)].

Depending on the skill of the experimenter, the patch will contain one or a few channels. When voltage is applied across this patch, the current rises and falls in a series of steps, all of which are multiples of unit current flow through a single channel. The smallest step current observed represents current flow through a single channel. From the applied voltage and the single-channel current flow observed, the single-channel conductance can be calculated. While both intact and excised patch configurations provide information on the single-channel conductance and rate of opening, the excised patch configuration is useful only to test certain agents that directly modulate the channels. That is, if Cl^- channels sensitive to cAMP-mediated phosphorylation are present in a patch membrane, cAMP will not be able to elevate single-channel current in an excised patch configuration, while it may do so in an intact patch configuration. This is because the cAMP-sensitive protein kinase A responsible for Cl⁻ channel activation through phosphorylation will not be present in an excised patch configuration. A combination of approaches is more useful to evaluate both direct and indirect regulation of single-channel properties. For instance, if phosphorylation of the Cl⁻ channel protein is responsible for activation, cAMP-induced stimulus should be obtained in an intact patch, which when pulled to form an excised inside-out patch should retain high Cl⁻ conductance as the channels were already phosphorylated.

Using whole cell and inside-out patches, the Cl⁻ channels of rat and rabbit corneal epithelium have been characterized [96]. Marshall and Hanrahan [96] observed ion channels with a single-channel conductance of 29 pS in the apical membranes of 1–4 day old corneal epithelial cells in culture. When bathed in identical electrolyte solutions, ion channels exhibit a linear current–voltage relationship with a reversal potential (potential at which the current changes its polarity) of 0 mV. A shift in reversal potential following dilution of the bathing electrolytes is often used to determine the ion selectivity of the channel in question. Thus, Marshall and Hanrahan [96] observed a shift in the reversal potential from 0 mV to -17.1 mV, when the bathing fluid was changed from 150 mM NaCl to 75 mM NaCl plus 75 mM sucrose, suggesting a Cl⁻-to-Na⁺ selectivity of $1:0.017$. The single-channel conductance as well as the Cl⁻ selectivity of the channels were identical in both rat and rabbit corneal epithelial cells. Fluctuation analysis, an indirect and more cumbersome approach, is also useful to obtain the single ion channel properties [97].

In the following discussion, the ion transport processes of the cornea and the conjunctiva obtained using one or more of the aforementioned techniques are summarized.

1. Cornea

Ion transport processes of the cornea and the conjunctiva can play an important role in maintaining intra- and extracellular fluid homeostasis, signal transduction, and intercellular communication. As all these functions may contribute to the modulation of drug transport (see Section IV.B), it is essential that the baseline ion transport processes in the cornea and the conjunctiva be understood.

a. *Origin of Potential Difference and Electrical Resistance.* The rabbit cornea is a tight epithelium with a mean transcorneal resistance (R_t) of 7.5 kohm · cm², a mean transepithelial potential difference (PD) of 21.4 mV (measured in vitro as the difference between the tear potential and the stromal potential [98]), and a short-circuit current (I_{sc}) of 2.85 µA/cm². By placing microelectrodes at different depths in the epithelium of the isolated cornea, Klyce [76] discerned three major steps in the corneal PD. When the superficial cells were first penetrated, the microelectrode recorded a negative potential of 30 mV. The underlying wing cells had the same potential difference as the superficial cells, possibly due to the presence of numerous ion-conducting desmosomes between superficial and wing cells. The potential became 5–10 mV more negative when the microelectrodes penetrated the basal cells, indicating less electrical coupling between cells at this level. Finally, penetration of the microelectrodes beyond the basolateral membrane of basal cells revealed a more positive potential that was similar to the electrical potential in the aqueous humor, since the transendothelial potential was small.

Using the same microelectrode technique, Klyce [76] measured the contribution of the cell membranes to the overall corneal resistance. He found that the outer membranes of the superficial cells contributed as much as 60% to the overall resistance, while stroma and endothelium together contributed about 3%. The remaining resistance was mainly contributed by the tight junctions of superficial cells. Therefore, it is not surprising that the superficial cells of the corneal epithelium form the most resistant barrier in the transport of timolol and levobunolol, two clinically used antiglaucoma agents [99]. Much of this resistance originates from the apical membrane and tight junctions. Marshall and Klyce [98] estimated that the electrical resistances of the apical membrane, the basolateral membrane, and the shunt pathway of rabbit corneal epithelium were 23.4, 3.8, and 15.4 kohm · cm^2, respectively. Thus, the paracellular shunt resistance is about threefold higher than that of the basolateral membrane and less than that of the apical membrane. While the shunt resistance of corneal epithelium is comparable to that (10–20 kohm · cm^2) of toad urinary bladder, another tight epithelium, the apical membrane resistance of the cornea is as much as seven times greater than that reported for toad urinary bladder (3–5 kohm · cm^2). Similar shunt resistance for cornea and toad urinary bladder suggests that these two epithelia are equally impermeable to hydrophilic drugs such as peptides and oligonucleotides, which are transported via the paracellular route. Higher apical membrane resistance implies that membrane-perturbing agents or stimulators of active ion transport across the apical membrane will lead to a significant decrease in transcorneal resistance. For this reason, a decrease in transcorneal R_t is not a good index for judging alterations in tight junctional permeability in the corneal epithelium.

In order to obtain accurate bioelectric parameters of the cornea, careful tissue handling and the use of salt bridges are highly recommended. Maurice [100] emphasized the importance of preserving the corneal curvature in order to obtain high values of PD and R_t. Any folding or mechanical damage induced by a wick can lead to reduced corneal PD and R_t. Improper securement of cornea in corneal rings used to mount the tissue can impair the epithelium along the edges of the ring, allowing leakage of ions along the edges, which, in turn, may reduce the corneal resistance. Rojanasakul and Robinson [101] observed a surprisingly low initial R_t (600–800 ohm · cm^2) of the cornea compared to that (6–10 kohm · cm^2) reported by Klyce [76]. The most likely explanation for such a low initial R_t is an increase in the cationic permeability of the apical membrane of the corneal epithelium induced by Ag/AgCl electrodes that were connected directly to the bathing fluid without salt bridges. This arrangement would allow Ag^+ ions to migrate to the epithelium and induce cationic permeability across the cornea as has been shown by Klyce and Marshall [102], thereby loosening the tight junctions (see Section IV.B.2).

The cornea has a remarkable capacity to regenerate its resistance within minutes of surface cell exfoliation. Evidence for this comes from the work of Wolosin [103]. Following 4–8 min exposure to 20 µM digitonin, the superficial cells of the corneal epithelium of the albino rabbit are permeabilized, resulting in complete loss of R_t. After the damaged monolayer of cells was removed by gentle prodding with a smooth glass rod and digitonin exposure discontinued, there was an immediate recovery of R_t that was >90% complete by the end of 70 min. A second exposure to digitonin resulted in further permeabilization and complete loss of R_t. One more layer of cells could be exfoliated by gentle prodding at the end of the second treatment. Although recovery of R_t started immediately after removal of digitonin, it took about 4–5 hr for 73% recovery. This recovery of R_t after surface cell permeabilization with digitonin was attributed to the reestablishment of tight junctions in the lower layers of the epithelium. From these studies, it can be inferred that by choosing an optimal concentration of a barrier-perturbing penetration enhancer, the rate of barrier regeneration can be optimized.

b. *Electrogenic Absorption of* Na^+. The lumen negative PD of the cornea suggests that it is capable of achieving (1) net anion transport in the endothelial-to-epithelial direction or (2) net cation transport in the epithelial-to-endothelial direction. The short-circuit current, therefore, will be the algebraic sum of currents associated with the bidirectional fluxes of cations and anions. Figure 2 summarizes the various ion transport mechanisms in the rabbit cornea.

Using the short-circuit current technique, Donn et al. [104] demonstrated that Na^+ was actively transported from tears to stroma, a finding that was confirmed by Green [105]. Klyce et al. [106] further demonstrated that net Na^+ absorption occurred at a rate of 0.039 µeq/(cm$^2 \cdot$ hr), accounting for about half of the short-circuit current. Sodium ion absorption, therefore, contributes about 50% to corneal active ion transport. Na^+ enters the superficial cells either through low conductance Na^+ channels that are insensitive to amiloride [54,107] or together with polar solutes such as amino acids [108]. Indeed, Liaw et al. [108] demonstrated preferential L-lysine transport in the epithelial-to-endothelial direction in an Na^+-dependent and temperature-sensitive manner in the albino rabbit cornea. Furthermore, inhibition of Na^+/K^+-ATPase with endothelial application of ouabain inhibited epithelial-to-endothelial permeability of lysine by 75%, suggesting that lysine transport was energized by Na^+/K^+-ATPase. This was not the case

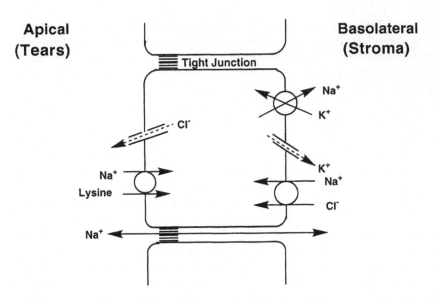

Figure 2 Schematic representation of the active ion transport processes in the rabbit corneal epithelium.

for L-glutamic acid. An Na^+ glucose cotransporter may also be absent from the pigmented rabbit cornea, since neither the epithelial nor the endothelial application of phlorizin (a Na^+-glucose cotransport inhibitor) up to 1 mM altered the I_{sc} in corneas bathed in glutathione bicarbonate Ringer's (GBR) solution containing 10 mM glucose (Kompella, Kim, and Lee, unpublished observation).

c. *Modulation of Na^+ Absorption.*   Amphotericin B and Ag^+ ions are two agents that can elevate the absorption of Na^+ from the tear side to the stromal side of the cornea [102,109]. Candia et al. [109] demonstrated that 10 μM amphotericin B elevated the corneal I_{sc} from 0.95 to 19.85 μA/cm². Since this effect could be totally reversed by 0.2 mM ouabain, the elevation in active transport due to amphotericin B must be energized by Na^+/K^+-ATPase. Furthermore, reduction in the tear side Na^+ from 104 to 5 mM reduced the I_{sc} to 7.65 μA/cm², suggesting that at least 65% of the elevation in I_{sc} is contributed by enhanced active absorption of Na^+ from the tear side. Klyce and Marshall [102] demonstrated that addition of 10 μM $AgNO_3$ to the tear side of the cornea elevated I_{sc} from 3–5 μA/cm² to about 35–55 μA/cm² within seconds. Because this effect was inhibited by at least 80% when the tear side was bathed in Na^+-free Ringer's solution, enhanced Na^+ absorption predominantly accounts for the observed increase in active ion transport.

d. *Electrogenic Secretion of Cl⁻.* The cornea actively secretes Cl^- from the aqueous to the tear side at a rate of 0.042 μeq/(cm² · hr), accounting for the remaining half of the active ion transport [106]. While an electrochemical potential gradient exists in the cells to facilitate the passive transport of Cl^- across the apical membrane, the entry of Cl^- across the basolateral membrane is in opposition to an electrochemical potential. Consequently, an active transport mechanism in the basolateral membrane would be required to maintain passive diffusion across the apical membrane. When bathed in a Cl^--free medium on the endothelial side, the corneal PD decreases by 90%. This drop in PD could be reversed by replacing the endothelial medium with Cl^--containing medium, but not with 10 μM ouabain in the medium. This suggests that the reentry of Cl^- into the epithelium from the stromal side is energized by the Na^+/K^+-ATPase pump [88].

The work of Bonanno et al. [110] revealed the existence of a loop diuretic–sensitive Cl^- uptake mechanism on the stromal side of corneal epithelium. Furosemide (0.1 mM), a loop diuretic capable of inhibiting Na^+-Cl^- and Na^+-K^+-$2Cl^-$ cotransport processes, reduced the rabbit corneal I_{sc} by 60% when exposed to the stromal side for 90 min. Removal of Cl^- from the solutions bathing a furosemide-inhibited preparation did not further reduce I_{sc}, suggesting that furosemide inhibited almost all of Cl^- transport. Bumetanide (0.01 mM), a more potent loop diuretic, also reduced I_{sc} by 65% in 90 min. Increasing the concentrations of Na^+ or Cl^- in both the bathing solutions yielded Hill coefficients of 0.99 and 1.04, suggesting a 1:1 stoichiometry in the uptake of these ions. The entry of Cl^- into the cells is independent of K^+ since neither a 50% reduction nor a twofold increase in K^+ concentration in both bathing fluids altered I_{sc}. The complementary experiment involving a K^+-free medium was not attempted, because the absence of K^+ in the serosal medium will impair the function of the Na^+/K^+-ATPase pump of the basolateral membrane, thereby reducing I_{sc}. Although this still leaves some room for involvement by K^+ in Cl^- entry from the basolateral side, the Hill coefficient of 1 for Cl^- and 1:1 stoichiometry of Na^+ and Cl^- strongly suggest that the basolateral entry of Cl^- is driven by the Na^+-Cl^- but not the Na^+-K^+-$2Cl^-$ cotransport process. A Hill coefficient of 2 would be expected if the latter process were involved.

e. *Modulation of Cl⁻ Secretion.* Three principal types of Cl^- channels exist in epithelial tissues and may coexist in some epithelial cells: cAMP-, Ca^{2+}-, and cell volume–regulated Cl^- channels [111,112]. All three types were identified in the apical membrane of T84 colonic epithelial cells. Using the whole cell patch-clamp technique, Cliff and Frizzell [113] detected two biophysically distinct secretagogue-induced Cl^- conductance activities in T84 cells. The Cl^- conductance stimulated by 0.4 mM 8-(4-chlorophenylthio)cAMP or 5 μM forskolin, an activator of adenylate cyclase, exhibited a linear current–voltage relationship in the range of ±100 mV. The permselectivity of this conductance was

$Br^- > Cl^- > I^-$. This conductance was not time-dependent. On the other hand, Cl^- conductance stimulated by the Ca^{2+} ionophore A23187 (2.5 μM) or elevation in the free Ca^{2+} levels in the pipet from 100 to 500 nM was time-dependent and exhibited a nonlinear current–voltage relationship that was outwardly rectifying. The permselectivity of this Ca^{2+}-stimulated conductance was $I^- > Br^- > Cl^-$, distinguishing it from the cAMP-stimulated Cl^- conductance. As both calcium ionophore A23187 and cAMP were shown to elevate the Cl^- conductance of the cornea, it is quite likely that these two types of Cl^- channels are present in the corneal epithelium [96,106,114].

Worrell et al. [115], using whole cell patch-clamp studies, identified yet another Cl^- conductance that is regulated by cell volume. With Cl^- as the permeable ion in the patch pipet and bath, the initial conductance of this channel was about 50 pA at 100 mV, increasing to about 1–3 nA within 10 min. This increase was associated with visible swelling of the cells and could be reversed by the addition of 50–75 mM sucrose to the bath, suggesting their cell volume–dependent regulation. These channels were similar to Ca^{2+}-activated Cl^- conductance in their outwardly rectifying current–voltage relationship and the permselectivity of the channel, which was $I^- > Br^-, Cl^-$. This conductance, however, was inactivated during depolarizing voltages and activated during hyperpolarizing voltages, a behavior kinetically opposite to that of Ca^{2+}-regulated Cl^- channels. Thus far, no evidence exists for the presence of a volume-regulated Cl^- channel in the corneal epithelium.

In order to elevate corneal drug transport by modulating Cl^- secretion, it is important that the stimulatory and inhibitory factors be known. Table 1 lists various such factors and the possible mechanisms by which they regulate corneal chloride transport. Chloride ion secretion across the rabbit cornea has been shown to be increased up to threefold by alkaline pH [116], theophylline [106], epinephrine [106], and serotonin [117]. pH was tested in the range of 7.2–8.6, and theophylline and serotonin were tested at concentrations of 1 and 0.1 mM, respectively. The concentration range employed for epinephrine was 0.001–1 μM. While the first three factors were effective when applied to the epithelial side, serotonin was effective only when applied to the endothelial side. Candia [116] observed a 65% elevation in the net secretory flux of Cl^- and I_{sc} across the bullfrog cornea when the pH was raised from 7.2 to 8.6. Although the exact role of pH on the corneal secretion of chloride is not clear, it has been suggested that certain proteins are capable of binding chloride with an affinity that is enhanced by protonation.

Of the above four factors, epinephrine is the most potent ($EC_{50} = 5$ nM), elevating the corneal I_{sc} within 15 sec at a concentration as low as 1 nM [106]. All three compounds probably elevate Cl^- secretion by increasing intracellular cAMP. Indeed, Klyce et al. [106] demonstrated that exposure of the mucosal side of the cornea to 1 mM dibutyryl cAMP elevated the I_{sc} by 144%, which correlated with a 200% change in the net Cl^- secretion measured as $^{36}Cl^-$ flux. This com-

Table 1 Factors Modulating Chloride Secretion in the Cornea

Modulator[a]	Mechanism	Animal model
Activators		
Adenosine (M)	↑ $[cAMP]_i$	Bullfrog [239,240]
Dibutyryl cAMP (M)	↑ $[cAMP]_i$	Rabbit [106]; bull-frog [93,241]
Dopamine (M&S)	↑ $[cAMP]_i$	Rabbit [242]
Epinephrine (M)	↑ $[cAMP]_i$	Rabbit [106]; bull-frog [93,241]
Fatty acids (M&S)	↑ $[cAMP]_i$?	Rabbit [243]; bull-frog [243]
Forskolin (M/S)	↑ $[cAMP]_i$	Rabbit [117]; bull-frog [117]
Prostaglandins (M)	↑ $[cAMP]_i$?	Bullfrog [244]
Serotonin (S)	↑ $[cAMP]_i$	Rabbit [98,245]
Aminophylline (M)	↓ cAMP breakdown	Rabbit [106]; bull-frog [93]
Ascorbic acid (M&S)	↓ cAMP breakdown	Bullfrog [246]
Theophylline (M)	↓ cAMP breakdown	Rabbit [106]
A23187 (M)	↑ $[Ca^{2+}]_i$	Bullfrog [114]
Alkaline pH (M)	↓ Cl^- holding by proteins	Bullfrog [116]
Amphotericin B (M)	↑ Cl^- permeability	Bullfrog [109]
Inhibitors		
Amytal (M&S)	↓ Respiration	Bullfrog [240,247]
Antimycin A (M&S)	↓ Respiration	Bullfrog [240,247]
Ouabain (S)	Inhibition of Na^+/K^+-ATPase	Rabbit [88]; bullfrog [248]
Bumetanide (S)	Inhibition of serosal Cl^- entry	Rabbit [110]; bull-frog [249,250]
Ethacrynic acid (S)	Inhibition of serosal Cl^- entry	Bullfrog [249]
Furosemide (S)	Inhibition of serosal Cl^- entry	Rabbit [110]; bull-frog [249,250]
Diltiazem (M)	↓ Serosal K^+ conductance	Bullfrog [121]
Dexamethasone (M&S)	↓ Adenylate cyclase activity	Bovine cornea [20]
Hydrocortisone (M&S)	↓ Adenylate cyclase activity	Bovine cornea [20]
Prednisolone (M&S)	↓ Adenylate cyclase activity	Bovine cornea [20]
Beta-blockers (M&S)	↓ Adenylate cyclase activity Inhibition of Na^+/K^+-ATPase	Bovine cornea [118] Rabbit [251]

[a] M = mucosal application; S = serosal application; M&S = simultaneous application to both mucosal and serosal sides; M/S = application to either mucosal or serosal side.

pound is specific in elevating Cl$^-$ secretion, because elevation in I_{sc} was not observed in the absence of Cl$^-$ in the bathing fluid. Exposure of the corneal epithelium to 1 mM theophylline resulted in a doubling of the I_{sc} and net Cl$^-$ flux [106], a response that was not observed in Cl$^-$-free medium. Similarly, epinephrine produced a maximal threefold elevation in the corneal I_{sc} in a Cl$^-$-containing but not a Cl$^-$-free medium. Since the effects of epinephrine on Cl$^-$ secretion could be blocked by the mixed beta-adrenergic antagonist timolol, epinephrine may act through beta-adrenergic receptors. Toward that end, Reinach and Holmberg [118] demonstrated in bovine corneal epithelial cells that propranolol completely inhibited the maximal 16-, 6.6-, and 4.7-fold elevation in cAMP in bovine corneal epithelial cells by isoproterenol (beta-selective agonist), norepinephrine, and epinephrine (nonselective adrenergic agonists), respectively. There is, therefore, evidence for beta-adrenergic modulation of corneal Cl$^-$ secretion. Figure 3 depicts the modulation of Cl$^-$ secretion in the rabbit cornea by various biogenic amines.

Serosal application of serotonin, a neurotransmitter localized histochemically in the corneal nerves, increases the intracellular levels of cAMP, thereby elevating secretory flux of Cl$^-$, which accounts for all the elevation in the corneal I_{sc} [119]. The effect of serotonin on I_{sc} is slower than that of epinephrine, oc-

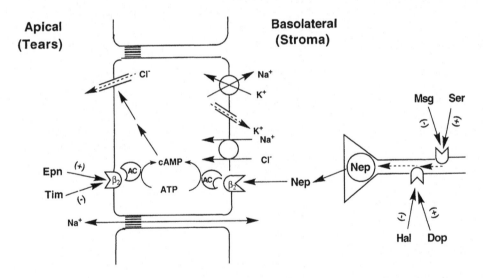

Figure 3 Putative model for the mechanism by which biogenic amines stimulate Cl$^-$ secretion across the rabbit corneal epithelium. Epn = epinephrine; Nep = norepinephrine; Tim = Timolol; Ser = serotonin; Msg = methysergide; Dop = dopamine; Hal = haloperidol; β_2 = β_2-adrenoceptor; AC = adenylate cyclase. The scheme is consistent with the observation that epithelial responsiveness to serotonin and dopamine can be blocked by their receptor antagonists haloperidol and methysergide, respectively, and by both timolol treatment and sympathectomy. The probable source of serotonin or dopamine is the sympathetic fibers that innervate the cornea. (From Ref. 284.)

curring in 3–7 min. Application of methysergide (a serotonin receptor antagonist) and substitution of SO_4^{2-} for Cl^- in the medium inhibited the I_{sc} response to serotonin. Since serotonin failed to alter net Na^+ absorption, its cAMP-mediated effects must be specific for Cl^- secretion. Using microelectrodes, Marshall and Klyce [98] demonstrated that 0.1 mM serotonin when applied with 0.1 mM nialamide, an inhibitor of monoamine oxidase, elevated Cl^- secretion by reducing the apical membrane resistance of the squamous superficial cells from 21.3 to 3.9 kohm · cm^2.

In addition to timolol, propranolol, and methysergide, anti-inflammatory steroids and Ca^{2+} channel blockers also reduce corneal Cl^- secretion. Walkenbach and Legrand [120] demonstrated that the phosphate salts of dexamethasone, prednisolone, and hydrocortisone inhibited adenylate cyclase activity in bovine corneal epithelial cells. The rank order of potency was dexamethasone (IC_{50} = 2 mM) > prednisolone (IC_{50} = 6 mM) > hydrocortisone (IC_{50} = 10 mM). At 5 mM, the extent of adenylate cyclase inhibition was 75%, 35%, and 20%, respectively. Such an inhibition of adenylate cyclase and hence synthesis of cAMP by anti-inflammatory steroids may reduce corneal Cl^- secretion. Huff and Reinach [121] observed a 98% inhibition in the bullfrog corneal I_{sc} following the mucosal application of 1 mM diltiazem. This effect was associated with a fall in the cAMP content and an elevation in the basolateral membrane resistance measured with microelectrodes. Nevertheless, the fall in I_{sc} induced by diltiazem could not be reversed even after elevating the intracellular levels of cAMP fourfold with isoproterenol. Suppression of cAMP levels by diltiazem cannot, therefore, account for its effect on I_{sc}. Moreover, since elevation in the intracellular levels of Ca^{2+} with 1 μM A23187, a calcium ionophore, failed to reverse the effects of diltiazem, there is a lack of involvement of altered Ca^{2+} levels in the I_{sc} reduction caused by it. Finally, while reducing basolateral K^+ conductance, diltiazem failed to inhibit Na^+/K^+-ATPase activity in the basolateral membrane fractions of bullfrog corneal epithelium, suggesting the involvement of K^+ channels in the inhibition of corneal active ion transport.

As regulation of Cl^- channel activity can modulate drug transport (see Section IV.B), it is quite possible that some modulators of Cl^- channels such as epinephrine, theophylline, serotonin, beta-adrenergic antagonists, and anti-inflammatory steroids may have permeability coefficients that do not obey the pH-partition hypothesis. Furthermore, it can be anticipated that these agents may affect the transport of coadministered hydrophilic drugs. For instance, elevation of Cl^- secretion by epinephrine can be expected to decrease the permeability of coadministered hydrophilic drugs. Effects opposite to those achieved with stimulants of Cl^- secretion would be expected following inhibition of Cl^- secretion with Cl^- channel blocker N-phenylanthranilic acid (NPAA) and loop diuretics such as bumetanide and furosemide [122]. Indeed, Kompella et al. [123] demonstrated that the inhibition of apically localized Cl^- channels of rabbit cornea with 1 mM NPAA elevates the paracellular permeability of mannitol by 60%.

f. *Modulation of Ca^{2+} Influx.* Although the total concentration of Ca^{2+} within animal cells approaches 1 mM, most of this is compartmentalized or bound to proteins, thereby limiting the free cytosolic Ca^{2+} to between 50 and 200 nM [124]. Given the large amount of cellular Ca^{2+} that is not part of the free cytosolic compartment and the large gradient of free Ca^{2+} across the cell membrane, even minor changes in the Ca^{2+} permeability of the plasma membrane or release of stored Ca^{2+} triggered by physiological stimuli such as hormones activating phospholipase C can significantly elevate intracellular levels of free Ca^{2+}. The end result will be activation of several biological responses, including contraction of the cytoskeletal proteins. Cytoskeletal contraction induced by intracellular Ca^{2+} may promote the paracellular transport of hydrophilic drugs (see Section IV.B.2.b). Various putative transport and homeostatic mechanisms of intracellular Ca^{2+} and their modulatory mechanisms are represented in Figure 4, and the means to elevate intracellular levels of Ca^{2+} are listed in Table 2.

Mammalian cells are equipped with Ca^{2+} channels in their plasma mem-

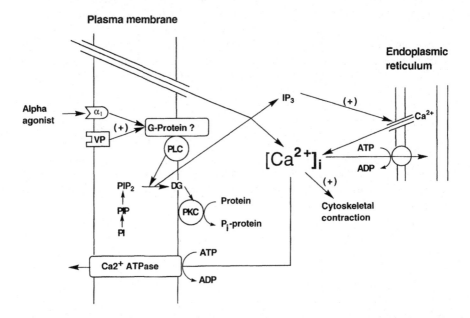

Figure 4 Schematic representation of the Ca^{2+}-transporting systems affecting cellular calcium homeostasis during hormonal stimulation. $\alpha_1 = \alpha_1$-adrenergic receptor; VP = vasopressin receptor; PLC = phospholipase C; PI = phosphatidylinositol; PIP = phosphatidylinositol-4-phosphate; PIP_2 = phosphatidylinositol-4,5-biphosphate; IP_3 = inositol-1,4,5-triphosphate; DG = diacylglycerol; PKC = protein kinase C. (Modified from Refs. 125 and 285.)

Table 2 Approaches to Elevate Intracellular Levels of Ca^{2+}

Approaches	Example	Ref.
Increased cellular entry		
Direct activation of Ca^{2+} channels in plasma membrane	Bay K[a]	252
Hyperpolarization of plasma membrane	Carbachol	130
Use of Ca^{2+} ionophore	Ionomycin	118, 253
Release from intracellular pools		
Hormonal stimulation of phospholipase C	Phenylephrine	254
	Vasopressin	254, 255
	Angiotensin II	256, 257
Direct stimulation of inositol triphosphate–sensitive Ca^{2+} pools	2,5-Di(tert-butyl)-1,4-benzohydroquinone	220

[a] Bay K = methyl-1,4-dihydro-2,6-dimethyl-3-nitro-4-(2-trifluoromethylphenyl) pyridine-5-carboxylate.

brane to facilitate Ca^{2+} influx into the cytosol [125,126]. Reinach and Holmberg [118] observed that treatment of bovine corneal epithelial cells with 1 mM diltiazem, a Ca^{2+} channel blocker, reduced the intracellular levels of Ca^{2+}, suggesting the possible existence of Ca^{2+} channels in the corneal epithelium. Rich and Rae [127] observed a non-voltage-gated Ca^{2+} channel activity in the rabbit corneal epithelium. Some Ca^{2+} channels may be activated with Bay K (see footnote to Table 2) to promote the cellular entry of Ca^{2+} [128,129]. However, application of 1 mM Bay K did not affect intracellular Ca^{2+} levels in the rabbit corneal epithelial cells [127].

Another way to increase the entry of Ca^{2+} across the plasma membrane is to hyperpolarize the plasma membrane by elevating active ion transport. Fischer et al. [130] demonstrated that hyperpolarization of colonic epithelial cells (HT-29) with carbachol elevates the intracellular levels of Ca^{2+}, $[Ca^{2+}]_i$ while depolarization with gramicidin D or elevation of K^+ in the bathing fluid reverses it. Treatment with 0.1 mM carbachol produced a spontaneous increase in $[Ca^{2+}]_i$ from 63 nM to 901 nM. This lasted for about 3 min, beyond which a plateau level of 309 nM was maintained. While the initial Ca^{2+} transient was present in Ca^{2+}-free medium containing 0.1 mM EGTA, the plateau phase was suppressed to baseline levels, suggesting that carbachol initially releases Ca^{2+} from the intracellular stores and subsequently increases the Ca^{2+} entry across the plasma membrane. In cells hyperpolarized with carbachol, induction of depolarization by elevating the extracellular levels of K^+ from 5 to 95 mM abolished the plateau phase. Furthermore, depolarization with 2 μM gramicidin D produced similar results, confirming that $[Ca^{2+}]_i$ elevation achieved by carbachol is due to elevated cellular Ca^{2+} entry following hyperpolarization of the membrane potential. Indeed, hyperpolarization of rabbit corneal epithelial cells with flufenamic acid increased the intracellular levels of Ca^{2+} [127]. Yet a third way to increase the entry of Ca^{2+} from the bathing fluid is to use A23187, a calcium ionophore [118].

g. *Modulation of Ca^{2+} Release from Intracellular Pools.* Calcium ion channels are present in the endoplasmic reticulum membrane of mammalian cells to facilitate the release of Ca^{2+} into the cytosol [125,126]. Certain hormones like α1-adrenergic agonists and vasopressin are coupled to plasma membrane–bound phospholipase C (PLC). Immediately following receptor occupation by these hormones, PLC hydrolyzes phosphatidylinositol 4,5-biphosphate (PIP_2) of the plasma membrane to inositol 1,4,5-triphosphate (IP_3) and diacylglycerol (DAG) [51]. The IP_3 generated in this process can stimulate the release of Ca^{2+} from the endoplasmic reticulum [131].

In addition to the aforementioned Ca^{2+} channels capable of elevating the intracellular levels of Ca^{2+}, epithelial cells can be equipped with ion transport processes to reduce the $[Ca^{2+}]_i$ [124]. Indeed, the Ca^{2+} pump, an ATPase that pumps Ca^{2+} out of the cell in exchange for extracellular protons, is known to exist in the plasma membrane of the bovine corneal epithelial cells [132]. In addition, intracellular Ca^{2+} can be depleted in exchange for extracellular Na^+ by the Na^+/Ca^{2+} exchanger located in the plasma membrane. Using rabbit corneal epithelial cells, Rich and Rae [127] demonstrated that the replacement of extracellular Na^+ with Li^+ results in an elevation in the intracellular levels of Ca^{2+} if the cells were initially loaded with Na^+ using monensin, a sodium ionophore, and ouabain, an Na^+/K^+-ATPase inhibitor. These findings suggested that Na^+/Ca^{2+} exchange may also regulate intracellular Ca^{2+} levels in the corneal epithelium. As there is evidence that Ca^{2+} channels [121,127], Na^+/Ca^{2+} exchange [127], and Ca^{2+} pumps [132] exist in the corneal epithelium, both of them represent a potential target for pharmacological manipulation for the purpose of increasing drug transport (see Section IV.B.2.b).

h. *Electroneutral Transporters.* In addition to the aforementioned electrogenic ion transport processes, electroneutral exchangers or cotransporters are also present in the cornea. These transporters either transport ions of equal charge in opposite directions in a neutral stoichiometry (electroneutral exchangers, e.g., K^+/H^+, Na^+/H^+, and Cl^-/HCO_3^-) or cotransport two or more ions of opposite charge in the same direction in a neutral stoichiometry (electroneutral cotransporters, e.g., Na^+-K^+-$2Cl^-$ cotransporter). The majority of these neutral transport systems are involved in the regulation of intracellular pH, which can be monitored using fluorescent probes. Examples of pH-sensitive probes include pyranine, 4-

methylumbelliferone (4-MU), and derivatives of fluorescein such as carboxy-fluorescein (CF), dimethylcarboxyfluorescein (DMCF), and biscarboxyethylcar-boxyfluorescein (BCECF) [133]. These probes are either impermeable in their native forms or converted to their impermeable forms after cellular entry. Pyranine, for instance, is permanently charged and can be trapped in membrane vesicles to study intravesicular pH regulation. On the other hand, BCECF is used in its membrane-permeable acetoxymethyl ester form (BCECF/AM). Upon entry into the cytosol, BCECF/AM is converted by cytosolic esterases to its membrane-impermeable BCECF form. These properties make it suitable for monitoring intracellular pH of intact cells or epithelia. Of all the pH-indicating dyes, BCECF is used the most, since it yields a better linear relationship between fluorescence ratio and pH in the range of 6.4–7.4 [134].

Intracellular Ca^{2+} can be regulated by Na^+/Ca^{2+} exchanger. As this exchanger is slightly electrogenic and contributes little to the transepithelial I_{sc}, it is best characterized by fluorescent probes rather than the short-circuit current approach. Fluorescent probes useful for intracellular Ca^{2+} measurement include Quin-2, Fura-2, and Indo-1 [135,136]. Like BCECF, these probes are used in their membrane-permeable ester forms. Once inside the cell, esterases produce the membrane-impermeable probes for measuring intracellular levels of Ca^{2+}. The fluorescent probes used for intracellular pH or Ca^{2+} measurements can be monitored in cell populations by using spectrofluorometry and in single cells by using fluorescence microscopy. pH or Ca^{2+} levels within epithelial cells can also be monitored by impaling cells with microelectrodes sensitive to pH or Ca^{2+}. In association with fluorescent probes or microelectrodes, suitable inhibitors, activators, and specific ion-free solutions can be used to characterize electroneutral transporters.

Using the fluorescent probe technique, Bonanno [137] identified a lactate-proton cotransport process and a K^+/H^+ exchanger in the basal cells of the rabbit corneal epithelium. To probe the lactate-proton cotransport process, BCECF/AM was loaded into epithelial cells for measurement of variations in intracellular pH (estimated using a ratio of fluorescence emission output at 520 nm following excitation at 490 and 440 nm) when exposed to Ringer's solution containing increasing concentrations of lactate. There was a saturable influx of protons into the epithelial cells. The maximal rate of proton influx was 10.2 mM/min; the half-maximal rate was reached at 10.7 mM lactate. This proton influx induced by 10 mM lactate was inhibited by cyanohydroxycinnamic acid (1 mM), 4,4′-diisothiocyanodihydrostilbene-2,2′-disulfonic acid (0.5 mM), and lactic acid isobutyl ester (1 mM) to the extent of 36%, 60%, and 47%, respectively, suggesting the presence of a lactate-H^+ symporter in the basal cells of the corneal epithelium.

Evidence for the presence of a K^+/H^+ exchanger in the basal cells of corneal epithelium was also obtained by Bonanno [137]. Elevated levels of intracellular K^+ maintained by the basolateral Na^+/K^+-ATPase are utilized by K^+/H^+

exchangers of epithelial cells to exchange extracellular H^+ for intracellular K^+. That is, if the K^+ gradient is offset by exposing high concentrations of K^+ to the plasma membrane exterior, the activity of the K^+/H^+ exchanger can be inhibited, thereby alkalinizing the intracellular pH. When K^+ in the bathing fluid is elevated to its level within the rabbit corneal epithelial cells (75 mM), the activity of K^+/H^+ exchanger can be inhibited [138]. Indeed, a 0.28 unit rise in the intracellular pH of the corneal epithelial basal cells was observed when the K^+ concentration was increased from 4.8 to 77 mM in a bicarbonate-free Ringer's solution [137]. In order to rule out membrane potential–directed passive influx of H^+, depolarization of the membrane with 0.2 μM gramicidin D (a Na^+- and K^+-permeable ionophore), with 5 mM $BaCl_2$ (K^+ channel blocker), and with 0.1 μM ouabain (Na^+/K^+-ATPase inhibitor) was tested. These agents acidified the cells by only 0.05, 0.09, and 0.04 pH unit, respectively; thus, passive flux of H^+ contributes little to the intracellular pH regulation of corneal epithelial cells. Since high K^+-induced alkalinization was observed even in the presence of 0.1 mM SCH28080 (a K^+/H^+-ATPase blocker), it is the K^+/H^+ exchanger, not K^+/H^+-ATPase, that contributes to alkalinization.

While the lactate-H^+ symporter and the K^+/H^+ exchanger are involved in acidification of the cell, the Na^+/H^+ exchanger present in the basal cells exports protons out of the cell in exchange for Na^+ [139]. It was observed that removal of Na^+ from the Ringer's solution decreased intracellular pH by 0.5 unit in basal cells, possibly due to inhibition of the Na^+/H^+ exchanger. As the basal cells are the precursors for the superficial cells of the corneal epithelium, it is quite likely that similar exchange processes are also present in the superficial layer, the principal barrier to ion and drug transport [99,103].

2. Conjunctiva

Ion transport processes in the conjunctiva have only been characterized recently (Fig. 5). Recent investigations in our laboratory have demonstrated, for the first time, that the pigmented rabbit conjunctiva is a moderately tight epithelium capable of active ion transport [140–143]. These findings were subsequently confirmed by Shi and Candia in the albino rabbit [144]. The bioelectrical properties of the albino rabbit cornea and the pigmented rabbit conjunctiva are compared in Table 3. The potential difference of the conjunctiva is lumen-negative as in the cornea, suggesting a net secretion of anions and/or a net absorption of cations. The I_{sc} is 14 μA/cm^2, and the R_t is 1.3 kohm · cm^2 in the conjunctiva, compared with a mean I_{sc} and R_t of 2.85 μA/cm^2 and 7.5 kohm · cm^2 in the cornea. Thus, the I_{sc} and the overall conductance of conjunctiva are, respectively, 3.9 and 4.8 times higher than those in the cornea. Both active and passive ionic conductances of the conjunctiva are therefore higher than those of the cornea.

Amiloride-sensitive Na^+ channels are probably absent from the pigmented rabbit conjunctival epithelium, since the conjunctival I_{sc} is not sensitive to either mucosal or serosal amiloride up to 1 mM. Conjunctival I_{sc} is, however, inhibited

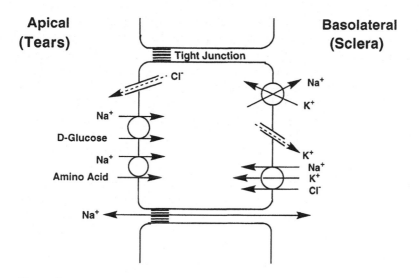

Figure 5 Proposed model depicting active ion transport processes in the rabbit conjunctiva.

by about 80% by 1 mM N-phenylanthranilic acid (NPAA), a Cl^- channel blocker, suggesting that about 80% of the conjunctival active ion transport is contributed by Cl^- secretion through apically localized channels. These Cl^- channels appear to be subject to cAMP, Ca^{2+}, and protein kinase C (PKC) modulation [143]. There is also evidence for the entry of Cl^- from the basolateral side into the conjunctival cells via Na^+-(K^+)-$2Cl^-$ cotransport, in part energized by the Na^+ gradient set up by basolateral Na^+/K^+-ATPase activity [141].

Ion-dependent solute transport processes such as Na^+-glucose and Na^+-amino acid cotransporters can be identified in epithelial tissues by observing an elevation in I_{sc} following solute addition in Na^+-containing but not Na^+-free

Table 3 Bioelectrical Properties of the Albino Rabbit Cornea and the Pigmented Rabbit Conjunctiva

Parameter	Cornea[a]	Conjunctiva[b]
PD (mV)	21.4 ± 0.8	17.73 ± 0.79
I_{sc} ($\mu A/cm^2$)	2.85^c	14.48 ± 0.70
R_t (ohm · cm^2)	7500 ± 200	1323 ± 83
n	121	45

Entries are mean $\pm$ SEM.
[a] From Ref. 98.
[b] From Ref. 258.
[c] Calculated as a ratio of PD to R_t.

media. After bathing the conjunctiva in a glucose-free GBR solution, mucosal addition of D-glucose elevated I_{sc} by a maximum of 20% (at 10 mM) in a dose-dependent manner. On the other hand, phloridzin, a specific inhibitor of Na^+-glucose cotransporter, reduced the I_{sc} in a dose-dependent manner. An optimal I_{sc} reduction of 16% was observed at 1 mM phloridzin. These findings based on electrophysiology [145], which were subsequently verified by [^{14}C]methylglucose flux determinations [146], suggest the possible apical localization of Na^+-glucose cotransporter in the pigmented rabbit conjunctiva. Furthermore, mucosal but not serosal application of 5 mM glycine, 1 mM arginine, and 1 mM glutamic acid produced maximal I_{sc} elevations of 35%, 45%, and 7.5%, respectively. Such I_{sc} elevation was absent from Na^+-free GBR, suggesting the possible apical localization of Na^+-amino acid cotransporters in the pigmented rabbit conjunctiva. This was further confirmed by the competitive inhibition of the effects of 5 mM

Table 4 Comparison of the Ion Transport Processes of the Cornea and the Conjunctiva

	Location	
Ion transporter	Cornea	Conjunctiva
Channels		
Cl⁻ channel	Apical [259]	Apical [258]
Amiloride-sensitive		
Na^+ channel	Absent[a]	Absent[a]
K^+ channel	Basolateral [196]	Basolateral [140]
Ca^{2+} channel	Present[b] [121,127]	N.D.
Cotransporters		
Na^+-glucose contransporter	Absent[a]	Apical [140]
Na^+-amino acid contransporter	Apical [108]	Apical [147]
	Basolateral [260]	
Na^+-(K^+)-Cl⁻ cotransporter	Basolateral [157]	Basolateral [258]
H^+-lactate symporter	Present [137]	N.D.
Exchangers		
Na^+/H^+ exchanger	Basolateral [139,261–263]	N.D.
Na^+/Ca^{2+} exchanger	Present [127]	N.D.
Cl⁻/HCO_3^- exchanger	Present [263]	N.D.
K^+/H^+ exchanger	Present [137]	N.D.
Pumps		
Na^+/K^+-ATPase	Basolateral [88,264]	Basolateral [258]
Ca^{2+}-ATPase	Present [132]	N.D.

N.D. = not determined.
[a] Kompella, Kim, and Lee, unpublished observation.
[b] Present = detected, but polarity was not determined.

glycine, 1 mM L-arginine, and 1 mM glutamic acid by conjunctival pretreatment with 5 mM serine, 1 mM aspartic acid, or 1 mM L-arginine, in that order [147]. Followup experiments by Hosoya et al. [148] based on [³H]L-arginine flux measurements have revealed the possible existence of the $B^{0,+}$ amino acid transport system in the pigmented rabbit conjunctiva.

Table 4 compares the ion transport processes in the pigmented rabbit conjunctiva to those of the albino rabbit cornea. Both the conjunctiva and the cornea are capable of secreting Cl^- through apically localized channels. In both epithelia, Na^+/K^+-ATPase energizes the cellular entry of Cl^- via the basolaterally localized Na^+-(K^+)-$2Cl^-$ cotransporter. While Na^+-glucose cotransporter is present in the conjunctiva, it may be absent from the cornea. Work is in progress in our laboratory to identify Ca^{2+} channels and Ca^{2+} pumps that are known to regulate intracellular levels of Ca^{2+} in the cornea and Na^+/H^+, K^+/H^+, and Cl^-/HCO_3^- exchangers that regulate intracellular pH in the corneal epithelium.

III. PERMEABILITY BARRIERS IN THE CORNEA AND THE CONJUNCTIVA

There are two principal routes of drug transport across any epithelium: transcellular and paracellular (Fig. 6). In the transcellular route, drugs are transported

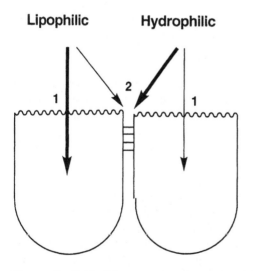

Figure 6 Epithelial penetration routes for topically applied drugs. The transcellular route (1) is preferred by lipophilic drugs, while the paracellular route (2) is preferred by hydrophilic drugs.

through the apical membrane, cytoplasm, and the basolateral membrane, in that order, whereas in the paracellular route drugs are transported through fluid-filled intercellular junctions. In both the cornea and the conjunctiva, while the transcellular route is preferred by lipophilic drugs, the paracellular route is the route preferred by hydrophilic drugs for which no specialized transport mechanisms exist. Ashton et al. [149] showed that opening of the paracellular pathway with 0.1% EDTA caused about 21- and 1.5-fold increases, respectively, in the corneal and conjunctival permeabilities of atenolol (log PC = 0.16), a relatively hydrophilic drug, without affecting the permeability of highly lipophilic levobunolol (log PC = 2.4) and betaxolol (log PC = 3.44). Furthermore, the permeability of atenolol was about 30 times higher across the leakier conjunctiva, while the permeabilities of levobunolol and betaxolol were very similar in these two epithelia.

Transport of drugs via the transcellular route requires either preferential membrane partitioning of the drug [estimated as log (octanol/water partition coefficient), log PC] or the presence of specialized transport mechanisms such as facilitated diffusion, carrier-mediated transport, receptor-mediated endocytosis, adsorptive endocytosis, fluid-phase endocytosis, and potocytosis. There is preliminary evidence that carrier-mediated transport exists to transport di- and tripeptides across the cornea [150,151] and the conjunctiva of the rabbit [152,153] and that fluid-phase endocytosis exists in the conjunctiva but not the cornea to transport horseradish peroxidase [154,155]. Potocytosis is a novel membrane transport mechanism for small molecules that has not yet been demonstrated in epithelial tissues including the cornea and the conjunctiva [67]. It is analogous to receptor-mediated endocytosis (RME) for the uptake and transport of macromolecular drugs. Unlike RME, potocytosis is temperature-insensitive and does not require coated pits for drug internalization. Instead, potocytosis employs plasma membrane vesicles (caveolae), 50 nm in diameter, which may act in concert with glycosylphosphatidylinositol (GPI)–anchored membrane proteins to sequester small molecules for internalization. Receptors may also be part of the caveolae [156]. Folate, for instance, is captured by potocytosis in MA104 cells via the receptor-mediated mechanism. It is thought that after folate binds to its receptors in the caveolae, the vesicles pinch off from the membrane. Detachment of folate from its receptors is facilitated by acidification of the vesicle, whereas export of folate across the vesicular membrane by an anion carrier is facilitated by the concentration gradient of folate accumulated within the caveolae. Following liberation of their contents, the caveolae recycle.

The cornea is highly impermeable to small electrolytes. Using beta-adrenergic antagonists of diverse log PC values, Huang et al. [157] delineated the resistance offered by the corneal epithelium, stroma, and endothelium to the overall permeability of these drugs. For hydrophilic beta-adrenergic antagonists with log

PC values in the range of -2 to 0.2, more than 90% of the resistance resides in the epithelium, with stroma and endothelium being highly permeable. For lipophilic beta-adrenergic antagonists with log PC values in the range of 1.62–2.19, stroma and endothelium together contribute 80% or more to the resistance. For beta-adrenergic antagonists with intermediate lipophilicity (log PC values in the range of 0.28–0.72), the epithelium is the major permeability barrier contributing 45–72% of the resistance, with stromal and endothelial resistances contributing less than 21% and 34%, respectively.

Shih and Lee [99], using a digitonin permeabilization technique developed by Wolosin [103], further localized the rate-limiting barriers to drug transport in the pigmented rabbit corneal epithelium. Exposure of the corneal epithelium to 40, 60, or 80 µM of digitonin for 15 min resulted in stepwise exfoliation of the corneal epithelium up to the wing cells, basement membrane, and stroma, respectively. The transcorneal flux of atenolol (log PC = 0.16) was gradually increased over the above concentration range of digitonin, suggesting that resistance to paracellular transport of hydrophilic drugs probably resides throughout the epithelium. On the other hand, the permeabilities of timolol (log PC = 1.91) and levobunolol (log PC = 2.4) were increased up to 40 µM digitonin. The resistance to the transport of these drugs is therefore confined to superficial and wing cell layers of the epithelium. Permeability of betaxolol (log PC = 3.44), the most lipophilic beta-adrenergic antagonist tested, was not improved even after total epithelial permeabilization (i.e., deepithelialization). The corneal epithelium, therefore, offers the least impedance to the transport of lipophilic drugs.

Compared with the cornea, the conjunctiva is more permeable to drugs. The ratio between the permeability coefficients of a series of beta-adrenergic antagonists in the cornea and those in the conjunctiva exhibits a sigmoidal correlation with log PC (Fig. 7) [158]. The ratio is the highest for the very lipophilic drug alprenolol (log PC = 2.61), with a value of 0.54. As expected, the inverse ratio (conjunctival permeability/corneal permeability) increases with decreasing drug lipophilicity, attaining a value of 24 with sotalol (log PC = -0.62), the most hydrophilic beta-adrenergic antagonist tested by these investigators. The ratio is even higher for some peptide drugs (Table 5). Given that the corneal and conjunctival permeabilities are similar for lipophilic drugs but cover a wide range for hydrophilic drugs, it is the paracellular route that differentiates these epithelia from each other.

The permselectivity of the corneal and conjunctival paracellular routes was investigated by Huang et al. [159] in an attempt to show that nutrients can be extracted from the blood by the conjunctiva. Neither the blood vessels supplying the conjunctiva nor its basement membrane are rate-limiting to the transport of horseradish peroxidase. This 40 kDa tracer is restricted underneath the conjuncti-

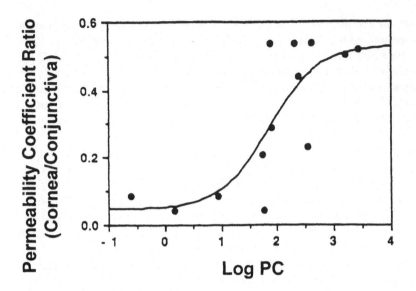

Figure 7   Influence of drug lipophilicity (log PC) on the ratio of corneal to conjunctival permeability coefficients of beta-blockers. (From Ref. 158.)

val (and corneal) epithelia at 45 min following intravenous administration in the rabbit. In vitro permeability studies from the subepithelial space to the apical side with paracellular markers such as mannitol (182 Da), inulin (5 kDa), and FITC-dextrans (20, 40, and 70 kDa) revealed that the intact cornea was permeable to mannitol but not to inulin or dextrans, whereas the conjunctiva was permeable to mannitol, inulin, and dextrans of 20 kDa but not to those of 40 or 70 kDa. The conclusion is that the molecular weight cutoff for corneal paracellular transport is below 5 kDa, whereas that for conjunctival paracellular transport is between 20 and 40 kDa [160]. Indeed, arginine vasopressin, a 1084 Da nonapeptide, appears to permeate the conjunctiva via the paracellular pathway [161].

Although the cornea is less permeable than the conjunctiva, its proximity to the anterior chamber tissues makes it a more direct route of absorption in the treatment of diseases related to the anterior chamber such as glaucoma. However, the conjunctiva can also contribute to drug concentations in the anterior eye, albeit to a lesser extent [1–3]. The attractiveness of the conjunctival route of drug transport is the higher likelihood of drug access to the posterior segment tissues for treatment of diseases there [162]. It is conceivable that such a therapeutic goal may be facilitated by the existence of a variety of drug transporters in the conjunctiva, including those for arginine derivatives [163], nucleosides [164], monocarboxylates [165], dipeptides [166], and organic cations [167]. A possible

Table 5 Molecular Weights (MW), n-Octanol/pH 7.4 Buffer Partition Coefficients (log PC), Apparent Permeability Coefficients, and Permeability Ratios of Selected Peptide and Nonpeptide Drugs in the Cornea and Conjunctiva

Compound	MW	Log PC	Permeability coefficient[a] (µm/sec)		Permeability ratio (conjunctiva/cornea)
			Conjuctiva	Cornea	
Peptides					
Diglycine	132	−1.89[b]	0.406 ± 0.057	0.040 ± 0.005[b]	10
Triglycine	189	−1.74[b]	0.195 ± 0.054	0.002 ± 0.001[b]	93
Tetraglycine	246	−1.70[b]	0.238 ± 0.049	0.051 ± 0.019[b]	5
Pentaglycine	303	−0.77[b]	0.522 ± 0.128	0.061 ± 0.007[b]	9
TRH	362	−1.59[b]	0.286 ± 0.021	0.013 ± 0.004[b]	22
Nonpeptides					
Atenolol	266	0.164[c]	0.295 ± 0.065	0.008 ± 0.002[d]	37
Timolol	316	1.914[c]	0.486 ± 0.049	0.123 ± 0.121[d]	4

[a] Values expressed as mean ± SEM.
[b] From Ref. 151.
[c] From Ref. 265.
[d] From Ref. 158.
Source: Ref. 168, except as noted.

deterrent to drug penetration across the conjunctiva may be the p-glycoprotein gp170 (P-gp) drug efflux pump [168].

IV. MEANS TO ENHANCE DRUG TRANSPORT ACROSS THE CORNEA AND THE CONJUNCTIVA

A. Conventional Means

The poor ocular bioavailability of topically applied drugs leaves considerable room for improvement. Aside from the benefit of improved patient compliance, an increase in ocular drug bioavailability, which then allows the use of a smaller dose at less frequent intervals, is essential in minimizing the systemic side effects of such drugs as timolol [169], betaxolol [170], levobunolol [171], and metiprano-lol [172]. As mentioned earlier, the principal reasons for poor ocular drug absorption are precorneal drug loss, resistance of the cornea to penetration, and possible metabolic breakdown in the eye. Because of the efficiency of the first two processes, it will be necessary to reduce precorneal drug clearance or increase corneal drug absorption by one to two orders of magnitude in order to significantly improve ocular drug bioavailability [173]. This is a formidable task. Table 6 lists various factors constituting each of these problems along with attempts to overcome these barriers.

One attempt to reduce precorneal drug clearance is based on mucoadhesive polymers and was pioneered by Hui and Robinson [174] in 1985. These investigators demonstrated that polymers of acrylic acid cross-linked (0.3% w/w) with divinyl glycol and 2,5-dimethyl-1,5-hexadiene were retained in the conjunctival sac of the rabbit eye for as long as 12 hr, resulting in a 4.2-fold improvement in the ocular bioavailability of progesterone compared with the suspension control. Since then, other mucoadhesive polymers have been investigated, including sodium hyaluronate [12] and poly(acrylic acid) derivatives [11]. One such poly(acrylic acid) derivative, Carbopol 934P, was investigated by Davies et al. [13]. Compared with poly(vinyl alcohol) of the same viscosity (60 mPa · sec), this polymer afforded a threefold improvement in precorneal retention and a 2.5-fold improvement in the ocular bioavailability of pilocarpine based on the miotic response. A similar favorable precorneal retention response was observed when the above mucoadhesive polymer was coated on multilamellar liposomes made of egg phosphatidylcholine [13].

An alternative to improving precorneal drug retention is to improve corneal and conjunctival permeability so as to improve drug absorption by the corneal and non-corneal pathways, respectively. These permeability barriers can be overcome either by modifying the chemical structure and physicochemical properties of the drug or by modulating the epithelial membrane itself. These approaches are listed with examples in Table 6. Prodrugs and ion-pairing agents, which are means to

Table 6 Some Limiting Factors in Ocular Drug Delivery and Their Possible Solutions

Limiting factor	Solution	Ref.
Short precorneal residence time	Viscous vehicles.	8–10
	In situ gels.	20, 266
	Inserts.	16, 17
	Bioadhesives.	11–13, 174, 267, 268
	Liposomes coated with bioadhesives.	269
	Liposomes in collagen shields.	23
	Collagen shields.	21, 22, 24, 270
	Ophthalmic rods.	25, 26
	Reduce instillation drop volume.	271, 272
	Allow 5 min interval between consecutive drops.	272
	Reduce tear flow rate with drugs.	269, 273
	Reduce tear flow rate by diluting buffer strength.	274
Drug binding to tear proteins	Adjuvants competing for binding sites of proteins.	275
Loss of drug to nasal and conjunctival vasculature	Prodrugs.	276
	Vasoconstrictors.	277
	Nanocapsules.	172
Permeability barrier	Penetration enhancers.	33–37, 149, 179, 201, 278, 279
	Prodrugs.	10, 27, 175–177, 179–183, 280, 281
Enzymatic barrier	Ion-pairing agents.	30–32, 184, 185
	Protease inhibitors.	282

modify drug properties, and penetration enhancers, which are agents capable of altering membrane properties, are discussed below. Finally, novel, promising means based on ion transport modulation are also described.

1. Prodrugs

The prodrug approach is the best developed of the aforementioned approaches. Prodrugs are bioreversible derivatives of existing drugs designed to alter their absorption, decrease their side effects, or prolong their duration of action. Several ophthalmic drugs have been considered for prodrug derivatization, including anti-glaucoma drugs [4,27], antiviral drugs [175,176], anti-inflammatory drugs [177,178], and mydriatics [179,180]. The effectiveness of prodrugs as a means to improve ocular drug bioavailability has been reviewed by Lee and Li [28] and by Lee [29].

All the ophthalmic prodrugs that have been investigated share the characteristic of higher lipophilicity as a result of derivatization of (1) their amino groups forming *N*-alkylated derivatives or (2) their hydroxyl or carboxyl functional groups forming esters. For instance, Wang et al. [181] tested the corneal permeability of various 1-alkoxycarbonyl, 1-acyloxymethyl, and 3-acyl prodrug derivatives of 5-fluorouracil, useful as an adjunct in glaucoma filtration surgery, in an attempt to identify the most permeable prodrug. The log PC of 5-fluorouracil was -0.96, while that of the prodrugs ranged from -0.79 to 1.34. There was an 8.5–53-fold improvement in the corneal permeability of 5-fluorouracil by prodrugs, with the highest permeability (1.09 µm/sec) for 1-hexyloxycarbonyl-5-fluorouracil (log PC = 1.34).

As another example, Chang et al. [27] tested lipophilic *O*-acetyl, propionyl, butyryl, and pivalyl ester prodrugs for their corneal permeability and extent of absorption into the aqueous humor. All the prodrugs tested had a higher corneal permeability than timolol. Peak corneal permeability was reached for *O*-butyryl timolol at a log PC value of 2.08. This improved corneal permeability led to a 4.5-fold increase in ocular drug absorption into the aqueous humor. When the topically applied dose was reduced proportionally, ocular drug absorption was the same as a fourfold higher dose of timolol while its systemic drug absorption was reduced by a factor of 10. The net result was a 15-fold increase in the ratio of ocular to systemic absorption [182]. Similar to *O*-butyryl timolol, diacetyl nadolol, an ester prodrug of nadolol, is 20 times more lipophilic and 10 times more readily permeable than the parent drug [183].

Not all drugs contain functional groups that lend themselves readily to prodrug derivatization. A case in point is the carbonic anhydrase inhibitors such as acetazolamide, ethoxzolamide, and methazolamide. Although the amino functional group of their sulfonamide moiety can be methylated, the resulting analogs are inactive due to lack of demethylation [28]. In this case, means such as penetration enhancers may have to be considered.

2. Ion-Pairing Agents

"Ion pairing" refers to the interaction between two oppositely charged species resulting in the formation of electrically neutral species with increased lipid solubility and hence improved membrane permeability. The utility of this approach in increasing intraocular absorption is controversial [32]. Ion pairing was studied by Ashton et al. [31] to improve the corneal penetration of phenylephrine, a positively charged hydrophilic drug with a log PC of 0.021. Flurbiprofen, a negatively charged drug capable of exerting mydriatic and anti-inflammatory effects like phenylephrine, was used as an ion-pairing agent. The corneal permeability of 1 mM phenylephrine was increased in a dose-dependent manner over a flurbiprofen concentration range of 0–16 mM; there was a ninefold increase at 16 mM flurbiprofen. Permeability enhancement obtained in the presence of ion-pairing counterions must be interpreted with caution, however, as the counterions themselves may increase membrane permeability. This is certainly the case for flurbiprofen, which at 10 mM caused a 4.2- and 2.4-fold increase in the corneal permeabilities of benzoic acid and sorbitol, respectively, two solutes incapable of ion pairing with flurbiprofen. This may also be the case for the enhancement in ocular absorption of sodium chromoglycate by benzalkonium chloride [184] and of bunazosin by the long-chain fatty acid caprylic acid [185], since both ion-pairing agents are capable of increasing membrane permeability [186].

3. Penetration Enhancers

Penetration enhancers are low molecular weight compounds that can increase the absorption of poorly absorbed hydrophilic drugs such as peptides and proteins from the nasal, buccal, oral, rectal, and vaginal routes of administration [186]. Chelators, bile salts, surfactants, and fatty acids are some examples of penetration enhancers that have been widely tested [186]. The precise mechanisms by which these enhancers increase drug penetration are largely unknown. Bile salts, for instance, have been shown to increase the transport of lipophilic cholesterol [187] as well as the pore size of the epithelium [188], indicating enhancement in both transcellular and paracellular transport. Bile salts are known to break down mucus [189], form micelles [190], extract membrane proteins [191], and chelate ions [192]. While breakdown of mucus, formation of micelles, and lipid extraction may have contributed predominantly to the bile salt–induced enhancement of transcellular transport, chelation of ions possibly accounts for their effect on the paracellular pathway. In addition to their lack of specificity in enhancing membrane transport, penetration enhancers may cause nonspecific forms of damage such as release of membrane lipids [193], lysis of cells [194], impairment of ciliary motility [195], and the release of inflammatory mediators [186], thereby limiting their utility as therapeutic adjuvants.

Although the ocular absorption of peptide as well as nonpeptide drugs is poor [96,196–198], the ocular route is by far the least studied for the usefulness of penetration enhancers. This is in part due to the perceived sensitivity of ocular tissues to irritation and the fear of corneal and conjunctival damage caused by the enhancers. Whereas the rat nasal epithelium may tolerate up to 5% sodium glycocholate [199], ocular administration of sodium glycocholate at a concentration of 2% and beyond induces reddening of the eye and tear production in rabbits (Kompella and Lee, unpublished observation).

Nevertheless, there are reports on enhancement of ocular drug absorption by bile salts [33], surfactants [200], and chelators [149]. Newton et al. [35] demonstrated that Azone, an enhancer widely tested in transdermal drug delivery [201], increased the ocular absorption of cyclosporine, an immunosuppressant, by a factor of 3, thereby prolonging the survival of a corneal allograft. In 1986, Lee et al. [34] reported that 10 µg/mL cytochalasin B, an agent capable of condensing the actin microfilaments, increased the aqueous humor and iris-ciliary body concentrations of topically applied inulin (5 kDa) by about 70% and 700%, respectively, in the albino rabbit.

Morimoto et al. [33] demonstrated that the ocular absorption of hydrophilic compounds over a wide range of molecular weights could be increased by 2 and 10 mM sodium taurocholate and sodium taurodeoxycholate in a dose-dependent manner. The compounds were glutathione (307 Da), 6-carboxyfluorescein (376 Da), FITC-dextran (4 kDa), and insulin (5.7 kDa). Of the two bile salts, sodium taurodeoxycholate was more effective. At 10 mM, this bile salt increased the permeability of 6-carboxyfluorescein from 0.02% to 11%, glutathione from 0.08% to 6%, FITC-dextran from 0% to 0.07%, and insulin from 0.06% to 3.8%. Sodium taurocholate, on the other hand, increased the permeability to 0.13%, 0.38%, 0.0011%, and 0.14%, respectively. Taurodeoxycholate was more effective than taurocholate in the nasal epithelium as well [202]. This difference in activities can possibly be attributed to their micelle-forming capability, which is higher for taurodeoxycholate, a dihydroxy bile salt [190].

Conjunctival insulin absorption in rabbits estimated as plasma insulin levels after punctal occlusion was also shown to be increased by bile salts (sodium deoxycholate, glycocholate, and taurocholate) and a surfactant (polyoxyethylene-9-lauryl ether) [200]. Their rank order of effectiveness at 1% was sodium deoxycholate > polyoxyethylene-9-lauryl ether > sodium glycocholate = sodium taurocholate. There was an 18-, 29-, 3-, and 3-fold increase, respectively, in conjunctival absorption. Sodium deoxycholate, a dihydroxy bile salt, was more effective than the trihydroxy bile salts sodium glycocholate and taurocholate, due possibly to its greater surface activity.

B. Novel Means Based on Modulation of Ion Transport Processes

There is growing evidence implicating Na^+-dependent solute transporters and intracellular as well as extracellular Ca^{2+} in the physiological regulation of the paracellular pathway [81,203,204]. Such modulation of paracellular permeability is especially important for drugs such as peptides and oligonucleotides that exhibit poor permeability characteristics across both the cornea and the conjunctiva [150,152,154,155]. In addition, ion transporters such as Cl^- and Ca^{2+} channels have been implicated in macromolecular transport (see Sections IV.B.2 and IV.B.4). In the following discussion, some key ion transport processes and their possible roles in solute transport across epithelial tissues are summarized.

1. Activation of Ion-Dependent Cotransport Processes

Several types of cells are equipped with carrier proteins to transport essential nutrients such as glucose and amino acids that cannot cross the plasma membrane freely because of their hydrophilicity. Intestinal and renal epithelia have long been known to possess specialized Na^+ cotransport processes for glucose [205], amino acids [206], and di- and tripeptides [207].

Glucose and amino acids can be transported preferentially via the paracellular route when the transcellular route is saturated beyond 10 mM. Using flat sheets of jejunum, Thomson et al. [208] demonstrated that passive transport of glucose began to exceed the active component at concentrations ranging from 10 to 40 mM in rats, hamsters, guinea pigs, and rabbits. Madara and Pappenheimer [46] attributed this increase in the passive transport of glucose to the opening of tight junctions at high lumenal glucose concentrations. They demonstrated that 25 mM glucose widened the intestinal tight junctions by inducing contraction of the perijunctional actomyosin ring (PAMR), as seen in electron micrographs. Accumulation of water in the lateral intercellular spaces following deposition of osmotically active glucose in this region is thought to induce PAMR contraction and opening of tight junctions (Fig. 8). Such glucose-induced opening of tight junctions has been shown to enhance the paracellular transport of creatinine, polyethylene glycol 4000, and inulin. In addition to glucose, other nutrients such as alanine and leucine, which are also transported in a Na^+-coupled manner, also induced PAMR contraction and opening of tight junctions in the hamster intestine [46,47]. There may, therefore, be a common role for Na^+-dependent cotransporters.

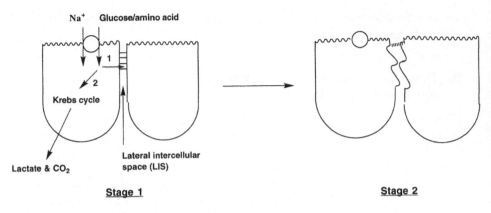

Figure 8 Schematic illustration of the cytoskeletal contraction and tight junctional opening achieved through activation of Na$^+$-glucose or Na$^+$-amino acid contransporters. Stage 1: Activation of the contransporter. Stage 2: Expansion of LIS and cytoskeletal contraction.

The utility of activating Na$^+$ cotransporters to improve paracellular drug permeability was tested by Lu et al. [209]. Thus, 112 mM D-glucose improved the jejunal uptake of less lipophilic acetaminophen (log PC = 0.3) to 233% and that of more lipophilic prednisolone (log PC = 1.6) to about 140%. Further penetration enhancement may be possible with further increases in glucose concentration, in light of the report by Magee and Reid [210] that the maximal rate of glucose absorption from the gastrointestinal tract of rats, cats, and rabbits occurred at 750 mM. It remains to be seen how such a high osmotic load of glucose may affect fluid and electrolyte homeostasis.

Glucose is taken up and metabolized by the cornea. By incubating isolated rabbit corneas with [6-^{14}C]glucose, Thies and Mandel [211] demonstrated that oxidation of exogenous glucose in the Krebs cycle accounts for 75% of the corneal oxygen consumption during oxidative phosphorylation. Tears and/or aqueous humor may be the exogenous source of glucose; the glucose levels in these two fluids in humans were estimated to be 0.45 and 2.7–3.7 mM, respectively [54,212]. No evidence exists, however, for the existence of glucose permease and Na$^+$-dependent glucose cotransporter in the cornea, the two principal entry mechanisms for glucose in the intestine [205]. Our investigations using the voltage-clamp technique (see Section II.B.2) also failed to demonstrate the existence of a Na$^+$-glucose cotransporter in the cornea. There is, however, evidence for the existence of Na$^+$-dependent cotransport processes for glucose and amino acids in the pigmented rabbit conjunctiva (see Section II.B.2) [140].

In 1986, Lee et al. [34] reported that 75 mM glucose increased the concentrations of topically applied inulin (log PC = -2.9) in the aqueous humor and iris-ciliary body four- and tenfold, respectively. Although glucose may exert its effect on paracellular permeability in both the cornea and the conjunctiva, its main effect may be at the conjunctiva for the following reason. First, the work of Chien et al. [4] demonstrated that the conjunctival/scleral route is the principal route of uptake of hydrophilic drugs by the iris-ciliary body. Second, inulin itself reaches the iris-ciliary body via the conjunctival/scleral route to the extent of 80% [3]. The higher concentration of inulin in the iris-ciliary body than in the aqueous humor at 30 min after ocular administration [34] indicates that the conjunctival/scleral route may have contributed to the intraocular absorption of inulin. Third, the absorption enhancement afforded by glucose was higher in the iris-ciliary body than that in the aqueous humor. Finally, while there is no evidence for the presence of a Na^+-glucose cotransporter in the cornea, evidence exists for its presence in the conjunctival epithelium [140].

The effects of D-glucose observed in vivo are not well reproduced in vitro. Madara [203] reported that cytoskeletal contraction and enhanced paracellular permeability were observed only in an in situ perfusion preparation and not in an isolated tissue preparation. Although its in vivo effect was not tested, 25 mM D-glucose, an effective concentration in the jejunum [47], failed to enhance the in vitro transport of sotalol (log PC = −0.62), atenolol (log PC = 0.16), or nadolol (log PC = 0.93) across the isolated conjunctiva [213]. For a similar reason and possibly due to the absence of a Na^+-glucose cotransporter in the cornea, 25 mM D-glucose was ineffective in increasing the corneal transport of these three drugs.

2. Modulation of Extra- and Intracellular Ca^{2+} Levels

a. *Role of Extracellular Ca^{2+}.* Calcium ion (Ca^{2+}) plays a unique role in maintaining an impermeable epithelial barrier through its effect on the tight junction and cell–cell adhesion mediated through uvomorulin. Cereijido et al. [214] showed that chelation of extracellular Ca^{2+} with 2.5 mM EDTA opened the tight junctions and impaired polarity in a Madin–Darby canine kidney (MDCK) monolayer. Upon replating, the cells attached and formed a monolayer in 10–30 min and started to repolarize and reestablish tight junctions and transepithelial electrical resistance (TER) in 12–15 hr. One locus for the action of extracellular Ca^{2+} in establishing intercellular junctions is the protein uvomorulin, also known as epithelial cell adhesion molecule (ECAM), located below the tight junctions [79].

Human tears contain 0.5–1.1 mM Ca^{2+} [215], which may be the source of extracellular calcium for maintaining the integrity of the corneal and conjunctival intercellular junctions. Indeed, chelation of extracellular Ca^{2+} with 0.5% EDTA reduced transcorneal and transconjunctival resistance by 80 and 65%, respectively (Kompella, Kim, and Lee, unpublished observation). In 1991, Rojanasakul and Robinson [101] also demonstrated that exposure of the rabbit cornea to a Ca^{2+}-free GBR or 0.1 mM EDTA for 45 min reduced the corneal resistance by 33%, an effect that was fully reversible within 1 hr of exposure of the cornea to Ca^{2+}-containing GBR. These findings suggest that chelation of extracellular Ca^{2+} possibly disrupts the intercellular junctions, thereby reducing transepithelial resistance. As expected, chelation of Ca^{2+} by 0.05% and 0.1% EDTA led to an increase in the corneal permeability to glycerol (log PC = −1.79), a hydrophilic marker, by 88% and 307%, respectively [216]. Chelation of Ca^{2+} by 0.1% and 0.5% EDTA led to an increase in the conjunctival permeability to sotalol (log PC = −0.62), a hydrophilic beta-adrenergic antagonist, by two- to threefold [213].

In addition to the aforementioned effects on paracellular drug transport, Ca^{2+} also plays an important role in the transcytosis of macromolecules. The entry of the plant lectins abrin, modeccin, and viscumin into Vero cells was inhibited in a Ca^{2+}-free medium a well as in a Ca^{2+}-containing medium containing verapamil and Co^{2+}, both inhibitors of Ca^{2+} [217]. Ca^{2+} is therefore required at a stage after the binding of the above lectins, perhaps in the fusion and exocytosis of membrane vesicles.

   b. *Role of Intracellular Ca^{2+}.* In addition to depleting extracellular Ca^{2+}, elevating intracellular Ca^{2+} via either increased entry from the bathing fluid or release from intracellular pools would also lead to opening of tight junctions triggered by cytoskeletal contraction, an event that typically occurs within 30 min of pharmacological manipulation [204]. Following the increase of intracellular levels of Ca^{2+} using A23187, Rojanasakul et al. [218] reported a substantial drop in the TER of rabbit tracheal epithelium, an effect that was fully reversible upon replacing the A23187-containing medium with ionophore-free medium. Similar observations were made in the rat liver by Kan and Coleman [219]. The work of Candia et al. [114] suggests that A23187 may also increase the passive permeability of the frog cornea, which, like the rabbit cornea, secretes Cl^- in the endothelial-to-epithelial direction and lacks specialized uptake mechanisms for Cl^- in the apical membrane. Therefore, $^{36}Cl^-$ fluxes in the epithelial-to-endothelial direction represent a passive paracellular permeability of this ion. A 50% increase in the epithelial-to-endothelial flux of $^{36}Cl^-$ was observed following epithelial application of 10 µM A23187, suggesting that an increase in intracellular Ca^{2+} levels could enhance paracellular permeability of the cornea. Similar results are expected in the conjunctival epithelium.

Mobilization of Ca^{2+} from intracellular pools can be achieved with 2,5-di(*tert*-butyl)-1,4-benzohydroquinone (tBuBHQ), an agent that elevates intracellular levels of Ca^{2+} by mobilizing the inositol 1,4,5-triphosphate-sensitive Ca^{2+} pool without increasing inositol phosphate itself [220]. Llopis et al. [204] showed that the paracellular component of HRP transport in the perfused rat liver was increased fivefold when 25 µM tBuBHQ was infused for 10 min before 0.5 mg of HRP was infused as a bolus.

3. Modulation of Na^+/H^+ Exchanger

Protons are continuously generated during glycolysis, glycogen synthesis and degradation, lipolysis, triglyceride synthesis and degradation, and ATP hydrolysis [221]. As virtually all intracellular processes are pH-sensitive, the intracellular pH is usually kept within a very narrow range through specialized proton export proteins such as the Na^+/H^+ exchanger, the ATP-driven proton pump, and the lactate-H^+ symporter [222] as well as through the Cl^-/HCO_3^- exchangers that indirectly influence the intracellular pH by altering intracellular HCO_3^- concentrations. In this section, the Na^+/H^+ exchanger is discussed relative to its role in the modulation of drug transport.

Dowty and Braquet [223] noticed a 0.25–0.3 unit decrease in the intracellular pH in cultured corneal endothelial cells upon inhibiting the Na^+/H^+ exchanger by 1 mM amiloride or in a Na^+-free Ringer's buffer. Intracellular acidification often leads to increased intracellular Ca^{2+} concentrations [224,225], thereby opening tight junctions. In order to determine whether inhibition of Na^+/H^+ exchange would enhance paracellular permeability, Kompella, Kim, and Lee tested the effect of 1 mM 5-(N-hexamethylene)amiloride (HA) on the corneal and conjunctival permeability of atenolol (log PC = 0.16) and mannitol (a paracellular marker). HA enhanced the corneal transport of atenolol and mannitol by 150% and 140% and the conjunctival transport by 110% and 150%, respectively (Fig. 9). Thus, the permeability of mannitol was enhanced the most in both tissues, suggesting that HA elevates the paracellular permeability in the cornea and conjunctiva. This enhancement in permeability correlated with a 70–75% decrease in the transepithelial resistance of the cornea and conjunctiva. This penetration enhancement effect of HA is consistent with the enhancement of the ocular hypotensive effects of p-aminoclonidine by coadministered Na^+/H^+ exchange inhibitors (1%) [226]. The rank order of potency—5-(N-hexamethylene)amiloride > 5-(N-ethyl-N-isopropyl)amiloride > 5-(N,N-dimethyl)amiloride ≫ amiloride—correlated very well with the affinities of the above inhibitors for the Na^+/H^+ exchanger.

4. Modulation of Cl^- Channels

Modulation of Cl^- channels is expected to alter paracellular drug transport across the cornea and the conjunctiva by modulating either net water transport or intracellular levels of Ca^{2+}. In general, epithelia with Cl^- channels secrete Cl^- along

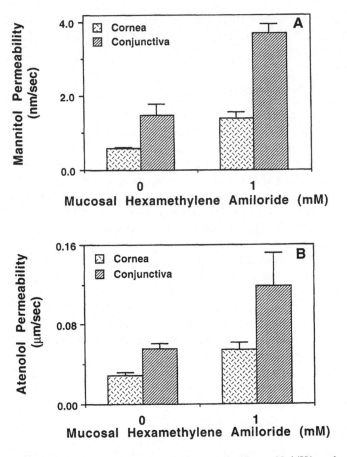

Figure 9 Influence of hexamethylene amiloride, an Na^+/H^+ exchange inhibitor, on the corneal and conjunctival permeability of (A) mannitol and (B) atenolol. Error bars denote mean $\pm$ SEM for $n = 4$. (Kompella, Kim, and Lee, unpublished data.)

with water in the serosal-to-mucosal direction. Inhibition of Cl^- channels with specific inhibitors such as N-phenylanthranilic acid (NPAA) and 5-nitro-2-(3-phenylpropylamino) benzoic acid (NPPB) can decrease water efflux, offering less impedance to the transport of polar drugs.

Chloride channels themselves may directly export drugs, particularly since certain chloride channels appear to act in the same manner as P-glycoprotein (multiple drug resistance protein) [227], which is involved in the export of vincristine [228,229], daunomycin [228,229], gramicidin D [230], and cyclosporin

A [228,229] by an as yet unknown mechanism. Owing to their lipophilicity, these drugs tend to accumulate rapidly within the cell to reach concentrations high enough to elicit toxic effects. The expression of *P*-glycoprotein in the plasma membrane may be up-regulated in response, in order to export the cytotoxic agents. Indeed, Riordan and Ling [231] demonstrated that the plasma membrane vesicles of colchicine-resistant but not colchicine-sensitive Chinese hamster ovary (CHO) cells expressed high levels of *P*-glycoprotein.

Valverde et al. [227] demonstrated that the expression of *P*-glycoprotein in fibroblasts generated volume-regulated ATP-dependent Cl⁻-selective channels with properties similar to those characterized previously in colonic epithelium [115]. At 50 µM, compounds known to inhibit *P*-glycoprotein-mediated drug transport also inhibited the *P*-glycoprotein-mediated volume-activated chloride current in fibroblast cells transfected with the multiple drug resistance gene (MDR1) [227]. The extent of inhibition was 15% for forskolin, 90% for 1,9-dideoxyforskolin, 75% for verapamil, 20% for nifedepine, 25% for quinine, and 60% for quinidine.

There is mounting evidence that *P*-glycoprotein belongs to the same membrane transporter superfamily as the cystic fibrosis transmembrane receptor (CFTR) involved in anion transport and the transport of other solutes such as anticancer drugs. Indeed, CFTR has been shown to export vincristine and daunomycin in Chinese hamster ovary cells [194]. The above two proteins are structurally similar, with a membrane-associated domain comprising six membrane-spanning segments as well as restricted regions of amino acid sequence similarity within the nucleotide-binding folds. The drugs exported by these ion transporters are lipophilic with high affinity for membranes. It remains to be seen whether these Cl⁻ channels can import drugs into the cells.

The entry of Cl⁻ into cells may be essential for the cellular entry [232] or secretion [233] of some macromolecules such as diphtheria toxin and modeccin. Sandvig and Olsnes [232] studied the entry of diphtheria toxin and modeccin into Vero cells in pH 7.2 media containing 20 mM Hepes, 1 mM Ca(OH)₂, 5 mM glucose,a sufficient amount of mannitol to ensure isotonicity, and varying concentrations of NaCl. The cellular uptake of 0.1 nM diphtheria toxin at the end of 50 min was strongly dependent on Cl⁻ concentration. It was 0% at 0 mM NaCl, 25% of the 140 mM NaCl control at 2 mM NaCl, and 60% of the control at 70 mM NaCl. A similar trend was observed for modeccin, i.e., no transport at 0 mM NaCl, 20% of control at 0.05 mM NaCl, 60% of control at 0.1 mM NaCl, 80% of control at 0.5 mM NaCl, and 100% of control at 2 mM NaCl.

The use of Cl⁻-free media required 100 times higher concentrations of the toxins for any cellular uptake. Furthermore, the entry of both toxins was denied when Vero cells were incubated in normal medium with inhibitors of Cl⁻ entry, including 4-acetamido-4'-isothiocyanostilbene-2,2'-disulfonate (SITS), pyridoxal phosphate, SCN^-, or SO_4^{2-}. The exact role of Cl^- entry in the internalization of diphtheria toxin and modeccin is not known.

Chloride ion conductive pathways may be involved in the exocytosis of vesicles containing macromolecules [234]. Gasser et al. [234] demonstrated in rat pancreatic zymogen granules the presence of Cl^- channels and Cl^-/HCO_3^- exchangers whose activity was increased in cholecystokinin- and secretin-treated rats. It was postulated that increased Cl^- conductance may aid in (1) the accumulation and storage of secretory proteins, (2) fluid secretion along with macromolecules, (3) osmotic swelling of granules and subsequent membrane fusion or fission, and (4) increase of Cl^- conductance and fluidity of the plasma membrane after fusion. In addition, Cl^- conductive pathways also seem to be essential in the acidification of various intracellular vesicles such as endosomes, lysosomes, Golgi vesicles, and secretory granules [235]. The acidification of these vesicles is achieved by H^+-ATPase, which, unlike H^+/K^+-ATPase of the stomach, does not exchange K^+ or any other cation for H^+. However, it is sensitive to chloride. It was speculated that Cl^-, a relatively permeable anion, was essential in nullifying the potential difference developed during the inward pumping of H^+ by the electrogenic H^+-ATPase of the intracellular vesicles.

More recent reports further support the role of Cl^- channels in the transport of macromolecules [236,237]. Bradbury et al. [237] demonstrated that stimulation of plasma membrane recycling during cAMP-mediated regulation is dependent on the presence of CFTR, a cAMP-dependent Cl^- conductive pathway. Following the loading of biotinylated wheat germ agglutinin (bWGA), 8-(4-chlorophenyl-thio)cAMP failed to stimulate the exocytosis of this lectin in cystic fibrosis pancreatic cells with a defective CFTR. On the other hand, cystic fibrosis pancreatic cells transfected with a retroviral vector containing the full-length cDNA encoding CFTR responded to cAMP stimulus and facilitated the exocytosis of bWGA. Defective pancreatic cells transfected with vector alone failed to respond to cAMP stimulation, confirming the role of CFTR in facilitating cAMP-mediated exocytosis. A schematic representation of cAMP-stimulus-coupled exocytosis of bWGA in pancreatic cells is given in Figure 10. It is speculated that protein kinase A activated by cAMP phosphorylates the Cl^- channels of the bWGA-containing vesicles, thereby allowing the osmotic swelling and exocytosis of bWGA.

Recent work in our laboratory (Kompella, Mathias, and Lee, unpublished observation) has revealed that activation of the cAMP-regulated Cl^- channels in the conjunctiva also enhances the transcytosis of horseradish peroxidase. 8-Bromo-cAMP (a membrane-permeable analog of cAMP) and terbutaline (a β2-adrenergic agonist known to increase intracellular levels of cAMP in other epithelial tissues [238]), at 0.5 mM, were found to enhance the transport of 100 μg/mL HRP from the mucosal side to the serosal side of the pigmented rabbit conjunctiva by a factor of 4 (Fig. 11).

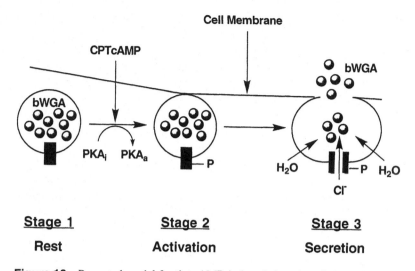

Figure 10 Proposed model for the cAMP-induced elevation observed in the exocytosis of biotinylated wheat germ agglutinin containing vesicles in pancreatic cells. CPTcAMP = 8-(4-chlorophenylthio)cAMP; PKA_i = inactive form of protein kinase A; PKA_a = active form of protein kinase A; P = inorganic phosphate; bWGA = biotinylated wheat germ agglutinin. (Based on the findings of Ref. 237.)

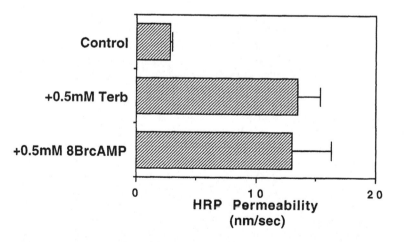

Figure 11 Influence of intracellular elevation of cAMP on the fluid-phase transport of HRP in the rabbit conjunctiva. Terb = terbutaline; 8BrcAMP = 8-bromo cAMP. Error bars denote mean ± SEM. for $n = 4$. (Kompella, Mathias, and Lee, unpublished data.)

V. CONCLUSION

Improving the corneal and conjunctival permeability to drugs is a major challenge in ocular drug delivery. Thus far, the main focus has been on improving drug lipophilicity through the use of prodrugs or analogs, although the use of ion pairs and penetration enhancers has also been considered. Each approach is not without drawbacks. Prodrugs may suffer from the lack of suitable functional groups for derivatization or the lack of a mechanism to regenerate the active drug in the eye. While there is no conclusive evidence for the usefulness of ion pairing in enhancing membrane transport, this approach is confined to the transcellular uptake of drugs. Penetration enhancers, on the other hand, are nonspecific with respect to the transcellular and paracellular routes of transport and would cause cell damage ranging from lipid extraction to cell lysis. All of the above approaches are based on the premise that the cornea and the conjunctiva play a passive role in drug transport, when in fact both are very active ion-transporting tissues. Elucidating how these ion transport processes may be primed to improve drug transport is an area that deserves more attention in the future.

REFERENCES

1. I Ahmed, TF Patton. (1987). Disposition of timolol and inulin in the rabbit eye following corneal versus non-corneal absorption. Int J Pharm 38:9–21.
2. MG Doane, AD Jenson, CH Dohlman. (1978). Penetration routes of topically applied eye medications. Am J Ophthalmol 85:383–386.
3. I Ahmed, TF Patton. (1985). Importance of the noncorneal absorption route in topical ophthalmic drug delivery. Invest Ophthalmol Vis Sci 26:584–587.
4. DS Chien, JJ Homsy, C Gluchowski, DD Tang-Liu. (1990). Corneal and conjunctival/scleral penetration of p-aminoclonidine. Curr Eye Res 9:1051–1059.
5. TF Patton, JR Robinson. (1976). Quantitative precorneal disposition of topically applied pilocarpine nitrate in rabbit eyes. J Pharm Sci 65:1295–1301.
6. CH Chiang, RD Schoenwald. (1986). Ocular pharmacokinetic models of clonidine-^{3}H hydrochloride. J Pharmacokin Biopharm 14:175–211.
7. DDS Tang-Liu, SS Liu, RJ Weinkam. (1984). Ocular and systemic bioavailability of ophthalmic flurbiprofen. J Pharmacokin Biopharm 12:611–626.
8. TF Patton, JR Robinson. (1974). Ocular evaluation of polyvinyl alcohol vehicle in rabbits. J Pharm Sci 64:1312–1316.
9. SS Chrai, JR Robinson. (1974). Ocular evaluation of methyl cellulose vehicle in albino rabbits. J Pharm Sci 63:1218–1223.
10. SC Chang, DS Chien, H Bundgaard, VHL Lee. (1988). Relative effectiveness of prodrug and viscous solution approaches in maximizing the ratio of ocular to systemic absorption of topically applied timolol. Exp Eye Res 46:59–69.
11. MF Saettone, D Monti, MT Torracca, P Chetoni, B Giannaccini. (1989). Muco-

adhesive liquid ophthalmic vehicles—Evaluation of macromolecular ionic complexes of pilocarpine. Drug Dev Ind Pharm 15:2475–2489.

12. O Camber, P Edman, R Gurny. (1987). Influence of sodium hyaluronate on the meiotic effect of pilocarpine in rabbits. Curr Eye Res 6:779–784.

13. NM Davies, SJ Farr, J Hadgraft, IW Kellaway. (1991). Evaluation of mucoadhesive polymers in ocular drug delivery. I. Viscous solutions. Pharm Res 8:1039–1043.

14. MF Saettone, D Monti, MT Torracca, P Chetoni. (1994). Mucoadhesive ophthalmic vesicles: Evaluation of polymeric low-viscosity formulations. J Ocular Pharmacol 10:83–92.

15. DS Chien, H Sasaki, H Bundgaard, A Buur, VHL Lee. (1991). Role of enzymatic lability in the corneal and conjunctival penetration of timolol ester prodrugs in the pigmented rabbit. Pharm Res 8:728–733.

16. MF Saettone, B Giannaccini, P Chetoni, G Galli, E Chiellini. (1984). Vehicle effects in ophthalmic bioavailability: An evaluation of polymeric inserts containing pilocarpine. J Pharm Pharmacol 36:229–234.

17. VHL Lee, S Li, MF Saettone, P Chetoni, H Bundgaard. (1991). Systemic and ocular absorption of timolol prodrugs from erodible inserts. Proc Int Symp Controlled Release Bioact Mater 18:291–292.

18. B Hoffmann, I Nagel, W Clauss. (1990). Aldosterone regulates paracellular pathway resistance in rabbit distal colon. J Comp Physiol B 160:381–388.

19. RA Lewis, RD Schoenwald, MG Eller, CF Barknecht, DC Phelps. (1984). Ethoxzolamide analogue gel, a topical carbonic anhydrase inhibitor. Arch Ophthalmol 102:1821–1824.

20. SC Miller, MD Donovan. (1982). Effect of poloxamer 407 gel on the mitotic activity of pilocarpine nitrate in rabbits. Int J Pharm 12:147–152.

21. JA Hobden, JJ Reidy, RJ O'Callaghan, MS Insler, JM Hill. (1990). Quinolones in collagen shields to treat aminoglycoside-resistant pseudomonas keratitis. Invest Ophthalmol Vis Sci 31:2241–2243.

22. JA Hobden, JJ Reidy, RJ O'Callaghan, MS Insler, JM Hill. (1990). Ciprofloxacin iontophoresis for aminoglycoside-resistant pseudomonas keratitis. Invest Ophthalmol Vis Sci 31:1940–1944.

23. U Pleyer, B Elkins, D Ruckert, S Lutz, J Grammer, J Chou, KH Schmidt, BJ Mondino. (1994). Ocular absorption of cyclosporine A from liposomes incorporated into collagen shields. Curr Eye Res 13:177–181.

24. JK Milani, I Verbukh, U Pleyer, H Sumner, S Adamu, HP Halabi, HJ Chou, DA Lee, BJ Mondino. (1993). Collagen shields impregnated with gentamicin-dexamethasone as a potential drug delivery device. Am J Opthalmol 116:622–627.

25. SD Alani. (1990). The ophthalmic rod—A new ophthalmic drug delivery system. I. Graefes' Arch Clin Exp Ophthalmol 228:297–301.

26. SD Alani, W Hammerstein. (1990). The ophthalmic rod—A new drug delivery system. II. Graefes' Arch Clin Exp Ophthalmol 228:302–304.

27. SC Chang, H Bundgaard, A Buur, VHL Lee. (1987). Improved corneal penetration of timolol by prodrugs as a means to reduce systemic drug load. Invest Ophthalmol Vis Sci 28:487–491.

28. VHL Lee, VHK Li. (1989). Prodrugs for improved ocular drug delivery. Adv Drug Delivery Rev 3:1–38.

29. VHL Lee. (1993). Improved ocular drug delivery by use of chemical modification (prodrugs). In: P Edman, ed. Biopharmaceutics of Ocular Drug Delivery. Boca Raton, FL, CRC Press, pp 121–143.

30. PM Hughes, AK Mitra. (1993). Effect of acylation on the ocular disposition of acyclovir. II: Corneal permeability and anti-HSV 1 activity of 2'-esters in rabbit epithelial keratitis. J Ocular Pharmacol 9:299–309.

31. P Ashton, DS Clark, VHL Lee. (1992). A mechanistic study on the enhancement of corneal penetration of phenylephrine by flurbiprofen in the rabbit. Curr Eye Res 11:85–90.

32. CW Conroy, RH Buck. (1992). Influence of ion pairing salts on the transcorneal permeability of ionized sulfonamides. J Ocular Pharmacol 8:233–240.

33. KM Morimoto, T Nakai, K Morisaka. (1987). Evaluation of permeability enhancement of hydrophilic compounds and macromolecular compounds by bile salts through rabbit corneas in vitro. J Pharm Pharmacol 39:124–126.

34. VHL Lee, LW Carson, KA Takemoto. (1986). Macromolecular drug absorption in the albino rabbit eye. Int J Pharm 29:43–51.

35. C Newton, BM Gebhardt, HE Kaufman. (1988). Topically applied cyclosporine in azone prolongs corneal allograft survival. Invest Ophthalmol Vis Sci 29:208–215.

36. D BenEzra, G Maftzir. (1990). Ocular penetration of cyclosporine A in the rat eye. Arch Ophthalmol 108:584–587.

37. DD Tang-Liu, JB Richman, RJ Weinkam, H Takruri. (1994). Effects of four penetration enhancers on corneal permeability of drugs in vitro. J Pharm Sci 83:85–90.

38. VHL Lee, LW Carson, S Dodda Kashi, RE Stratford Jr. (1986). Metabolic and permeation barriers to the ocular absorption of topically applied enkephalins in albino rabbits. J Ocular Pharmacol 2:345–352.

39. VHL Lee, KW Morimoto, RE Stratford Jr. (1982). Esterase distribution in the rabbit cornea and its implications in ocular drug bioavailability. Biopharm Drug Dispos 3:291–300.

40. VHL Lee. (1983). Esterase activities in adult rabbit eyes. J Pharm Sci 72:239–244.

41. VHL Lee, RE Stratford Jr, KW Morimoto. (1983). Age related changes in esterase activity in rabbit eyes. Int J Pharm 13:183–195.

42. RE Stratford Jr, VHL Lee. (1985). Aminopeptidase activity in albino rabbit extraocular tissues relative to the small intestine. J Pharm Sci 74:731–734.

43. VHL Lee, DS Chien, H Sasaki. (1988). Ocular ketone reductase distribution and its role in the metabolism of ocularly applied levobunolol in the pigmented rabbit. J Pharmacol Exp Therap 246:871–878.

44. DA Campbell, RD Schoenwald, MW Duffel, CF Barknecht. (1991). Characterization of arylamine acetyltransferases in the rabbit eye. Invest Ophthalmol Vis Sci 32:2190–2200.

45. RE Stratford Jr, LW Carson, S Dodda Kashi, VHL Lee. (1988). Systemic absorption of ocularly administered enkephalinamide and inulin in the albino rabbit: Extent, pathways, and vehicle effects. J Pharm Sci 77:838–842.

46. JL Madara, JR Pappenheimer. (1987). Structural basis for physiological regulation of paracellular pathways in intestinal epithelium. J Membr Biol 100:149–164.

47. JR Pappenheimer, KZ Reiss. (1987). Contribution of solvent drag through intercel-

lular junctions to the absorption of nutrients by the small intestine of the rat. J Membr Biol 100:123–136.

48. MS Balda, GL Mariscal, RG Contreras, MM Silva, TME Marquez, GJA Sainz, M Cereijido. (1991). Assembly and sealing of tight junctions: Possible participation of G-proteins, phospholipase C, protein kinase C, and calmodulin. J Membr Biol 122:193–202.

49. PJ Lowe, K Miyal, JH Steinbach, WGM Hardison. (1988). Hormonal regulation of hepatocyte tight junctional permeability. Am J Physiol 255:G454–G461.

50. EE Schneeberger, RD Lynch. (1992). Structure, function, and regulation of cellular tight junctions. Am J Physiol 262:L647–L661.

51. MJ Berridge. (1987). Inositol triphosphate and diacylglycerol: Two interacting second messengers. Annu Rev Biochem 56:159–193.

52. VHL Lee. (1993). Precorneal, corneal, and postcorneal factors. In: AK Mitra, ed. Ophthalmic Drug Delivery Systems. New York: Marcel Dekker, pp 59–81.

53. SD Klyce, RW Beuerman. (1988). Structure and function of the cornea. In: HE Kaufman, BA Barron, MB McDonald, SR Waltman, eds. The Cornea. New York: Churchill Livingstone, pp 3–54.

54. JS Pepose, JL Ubels. (1992). The cornea. In: W Hart Jr, ed. Adler's Physiology of the Eye. St. Louis; Mosby Year Book, pp 29–70.

55. DM Maurice. (1951). The permeability to sodium ions of the living rabbit cornea. J Physiol 112:367–391.

56. DM Maurice. (1984). The cornea and sclera. In: H Davson, ed. The Eye. New York: Academic Press, pp 1–158.

57. S Mishima, BO Hedbys. (1967). The permeability of the corneal epithelium and endothelium to water. Exp Eye Res 6:10–32.

58. JC Moore, JM Tiffany. (1979). Human ocular mucus. Origins and preliminary characterization. Exp Eye Res 29:1–11.

59. JV Greiner, DR Korb, HI Covington, DG Peace, MR Allansmith. (1982). Human ocular mucus: Scanning electron microscopy study. Arch Ophthalmol 100:1614–1617.

60. RA Thoft, J Friend. (1983). The XYZ hypothesis of corneal epithelial maintenance. Invest Ophthalmol Vis Sci 24:1442–1443.

61. M Hirsch, G Renard, JP Faure, Y Pouliquen. (1977). Study of the ultrastructure of the rabbit corneal endothelium by the freeze-fracture technique: Apical and lateral junctions. Exp Eye Res 25:277–288.

62. MA Watsky, MM Jablonski, HF Edelhauser. (1988). Comparison of conjunctival and corneal surface areas in rabbit and human. Curr Eye Res 7:483–486.

63. FJ Holly. (1973). Formation and rupture of the tear film. Exp Eye Res 15:515–525.

64. JV Greiner, KR Kenyon, AS Henriquez, DR Korb, TA Weidman, MR Allansmith. (1980). Mucus secretory vesicles in conjunctival epithelial cells of wearers of contact lenses. Arch Ophthalmol 98:1843–1846.

65. VHL Lee, DJ Schanzlin, RE Smith. (1986). Interaction of rabbit conjunctival mucin with tear protein and peptide analogs. In: FJ Holly, ed. The Preocular Tear Film in Health, Disease, and Contact Lens Wear. Lubbock, TX: Dry Eye Institute, pp 341–355.

66. AJ Bron, LS Mengher, CC Davey. (1985). The normal conjunctiva and its response to inflammation. Trans Ophthalmol Soc UK 104:424–435.

67. MJ Hogan, JA Alvarado, JE Weddell. (1971) The limbus. In: MJ Hogan, JA Alvarado, JE Weddel, eds. Histology of the Human Eye. Philadelphia: WB Saunders, pp 112–182.

68. HS Geggel, J Friend, RA Thoft. (1984). Conjunctival epithelial wound healing. Invest Ophthalmol Vis Sci 25:860–863.

69. MS Shapiro, J Friend, RA Thoft. (1981). Corneal reepithelialization from the conjunctiva. Invest Ophthalmol Vis Sci 21:135–142.

70. D Aitken, J Friend, RA Thoft, WR Lee. (1988). An ultrastructural study of rabbit ocular surface transdifferentiation. Invest Ophthalmol Vis Sci 29:224–231.

71. AJW Huang, SCG Tseng. (1991). Corneal epithelial wound healing in the absence of limbal epithelium. Invest Ophthalmol Vis Sci 32:96–105.

72. B Nichols, CR Davson, B Togni. (1983). Surface features of the conjunctiva and cornea. Invest Ophthalmol Vis Sci 24:570–576.

73. BJ McLaughlin, RB Caldwell, Y Sasaki, T Wood. (1985). Freeze-fracture quantitative comparison of rabbit corneal epithelial and endothelial membranes. Curr Eye Res 4:951–961.

74. JM Wolosin, M Chen. (1993). Ontogeny of corneal epithelial tight junctions: Stratal locale of biosynthetic activities. Invest Ophthalmol Vis Sci 34:2655–2664.

75. DM Maurice. (1967). Epithelial potential of the cornea. Exp Eye Res 6:138–140.

76. SD Klyce. (1972). Electrical profiles in the corneal epithelium. J Physiol 226:407–429.

77. J Behrens, W Birchmeir, SL Goodman, BA Imhof. (1985). Dissociation of Madin–Darby canine kidney epithelial cells by the monoclonal antibody anti-arc-1: Mechanistic aspects and identification of the antigen as a component related uvomorulin. J Cell Biol 101:1307–1315.

78. WJ Nelson, EM Shore, AZ Wang, RW Hammerton. (1990). Identification of a membrane-cytoskeletal complex containing the cell adhesion molecule uvomorulin (E-cadherin), ankyrin, and fodrin in Madin–Darby canine kidney epithelial cells. J Cell Biol 110:349–357.

79. B Gumbiner, K Simons. (1987). The role of uvomorulin in the formation of epithelial occluding junctions. In: Anonymous. Junctional Complexes of Epithelial Cells, CIBA Foundation Symposium 125, New York: Wiley, pp 168–186.

80. JM Anderson, BR Stevenson, LA Jesaitis, DA Goodenough, MS Mooseker. (1989). Characterization of ZO-1, a protein component of the tight junction from mouse-liver and Madin–Darby canine kidney cells. J Cell Biol 106:1141–1149.

81. JM Anderson, CM Van Itallie, MD Peterson, BR Stevenson, EA Carew, MS Mooseker. (1989). ZO-1 mRNA and protein expression during tight junction assembly in Caco-2 cells. J Cell Biol 109:1047–1056.

82. RA Conradi, AR Hilgers, NFH Ho, PS Burton. (1991). The influence of peptide structure on transport across Caco-2 cells. Pharm Res 8:1453–1460.

83. Y Wang, M Chen, JM Wolosin. (1993). ZO-1 in corneal epithelium: Stratal distribution and synthesis induction by outer cell removal. Exp Eye Res 57:283–292.

84. A Smulders, EM Wright. (1971). Galactose transport across the hamster small intestine: The effect of sodium electrochemical gradients. J Physiol 212:277–286.

85. KR Page. (1980). The use of KCl salt bridges in electrophysiological circuits. Comp Biochem Physiol A67:637–642.

86. SD Klyce. (1973). Relationship of epithelial membrane potentials to corneal potential. Exp Eye Res 15:567–575.

87. SD Klyce. (1975). Transport of Na^+, Cl^-, and water by the rabbit corneal epithelium at resting potential. Am J Physiol 228:1446–1452.

88. SD Klyce, RKS Wong. (1977). Site and mode of adrenaline action on chloride transport across the rabbit corneal epithelium. J Physiol 266:777–799.

89. E Fromter. (1972). The route of passive ion movement through the epithelium of *Necturus* gallbladder. J Membr Biol 8:259–301.

90. D Amman. (1986). Ion-Selective Microelectrodes: Principles, Design, and Application. Springer Verlag, New York.

91. J DeLong, MM Civan. (1983). Microelectrode study of K^+ accumulation by tight epithelia. I. Baseline values of split frog skin and toad urinary bladder. J Membr Biol 72:183–193.

92. RD Gunther, EM Wright. (1983). Na^+, Li^+, and Cl^- transport by brush border membranes from rabbit jejunum. J Membr Biol 74:85–94.

93. M Chalfie, AH Neufeld, JA Zadunaisky. (1972). Action of epinephrine and other cyclic AMP-mediated agents on the chloride transport of the frog cornea. Invest Ophthalmol 11:644–650.

94. E Neher, B Sakman, JH Steinbach. (1978). The extracellular patch clamp: A method for resolving currents through individual open channels in biological membranes. Pfluegers Arch 375:219–228.

95. OP Hamill, A Marty, E Neher, B Sakman, FJ Sigworth. (1981). Improved patch-clamp techniques for high-resolution current recording from cells and cell-free membrane patches. Pfluegers Arch 391:85–100.

96. WS Marshall, JW Hanrahan. (1991). Anion channels in the apical membrane of mammalian corneal epithelium primary cultures. Invest Ophthalmol Vis Sci 32:1562–1568.

97. W Van Driessche, D Erlij. (1983). Noise analysis of inward and outward Na^+ currents across the apical border of ouabain-treated frog skin. Pfluegers Arch 398:179–188.

98. WS Marshall, SD Klyce. (1984). Cellular mode of serotonin action on Cl transport in the rabbit corneal epithelium. Biochim Biophys Acta 778:139–143.

99. RL Shih, VHL Lee. (1990). Rate limiting barrier to the penetration of ocular hypotensive beta-blockers across the corneal epithelium in the p.gmented rabbit. J Ocular Pharmacol 6:329–336.

100. DM Maurice. (1976). Techniques of investigation of the cornea. In: S Dikstein, ed. Drugs and Ocular Tissues. Basel: Krager, pp. 90–101.

101. Y Rojanasakul, JR Robinson. (1991). The cytoskeleton of the cornea and its role in tight junction permeability. Int J Pharm 68:135–149.

102. SD Klyce, WS Marshall. (1982). Effects of Ag^+ on ion transport by the corneal epithelium of the rabbit. J Membr Biol 66:133–144.

103. JM Wolosin. (1988). Regeneration of resistance and ion transport in rabbit corneal epithelium after induced surface cell exfoliation. J Membr Biol 104:45–55.

104. A Donn, DM Maurice, NL Mills. (1959). Studies in the living cornea in vitro. II.

The active transport of Na^+ across the epithelium. Arch Ophthalmol 62:748–757.

105. K Green. (1965). Ion transport in the isolated cornea of the rabbit. Am J Physiol 209:1311–1316.

106. SD Klyce, AH Neufeld, JA Zadunaisky. (1973). The activation of chloride transport by epinephrine and Db cyclic-AMP in the cornea of the rabbit. Ophthalmol 12: 127–138.

107. MA Watsky, K Cooper, JL Rae. (1991). Sodium channels in ocular epithelia. Pfluegers Arch Eur J Physiol 419:454–459.

108. J Liaw, Y Rojanasakul, JR Robinson. (1991). The effect of charge type and charge density on corneal transport of lysine and glutamic acid—Evidence for active transport. Pharm Res 8(Suppl):S-130.

109. OA Candia, PJ Bently, PI Cook. (1974). Stimulation by amphotericin B of active sodium transport across amphibian cornea. Am J Physiol 222:1438.

110. JA Bonanno, SD Klyce, EJ Cragoe Jr. (1989). Mechanism of chloride uptake in rabbit corneal epithelium. Am J Physiol 257:C290–C296.

111. N Ehlers. (1973). In vitro studies of trans- and intraepithelial potentials of the cornea. Exp Eye Res 15:553–565.

112. WS Rehm, RL Shoemaker, SS Sanders, JT Tarvin, JA Wright Jr, EA Friday. (1973). Conductance of epithelial tissues with particular reference to the frog's cornea and gastric mucosa. Exp Eye Res 15:533–552.

113. WH Cliff, RA Frizzell. (1990). Separate Cl^- conductances activated by cAMP and Ca^{2+} in Cl^- secreting epithelial cells. Proc Natl Acad Sci USA 87:4956–4960.

114. OA Candia, R Montoreano, SM Podos. (1977). Effect of the ionophore A23187 on chloride transport across isolated frog cornea. Am J Physiol 233:F94–F101.

115. RT Worrell, AG Butt, WH Cliff, RA Frizzell. (1989). A volume-sensitive chloride conductance in human colonic cell line T84. Am J Physiol 256:C1111–C1119.

116. OA Candia. (1973). Effect of pH on chloride transport across the isolated bullfrog cornea. Exp Eye Res 15:375–382.

117. OA Candia, L Grillone, TC Chu. (1986). Forskolin effects on frog and rabbit corneal epithelium ion transport. Am J Physiol 251:C448–C454.

118. P Reinach, N Holmberg. (1989). Inhibition of calcium of beta adrenoceptor mediated cAMP responses in isolated bovine corneal epithelial cells. Curr Eye Res 8: 85–90.

119. SD Klyce, KA Palkama, M Harkonen, WS Marshall, S Huhtaniitty, KP Mann, AH Neufeld. (1982). Neural serotonin stimulates chloride transport in the rabbit corneal epithelium. Invest Ophthalmol Vis Sci 23:181–192.

120. RJ Walkenbach, RD LeGrand. (1982). Inhibition of adenylate cyclase activity in the corneal epithelium by anti-inflammatory steroids. Exp Eye Res 34:161–168.

121. JW Huff, PS Reinach. (1985). Mechanism of inhibition of net ion transport across frog corneal epithelium by calcium channel antagonists. J Membr Biol 85:215–223.

122. O Eriksson, PJ Wistrand. (1986). Inhibitory effects of chemically-different "loop" diuretics on chloride transport across the bullfrog cornea. Acta Physiol Scand 127: 137–144.

123. UB Kompella, KJ Kim, VHL Lee. (1992). Paracellular permeability of a chloride secreting epithelium. Proc Int Symp Controlled Release 19:425–426.
124. E Carafoli. (1987). Intracellular calcium homeostasis. Annu Rev Biochem 6:395–433.
125. JA Williamson, JR Monack. (1989). Hormone effects on cellular Ca^{2+} fluxes. Annu Rev Physiol 51:107–124.
126. S Muallem. (1989). Calcium transport pathways of pancreatic acinar cells. Annu Rev Physiol 51:83–105.
127. A Rich, JL Rae. (1995). Calcium entry in rabbit corneal epithelial cells: Evidence for a non-voltage dependent pathway. J Membr Biol 144:177–184.
128. P Hess, JB Lansman, RW Tsien. (1984). Different modes of Ca channel gating behaviour favoured by dihydropyridine Ca agonists and antagonists. Nature 311: 538–544.
129. M Schramm, G Thomas, R Towart, G Frackowiak. (1983). Activation of calcium channels by novel 1,4-dihydropyridines. A new mechanism for positive inotropies of smooth muscle stimulants. Arzneim Forsch 33:1268–1272.
130. H Fischer, B Illek, PA Negulescu, W Clauss, TE Machen. (1992). Carbachol-activated calcium entry into HT-29 cells is regulated by both membrane potential and cell volume. Proc Natl Acad Sci USA 89:1438–1442.
131. H Sterb, RF Irvine, MJ Berridge, I Schulz. (1983). Release of Ca^{2+} from a nonmitochondrial intracellular store in pancreatic acinar cells by inositol-4,5-triphosphate. Annu Rev Physiol 306:67–69.
132. PS Reinach, N Holmbeg, R Chiesa. (1991). Identification of calmodulin-sensitive Ca^{2+}-transporting ATPase in the plasma membrane of bovine corneal epithelial cell. Biochim Biophys Acta 1068:1–8.
133. ML Graber, DC DiLillo, BL Friedman, EP Munoz. (1986). Characteristics of fluoroprobes for measuring intracellular pH. Anal Biochem 156:202–212.
134. RJ Alpern. (1985). Mechanism of basolateral membrane H^+-OH^-/HCO_3^- transport in the rat proximal convoluted tubule. A sodium-coupled electrogenic process. J Gen Physiol 86:613–636.
135. RY Tsien. (1980). New calcium indicators and buffers with high selectivity against magnesium and protons: Design, synthesis, and properties of prototype structures. Biochemistry 19:2396–2404.
136. G Grynkiewioz, M Poenie, RY Tsien. (1985). A new generation of Ca^{2+} indicators with greatly improved fluorescence properties. J Biol Chem 260:3440–3450.
137. JA Bonanno. (1991). K^+-H^+ exchange, a fundamental cell acidifier in corneal epithelium. Am J Physiol 260:C618–C625.
138. CMAW Festen, JFG Slegers, CH Van Os. (1983). Intracellular activities of chloride, potassium, and sodium ions in rabbit corneal epithelium. Biochim Biophys Acta 732:394–404.
139. JA Bonanno, TE Machen. (1989). Intracellular pH regulation in basal corneal epithelial cells measured in corneal explants: Characterization of Na/H exchange. Exp Eye Res 49:129–142.
140. U Kompella, KJ Kim, VHL Lee. (1992). Active ion and nutrient transport mechanisms of the pigmented rabbit conjunctiva. Pharm Res 9(Suppl):3.
141. U Kompella, KJ Kim, VHL Lee. (1993). Active chloride transport in the pigmented rabbit conjunctiva. Curr Eye Res 12:1041–1048.

142. U Kompella, KJ Kim, MHI Shiue, VHL Lee. (1995). Possible existence of Na$^+$-coupled amino acid transport in the pigmented rabbit conjunctiva. Life Sci 57: 1427–1431.

143. MHI Shiue, KJ Kim, VHL Lee. (1998). Modulation of chloride secretion across the pigmented rabbit conjunctiva. Exp Eye Res 66:275–282.

144. XP Shi, OA Candia. (1995). Active sodium and chloride transport across the isolated rabbit conjunctiva. Curr Eye Res 14:927–935.

145. K Hosoya, UB Kompella, KJ Kim, VHL Lee. (1996). Contribution of Na$^+$-glucose cotransport to the short-circuit current in the pigmented rabbit conjunctiva. Curr Eye Res 15:447–451.

146. Y Horibe, K Hosoya, KJ Kim, VHL Lee. (1997). Kinetic evidence for Na$^+$-glucose co-transport in the pigmented rabbit conjunctiva. Curr Eye Res 16:1050–1055.

147. UB Kompella, KJ Kim, MHI Shiue, VHL Lee. (1995). Possible existence of Na$^+$-coupled amino acid transport in the pigmented rabbit conjunctiva. Life Sci 57: 1427–1431.

148. K Hosoya, Y Horibe, KJ Kim, VHL Lee. (1997). Na$^+$-dependent L-arginine transport in the pigmented rabbit conjunctiva. Exp Eye Res 65:547–553.

149. P Ashton, SK Podder, VHL Lee. (1991). Formulation influence on the conjunctival penetration of four beta-blockers in the pigmented rabbit: Comparison with corneal penetration. Pharm Res 8:1166–1174.

150. S Dodda-Kashi, W Wang, VHL Lee. (1989). Corneal penetration of angiotensin converting enzyme inhibitors and related peptides in the albino rabbit. Invest Ophthalmol Vis Sci 30(Suppl):21.

151. S Dodda-Kashi. (1989). Mechanistic studies on the corneal penetration of peptides in the albino rabbit. PhD Thesis, University of Southern California.

152. W Wang, VHL Lee. (1991). Carrier-mediated transport of peptides in the rabbit conjunctiva. Pharm Res 8(Suppl):S-129.

153. W Wang. (1992). Peptide transport in the pigmented rabbit conjunctiva. PhD Thesis, University of Southern California, Los Angeles.

154. P Ashton, WC Shen, VHL Lee. (1989). Penetration mechanism of a protein across the conjunctiva of the pigmented rabbit. Pharm Res 6:S116.

155. M Narawane, VHL Lee. (1992). Hormonal and oxidative control of protein uptake by the conjunctiva of the pigmented rabbit. Invest Ophthalmol Vis Sci 33(Suppl):733.

156. BA Kamen, A Capdevile. (1986). Receptor-mediated folate accumulation is regulated by the cellular folate content. Proc Natl Acad Sci USA 83:5983–5987.

157. HS Huang, RD Schoenwald, JL Lach. (1983). Corneal penetration behavior of beta-blocking agents. II: Assessment of barrier contributions. J Pharm Sci 72:1272–1279.

158. W Wang, H Sasaki, DS Chien, VHL Lee. (1991). Lipophilicity influence on conjunctival drug penetration in the pigmented rabbit: A comparison with corneal penetration. Curr Eye Res 10:571–579.

159. AJW Huang, SCG Tseng, R Kenyon. (1990). Paracellular permeability of corneal and conjunctival epithelia. Invest Ophthalmol Vis Sci 30:684–689.

160. Y Horibe, K Hosoya, KJ Kim, T Ogiso, VHL Lee. (1997). Polar solute transport across the pigmented rabbit conjunctiva: Size dependence and the influence of 8-bromo cyclic adenosine monophosphate. Pharm Res 14:1246–1251.

161. L Sun, SK Basu, KJ Kim, VHL Lee. (1998). Arginine vasopressin transport and metabolism in the pigmented rabbit conjunctiva. Eur J Pharm Sci 6:47–52.

162. K Hosoya, VHL Lee. (1997). Cidofovir transport in the pigmented rabbit conjunctiva. Curr Eye Res 16:693–697.

163. K Hosoya, Y Horibe, KJ Kim, VHL Lee. (1998). Carrier-mediated N^G-nitro-L-arginine absorption across the pigmented rabbit conjunctiva. J Pharmacol Exp Therap 285:223–227.

164. K Hosoya, Y Horibe, KJ Kim, VHL Lee. (1998). Nucleoside transport mechanisms in the pigmented rabbit conjunctiva. Invest Ophthalmol Vis Sci 39:372–377.

165. Y Horibe, KJ Kim, VHL Lee. (1998). Carrier-mediated transport of monocarboxylate drugs in the pigmented rabbit conjunctiva. Invest Ophthalmol Vis Sci 39:1436–1443.

166. SK Basu, IS Haworth, MB Bolger, VHL Lee. (1998). Proton-driven dipeptide uptake in primary cultured rabbit conjunctival epithelial cells. Invest Ophthalmol Vis Sci 39:2365–2373.

167. Y Horibe, H Ueda, KJ Kim, VHL Lee. (1998). Organic cation drug transport in the pigmented rabbit conjunctiva. In preparation.

168. P Saha, J Yang, VHL Lee. (1998). Existence of a p-glycoprotein drug efflux pump in cultured rabbit conjunctival epithelial cells. Invest Opthalmol Vis Sci 39:1221–1226.

169. WL Nelson, FT Fraunfelder, JM Sills, JB Arrowsmith, JN Kuritsky. (1986). Adverse respiratory and cardiovascular events attributed to timolol ophthalmic solution. Am J Ophthalmol 102:606–611.

170. TS Lesar. (1987). Comparison of ophthalmic beta-blocking agents. Clin Pharm 6:451–463.

171. DE Silverstone, D Arkfeld, G Cowarn. (1985). Long term diurnal control of intraocular pressure with levobunolol and with timolol. Glaucoma 7:138–140.

172. C Losa, MJ Alonso, JL Vila, F Orallo, J Martinez, JA Saavedra, JC Pastor. (1992). Reduction of cardiovascular side effects associated with ocular administration of metipranolol by inclusion in polymeric nanocapsules. J Ocular Pharmacol 8:191–198.

173. GM Grass, JR Robinson. (1984). Relationship of chemical structure to corneal penetration and influence of low viscosity solution on ocular bioavailability. J Pharm Sci 73:1021–1027.

174. HW Hui, JR Robinson. (1985). Ocular delivery of progesterone using a bioadhesive polymer. Int J Pharm 26:203–213.

175. PC Madugal, KD Clercq, J Descamps, L Missotten. (1984). Topical treatment of experimental *Herpes simplex* keratouveitis with 2′-O-glycylacyclovir. Arch Ophthalmol 102:140–142.

176. HJ Schaeffer, L Beauchamp, P De Miranda, GD Elison, DJ Bauer, P Collins. (1978). 9-(2-Hydroxyethoxymethyl)guanine activity against viruses of herpes group. Nature 272:583–585.

177. A Kupferman, MV Pratt, K Suckewer, HM Leibowitz. (1974). Topically applied steroids in corneal disease. III. The role of drug derivative in stromal absorption of dexamethasone. Arch Ophthalmol 91:373–376.

178. D Burston, DM Matthews. (1972). Intestinal transport of dipeptides containing acidic and basic L-amino acids and a neutral D-amino acid. Clin Sci Mol Med 42: 4P.

179. DS Chien, RD Schoenwald. (1986). Improving the ocular absorption of phenylephrine. Biopharm Drug Disp 7:453–462.

180. MVW Bergamini, DL Murray, PD Krause. (1979). Pivalyl phenylephrine (PPE), a mydriatic prodrug of phenylephrine with reduced cardiovascular effects. Invest Ophthalmol Vis Sci 20:187.

181. W Wang, H Bundgaard, A Buur, VHL Lee. (1991). Corneal penetration of 5-fluorouracil and its improvement by prodrug derivatization in the albino rabbit: Implications in glaucoma filtration surgery. Curr Eye Res 10:87–97.

182. SC Chang, H Bundgaard, A Buur, VHL Lee. (1988). Low dose O-butyryl timolol improves the therapeutic index of timolol in the pigmented rabbit. Invest Ophthalmol Vis Sci 29:626–629.

183. E Duzman, CC Chen, J Anderson, M Blumenthal, H Twizer. (1982). Diacetyl derivative of nadolol, I. Ocular pharmacology and short-term ocular hypotensive effect in glaucomatous eyes. Arch Ophthalmol 100:1916–1919.

184. CG Wilson, E Tomlinson, SS Davis, O Olejnik. (1981). Altered ocular absorption and disposition of sodium cromoglycate upon ion pair and complex coacervate formation with dodecylbenzyldimethylammonium chloride. J Pharm Pharmacol 31: 749–753.

185. A Kato, S Iwata. (1988). Studies on improved corneal permeability to bunazosin. J Pharmacobio Dyn 11:330–334.

186. VHL Lee, A Yamamoto, U Kompella. (1991). Mucosal penetration enhancers for facilitation of peptide and protein drug absorption. CRC Crit Rev Drug Carrier Sys 8:91–192.

187. PE Ross, AN Butt, G Gallacher. (1990). Cholesterol absorption by the gallbladder. J Clin Physiol 43:572–575.

188. M Tomita, M Shiga, M Hayashi, S Awazu. (1988). Enhancement of colonic drug absorption by the paracellular permeation route. Pharm Res 5:341–351.

189. GP Martin, C Marriott, I Kellaway. (1976). The effect of natural surfactants on the rheological properties of mucus. J Pharm Pharmacol 28:76P.

190. ABR Thomson, JM Dietschy. (1981). Intestinal lipid absorption: Major extracellular and intracellular events. In: LR Johnson, J Christensen, SL Schultz, eds. Physiology of the Gastrointestinal Tract. New York: Raven Press, p 1146.

191. S Hirai, T Yashiki, H Mima. (1981). Mechanisms for the enhancement of the nasal absorption of insulin by surfactants. Int J Pharm 9:173–184.

192. T Murakami, Y Sasaki, R Yamajo, N Yata. (1984). Effect of bile salt on the rectal absorption of sodium ampicillin in rats. Chem Pharm Bull 32:1948–1955.

193. DA Whitmore, LG Brookes, KP Wheeler. (1979). Relative effects of different surfactants on intestinal absorption and the release of proteins and phospholipids from the tissue. J Pharm Pharmacol 31:277–283.

194. JP Longenecker, AC Moses, JS Flier, RD Silver, MC Carey, EJ Dubovi. (1987). Effects of sodium taurodihydrofusidate on nasal absorption of insulin in sheep. J Pharm Sci 76:351–355.

195. WAJJ Hermens, PM Hooymans, JC Verhoef, FWHM Merkus. (1990). Effects of

absorption enhancers on human nasal tissue ciliary movement in vitro. Pharm Res 7:144–146.

196. JL Rae. (1990). Single K$^+$ channels in corneal epithelium. Invest Ophthalmol Vis Sci 31:1799–1809.

197. S Dodda Kashi, VHL Lee. (1986). Enkephalin hydrolysis in homogenates of various absorptive mucosae of the albino rabbit: Similarities in rates and involvement of aminopeptidases. Life Sci 38:2019–2028.

198. RE Stratford Jr, VHL Lee. (1985). Ocular aminopeptidase activity and distribution in the albino rabbit. Curr Eye Res 4:995–999.

199. BJ Aungst, NJ Rogers, E Shefter. (1988). Comparison of nasal, rectal, buccal, sublingual, and intramuscular insulin efficacy and the effects of a bile salt absorption promoter. J Pharmacol Exp Therap 244:23–27.

200. A Yamamoto, AM Luo, S Doddakashi, VHL Lee. (1989). The ocular route for systemic insulin delivery in the albino rabbit. J Pharmacol Exp Ther 249:249–255.

201. PS Banerjee, WA Ritschel. (1989). Transdermal permeation of vasopressin. II. Influence of azone on in vitro and in vivo permeation. Int J Pharm 49:199–204.

202. GSMJE Duchateau, J Zuidema, FWHM Merkus. (1986). Bile salts and intranasal drug absorption. Int J Pharm 31:193–199.

203. JL Madara. (1989). Loosening tight junctions: Lessons from the intestine. J Clin Invest 83:1089–1094.

204. J Llopis, GEN Kass, SK Duddy, GA Moore, S Orrenius. (1991). Mobilization of hormone sensitive calcium pool increases hepatocyte tight junctional permeability in the perfused rat liver. FEBS Lett 280:84–86.

205. RK Crane, FC Dorando. (1982). The kinetics and mechanism of Na$^+$ gradient-coupled glucose transport. In: AN Martonosi, ed. Membranes and Transport. New York: Plenum Press, pp 153–160.

206. GG Munck. (1972). Effects of sugar and amino-acid transport on transepithelial fluxes of sodium and chloride of short circuited rat jejunum. J Physiol 223:699–717.

207. V Ganapathy, FH Leibach. (1990). Peptide transport in intestinal and renal brush border membrane vesicles. Life Sci 30:2137–2146.

208. ABR Thomson, CA Hotke, WM Weinstein. (1982). Comparison of kinetic constants of hexose uptake in four animal species and man. Comp Biochem Physiol 72A:225–236.

209. HH Lu, J Thomas, D Fleisher. (1991). Influence of D-glucose induced water absorption on rat jejunal uptake of two passively absorbed drugs. J Pharm Sci 80:1–5.

210. HE Magee, E Reid. (1931). The absorption of glucose from the alimentary canal. J Physiol 73:163–183.

211. RS Thies, LJ Mandel. (1985). Role of glucose in corneal metabolism. Am J Physiol 249:C409–C416.

212. KM Daum, RM Hill. (1984). Human tears: Glucose instabilities. Acta Ophthalmol 62:472–478.

213. UB Kompella, P Ashton, VHL Lee. (1992). Paracellular drug transport in the pigmented rabbit conjunctiva and cornea. Invest Ophthalmol Vis Sci 33:1015.

214. M Cereijido, I Meza, A Martinez-Palomo. (1981). Occluding junctions in cultured epithelial monolayers. Am J Physiol 240:C96–C102.

215. I Tapaszto. (1973). Pathophysiology of human tears. The preocular tear film and dry eye syndromes. Int Ophthalmol Clin 13:119–122.

216. GM Grass, JR Robinson. (1988). Mechanisms of corneal drug penetration. I: In vivo and in vitro kinetics. J Pharm Sci 77:3–14.

217. K Sandvig, S Olsnes. (1982). Entry of toxic proteins abrin, modeccin, ricin, and diphtheria toxin into cells. I. Requirement for Ca^{2+}. J Biol Chem 257:7495–7503.

218. Y Rojanasakul, M Bhat, L Wang, CJ Malanga, DD Glover, JKH Ma. (1991). Regulation of tight junction permeability by calcium mediators and cell cytoskeleton in tracheal epithelium. Pharm Res 8(Suppl):S129.

219. KS Kan, R Coleman. (1988). The calcium ionophore A23187 increases the tight-junctional permeability in rat liver. Biochem J 256:1039–1041.

220. GEN Kass, SK Duddy, GA Moore, S Orrenius. (1992). 2,5-Di(*tert*-butyl)-1,4-benzohydroquinone rapidly elevates cytosolic Ca^{2+} concentration by mobilizing the inositol 1,4,5-trisphosphate-sensitive Ca^{2+} pool. J Biol Chem 264:15192–15198.

221. W Gevers. (1977). Generation of protons by metabolic processes in heart cells. J Mol Cardiol 9:867–874.

222. C Frelin, P Vigne, A Ladoux, M Lazdunski. (1988). The regulation of intracellular pH in cells from vertebrates. Eur J Biochem 174:3–14.

223. ME Dowty, P Braquet. (1991). Effect of extracellular pH on cytoplasmic pH and mechanism of pH regulation in cultured bovine corneal endothelium: Possible importance in drug transport studies. Int J Pharm 68:231–238.

224. D Chang, NL Kushman, DC Dawson. (1991). Intracellular pH regulates basolateral K^+ and Cl^- conductances in colonic epithelial cells by modulating Ca^{2+} activation. J Gen Physiol 98:183–196.

225. CJ Swallow, S Grinstein, OD Rotstein. (1990). A vacuolar type H^+-ATPase regulates cytoplasmic pH in murine macrophages. J Biol Chem 265:7645–7654.

226. J Burke, E Padillo, G Sachs, E Cragoe. (1992). Inhibitors of Na^+/H^+ exchange modify IOP responses to alpha adrenoceptor agonists. Invest Ophthalmol Vis Sci 33:1116.

227. MA Valverde, M Diaz, FV Sepulveda, DR Gill, SC Hyde, CF Higgins. (1992). Volume-regulated chloride channels associated with the human multidrug-resistance P-glycoprotein. Nature 355:830–833.

228. L Slater, P Sweet, M Stupecky, S Gupta. (1986). Cyclosporin A reverses vincristine and daunomycin resistance in acute lymphatic leukemia in vitro. J Clin Invest 77:1405–1408.

229. K Osann, P Sweet, LM Slater. (1992). Synergistic action of cyclosporin A and verapamil on vincristine and daunorubicin resistance in multidrug-resistant human leukemia cells in vitro. Cancer Chemother Pharmacol 30:152–154.

230. JR Riordan, V Ling. (1985). Genetic and biochemical characterization of multidrug resistance. Pharmacol Ther 28:51–75.

231. J Riordan, V Ling. (1979). Purification of P-glycoprotein from plasma membrane vesicles of Chinese hamster ovary cell mutants with reduced colchicine permeability. J Biol Chem 254:12701–12705.

232. K Sandvig, S Olsnes. (1984). Receptor-mediated entry of protein toxins into cells. Acta Histochem 29:979–994.

233. RC Delisle, U Hopfer. (1986). Electrolyte permeabilities of pancreatic zymogen granules: Implications for pancreatic secretion. Am J Physiol 250:G489–G496.

234. KW Gasser, J DiDomineco, UH Hopfer. (1988). Secretagogues activate chloride transport pathways in pancreatic zymogen granules. Am J Physiol 254:G93–G99.

235. RW Van Dyke. (1990). Acid transport by intracellular vesicles. J Internal Med 228(Suppl 1):41–46.

236. NA Bradbury, T Jilling, KL Kirk, RJ Bridges. (1992). Regulated endocytosis in a chloride secretory epithelial cell line. Am J Physiol 262:C752–C759.

237. NA Bradbury, T Jilling, G Berta, EJ Sorscher, RJ Bridges, KL Kirk. (1992). Regulation of plasma membrane recycling by CFTR. Science 256:530–532.

238. TO Mjorndal, SE Chesrown, MJ Frey, BR Reed, WM Gold. (1978). Effect of increased cyclic AMP on antigen-induced mediator release and function in canine lung in vivo. Clin Res 26:539A.

239. BS Spinowitz, JA Zadunaisky. (1979). Action of adenosine on chloride active transport of isolated frog cornea. Am J Physiol 237:F121–F127.

240. JA Zadunaisky, B Spinowitz. (1977). Drugs affecting the transport and permeability of the corneal epithelium. In: S Dikstein, ed. Drugs and Ocular Tissues. Basel: Krager, pp 57–78.

241. JA Zadunaisky, MA Lande, M Chalfie, AH Neufeld. (1973). Ion pumps in the cornea and their stimulation by epinephrine and cyclic-AMP. Exp Eye Res 15:577–584.

242. CE Crosson, RW Beuerman, SD Klyce. (1984). Dopamine modulation of active ion transport in rabbit corneal epithelium. Invest Ophthalmol Vis Sci 25:1240–1245.

243. BE Schaeffer, JA Zadunaisky. (1979). Stimulation of chloride transport by fatty acids in corneal epithelium and relation to changes in membrane fluidity. Biochim Biophys Acta 556:131–143.

244. BR Beitch, I Beitch, JA Zadunaisky. (1975). The stimulation of chloride transport by prostaglandins and their interaction with epinephrine, theophylline and cyclic AMP in the corneal epithelium. J Membr Biol 19:381–396.

245. SD Klyce, KA Palkama, M Harkonen, WS Marshall, S Huhtaniitty, KP Mann, AH Neufeld. (1982). Neural serotonin stimulates chloride transport in the rabbit corneal epithelium. Invest Ophthalmol Vis Sci 23:181–192.

246. M Gilmour Buck, JA Zadunaisky. (1975). Stimulation of ion transport by ascorbic acid through inhibition of 3',5'-cyclic AMP phosphodiesterase in the corneal epithelium and other tissue. Biochim Biophys Acta 376:82–88.

247. JA Zadunaisky, MA Lande, J Hafner. (1971). Further studies on chloride transport in the frog cornea. Am J Physiol 221:1832–1836.

248. OA Candia. (1972). Ouabain and sodium effects on chloride fluxes across the isolated bullfrog cornea. Am J Physiol 223:1053–1057.

249. OA Candia. (1972). Short-circuit current related to active transport of chloride in the frog cornea: Effects of furosemide and ethacrynic acid. Biochim Biophys Acta 258:1011.

250. W Nagel, G Carrasquer. (1989). Effect of loop diuretics on bullfrog corneal epithelium. Am J Physiol 256:C750–C755.

251. DR Whikehart, B Montgomery, DH Sorna, JD Wells. (1991). Beta-blocking agents

inhibit Na+-K+-ATPase in cultured corneal endothelial and epithelial cells. J Ocular Pharmacol 7:195–200.

252. JH Kinoshita, T Masurat, M Herfaut. (1955). Pathway of glucose metabolism in corneal epithelium. Science 122:72.

253. BK Toeplitz, AI Cohen, PT Funke, WL Parker, JZ Gougoutas. (1979). Structure of ionomycin—A novel diacidic polyether antibiotic having high affinity for calcium ions. J Am Chem Soc 101:3344–3353.

254. SK Joseph, JR Williamson. (1983). The origin, quantitation, and kinetics of intracellular calcium mobilization by vasopressin and phenylephrine in hepatocytes. J Biol Chem 258:10425–10432.

255. BP Hughes, JN Crofts, AM Auld, LC Read, GJ Barritt. (1987). Evidence that a pertussis-toxin-sensitive substrate is involved in the stimulation by epidermal growth factor and vasopressin of plasma-membrane Ca^{2+} inflow in hepatocytes. Biochem J 248:911–918.

256. I Kojima, H Shibata, E Ogata. (1986). Pertussis toxin blocks angiotensin II-induced calcium influx but not inositol trisphosphate production in adrenal glomerulosa cell. FEBS Lett 204:347–351.

257. JD Johnson, JC Garrison. (1987). Epidermal growth factor and angiotensin II stimulates formation of inositol 1,4,5 and inositol 1,3,4-trisphosphate in hepatocytes. J Biol Chem 262:17285–17293.

258. UB Kompella, KJ Kim, VHL Lee. (1993). Active chloride transport in the pigmented rabbit conjunctiva. Curr Eye Res 12:1041–1048.

259. WS Marshall, JW Hanrahan. (1991). Anion channels in the apical membrane of mammalian corneal epithelium primary cultures. Invest Ophthalmol Vis Sci 32:1562–1568.

260. WN Scott, F Friedenthal. (1973). A proposed role for ascorbate in the transport of amino acids and ions in the cornea. Exp Eye Res 15:683–692.

261. C Korbmacher, H Helbig, C Forster, M Wiederholt. (1988). Characterization of Na+/H+ exchange in a rabbit corneal epithelial cell line (SIRC). Biochim Biophys Acta 943:405–410.

262. C Korbmacher, H Helbig, C Forster, M Wiederholt. (1988). Evidence for Na+/H+ exchange and pH sensitive membrane voltage in cultured bovine corneal epithelial cells. Curr Eye Res 7:619–626.

263. JA Bonanno, C Giasson. (1992). Intracellular pH regulation in fresh and cultured bovine corneal epithelium. I. Na+/H+ exchange in the absence and presence of HCO_3^-. Invest Ophthalmol Vis Sci 33:3058–3067.

264. DF Cooperstein. (1987). Na+-K+-ATPase activity and transport process in toad corneal epithelium. Comp Biochem Physiol 87A:1119–1121.

265. RD Schoenwald, HS Huang. (1983). Corneal penetration behavior of beta-blocking agents. I: Physicochemical factors. J Pharm Sci 72:1266–1272.

266. R Vogel, SF Kulaga, JK Laurence, RL Gross, BG Haik, D Karp, M Koby, TJ Zimmerman. (1990). The effects of a Gerlite (G) vehicle on the efficacy of low concentrations of timolol (T). Invest Ophthalmol Vis Sci 31(Suppl):404.

267. MF Saettone, D Monti, MT Torracca, P Chetoni, B Giannaccini. (1989). Mucoadhesive liquid ophthalmic vehicles—Evaluation of macromolecular ionic complexes of pilocarpine. Drug Dev Ind Pharm 15:2475–2489.

268. CM Lehr, YH Lee, VHL Lee. (1994). Improved ocular penetration of gentamicin by mucoadhesive polymer polycarbophil in the pigmented rabbit. Invest Ophthalmol Vis Sci 35:2809–2814.

269. NM Davies, SJ Farr, J Hadgraft, IW Kellaway. (1992). Evaluation of mucoadhesive polymers in ocular drug delivery. II. Polymer-coated vesicles. Pharm Res 9:1137–1144.

270. JT Jacob-LaBarre, HE Kaufman. (1990). Investigation of pilocarpine loaded poly-butylcyanoacrylate nanocapsules in collagen shields as a drug delivery system. Invest Ophthalmol Vis Sci 31(Suppl):485–488.

271. SS Chrai, TF Patton, A Mehta, JR Robinson. (1973). Lacrimal and instilled fluid dynamics in rabbit eyes. J Pharm Sci 62:1112–1121.

272. SS Chrai, MC Makoid, SP Eriksen, JR Robinson. (1974). Drop size and initial dosing frequency problems of topically applied ophthalmic drugs. J Pharm Sci 63:333–338.

273. G Aberg, G Adler, J Wikberg. (1978). Inhibition and facilitation of lacrimal flow by beta-adrenergic drugs. Ophthalmol Acta 57:225.

274. AK Mitra, TJ Mikkelson. (1982). Ophthalmic solution buffer systems. I. The effect of buffer concentration on the ocular absorption of pilocarpine. Int J Pharm 10:219–229.

275. TJ Mikkelson, SS Chrai, JR Robinson. (1973). Competitive inhibition of drug–protein interaction in the eye fluids and tissues. J Pharm Sci 62:1942–1945.

276. H Sasaki, H Bundgaard, VHL Lee. (1989). Design of prodrugs to selectively reduce systemic timolol absorption on the basis of the differential lipophilic characteristics of the cornea and the conjunctiva. Invest Ophthalmol Vis Sci 30(Suppl):25.

277. AM Luo, H Sasaki, VHL Lee. (1991). Ocular drug interactions involving topically applied timolol in the pigmented rabbit. Curr Eye Res 10:231–240.

278. A Yamamoto, AM Luo, S Doddakashi, VHL Lee. (1989). The ocular route for systemic insulin delivery in the albino rabbit. J Pharmacol Exp Therap 249:249–255.

279. K Morimoto, T Nakamura, K Morisaka. (1989). Effect of medium-chain fatty acid salts on penetration of a hydrophilic compound and a macromolecular compound across rabbit corneas. Arch Int Pharmacodyn 302:18–26.

280. DS Chien, JJ Homsky, C Gluchowski, DDS Tang-Liu. (1990). Corneal and conjunctival/scleral penetration of p-aminoclonidine, AGN 190342, and clonidine in rabbit eyes. Curr Eye Res 9:1051–1059.

281. A Kupferman, HM Leibowitz. (1974). Topically applied steroids in corneal disease. IV. The role of drug concentration in stromal absorption of prednisolone acetate. Arch Ophthalmol 91:377–380.

282. VHL Lee, LW Carson, S Dodda-Kashi, RE Stratford Jr. (1986). Metabolic and permeation barriers to the ocular absorption of topically applied enkephalins in albino rabbits. J. Ocular Pharmacol 2:345–352.

283. OH Peterson, I Findlay. (1987). Electrophysiology of the pancreas. Physiol Rev 67:1054–1116.

284. SD Klyce, CE Crosson. (1985). Transport processes across the rabbit corneal epithelium: A review. Curr Eye Res 4:323–331.

285. JR Williamson, SK Joseph, KE Coll, AP Thomas, A Verhoeven, M Prentki. (1986). Hormone-induced inositol lipid breakdown and calcium-mediated cellular responses in liver. In: G Poste, ST Crooke, eds. New Insights into Cell and Membrane Transport Processes. New York: Plenum Press, pp 217–247.

10

Predicting Oral Drug Absorption in Humans

Lawrence X. Yu*
Glaxo Wellcome Research and Development, Research Triangle Park, North Carolina

Larry Gatlin
Biogen, Inc., Cambridge, Massachusetts

Gordon L. Amidon
The University of Michigan, Ann Arbor, Michigan

I. INTRODUCTION

Drugs are most commonly given into the body by the oral route of administration. In fact, the vast majority of pharmaceutical dosage forms are designed for oral introduction, primarily for ease of administration [1]. The prediction of oral drug absorption is of pharmaceutical interest [2]. If used appropriately, it can greatly accelerate the screening of new molecular entities and the development of new products. In a broad sense, there exist qualitative and quantitative approaches to the estimatation of oral drug absorption [3]. Qualitative models include the pH-partition hypothesis and the absorption potential concept. Quantitative models include dispersion models, mass balance approaches, and compartmental absorption and transit models. This chapter is aimed at reviewing these absorption models, with an emphasis on recent developments.

For convenience, these qualitative and quantitative absorption models have been classified into three categories based on their dependence on spatial and temporal variables [2]. The first category is referred to as "quasi-equilibrium models." The quasi-equilibrium model, including the pH-partition hypothesis

---

* Current affiliation: Food and Drug Administration, Rockville, Maryland.

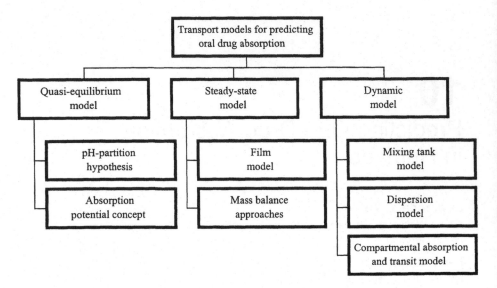

Figure 1 Overview of transport models for predicting oral drug absorption. Absorption models are classified into three categories based on their dependence on the spatial and temporal variables. These three categories are quasi-equilibrium, steady-state, and dynamic models.

and the absorption potential concept, is independent of the spatial and temporal variables. The quasi-equilibrium model is generally qualitative and provides a basic guideline for understanding drug absorption trends. The second category is referred to as "steady-state models" and includes the film model and the mass balance approaches. The steady-state model is independent of the temporal variable but dependent on the spatial variable. The steady-state model can be employed to estimate the extent of drug absorption, but not the rate of drug absorption. The third category is referred to as "dynamic models." The dynamic model, including the dispersion model, the mixing tank, and the compartmental absorption and transit (CAT) model, is dependent on the temporal variable. The dynamic model can be used to predict the rate and extent of drug absorption. Figure 1 shows an overview of these absorption models.

II. QUASI-EQUILIBRIUM MODEL

A. pH-Partition Hypothesis

In 1940, Jacobs [4] described the theory of nonionic membrane permeation of organic compounds in quantitative terms. The influence of pH and pK_a on drug absorption from the gastrointestinal tract was then extensively investigated in the 1950s and 1960s [4–10]. These studies resulted in the development of the pH-partition hypothesis. According to this hypothesis, ionizable compounds diffuse through biological membranes primarily in the un-ionized forms. Therefore, the extent of absorption of compounds across lipid membranes depends on the degree of ionization. The pH and pK_a of compounds control the fraction of the un-ionized species. Thus, for acids,

$$pH = pK_a + \log_{10}\left(\frac{\text{ionized concentration}}{\text{un-ionized concentration}}\right) \tag{1}$$

and for bases,

$$pH = pK_a + \log_{10}\left(\frac{\text{un-ionized concentration}}{\text{ionized concentration}}\right) \tag{2}$$

The majority of evidence supporting the pH-partition hypothesis is from studies of gastrointestinal absorption, renal excretion, and gastric secretion of drugs [11]. While correlation between absorption rate and pK_a was found to be consistent with the pH-partition hypothesis, deviations from this hypothesis were often reported [12]. Such deviations were explained by the existence of a mucosal unstirred layer [13,14] and/or a microclimate pH [15].

The un-ionized form is assumed to be sufficiently lipophilic to traverse membranes in the pH-partition hypothesis. If it were not, no transfer could be predicted, irrespective of pH. The lipophilicity of compounds is experimentally determined as the ''partition coefficient (log P)'' or ''distribution coefficient (log D)'' [16]. The partition coefficient is the ratio of concentrations of the neutral species between aqueous and nonpolar phases, while the distribution coefficient is the ratio of all species between aqueous and nonpolar phases [17,18].

The octanol–water system is one of the most commonly used systems for predicting the ability of drugs to enter into and diffuse across a cell membrane [19–22]. When comparing permeability for a series of drugs with the octanol–water distribution coefficient, a sigmoidal relationship is often obtained, as shown in Figure 2 [23–26]. The phenomena were recently interpreted by Burton et al. [27] on the basis of the Caco-2 cell model. For very polar compounds with small partition coefficients, entry into the cell membrane is unfavorable and transport across the epithelium is correspondingly slow and occurs to a significant extent by diffusion via the aqueous, paracellular pathway that is dependent on molecular size and charge. As the partition coefficient increases, the ability to enter the cell membrane increases, resulting in increasing permeability until a plateau is

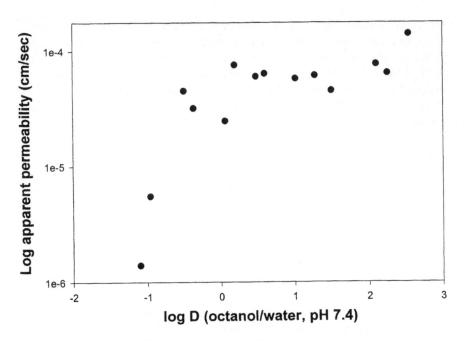

Figure 2 The permeability in Caco-2 cell monolayers as a function of distribution coefficient (octonol/water, pH 7.4) for 5-fluorouracil and prodrugs. A sigmoidal relationship can be observed. (Replot from Ref. 26, with kind permission from Elsevier Science-NL, Amsterdam, The Netherlands.)

reached that is representative of diffusion through the unstirred aqueous boundary layer adjacent to the cell membrane.

B. Absorption Potential Concept

In reviewing the pH-partition hypothesis, it is apparent that it is an oversimplification of a very complex process. It does not consider one of the critical physicochemical factors, solubility. Low aqueous solubility is often the cause of the low bioavailability. To address this issue, Dressman et al. [28] developed an absorption potential concept that takes into account not only the partition coefficient but also solubility and dose. Using a dimensional analysis approach, the following simple equation was proposed:

$$AP = \log\left(\frac{PF_{un}}{Do}\right) \tag{3}$$

where AP is the absorption potential as a predictor of the extent of drug absorption, P is the partition coefficient, and F_{un} is the fraction in un-ionized form at pH 6.5. Do in Eq. (3) is the dimensionless dose number defined as the ratio of dose concentration to solubility:

$$Do = \frac{C_0}{C_s} = \frac{M_0/V_0}{C_s} \tag{4}$$

where C_s is the physiological solubility, M_0 is the dose, and V_0 is the volume of water taken with the dose, which is generally set to be 250 mL [28]. The quantitative absorption potential concept was proposed by Macheras and Symillides [29]:

$$F_a = \frac{(10^{AP})^2}{(10^{AP})^2 + F_{un}(1 - F_{un})} \tag{5}$$

with constraints that P be set equal to 1000 when $P > 1000$ and Do be set equal to 1 when Do < 1.

Several drugs covering a wide range of absorption characteristics, from poorly absorbed compounds to those with virtually complete absorption, were selected to evaluate the ability of the absorption potential concept to predict the fraction of dose absorbed (Fig. 3). The drugs varied widely in their physicochemical characteristics. It was demonstrated that the absorption potential strongly correlated with the fraction of dose absorbed. The advantage of the absorption potential concept is its simplicity. It is based solely on the physicochemical properties of drugs. The absorption potential concept provides an alternative to the pH-partition hypothesis to forecast the absorption trends and to identify the critical limiting physicochemical properties, particularly for poorly soluble drugs. With more data available, the concept may need to be refined to incorporate other important absorption variables, such as gastrointestinal transit time.

III. STEADY-STATE MODEL

A. Macroscopic Mass Balance Approach

Although the pH-partition hypothesis and the absorption potential concept are useful indicators of oral drug absorption, physiologically based quantitative approaches need to be developed to estimate the fraction of dose absorbed in humans. We can reasonably assume that a direct measure of tissue permeability, either in situ or in vitro, will be more likely to yield successful predictions of drug absorption. Amidon et al. [30] developed a simplified film model to correlate the extent of absorption with membrane permeability. Sinko et al. [31] extended this approach by including the effect of solubility and proposed a macroscopic mass balance approach. That approach was then further extended to include facili-

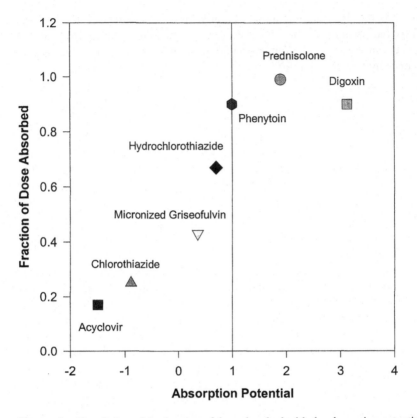

Figure 3 Correlation of the fraction of dose absorbed with the absorption potential. The calculation of the absorption potential is detailed in Dressman et al. [28]. The quantitative form of the fraction of dose absorbed with the absorption potential was derived by Macheras and Symillides [29]. (Replot from Ref. 28 with kind permission from APhA, Washington, DC.)

tated drug absorption and degradation [32]. Since it includes the results of the film model, we focus on the macroscopic mass balance approach.

The small intestine is assumed to be a cylindrical tube with a surface area of $2\pi RL$, where R is the radius and L is the length of the tube (Fig. 4). The rate at which the drug enters the tube is the product of the inlet concentration, C_0, and the volumetric flow rate, Q. The rate at which it exits the tube is the product of the outlet concentration, C_{out}, and the volumetric flow rate, Q. The absorption flux across the small intestinal membrane, j, is the product of the effective permeability, P_{eff}, and concentration, C. The total drug loss by absorption from the

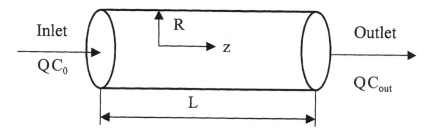

Figure 4 Schematic of macroscopic mass balance approach. The small intestine is assumed to be a cylindrical tube with a radius of R and a length of L. The inlet and outlet concentrations are C_0 and C_{out}. The inlet or outlet volumetric flow rate is Q.

small intestine is the integral of the absorption flux over the surface area of the tube. From mass balance, we have

$$Q(C_0 - C_{out}) = 2\pi R P_{eff} \int_0^L C \, dz \tag{6}$$

Under the steady-state assumption, the fraction of dose absorbed, F_a, is $1 - C_{out}/C_0$; so

$$F_a = 1 - \frac{C_{out}}{C_0} = \frac{2\pi R P_{eff}}{Q C_0} \int_0^L C \, dz = 2\,\mathrm{An} \int_0^1 C^* \, dz^* \tag{7}$$

where C^* and z^* are dimensionless variables, $C^* = C/C_0$, $z^* = z/L$. An is the dimensionless absorption number and is defined as the ratio of the mean small intestinal transit time $\langle T_{si} \rangle$ to absorption time (R/P_{eff}):

$$\mathrm{An} = \frac{\langle T_{si} \rangle}{R/P_{eff}} = \frac{\pi R L P_{eff}}{Q} \tag{8}$$

The evaluation of the integral in Eq. (8) requires the knowledge of how the concentration of drug varies along the tube. The concentration profile in the tube depends on effective permeability, solubility, and flow pattern. Three cases have been considered separately with regard to the inlet and outlet concentrations and the solubility of the drug:

Case I: $C_0 \leq C_s$ and $C_{out} \leq C_s$
Case II: $C_0 > C_s$ and $C_{out} < C_s$
Case III: $C_0 > C_s$ and $C_{out} > C_s$

It can be seen that case II is intermediate between case I and case III. The results of case I and case III will facilitate the derivation for case II. We will discuss case I and case III first, and case II last.

1. Case I

In case I, the drug is highly soluble and the dissolution is not a limiting step. Assuming a complete radial mixing model, the drug concentration profile in the intestine is [23,53]

$$C^* = \frac{C_{out}}{C_0} = \exp\left[- \frac{2\pi R P_{eff} z}{Q} \right] = e^{-2Anz^*} \tag{9}$$

Substituting Eq. (9) into Eq. (7) and integrating the resulting equation, we have

$$F_a = 1 - e^{-2An} \tag{10}$$

Thus, the fraction of dose absorbed is exponentially related to the absorption number. Equation (10) shows that the absorption number (and therefore the membrane permeability) is a fundamental parameter while other parameters such as the partition coefficient and pK_a are useful guides but not fundamental parameters. For highly soluble drugs with linear absorption kinetics, dose and dissolution have no effect on the fraction of dose absorbed. In the case of drugs that are absorbed by a carrier-mediated process, a mean permeability should be used [30].

2. Case III

In case III, solid drug exists at both inlet and outlet. The concentration in the intestinal lumen can be assumed to be equal to the solubility:

$$C^* = \frac{C_s}{C_0} = \frac{1}{Do} \tag{11}$$

Substituting Eq. (11) into Eq. (7) and integrating the resulting equation, we have

$$F_a = \frac{2An}{Do} \tag{12}$$

From Eq. (12), the dependence of F_a on both An and Do is apparent. The relationship suggests that a lower fraction of dose will be absorbed at a higher dose. However, for compounds with low solubility and/or high dose, the concentration in the intestinal lumen may not be the same as the solubility, owing to slow dissolution. Dissolution-limited absorption is discussed in Section III.B.

3. Case II

In case II, the inlet drug concentration is above the solubility and the outlet concentration of the drug is below the solubility. There exists a point where the concentration is equal to the solubility, and the whole intestine can be divided into two regions. The first region, where solid drug exists, is equivalent to case III. The second region, where no solid drug exists, is equivalent to case I. The fraction of dose absorbed in the first region is

$$F_a^1 = 1 - \frac{1}{Do} \tag{13}$$

The fraction of dose absorbed in the second region is

$$F_a^2 = \frac{1}{Do}(1 - e^{-2An+Do-1}) \tag{14}$$

The total fraction of dose absorbed is

$$F_a = 1 - \frac{1}{Do}e^{-2An+Do-1} \tag{15}$$

The mass balance approach has been used to correlate the fraction of dose absorbed and the rat intestinal permeability for 19 drugs covering a wide range of physicochemical properties [34]. Intestinal permeability was determined by using the rat single-pass perfusion model [35–37]. Figure 5 shows the good correlation achieved by the macroscopic mass balance or film model. Since these drugs include acidic, basic, and zwitterionic compounds, a simple relationship between absorption and partition coefficient is not expected. Furthermore, some of these drugs involve a facilitated drug absorption mechanism. They certainly will not follow the pH-partition hypothesis and the absorption potential concept.

Artursson et al. [38] examined the use of Caco-2 monolayers in the prediction of intestinal drug absorption. They compared the Caco-2 cell permeability data from different laboratories, as illustrated in Figure 6. Figure 6 shows that the permeability corresponding to incomplete or complete drug absorption varies considerably among laboratories. However, qualitatively similar correlations with the fraction of dose absorbed in humans were established in all laboratories [39–42]. It was concluded that the Caco-2 monolayers could be used as a simple model in predicting passive drug absorption.

The mass balance approach was also applied to amoxicillin data. Amoxicillin is a broad-spectrum bacterial antibiotic administered orally for the treatment of various gram-positive and gram-negative infections. The dose ranged from 250 to 3000 mg with 200 mL of water. The corresponding dose concentration varied from 1.25 to 15 mg/mL. Low solubility (6 mg/mL) and a nonpassive absorption mechanism makes estimation or prediction of absorption more diffi-

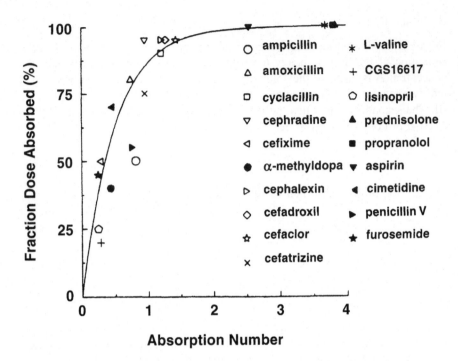

Figure 5 Correlation of the fraction of dose absorbed with rat permeability. The rat permeability data were determined using the rat perfusion model. Considering the difference between the human permeability and the rat permeability, a scale factor was used to calculate the absorption number in Eq. (8). (Ref. 34 with kind permission from Plenum Publishing Corporation, New York.)

cult. However, Eqs. (10), (14), and (15) were able to predict the experimental fraction of dose absorbed for doses from 250 to 4000 mg.

Sinko et al. [32] extended the macroscopic mass balance approach to include chemical and enzymatic degradation, where only highly soluble drugs were taken into account. When the effect of degradation is considered, the total loss in the intestinal tube becomes

$$F_{tot} = F_a + F_d = (2\,An + Da) \int_0^1 C^* dz^* \qquad (16)$$

where Fd is the fraction of dose degraded and Da is the dimensionless Damköhler number, defined as

$$Da = K_d \langle T_{si} \rangle \qquad (17)$$

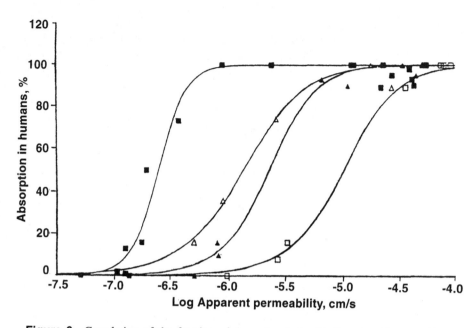

Figure 6 Correlation of the fraction of dose absorbed with Caco-2 cell permeability obtained in four different laboratories (■, □, ▲, and △). Qualitatively similar correlations were established in all four laboratories, but the data are not directly comparable due to quantitative differences in the permeability of the Caco-2 monolayers. (From Ref. 38 with kind permission from Elsevier Science-NL, Amsterdam.)

where K_d is the first-order degradation constant. In the case of the complete radial mixing model, the drug concentration profile is

$$C^* = e^{-(2An+Da)z^*} \qquad (18)$$

Therefore,

$$F_{tot} = 1 - e^{-(2An+Da)} \qquad (19)$$

The fractions of dose absorbed and degraded are

$$F_a = \frac{2\,An}{2\,An + Da}(1 - e^{-(2An+Da)}) \qquad (20)$$

and

$$F_d = \frac{Da}{2\,An + Da}(1 - e^{-(2An+Da)}) \qquad (21)$$

Equations (20) and (21) have been used to estimate the oral absorption of cefaclor, cefatrizine, and insulin. The simulated results compare favorably to the reported literature values in humans. The macroscopic mass balance approach provides a quick approximation to the fraction of dose absorbed and degraded for both passively and nonpassively absorbed drugs.

B. Microscopic Mass Balance Approach

Sinko et al. [31] also used the microscopic approach to predict the fraction of dose absorbed for highly soluble drugs. The results from the microscopic approach are similar to those from the macroscopic approach. The macroscopic approach, therefore, is recommended for such a purpose because of its simplicity. However, when the dissolution is limiting, the microscopic approach should be used. Oh et al. [43] developed a mathematical model to estimate the fraction of dose absorbed from suspensions of poorly soluble drugs. Again, steady state is assumed. Considering drug absorption from a cylindrical tube, we have the following equation to describe the rate of change of particle radius:

$$\frac{dr_p}{dz} = -\frac{D\pi R^2}{QP}\left(\frac{C_s - C}{r_p}\right) \tag{22}$$

From a mass balance for the solution phase, we have

$$\frac{dC}{dz} = \frac{4\pi^2 R^2 D(N_0/V_0)}{Q} r_p(C_s - C) - \frac{2\pi R P_{eff}}{Q} C \tag{23}$$

where r_p is the particle radius, ρ is the particle density, N_0/V_0 is the particle number density, and D is the diffusion coefficient. Although it was not pointed out by the authors, Eqs. (22) and (23) imply that the volume of particles is negligible compared to the volume of the solution phase. Let

$$z^* = \frac{z}{L}, \qquad r^* = \frac{r_p}{r_0}, \qquad C^* = \frac{C}{C_0} \tag{24}$$

Equations (22) and (23) then become

$$\frac{dr^*}{dz^*} = -\frac{Dn}{3}\left(\frac{1 - C^*}{r^*}\right) \tag{25}$$

$$\frac{dC^*}{dz^*} = Dn\ Do\ r^*(1 - C^*) - 2\ An\ C^* \tag{26}$$

where dose number Do and absorption number An are as defined by Eqs. (4) and (8). Dn is the dimensionless dissolution number, defined as the ratio of the dissolution rate to the flow rate along the intestine:

$$
\mathrm{Dn} = \frac{\pi R^2 L}{Q}\left(\frac{3DC_0}{\rho r_0^2}\right) = \frac{\langle T_{si}\rangle}{t_{\mathrm{diss}}}
\tag{27}
$$

Equations (25) and (26) show that the dissolution number will also influence the drug absorption in addition to the absorption and dose numbers. Assuming that the initial amount of drug in solution is insignificant compared to the amount of solid drug, the fraction of dose absorbed can be estimated by

$$
F_a = 1 - ((r^*)_{z^*=1})^3 - \left(\frac{C^*}{\mathrm{Do}}\right)_{z^*=1}
\tag{28}
$$

Equation (28) is valid only for monodisperse drugs. For polydisperse drugs, the overall fraction of dose absorbed may be estimated on the basis of the particle size distribution [44].

Figure 7 gives a typical profile of the fraction of dose absorbed as a function of the dissolution number and dose number for highly permeable (large absorption number) drugs. It shows that the fraction of dose absorbed depends sharply on the dose and dissolution numbers when they are in critical ranges around 1 for highly permeable drugs. Figure 7 shows experimental results for griseofulvin and digoxin as a further illustration of the significance of the dose and solubility of drugs. Griseofulvin and digoxin have similar solubilities, 15 and 24 mg/mL, respectively. Based on the solubility data, it can be assumed that both compounds should be equally absorbed. However, from the dose numbers of the two compounds (133 for griseofulvin and 0.08 for digoxin), the fraction of dose absorbed for digoxin is expected to be much greater than that for griseofulvin, as shown in Figure 7. In fact, an increase in the dissolution number via micronization for digoxin makes it completely absorbed [45]. The relative bioavailability of griseofulvin can be improved by only a factor of 1.7 via micronization, suggesting incomplete absorption due to its high dose number [46]. At high dose number, the dissolution number weakly influences the fraction of dose absorbed.

Crison and Amidon [47,48] recently used the mass balance approach to study the variability in absorption due to intestinal transit time for water-insoluble drugs. As expected, for low dose drugs, such as digoxin, the variability in absorption sharply decreases with the increase of dissolution number via micronization. For high dose drugs, such as griseofulvin, little effect was observed when the dissolution number was increased by micronization.

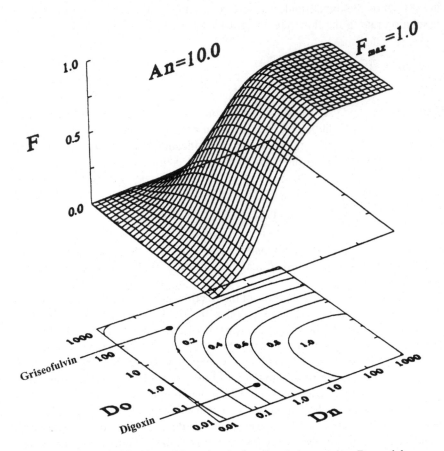

Figure 7 Estimated fraction of dose absorbed vs dissolution number, Dn, and dose number, Do, for a high permeability drug. An = 10 corresponds to a drug with a permeability approximately equal to that of glucose. Dn and Do for digoxin and griseofulvin were calculated from Eqs. (4) and (27) and the following physicochemical/physiological parameters.

Drug	S (mg/mL)	ρ (g/mL)	D (mg)	V_0 (mL)	r_0 (μm)	D (cm^2/sec)
Digoxin	0.024	1200	0.5	250	25	5×10^{-6}
Griseofulvin	0.015	1200	500	250	25	5×10^{-6}

(From Ref. 33 with kind permission from Plenum Publishing Corporation.)

IV. DYNAMIC MODEL

A. Dispersion Model

The dispersion model approach was first proposed to simulate dynamic absorption processes [49]. The dispersion model assumes that the small intestine can be considered as a uniform tube with constant axial velocity, constant dispersion behavior, and uniform concentration across the tube diameter. Then the absorption of highly soluble drugs in the small intestine can be delineated by the following dispersion model equation:

$$\frac{\partial C}{\partial t} = \alpha \frac{\partial^2 C}{\partial z^2} - v\frac{\partial C}{\partial z} - K_a C \tag{29}$$

where C is the concentration of the drug, z is the axial distance from the stomach, K_a is the (intrinsic) absorption rate constant, v is the velocity in the axial direction, and α is a longitudinal coefficient that accounts for mixing by both molecular diffusion and physiological effects, such as membrane surface solute binding, peristaltic and villous activities, and the multi-S course of the small intestine [50]. Equation (29) generally has to be solved numerically [47,48], but in some cases analytical solutions may be possible. The initial condition for Eq. (29) is

$$t = 0, \quad \text{all } z, \quad C = 0 \tag{30}$$

Two boundary conditions are required to solve Eq. (29). One is obtained by assuming that the concentration of the drug is zero at infinite distance:

$$z = \infty, \quad \text{all } t, \quad C = 0 \tag{31}$$

Another boundary condition is at $z = 0$. Since gastric emptying is expressed with respect to volume and the boundary condition with respect to concentration, we have to transform volume into concentration. A mechanistic way to accomplish this transformation has not yet been established. Consequently, various conditions have been used in the literature [50]. This results in various analytical solutions. Assuming that the stomach can be considered as an infinite reservoir with constant output rate with respect to concentration and volume, we have

$$z = 0, \quad \text{all } t, \quad C = C_0 \tag{32}$$

The analytical solution is then

$$\frac{C}{C_0} = \frac{e^{vz/2\alpha}}{2}\left(e^{-z\sqrt{\xi/\alpha}}\text{erf}\left[\frac{z}{\sqrt{4\alpha t}} - \sqrt{\xi t}\right] + e^{z\sqrt{\xi/\alpha}}\text{erf}\left[\frac{z}{\sqrt{4\alpha t}} + \sqrt{\xi t}\right]\right) \tag{33}$$

where

$$\xi = \frac{v^2}{4\alpha} + K_a \tag{34}$$

Obviously, even in the most simplified condition, the analytical solution is still very complex. The accurate calculation of Eq. (33) requires skillful numerical techniques because it involves the product of an infinite number and a value close to zero.

Given its complexity, the dispersion model has not been widely used despite the fact that it provides a rigorous potential framework for describing oral drug absorption. A concept extracted from the dispersion model, the anatomic reserve length, has instead been widely used to explain absorption phenomena [53,54]. The reserve length for absorption is the length of the intestine beyond that at which absorption is complete. If absorption from the stomach and colon is minor compared with that from the small intestine, the maximum reserve length is then the length of the small intestine. Thus, the reserve length is greater when absorption is efficient and correspondingly shorter for less efficient absorption. Mathematically,

$$RL = L - l \tag{35}$$

where RL is the anatomic reserve length, L is the length of the small intestine, and l is the intestinal length at which absorption is complete.

When $l < L$, the reserve length is positive and absorption is complete within the small intestine. When $l > L$, the reserve length is negative and absorption is incomplete within the small intestine. If the fraction of dose absorbed above 95% is defined as complete absorption, then l can be estimated by [53,54]

$$l = \frac{3Rv}{2P_{eff}} + \frac{3\alpha}{v} \tag{36}$$

and, in turn,

$$RL = L - \frac{3Rv}{2P_{eff}} + \frac{3\alpha}{v} \tag{37}$$

where R is the radius of the small intestine and P_{eff} is the effective membrane permeability. The third term on the right in Eq. (37) is relatively small, and Eq. (37) can be simplified into

$$RL = L - \frac{3Rv}{2P_{eff}} \tag{38}$$

R and α in Eq. (38) are physiologically related parameters; the reserve length therefore depends on the effective permeability only. In that regard, the reserve length concept and the macroscopic mass balance approach are quite similar. The reserve length concept provides qualitative information, whereas the macroscopic mass balance approach provides quantitative information.

B. Mixing Tank Model

The mixing tank model has been developed and utilized to simulate oral absorption phenomena [55]. This approach considers the gastrointestinal tract as one or more serial mixing tanks with linear transfer kinetics. Each tank is well mixed and has a uniform concentration. Dressman and coworkers [56,57] treated the gastrointestinal tract as one or two mixing tanks to investigate dose-dependent and dissolution rate–controlled drug absorption. Johnson and coworkers [58–60] extended Dressman's approach to include polydisperse drugs to show the effect of particle size distribution and shape on dissolution and absorption. Oberle and Amidon [61] employed four mixing tanks to explain plasma level double-peak phenomena. Leesman et al. [62] proposed a physiological flow model and have demonstrated its utilization in dosage form design and evaluation. Luner and Amidon [63] employed a four mixing tank model to study the effect of bile sequestrants on bile salt excretion.

A single mixing tank model is reviewed here for illustration (Fig. 8). The single mixing tank model was originally proposed by Dressman et al. [56] to

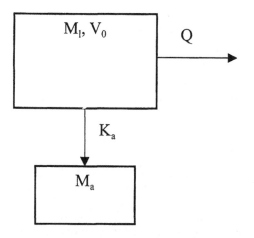

Figure 8 Schematic of mixing tank model with volume V_0 and flow rate Q. M_l is the amount of drug (liquid) and M_a is the amount of drug absorbed. K_a is the intrinsic absorption rate constant.

investigate dissolution-controlled drug absorption. However, here we consider soluble drugs only. The rate of change of drug mass in the mixing tank is

$$\frac{dM_l}{dt} = -\left(K_a + \frac{Q}{V_0}\right)M_l \tag{39}$$

The rate of drug absorption is

$$\frac{dM_a}{dt} = K_a M_l \tag{40}$$

where M_l is the amount of drug in the small intestine, M_a is the amount of drug absorbed, t is time, Q is flow rate, V_0 is the volume of the mixing tank, and K_a is the intrinsic absorption rate constant. Defining $Y_l = M_l/M$ and $Y_a = M_a/M$ and solving Eqs. (39) and (40) yields

$$Y_l = \exp\left[-\left(K_a + \frac{Q}{V_0}\right)t\right] \tag{41}$$

$$Y_a = \frac{K_a}{K_a + Q/V_0}\left\{1 - \exp\left[-\left(K_a + \frac{Q}{V_0}\right)t\right]\right\} \tag{42}$$

The fraction of dose absorbed is then given by

$$F_a = Y_a\Big|_{t\to\infty} = \frac{K_a}{K_a + Q/V_0} \tag{43}$$

The absorption rate constant K_a can be estimated from the effective permeability:

$$K_a = \frac{2P_{\text{eff}}}{R} \tag{44}$$

Rewriting Eq. (43) in terms of the absorption number according to Eq. (8) results in

$$F_a = \frac{2\,\text{An}}{1 + 2\,\text{An}} \tag{45}$$

Note that we define the fraction of dose absorbed as the upper limit of the percent of dose absorbed. Therefore, the percent of dose absorbed changes with time, but the fraction of dose absorbed does not.

The advantage of the mixing tank model approach is its relative simplicity, intuitive accessibility, and easy correlation with pharmacokinetic models. However, the physical basis for considering a segment of the small intestine as one or more serial mixing tanks is limited, although such an assumption has been commonly and successfully utilized in the physical and biological sciences.

C. Compartmental Absorption and Transit Model

As reviewed in the previous sections, different numbers of mixing tanks have been used to simulate and explain oral drug absorption in various publications. Obviously, the number of mixing tanks will affect simulation results. Therefore, there is a need to define the number of mixing tanks that is most appropriate to characterize the flow and absorption process in the human small intestine. A mathematical model was developed in such a way that it best describes the transit flow of drugs in the human small intestine [64,65]. To emphasize the importance of the transit, the model was named compartmental absorption and transit (CAT). The CAT model has been successfully utilized in predicting oral drug absorption, estimating oral plasma concentration profiles, and optimizing the release profiles from sustained release dosage forms [66–69].

1. Compartmental Transit Model

The process of a drug passing through the small intestine was viewed as flow through a series of segments. A single compartment can describe each segment with linear transfer kinetics from one to the next. All compartments may have different volumes and flow rates, but they have the same transit rate constant K_t. It is assumed that a drug is neither absorbable nor degradable (absorption is discussed in the next section). Therefore, from mass balance, we have

$$\frac{dM_n}{dt} = K_t M_{n-1} - K_t M_n, \qquad n = 1, 2, \ldots, N \tag{46}$$

where M_n is the amount of the drug in the nth compartment, t is the time, and N is the number of compartments. The rate at which the drug exits the small intestine or enters the colon is

$$\frac{dM_c}{dt} = K_t M_N \tag{47}$$

where M_c is the amount of drug in the colon. Coupling this with Eq. (46), the analytical solution of Eq. (47) is

$$F(t) = \frac{M_c}{M_0} = 1 - e^{-K_t t}\left(1 + K_t t + \frac{(K_t t)^2}{2} + \cdots + \frac{(K_t t)^{N-1}}{(N-1)!}\right) \tag{48}$$

By definition, the transit rate constant K_t is

$$K_t = \frac{N}{\langle T_{si} \rangle} \tag{49}$$

In order to determine the optimal number of compartments, literature information on small intestinal transit times was utilized. A total of over 400 human small intestinal transit time data were collected and compiled from various publications, since the small intestinal transit time is independent of dosage form, gender, age, body weight, and the presence of food [70]. Descriptive statistics showed that the mean small intestinal transit time was 199 min with a standard deviation of 78 min and a 95% confidence interval of 7 min. The data set was then analyzed by arranging the data into 14 classes, each with a width of 40 min. Figure 9 shows the distribution of this data set.

Figure 9 shows the results of population analysis. It does not distinguish the intra- and intersubject variability. In our subsequent analysis [65], we defined intradose, intrasubject, and intersubject variances. The intradose variance charac-

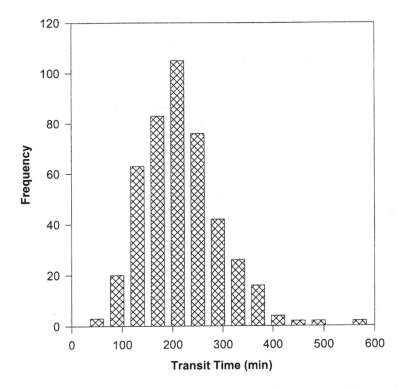

Figure 9 Distribution of small intestinal transit time in humans. The transit time was measured by γ-scintigraphy based on the difference in time between 50% of the drug arriving at the colon and 50% of the drug leaving the stomach. The distribution was constructed from over 400 literature data points. (From Ref. 64 with kind permission from Elsevier Science-NL, Amsterdam.)

terizes the spreading of a drug solution dose traveling along the small intestine. The intrasubject variance describes the dose-to-dose variability of the transit time in a subject, while the subject-to-subject variability of the transit time in a population was characterized by the intersubject variance. It was shown that either intrasubject or intersubject variance is significantly smaller than the intradose variance.

The cumulative curve obtained from the transit time distribution in Figure 9 was fitted by Eq. (48) to determine the number of compartments. An additional compartment was added until the reduction in residual (error) sum of squares (SSE) with an additional compartment becomes small. An F test was not used, because the compartmental model with a fixed number of compartments contains no parameters. SSE then became the only criterion to select the best compartmental model. The number of compartments generating the smallest SSE was seven. The seven-compartment model was thereafter referred to as the compartmental transit model.

The seven-compartment transit model may be physiologically sound. We may visualize that the first half of the first compartment represents the duodenum; the second half of the first compartment, along with the second and third compartments, the jejunum; and the rest of the compartments the ileum. The corresponding transit times in the duodenum, jejunum, and ileum are 14, 71, and 114 min, respectively. Considering the volumes and flow rates in these three segments [71,72], such an assignment seems reasonable.

2. Compartmental Absorption Model

The compartmental absorption and transit model was developed based on the transit model. The assumptions for the CAT model include the following.

1. Absorption from the stomach and colon is minor compared with that from the small intestine.
2. Transport across the small intestinal membrane is passive.
3. Dissolution is instantaneous.

Therefore, for a nondegradable drug dosed in immediate release dosage forms, the absorption and transit in the gastrointestinal tract can be depicted as follows. Stomach:

$$\frac{dM_s}{dt} = -K_s M_s \tag{50}$$

Small intestine:

$$\frac{dM_n}{dt} = K_t M_{n-1} - K_t M_n - K_a M_n, \qquad n = 1, 2, \ldots, 7 \tag{51}$$

Colon:

$$\frac{dM_c}{dt} = K_t M_N \tag{52}$$

where M_s is the amount of drug in the stomach; M_c is the amount of drug in the colon; M_n is the amount of drug in the nth compartment; t is the time; and K_s, K_t, and K_a are the rate constants of gastric emptying, small intestinal transit, and intrinsic absorption, respectively. In Eq. (51), when $n = 1$, the term $K_t M_0$ is replaced by $K_s M_s$. The rate of drug absorption from the small intestine into the plasma is calculated as

$$\frac{dM_a}{dt} = K_a \sum_{n=1}^{7} M_n \tag{53}$$

where M_a is the amount of drug absorbed. From mass balance, we have

$$M_s + \sum_{n=1}^{7} M_n + M_c + M_a = M_0 \tag{54}$$

At $t \rightarrow \infty$, M_s and the M_n's go to zero, so

$$M_c + M_a = M_0 \tag{55}$$

The fraction of dose absorbed, F_a, can then be estimated by

$$F_a = \frac{M_a}{M_0} = \frac{1}{M_0} \int K_a \sum_{n=1}^{7} M_n dt \tag{56}$$

Coupling with Eqs. (50) and (51), the analytical solution of Eq. (56) is

$$F_a = 1 - (1 + K_a/K_t)^{-7} \tag{57}$$

Substituting Eqs. (43) and (48) with $N = 7$ into Eq. (57), we have

$$F_a = 1 - \left(1 + \frac{2P_{eff}\langle T_{si}\rangle}{7R}\right)^{-7} \tag{58}$$

Substitution of $\langle T_{si}\rangle$ of 3.32 hr and the radius of 1.75 cm [50] into Eq. (57) yields

$$F_a = 1 - (1 + 0.54 P_{eff})^{-7} \tag{59}$$

The human effective permeability P_{eff} in Eq. (59) is expressed in centimeters per hour. In terms of the dimensionless absorption number An, Eq. (58) can be written as

$$F_a = 1 - (1 + 0.29 \text{An})^{-7} \tag{60}$$

Ten compounds covering a wide range of absorption characteristics, from the least permeable, enalaprilat, to the most permeable, ketoprofen, were used to evaluate the ability of the CAT model to predict the fraction of dose absorbed. Figure 10 shows the predicted values from Eq. (59) of the CAT model. Enalaprilat is an angiotension-converting enzyme inhibitor. The fraction of dose absorbed of enalaprilat in laboratory animals was estimated to be only 5–12%; in humans, oral absorption of radiolabeled enalaprilat was probably less than 10% [73]. The model predicted the fraction of dose absorbed of enalaprilat to be 23%, greater than the experimental value. Furosemide is a potent loop diuretic used in the treatment of edematous states associated with cardiac, renal, and hepatic failure and for the treatment of hypertension. Although furosemide is practically insoluble in water, absorption is limited by transport across the gastrointestinal mem-

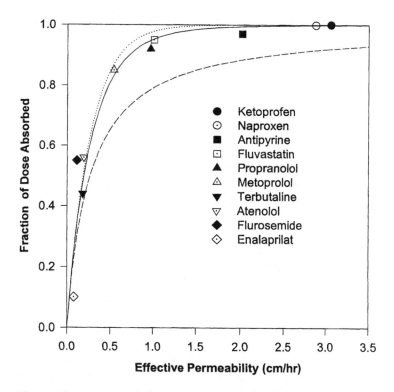

Figure 10 The fraction of dose absorbed as a function of the effective human permeability. (——) Compartmental absorption and transit model (Eqs. (59) or (60)); (---) single-compartment model (Eq. (45)); (· · ·) plug flow model (Eq. (10)). The human permeability was measured using Loc-I-Gut apparatus [85]. The correlations of human permeability with the permeability from preclinical in vitro absorption models are being established [86].

brane rather than by dissolution [74]. The fraction of dose absorbed varies from 37% to 83% in healthy volunteers, and the mean value is around 55% [70]. The predicted fraction of dose absorbed is 33%, slightly below the experimental observations. The model predicted the fraction of dose absorbed to be 48, 50, 83, and 95% for terbutaline, atenolol, metoprolol, and propranolol, respectively, based on the permeability data. The predicted results are in agreement with the experimental data [75–79]. Fluvastatin, antipyrine, naproxen, and ketoprofen are completely absorbed [77,80,81], as predicted by the CAT model.

Figure 10 also shows the fraction of dose absorbed calculated by the single mixing tank and the mass balance approach. Overall, the single-compartment model underestimates the fraction of dose absorbed, whereas the plug flow and the CAT models give a much closer fit to the data. The single-compartment model assumes that the drug will leave the small intestine and enter the colon as long as $t > 0$, contradicting the fact that traveling through the small intestine takes at least a few minutes [70]. Consequently, the fraction of dose absorbed is underestimated.

In the case of the plug flow model, it is assumed that the drug will not leave the small intestine and enter the colon until the mean small intestinal transit time has been reached. This contradicts the fact that some of the drug leaves the small intestine in a much shorter time due to rapid longitudinal dispersion. Both CAT and dispersion models characterized the human small intestinal transit flow well [64]. We can reasonably assume that the results of the CAT model and the dispersion model would be very close. The plug flow model is a simplification of the dispersion model which assumes no longitudinal dispersion. (The dispersion model is also called the dispersed plug flow model.) Since the effect of longitudinal dispersion on the anatomic reserve length (the fraction of dose absorbed) is negligible [50], the results of the dispersion and plug flow models are expected to be similar. Consequently, the results of the CAT and plug flow models are essentially the same; nevertheless, since dispersion is not considered in the plug flow model, it is unlikely that this model will be able to estimate the rate of drug absorption.

3. Application to Pharmacokinetic Modeling

The CAT model estimates not only the extent of drug absorption, but also the rate of drug absorption; that makes it possible to couple the CAT model to pharmacokinetic models to estimate plasma concentration profiles. The CAT model has been used to estimate the rate of absorption for saturable and region-dependent drugs, such as cefatrizine [67]. In this case, the model simultaneously considers passive diffusion, saturable absorption, GI degradation, and transit. The mass balance equation, Eq. (51), needs to be rewritten to include all these processes:

$$\frac{dM_n}{dt} = K_t M_{n-1} - K_t M_n - K_{an} M_n - K_{dn} M_n, \qquad n = 1, 2, \ldots, 7 \quad (61)$$

where K_{an} and K_{dn} ($n = 1, 2, \ldots, 7$) are the absorption and degradation rate constants. The rate constants of drug absorption K_{an} ($n = 1, 2, \ldots, 7$) consist of saturable absorption that is compartment-dependent and passive absorption that is compartment-independent. K_{an} ($n = 1, 2, \ldots, 7$) may be estimated by

$$K_{an} = \frac{V_{\max n}}{K_{mn} + C_{mn}} + K_{a1}, \qquad n = 1, 2, \ldots, 7 \quad (62)$$

where $V_{\max n}$, K_{mn}, and C_{mn} ($n = 1, 2, \ldots, 7$) are the maximum rate of absorption, Michaelis constant of saturable absorption, and the concentration of the drug [50,51]. K_{a1} is the passive absorption rate constant. The degradation rate constant K_{dn} ($n = 1, 2, \ldots, 7$) may also be compartment-dependent and can be estimated as

$$K_{dn} = K_{d1} + K_{d2} C_{mn}, \qquad n = 1, 2, \ldots, 7 \quad (63)$$

where K_{d1} and K_{d2} are the first-order and second-order rate constants of degradation. Clearly, we could incorporate the other forms of degradation kinetics in Eq. (63).

The rates of drug absorption and entry into the colon are the same as in Eq. (53) and Eq. (52). The rate of degradation is

$$\frac{dM_d}{dt} = \sum_{n=1}^{7} K_{dn} M_n \quad (64)$$

To predict oral plasma concentration–time profiles, the rate of drug absorption (Eq. (53)) needs to be related to intravenous kinetics. For example, in the case of the one-compartment model with first-order elimination, the rate of plasma concentration change is estimated as

$$\frac{dC}{dt} = \frac{1}{V}\frac{dM_a}{dt} - k_e C \quad (65)$$

where C is the plasma concentration in the central compartment, V is the volume of plasma (volume of distribution), and k_e is the elimination rate constant.

Coupling with its intravenous pharmacokinetic parameters, the extended CAT model was used to predict the observed plasma concentration–time profiles of cefatrizine at doses of 250, 500, and 1000 mg. The human experimental data from Pfeffer et al. [82] were used for comparison. The predicted peak plasma concentrations and peak times were 4.3, 7.9, and 9.3 µg/mL at 1.6, 1.8, and 2.0 hr, in agreement with the experimental mean peak plasma concentrations of

4.9 ± 1.2, 8.6 ± 1.0, and 10.2 ± 2.1 μg/mL at the peak times 1.4 ± 0.4, 1.6 ± 0.2, and 2.0 ± 0.6 hr. The reported absolute bioavailability was 75% and 50% at 250 and 1000 oral doses, which compared favorably with the theoretical fraction of dose absorbed (74% and 48%). The calculated fraction of dose absorbed at a 500 mg dose is 61%, lower than the experimental bioavailability of 75%. With an increase of the dose from 250 to 1000 mg, the overall fraction of dose absorbed decreases from 74% to 48%, while the fraction of dose in the colon increases from 16% to 33%, and the fraction of dose degraded in the small intestine increases from 10% to 19%. Although the overall fraction of dose absorbed decreases with increasing dose, the fraction of dose absorbed by passive transport increases slightly. The decrease in the overall fraction of dose absorbed is mainly due to saturable absorption mechanisms. Sixty-four percent of a 250 mg dose was absorbed by a saturable route, compared to only 34% of a 1000 mg dose. Meanwhile, with a 1000 mg dose, 14% of the dose was absorbed by passive diffusion, compared to 34% by saturable absorption, suggesting the importance of passive diffusion in the simulation of cefatrizine absorption.

The CAT model considers passive absorption, saturable absorption, degradation, and transit in the human small intestine. However, the absorption and degradation kinetics are the only model parameters that need to be determined to estimate the fraction of dose absorbed and to simulate intestinal absorption kinetics. Degradation kinetics may be determined in vitro and absorption parameters can also be determined using human intestinal perfusion techniques [85]; therefore, it may be feasible to predict intestinal absorption kinetics based on in vitro degradation and in vivo perfusion data. Nevertheless, considering the complexity of oral drug absorption, such a prediction is only an approximation.

V. SUMMARY

Numerous models have been developed ranging from the pH-partition hypothesis to the CAT model throughout the last 50 years. The pH-partition hypothesis and the absorption potential concept provide basic guidelines for understanding drug absorption. These quasiequilibrium models are entirely based on the physicochemical properties of drugs. Such approaches may be useful in the qualitative assessment of drug absorption for new drugs. The dispersion model has the advantage of being more physically realistic. Both the mass balance approach and the CAT model gave simple equations for quantitative estimation of the fraction of dose absorbed, while the CAT model can also be directly related to pharmacokinetic models to predict plasma concentration profiles.

It should be noted, however, that any of the absorption models discussed above works well only when the situation is close to what has been assumed. In balance, we believe that the absorption models are best applied with scientific judgment, bearing in mind that many other factors, such as crystal form, blood flow, gastrointestinal pH, and other dosage form factors, have not been fully considered. The consideration of these factors will advance our ability to predict oral drug absorption. This may occur through extending existing models and/or developing novel approaches.

VI. NOTATION

A	Amount of drug absorbed
An	Absorption number $= P_{\text{eff}}\langle T_{\text{si}}\rangle/R$
AP	Absorption potential
C	Lumenal drug concentration or plasma concentration
C_0	Dose concentration
C_{out}	Drug concentration exiting the small intestine
D	Diffusivity
Da	Damkohler number $= K_d\langle T_{\text{si}}\rangle$
Dn	Dissolution number $= 3DC_0\langle T_{\text{si}}\rangle/\rho r_0$
Do	Dose number $= M_0V_0/C_s$
F_a	Fraction of dose absorbed
F_d	Fraction of dose degraded
$F(t)$	Cumulative distribution of the small intestinal transit time
F_{tot}	Total drug loss (clearance) in the intestine
F_{un}	Fraction in un-ionized form at pH 6.5
f	Proportional constant
k_e	Elimination rate constant
K_a	Absorption rate constant
K_d	Degradation rate constant
K_*	Transit rate constant
L	Length of the small intestine
l	Intestinal length at which absorption is complete
M_a	Amount of drug absorbed
M_c	Amount of drug in the colon
M_d	Amount of drug degraded
M_n	Amount of drug in the nth compartment
M_0	Dose
M_s	Amount of drug in the stomach
N	Number of compartments
N_0	Number of particles
P	Partition coefficient
P_{sq}	Aqueous permeability
P_{eff}	Effective permeability

P_w Wall permeability
Q Flow rate
R Radius of the small intestine
RL   Anatomic reserve length
r_o Initial radius of particle
r_p Radius of particle
$\langle T_{si} \rangle$ Mean small intestinal transit time
t Time
V Volume of distribution
V_{max} Maximum rate of absorption
V_0 Volume of water taken with dose
z Axial distance
α Longitudinal coefficient
ρ Density of particles
v Velocity in the axial direction

REFERENCES

1. M Mayersohn. Principles of drug absorption. In: GS Banker, CT Rhodes, eds. Modern Pharmaceutics. 2nd ed. New York: Marcel Dekker, 1990, pp 23–90.
2. LX Yu, JR Crison, E Lipka, GL Amidon. Transport approaches to the biopharmaceutical design of oral drug delivery systems: Prediction of intestinal absorption. Adv Drug Delivery Rev 19:359–376, 1996.
3. LX Yu, LA Gatlin. Dosage form development aid: Oral drug absorption. Glaxo Wellcome Dosage Form Development Guidelines, 1997.
4. MH Jacobs. Some aspects of cell permeability to weak electrolytes. Cold Spring Harbor Symp Quant Biol 8:30–39, 1940.
5. PA Shore, BB Brodie, CAM Hogben. A mathematical model of drug absorption. J Pharmacol Exp Ther 119:361–369, 1957.
6. LS Schanker, PA Shore, BB Brodie, CAM Hogben. Absorption of drugs from the stomach. I. The rat. J Pharmacol Exp Ther 120:528–539, 1957.
7. CAM Hogben, LS Schanker, DJ Tocco, BB Brodie. Absorption of drugs from the stomach. II. The human. J Pharmacol Exp Ther 120:540–545, 1957.
8. LS Schanker, DJ Tocco, BB Brodie, CAM Hogben. Absorption of drugs from the rat small intestine. J Pharmacol Exp Ther 123:81–88, 1958.
9. CAM Hogben, DJ Tocco, BB Brodie, LS Schanker. On the mechanism of intestinal absorption of drugs. J Pharmacol Exp Ther 125:275–282, 1959.
10. LS Schanker. On the mechanism of absorption from the gastrointestinal tract. J Med Pharm Chem 2:343, 1960.
11. M Rowland, TN Tozer. Clinical Pharmacokinetics. 3rd ed. Media, PA: Williams & Wilkins, 1995, pp 114–116.
12. M Gibaldi. Biopharmaceutics and Clinical Pharmacokinetics. 4th ed. Philadelphia: Lea & Lea Febiger, 1991.

13. S Suzuki, WI Higuchi, NFH Ho. Theoretical model studies of drug absorption and transport in the gastrointestinal tract. I. J Pharm Sci 59:644–651, 1970.
14. A Suzuki, WI Higuchi, NFH Ho. Theoretical model studies of drug absorption and transport in the gastrointestinal tract. II. J Pharm Sci 59:651–659, 1970.
15. D Winne. Shift of pH-absorption curves. J Pharmacok Biopharm 5:53–94, 1977.
16. B Testa, P-A Carrupt, P Gaillard, F Billois, P Weber. Lipophilicity in molecular modeling. Pharm Res 13:335–343, 1996.
17. H van de Waterbeemd, B Testa. The parameterization of lipophilicity and other structural properties in drug design. In: B. Testa, ed. Advances in Drug Research, Vol. 16. London: Academic Press, 1987, pp 87–227.
18. BH Stewart, OH Chan, N Jezyk, D Fleisher. Discrimination between drug candidate using models for evaluation of intestinal absorption. Adv Drug Deliv Rev 23:27–45, 1997.
19. RN Smith, C Hansch, MM Ames. Selection of a reference partitioning system for drug design work. J Pharm Sci 64:599–606, 1975.
20. V Austel, E Kutter. Absorption, distribution and metabolism of drugs. In: JG Topless, ed. Quantitative Structure–Activity Relationships of Drugs. New York: Academic Press, 1980, pp 437–496.
21. A Leo, C Hansch, D Elkins. Partition coefficients and their uses. Chem Rev 71:525–616, 1977.
22. SH Yalkowsky, W Morozowich. A physical chemical basis of the design of orally active prodrugs. Drug Des 9:122–185, 1980.
23. I Komiya, JY Park, A Kamani, NFH Ho, WI Higuchi. Physical model studies on the simultaneous fluid flow and absorption of steroids in the rat intestines. Int J Pharm 4:249–262, 1980.
24. NFH Ho, JY Park, W Morozowich, WI Higuchi. Physical model approach to the design of drugs with improved intestinal absorption. In: EB Roche, ed. Design of Biopharmaceutical Properties Through Prodrugs and Analogs. Washington DC: American Pharmaceutical Association, Acad Pharm Sci, 1977, pp 136–227.
25. JBM Van Bree, AG De Boer, M Danhof, L Gisel, DD Breimer. Characterization of an in vitro blood-brain barrier: Effects of molecular size and lipophilicity on cerebrovascular endothelial transport rates of drugs. J Pharmacol Exp Ther 247:1233–1239, 1988.
26. A Buur, L Trier, C Magnusson, P Artursson. Permeability of 5-fluorouracil and prodrugs in Caco-2 cell monolayers. Int J Pharm 129:223–231, 1996.
27. PS Burton, RA Conradi, NFH Ho, AR Hilgers, RT Borchard. How structure features influence the biomembrane permeability of peptides. J Pharm Sci 85:1336–1340, 1996.
28. JB Dressman, GL Amidon, D Fleisher. Absorption potential: Estimating the fraction absorbed for orally administered compounds. J Pharm Sci 74:588–589, 1985.
29. P Macheras, MY Symillides. Toward a quantitative approach for the prediction of the fraction of dose absorbed using the absorption potential concept. Biopharm Drug Dispos 10:43–53, 1989.
30. GL Amidon, PJ Sinko, D Fleisher. Estimating human oral fraction dose absorbed: A correlation using rat intestinal membrane permeability for passive and carrier-mediated compounds. Pharm Res 5:651–654, 1988.

31. PJ Sinko, GD Leesman, GL Amidon. Predicting fraction dose absorbed in humans using a macroscopic mass balance approach. Pharm Res 8:979–988, 1991.

32. PJ Sinko, GD Leesman, GL Amidon. Mass balance approaches for estimating the intestinal absorption and metabolism of peptides and analogues: Theoretical development and applications. Pharm Res 10:271–275, 1993.

33. GL Amidon, H Lennernas, VP Shah, JR Crison. A theoretical basis for a biopharmaceutic drug classification: The correlation of in vitro drug product dissolution and in vivo bioavailability. Pharm Res 12:413–420, 1995.

34. DM Oh, PJ Sinko, GL Amidon. Predicting oral drug absorption in humans: A macroscopic mass balance approach for passive and carrier-mediated compounds. In:DZ D'Argenio, ed. Advanced Methods of Pharmacokinetic and Pharmacodynamic Systems Analysis. New York: Plenum Press, 1990, pp 3–11.

35. PJ Sinko, GL Amidon. Characterization of the oral absorption of β-lactam antibiotics. I. Cephalosporins: Determination of intrinsic membrane absorption parameters in the rat intestine in situ. Pharm Res 5:645–650, 1988.

36. PJ Sinko, GL Amidon. Characterization of the oral absorption of β-lactam antibiotics. II. Competitive absorption and peptide carrier specificity. J Pharm Sci 78:723–727, 1989.

37. DM Oh, PJ Sinko, GL Amidon. Characterization of the oral absorption of β-lactam: Effect of the α-amino side chain group. J Pharm Sci 82:897–900, 1993.

38. P Artursson, K Palm, K Luthman. Caco-2 monolayers in experimental and theoretical predictions of drug transport. Adv Drug Deliv Rev 22:67–84, 1996.

39. P Artursson, J Karlsson. Correlation between oral drug absorption in humans and apparent drug permeability coefficients in human intestinal epithelial (Caco-2) cells. Biochem Biophys Res Commun 175:880–885, 1991.

40. BH Stewart, OH Chan, RH Lu, EL Reyner, HL Schmid, HW Hamilton, BA Steinbaugh, MD Taylor. Comparison of intestinal permeabilities determined in multiple in vitro and in situ model: Relationship to absorption in humans. Pharm Res 12:693–699, 1995.

41. JN Cogburn, MG Donovan, CS Schasteen. A model of human small intestinal absorptive cells. 1. Transport barrier. Pharm Res 9:210–216, 1991.

42. W Rubas, N Jezyk, GM Grass. Comparison of the permeability characteristics of a human colonic epithelial (Caco-2) cell line to colon of rabbit, monkey, and dog intestine and human drug absorption. Pharm Res 10:113–118, 1993.

43. DM Oh, RL Curl, GL Amidon. Estimating the fraction dose absorbed from suspensions of poorly soluble compounds in humans: A mathematical model. Pharm Res 10:264–270, 1993.

44. JR Crison. Estimating the dissolution and absorption of water insoluble drugs in the small intestine. PhD dissertation, The University of Michigan, Ann Arbor, MI, 1993.

45. AJ Jounela, PJ Pentikainen, A Sothman. Effect of particle size on the bioavailability of digoxin. Eur J Clin Pharmacol 8:365–370, 1975.

46. M Kraml, J Dubuc, R Gaudry. Gastrointestinal absorption of griseofulvin: II. Influence of particle size in man. Antibiot Chemother 12:239–242, 1962.

47. JR Crison, GL Amidon. Expected variation in bioavailability for water insoluble drugs. In: HH Blume, KK Midha, eds. Bio-International 2, Bioavailability, Bioequi-

valence and Pharmacokinetic Studies. Stuttgart:Medpharm Scientific Publishers, 1995, pp 301–309.

48. JR Crison, GL Amidon. The effect of particle size distribution on drug dissolution: A mathematical model for predicting dissolution and absorption of suspensions in the small intestine. Pharm Res 10:S170, 1992.

49. PF Ni, NFH Ho, JF Fox, H Leuenberger, WI Higuchi. Theoretical model studies of intestinal drug absorption. V. Nonsteady-state fluid flow and absorption. Int J Pharm 5:33–47, 1980.

50. NFH Ho, JY Park, PF Ni, WI Higuchi. Advanced quantitative and mechanistic approaches in interfacing gastrointestinal drug absorption studies in animals and humans. In: W Crouthanel, AC Sarapu, eds. Animal Models for Oral Drug Delivery in Man: In Situ and In Vivo Approaches. Washington DC: American Pharmaceutical Association, 1983, pp 27–106.

51. R Zipp, NFH Ho. Nonsteady state model of absorption of suspensions in the GI tract: Coupling multi-phase intestinal flow with blood level kinetics. Pharm Res 10: S210, 1993.

52. RJ Leipold. Description and simulation of a tubular, plug-flow model to predict the effect of bile sequestrants on human bile salt excretion. J Pharm Sci 84:670–672, 1995.

53. GE Amidon, NFH Ho, AB Frech, WI Higuchi. Predicted absorption rates with simultaneous bulk fluid flow in the intestinal tract. J Theor Biol 89:195–210, 1981.

54. NFH Ho, HP Merkle, WI Higuchi. Quantitative, mechanistic and physiologically realistic approach to the biopharmaceutical design of oral drug delivery systems. Drug Deliv Ind Pharm 9:1111–1184, 1983.

55. BC Goodacre, RJ Murry. Quantitative, mechanistic and physiologically realistic approach to the biopharmaceutical design of oral drug delivery systems. J Clin Hosp Pharm 6:117–133, 1981.

56. JB Dressman, D Fleisher, GL Amidon. Physicochemical model for dose-dependent drug absorption. J Pharm Sci 73:1274–1279, 1984.

57. JB Dressman, D Fleisher. Mixing-tank model for predicting dissolution rate control of oral absorption. J Pharm Sci 75:109–116, 1986.

58. RJ Hintz, KC Johnson. The effect of particle size distribution on the dissolution rate and oral absorption. Int J Pharm 51:9–17, 1989.

59. ATK Lu, ME Frisella, KC Johnson. Dissolution modeling: Factors affecting the dissolution rates of polydisperse powders. Pharm Res 10:1308–1314, 1993.

60. KC Johnson, AC Swindell. Guidance in the setting of drug particle size specifications to minimize variability in absorption. Pharm Res 13:1795–1798, 1996.

61. RL Oberle, GL Amidon. The influence of variable gastric emptying and intestinal transit rates on the plasma level curve of cimetidine: An explanation for the double peak phenomenon. J Pharmacok Biopharm 15:529–544, 1987.

62. GD Leesman, PJ Sinko, GL Amidon. Simulation of oral drug absorption: Gastric emptying and gastrointestinal motility. In: PG Welling, FLS Tse, eds. Pharmacokinetics. 2nd ed. New York: Marcel Dekker, 1989, pp 267–284.

63. P Luner, GL Amidon. Description and simulation of a multiple mixing tank model to predict the effect of bile sequestrants on bile salt excretion. J Pharm Sci 82:311–318, 1993.

64. LX Yu, JR Crison, GL Amidon. Compartmental transit and dispersion model analysis of small intestinal transit flow in humans. Int J Pharm 140:111–118, 1996.

65. LX Yu, GL Amidon. Characterization of small intestinal transit time distribution in humans. Int J Pharm 171:157–163, 1998.

66. LX Yu, JR Crison, GL Amidon. A strategic approach for predicting oral drug absorption in humans. Pharm Res 12:S8, 1995.

67. LX Yu, GL Amidon. Saturable small intestinal drug absorption in humans: Modeling and explanation of the cefatrizine data. Eur J Pharm Biopharm 45:199–203, 1998.

68. SY Choe, E Lipka, LX Yu, JR Crison, GL Amidon. Optimization of dosage form release parameters based on pharmacokinetic and gastrointestinal transit time considerations. Pharm Res 13:S292, 1996.

69. E Lipka, LX Yu, D Liu, JR Crison, GL Amidon. Evaluation of the intestinal permeability of β-blocker drugs and their potential for oral controlled release. Proc Int Symp Controlled Release Bioact Mater 22:366–367, 1995.

70. SS Davis, JG Hardy, JW Fara. Transit of pharmaceutical dosage forms through the small intestine. Gut 27:886–892, 1986.

71. NW Weisbrodt. Motility of the small intestine. In: LR Johnson, ed. Physiology of the Gastrointestinal Tract. New York: Raven Press, 1989, pp 631–664.

72. P Kerlin, A Zinsmeister, S Phillips. Relationship of motility to flow of contents in the human small intestine. Gastroenterology 82:701–706, 1982.

73. SH Kubo, RJ Cody. Clinical pharmacokinetics of the angiotension converting enzyme inhibitors. Clin Pharmacokinet 10:377–391, 1985.

74. LLB Ponto, RD Schoenwald. Furosemide: A pharmacokinetics/pharmacodynamics review. Clin Pharmacok 18:381–408, 1993.

75. DS Davies. Pharmacokinetics of terbutaline after oral administration. Eur J Resp Dis Suppl 65:111–117, 1984.

76. WD Mason, N Winer, G Kochak, I Cohen, R Bell. Kinetics and absolute bioavailability of atenolol. Clin Pharmacol Ther 25:408–415, 1979.

77. LZ Benet, S Øie, JB Schwartz. Design and optimization of dosage regiments; pharmacokinetic data. In: JG Hardman, LE Limbird, AG Gilman, eds. The Pharmacological Basis of Therapeutics. 9th ed. New York: McGraw-Hill; 1996, pp 1707–1793.

78. American Hospital Formulary Service. 1996 Edition, p 1982.

79. JW Paterson, ME Conolly, CT Dollery, A Hayes, RG Cooper. The pharmacodynamics and metabolism of propranolol in man. Pharmacol Clin 2:127–133, 1970.

80. FLS Tse, JM Jaffe, AJ Troendle. Pharmacokinetics of fluvastatin after single and multiple doses in normal volunteers. J Clin Pharmacol 32:630–638, 1992.

81. E Eichelbaum, HR Ochs, G Roberts, A Somogyi. Pharmacokinetics and metabolism of antipyrine after intravenous and oral administration. Arzneim-Forsch 32:575–578, 1982.

82. M Pfeffer, RC Gaver, J Ximenez. Human intravenous pharmacokinetics and absolute oral bioavailability of cefatrizine. Antimicrob Agents Chemother 24:915–920, 1983.

83. AM Metcalf, SF Phillips, AR Zinsmeister, RL MacCarty, RW Beart, BG Wolff. Simplified assessment of segmental colonic transit. Gastroenterology 92:40–47, 1987.

84. RC Gaver, G Deeb. Disposition of ^{14}C-cefatrazine in man. Drug Dis Metab Dispos 8:157–162, 1980.

85. H Lennernas, O Ahrenstedt, R Hallgren, L Knutson, M Ryde, LK Paalzow. Regional jejunal perfusion, a new in vivo approach to study oral drug absorption in man. Pharm Res 9:1243–1251, 1992.

86. H Lennernas. Human jejunal effective permeability and its correlation with preclinical drug absorption models. J Pharm Pharmacol 49:627–638, 1997.

11

Controlled Release Osmotic Drug Delivery Systems for Oral Applications

John R. Cardinal*
Oakmont Pharmaceuticals, North Wales, Pennsylvania

I. INTRODUCTION

One of the strengths of the drug delivery field is the number and variety of drug delivery systems available to the formulation scientist. Systems are available for delivery of drug to essentially every site and orifice for both humans [1,2] and animals [3,4]. Such systems can be classified in many ways, but one would be to subdivide the systems into those intended for immediate release of drug (<1 hr) such as solutions or conventional tablets and capsules, and those intended for long periods (>4 hr) such as coated multiparticulates or dispersions of drug in biodegradable matrices. For the latter, some means must be incorporated into the system to prolong the release of drug beyond what would arise from the normal dissolution characteristics of the solid drug particles in the dosage form. Almost invariably, this long-acting delivery is achieved through the use of polymeric materials that in some way affect the rate of diffusion of the drug to the surface of the delivery system.

While this control can be (and has been) built into the system in many ways, it is useful to first consider the characteristics and attributes desired for the final product. For systems intended for long-acting delivery of drug via the peroral route in humans, most products would require some or all of the following characteristics:

---

* Current affiliation: Applied Analytical Industries, Inc., Wilmington, North Carolina.

A definable duration of delivery

A predictable delivery rate independent of environmental factors such as viscosity, pH, water content, and stirring rate (i.e., GI motility)

Controllable release rates during the delivery period

Delivery characteristics independent of the properties of the drug (e.g., aqueous solubility)

Delivery of either a small or large dose of the drug at the same rate

Delivery of drug at the intended site of absorption irrespective of the region of the gastrointestinal tract where that process occurs

While no system has been designed that will meet all of these characteristics, the ones that come closest are those that utilize coatings of semipermeable polymers on solid cores leading to release rates that are wholly or largely controlled by transport of water across the coatings in response to an osmotic pressure gradient. The reasons for this success are due, in large part, to the fundamental nature of the processes involved.

In this chapter the fundamental characteristics of osmotic systems are explored. The basic transport equations are developed. Examples of the various types of systems available are discussed and related to these transport equations. The relative advantages for each approach are delineated. Also, various aspects of the manufacture of these systems are reviewed. It is important to emphasize that much of the relevant work in this area has been published only in the patent literature. An excellent review of this work for patents published through 1993 is available [5].

II. FUNDAMENTAL CONCEPTS

Most pharmaceutical scientists receive extensive training and are well versed in the fundamental concepts and applications of Fick's laws of diffusion as they relate to drug transport in fluid media and/or polymeric films. Indeed, such issues are the fundamental focus of most of this volume and are critical to the understanding of the rate of drug release from most controlled release drug delivery systems. It is important to recognize, however, that in osmotic systems it is the transport of *water*, not *drug*, that is the defining event. Drug release occurs only as a consequence of water absorption by the system. Thus, a review of the laws that describe water transport through semipermeable films is a reasonable place to begin.

In conventional analyses of transport based on Fick's laws, the fundamental parameters that define the transport process are the solute diffusion coefficient in the polymer film, D_M, and the partition coefficient, K_P. Essentially, the diffusion coefficient defines how fast a solute molecule moves, and the partition coefficient describes how many molecules can move. Taken together they define the total amount of solute transported in a given time (dM/dt) by equations such as the following, which describes solute transport across a polymer film:

$$\frac{dM}{dt} = \frac{AD_M K_P C_A}{h} \tag{1}$$

where C_A is the aqueous solubility of the solute, K_P is the partition coefficient of the solute between the film and the aqueous solution, A is the surface area, and h is the thickness of the film. Note that a fundamental assumption of this approach is that only solute is transported in response to the concentration gradient.

The transport of both solute and solvent can be described by an alternative approach that is based on the laws of irreversible thermodynamics. The fundamental concepts and equations for biological systems were described by Kedem and Katchalsky [6] and those for artificial membranes by Ginsburg and Katchalsky [7]. In this approach the transport process is defined in terms of three phenomenological coefficients, namely, the filtration coefficient L_P, the reflection coefficient σ, and the solute permeability coefficient ω.

Experimentally, these coefficients can be determined from transport experiments designed to measure the total volume flow per unit area, J_V, and the total solute flow per unit area, J_S, through a polymer film. The latter value can be obtained by using conventional diffusion cells wherein a film is placed between two well-stirred solutions and the change in concentration is determined over time. The value of J_V is determined from measurements of the volume flow across the film using an appropriately designed cell. One example used previously by Song et al. [8] is given by Fritzinger et al. [9]. This cell is designed such that a hydrostatic pressure can be applied to one side of the diffusion cell by placing a volume of fluid at an elevated height above the cell, which in turn is connected to the cell via flexible tubing. The second cell is equipped with a calibrated capillary. The volume of solvent transported is measured as an increase in volume in the capillary via an instrument such as a cathetometer. For the case where a hydrostatic pressure ΔP is applied to the compartment of lower concentration, the value of J_v is obtained from

$$J_v = \frac{A}{h} L_P (\sigma \Delta \pi - \Delta P) \tag{2}$$

and

$$J_s = \Delta c_s (1 - \sigma) J_v + \omega \Delta \pi \tag{3}$$

where $\Delta \pi$ is the osmotic pressure difference across the film and Δc_s is the concentration gradient across the membrane. Rearrangement and substitution of the van't Hoff law leads to equations that define L_P, σ, and ω. For the case where equal solute concentrations exist on both sides of the film, L_P can be defined as

$$L_P = -\left(\frac{hJ_v}{A\,\Delta P}\right)_{\Delta\pi=0} \tag{4}$$

From Eq. (2), σ can be defined as

$$\sigma = \frac{(hJ_v - L_P\,\Delta PA)}{AL_P\,\Delta\pi} \tag{5}$$

and when $\Delta P = 0$, $J_s = AC_aD_mK_P/h$ and $J_v = \sigma\,\Delta\,\pi L_P A/h$. Substitution of these values into Eq. (5) gives the following equation for ω:

$$\omega = \frac{AD_mK_P}{hRT} + \frac{A}{h}\Delta c_s(1 - \sigma)\sigma L_P \tag{6}$$

From the above it can be seen that L_P provides a direct measure of the volume flow through the film. The reflection coefficient σ is a measure of the degree to which the membrane is semipermeable. It has the value of 1.0 for a perfectly semipermeable film and 0 for a film which does not differentiate between solvent and solute. Some representative values of σ, ω, and L_P for urea, glucose, and sucrose in polyhydroxyethyl methacrylate (pHEMA) and several cellulosic membranes are shown in Table 1 [8]. Several points are of interest. Note the significant decrease in σ with the decrease in solute size for the pHEMA type films whereas the cellulosic films are much less discriminating. The data also help make the point that a membrane that is essentially semipermeable for a large solute like sucrose can be nondiscriminating for a solute such as urea. The solute permeability coefficients are strongly dependent on solute molecular weight, but the values of L_P are only mildly so.

For osmotic drug delivery systems, Eq. (2) is of critical importance. This equation demonstrates that the quantity of water that can pass a semipermeable film is directly proportional to the pressure differential across the film as measured by the difference between the hydrostatic and osmotic pressures. Osmotic delivery systems are generally composed of a solid core formulation coated with a semipermeable film. Included in the core formulation is a quantity of material capable of generating an osmotic pressure differential across the film. When placed in an aqueous environment, water is transported across the film. This transported water in turn builds up a hydrostatic pressure within the device which leads to expulsion of the core material through a suitably placed exit port.

Table 1 Phenomenological Coefficients and Partition Coefficients for Urea, Glucose, and Sucrose in Various Membranes

Membrane	Solute	$L_p h$ [cm⁴/(dyn · sec)]	σ	ωh $\left(\dfrac{mol \cdot cm}{dyn \cdot sec}\right)$	K_P
pHEMA	Urea	8.56×10^{-15}	0.071	3.09×10^{-17}	0.53
	Glucose	8.75×10^{-15}	0.752	1.87×10^{-18}	0.23
	Sucrose	5.95×10^{-15}	0.940	4.63×10^{-19}	0.23
pHEMA + 1 mol %	Urea	6.34×10^{-15}	0.063	2.48×10^{-17}	0.48
EGDMA	Glucose	3.05×10^{-15}		1.52×10^{-18}	0.24
	Sucrose			3.58×10^{-19}	0.25
Cupraphane	Urea	1.08×10^{-13}	0.026	1.07×10^{-16}	0.54
	Glucose	9.09×10^{-14}	0.088	2.41×10^{-17}	0.29
	Sucrose	7.74×10^{-14}	0.174	1.72×10^{-17}	0.30
Cellophane	Urea				
	Glucose	1.70×10^{-13}	0.088	3.39×10^{-17}	0.70
	Sucrose	1.62×10^{-13}	0.105	2.12×10^{-17}	0.73
Wet gel	Urea	8.21×10^{-13}	0.002	2.66×10^{-16}	
	Glucose	8.21×10^{-13}	0.024	9.51×10^{-17}	
	Sucrose	8.21×10^{-13}	0.036	6.44×10^{-17}	

Source: Ref. 8.

III. OSMOTIC DRUG DELIVERY SYSTEMS

A. One-Compartment Systems

From a historical and practical perspective, the most important event in the development of osmotic system technologies was the introduction of the elementary osmotic pump technology by Felix Theeuwes of the ALZA Corporation in the mid-1970s [10]. This system was not the first, as several other systems had been described previously including those of Rose and Nelson [11] and Polli et al. [12] and several earlier designs at ALZA commonly called the Higuchi–Leeper systems [13–15]. Of these only the Theeuwes system was commercialized for oral drug delivery. It was given the trade name OROS, which is an acronym for oral osmotic system.

The basic design of an OROS is shown in Figure 1. The system is composed of a tablet containing the drug plus excipients. The tablet is coated with a semipermeable film, usually cellulose acetate (CA). A hole is drilled through the coating using a laser beam. The dimensions of the hole are not critical provided that it is large enough to permit efflux of the imbibed solvent without back pressure

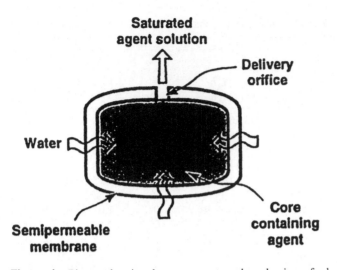

Figure 1   Picture showing the components and mechanism of release for the elementary osmotic pump. (From Ref. 3.)

and small enough to ensure that conventional drug diffusion through the port is not significant. The coating must be of sufficient thickness to ensure that the mechanical integrity of the film is maintained. However, increases in the film thickness to provide sufficient mechanical strength will lead to decreasing release rates. This is clearly a limiting factor with respect to the total dose that can be delivered over a given time period.

Figures 2 and 3, taken from the work of Theeuwes [10], demonstrate several of the unique features of osmotic systems. Figure 2 shows the average delivery rate over time with and without stirring of the dissolution medium. Several points are noteworthy: (1) The tight error bars demonstrate the high reproducibility from tablet to tablet within a given lot; (2) after the initial start-up, the release rate is essentially constant over time; and (3) the rate of stirring has essentially no effect on the release rate. It should be noted that the constancy of release over time is maintained only during the period that solid drug particles remain within the device. Once solid drug disappears, the release pattern becomes first-order [10]. Figure 3 shows the same in vitro data compared with release rates for tablets removed from dogs at various time intervals. The in vitro–in vivo correlation is excellent; this feature is becoming critically important with the increasing role of such correlations in the regulatory review of controlled release delivery systems.

The release rate of drug from these devices is obtained by suitable modifi-

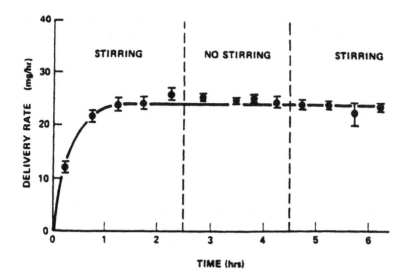

Figure 2 In vitro release of KCl from the elementary osmotic pump with and without stirring of the dissolution medium. The brackets represent the range of the data from five systems. (From Ref. 10.)

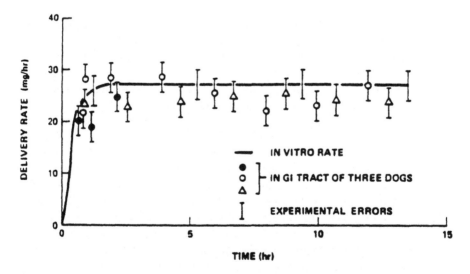

Figure 3 In vitro and in vivo release of KCl from the elementary osmotic pump. The devices were removed from the Gl tract of dogs at the time indicated. (From Ref. 10.)

cation of Eq. (2) [10]. If it is assumed that ΔP is small relative to the osmotic pressure gradient and that

$$\frac{dm}{dt} = \frac{dV}{dt}C_s \tag{7}$$

where m is the mass of drug released and C_s is the saturation solubility of the drug in the tablet, then Eq. (2) reduces to

$$\frac{dm}{dt} = \frac{A}{dt}L_p\sigma\pi_sC_s \tag{8}$$

where π_s is the osmotic pressure of core formulation at saturation. This equation is valid provided that solid drug particles remain within the tablet.

Equation (8) contains the key information required to understand the functionality of osmotic systems. Therefore, an examination of each of the key variables will provide the basis for an understanding of osmotic systems in general.

A, the surface area of the system, is essentially a dependent variable whose value is set by the choices made in the formulation of the core. The value of h, the membrane thickness, is a critical parameter in many ways. The lower limit on h is set by the fact that the membrane coating must have sufficient mechanical integrity to maintain the film as an intact barrier throughout its journey down the GI tract. Zentner et al. [16] measured the mechanical integrity of the applied films using an Instron tensile tester and showed that films made with CA containing varying levels of porosigens have sufficient mechanical integrity to withstand the typical pressures found within the gastrointestinal tract [17]. The wall thicknesses they evaluated were 0.12 mm or greater. It is important to recognize that the system geometry will play a critical role in this regard since the weak points in the films will generally be found at points such as the tablet edges that are subject to abrasion during the coating operation.

For osmotic systems the values of L_P and σ are generally not measured independently. Rather the values are combined as a new constant k, which is termed the fluid permeability of the membrane [18]. Its value for cellulose acetate (CA), calculated from the work of Theeuwes [10], is about 7.7×10^{-16} cm$^3 \cdot$ sec/g. Zentner et al. [16] measured $L_P\sigma$ for two CA polymers containing sorbitol as a porosigen and with and without diethylphthalate as a plasticizer. The results are shown in Figure 4. The value for dense CA taken from Ref. 10 is indicated by an arrow. Note that the incorporation of porosigens and/or the use of a CA with a higher hydroxyl content increases the value of $L_P\sigma$ whereas the incorporation of both a plasticizer and a porosigen decreases the value back toward that found with the dense CA film alone. The value of $L_P\sigma$ and hence the release rate can also be varied by changes in the polymer per se.

The value of the osmotic pressure π_s, as defined in Eq. (8), is a function

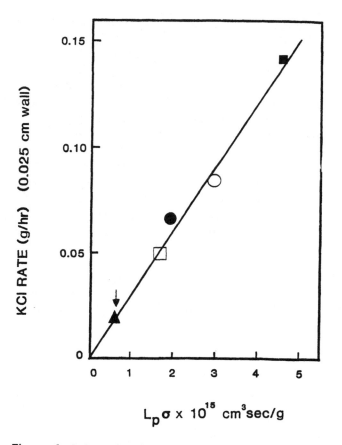

Figure 4 Release rate for KCl plotted versus the fluid permeability of the membrane. (○) CA-398-10 + 50% sorbitol; (■) CA-320S:398-30 (50:50) + 50% sorbitol; (●, □, ▲) Same as (■) but contains 5, 10, 25% diethylphthalate respectively. (From Ref 16.)

of the composition of the core formulation. Its maximum value is on the order of several hundred atmospheres for highly soluble, low molecular weight solutes such as NaCl or KCl [19]. Furthermore, its value will be set by the sum of the contributions of the individual components of the core including both excipients and the drug. This point is critical in the design of the core formulation for an osmotic system. Clearly the overall release rate from the system can be maximized through the incorporation of highly soluble, low molecular weight excipients. However, excipients and drug will be dispensed from the exit port at rates consistent with their relative solubilities. Therefore, the relative composition of the drug and excipient in the core must also be consistent with their relative solubilities in water.

An example of the effects of the osmotic pressure of the core formulation on the overall release rate from an osmotic tablet, taken from the work of Herbig et al. [20], is shown in Figure 5. For this work, various core formulations containing 1 wt% doxazosin were prepared such that the total osmotic pressure of the core increased while maintaining the same solubility of the drug (27 mg/mL). The osmotic pressure was varied through the incorporation of organic acids of varying solubility. From Figure 5 it is apparent that as the osmotic pressure of the core formulation increases, the total release rate increases proportionately.

The value of C_s is the most critical parameter in determining the overall release rate from a given osmotic system. Indeed, its value will determine whether or not it is feasible to utilize an osmotic system to deliver a particular drug for a specified duration. The maximum release rate achievable is likely that seen with KCl. The relevant values for the parameters in Eq. (6) for OROS [10] are as follows: $A = 2.2$ cm^2, $h = 0.025$ cm, $L_p\sigma = 2.8 \times 10^{-6}$ cm^2/(atm · hr), $\pi_s = 245$ atm, and $C_s = 330$ mg/mL. This translates to about 20 mg/hr or about 250 mg over a 24 hr period. This is for a highly water soluble drug with a high osmotic pressure differential. For drugs of moderate solubility—for example,

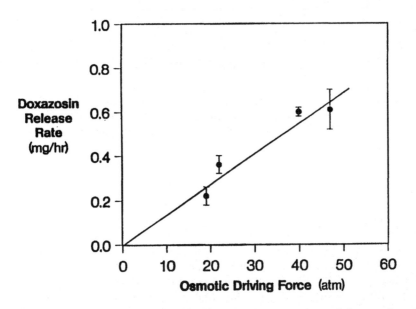

Figure 5 Doxazosin release rate versus the osmotic pressure of the core formulation. (From Ref. 20.)

those with solubilities in the range of 1–10 mg/mL—the maximum dose that can be delivered over 24 hr will be less than 20 mg [5].

For this reason much work has been done at the ALZA Corporation and elsewhere to increase the water permeation rates by various technologies. For example, ALZA scientists utilized a composite membrane in the development of their first commercial product with this technology [21,22]. In this system they first applied a CA membrane containing a high concentration of porosigens. A second dense membrane containing only CA was added. In this way the overall fluid permeability was increased, since the thickness of the dense portion of the film could be proportionately reduced.

At Merck, Zentner and his colleagues [16,23–26] developed a new osmotic system, termed the controlled porosity osmotic pump, that utilized more permeable CA films based on the incorporation of porosigens in the polymer film. As noted above, the value of $L_P\sigma$ increases significantly with the incorporation of the porosigen. This change also leads to several other changes in the performance of their systems relative to OROS. First, because of the porous nature of these films, it is not necessary to incorporate the exit port as part of the manufacturing process. Rather, exit passageways develop spontaneously during the release process. Second, also because of the use of porosigens, the films are not strictly defined as semipermeable. Because of this a significant contribution to the total transport arises from conventional diffusive transport processes. This is seen, for example, in their plot of the release rate for KCl versus the osmotic pressure difference for systems made from CA containing 50 wt% sorbitol (Fig. 6). The release rate at zero total osmotic pressure difference is not zero. This transport arises due to conventional solute transport across the porous membrane coating. A detailed analysis of the relative contributions of diffusive and osmotic transport for their systems is provided by Zentner et al. [16].

As noted above, the value of $L_P\sigma$ increases significantly with the incorporation of porosigens into the polymer film. This leads to significant increases in the total solute that can be delivered with this system. From the data in Figure 3 it is apparent that the release rate increases by at least a factor of 5 through the incorporation of the porosigen. Thus the minimum total solute that can be delivered over 24 hr increases to at least 100 mg for moderately soluble solutes.

Another approach to increasing the permeability of the films was taken by scientists at Pfizer [20,27–32]. Their approach was to utilize asymmetrical membranes that surround a core formulation containing drug and excipient. These types of films are commonly used in applications such as reverse osmosis where very high solvent flows are required for various separation processes. Figure 7 is a picture of an asymmetrical membrane made with CA taken from Herbig et al. [20]. It shows the typical structure for this type of film where it can be seen that there is a relatively thin dense region at the surface supported by a much thicker porous substructure. This structure provides strength and yet is highly

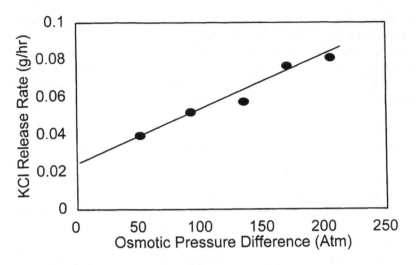

Figure 6 KCl release rate versus the osmotic pressure difference between the core formulation and the release medium. The osmotic pressure of the release medium was varied by addition of urea. (Redrawn from data in Ref. 16.)

permeable to solvent. The membranes are prepared by either a "wet" or a "dry" process. In both cases the film-forming polymer is dissolved in a solvent containing both good and poor solvents for the polymer. Upon exposure to either air for the dry process or a poor solvent for the polymer in the wet process, the polymer precipitates at the surface, creating the thin skin seen in the picture. At the same time, evaporation of the more volatile good solvent enhances the relative concentration of the poor solvent within the bulk solution, which in turn leads to phase separation of the polymer, creating the porous substructure seen within the inner regions of the film.

Asymmetrical membranes are highly permeable to solvent relative to dense films of the same thickness and total weight. The relative effects on the release of solute are shown in Figure 8 for osmotic tablets prepared with the same core formulation containing trimazosin but coated with either a dense membrane or an asymmetrical membrane. An exit hole was drilled in the dense membrane prior to the release studies, but not in the asymmetrical membrane coatings. In the latter systems, the local buildup of hydrostatic pressure leads to rupture of the thin dense film at one or more locations in the coating, obviating the need for the hole. The release rate for the asymmetrical membrane–coated tablet is about seven times that seen with the dense film. Furthermore, as can be seen in Figure 9, the release rate from devices with asymmetrical membrane coatings decreases only slightly as the overall thickness increases. These results arise be-

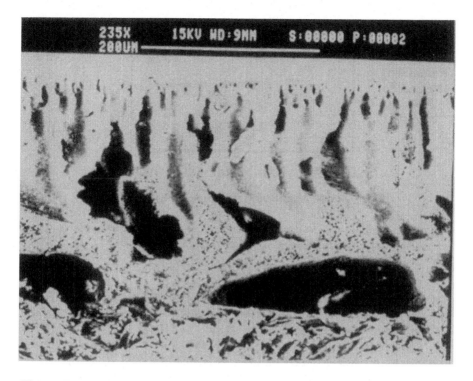

Figure 7 Picture of an asymmetrical membrane coating layered onto a tablet core. (From Ref. 20.)

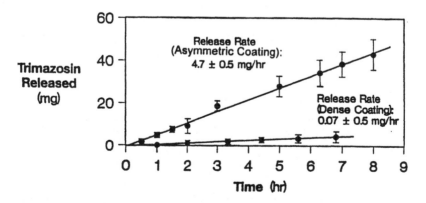

Figure 8 The release of trimazosin versus time for a tablet coated with either a dense or an asymmetrical membrane. (From Ref. 20.)

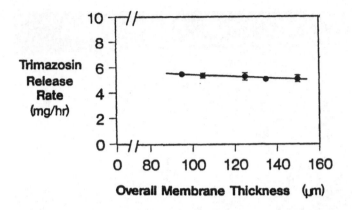

Figure 9   The trimazozin release rate plotted versus the overall membrane thickness for an asymmetrical membrane coating on tablets. (From Ref. 20.)

cause the limiting barrier for transport is the thin dense region at the surface of the device, the thickness of which does not increase with overall increases in the full thickness of the coating.

As with the controlled porosity osmotic pump, the asymmetrical membrane coatings are not strictly semipermeable, as is seen in Figure 10 for the release of trimazosin from tablets with asymmetrical membrane coatings. In this figure the release of trimazosin is plotted as a function of time and as a function of the composition of the receptor solution. At early times the receptor solution was 2.4 wt% $MgSO_4$. The osmotic pressure of this solution is about 6 atm compared with the total osmotic pressure of the core formulation, which is about 3 atm. Under this condition, a small but detectable rate is seen. This rate is attributable to conventional diffusive transport. When the receptor solution is changed to water, the release rate increases significantly, showing the importance of the osmotic pressure differential to the total release from this tablet.

The thin film at the surface can be made to be continuous or porous depending on the composition of the coating solvent and, in particular, on the level of pore former added to the coating solutions [20,27]. As noted earlier, pore formers are hydrophilic solutes added to the coating solvent which are poor solvents for the polymer and enhance the overall porosity of the resultant films. Depending on the levels of the pore formers, the thin surface films may be intact or perforated. For example, with glycerol as the pore former, the surface film obtained with concentrations of less than about 1% are imperforate, but those made at 5% or greater contain clear evidence of pores as viewed with a scanning electron microscope. Again depending on the solute, conventional diffusion processes will contribute to the total transport. This is shown in Figure 11 for the

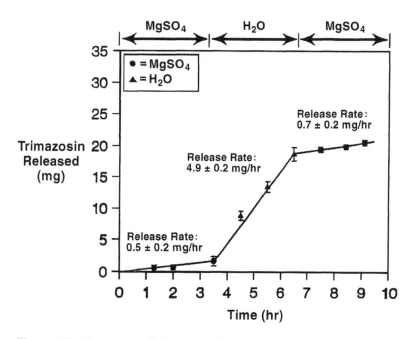

Figure 10 The amount of trimazosin released versus time with changes of the release medium from 2.4 wt% $MgSO_4$ at the times indicated. (From Ref. 20.)

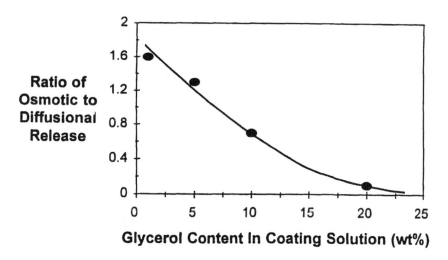

Figure 11 Ratio of osmotic to diffusional release rates plotted versus the glycerol content in the coating solution. (From Ref. 20.)

release rates of trimazosin from tablets made with asymmetrical films of CA made from a coating solvent containing varying levels of glycerol as the pore former. In this figure the ratio of osmotic to diffusive contributions is plotted versus the level of glycerol in the coating solvent. As above, the ratio was obtained from the relative release rates for trimazosin into water and 2.4% $MgSO_4$. A secondary aspect of the relative release rates obtained via diffusional or osmotic processes for these systems is seen in Table 2, where the release rates obtained with either water or 2.4 wt% $MgSO_4$ are listed. As the concentration of the pore former increases, the release rate increases up to about 10 wt% glycerol. However, at 20 wt% glycerol the total release rate is lower and is the same value in both release media. Thus for this solute the release that can be achieved from the system operating via an osmotic mechanism is significantly greater than that found with the diffusive mechanism through a highly porous film.

In addition to coating tablets, asymmetrical membrane systems can also be prepared as capsules or multiparticulates.

As described in the patent literature [27,28] and by Thombre et al. [30,31], osmotic systems can also be made in the form of a capsule shell. These systems are manufactured in a manner analogous to conventional manufacturing processes for gelatin capsules with the exceptions that CA is used as the polymer and the coating solution is dipped into a quench bath to precipitate the polymer to form the asymmetrical membrane coating. The resultant shell has the same general characteristics and utility as conventional gelatin capsules, but now it can be used as a generic system for the controlled release of drugs. The shells can be filled on conventional capsule-filling equipment. The core formulations are composed of drug with or without excipient. As with the asymmetrical membrane tablets, it is not necessary to incorporate an exit port in the systems prior to release studies. Because of the generic nature and simplicity of their manufacturing process, the shells can be used to test various formulations both in vitro and in vivo with limited quantities of bulk drug and with limited formulation work. This greatly simplifies the effort required to test the feasibility of controlled release

Table 2 Release of Trimazosin into Water and 2.4% $MgSO_4$

Tablet coating	Release rate (mg/hr) into 2.4 wt% $MgSO_4$	Release rate (mg/hr) into water
15 wt% CA; 1 wt% glycerol; 84 wt% acetone	2.41 ± 0.43	6.30 ± 0.27
15 wt% CA; 10 wt% glycerol; 75 wt% acetone	4.52 ± 0.54	7.65 ± 1.05
15 wt% CA; 20 wt% glycerol; 65 wt% acetone	3.03 ± 2.22	3.39 ± 0.35

Source: Ref. 27.

formulations for new chemical entities in the development setting, where the quantities of bulk drug available for development work are often severely limited.

Asymmetrical membrane capsules were prepared, for example, by dip coating mandrels with solutions containing 15 wt% CA and 33 wt% ethanol in acetone. After the mandrels were withdrawn from the coating solution they were immersed in water to precipitate the polymer and create the asymmetrical membrane capsule shell. This process was used to create both the body and the cap for the capsules. The capsules were sealed at the juncture between the cap and body by banding with a solution of CA in acetone.

Figure 12 [28] is a plot of the release rate for doxazosin versus the osmotic pressure of the receptor solution. As with tablets, this plot shows that the release from these capsules occurs via an osmotic mechanism, i.e., the release rate decreases with an increase in the osmotic pressure of the receptor solution. The interesting aspect of these results, however, is the fact that the solubility of doxazosin in the two high osmotic pressure buffers (10 mg/mL) is significantly greater than that of doxazosin in the gastric buffer (250 ppm). Thus in these systems the release rate decreases as the solubility of the drug in the release medium increases. This is another example of how osmotic pressure driven systems differ from classical diffusion-based systems where the concentration gradient, and hence

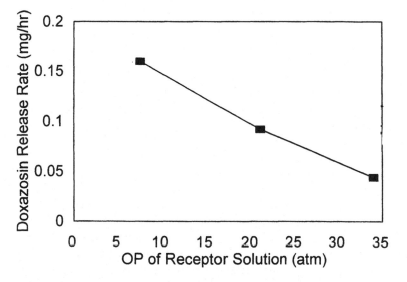

Figure 12 Release of doxazosin from asymmetrical capsule shells versus the osmotic pressure of the receptor solution (gastric buffer, 7.5 atm, and dextrose solutions, 21 and 34 atm). (From Ref. 28.)

the solubility of drug in the receptor solution, is a critical determinant of the overall release rate.

Osmotic systems can also be prepared from multiparticulates by using the asymmetrical membrane coating process. As described by Korsmeyer et al. [30], a number of processes were evaluated to manufacture such systems, and the process is not straightforward because of the complexities of handling these solvent systems where the polymer is present at concentrations near saturation. The simplest and most efficient process was one in which the polymeric solution and multiparticulate beads containing drug are separately forced through a nozzle where mixing of the beads and coating solution occurs only as the mixture exits the nozzle. In this way the contact time between solvent and bead is minimized. An example of the release rates seen for doxazosin from multiparticulate beads with one to three coats of an asymmetrical membrane is shown in Figure 13 [28].

As noted earlier, perhaps the key determinant of the overall release characteristics from an osmotic pump system is the solubility of the drug. If the solubility of the drug is too low, then the total dose that can be delivered is limited. If the solubility is too high, then the percent of the total dose that can be delivered at a zero-order rate can be relatively small. For example, as shown by Theeuwes [10], the fraction of the total dose (F) that can be delivered at a zero-order rate is given by

$$F = (1 - C_s/\rho) \qquad (9)$$

where ρ is the density of the core tablet. Table 3 shows some results of this calculation for the case of diltiazem HCl taken from McClelland et al. [24]. This drug has an aqueous solubility that is very high (>590 mg/mL). Its solubility can be decreased through the common ion effect. Table 3 shows the approximate solubility of diltiazem as a function of the concentration of NaCl in the medium and the projected percent released at zero order calculated according to Eq. (9). Obviously, the percent delivered at zero order increases as the solubility decreases, but the release rate also decreases.

Zentner and coworkers [24,26] utilized this information in their development of a system that releases this drug over a 24 hr period. The use of NaCl to modulate the release of diltiazem presents an interesting problem in that the concentration of the solubility modifier must be maintained within certain limits and below its saturation solubility within the device. To solve this problem, core formulations were developed that contained both free and encapsulated NaCl. The encapsulated NaCl was prepared by placing a microporous coating of cellulose acetate butyrate containing 20 wt% sorbitol onto sieved NaCl crystals. The coated granules released NaCl over 12–14 hr period via an osmotic mechanism into either water or the core tablet formulation. The in vitro release profile for tablets (core I devices) containing 360 mg of diltiazem HCl and 100 mg of NaCl equally divided between the immediate release and controlled release fractions

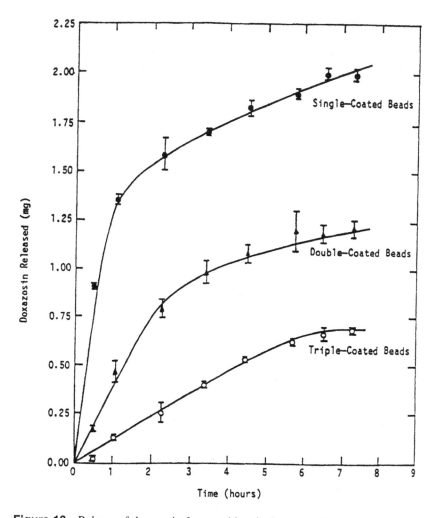

Figure 13 Release of doxazosin from multiparticulates coated with one to three coats of an asymmetrical membrane. (From Ref. 28.)

is shown in Figure 14. With this formulation, greater than 70% of the drug is released at a zero-order rate. Figure 15 shows the results for a similar formulation containing a total of 90 mg of the NaCl mix but only 72 mg of drug (core II). Clearly, the relative concentration of NaCl within this device would be greater throughout the duration of delivery than with the device containing 360 mg of drug. This is verified by the fact that the percent released at zero order is signifi-

Table 3 Diltiazem HCl Solubility in Sodium Chloride Solutions

Diltiazem HCl solubility (mg/mL)	NaCl concentration (M)	Theoretical profile (% zero-order)
>590	0.00	51
545	0.25	55
395	0.50	67
278	0.75	77
155	1.00	87
40	1.20	99

Source: Ref. 24.

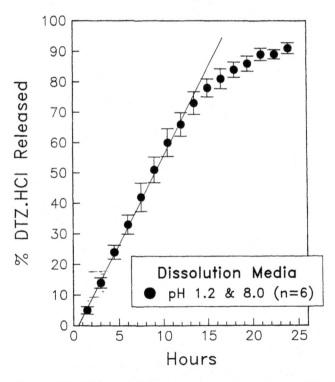

Figure 14 Diltiazem HCl release versus time from core I devices containing 100 mL of NaCl 360 mg of diltiazem HCl with dissolution media of pH 1.2 and 8.0. The data from the separate media are superimposable. (From Ref. 24.)

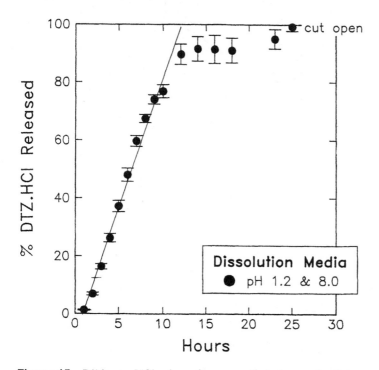

Figure 15   Diltiazem HCl release from core II devices under the same conditions as Figure 14. The tablets contain 72 mg of diltiazem HCl and 90 mg of a 50:50 mix of immediate and controlled release NaCl. (From Ref. 24.)

cantly greater than that seen with the core I device. While the results are consistent with expected theories, the overall mechanism of release from these tablets is undoubtedly very complicated given the multiple components of this formulation and their relative concentrations in the fluids that evolve from the device during their delivery.

As noted earlier, osmotic systems have been shown to provide good in vitro–in vivo correlations between the observed release rates. This has been shown explicitly for the core I devices described above [33]. The data are shown in Figure 16, where the in vitro data are plotted along with the release curves from six devices administered to dogs. The animals were in the fed state at the time of administration and maintained that way throughout the duration of the experiment via the administration of ~50 g of dog chow before device administration and every hour thereafter. The individual release curves shown in Figure 16 were obtained by numerical deconvolution of the plasma data with an oral solution dose given to the same dog. Clearly the in vivo and in vitro data are

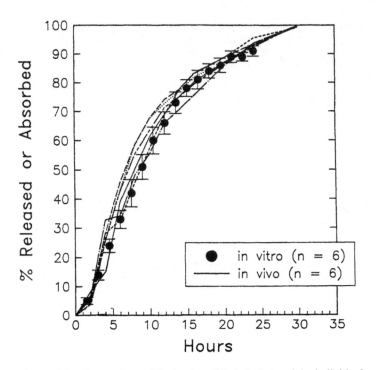

Figure 16 Comparison of the in vitro (filled circles) and the individual percent absorbed (curves) following the administration of core I devices to dogs. (From Ref. 24.)

superimposable, showing the remarkable correlation obtained between the performance of these types of devices in these widely divergent environments.

Another interesting example of the use of the common ion effect to modulate the release of drug from the elementary osmotic pump is that summarized in a patent by Magruder et al. [34] for the delivery of salbutamol. This drug has a high solubility in water (270 mg/mL), which decreases to only about 11 mg/mL in saturated solutions of NaCl. In their formulation sufficient NaCl is added to control the release of salbutomal at the rate dictated by the presence of saturated solutions of NaCl. The quantity of NaCl present is not sufficient, however, to maintain saturation of the salt throughout the duration of delivery of the drug. Upon depletion of the salt, the drug release rate will initially decline due to a decrease in the osmotic pressure. However, upon further depletion of the salt, the solubility of the salbutomal increases dramatically, leading to a dramatic increase in its release rate. In effect the system provides constant release followed by a dose dumping. This type of delivery profile is desirable to control nocturnal asthma.

Numerous other examples of the use of solubility to control the delivery profile of drugs from the elementary osmotic pump can be found in the literature, especially the patent literature [35–40]. These systems apply to drugs of moderate to high aqueous solubility where either the excipient or the salt form of the drug was used to control the drug solubility within the core formulation.

B. Two-Compartment Systems

As noted in the previous section, one of the major drawbacks of OROS and the other elementary osmotic pump designs reviewed is the limitation in their ability to deliver drugs which are of low solubility in water. As noted earlier, the controlled porosity and the asymmetrical membrane systems expand this range somewhat over OROS because of the increased permeability of their membrane coatings to water. Nevertheless there was a crying need for a system that could be manufactured using conventional pharmaceutical manufacturing equipment that could deliver water-insoluble compounds at controlled rates. This goal was the subject of numerous patents issued by the ALZA Corporation beginning in the late 1970s [41–45]. The focus of this work was to identify a system that could effectively deliver a dispersion of solid drug particles using the fundamental principles of osmotic pressure as outlined above for the OROS system. The primary problem was the development of a composition that would have sufficient osmotic pressure to draw in water through a semipermeable membrane and sufficient fluidity to flow out through the orifice of the tablet without significant change in the relative composition of the ingredients in the formulation.

An elegant solution to this problem was ultimately developed by Theeuwes and coworkers [46,47]. This system is termed the GITS (gastrointestinal therapeutic system) or the "push–pull" system. The system is composed of a bilayer tablet coated with a semipermeable membrane. An orifice to serve as the drug release port is placed with a laser beam in the membrane such that the upper layer of the tablet is contiguous with the exit port. The upper layer of the tablet is composed of drug plus hydrogel plus an osmotic agent such as NaCl. The lower tablet layer is also composed of a hydrogel. In operation, water is imbibed through the semipermeable membrane into both tablet layers. The imbibed water combines with the hydrogels to create a gel with sufficient fluidity to flow from the device through the exit port.

The equations required to describe the release rate from this system are fundamentally the same as those utilized for the elementary osmotic pump. The basic equation is

$$\frac{dM}{dt} = (Q + F)F_D C_0 \tag{10}$$

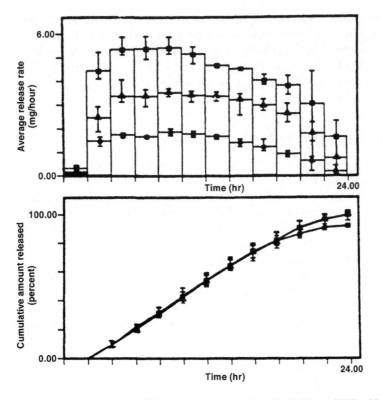

Figure 17 Comparison of in vitro release profiles of nifedipine GITS tablets containing 30 (●; $n = 10$), 60 (△; $n = 10$), or 90 (■; $n = 5$) mg of drug. The error bars indicate the range of data. (From Ref. 47.)

where C_0 is the concentration of solids, F_D is the fraction of drug in the upper layer, and Q and F are defined as follows:

$$Q = \frac{L_P \sigma}{h} A_P(H) \pi_P(H) \tag{11}$$

$$F = \frac{L_P \sigma}{h} [A - A_P(H)] \pi_D(H) \tag{12}$$

where A_P is the membrane area of the lower or "push" compartment, π_P and π_D are the osmotic pressures of the push and drug compartments, respectively, and the other symbols are as previously defined. The parenthetical H is added to emphasize that the associated values are a function of the degree of hydration of the core components. While the equations are somewhat more complex, the fundamental concepts are the same as with the elementary osmotic pump; that is, the systems imbibe water at a rate that is dependent on the osmotic pressure difference and the fluid permeability coefficient of the coating. The total drug release rate is a function of the concentration of the drug in the solid dispersion that exits from the device.

The manufacture of this system utilizes essentially the same process as the one inherent in the manufacture of OROS. The complicating factors are that a bilayer tablet press is required and the tablet must be oriented during the laser drilling operation to ensure that the drug-containing layer is directed toward the laser beam.

The performance of this system is shown in Figure 17 for the release of nifedipine from the GITS system [47]. The reproducibility of the release rates is remarkable. Also note that the fractions released over time from three separate doses are basically superimposable. It should also be noted that these systems have an inherent delay in the onset of drug delivery which arises from the time required to build up a sufficient hydrostatic pressure to permit release of the gel that is formed within the tablet during delivery. Figure 18 shows the comparison of the in vitro and in vivo cumulative fraction released for the 30 mg system. Clearly, the in vitro performance is mirrored in the in vivo data.

This system is the only osmotic system developed commercially at this time that is suitable for the oral administration of insoluble drugs to humans. It has been utilized in the development of several other drugs including isradipine, doxazosin, diltiazem, contraceptive steroids, glipizide, and verapamil [48–53]. The system has also been utilized to codeliver the free bases of compounds normally administered as water-soluble salts such as pseudoephedrine and bromopheniramine [54]. The latter system includes both a loading dose and a controlled release dose and is intended for applications in the over-the-counter market.

More recently this push–pull concept has been reconfigured to permit the delivery of liquids. In these systems [55,56], drug as a hydrophobic neat liquid such as vitamin E or drug dispersed in a hydrophobic oil is placed within a gelatin capsule. This capsule is coated first with an osmopolymer such as carbopol or sodium carboxymethyl cellulose and then coated with a semipermeable film such as CA with or without porosigens. An exit port is drilled to be contiguous with the oil compartment. In use, water is drawn in through the semipermeable film to create a pressure which forces the oil from the device through the exit port. This system is likely to be of particular value in delivery of highly insoluble drugs.

Using fundamentally the same concepts as those embodied in the push–pull system, ALZA scientists along with those at Merck have commercialized

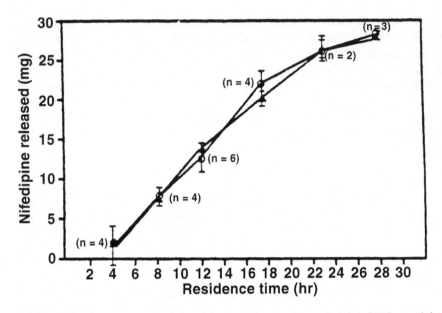

Figure 18 Comparison of the in vitro and in vivo release for nifedipine GITS containing 30 mg ($\bigcirc$ = in vivo; n as indicated); ($\triangle$ = in vitro release; n = 4). Error bars indicate the standard deviation of the data. (From Ref. 47.)

an osmotic technology for the delivery of drugs to grazing animals. The system is known as the IVOMEC SR bolus [57–62] and is intended for the very long term delivery of ivermectin to ruminants. The goal is to provide control of a variety of both ecto- and endocytes for the full grazing season. The system has also been developed by ALZA for the delivery of trace elements such as selenium to grazing cattle [63].

This system is shown in Figure 19. It is composed of a rigid membrane cup made from CA containing plasticizers. At the bottom of the cup is placed an osmotic tablet containing a hydrogel and salt. Placed above this is a partition layer made from a high melting paraffin wax. The drug-containing layer is placed above this. It is composed of drug plus a lower melting wax which softens in the range of 35–40°C. The wax layer is maintained at a viscosity which is low enough to permit flow under the generated osmotic pressure but high enough to ensure that the drug in suspension does not settle during the very long delivery period (135 days) for this product. Placed above the drug layer is a densifier which is incorporated to prevent regurgitation of the device following administration to the rumen of cattle. An exit port is placed at the center of this densifier to permit delivery of material through the densifier. Finally an exit port screen is placed

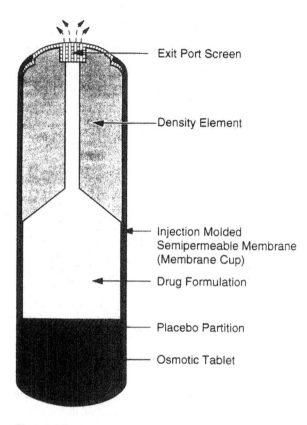

Exit Port Screen

Density Element

Injection Molded
Semipermeable Membrane
(Membrane Cup)

Drug Formulation

Placebo Partition

Osmotic Tablet

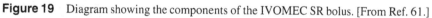

Figure 19 Diagram showing the components of the IVOMEC SR bolus. [From Ref. 61.]

within the exit port to ensure that extraneous matter from the rumen cannot enter the device.

This overall design has the same essential components as the GITS system for human oral applications described in the previous paragraphs. In operation, water is imbibed through the membrane in the region of the osmotic tablet. The partition layer is added to essentially act as a piston to ensure complete and steady delivery of the drug layer throughout the duration of delivery. Expansion of the tablet pushes the drug layer through the exit port and exit port screen.

An example of the delivery profile for this device is shown in Figure 20, where the in vitro and in vivo release rates for the drug are plotted over time. Several points are of interest regarding these data.

1. Two sets of curves are shown, one for anhydrous and one for hydrated boluses. As with the GITS system, there is a lag time associated with the onset

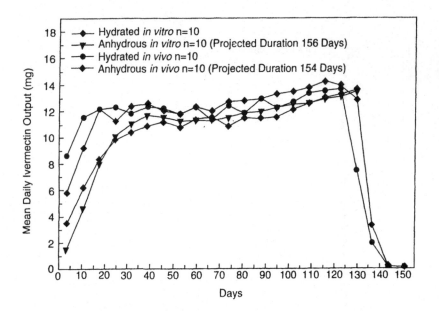

Figure 20 Mean daily ivermectin output of hydrated (1.3 g) and anhydrous boluses in vitro and in vivo. (From Ref. 61.)

of delivery from this system, but in this case the lag time is unusually long due to the significant pressures required to move this viscous material through the exit port screen [60]. To help decrease the lag time, the boluses are hydrated prior to administration by incorporating water into the package. This water is absorbed during storage and produces a buildup of hydrostatic pressure prior to administration. It is apparent that hydrated boluses start up much more rapidly (within 7–10 days) than do the anhydrous boluses (~20 days).

2. The duration of delivery for these systems is very long, about 135 days, followed by an abrupt cessation of delivery upon exhaustion of the drug formulation. All boluses within a given lot shut down within a 14 day period. This abrupt decline in delivery is critical for a product intended for delivery to food-producing animals where it is important to know the last day of drug delivery. This information is critical to help set a withdrawal period, i.e., the time before the animal can be slaughtered following administration of the drug.

3. Within the hydrated or the anhydrous lots, the in vitro–in vivo correlation is excellent in terms of both the average steady-state release rate and the total duration of delivery [60].

4. The average daily release rate of ivermectin increases modestly during the steady-state period. This is most likely the result of a combination of factors

including an increase in the effective area for water transport as the tablet moves up during delivery and a decrease in the effective osmotic pressure due to the dilution effects arising from the incorporation of water in the tablet.

Overall this system provides remarkable reproducibility in performance given the fact that the delivery period extends for a full grazing season.

C. Pulsatile Systems

In general, osmotic delivery systems have achieved wide acceptance because of their capacity to deliver drugs at a constant rate that is largely independent of the environment. Recently, however, several groups have developed osmotic systems that are designed to release bolus doses of drug at predetermined times. Such systems are expected to be of particular utility for site-specific delivery of drugs in the gastrointestinal tract.

For example, ALZA scientists have modified their push–pull system to provide an immediate dose and then a controlled release dose that begins to release at some point well delayed from the time of the bolus dose [64,65]. In effect, these systems have a core tablet of the same type of design as the push–pull tablet; however, prior to application of the semipermeable CA coating, an intermediate drug-free coating is applied to the tablet. This layer creates a delay in the delivery of the drug from the core formulation. Also subsequent to the application of the CA coating a second outer coating is applied that contains a dose of drug that is released immediately upon contact of the tablet with aqueous fluids. Also it is likely that an outer aqueous film coat would be applied to provide some masking of the taste of the drug layer. An example of the development of a specific system for the release of doxylamine to control sleep patterns is given in Ref. 64.

Another approach to the pulsatile delivery of drugs with an osmotic pressure driven system has been suggested by Amidon et al. [66]. This system provides the option of an immediate bolus dose and a second dose that can be timed to be released at some subsequent point following the administration of the dosage form. One or both of the doses can be composed of multiparticulates, for example, that would themselves be sustained release systems.

The design of their system is as follows. A conventional gelatin capsule shell bottom is coated with a CA membrane. The pulsed dose is placed in the bottom of this coated gelatin capsule. This layer is composed of drug plus an osmotic agent such as NaCl and a dispersing agent such as Explotab. The latter material helps to ensure dispersal of the drug from the capsule bottom. Layered above this pulsed dose is a fluid-impermeable plug made from a wax such as bees wax, or paraffin, white, or carnauba wax. The plug is placed in the capsule shell in such a manner as to ensure that a frictional resistance exists between the capsule shell wall and the wax. The cap of the shell is made from uncoated gelatin so that upon placement into an aqueous environment the cap dissolves, making available an immediate release dose that has been placed in the capsule shell above the plug.

In operation, the first dose is released following dissolution of the gelatin cap of the shell. The bottom of the shell remains intact due to the CA coating, which in turn serves as a semipermeable film that controls the rate of water permeation into the bottom of the capsule shell. This imbibed water in turn causes a buildup of hydrostatic pressure which ultimately leads to ejection of the wax plug and release of the entrapped second dose. The duration before release of this dose is a function of the composition of the wax plug and the thickness and composition of the CA coating. An example of the control of the timing of the release of the second dose is shown in Figure 21.

As noted in the patent for this system, pulsatile systems are expected to have numerous applications in drug delivery, including site-specific delivery in the GI tract and bid (twice-a-day) dosing of two pulsatile systems that simulates qid (four times a day) dosing of immediately available products. The latter concept is expected to have a particular advantage for those drugs that have high first-pass metabolism.

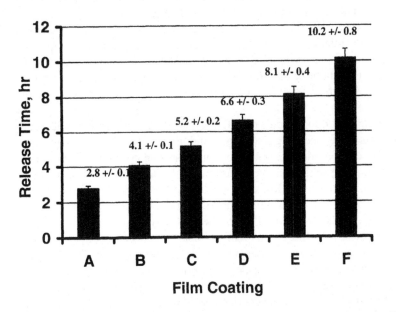

Figure 21 Effect of film thickness on the in vitro release time for pulsatile capsules. (From Ref. 66.)

IV. CONCLUSIONS

This chapter provides an overview of many of the approaches that have been developed utilizing osmotic pressure to control the overall rate of delivery of a drug to the gastrointestinal tract. It has been demonstrated that the overall release rates are functions of the properties and design of the system per se and largely independent of the characteristics of the release medium. The systems demonstrate good in vitro–in vivo correlations. The aqueous solubility of the drug is the most critical factor in determining the utility of any given osmotic system design for the oral delivery of that drug. With the introduction of the push–pull concept in the 1980s, systems can be designed to deliver most drugs at a controlled rate over a 12–24 hr period.

REFERENCES

1. JR Robinson, VHL Lee, eds. Controlled Drug Delivery, Fundamentals and Applications. New York: Marcel Dekker, 1987.
2. A Kydonius, ed. Treatise on Controlled Drug Delivery. New York: Marcel Dekker, 1992.
3. JR Cardinal, L Witchey-Lakshmann. Drug delivery in veterinary medicine. In: A Kydonius, ed. Treatise on Controlled Drug Delivery. New York: Marcel Dekker, 1992, pp 465–490.
4. M Rathbone, ed. Veterinary drug delivery. Part I. Adv Drug Delivery Rev 28:301–392, 1997.
5. G Santus, RW Baker. Osmotic drug delivery: A review of the patent literature. J Controlled Release 35:1–21, 1995.
6. O Kedem, A Katchalsky. Thermodynamic analysis of the permeability of biological membranes to nonelectrolytes. Biochim Biophys Acta 27:229–246, 1958.
7. BZ Ginsburg, A Katchalsky. The frictional coefficients of the flows of non-electrolytes through artificial membranes. J Gen Physiol 47:403–420, 1963.
8. SZ Song, JR Cardinal, SJ Wisniewski, SW Kim. Mechanisms of solute permeation through hydrogel films: An irreversible thermodynamic approach. Abstracts of Papers Presented at the 126th National Meeting of the American Pharmaceutical Association, 1979.
9. BK Fritzinger, SK Bauman, DJ Lyman. Membrane characteristics, permeability parameters, and frictional coefficients for cupraphane. J Biomed Mater Res 5:3–16, 1971.
10. F Theeuwes. Elementary osmotic pump. J Pharm Sci 64:1987–1991, 1975.
11. S Rose, JF Nelson. A continuous long-term injector. Aust J Exp Biol 33:415–420, 1955.
12. GP Polli, CE Shoop, WM Grim. Controlled release medicinal tablets. US Patent 3,538,214, 1970.

13. T Higuchi, HM Leeper. Osmotic dispenser. US Patent 3,732,865, 1973.

14. T Higuchi, HM Leeper. Improved osmotic dispenser employing magnesium sulfate and magnesium chloride. US Patent 3,760,804, 1973.

15. T Higuchi. Osmotic dispenser with collapsible supply container. US Patent 3,760,805, 1973.

16. GM Zentner, GS Rork, KJ Himmelstein. Osmotic flow through controlled porosity films: An approach to the delivery of water soluble compounds. In: JM Anderson, SW Kim, eds. Advances in Drug Delivery Systems. New York: Elsevier, 1986, pp 217–230.

17. KA Kelly. Motility of the stomach and gastroduodenal junction. In: LR Johnson, ed. Physiology of the Gastrointestinal Tract. New York: Raven Press, 1981, pp 393–410.

18. A Katchalsky, PF Curran. Nonequilibrium Thermodynamics in Biophysics. Cambridge, MA: Harvard University Press, 1965.

19. RA Robinson, RH Stokes. Electrolyte Solutions, 2nd ed (revised). London: Butterworths, 1970.

20. SM Herbig, JR Cardinal, RW Korsmeyer, KL Smith. Asymmetric-membrane tablet coatings for osmotic drug delivery. J Controlled Release 35:127–136, 1995.

21. F Theeuwes, AD Ayer. Laminated osmotic system for dispensing beneficial agent. US Patent 4,014,334, 1976.

22. F Theeuwes, D Swanson, P Wong, P Bonson, V Place, K Heimlich, KC Kwan. Elementary osmotic pump for indomethacin. J Pharm Sci 72:253–258, 1983.

23. AG Thombre, GM Zentner, KJ Himmelstein. Mechanism of water transport in controlled porosity osmotic devices. J Membrane Sci 40:279–310, 1989.

24. GA McClelland, SC Sutton, K Engle, GM Zentner. The solubility-modulated osmotic pump: In vitro/in vivo release of diltiazem hydrochloride. Pharm Res 8:88–92, 1991.

25. GA McClelland, RJ Stubbs, JA Fix, SA Pogany, GM Zentner. Pharm Res 8:873–876, 1991.

26. GM Zentner, GA McClelland, SC Sutton. Controlled porosity solubility- and resin-modulated osmotic delivery systems for release of diltiazem hydrochloride. J Controlled Release 16:237–244, 1991.

27. JR Cardinal, SM Herbig, RW Korsmeyer, J Lo, KL Smith, AG Thombre. The use of asymmetric membranes in delivery devices. EPO 0357369, 1990.

28. JR Cardinal, SM Herbig, RW Korsmeyer, J Lo, KL Smith, AG Thombre. Asymmetric membranes in delivery devices. US Patent 5,698,220, 1997.

29. RW Korsmeyer, KD Wilner, WE Ballinger, FC Falkner. Asymmetric membrane tablets for delivery of glipizide. Proceedings of the International Symposium on the Controlled Release of Bioactive Materials, Stockholm, 1997, pp 239–240.

30. RW Korsmeyer, SM Herbig, KL Smith, JR Cardinal, AC Curtiss, MB Fergione, RA Wilson. Single-step process for coating multiparticulates with asymmetric membranes. Proceedings of the International Symposium on the Controlled Release of Bioactive Materials, Stockholm, 1997, pp 531–532.

31. AG Thombre, JR Cardinal, AR DeNoto, SM Herbig, KL Smith. Asymmetric membrane capsules for osmotic drug delivery. I. Development of manufacturing process. J Controlled Release 57:55–64, 1999.

32. AG Thombre, JR Cardinal, AR DeNoto, DC Gibbes. Asymmetric membrane capsules for osmotic drug delivery. II. In vitro and in vivo performance. J Controlled Release 57:65–73, 1999.

33. SC Sutton, K Engle, RA Deeken, KE Plute, RD Shaffer. Performance of diltiazem tablet and multiparticulate osmotic formulations in the dog. Pharm Res 7:874–878, 1990.

34. PR Magruder, B Barclay, PSL Wong, F Theeuwes. Composition comprising salbutomal. US Patent 4,751,071, 1988.

35. F Theeuwes. Osmotic device having microporous reservoir. US Patent 3,977,404, 1977.

36. F Theeuwes. Osmotic dispenser with gas generating means. US Patent 4,036,228, 1977.

37. D Swanson, D Edgren. Osmotic device that improves delivery properties in situ. US Patent 4,326,525, 1982.

38. P Bonson, PSL Wong, F Theeuwes. Method of delivering drug with aid of effervescent activity generated in the environment of use. US Patent 4,344,929, 1982.

39. AD Ayer, PSL Wong. Dosage form comprising solubility regulating member. US Patent 4,755,180, 1988.

40. S Khanna. Therapeutic system for sparing soluble active ingredients. US Patent 4,992,278, 1991.

41. F Theeuwes. Osmotic system for delivery of selected beneficial agents having varying degrees of solubility. US Patent 4,111,201, 1978.

42. F Theeuwes. Osmotic system for the controlled delivery of agent over time. US Patent 4,111,202, 1978.

43. F Theeuwes. Osmotic system with means for improving delivery kinetics of system. US Patent 4,111,203, 1978.

44. F Theeuwes. System with microporous releasing diffusor. US Patent 4,309,996, 1982.

45. R Cortese, F Theeuwes. Osmotic device with hydrogel driving member. US Patent 4,327,725, 1982.

46. PSL Wong, BL Barclay, JC Deters, F Theeuwes. Osmotic device for administering certain drugs. US Patent 4,765,989, 1988.

47. D Swanson, B Barclay, P Wong, F Theeuwes. Nifedipine gastrointestinal system. Am J Med 83(Suppl 6B):3–9, 1987.

48. AD Ayer, DR Swanson, AL Kuczynski. Dosage form for treating cardiovascular diseases. US Patent 4,950,486, 1990.

49. JC Deters, PSL Wong, BL Barclay, F Theeuwes, DR Swanson. Dosage form for dispensing drug for human therapy. US Patent 4,837,111, 1989.

50. GV Guittard, PSL Wong, F Theeuwes, R Cortese. Dosage form for delivering diltiazem. US Patent 4,859,470, 1989.

51. J Wright, JD Childers, BL Barclay, PSL Wong, LE Atkinson. Osmotic dosage form comprising an estrogen and a progestogen. US Patent 4,948,593, 1990.

52. AL Kuczynski, AD Ayer, PSL Wong. Oral hypoglycemic glipizide granulation. US Patent 5,024,843, 1991.

53. F Jao, PS Wong, HT Huynh, K McCheeney, PK Wat. Verapamil therapy. US Patent 5,160,744, 1992.

54. AD Ayer, LG Lawrence. Pseudoephedrine brompheniramine therapy. US Patent 4,810,502, 1989.

55. PSL wong, F Theeuwes, BL Bailey, MH Desley. Osmotic dosage system for delivering a formulation comprising liquid carrier and drug. US Patent 5,324,280, 1995.

56. PSL Wong, F Theeuwes, BL Bailey, MH Desley. Osmotic dosage form for liquid delivery. US Patent 5,413,572, 1995.

57. DG Pope, PK Wilkinson, JR Egerton, J Conroy. Oral controlled release delivery of ivermectin in cattle via an osmotic pump. J Pharm Sci 74:1108–1110, 1985.

58. JR Egerton, D Suhayda, CH Eary. Prophylaxis of nematode infections in cattle with an indwelling, ruminoreticular ivermectin sustained release bolus. Vet Parasitol 74: 614–617, 1986.

59. B Eckenhoff, R Cortese, JC Wright, PK Wilkinson, JR Zingerman, DG Pope. Osmotically-driven ruminal delivery system for the long term rate controlled delivery of ivermectin. Pharm Res (Suppl)4:46, 1987.

60. B Eckenhoff, R Cortese, FA Landrau. Delivery system for controlled administration to ruminants, US Patent 4,595,583, 1986.

61. JL Zingerman, JR Cardinal, RT Chern, J Holste, JB Williams, B Eckenhoff, JT Wright. The in vitro and in vivo performance of an osmotically controlled delivery system-IVOMEC SR Bolus®. J Controlled Release 47:1–11, 1997.

62. JR Cardinal. Intraruminal boluses. Adv Drug Deliv Rev 28:303–322 (1997).

63. JB Eckenhoff, R Cortese, FA Landrau. Dispenser for delivering drug to livestock. US Patent 4,927,633, 1990.

64. AD Ayer, F Theeuwes, PS Wong. Pulsed drug delivery of doxylamine. US Patent 4,842,867, 1989.

65. AD Ayer, F Theeuwes, PS Wong. Pulsed drug delivery. US Patent 4,848,592, 1990.

66. GL Amidon, GD Leesman, LB Sherman. Multi stage delivery system. US Patent 5,387,421, 1995.

12

Transport in Polymer Systems

Jim H. Kou
Schering–Plough Research Institute, Kenilworth, New Jersey

I. INTRODUCTION

Recent advances in controlled drug delivery place a heavy emphasis on the application of polymers. They are employed as release rate controlling matrices or membranes. In either case, diffusion of solvent or drug solute in polymers becomes a subject which concerns all researchers in the field. The purpose of the present chapter is to provide a basic conceptual framework for mass transport in polymers. In Section II, general concepts of Fickian transport are introduced, including terminology and definitions, basic methods for determining diffusion coefficients, and characteristics of Fickian transport. In Section III, the theoretical basis of Fickian transport on free volume is introduced. A cardinal feature of diffusion in amorphous polymers is the dependence of the diffusion coefficient on temperature and concentration. It is shown that this dependence can be accounted for via free volume theory. A survey of diffusion anomalies in glassy polymers is also summarized. In Section IV, features particular to transport in swellable polymers such as hydrogels are described. Hydrogels represent a category that is most attractive for biomedical and drug delivery applications [1]. Various models developed to understand the diffusion in the swollen matrix are presented.

In this chapter, the subscript 1 denotes the penetrant and subscript 2, the polymer. The term "penetrant" refers to solvents which have sufficient thermodynamic affinity for and interaction with the polymer. It is because of this interaction that penetrant diffusion exhibits a significant concentration dependence. This orientation excludes consideration of the permeation of small gaseous molecules.

Where hydrogels are concerned, the subscript 3 denotes a drug solute for which diffusional transport is of interest.

II. FUNDAMENTALS

A. Frames of Reference

Transport of component i in a binary system is described by the equation of continuity [2], which is an expression for mass conservation of the subject component in the system, i.e.,

$$\frac{\partial c_i}{\partial t} + \Delta \cdot (c_i v_i) + R_i = 0 \tag{1}$$

where c_i, v_i, and R_i are, respectively, the concentration, velocity relative to an external stationary reference, and generation rate of component i. In the absence of any chemical reaction, the absolute mass flux, $c_i v_i$, is generally expressed as

$$c_i v_i = c_i \bar{v}_X + J_{X,i} \tag{2}$$

where $\bar{v}_X$ is an average velocity needing definition and $J_{X,i}$ is the diffusional flux of component i relative to $\bar{v}_X$. It is clear that the form of $J_{X,i}$ depends on the definition of $\bar{v}_X$.

The choice of $\bar{v}_X$ is a matter of convenience for the system of interest. Table 1 summarizes the various definitions of $\bar{v}_X$ and corresponding $J_{X,i}$ commonly in use [3]. The various diffusion coefficients listed in Table 1 are interconvertible, and formulas have been derived. For polymer–solvent systems, the volume average velocity, $\bar{v}_v$, is generally used, resulting in the simplest form of $J_{X,i}$. Assuming that this $\bar{v}_v = 0$, implying that the volume of the system does not change, the equation of continuity reduces to the common form of Fick's second law. In one dimension, this is

Table 1 Definition of Velocities, $\bar{v}$, in Eq. (2) and Corresponding Diffusive Fluxes in Binary Mixtures[a]

Average velocity, $\bar{v}_x$	Definition of $\bar{v}_x$	Constitutive equation for $J_{x,i}$
Mass average velocity	$\bar{v}_m = \sum_{i=1}^{2} \omega_i v_i$	$J_{m,i} = \rho D_m \, \Delta\omega_i$
Molar average velocity	$\bar{v}_M = \sum_{i=1}^{2} x_i v_i$	$J_{M,i} = -M_i c^* D_M \, \Delta x_i$
Volume average velocity	$\bar{v}_v = \sum_{i=1}^{2} \hat{V}_i c_i v_i$	$J_{v,i} = -D_v \, \Delta x_i$

[a] ω_i = mass fraction of i; x_i = mole fraction of i; $\hat{V}_i$ = partial specific volume of i; c^* = molar concentration of mixture; c_i = mass concentration of i; M_i = molecular weight of i.

$$\frac{\partial c_i}{\partial t} = \frac{\partial}{\partial x}\left(D\frac{\partial c_i}{\partial x}\right) \tag{3}$$

where D is the diffusion coefficient, which, in general, is a function of concentration, temperature, and pressure. This equation is applicable to both polymer and penetrant, and the same diffusion coefficient will appear in both. The subscript v is dropped from D in Eq. (3) with the understanding that the volume of the system is constant throughout the diffusion process.

B. Diffusion Coefficients

Various diffusion coefficients have appeared in the polymer literature. The diffusion coefficient D that appears in Eq. (3) is termed the mutual diffusion coefficient in the mixture. By its very nature, it is a measure of the ability of the system to dissipate a concentration gradient rather than a measure of the intrinsic mobility of the diffusing molecules. In fact, it has been demonstrated that there is a bulk flow of the more slowly diffusing component during the diffusion process [4]. The mutual diffusion coefficient thus includes the effect of this bulk flow. An intrinsic diffusion coefficient, D_i^*, also has been defined in terms of the rate of transport across a section where no bulk flow occurs. It can be shown that these quantities are related to the mutual diffusion coefficient by

$$D = \hat{V}_1 c_1(D_2^* - D_1^*) + D_1^* \tag{4}$$

where $\hat{V}_1$ and c_1 are, respectively, the specific volume and concentration of component 1. In polymer–penetrant systems where the diffusive mobility of polymer is sufficiently small compared to that of the penetrant, D_1^* can be determined from D by

$$D_1^* = D/\phi_2 \tag{5}$$

where ϕ_2 is the polymer volume fraction.

The self-diffusion coefficient is determined by measuring the diffusion rate of the labeled molecules in systems of uniform chemical composition. This is a true measure of the diffusional mobility of the subject species and is not complicated by bulk flow. It should be pointed out that this quantity differs from the intrinsic diffusion coefficient in that a chemical potential gradient exists in systems where diffusion takes place. It can be shown that the self-diffusion coefficient, D_1, is related to the intrinsic diffusion coefficient, D_1^*, by

$$D_1 = D_1^* RT\left(\frac{\partial \ln a_1}{\partial \ln c_1}\right)^{-1} \tag{6}$$

where a_1 is the thermodynamic activity of component 1 [4].

For practical purposes, the mutual diffusion coefficient is the quantity commonly reported to characterize diffusional transport in pharmaceutical systems. It is thus the purpose of investigators to determine this quantity experimentally. To this end, both sorption and permeation methods are commonly used.

C. Fickian Sorption

A typical sorption experiment involves exposing a polymer sample, initially at an equilibrium penetrant concentration of c_1^0, to a bathing penetrant concentration of c_1^∞. The weight gain or loss is then measured as a function of time. The term "sorption" used in this context includes both absorption and desorption. The sorption is of the integral type if $c_1^0 = 0$ in the case of absorption or if $c_1^\infty = 0$ in the case of desorption. Details of the experimental setup for the sorption measurement are discussed elsewhere [4].

For a constant D, the solution to the non-steady-state sorption problem gives the fraction sorbed, M_t/M_∞, as

$$\frac{M_t}{M_\infty} = \frac{4}{\sqrt{\pi}}\sqrt{\frac{Dt}{L^2}} + 8\sqrt{\frac{Dt}{L^2}}\sum_{n=0}^{\infty}(-1)^n \, ierfc\left(\frac{n}{2}\sqrt{\frac{L^2}{Dt}}\right) \tag{7}$$

where L is the polymer membrane thickness. In an absorption experiment, the mutual diffusion coefficient, D_a, can be calculated from the initial slope, s_a, of a small-time approximation of Eq. (7), i.e.,

$$D_a = \frac{\pi}{16}L^2 s_a^2 \tag{8}$$

where the subscript a indicates absorption. For most polymer–penetrant systems, D_a will be a function of c_1. In this case, the D_a determined is apparently an average over the concentration range from c_1^0 to c_1^∞. It can be shown that this average $\overline{D}_a$ is

$$\overline{D}_a = \frac{1}{c_1^\infty - c_1^0}\int_{c_1^0}^{c_1^\infty} D_a(c)dc \tag{9}$$

To evaluate the concentration dependence of D, a series of sorption experiments are performed at successive concentration intervals. A numerical differentiation is then performed on the plot of $\overline{D}_a(c_1^0)$ versus c_1^0 to obtain a first approximation for the relationship between D_a and c_1. Similarly, the c_1 dependence of D_d, where d denotes desorption, can be determined from desorption data. This estimation method works quite satisfactorily for cases where D has a mild dependence on c_1 and both $D_a(c_1)$ and $D_d(c_1)$ give good estimates of $D(c_1)$. It is to be emphasized that the sorption method for determining D is valid only when the transport strictly adheres to Fickian characteristics.

For a classical diffusion process, "Fickian" is often the term used to describe the kinetics of transport. In polymer–penetrant systems where the diffusion is concentration-dependent, the term "Fickian" warrants clarification. The result of a sorption experiment is usually presented on a normalized time scale, i.e., by plotting M_t/M_∞ versus $t^{1/2}/L$. This is called the reduced sorption curve. The features of the Fickian sorption process, based on Crank's extensive mathematical analysis of Eq. (3) with various functional dependencies of $D(c_1)$, are discussed in detail by Crank [5]. The major characteristics are

1. Both reduced absorption and desorption curves have a linear region extending up to a M_t/M_∞ value of 0.6. Should D be an increasing function of c_1, this linear region can be further extended.
2. The sorption curves eventually become concave against the time axis.
3. The reduced absorption and desorption curves do not depend on the thickness of the polymer film.
4. If D is an increasing function of c_1, the reduced absorption curve always lies above the desorption curve. The divergence becomes more significant when the dependence of D on c_1 becomes stronger; the difference will disappear if D is a constant.

Criteria 1–3 are the cardinal characteristics of Fickian diffusion and disregard the functional form of $D(c_1)$. Violation of any of these is indicative of non-Fickian mechanisms. Criterion 4 can serve as a check if the $D(c_1)$ dependence is known. As mentioned, it is crucial that the sorption curve fully adhere to Fickian characteristics for a valid determination of D from the experimental data. At temperatures well above the glass transition temperature, T_g, Fickian behavior is normally observed. However, caution should be exercised when the experimental temperature is either below or slightly above T_g, where anomalous diffusion behavior often occurs.

D. Fickian Permeation

The permeation technique is another commonly employed method for determining the mutual diffusion coefficient of a polymer–penetrant system. This technique involves a diffusion apparatus with the polymer membrane placed between two chambers. At time zero, the reservoir chamber is filled with the penetrant at a constant activity while the receptor chamber is maintained at zero activity. Therefore, the upstream surface of the polymer membrane is maintained at a concentration of c_1^∞. It is noted that c_1^∞ is the concentration within the polymer surface layer, and this concentration can be related to the bulk concentration or vapor pressure through a partition coefficient or solubility constant. The amount of penetrant appearing in the receptor, $q(t)$, is measured as a function of time. D is then calculated at steady state by using

$$\overline{D} = \left[\frac{dq}{dt}\right]_{ss} \frac{L}{Kc_r} \tag{10}$$

where c_r is the penetrant concentration in the reservoir chamber and K is the partition coefficient defined as c_1^{∞}/c_r. The diffusion coefficient obtained is an average over the concentration range from 0 to c_1^{∞}. The concentration dependence of D can be determined by numerical differentiation of the plot of $c_1^{\infty} \overline{D}(c_1^{\infty})$ versus c_1^{∞}. It is noted that this method does not contain any approximations other than ignoring the change in thickness due to swelling by the penetrant and is applicable even when Fickian characteristics are not strictly followed.

The use of the time lag method for extracting the diffusion coefficient is straight-forward with constant D. It can be shown that the steady-state portion of the $q(t)$ vs. t plot can be expressed by

$$q_{ss}(t) = \frac{Dc_1^{\infty}}{L}\left(t - \frac{L^2}{6D}\right) \tag{11}$$

D can then be calculated from the measured time lag, t_{lag}, which is the intercept on the time axis, by

$$D = \frac{L^2}{6t_{lag}} \tag{12}$$

The advantage of using the time lag method is that the partition coefficient K can be determined simultaneously. However, the accuracy of this approach may be limited if the membrane swells. With D determined by Eq. (12) and the steady-state permeation rate measured experimentally, K can be calculated by Eq. (10). In the case of a variable $D(c_1)$, equations have been derived for the time lag [6,7]. However, this requires that the functional dependence of D on c_1 be known. Details of this approach have been discussed by Meares [7]. The characteristics of systems in which permeation occurs only by diffusion can be summarized as follows:

1. The plot of $q(t)$ versus t is always convex toward the t axis and approaches a straight line as t increases. This behavior is always observed, regardless of the functional form of $D(c_1)$.
2. The linear portion of the $q(t)$ versus t plot represents a steady state during which the concentration distribution in the polymer film is independent of time.

III. DIFFUSION IN POLYMER–PENETRANT SYSTEMS

Diffusion of small molecular penetrants in polymers often assumes Fickian characteristics at temperatures above T_g of the system. As such, classical diffusion theory is sufficient for describing the mass transport, and a mutual diffusion coefficient can be determined unambiguously by sorption and permeation methods. For a penetrant molecule of a size comparable to that of the monomeric unit of a polymer, diffusion requires cooperative movement of several monomeric units. The mobility of the polymer chains thus controls the rate of diffusion, and factors affecting the chain mobility will also influence the diffusion coefficient. The key factors here are temperature and concentration. Increasing temperature enhances the Brownian motion of the polymer segments; the effect is to weaken the interaction between chains and thus increase the interchain distance. A similar effect can be expected upon the addition of a small molecular penetrant.

Figure 1 is a schematic diagram illustrating a typical composition dependence of the mutual diffusion coefficient for a polymer–penetrant system [8]. Here the penetrant is apparently a good solvent for the polymer since the entire composition range is realized. Note that four regions can be distinguished. In the

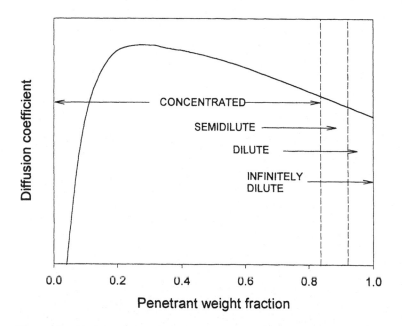

Figure 1 A schematic diagram illustrating a typical concentration dependence of mutual diffusion coefficient in a polymer–penetrant system. (From Ref. 8.)

infinitely dilute solution limit, the polymer chains are widely separated and do not interact with each other. In the dilute region, the polymer molecules begin to interact with each other hydrodynamically, leading to a concentration dependence of the diffusion coefficient. The dependence observed in this region is often weak and can be approximated by a linear function. The semidilute region is characterized by the overlap of polymer chains. Intermolecular entanglement leads to a dynamic network structure. Due to their complexity, diffusion in these systems is often analyzed by the use of scaling laws which give a power law correlation between the diffusion coefficient and system property [9]. Of interest to the present discussion is the concentrated region. It is clear from Figure 1 that a steep concentration dependence occurs in this region. This concentration dependence has been observed experimentally in numerous polymer–penetrant systems. An excellent summary can be found in Ref. 4. For an example, over the range of $0 < \omega_1 < 0.1$, the diffusion coefficient varies anywhere from tenfold in the case of polyisobutylene–alkane systems [10] to over a thousandfold in the case of the poly(vinyl acetate)–acetone system [11].

In addition to temperature and concentration, diffusion in polymers can be influenced by the penetrant size, polymer molecular weight, and polymer morphology factors such as crystallinity and cross-linking density. These factors render the prediction of the penetrant diffusion coefficient a rather complex task. However, in simpler systems such as non-cross-linked amorphous polymers, theories have been developed to predict the mutual diffusion coefficient with various degrees of success [12–19]. Among these, the most notable are the free volume theories [12,17]. In the following subsection, these free volume based theories are introduced to illustrate the principles involved.

A. Free Volume Theories

Cohen and Turnbull [20,21] laid down the foundation for the free volume concept in modeling self-diffusion in simple van der Waals liquids. They considered that the volume in a liquid is composed of two parts, the actual volume occupied by the liquid molecules and the free volume surrounding these molecules opened up by thermal fluctuation. Increasing temperature increases only the free volume and not the occupied volume. The average free volume per molecule, v_f, can be defined as

$$v_f = \bar{v} - v_o \tag{13}$$

where $\bar{v}$ and v_o are the average molecular volume in the liquid and the occupied volume, respectively. v_f is constantly being redistributed in the liquid space since no energy is required. An effective diffusive step will take place only if (1) a free volume hole of sufficient size is available and (2) the void left behind by the diffusing molecule is filled by another neighboring molecule. Assuming that an average liquid molecule confined in a cage created by the thermal fluctuation is moving with a gas kinetic velocity u, the contribution of this molecule to the overall diffusion coefficient is

$$D(v) = ga(v)u \tag{14}$$

where g is a geometrical factor and $a(v)$ is the cage diameter. The average diffusion coefficient is thus

$$D = \int_{v^*}^{\infty} D(v)p(v) \, dv \tag{15}$$

where $p(v)$ is the probability of finding a free volume of size v and v^* is the critical volume necessary for a diffusional jump. Using a probablistic approach, $p(v)$ is derived as

$$p(v) = \frac{\gamma}{v_f} e^{-\gamma v / v_f} \tag{16}$$

where γ is a factor correcting for the overlap of free volume and has a value between 0.5 and 1. The final expression for D is of the form

$$D = k e^{-\gamma v^* / v_f} \tag{17}$$

where k is related to g, u, and molecular size. Therefore, an exponential dependence on the free volume is derived for the self-diffusion coefficient. The temperature and pressure dependence of v_f can be derived from the definitions of thermal expansion coefficient and compressibility as

$$v_f = \alpha \bar{v}_m (T - T_o) - \beta \bar{v}_p \, \Delta P \tag{18}$$

where α is the mean thermal expansion coefficient, β is the mean compressibility, T_o is the temperature at which the free volume disappears, $\bar{v}_m$ is the mean molecular volume which is approximated by v_o, and $\bar{v}_p$ is the mean molecular volume for the pressure increment. The pressure dependence of v_f is usually small compared to the temperature effect [25]. It is clear from Eqs. (17) and (18) that the self-diffusion coefficient will decrease dramatically when temperature decreases, especially when approaching T_o [22–24]. If T_o is taken to be the glass transition temperature T_g, then Eqs. (17) and (18) serve as the basis for explaining the marked decrease in diffusivity in the neighborhood of T_g. Figure 2 illustrates the temperature dependence of the self-diffusion coefficient of 2,3-dimethylbutane based on the experimental data of McCall et al. [25] extrapolated according to Eqs. (17) and (18). It is clear from this graph that the diffusion coefficient decreases drastically when temperature approaches T_g at approximately 65 K.

Fujita [12] extended Eq. (17) to polymer–penetrant systems by a proper redefinition. For any polymer–penetrant binary system, the probability of finding

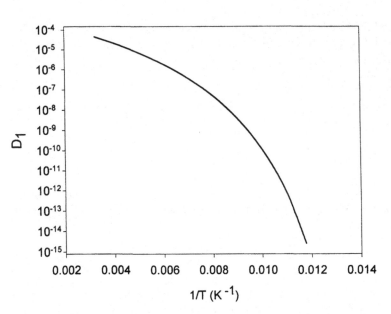

Figure 2   The temperature dependence of the self-diffusion coefficient of 2,3-dimethyl-butane predicted by Cohen and Turnbull's free volume model. (From Ref. 25.)

a hole larger than v is assumed to be of the same functional form as Eq. (16), i.e.,

$$p(v) = e^{-v/f} \qquad (19)$$

where f is now defined as the average fractional free volume of the system. Since the mobility of a penetrant molecule, m_1, is proportional to this probability, it is expressed by

$$m_1 = A_1 e^{-B_1/f} \qquad (20)$$

where B_1 is now the critical hole size for the diffusional jump of a diluent molecule to take place. Both A_1 and B_1 are independent of temperature and concentration but depend primarily on penetrant molecular size and shape. Following the definition of Hayes and Park [26], the thermodynamic diffusion coefficient, D_T, is usually related to m_1 by

$$D_T = RTm_1 \qquad (21)$$

D_T and the mutual diffusion coefficient, D, are interconvertible by correcting for the penetrant activity in the polymer [12]. For highly concentrated systems where the penetrant volume fraction, ϕ_1, is low, f can be approximated by

$$f(T, \phi_1) = f(T, 0) + \beta(T)\phi_1 \tag{22}$$

where $f(T, 0)$ is the average fractional free volume in the pure polymer at temperature T and $\beta(T)$ is the rate at which f increases with ϕ_1. Combining Eqs. (20)–(22), it can be shown that

$$\frac{1}{\ln(D_T/D_0)} = \frac{f(T, 0)}{B_1} + \frac{[f(T, 0)]^2}{[B_1\beta(T)]}\left(\frac{1}{\phi_1}\right) \tag{23}$$

where D_0 is the value of D_T at the limit of $\phi_1 = 0$, i.e.,

$$\ln\left(\frac{D_0}{RT}\right) = \ln A_1 - \frac{B_1}{f(T, 0)} \tag{24}$$

Equation (23) was found to be obeyed by a number of systems such as poly(ethyl acrylate)–benzene, rubber–benzene, and poly(methyl acrylate)–ethyl acetate [12]. According to the equation, a plot of $[\ln(D_T/D_o)]^{-1}$ versus ϕ_1^{-1} will yield a straight line, and from its slope the free volume parameter β can be determined. To construct this plot, $f(T, 0)$ is first calculated as [27]

$$f(T, 0) = f_g + \alpha(T - T_g) \tag{25}$$

where f_g is the fractional free volume at T_g with an universal value of 0.025; α assumes a value of 4.8×10^{-4} deg^{-1}. B_1 can then be determined from the value of the slope of the plot of $\ln(D_0/RT)$ versus $1/f(T, 0)$ according to Eq. (24). With the values of $f(T, 0)$ and B_1, $\beta(T)$ can be determined from the slope of a $[\ln(D_T/D_o)]^{-1}$ versus ϕ_1^{-1} plot using Eq. (23). The success of this theory in correlating the diffusion data for various polymer–penetrant systems implies that the marked concentration dependence of the diffusion coefficient is the result of the sensitivity of penetrant and polymer segmental mobility to a change in the free volume of the system. However, a shortcoming of the theory is that it is correlative rather than predictive in nature. For example, the parameter β can be determined only by using diffusion data and cannot be estimated from material properties. In the following, we will see that this shortcoming has been overcome by Vrentas and Duda's formulation of free volume.

Vrentas and Duda [17] and Vrentas et al. [28] derived a generalized free volume theory by redefining the free volume in polymer–penetrant binary systems. First, as shown in Figure 3, the interstitial free volume is that part of the unoccupied volume that requires significant energy for redistribution [8]. Therefore, diffusion does not take place through this space, since it is not readily available for cooperative movements. The remainder of the unoccupied volume is called the hole free volume. It is this part of the free volume which requires no energy for redistribution; therefore, it is assumed that diffusion takes place through this space [29]. Vrentas and Duda showed that the specific hole free

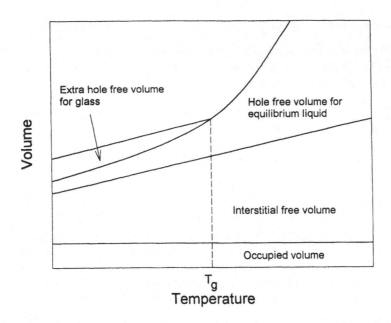

Figure 3 A schematic showing the temperature dependence of the specific volume of a polymer. (From Ref. 8.)

volume, $\hat{V}_{fh}$, can be estimated from the properties of the pure polymer and solvent according to the relationship

$$\hat{V}_{fh} = K_{11}\omega_1(K_{21} + T - T_{g1}) + K_{12}\omega_2(K_{22} + T - T_{g2}) \tag{26}$$

where ω_i is the mass fraction of component i, $K_{j,i}$ is the jth free volume parameter of component i, and T_{gi} is the glass transition temperature of the pure component i. The self-diffusion coefficient of penetrant, D_1, was then expressed as

$$D_1 = D_0 e^{-E'/RT} \exp\left[-\frac{\gamma(\omega_1 \hat{V}_1^* + \omega_2 \xi \hat{V}_2^*)}{\hat{V}_{fh}} \right] \tag{27}$$

where γ is the same overlap factor that appears in Eq. (16), $\hat{V}_i^*$ is the critical volume required for the diffusional jumps of component i, E' is the molar energy required to overcome the attractive forces for the diffusional jumps to take place, and ξ is a constant defined by $\hat{V}_1^* M_1 / \hat{V}_2^* M_j$. M_1 and M_j are the molecular weights of the penetrant and a jumping unit of the polymer chain, respectively. The form of Eq. (27) indicates that diffusion depends on the probability of a molecule having sufficient energy to overcome the attractive force field and the probability of the density fluctuation producing a sufficiently large hole.

In general, mass transfer processes involving polymer–penetrant mixtures are generally analyzed by using a mutual diffusion coefficient. Therefore, a relationship between the mutual diffusion coefficient, D, and self-diffusion coefficients, D_1's is needed. Vrentas et al. [30] proposed an equation relating D to D_1 for polymer–penetrant systems in which D_1 is much larger than D_2:

$$D = D_1(1 - \phi_1)^2(1 - 2\phi_1\chi) \qquad (28)$$

Here χ is the Flory–Huggins interaction parameter and ϕ_1 is the penetrant volume fraction. In order to use Eqs. (26)–(28) for the prediction of D, one needs a great deal of data. However, much of it is readily available. For example, $\hat{V}_1^*$ and $\hat{V}_2^*$ can be estimated by equating them to equilibrium liquid volume at 0 K, and K_{12}/γ and $K_{22} - T_{g2}$ can be computed from WLF constants which are available for a large number of polymers [31]. K_{11}/γ and $K_{11} - T_{g1}$ can be evaluated by using solvent viscosity–temperature data [28]. The interaction parameters, χ, can be determined experimentally and, for many polymer–penetrant systems, are available in the literature.

B. Glassy State and Anomalous Diffusion

From the volume–temperature relationship of an amorphous polymer shown in Figure 3, the specific volume will decrease with decreasing temperature. At high temperatures where the polymer chains are relatively mobile, the rate of densification is rapid, and a true equilibrium state can be reached instantaneously. However, when temperature continues to decrease, it will eventually reach a critical point below which chain movement becomes very sluggish. Consequently, the rapid cooling and slow densification result in a nonequilibrium state. This critical temperature is called the glass transition temperature, T_g, below which the polymer is said to be in a glassy state. The glass transition is characterized by a discontinuity in the second derivatives of free energy with respect to temperature such as heat capacity. Although this behavior resembles a second-order thermodynamic transition, it is cautioned that the portion of the curve in Figure 3 below T_g does not represent a true equilibrium state. Rather, it depends on the history of cooling. As such, the glass transition cannot be regarded as a true second-order thermodynamic transition. Since the glassy state is not in true equilibrium, an aging effect is expected.

When a penetrant diffuses into a polymer, the perturbation will cause the polymer molecules to rearrange to a new conformational state. The rate at which this conformational adaptation occurs depends on the mobility of the polymer chains. At temperatures well above the glass transition, this occurs quite rapidly and the diffusive process resembles that in the liquid state. At temperatures near or below the glass transition, the conformational change does not take place instantaneously. Instead, there is a finite rate of polymer relaxation induced by the perturbation due to the penetrating diffusant [32]. The relative magnitudes of the diffusional and relaxational rates have a major bearing on the kinetics of penetrant transport. When these kinetics are at variance with Fickian characteristics, they are generally called anomalous diffusion. Although a comprehensive theory is not yet available to account for all the anomalous effects, reasonable predictions of the occurrence can be made by using the dimensionless diffusion Deborah number, $(DEB)_D$, defined by

$$(DEB)_D = \frac{\lambda}{\theta} \tag{29}$$

where λ and θ are the characteristic times for relaxation and diffusion, respectively [33]. In general, λ can be estimated from the shear relaxation modulus, $G(t)$, of a polymer–penetrant system, i.e.,

$$\lambda = \frac{\displaystyle\int_0^\infty tG(t)\,dt}{\displaystyle\int_0^\infty G(t)\,dt} \tag{30}$$

and θ can be calculated as

$$\theta = \frac{L^2}{D^\dagger} \tag{31}$$

Here $D^\dagger$ is calculated from the self-diffusion coefficients of the polymer–penetrant pair, i.e.,

$$D^\dagger = x_2 D_1 + x_1 D_2 \tag{32}$$

where the x_i's are the mole fractions of the components. It should be noted that the value of $(DEB)_D$ is not unique, since other definitions of λ and θ are possible [34]. Nonetheless, the Deborah number diagrams constructed should give similar predictions of the temperature and composition range where anomalous diffusion will occur.

When $(DEB)_D$ is much smaller than unity, the polymer relaxation is relatively rapid compared to diffusion. In this case, conformational changes take place instantaneously and equilibrium is attained after each diffusional jump. This is the type of diffusion encountered ordinarily and is called viscous diffusion. Therefore, the transport will obey classical theories of diffusion. When $(DEB)_D$ is much larger than unity, the molecular relaxation is very slow compared to diffusion and there are no conformational changes of the medium within the diffusion time scale. In this case, Fick's law is generally valid, but no concentration dependence of the diffusion coefficient is expected. This is termed elastic diffusion. When $(DEB)_D$ is in the neighborhood of unity, molecular rearrangment and penetrant diffusion occur on the same time scale. This leads to a mixed transport mechanism in which both mechanical relaxation and diffusion contribute to transport kinetics. Anomalous diffusion is usually expected in these systems, and the transport process is called viscoelastic diffusion.

A Deborah number diagram is essentially a temperature–composition diagram with lines of constant Deborah number delineating various regions of transport. Figure 4 is a typical Deborah number diagram for a polystyrene–ethylbenzene system with polymer molecular weight of 3×10^5 and a sample thickness, L, of 10^{-3} cm [33]. It is clear from this diagram that anomalous transport can be expected at the temperatures and concentrations enclosed between the lines of $(DEB)_D$ values of 0.1 and 10. A line which represents the glass transition is also indicated in the diagram to emphasize the fact that anomaly can occur even at temperatures above T_g.

Various anomalous diffusional behaviors have been observed and documented for both sorption and permeation experiments. Detailed discussions of these anomalies can be found elsewhere [12,35,36]. Here only a brief summary of major findings is given. First, for the sorption anomaly, it has been observed that the reduced sorption curve has a distinctive thickness dependence. In this case, the reduced absorption and desorption curves obtained at various thick-

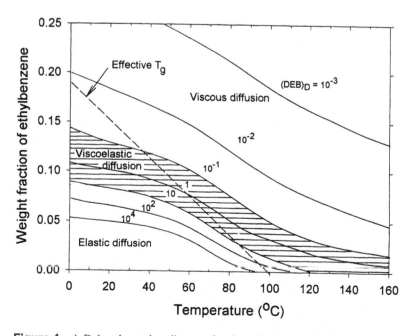

Figure 4 A Deborah number diagram for the polystyrene–ethylbenzene system showing the diffusion behavior as a function of weight fraction and temperature. (From Ref. 33.)

nesses do not coincide. Second, the sorption curves, when plotted as M_t/M_∞ versus $t^{1/2}$, are not linear even in the early time range. Third, the absorption curve often has an inflection point. Fourth, the paired absorption and desorption curves tend to intersect. Various theoretical interpretations of the observed anomalies have been attempted [12]. They are able to explain the anomalies partially, but none appears completely satisfactory. For the permeation anomaly, it has been shown that the permeation curve will go through a local maximum before converging to a linear steady-state curve. In this case, the curve is concave rather than convex toward the time axis. The explanations proposed include (1) the presence of an initial convective flow, (2) a time-dependent diffusion coefficient due to slow relaxation kinetics, and (3) a stress effect on the diffusion coefficient [12]. Although these offer qualitative explanations, quantitative models for prediction have yet to be developed.

IV. TRANSPORT IN SWELLABLE POLYMERS

A. Swelling and Case II Transport

When a glassy polymer is brought into contact with a thermodynamically compatible solvent, the solvent penetrates the polymer until a sorption equilibrium is reached. During the swelling process, a sharp penetrating front is often observed within the polymer matrix. This front separates the swollen rubbery region and the unpenetrated glassy core. Therefore, glass transition occurs at this glassy/rubbery front. In extreme cases, the swelling exhibits zero-order weight gain kinetics [37–39]. It is often observed that, on a fractional basis, the weight gain and front penetration time profiles coincide with each other. Mechanistically, the sorption process is said to be controlled by the moving glassy/rubbery front [40–43]. Thomas and Windle [37] showed that for methanol penetration into poly(methyl methacrylate) at 24°C, the concentration gradient of methanol is essentially zero in the rubbery region. Alfrey et al. [43] called this mode of transport case II in contrast to the case I (Fickian) mode. The intermediate cases are collectively termed anomalous diffusion. If the sorption kinetics are characterized according to Eq. (33),

$$\frac{M}{M_\infty} = kt^n \tag{33}$$

where k and n are constants, case I transport will be characterized by $n = 0.5$, case II by $n = 1.0$, and anomalous diffusion by $0.5 < n < 1.0$.

Investigations have been made to explore the potential of case II kinetics in the design of constant rate delivery systems [44–47]. The idea is that if the preloaded drug molecules have a high diffusivity in the swollen region and the swelling is limited by the glass/rubber relaxation front, then drug release will also be controlled by the zero-order front movement. The scheme which Alfrey et al. suggested for characterizing sorption kinetics can also be applied to the case of solute release from the swelling matrix. Hopfenberg and Hsu [39] were able to show a zero-order ($n = 1$) release of Sudan Red IV dye (MW 380) from initially penetrant-free polystyrene films swelling in hexane. However, subsequent studies [48] on sodium chloride and malachite green dye release from ethylene–vinyl alcohol copolymers did not reveal case II characteristics. Apparently the system is complicated by the osmotic effect of the loaded drug in the case of sodium chloride and by a diffusive contribution in the rubbery phase in the case of malachite green. Korsmeyer and Peppas [49] studied theophylline release from a poly(vinyl alcohol) gel cross-linked with glutaraldehyde and found that the highest value of n that the system can offer is around 0.76. This is again attributable to the contribution of the rubbery phase to the overall transport resistance. Nonetheless, an extended region of near zero-order release is often observed in these swelling systems due to the strong concentration dependence of the diffusion coefficient of the drug molecules. It was also pointed out that the drug loading level can significantly alter both the swelling and release kinetics [50–52]. This factor will certainly complicate the interpretation of the release mechanism. A useful method of predicting case II release from an initially dry glassy matrix was proposed by Peppas and Franson [53]. They defined a dimensionless swelling interface number, Sw, as

$$Sw = \frac{v}{D_s/\delta(t)} \tag{34}$$

where v is the velocity of the penetration front, $\delta(t)$ is the time-dependent thickness of the swollen rubbery phase, and D_s is the solute diffusion coefficient in the swollen phase. As written in Eq. (34), Sw is the ratio of the glassy/rubbery relaxation front velocity to the solute diffusive velocity in the rubbery phase. Sw is thus similar to the Peclet number in convective–diffusive problems as an indicator of the relative contribution of each process. It follows that if Sw $\gg$ 1, solute release will exhibit Fickian characteristics, i.e., $n = 0.5$, because diffusion in the swollen phase is rate-limiting. If Sw $\ll$ 1, then case II zero-order release will take place due to the slow relaxation front kinetics. When Sw $\approx$ 1, the release will be anomalous.

B. Diffusion in Swollen Gels

Diffusion in highly swollen matrices such as hydrogels involves a three-component system in which the polymer–solvent matrix constitutes a continuum through which the solute diffuses. The mode and rate of transport through the swollen matrices are determined by the physicochemical properties of the solutes and the structural characteristics of the gels. Two mechanisms have been described for the solute transport in swollen gel membranes [54]. The first is the pore mechanism, in which solutes permeate the membrane by diffusion through the solvent-filled pores. The sizes of the solute and pore are important for determining the transport rate. The second is the partition mechanism, in which solute permeation is achieved by solute partitioning into the membrane structure followed by diffusion along the polymer segments. The physicochemical properties of the solute and polymer thus play a dominant role in determining the membrane permeability, while the size of the solute is less critical. In a study of steroid permeation through poly(hydroxyethyl methacrylate), p(HEMA), membranes, Zentner et al. [55] demonstrated that the pore mechanism dominated in the uncross-linked membrane, while the partition mechanism dominated in the ethylene glycol dimethacrylate cross-linked membrane. It is important to note that the analysis is based on the differentiation of three types of water within the p(HEMA) gels, namely, bulk, interfacial, and bound water. The pore mechanism of transport is only associated with bulk water, while the partition mechanism has been shown for solute permeation through membranes devoid of bulk water.

For hydrophilic and ionic solutes, diffusion mainly takes place via a pore mechanism in the solvent-filled pores. In a simplistic view, the polymer chains in a highly swollen gel can be viewed as obstacles to solute transport. Applying this obstruction model to the diffusion of small ions in a water-swollen resin, Mackie and Meares [56] considered that the effect of the obstruction is to increase the diffusion path length by a tortuosity factor, θ. The diffusion coefficient in the gel, $D_{3,12}$, normalized by the diffusivity in free water, $D_{3,1}$, is related to θ by

$$\frac{D_{3,12}}{D_{3,1}} = \frac{1}{\theta^2} \tag{35}$$

The tortuosity factor appears as a squared term because it decreases the concentration gradient and increases the diffusive path length. Using a cubic lattice model and inquiring how many steps a diffusing molecule needs to take to get around an obstacle, θ was derived to be

$$\theta = \frac{1 + \phi_2}{1 - \phi_2} \tag{36}$$

where ϕ_2 is the polymer volume fraction. This model is interesting in that it is not specific to any particular polymer or solute. This model was successful in describing the diffusion of coions through ion-exchange membranes [57]. Osterhoudt [58] showed that Mackie and Meares' model gave an excellent correlation between the diffusion coefficient and the polymer volume fraction for ethanol and ethylene glycol permeation in water-swollen gelatin and poly(vinyl alcohol) membranes. However, the same model did not give a good prediction of D-fructose diffusivity in diethylene glycol swollen poly(2-hydroxyethyl methacrylate) gels when ϕ_2 is greater than 0.3. Paul et al. [59] also found that the Mackie–Meares model gave a good prediction of the diffusion coefficient of Sudan IV dye molecules in cross-linked natural rubbers swollen by various organic solvents up to a ϕ_2 value of 0.5. However, this model failed in other similar systems with no apparent explanations. Muhr and Blanshard [13] summarized and compared the applicability of the obstruction model to describe the diffusion data of a large variety of solutes and hydrogels and showed that the model was not universally obeyed. In particular, it tends to break down in the case of larger solutes. Since the model was originally derived for small ions, it can reasonably be expected that it will break down at a sufficiently high ϕ_2. In this case, the molecular sieving effect should be considered.

Burczak et al. [60] studied the diffusion of glucose, insulin, and immunoglobulin (IgG) through swollen cross-linked poly(vinyl alcohol) membranes. They found that, in addition to the water content of the membrane, the mesh size of the hydrogel network is also an important factor for a high molecular weight solute such as IgG. A positive correlation was found between the diffusion coefficient and the mesh size. Renkin [61] was among the earliest to postulate a model based on solvent drag to describe the effect of molecular size on the diffusion coefficient. The model is expressed as

$$\frac{D_{3,12}}{D_{3,1}} = \left(1 - \frac{r_s}{r_{cp}}\right)^2 \left[1 - 2.104\left(\frac{r_s}{r_{cp}}\right) + 2.09\left(\frac{r_s}{r_{cp}}\right)^3 - 0.95\left(\frac{r_s}{r_{cp}}\right)^5\right] \quad (37)$$

where r_s is the molecular radius of a spherical solute and r_{cp} is the radius of the cylindrical pores. Ratner and Miller [62] found that the diffusion coefficient data for small solutes such as sodium chloride, urea, and glycine in p(HEMA) membranes gave good correlation with molecular radius r_s using an r_{cp} value of 5.3 Å. The model, however, underestimates the diffusion coefficients of larger solutes such as glucose, sucrose, and raffinose in these membranes. A similar deviation has been observed in the diffusion of fluorescein-labeled dextran in poly(N-isopropylacrylamide-co-vinyl-terminated polydimethylsiloxane) gels [63]. Obviously, Renkin's model is a highly simplistic one in which no concentration dependence or structural feature of the membrane is considered. In addition, the higher values of the diffusion coefficients observed may be due to the involvement of mechanisms other than the simple solvent drag effect considered.

Yasuda et al. [64] developed a free volume theory describing the diffusion of sodium chloride through methacrylate copolymer gels. In their formulation, Yasuda et al. assumed that the free volume in the gels is a linear sum of those of the individual components, i.e.,

$$v_f = Hv_{f,1} + (1 - H)v_{f,2} \tag{38}$$

where H is the volume fraction of water and $v_{f,1}$ and $v_{f,2}$ are free volumes of pure water and polymer, respectively. For highly hydrophilic solutes which can only permeate through the aqueous phase, the polymer contribution to the free volume is negligible and the second term in Eq. (38) is thus dropped. Substituting this expression into Eq. (17) gives

$$\frac{D_{3,12}}{D_{3,1}} = \exp\left[-K\left(\frac{1}{H} - 1\right) \right] \tag{39}$$

where K is a constant related to v_3^* and $v_{f,1}$, and $D_{3,1}$ is the diffusivity of solute in free water. Yasuda et al. found that this relationship was obeyed by sodium chloride diffusion in methacrylate copolymer gels with varying water content. As shown in Figure 5, the experimental values of $\log(D_{3,12})$, varying over five orders of magnitude, were found to correlate well with $1/H$. A similar correlation was found with other solutes of varying sizes in hydroxyethyl methacrylate based gels [65–67]. An extension of this theory was made to include the sieving effect of the macromolecular network for larger solutes. This results in the following expression for $D_{3,12}$ [68]:

$$\frac{D_{3,12}}{D_{3,1}} = \psi(r_2^2)\exp\left[\frac{Br_3^2}{v_{f,1}}\left(\frac{1}{H} - 1\right) \right] \tag{40}$$

where B is a constant, r_3^2 is the effective cross section of the solute, and $\psi(r_3^2)$ is a probability term describing the sieving property of the polymer network. For small values of r_3^2, Ψ is essentially 1, but when the solute size approaches the mesh size of the network, Ψ decreases rapidly. According to this model, a plot of $\ln(D_{3,12}/D_{3,1})$ versus r_3^2 should initially decrease linearly, followed by a region of downward curvature when the sieving effect occurs. The linear dependence of $\ln(D_{3,12}/D_{3,1})$ on r_3^2 was demonstrated in both non-cross-linked and cross-linked p(HEMA) gels using hydrophilic solutes of varying sizes such as inorganic chlorides, sugars, and urea [69]. The same correlation was observed in the cellulosic membranes for molecules up to the size of albumin (MW 66,000) [68]. From the downward curvature, the distribution of the mesh size of the sieving network can be readily determined.

Another related theory expresses the screening effect of the macromolecu-

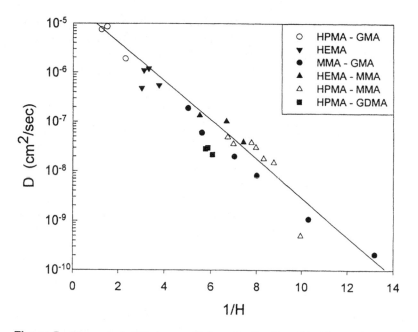

Figure 5 The mutual diffusion coefficient, D, of sodium chloride as a function of reciprocal matrix hydration, H, in various methacrylate gels. HPMA-GMA: poly(hydroxypropyl methacrylate-*co*-glyceryl methacrylate); HEMA: poly(hydroxyethyl methacrylate); MMA-GMA: poly(methyl methacrylate-*co*-glyceryl methacrylate); HEMA-MMA: poly-(hydroxyethyl methacrylate-*co*-methyl methacrylate); HPMA-MMA: poly(hydroxypropyl methacrylate-*co*-methyl methacrylate); HPMA-GDMA: poly(hydroxypropyl methacrylate-*co*-glyceryl dimethacrylate). (From Ref. 64.)

lar mesh on solute diffusion [70]. Using a stochastic approach, the mesh effect is expressed in terms of the molecular weight between cross-links, $\overline{M}_c$:

$$\frac{D_{3,12}}{D_{3,1}} = k_1 \left(\frac{\overline{M}_c - \overline{M}_c^*}{\overline{M}_n - \overline{M}_c^*} \right) \exp\left[\frac{k_2 r_3^2}{Q - 1} \right] \tag{41}$$

where k_1 and k_2 are constants, $\overline{M}_n$ is the number-average molecular weight before cross-linking, $\overline{M}_c^*$ is the threshold value of $\overline{M}_c$ below which no diffusion occurs, and Q is the equilibrium degree of swelling defined as $1/\phi_2$. It is interesting to note that this model has the same general form as Eq. (40), including a preexponential term for the sieving mechanism and an exponential term incorporating the effects of free volume and solute size. In general, the correlations between $\ln(D_{3,12}/D_{3,1})$ and r_3^2 or $1/(Q - 1)$ are excellent; examples include the diffusion of solutes of varying sizes in cellulosic membranes and of vitamin B_{12} in cross-linked polyacrylamide gels [70,71]. However, the model fails to give a quantitative prediction in cross-linked poly(vinyl alcohol) membranes with independently determined $\overline{M}_c$ and $\overline{M}_c^*$ values [72]. Nonetheless, the model represents an attempt to describe diffusion in terms of structural characteristics of the polymer. An unproven variation of this model was derived for diffusional transport through the mucus layer [73].

Since the prediction of the solute diffusion coefficient in a swollen matrix is complex and no quantitative theory is yet possible, Lustig and Peppas [74] made use of the scaling concept, arriving at a functional dependence of the solute diffusion coefficient on structural characteristics of the network. The resulting scaling law thus avoids a detailed description of the polymer structure and yet provides a dependence on the parameters involved. The final form of the scaling law for description of the solute diffusion in gels is

$$\frac{D_{3,12}}{D_{3,1}} = \left(1 - \frac{r_2}{\xi}\right) e^{-Y/(Q-1)} \tag{42}$$

where ξ is the mesh size of the network and Y is a factor related to the ratio of the critical volume for solute diffusion and the molecular free volume of the solvent. Similar to Eqs. (40) and (41), the preexponential term is the probability of moving through the gel mesh (representing the sieving mechanism), and the exponential term is the probability of moving through the free volume. In this expression, ξ is a parameter which is not always easy to determine experimentally. Therefore, a simple scaling approach again is utilized for correlating ξ and Q because Q is a parameter that is readily accessible experimentally. This correlation was demonstrated in cross-linked poly(vinyl alcohol) and p(HEMA) gels [75]. Writing the general correlation function in the form

$$\xi = k' + k''Q^m \tag{43}$$

and fitting experimental ξ to Q, m was determined to be 0.5 for $\phi_2 < 0.1$ and 1 for $\phi_2 > 0.1$. However, it should be noted that the initial polymer concentration before cross-linking was found to influence the correlation. If ξ is not determinable, a scaling law based on a semidilute polymer solution in the form of Eq. (44) can be used [9,74], i.e.,

$$\xi \approx Q^{3/4} \tag{44}$$

Substituting this into Eq. (42) gives

$$\frac{D_{3,12}}{D_{3,1}} = r_3 Q^{-3/4} e^{-Y/(Q-1)} \tag{45}$$

This correlation was found to be valid for solute diffusion in heparinized poly(vinyl alcohol) and cellulosic gels [74]. It was also found that the model is quite insensitive to the value of the parameter Y. As such, a value of unity was found to be a good approximation.

V. SUMMARY

In conclusion, this chapter has covered the basics of mass transport phenomena in polymeric systems. Many problems encountered in engineering the polymeric materials for applications in pharmaceutical systems are of this nature. The discussion in Section III will be essential for treating diffusion problems in dense hydrophobic polymers. Both Fickian and anomalous diffusion were described in detail. Where Fickian diffusion is concerned, free volume based theories are the foundation for understanding the transport. A reasonable prediction can be made by using Eqs. (27) and (28). Although no adequate theories have yet been developed for dealing with anomalous diffusion, a guideline for predicting the occurrence is available. The use of the Deborah number and the swelling interface number are useful for the purpose. The discussion in Section IV is critical for describing the diffusion in hydrophilic polymers such as hydrogels. In most cases, a free volume theory such as Eq. (39) is able to provide a good correlation between the diffusion coefficient and matrix hydration. A further refinement can be made by incorporating a polymer structural factor such as the mesh size, as shown in Eq. (42).

The subject matter covered in this chapter is a complex science. Active research in the field continues to develop new understanding and theories to replace the old. It is important for the interested reader to consult the primary literature in the polymer and materials sciences in order to build on the foundation provided in this chapter.

REFERENCES

1. JD Andrade. Hydrogels for Medical and Related Applications. ACS Symp Ser 31. Washington DC: Am Chem Soc, 1976.
2. RB Bird, WE Stewart, EN Lightfoot. Transport Phenomena. New York: John Wiley, 1960, p 559.
3. JS Vrentas, JL Duda. Diffusion. In: J. Kroschwitz, ed. Encyclopedia of Polymer Science and Engineering. New York: Wiley, 1986, pp 36–68.
4. J Crank, GS Park. Diffusion in Polymers. New York: Academic Press, 1968.
5. J Crank. The Mathematics of Diffusion. 2nd ed. Oxford: Clarendon Press, 1975, p 160.

6. HL Frisch. The time lag in diffusion. J Phys Chem 61:93–95, 1957.

7. P Meares. Transient permeation of organic vapors through polymer membranes. J Appl Polym Sci 9:917–932, 1965.

8. JL Duda. Molecular diffusion in polymeric systems. Pure Appl Chem 57:1681–1690, 1985.

9. PG deGennes. Scaling Concepts in Polymer Plastics. Ithaca, NY: Cornell University Press, 1979.

10. S Prager, FA Long. Diffusion of hydrocarbons in polyisobutylene. J Am Chem Soc 73:4072–4075, 1951.

11. RJ Kokes, FA Long. Diffusion of organic vapors into polyvinyl acetate. J Am Chem Soc 75:6142–6149, 1953.

12. H Fujita. Diffusion in polymer–diluent systems. Fortschr Hochpolym-Forsch 3:1–21, 1961.

13. AH Muhr, MV Blanshard. Diffusion in gels. Polymer 23:1012–1026, 1982.

14. F Bueche. Segmental mobility of polymers near their glass temperature. J Chem Phys 21:1850–1855, 1953.

15. WW Brandt. Model calculation of the temperature dependence of small molecule diffusion in high polymers. J Phys Chem 63:1080–1084, 1959.

16. AT DiBenedetto, DR Paul. Interpretation of gaseous diffusion through polymers by using fluctuation theory. J Polym Sci Part A 2:1001–1015, 1964.

17. JS Vrentas, JL Duda. Diffusion in polymer–solvent systems. I. Reexamination of the free volume theory. J Polym Sci, Polym Phys Ed 15:403–416, 1977.

18. RJ Pace, A Datyner. Statistical mechanical model of diffusion of complex penetrants in polymers. I. Theory. J Polym Sci, Polym Phys Ed 17:1675–1692, 1979.

19. CW Paul. A model for predicting solvent self-diffusion coefficients in nonglassy polymer/solvent solutions. J Polym Sci, Polym Phys Ed 21:425–439, 1983.

20. MH Cohen, D Turnbull. Molecular transport in liquids and glasses. J Chem Phys 31:1164–1169, 1959.

21. D Turnbull, MH Cohen. Free-volume model of the amorphous phase: Glass transition. J Chem Phys 34:120–125, 1961.

22. JL Duda, YC Ni, JS Vrentas. An equation relating self-diffusion and mutual diffusion coefficients in polymer–solvent systems. Macromolecules 12:459–462, 1979.

23. JG Kirkwood, J Riseman. The intrinsic viscosities and diffusion constants of flexible macromolecules in solution. J Chem Phys 16:565–573, 1948.

24. M Kurata, WH Stockmayer. Intrinsic viscosities and unperturbed dimensions of long chain molecules. Fortschr Hochpolym-Forsch 3:196–312, 1963.

25. DW McCall, DC Douglas, EW Anderson. Self diffusion in liquids: Paraffin hydrocarbons. Phys Fluids 2:87–91, 1959.

26. MJ Hayes, GS Park. The diffusion of benzene in rubber. Trans Faraday Soc 52:949–955, 1956.

27. ML Williams, RF Landel, JD Ferry. The temperature dependence of relaxation mechanisms in amorphous polymers and other glass-forming liquids. J Am Chem Soc 77:3701–3707, 1955.

28. JS Vrentas, JL Duda, HC Ling, AC Hou. Free volume theories for self-diffusion in polymer–solvent systems. II. Predictive capabilities. J Polym Sci, Polym Phys Ed 23:289–304, 1985.

29. IC Sanchez. Theory of polymer crystal thickening during annealing. J Appl Phys 45:4216–4219, 1974.

30. JS Vrentas, JL Duda, MK Lau. Solvent diffusion in molten polyethylene. J Appl Polym Sci 27:3987–3997, 1982.

31. JD Ferry. Viscoelastic Property of Polymers. 2nd ed. New York: John Wiley, 1970.

32. JS Vrentas, JL Duda. Molecular diffusion in polymer solutions, AIChE J 25:1–24, 1979.

33. JS Vrentas, JL Duda. Diffusion in polymer–solvent systems. III. Construction of Deborah number diagrams. J Polym Sci, Polym Phys Ed 15:441–453, 1977.

34. JC Wu, NA Peppas. Numerical simulation of anomalous penetrant diffusion in polymers. J Appl Polym Sci 49:1845–1856, 1993.

35. J Crank, GS Park. Diffusion in Polymers. New York: Academic Press, 1968, Chapter 3.

36. HB Hopfenberg, V Stannett. Diffusion and sorption of gases and vapors in glassy polymers. In: RN Haward, ed. The Physics of Glassy Polymers. New York: Wiley, 1973, pp 504–547.

37. N Thomas, AH Windle. Transport of methanol in poly(methyl methacrylate). Polymer 19:255–265, 1978.

38. HB Hopfenberg, L Nicolais, E Drioli. Relaxation controlled (case II) transport of lower alcohols in poly(methyl methacrylate). Polymer 17:195–198, 1976.

39. HB Hopfenberg, KC Hsu. Swelling controlled, constant rate delivery systems. Polym Eng Sci 18:1186–1191, 1978.

40. NL Thomas, AH Windle. A deformation model for case II diffusion. Polymer 21: 613–619, 1980.

41. NL Thomas, AH Windle. A theory of case II diffusion. Polymer 23:529–542, 1982.

42. DS Cohen, AB White Jr. Sharp fronts due to diffusion and stress at the glass transition in polymers. J Polym Sci: Part B 27:1731–1745, 1989.

43. T Alfrey, EF Gurnee, WG Lloyd. Diffusion in glassy polymers. J Polym Sci: Part C 12:249–261, 1966.

44. VS Vadalkar, MG Kulkarni, SS Bhagwat. Anomalous sorption of binary solvents in glassy polymers: Interpretation of solute release at constant rates. Polymer 34: 4300–4306, 1993.

45. NR Vyavahare, MG Kulkarni, RA Mashelkar. Zero order release from glassy hydrogels. II. Matrix effects. J Membrane Sci 54:205–220, 1990.

46. AY Polishchuk, GE Zaikov. General model of transport of water and low-molecular weight solute in swelling polymer. Int J Polym Mater 25:1–12, 1994.

47. P Colombo, R Bettini, P Santi, A De Ascentiis, NA Peppas. Analysis of the swelling and release mechanisms from drug delivery systems with emphasis on drug solubility and water transport. J Controlled Release 39:231–237, 1996.

48. HB Hopfenberg, A Apicella, DE Saleeby. Factors affecting water sorption in and solute release from glassy ethylene-vinyl alcohol copolymers. J Membrane Sci 8: 273–281, 1981.

49. RW Korsmeyer, NA Peppas. Effect of the morphology of hydrophilic polymeric matrices on the diffusion and release of water soluble drugs. J Membrane Sci 9: 211–227, 1981.

50. PI Lee. Dimensional changes during drug release from a glassy hydrogel matrix. Polym Commun 24:45–50, 1981.

51. C-J Kim, PI Lee. Hydrophobic anionic gel beads for swelling-controlled drug delivery. Pharm Res 9:195–201, 1991.

52. CM Klech, X Li. Consideration of drug load on the swelling kinetics of glassy gelatin matrices. J Pharm Sci 79:999–1007, 1990.

53. NA Peppas, NM Franson. The swelling interface number as a criterion for prediction of diffusional solute release mechanisms in swellable polymers. J Polym Sci: Polym Phys Ed 21:983–997, 1983.

54. DJ Lyman, SW Kim. Aqueous diffusion through partition membranes. J Polym Sci, Polym Symp 41:139–144, 1973.

55. GM Zentner, JR Cardinal, J Feijen, S-Z Song. Progestin permeation through polymer membranes. IV. Mechanism of steroid permeation and functional group contributions to diffusion through hydrogel films. J Pharm Sci 68:970–975, 1979.

56. JS Mackie, P Meares. The diffusion of electrolytes in a cation-exchange resin membrane: I. Theoretical. Proc Roy Soc Lond A232:498–509, 1955.

57. JS Mackie, P Meares. The diffusion of electrolytes in a cation-exchange resin membrane: II. Experimental. Proc Roy Soc Lond A232:510–518, 1955.

58. HW Osterhoudt. Transport properties of hydrophilic polymer membranes. The influence of volume fraction polymer and tortuosity on permeability. J Phys Chem 78:408–411, 1974.

59. DR Paul, M Garcin, WE Garmon. Solute diffusion through swollen polymer membranes. J Appl Polym Sci 20:609–625, 1976.

60. K Burczak, T Fujisato, M Hatada, Y Ikada. Protein permeation through polymer membranes for hybrid-type artificial pancreas. Proc Jpn Acad B 67:83–88, 1991.

61. EM Renkin. Multiple pathway of capillary permeability. Circ Res 41:735–743, 1977.

62. BD Ratner, IF Miller. Transport through crosslinked poly(2-hydroxyethyl methacrylate) hydrogel membranes. J Biomed Mater Res 7:353–367, 1973.

63. L-C Dong, Q Yan, AS Hoffman. Macromolecular permeation through hydrogels. Proc Int Symp Controlled Release Bioact Mater Amsterdam 18:261–262, 1991.

64. H Yasuda, CE Lamaze, LD Ikenberry. Permeability of solutes through hydrated polymer membranes: I. Diffusion of sodium chloride. Makromol Chem 118:196–206, 1968.

65. JH Kou, GL Amidon, PI Lee. pH dependent swelling and solute diffusion characteristics of poly(hydroxyethyl methacrylate-co-methacrylic acid) hydrogels. Pharm Res 5:592–597, 1988.

66. K Ishihara, M Kobayashi, I Shinohara. Insulin permeation through amphiphilic polymer membranes having 2-hydroxyethyl methacrylate moiety. Polymer 16:647–651, 1984.

67. A Domb, GW Davidson III, LM Sanders. Diffusion of peptides through hydrogel membranes. J Controlled Release 14:133–139, 1990.

68. H Yasuda, A Peterlin, CK Colton, KA Smith, EW Merrill. Permeability of solutes through hydrated polymer membranes: III. Theoretical background for the selectivity of dialysis membranes. Makromol Chem 126:177–186, 1969.

69. S Wisniewski, SW Kim. Permeation of water soluble solutes through poly(hydroxy-

ethyl methacrylate) and poly(hydroxyethyl methacrylate) crosslinked with ethylene glycol dimethacrylate. J Membrane Sci 6:299–308, 1980.

70. NA Peppas, CT Reinhart. Solute diffusion in swollen membranes. Part I. A new theory. J Membrane Sci 15:275–287, 1983.

71. SH Gehrke, EL Cussler. Mass transfer in pH-sensitive hydrogels. Chem Eng Sci 44:559–566, 1989.

72. CT Reinhart, NA Peppas. Solute diffusion in swollen membranes. Part II. Influence of crosslinking on diffusive properties. J Membrane Sci 18:227–239, 1984.

73. NA Peppas, PJ Hansen, PA Buri. A theory of molecular diffusion in the intestinal mucus. Int J Pharm 20:107–118, 1984.

74. SR Lustig, NA Peppas. Solute diffusion in swollen membranes. IX. Scaling laws for solute diffusion in gels. J Appl Polym Sci 36:735–747, 1988.

75. T Canal, NA Peppas. Correlation between mesh size and equilibrium degree of swelling of polymeric networks. J Biomed Mater Res 23:1183–1193, 1989.

13

Synthesis and Properties of Hydrogels Used for Drug Delivery

Stevin H. Gehrke
Kansas State University, Manhattan, Kansas

I. INTRODUCTION

In colloquial definitions, the term "gel" applies to states of matter intermediate between solids and liquids in some sense, while a "hydrogel" is simply an aqueous gel. In fact, the latter word has its origins in the Latin *gelare*, meaning "to congeal." Over the decades, numerous authors have attempted to formulate more precise definitions; thus, there are probably as many definitions for the term as there are authors on the subject. A most astute statement on the subject was the pronouncement by D. Jordon-Lloyd in the 1920s that "the gel is . . . easier to recognize than to define" [1]. The basis of all definitions is that a gel possesses a continuous macroscopic structure giving the sense of solidity—that is, exhibiting an elastic response to stress, at least up to a certain yield stress—while having a significant liquid component as well. Opinions diverge on whether a dried gel can still be termed a gel.

P. J. Flory [1] suggested that gels, defined loosely along these lines, could be classified on the basis of structural characteristics as follows:

1. Highly ordered lamellar mesophases
2. Polymeric networks held together by covalent bonds yet completely disordered
3. Polymeric networks held together by noncovalent interactions, mostly disordered, but typically with regions of local order
4. Particulate solids dispersed in liquid and devoid of order

Type 1 gels are mesophases that are so highly ordered that they resist disruption of their structure and are thus extraordinarily viscous, to the point of appearing solid-like, even though no high molecular weight species need be present in the system. Surfactants, both synthetic (e.g., sodium dodecylsulfate) and natural (e.g., phospholipids), and clays are typical representatives of this class.

Type 2 gels are essentially infinite molecular weight molecules: Their three-dimensional macroscopic networks comprise structural components that are covalently linked through multifunctional units. This is a very broad class that includes linear polymers that have been chemically or radiochemically cross-linked into a permanent structure as well as networks that have been built up by the step or chain polymerization of difunctional and multifunctional monomers.

Type 3 gels are composed of linear polymers of finite molecular weight that have been bound into a macroscopic structure by noncovalent interactions. These interactions include hydrogen bonds, electrostatic interactions, van der Waals interactions (nonspecific or hydrophobic interactions), insoluble crystallites, and physical entanglements. In general, these gels have some regions of local order although they have no long-range order. Because no covalent bonds hold the network together, many of these gels are reversible; that is, they may form spontaneously or dissolve in response to environmental changes. Gelatin is an archetype of this class; the structure is held together by infrequently occurring triple helices bound together by extensive hydrogen bonding. Polysaccharides (e.g., carrageenan, agar, alginate), collagen, poly(vinyl alcohol), and high molecular weight linear polymers are other important examples of this class.

Type 4 gels are predominantly flocculated precipitates. These include aggregated proteins and needle-like precipitated inorganic crystals. Highly viscous polymer solutions such as cellulose ethers are probably best placed in this category. Type 4 gels differ from physically entangled Type 3 gels in that they have a finite yield stress and are viscoelastic, while Type 3 gels are elastic solids despite their liquid component.

All of these classes of gels prove useful in drug delivery [2]. However, it is the Type 2 and 3 gels that are truly solid-like, displaying elastic responses to shear and maintaining well-defined geometrical shapes. For these materials, their most important property is the equilibrium solvent uptake. In contrast, Type 1 and 4 gels are viscoelastic, and their most important property is their rheological behavior. Because of this difference in fundamental properties, Type 2 and 3 gels must be examined separately from Type 1 and 4 gels.

Thus, for the purposes of this chapter, a hydrogel is considered to be a polymeric material that can absorb more than 20% of its weight in water while maintaining a distinct three-dimensional structure. This definition includes dry polymers that will swell in aqueous environments in addition to the water-swollen materials which inspired the original definitions [3]. A hydrogel that dries without significant collapse of the macroscopic structure and which absorbs water into macropores without substantial macroscopic swelling is termed an "aerogel," while a nonporous gel that absorbs water by swelling is called a "xerogel." This definition does exclude the Type 1 and 4 gels used in pharmaceuticals, such as the various hydrophilic polymers that are used as binders, disintegrants, and so on, but is consistent with the understanding of gels accepted by the broadest range of polymer scientists. The chapter is also restricted to organic hydrogels, since inorganic gels meeting this definition are not often used in drug delivery (although the USP has defined gels as semisolids consisting of inorganic particles enclosed and penetrated by a liquid [2]).

Even with the restrictions imposed by this definition, hydrogels remain an exceptionally diverse group of materials. Any technique used to form polymeric networks can be used to produce hydrogels from hydrophilic polymers. Virtually all hydrophilic polymers can be cross-linked to produce hydrogels, whether the polymer is of biological origin, semisynthetic, or wholly synthetic. Hydrogels can be made biologically inert or biodegradable; they are easily derivatized with specialized groups like enzymes and can be grafted or bonded onto other materials, even living tissue [4]. They can be made in virtually any size or shape using any polymer production technique. Hydrogels can even be made such that their properties continually adjust in response to changes in their environment.

The key properties that make hydrogels thus defined valuable in drug delivery applications are their equilibrium swelling degree, sorption kinetics, solute permeability, and in vivo performance characteristics. The equilibrium swelling degree or sorption capacity (swollen volume per dry volume) is the most important property of a hydrogel and directly influences the other properties. Depending upon the formulation, the swelling degree can be varied virtually without limit. Although the sorption kinetics do depend upon the gel composition, the sorption rate is roughly proportional to the equilibrium swelling degree and thus is also widely variable. Permeability to water, drugs, proteins, and other solutes can also be varied over extraordinarily broad ranges, again depending primarily upon the swelling degree or water content. In vivo, hydrogels can be designed to be biocompatible and/or biodegradable. The high water content of hydrogels also enhances their biocompatibility, since impurities can be easily extracted, their soft, flexible nature mimics natural tissue, and their hydrophilic character minimizes interfacial tension. Unfortunately, the mechanical strength of a gel declines in rough proportion to the swelling degree, although strength is usually of lesser concern for drug delivery than the other four properties. However, when mechanical strength is important, a hydrogel can be bonded onto a support made of plastic, ceramic, or metal. The composite system then gains the mechanical strength of the substrate along with the useful drug delivery properties of the hydrogel.

These properties, either singly or in combination, have led to widespread interest in the use of hydrogels for drug delivery. These materials can be used to protect labile drugs from denaturants, control the release rate of the therapeutic agent, or help target release to a chosen site within the body. They can be made as oral dosage forms, suppositories, injectable micro- or nanoparticles, or implants at any site within the body, even when blood contact is required. This chapter is intended to make clear how the properties of gels can be controlled to deliver a therapeutic agent in a desired manner.

II. HYDROGEL SYNTHESIS

In the most succinct sense, a hydrogel is simply a hydrophilic polymeric network cross-linked in some fashion to produce an elastic structure. Thus any technique which can be used to create a cross-linked polymer can be used to produce a hydrogel. Copolymerization/cross-linking free radical polymerizations are commonly used to produce hydrogels by reacting hydrophilic monomers with multifunctional cross-linkers. Water-soluble linear polymers of both natural and synthetic origin are cross-linked to form hydrogels in a number of ways:

1. By linking polymer chains via chemical reaction
2. By using ionizing radiation to generate main-chain free radicals which can recombine as cross-link junctions
3. Through physical interactions such as entanglements, electrostatics, and crystallite formation

Almost any combination of synthetic, biological, or semisynthetic (modified biological polymers, such as cellulose ethers) materials can be used to form a hydrogel, as long as the combination is sufficiently hydrophilic to absorb water. Net hydrophilicity of the network arises through inclusion in the network of a sufficient concentration of amide, amine, carboxylate, ether, hydroxyl, sulfonate, or other groups which interact favorably with water. Hydrogels can even be produced by the chemical conversion of one type of gel into another. Any of the various polymerization techniques can be used to form gels, including bulk, solution, and suspension polymerization.

Thus hydrogels form a truly diverse class of materials. In fact, it is this diversity and flexibility in preparation that leads to their utility: Many different formulations can be tested for a given application, and the formulation can be readily fine-tuned for optimal performance. In the following subsection, the range of hydrogel types available and their preparation are described.

A. Monomer Polymerization

The most common and most versatile technique for the production of synthetic hydrogels is the free radical (chain) copolymerization of monofunctional and multifunctional vinyl monomers. Gels can also be created by the step (e.g., condensation) polymerization of suitable monomers. However, this approach is rarely used to produce the organic hydrogels useful in pharmaceutical applications, although it is very important for production of the inorganic gels used as ceramic precursors. Other methods such as ionic polymerization have also been used to produce hydrogels, but again, these are infrequently used. Thus this section focuses on monomer polymerization using free radical copolymerization/cross-linking reactions.

1. Types of Gels

Copolymerization/cross-linking synthesis of gels is widely used because of the virtually endless variety of gels which can be produced by mixing and reacting different combinations of cross-linkers and monomers. Tables 1–5 list some of the monomers which can be used to create hydrogels (adapted from Refs. 3 and 5). Commonly used nonionic, hydrophilic monomers are listed in Table 1; a cross-linked homogel of these monomers would yield a hydrogel. These can also be copolymerized with the nonionic, hydrophobic monomers of Table 2; a cross-linked homogel of these monomers would not swell in water. Tables 3 and 4 list examples of ionic monomers: Table 3 includes anionic, acidic monomers, and Table 4 lists cationic, basic monomers. Finally, Table 5 lists commonly used multifunctional cross-linking monomers.

Hydrogel properties can be varied widely by choosing different monomers and cross-linkers to form the gel. Properties can be fine-tuned by adjusting the ratio of monomer to cross-linker or by incorporating into the network varying ratios of comonomers. For example, the swelling of a given hydrogel can be reduced by adding a more hydrophobic monomer to the formulation or by increasing the amount of cross-linker used. Conversely, the swelling degree can be increased by copolymerization with a more hydrophilic monomer or an ionic monomer or by reducing the amount of cross-linker used. Although equilibrium swelling is the most important property of the gel, other properties can be subtly or dramatically modified by using different combinations of monomers in the polymerization mixture. In practice, the actual range of combinations which can be achieved will be limited by the need to find a compatible solvent for all monomers used and by differences in reaction rates, although the total number of combinations achievable remains virtually endless.

The gel properties will also be influenced by many other parameters including the nature of the cross-links, synthesis conditions, type and concentration of initiator used, phase separation, and the presence of noncovalent interactions such as hydrogen bonding and hydrophobic interactions. Nonetheless, the gel properties depend primarily upon the monomers used.

Monomers based upon acrylic and methacrylic acids, especially the methacrylics, form the most versatile and most important class of polymerizable mono-

Table 1 Nonionic, Hydrophilic Monomers Used for Hydrogel Synthesis

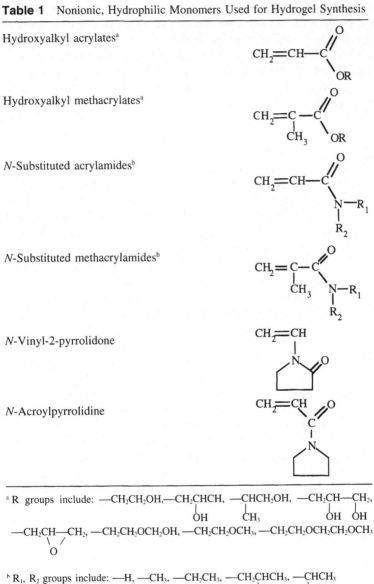

Hydroxyalkyl acrylates[a]

Hydroxyalkyl methacrylates[a]

N-Substituted acrylamides[b]

N-Substituted methacrylamides[b]

N-Vinyl-2-pyrrolidone

N-Acroylpyrrolidine

[a] R groups include: —CH₂CH₂OH, —CH₂CHCH, —CHCH₂OH, —CH₂CH—CH₂,

—CH₂CH—CH₂, —CH₂CH₂OCH₂OH, —CH₂CH₂OCH₃, —CH₂CH₂OCH₂CH₂OCH₃

[b] R₁, R₂ groups include: —H, —CH₃, —CH₂CH₃, —CH₂CHCH₃, —CHCH₃

Source: Ref. 5.

Table 2 Nonionic, Hydrophobic Monomers Used
for Hydrogel Synthesis

Acrylics[a]	$CH_2=CH-C\overset{\displaystyle O}{\underset{\displaystyle OR}{\diagup\diagdown}}$
Methacrylics[a]	$CH_2=\underset{\displaystyle CH_3}{C}-C\overset{\displaystyle O}{\underset{\displaystyle OR}{\diagup\diagdown}}$
Vinyl acetate	$CH_2=CH-C\overset{\displaystyle O}{\underset{\displaystyle CH_3}{\diagup\diagdown}}$
Acrylonitrile	$CH_2=\underset{\displaystyle CN}{CH}$
Styrene	$CH_2=CH$

[a] R groups include: $-CH_3$, $-CH_2CH_3$, $-CH_2CH_2CH_2CH_3$,
$-(CH_2)_5CH_3$
Source: Ref. 5.

mers for pharmaceutical hydrogels. A broad variety of esters can be prepared
from these acids, which can be either hydrophilic or hydrophobic, depending
upon the nature of the ester group, as indicated in Tables 1 and 2 [3,5]. Methacrylic acid (MAA) is also commonly derivatized with aminoalkyl groups to form
cationic monomers, as shown in Table 4. Even cross-linkers can be made from
MAA by condensing the carboxyls with the terminal hydroxyl groups of poly(ethylene glycol)s (Table 5). In fact, 2-hydroxyethyl methacrylate (HEMA) is perhaps
the most widely studied biomedical hydrogel of this class, while ethylene glycol
dimethacrylate (EGDMA) is one of the most common cross-linkers. A variety
of copolymers of methacrylates have been sold for pharmaceutical use as the
Eudragit series by Rohm Pharma. The many other monomers shown in these
tables are also quite valuable, either alone or in combination with other monomers. *N*-Vinylpyrrolidone is one of the most hydrophilic of monomers and is
quite useful for enhancing the swelling of a given gel, even though it doesn't
homopolymerize well or form good quality homopolymer gels. Acrylamide is
another strongly hydrophilic monomer, and *N*-substituted acrylamides and meth-

Table 3 Anionic Monomers Used for Hydrogel Synthesis

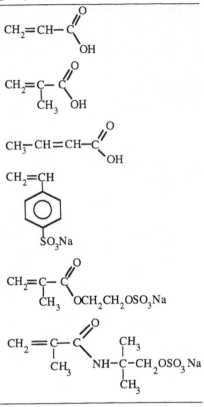

Acrylic acid	
Methacrylic acid	
Crotonic acid	
Sodium styrene sulfonate	
Sodium 2-sulfoxyethyl methacrylate	
2-Acrylamido-2-methylpropanesulfonic acid	

Source: Ref. 5.

acrylamides are usually hydrolytically stable if doubly substituted. In addition to the hydrophobic acrylate and methacrylate esters, vinyl acetate is a common hydrophobic monomer for hydrogel synthesis, especially as a precursor to poly(vinyl alcohol) (discussed later). And in addition to the weakly acidic methacrylic and acrylic acids, the strongly acidic sulfonated monomers and related species are valuable when pH-dependent ionization must be avoided. Of the cross-linkers, EGDMA, TEGDMA (tetramethylene glycol dimethacrylate), N,N'-methylenebisacrylamide, and divinylbenzene are the most commonly used, although many more cross-linkers are available than are shown in Table 5. The high molecular weight cross-linkers based on PEG terminated with reactive cross-linking groups promise improved biocompatibility over their low molecular weight analogues. Many proprietary cross-linkers promising improved mechanical properties of gels

Table 4 Cationic Monomers Used for Hydrogel Synthesis

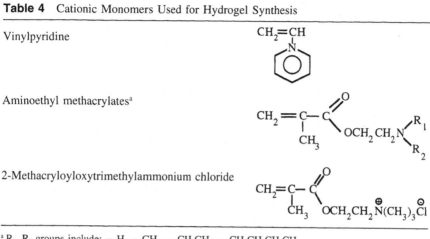

Vinylpyridine	
Aminoethyl methacrylates[a]	
2-Methacryloyloxytrimethylammonium chloride	

[a] R_1, R_2 groups include: —H, —CH_3, —CH_2CH_3, —$CH_2CH_2CH_2CH_3$
Source: Ref. 5.

Table 5 Multifunctional Cross-Linking Monomers Used for Hydrogel Synthesis

N,N'-Methylenebisacrylamide	
Poly(ethylene glycol) dimethacrylate	
2,2'-(p-Phenylenedioxy) diethyl dimethacrylate	
Divinylbenzene	
Triallylamine	

Source: Ref. 5.

are available; many of these were originally developed for use in soft contact lenses.

2. Synthesis Principles

In this section, the important concepts related to the formation of hydrogels by free radical copolymerization/cross-linking are examined. Greater depth beyond the scope of this chapter can be obtained from textbooks on polymer chemistry and the papers cited herein. As stated earlier, almost all gels produced from monomers for pharmaceutical applications are synthesized by free radical chain polymerizations.

Solution and bulk polymerization are the two basic routes to the production of a hydrogel. Bulk polymerization can be used if the monomer is a liquid in which suitable comonomers, cross-linkers, and initiators are also soluble. For solution polymerization, the monomers, cross-linkers, and initiators are dissolved in a common solvent, forming a single solution phase. The addition of a cross-linker may not be required for gels which do not swell enough to dissolve; poly(2-hydroxyethyl methacrylate) (PHEMA) is one prominent example. Because oxygen scavenges the free radicals which initiate and propagate the polymerization, oxygen is a potent inhibitor. Thus synthesis normally must be carried out in reduced-oxygen or oxygen-free environments.

The free radical sources which can be used include both chemical sources, including solvent-, thermally-, or light-activated chemicals, and ionizing radiation (γ-rays or high energy electrons). Chemical initiators which generate free radicals upon decomposition caused by thermal activation or dissolution are the most widely used initiators. A wide variety of such initiators can be used, including azo compounds, peroxides, redox couples, and photoinitiators [3,4]. Initiators commonly used for hydrogel synthesis include azobisisobutyronitrile (AIBN), benzoyl peroxide, diisopropyl percarbonate, and the ammonium persulfate (APS)–sodium metabisulfite or APS–N,N,N',N'-tetramethylethylenediamine (TEMED) redox couples. Since a chemical initiator is incorporated into the network as the first link in a chain, Gregonis et al. [6] used azo initiators modified to resemble the repeating units of the gel. However, since the total mass of initiator incorporated into the network is inconsequential in comparison to the total mass of the network, this is not usually a concern.

While subtle differences between gels synthesized with different initiators may exist, they are not easy to anticipate. The most important concern is usually the polymerization rate induced by a given initiator concentration [7]. Polymerization reactions are highly exothermic, so a fast initiation rate can lead to a rapid temperature increase; since the initiation and polymerization rate both increase rapidly with temperature, this process becomes autocatalytic. Poor quality, irreproducible gels result; on a production scale such a runaway reaction could become dangerous. However, a reaction rate that is very slow may be impractical and may also lead to poor quality gels; the solution may even fail to gel. Thus the choice of an initiator and its concentration level is usually made on an empirical basis. Other factors which enter into this decision include cost, safety, solubility of the initiator, the convenience of its mode of activation, and the ability to control the initiation rate. In terms of convenience and control, UV-initiated polymerizations using UV-sensitive AIBN or riboflavin initiators can be ideal, since the reaction can be readily modulated by the radiation intensity. Furthermore, UV initiation is often effective at lower concentrations than thermally initiated reactions, which can be desirable if the resulting gel cannot be easily leached of residuals. With most other initiators, free radicals begin to be generated as soon as the initiator dissolves, which may cause practical problems. Furthermore, thermally initiated reactions are inherently difficult to control because of the built-in time lags due to the thermal mass of the solution. Photoinitiation also provides a route for controlling gelation reactions in vivo [4].

Ionizing radiation can also be used to initiate free radical polymerization. The key advantage of this approach is that it leaves no residual initiator in the gel which must be removed prior to biomedical use. It is also conceivable that the system could be set up to simultaneously sterilize and synthesize the gel. However, the equipment and facilities needed to supply γ-radiation or high energy electrons are more expensive to use, less widely available, and more difficult to scale up for production. Also, some polymers tend to degrade under irradiation, including the methacrylates. Some differences in properties between gels formed by ionizing radiation and those formed with chemical initiators have been observed, possibly due to residual initiator effects.

The type and amount of solvent used, if any, when preparing a gel can also have significant effects upon the properties of the resulting gel. If the solvent is a good one for both the monomer and polymer, a transparent, homogeneous gel is usually formed. If the polymerization occurs in a solvent which is a poor one for the polymer, an opaque, heterogeneous, possibly spongelike gel can be formed due to the resulting phase separation. This can have a very dramatic influence on the properties of the gel produced, including the swelling degree, sorption/desorption rates, permeability, and mechanical properties [8]. Gels prepared in different solvents but otherwise identical in composition and synthesis conditions may also have different properties as a result [9].

Despite the importance of initiators, synthesis conditions, and diluents on the properties of a gel, composition is, of course, the most important variable. When growing polymeric chains are first initiated, they tend to grow independently. As the reaction proceeds, different chains become connected through cross-links. At a critical conversion threshold, called the "gel point" or the "sol–gel transition," enough growing chains become interconnected to form a macroscopic network. In other words, the solution gels. The reaction is typically far from complete at this point, however. A substantial amount of monomer or chains not yet attached to the network are still present; these species are termed the "sol," in contrast to the macroscopic network or "gel." As the reaction continues past the gel point, the amount of sol declines as it is incorporated into the network. In general, however, a nonnegligible sol fraction remains even upon cessation of reaction. The practical significance of all this is that reaction time becomes a significant variable affecting the properties of the resulting gels. Table 6 provides data which clearly indicate this [10]. In this case, sets of gels of different monomer concentrations and cross-linker/monomer ratios were prepared; one member of each set was allowed to proceed to completion (24 hr), while the other was quenched shortly after the gel point (0.75–1 hr). In all cases, the shear modulus and cross-linking efficiency were significantly lower, and thus the swelling degree was greater for the gels quenched at the gel point than for the gels allowed to reach equilibrium. Interestingly, the properties of the different gels are quite similar at the gel point, since at this point just enough network has formed to span the volume of the container; the more concentrated solutions seem to simply reach this point faster. The data also show that the cross-linking efficiency is greater for more concentrated monomer solutions and for larger ratios of monomer to cross-linker. Such observations have been examined from a theoretical point of view by a number of authors [11–14].

The composition of a copolymer gel may also turn out to be significantly different from that of the starting mixture if the reaction does not go to completion and if all the sol is not incorporated into the network. This results from the differences in the reactivities of different monomers; only in special cases are different

Table 6 Properties of N-Isopropylacrylamide Gels as a Function of Reaction Time[a]

Gel type	Reaction time (hr)	Equilibrium degree of swelling	Shear modulus (g/cm^2)	Effective cross-link density (10^{-5} g/cm^3)
10 × 1	1	25.7	30	3.6
	24	19.6	33	3.6
16 × 1	1	19.7	44	3.6
	24	14.8	87	6.5
10 × 4	0.75	18.2	28	3.0
	24	11.1	124	11.0
16 × 4	0.75	17.6	25	2.0
	24	7.8	220	13.0

[a] $M \times N$ notation indicates (M g monomer)/(100 g water) with (N g crosslinker)/(100 g monomer).
Source: Ref. 10.

monomers incorporated into a network at the same rates and in a random, evenly dispersed fashion. There are three basic extremes for the incorporation of different monomers into a copolymer: the formation of block, alternating, or random copolymers. Block copolymers tend to form if monomer A reacts with like monomers much more strongly than with unlike monomers, especially if the same holds true for monomer B. The resulting polymer thus develops long homopolymer segments: —AAAAAAAABBBBBBBB—. Such polymers tend to phase separate into A-rich and B-rich regions, especially if one segment is incompatible with the swelling solvent. Sometimes this phase separation can lead to improved mechanical strength or enhanced drug loading (e.g., a hydrophobic drug may be strongly absorbed by a hydrophobic region in the gel). Without special care, however, such gels may be of poor quality, especially if the reaction rates of the monomers are significantly different or if the comonomer is the cross-linker. Alternating copolymers form when a monomeric unit has a strong preference to add a monomer of the opposite kind, thus favoring development of a structure like —ABABABABABABAB—. If monomers have equal reactivities toward each other, random copolymers form: —ABBAAABABAABBBBAB—. In most situations, the copolymer formed will be something between these limiting cases. The type of copolymer which might be expected to form can be estimated with the simple, moderately successful Alfrey–Price "Q-e" scheme [15,16]. For example, this scheme predicts that monomers which are free acids or bases often will be incorporated into a network in much different fashions than their corresponding salts (e.g., methacrylic acid vs. sodium methacrylate). This could be a factor in developing a synthesis procedure for a given gel.

B. Cross-Linking Linear Polymers

Linear hydrophilic polymers, whether synthetic, biological, or semisynthetic (modified biological polymers), can be cross-linked into a permanent network in a number of ways. Since many hydrophilic polymers have free hydroxyl groups, the most versatile technique uses a reagent which can react with multiple hydroxyl groups and link different polymer chains together. Primary amines can be cross-linked similarly, and these groups are generally more reactive than the hydroxyls. Ionizing radiation can also covalently cross-link many classes of linear polymers by creating free radicals on the chains, which then recombine randomly, leading to network formation. Physical interactions of many different types can also lead to the formation of stable, elastic gels, even in the absence of covalent linkages.

1. Types of Gels

Table 7 lists in four categories some of the linear polymers which can be cross-linked to form hydrogels: nonionic, anionic, cationic, and ampholytic linear poly-

Table 7   Linear Polymers Which Can Be
Cross-Linked to Form Hydrogels[a]

Nonionic linear polymers
  Agarose (B)
  Amylose (B)
  Carrageenan (B)
  Dextran (B)
  Starch (B)
  Poly(ethylene glycol) (S)
  Poly(ethylene oxide) (S)
  Poly(propylene oxide) (S)
  Poly(vinyl alcohol) (S)
  Poly(vinyl methyl ether) (S)
  Cellulose ethers (SS)
Anionic linear polymers
  Sodium alginate (B)
  Sodium carrageenan (B)
  Sodium heparin (B)
  Sodium hyaluronate (B)
  Polygalacturonic acid (B)
  Polyglucuronic acid (B)
  Sodium carboxymethyl cellulose (SS)
  Sodium cellulose sulfate (SS)
  Sodium celluronate (SS)
  Sodium lignosulfonate (SS)
  Sodium polyribonucleate (SS)
Cationic linear polymers
  Chitosan (SS)
  Quaternized chitosan (SS)
  Poly(ethylene imine) (S)
  Aminoalkylated quaternized cellulose (SS)
  Poly(L-lysine) (S)
Polyampholytes
  Polypeptides (B,S)
  Gelatin (SS)

---

[a] B = biological; S = synthetic; SS = semisynthetic.

mers. This list includes a diverse set of materials, including synthetic polymers, biological polymers, and modified biological (semisynthetic) polymers; their only common feature is their hydrophilicity. Furthermore, any of the monomers in Tables 1–4 can be polymerized in the absence of a cross-linker to form linear polymers which can be subsequently cross-linked by techniques discussed in the following section. However, the techniques described here are most frequently used only for biologically derived polymers or synthetic polymers that cannot be formed directly by copolymerization or cross-linking reactions carried out by solution or bulk polymerization. But in pharmaceutical applications the formation of gels by purely physical interactions can be preferred over chemical techniques because gelation can occur in the presence of a drug without need for concern about side reactions or leachable impurities such as residual initiators or monomers.

The key synthetic gels for pharmaceutical applications which are not produced by copolymerization/cross-linking of monomers are made from poly(vinyl alcohol) (PVA), poly(ethylene glycol) (PEG), and poly(ethylene oxide) (PEO). PVA cannot be prepared by the polymerization of vinyl alcohol, since vinyl alcohol is much less stable than its tautomer, acetaldehyde. As a result, PVA must be prepared indirectly. Commercially, PVA is produced by polymerizing vinyl acetate followed by hydrolysis to PVA. Many different grades of PVA can be produced in terms of degree of hydrolysis and stereoregularity (isotactic, syndiotactic, atactic). The properties of PVA have been reviewed in depth by Peppas [17]. PVA gels are usually held together by crystallites, as discussed below. Although PEO and PEG have the same repeating units, namely $-(CH_2CH_2O)-$, gels are usually made from them in different ways, since PEG is usually of much smaller molecular weight than PEO and has hydroxyls as end groups. Thus PEG can be gelled by condensation reactions with cross-linkers such as triisocyanates and polyols. In contrast, PEO is usually cross-linked by γ-radiation, physical entanglements, or molecular complexes with other polymers such as poly(methacrylic acid). The formation of pharmaceutical gels from PEO and PEG has been reviewed by Graham [18].

There are many polymers of biological origin which form hydrogels, some spontaneously, while others are cross-linked synthetically. Perhaps the most familiar spontaneously gelling polymer is gelatin, a protein derived from collagen. Under appropriate conditions of pH, temperature, etc., the linear polymers form helices in regions which are stabilized by extensive hydrogen bonding. These helices function as cross-links which hold the amorphous regions together. Some polysaccharides such as carrageenan form similar gels [19]. These structures can be made permanent and stronger by reacting them with aldehydes to covalently cross-link the polymers through their hydroxyl groups. Starch gels can be synthesized in various ways, such as by cross-linking with phosphoryl chloride (POCl₃) [20–22].

Semisynthetic gels are also very useful for the creation of drug delivery systems. Cellulose ethers are particularly important in drug delivery. These compounds are made by derivatizing the cellulose hydroxyls with various groups such as hydroxypropyl, methyl, or carboxymethyl. This substitution breaks up the crystallinity of native cellulose and makes it water-soluble [23]. The degree of substitution and the nature of the substituent determine the hydrophobic/hydrophilic balance of the polymer and, together with the average molecular weight of the polymer, control solution properties. These polymers can be cross-linked with epichlorohydrin, diglycidyl ether, or divinylsulfone to form hydrogels with a very wide range of properties [24–28]. Cross-linked agarose and dextran gels and their derivatives (e.g., dextran sulfate) are extremely valuable in chromatography and also have biomedical value [29–31]. Gels of the polymers listed in Table 7 can also be made by a great variety of other techniques, as discussed below and by several authors [17–20,32].

2. Chemical Cross-Linking

Second only to copolymerization/cross-linking as a general technique for the production of a wide variety of hydrogels is the cross-linking of linear polymers with various cross-linking agents. As touched upon earlier, cross-links can be formed with many different multifunctional compounds capable of reacting with the labile groups available on most hydrophilic polymers [19]. Cross-linking through reaction with pendant hydroxyls is common, as illustrated in Figure 1, as well as cross-linking through amine groups, especially for proteinaceous gels such as gelatin [19,33]. Commonly used cross-linking agents include acetaldehyde, formaldehyde, glutaraldehyde, divinylsulfone, diisocyanates, dimethyl urea, epichlorohydrin, oxalic acid, phosphoryl chloride, trimetaphosphate, trimethylomelamine, polyacrolein, ceric redox systems, and so on [19,22].

Solution cross-linking is typical, although suspension cross-linking of linear polymers is used to produce cross-linked beads or particles [27]. While water is the usual solvent, other solvents can also be used, especially alcohols. The nominal cross-linking ratio X (ratio of moles of cross-linking agent to moles of polymer repeating units) is often used as an indication of the extent of cross-linking in the resulting gel. A nominal number-average molecular weight between cross-links, M_c, for difunctional cross-linkers can be defined as

$$M_c = M_r/2X \tag{1}$$

where M_r is the molecular weight of the repeating unit. However, it is critically important to recognize that X is really just a formulation specification and is not directly related to the actual number of cross-links present. Not all cross-linking molecules will form cross-links; some homopolymerize before cross-linking, some fail to form elastically effective cross-links, and some simply fail to react [22]. Since the number of effective cross-links which form cannot be predicted, direct measurement of the effective cross-link density is required; this is described in Section III.A.3.

Statistical theories which predict the gel point as a function of cross-linking have been developed. While the theoretical analysis of gelation caused by the

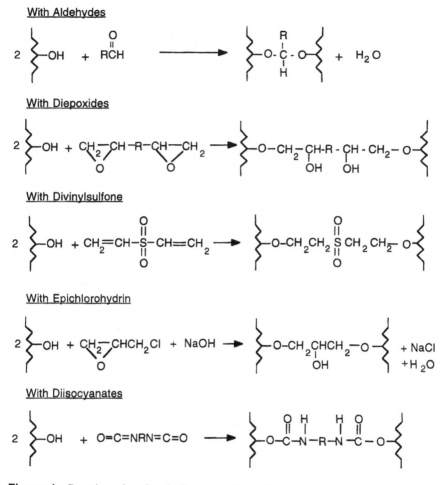

Figure 1 Reactions for chemically cross-linking linear polymers through hydroxyl groups.

cross-linking of linear polymers is quite different from copolymerization/cross-linking, the phenomena are qualitatively similar [34–38].

3. Radiation Cross-Linking

The use of ionizing radiation to induce cross-linking is another important technique for producing hydrogels from linear water-soluble polymers. When such polymers are bombarded by ionizing radiation, either in the bulk state or in solution, various covalent bonds are split, forming free radicals on the polymer chain. These radicals can cause the polymer chain to split into lower molecular weight fragments, or they can recombine in a random fashion, leading to the development of branched polymers. If the branching and cross-linking rate is faster than the degradation rate, a network will be formed. As a simple rule of thumb, vinyl polymers with structure 1 tend to form gels when irradiated, while polymers with structure 2 tend to degrade [35]:

$$
\begin{array}{cc}
\text{H} & \text{R'} \\
| & | \\
-\text{CH}_2-\text{C}- & -\text{CH}_2-\text{C}- \\
| & | \\
\text{R} & \text{R} \\
\mathbf{1} & \mathbf{2}
\end{array}
$$

Thus polyacrylates tend to cross-link, while polymethacrylates tend to degrade, for example. Cellulosics also tend to degrade when irradiated.

Typically, either γ-radiation from a ^{60}Co source or a high energy (2–10 MeV) electron beam from a Van de Graaff generator is used to cross-link polymers. Gamma rays have greater penetrating power, but otherwise the basic effects of the two types of radiation on polymers are similar. The primary variables used to control the extent of cross-linking are the dose rate and the total dose. The solvent type and degree of dilution also affect the nature of the gel formed. In any case, oxygen must be rigorously excluded from the system during irradiation, as oxygen promotes degradation of the polymer. This is especially critical for polymers cross-linked at low dose rates. Defects such as bubbles may be caused by the evolution of gases formed by the recombination of low molecular weight radicals, especially hydrogen, during irradiation. The gel point or extent of cross-linking must be determined by experiment; predictions cannot be made based simply upon the irradiation characteristics.

Despite the expense and inconvenience of working with radiation sources, radiation cross-linking is a valuable technique for the production of gels from a number of important water-soluble polymers, including PEO, PVA, poly(vinyl methyl ether) (PVME), and poly(N-vinylpyrrolidone) (PNVP). Radiation cross-linking also has the advantage of minimizing extractable low molecular weight impurities such as initiators, monomers, or cross-linkers, all of which tend to be toxic. Furthermore, it can be designed to simultaneously sterilize the gel. (It is also important to recognize that if radiation is used to sterilize a gel synthesized by a chemical or physical cross-linking technique, the properties of the gel may change as a result of unintentional cross-linking or degradation.)

4. Physical Cross-Linking

The two previous sections covered gels whose structure is formed through covalent cross-links; these gels are basically gigantic molecules of macroscopic dimension that fall into Flory's Type 2 gel category. In this section, gels whose structures are stabilized by noncovalent interactions are examined; these gels are conglomerations of finite molecular weight polymers—Flory's Type 3 gels. Because these gels are not formed by covalent bonds, many can be redissolved. Such gels are thus referred to as "reversible" gels. However, once formed, these gels can have strength and permanence comparable to those of Type 2 gels. By and large, gels in this category have the same basic properties and behavior as the previously discussed gels. The difference is simply that the cross-links are formed by physical interactions such as entanglements, charge complexes, hydrogen bonding, crystallite formation, or hydrophobic interactions rather than by covalent bonds.

The fact that no chemical reactions are required to form a Type 3 gel can be a significant advantage over Type 2 gels in pharmaceutical applications. First of all, the linear polymers used to make physical gels can be purified of the reagents used in their synthesis much more easily than covalently cross-linked gels can be purified. Second, active ingredients can be mixed directly with a polymer solution that gels physically. In contrast, regulatory approval is unlikely if covalent cross-linking reactions are carried out in the presence of the drug. In some applications, it could even be an advantage that the gel can be redissolved. The actual techniques used to create such gels vary widely; gelation may be induced by changes in temperature, ionic strength, cycles of freezing and thawing, or simply by mixing the appropriate components. Here the focus is on the different types of interactions and the gels that can be formed by using them, rather than on specific preparation techniques.

Simple physical entanglements can be sufficient to produce a structurally stable gel if the polymer has a sufficiently great molecular weight and if the polymer is of only modest hydrophilicity. In this case, the polymer will swell in water without dissolving, even in the absence of covalent cross-links. Poly(2-hydroxyethyl methacrylate) (PHEMA) is a prominent example of this type of hydrogel; when uncross-linked, it will dissolve in 1,2-propanediol but only swell in water.

An important class of gels are those stabilized by ionic bonds; these are solutions of polyelectrolytes and polyvalent counterions. Gels formed of a polyelectrolyte and a low molecular weight counterion, typically a multivalent metal ion, have been termed "ionotropic" gels [22], while gels formed of a complex of anionic and cationic polyelectrolytes have been termed "symplexes" [32]. In the former case, the small ion functions as a cross-linker, holding different chains together; calcium alginate is a prominent example of this type of gel. The structures of the symplex gels are somewhere between a highly ordered "ladder" structure, where two oppositely charged polymers couple in a one-to-one fashion, and the disordered "scrambled egg" formation, where specific cation–anion interactions form between polymers randomly. Such gels can be made from many combinations of natural and/or synthetic polyelectrolytes, such as the sodium carboxymethyl cellulose–poly(ethylene imine) symplex [32]. The formation of such gels has a complex dependence upon many parameters related to the solution environment and the chemical and physical structure of the polymer. Their sorption capacity also depends upon variables such as pH in a complex fashion, although there is not a great deal of sorption data available for such gels.

A related class of gels are those formed by extensive hydrogen bonding. An example is the poly(ethylene oxide)–poly(methacrylic acid) complex [18]. Spontaneously gelling natural polymer solutions are frequently of this type, including gelatin and native starch.

Semicrystalline hydrophilic polymers, most notably PVA, typically form hydrogels when swollen in water. The amorphous regions of these polymers will absorb water, but the crystalline regions (''crystallites'') normally will be insoluble and thus serve as polyfunctional cross-link junctions. The polymer may be inherently semicrystalline, but the crystallites may need to be grown to an appropriate level. This can be done by simple aging, but for PVA repeated cycles of freezing and thawing have been shown to enhance the crystallinity of PVA gels, leading to enhanced mechanical strength [39,40].

Hydrogel structure can also be stabilized by noncovalent hydrophobic interactions. Such gels can be prepared most readily from block or graft copolymers, where one sequence is hydrophilic and the other is hydrophobic. When immersed in water, the hydrophilic portion of the polymer will absorb water while the hydrophobic regions will tend to aggregate to avoid water contacts while maximizing polymer contacts. This interaction can be sufficient to form a strong, permanent gel, especially if the application temperature is below the glass transition temperature of the hydrophobic sections. Examples of these hydrogels include PEO-PMMA (poly(methyl methacrylate)) block copolymers [18], PHEMA-poly(tetramethylene oxide) graft copolymers, and hydrophobically modified cellulose ethers [41–43]. Such gels promise to be easy to form, extrude, and load with drug.

C. Other Techniques for Hydrogel Synthesis

While copolymerization/cross-linking of reactive monomers and cross-linking of linear polymers are the main methods for producing hydrogels, there are other techniques which also deserve mention, particularly chemical conversion and the formation of interpenetrating networks.

Production of hydrogels by chemical conversion uses chemical reactions to convert one type of gel to another. This can involve the conversion of a nonhydrogel to a hydrogel or the conversion of one type of hydrogel to another. Examples of the former include the hydrolysis of polyacrylonitrile to polyacrylamide or of poly(vinyl acetate) to poly(vinyl alcohol); an example of the latter includes the alkaline hydrolysis of polyacrylamide to poly(acrylic acid). As another example, PVA gels with superior mechanical properties can be produced by the bulk polymerization of vinyl trifluoroacetate (forming a non-hydrogel), followed by solvolysis of PVTFA to PVA [44].

An interpenetrating network (IPN) has two (or more) intertwined but structurally independent networks. They can be formed in several ways. One method swells an existing gel in a monomer solution, which is subsequently polymerized. Another technique carries out simultaneous but independent reactions, such as a step polymerization and a chain polymerization or a condensation cross-linking of linear polymers and copolymerization/cross-linking of monomers. Such IPNs can be made from two hydrogels or from one hydrogel and a hydrophobic network. IPNs may have superior mechanical properties compared to a homogel, or the technique may be a more effective way of modulating the swelling properties of a gel than copolymerization [45]. Another reason for making an IPN is to form a hydrogel from a desirable hydrophilic polymer that is difficult to cross-link. As an example, high molecular weight PEO can be admixed with vinyl monomers and cross-linkers. When the latter react, the PEO will be entangled with the vinyl gel, thus forming a physically entangled PEO gel within a chemically cross-linked vinyl monomer–based gel.

III. PROPERTIES OF HYDROGELS

The key properties of hydrogels which make them useful in drug delivery are their swelling degree, swelling kinetics, permeability, and in vivo properties such as biocompatibility or biodegradability. As Section II has shown, hydrogels are a diverse class of materials which can be prepared in many different ways. Their only common features are their hydrophilicity and their elastic structure, consistent with the definition given in the introduction. Despite this diversity, the analysis of their properties is for the most part independent of the specific type of gel, as long as the cross-links are permanent. Of the gels discussed in Section II, only the polymeric complexes held together by ionic or hydrogen bonds may not meet this requirement. Thus it is possible to take a generalized approach toward understanding the properties of gels despite their diversity. In the following sections, theoretical analysis and experimental measurement of these key gel properties are discussed in a general way, although important exceptions are identified.

A. Equilibrium Swelling Degree

The equilibrium swelling degree is the most important property of a hydrogel; it directly influences the rate of water sorption, the permeability to drugs, and the mechanical strength of the gel. It also affects the biocompatibility of the gel. Thus a close examination of equilibrium swelling properties of hydrogels is warranted. Regardless of the type of gel, the amount of water absorbed by a gel at equilibrium is primarily controlled by the hydrophilicity of the polymer, the number density of cross-link junctions, and the concentration of ionized groups in the network. Experimental data are usually reported as either the volume degree of swelling, Q, the mass degree of swelling, q, or the equilibrium water content, EWC, defined as

$$Q = \text{volume of swollen gel/volume of dry polymer} \qquad (2)$$

$$q = \text{mass of swollen gel/mass of dry polymer} \qquad (3)$$

$$\text{EWC} = \frac{\text{Mass of water absorbed by gel}}{\text{mass of swollen gel}} \times 100\% \qquad (4)$$

It is possible for Q or q to range from 1.2 to over 1000; this translates to an EWC range of 20% to over 99%. A commonly used hydrogel for drug delivery, poly(2-hydroxyethyl methacrylate) (PHEMA), has a q of about 1.7 or an EWC of about 40%.

1. Basic Theory

The theoretical principles for the swelling of a gel in equilibrium with a solvent were established in the 1940s [46,47]. But advances continue to be made, moving toward the goal of quantitative prediction of swelling degree in terms of independently measurable parameters. However, the basic approach described by Flory et al. [46] is still often used. In Flory's analysis, it is noted that a gel will absorb solvent until the chemical potentials of dissolved species in the gel and in the solution are equal. It is assumed that the swelling is the result of several independent free energy changes. This can also be written in terms of independent contributions to the osmotic swelling pressure, which is equal at equilibrium to the externally applied pressure, P_{ext} [47,48]:

$$P_{ext} = \Pi_{mix} + \Pi_{elas} + \Pi_{ion} + \Pi_{elec} \qquad (5)$$

In the case of free swelling, P_{ext} equals zero. The swelling pressure due to the tendency of the polymer to dissolve in the solvent (mixing of polymer and solvent) is given by Π_{mix}, while Π_{elas} accounts for the elastic response of the network due to cross-linking, which opposes dissolution. For gels with ionizable groups, the terms Π_{ion} and Π_{elec} are included; Π_{ion} represents osmotic pressure arising from a concentration difference of ions between the gel and solution, while Π_{elec} accounts for the electrostatic interactions of charges on the polymer chains and in solution. The equilibrium degree of water sorption generally cannot be quantitatively calculated a priori from theory and independently measured parameters despite substantial theoretical efforts in this area. However, the trends predicted by classical theory do agree qualitatively with experimental results in most cases. In this section, the classical expressions for the terms in Eq. (5) and their implications for hydrogels are examined.

Flory–Huggins theory for polymer solutions provides the classical expression for the polymer–solvent mixing term Π_{mix} [47]:

$$\Pi_{mix} = -(RT/\tilde{V}_1)[\ln(1 - \phi_2) + \phi_2 + \chi\phi_2^2] \tag{6}$$

where ϕ_2 = polymer volume fraction (note: $\phi_2 = 1/Q$); χ = polymer–solvent interaction parameter; R = gas constant; T = absolute temperature; and $\tilde{V}_1$ = solvent molar volume. The polymer–solvent interaction parameter accounts for free energy changes caused by mixing polymer and solvent, other than ideal mixing entropy. Interaction parameters are generally between 0 and 1: When $\chi < 1/2$, polymer–solvent interactions dominate, favoring dissolution; when $\chi > 1/2$, polymer–polymer interactions dominate, discouraging dissolution; for $\chi = 1/2$, termed the "theta state," there are no net interactions. While a theoretical basis has been developed for the interaction parameter, it is often used simply as an empirical parameter which depends most significantly upon concentration and temperature (see Section III.A.3).

The elastic contribution to Eq. (5) is a restraining force which opposes tendencies to swell. This constraint is entropic in nature; the number of configurations which can accommodate a given extension are reduced as the extension is increased; the minimum entropy state would be a fully extended chain, which has only a single configuration. While this picture of rubber elasticity is well established, the best model for use with swollen gels is not. Perhaps the most familiar model is still Flory's model for a network of freely jointed, random-walk chains, cross-linked in the bulk state by connecting four chains at a point [47]:

$$\Pi_{elas} = -\frac{RTv_e}{V_0}\left(\phi_2^{1/3} - \frac{1}{2}\phi_2\right) = -RT\rho_x\left(\phi_2^{1/3} - \frac{1}{2}\phi_2\right) \tag{7}$$

where v_e = number of elastically effective polymer chains (those which are deformed by stress), V_0 = unswollen gel volume, and $\rho_x = v_e/V_0$ = effective cross-link density relative to the solid state. The elastic retractive force is related to the state in which the cross-links were introduced, with the chains in the "relaxed" conformations. Thus, for networks prepared in solution, this equation can be modified as follows [49]:

$$\Pi_{elas} = -RT\phi_{2,r}\rho_x\left[\left(\frac{\phi_2}{\phi_{2,r}}\right)^{1/3} - 0.5\left(\frac{\phi_2}{\phi_{2,r}}\right)\right] \tag{8}$$

where $\phi_{2,r}$ = polymer volume fraction at network formation. Equation (8) reduces to (7) for the swelling of a network that was cross-linked in the absence of solvent; i.e., $\phi_{2,r} = 1$.

Equations (5), (6), and (8) can be combined to define the free swelling of a nonionic gel in equilibrium with a solvent:

$$\ln(1 - \phi_2) + \phi_2 + \chi\phi_2^2 + \tilde{V}_1\rho_x\phi_{2,r}\left[\left(\frac{\phi_2}{\phi_{2,r}}\right)^{1/3} - \frac{1}{2}\left(\frac{\phi_2}{\phi_{2,r}}\right)\right] = 0 \qquad (9)$$

Since the equilibrium volume degree of swelling Q equals $1/\phi_2$, this equation shows that the parameters which control swelling are the polymer–solvent interaction parameter χ and the effective cross-link density ρ_x. Figure 2 shows that Q falls as either χ or ρ_x increases, although the influence of the cross-link density becomes negligible in poor solvents.

The equation for the equilibrium swelling degree is more complicated for polyelectrolyte gels—those with ionizable groups—because of the need to include the additional ion-related terms in Eq. (5). The Π_{ion} term can be substantial

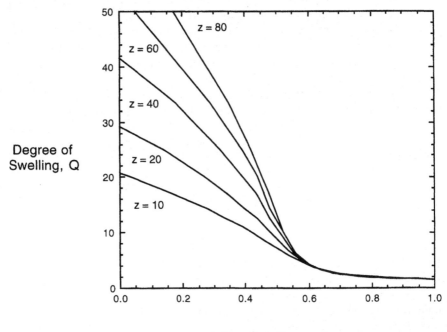

Figure 2   Equilibrium swelling degree of gels as predicted by Eq. (9). Z is the average number of repeat units between cross-links ($Z \approx 1/\rho_x\tilde{V}_m$).

even at rather low degrees of ionization and causes polyelectrolyte gels to swell to much greater extents than analogous nonionic gels. In contrast, theoretical calculations and experimental results [50–54] indicate that Π_{elec} is typically small in comparison to Π_{ion}, except possibly for rather hydrophobic gels with low degrees of swelling [53,54]. Qualitatively, this means that the increased swelling of polyelectrolyte gels is due to the high concentration of ions within the gel relative to the external solution and not to electrostatic repulsion of like charges on the polymeric network.

The Π_{ion} term is simply an expression for the osmotic pressure generated across a semipermeable membrane; effectively, the gel serves as a membrane which restricts the polyelectrolytes to one phase, while small ions can readily redistribute between phases. Assuming that the ions form an ideal solution, the expression for Π_{ion} becomes simply

$$\Pi_{ion} = RT(c - c^*) \tag{10}$$

where c = molar concentration of mobile ions in the solution inside the gel and c^* = molar concentration of mobile ions in the solution outside the gel. Since counterions arising from dissociation of the polymer's ionizable groups must remain in the gel to maintain electroneutrality, c is greater than c^*, often very much so. As with a semipermeable membrane, water flows into the gel, diluting the gel-phase ions until their chemical potentials become equal in the solution and gel phases. The resulting distribution of ions is governed by the principles of Donnan ion exclusion; in fact, measured ion distributions have been shown to match theory quite well [53–56]. The ions inside the gel can arise either from dissociation of ionic groups bound to the network or from mobile ions in the solution. For convenience, it is useful to rewrite (10) to note these two sources of ions explicitly, assuming that the gel's ionogenic groups are monovalent [47]:

$$\Pi_{ion} = RT\left[\frac{i\phi_2}{\tilde{V}_m} + \nu(c_s - c_s^*)\right] \tag{11}$$

where i = fraction of monomeric units on the gel which are ionized; $\tilde{V}_m$ = molar volume of a monomeric unit; ν = the number of ions into which a dissolved salt dissociates; c_s = molar concentration of electrolyte in the gel originating in the external solution; and c_s^* = molar concentration of electrolyte in the external solution.

Since the Π_{ion} term is frequently much greater than the Π_{mix} term for polyelectrolyte gels, the extent of ionization usually dominates the swelling degree of such gels. Thus changes in pH, which alter the degree of ionization, and changes in the ionic strength of the solution will directly affect the swelling degree of the gels. Such changes have been successfully correlated using Donnan exclusion theory [50,53–56]. By writing the degree of ionization in terms of the ionic group's dissociation constant, one can explicitly predict pH dependence of the gel swelling degree [57,58]. Weaknesses in theory for the mixing and elasticity terms are responsible for most of the theoretical failures of a priori prediction of swelling degrees of polyelectrolyte gels, discussed further in Section III.A.2. However, in polypeptide gels, the pK_a of the ionic group can also be a function of the neighboring groups on the polymer [59–62].

Substituting Eqs. (6), (9), and (11) into Eq. (5) yields an equation for the equilibrium swelling of polyelectrolyte gels:

$$
\frac{i\phi_2}{\tilde{V}_m} + v(c_s - c_s^*) - \frac{1}{\tilde{V}_1}\left[\ln(1 - \phi_2) + \phi_2 + \chi\phi_2^2\right]
$$
$$
- \rho_x\phi_{2,r}\left[\left(\frac{\phi_2}{\phi_{2,r}}\right)^{1/3} - \frac{1}{2}\left(\frac{\phi_2}{\phi_{2,r}}\right)\right] = 0 \tag{12}
$$

This predicts that the ionic swelling pressure can be great enough to cause substantial swelling even when the polymer–solvent interactions are unfavorable ($\chi > 0.5$), especially in the absence of added electrolyte, as shown in Figure 3. The swollen volume declines sharply as the concentration of the electrolyte is added, however [53,63]. Equation (12) also predicts a discontinuous change in the swollen volume of the gel as a function of the degree of ionization above a certain critical value of the polymer–solvent interaction parameter. The bends predicted are not thermodynamically stable and should be replaced by the straight line shown (indicating that the chemical potentials in the swollen, polymer-dilute phase are equal to the chemical potentials in the polymer-dense phase) [52,64]. This is thermodynamically analogous to predictions of phase transitions in fluids. That is, the swollen gels are analogous to vapors (low polymer density phase) and the shrunken gels to liquids (high polymer density phase); the discontinuous change in the degree of swelling at a specific set of interaction parameter and ionization is analogous to boiling or condensation at a specific set of temperature and pressure. Phase transitions in gels were first predicted by Dusek and Patterson in 1968 [65] and first observed by Tanaka in 1978 [66]. Since that time, phase transitions have been induced in both natural and synthetic gels by changes in the solvent [66,67], temperature [68,69], electric field [70], pH [52,64,71], ionic strength [71,72], pressure [73], and indirectly by light [74,75] and specific chemical stimuli such as glucose concentration [76]. Because of the potential for inducing large changes in the volume and/or permeability of gels which display phase transitions, interest has developed in their application to drug delivery. Comprehensive reviews of this subject are given in Refs. 77–79; reviews focused on drug delivery systems are in Refs. 80 and 81.

Thus, to summarize this section, the key parameters which determine the

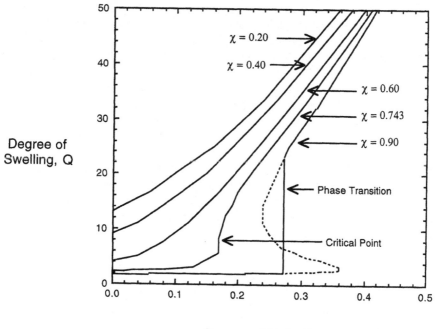

Figure 3   Equilibrium swelling degree of polyelectrolyte gels as predicted by Eq. (12).

degree of swelling of a gel are the polymer–solvent interaction parameter, the effective cross-linking density, and the degree of ionization of the network. The swelling degree of a gel increases with increasing affinity of the polymer for the solvent (decreasing values of χ), decreasing cross-link density, and increasing ionization. Abrupt changes in the degree of swelling can sometimes occur even with minimal changes in the ionization. While theoretical predictions correctly relate these variables qualitatively, there are some serious deficiencies in the predictions of the theory presented here which must be noted.

2.   Limitations of the Theory

The lack of quantitative success of the theory presented in the previous section is primarily the result of inadequacies in the mixing and elasticity terms. Sophisticated theoretical expressions are available which afford much improved predictions of swelling behavior. However, even these theories are not particularly successful when applied to hydrogels. Here the limitations of the theory presented in Section III.A.1 will be identified.

The quantitative failure of the swelling theory is largely due to difficulties in treating the nonideal entropic effects required in the mixing term ("nonideal entropy" relates to changes beyond the simple randomization of collections of objects and includes things such as reorientation of molecular interactions). As originally formulated, Flory–Huggins theory assumed an ideal combinatorial entropy change upon mixing, coupled with a heat of mixing term embedded into the polymer–solvent interaction parameter [in Eq. (6)]. In this form, the theory is unable to predict such important behavior as the phase separation of polymer solutions upon warming: The lower critical solution temperature (LCST) observed at near-ambient conditions in certain important water–polymer systems [e.g., poly(vinyl alcohol)–poly(vinyl acetate), cellulose ethers, poly(N-isopropyl acrylamide), poly(vinyl methyl ether)] [82–84]. Furthermore, according to the theory of Section III.A.1, the presence of ions in solution causes deswelling which depends only upon the concentration of ions and not their identity. But if the ions themselves can alter the interactions between solvent and polymer (for example, by disrupting hydrogen bonding), they will cause further deviation from Flory–Huggins theory [85,86].

Although more advanced theoretical interpretations of the interaction parameter have been developed within the framework of the Flory–Huggins lattice model, theoretical predictions of its value often deviate markedly from experimentally determined values. This is the result of complex dependences of χ upon numerous variables such as temperature, concentration, molecular weight, and the presence of any specific orientation-dependent interactions such as hydrogen bonding (Flory–Huggins is a mean field theory). Furthermore, when this theory is applied to gels, it must be assumed that χ for a network is the same as for the linear polymer. Although this is usually a reasonable assumption for lightly cross-linked networks, deviations may occur [1,87,88]. Deviations are often pronounced for hydrogels due to the complex interactions between water and polymers: Extensive hydrogen bonding exists, polar interactions are significant, and water structuring usually occurs around the network. While χ can be modified to include such effects, and methods of estimating of χ from solubility parameters are available, they rarely work well for hydrogels. Despite these problems, χ is widely used as an indicator of the quality of a solvent for a polymer, whether good ($\chi < 1/2$), moderate ($\chi \approx 1/2$), or poor ($\chi > 1/2$), even if it must be used as a purely empirical parameter. The theoretical formulation and estimation of χ and its experimental measurement have been reviewed in detail by Barton [89,90].

Prausnitz and coworkers [91,92] developed a model which accounts for nonideal entropic effects by deriving a partition function based on a lattice model with three categories of interaction sites: hydrogen bond donors, hydrogen bond acceptors, and dispersion force contact sites. A different approach was taken by Marchetti et al. [93,94] and others [95–98], who developed a mean field theory which accounts for the compressibility of the gel by combining Flory's network theory with the Sanchez and Lacombe polymer solution model. Both models have been at least partially successful in matching experimental data but require a number of parameters difficult to determine experimentally [95].

Weaknesses in the theory behind the elasticity expression used in Section III.A.1 also lead to deviations from observed swelling behavior. Both Eqs. (7) and (8) assume a Gaussian distribution of end-to-end chain lengths. However, this is not valid for highly swollen or highly cross-linked gels, since a Gaussian distribution exists only when the contour lengths of the chains are much greater than their end-to-end distances. Several models accounting for non-Gaussian distributions at large extensions have been proposed [47,99,100]. These models use power series in ϕ_2 and require knowledge of the number of effective links per network chain or, equivalently, the number of structural units in an effective chain. Thus they can account for finite chain extensibility and the resulting non-Hookean elasticity behavior (stress increases faster than strain at high extensions). As the solvent quality becomes increasingly good and the gel swells to greater extents, the use of non-Gaussian chain distributions becomes increasingly important. Because ionic gels have the potential to swell to truly extraordinary extents—over 1000 times the dry weight for some superabsorbent materials—it is particularly important to use a non-Gaussian elasticity expression for Π_{elas} for such gels. Figure 4 is analogous to Figure 3, except that the Galli and Brumage expression [100], modified for solution cross-linked gels, was used for the elasticity expression. The Galli–Brumage expression closely matches the exact distribution of freely jointed random walk chains even at high extensions. The result of this change in the assumed distribution is that the swelling degrees of the gels begin to level off at large values of swelling, as observed experimentally.

However, beyond the choice of the chain distribution assumed, more fundamental problems exist that have not been fully resolved. In the first place, the affine, real chain model of Flory and Rehner [Eq. (7)], while widely used, is not the most appropriate for highly swollen gels. In these systems, the cross-links are not necessarily firmly embedded in the polymer—they can fluctuate around an average position—and macroscopic deformation may not be exactly reflected at the microscopic level (nonaffine deformation). Thus swollen networks more closely match the phantom model of James and Guth [101]; this model explicitly allows cross-link junctions to fluctuate without hindrance from neighboring chains (hence the term "phantom"). Evidence of swelling through the topological rearrangement of chains rather than by their extension has also been presented [102]. The presence of elastically inactive structures or network imperfections—loops, dangling ends, etc.—should also be considered explicitly. Theoretical corrections for such imperfections are available in the literature, though their value in improving the prediction of swelling behavior for hydrogels has not been examined to the same extent as for rubbers [99,103].

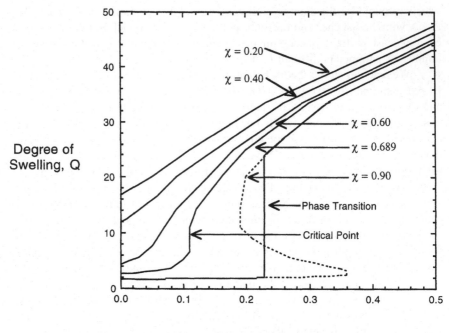

Figure 4 Equilibrium swelling degree of polyelectrolyte gels as predicted by Eq. (12), modified to account for non-Gaussian chain extensions using the Galli and Brumage [100] equation.

Furthermore, many gels are not homogeneous, especially on a microscopic level. Due to the nature of gel formation during solution copolymerization and cross-linking, the gel tends to form with polymer-dense regions clustered around an initiation site; these sites are then interconnected by less dense regions of polymer [104]. Phase separation due to polymer–solvent incompatibility during polymerization can lead to microporous, even spongy, structures to which standard elasticity expressions cannot be applied [8,25,26,105–107].

Thus quantitative analysis of elasticity is currently elusive, despite a great deal of work ongoing in this area. The extensive literature available on rubber elasticity by and large has not been adapted to hydrogels, and further work along these lines is necessary before quantitative predictions of swelling degree can be made from independent measurable polymer properties.

3. Empirical Analysis of Swelling Degree

The key property of a hydrogel is its swelling degree, since this directly influences other properties such as permeability. The theoretical description of hydrogel swelling, while quantitatively lacking, does interpret swelling qualitatively. Theory indicates that the parameters which control the swelling degree of a gel are the polymer–solvent interaction parameter χ, the effective cross-link density ρ_x, and the degree of ionization i. As described in Section II, hydrogels of many different types are available; the swelling degree for a given class of gels can be altered over an almost unlimited range by controlling these three parameters. These quantities are infrequently measured experimentally, however. Instead, an efficiency of reaction for the incorporation of comonomers into a network or for cross-linking linear polymers is typically assumed. However, this can lead to serious errors in the interpretation of experimental results because such reaction efficiencies are widely variable and cannot be reliably estimated. Thus these parameters should be measured directly; Harsh and Gehrke [63] describe the experimental measurement of cross-link density and degree of ionization of hydrogels in detail. Even when direct measurement is not feasible, it is important to obtain a firm empirical understanding of the effect of these variables.

The swelling degree of a nonionic polymer can be estimated from the solubility parameter δ, which is often more widely available than values of the interaction parameter; the references provided by Barton are particularly exhaustive on this subject [89,90,108]. For solvents, δ is defined as the square root of the cohesive energy density of the solvent; thus δ has units of $(cal/cm^3)^{1/2}$ or $MPa^{1/2}$. For polymers, δ is taken as equal to that of the solvent in which it displays maximum solubility. The effective value of δ for a mixed solvent or a copolymer can be estimated from mixing rules [89]. The degree of swelling of a polymer in a solvent is then inferred from the closeness of the match between the respective values of δ for the polymer and solvent, as illustrated in Figure 5 [109].

The polymer–solvent interaction parameter can be estimated by using the solubility parameters for the polymer and solvent. An approximate relationship between these parameters is

$$\chi = 0.34 + (\tilde{V}_1/RT)(\delta_1 - \delta_2)^2 \tag{13}$$

where the subscript 1 designates solvent and subscript 2 designates the polymer. The quantity 0.34 is an average contribution of all nonideal entropic effects, while the second term estimates the enthalpy of mixing.

The values of χ and δ are much less widely available for aqueous systems than for nonaqueous systems, however. This reflects the relative lack of success of the solution thermodynamic theory for aqueous systems. The concept of the solubility parameter has been modified to improve predictive capabilities by splitting the solubility parameter into several parameters which account for different contributions, e.g., nonpolar, polar, and hydrogen bonding interactions [89,90].

Nonetheless, in practice, it may prove more effective just to measure the interaction parameter directly and treat it as an empirical parameter independent of a specific theory. This can be done by measuring the solvent activity a_1 of the

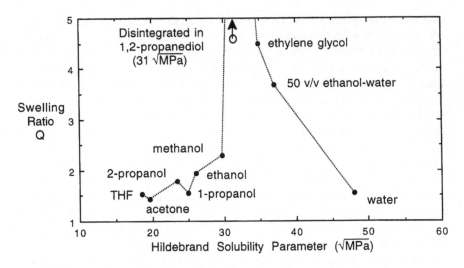

Figure 5 The solubility parameter for PHEMA gel is about 31 MPa^{1/2}. The swelling ratio decreases as the difference between the polymer and solvent solubility parameters increases. (From Ref. 109.)

polymer solution as a function of temperature and concentration and calculating χ as follows [89]:

$$\chi = \frac{\ln(a_1)}{\phi_2^2} - \frac{\ln(\phi_1)}{\phi_2^2} - \frac{1}{\phi_2} \tag{14}$$

This expression can be modified to apply directly to any of various techniques used to measure the interaction parameter, including membrane and vapor osmometry, freezing point depression, light scattering, viscometry, and inverse gas chromatography [89]. A polynomial curve fit is typically used for the concentration dependence of χ, while the temperature dependence can usually be fit over a limited temperature range to the form [47]

$$\chi = a + b/T \tag{15}$$

This form of the temperature dependence supports the idea that χ is fundamentally a Gibbs free energy parameter with entropic and enthalpic parameters.

However, the choice of a class of polymer for use in a given drug delivery system is often made for reasons unrelated to its swelling properties; the polymer might be chosen on the basis of cost, availability, supplier, biocompatibility, past use history, etc. Thus the hydrophilicity and χ will be fixed, and only the cross-link density and the ionic component can be readily adjusted to provide the swelling properties required for the problem at hand. However, the discussion above may also aid in the choice of a nonaqueous swelling solvent for leaching impurities from a gel or for drug loading.

As shown in Figure 2, the effective cross-link density has a significant effect on the equilibrium swelling degree of a polymer, except in poor solvents. The cross-link density is usually fixed at synthesis and depends upon the method of cross-linking, the cross-linking agent used (functionality, chemical composition, reactivity, etc.), and the physical structure of the gel (preexisting order in the polymer, inhomogeneities, network imperfections, etc.). Thus cross-link density must be measured rather than inferred from synthesis conditions. There are a number of ways to measure "effective" cross-link density, including swelling experiments and stress–strain measurements (either tensile or compressive). The swelling experiments are simple and direct: They only require measurement of the equilibrium swelling degree; ρ_x is then calculated from either (9) or (12). Use of this method thus requires knowledge of χ and is limited by the validity of the thermodynamic theory used for the calculation. Thus this technique is best restricted to comparisons within a family of gel formulations.

Determination of cross-link density from compression experiments is perhaps the most effective means of determining cross-link density as long as samples of the appropriate geometry can be prepared. When a hydrogel is subjected to an external force, it undergoes elastic deformation which can be related to the effective cross-link density of the network [63,99]. Here the measurements made to extract cross-link density from polymer deformation are briefly discussed.

The stress–strain response of ideal networks under uniaxial compression or extension is characterized as follows:

$$\tau = F/A_0 = G(\lambda - \lambda^{-2}) \tag{16}$$

where τ = engineering stress; F = applied force; A_0 = undeformed cross-sectional area of swollen polymer; G = shear modulus; $\lambda = L/L_0$; L = sample length under stress; and L_0 = undeformed sample length. At low strains, a plot of stress versus $\lambda - \lambda^{-2}$ will yield a straight line whose slope is the modulus; an example is given in Figure 6. The effective cross-link density may then be calculated from the modulus as

$$\rho_x' = G/(\phi_2^{1/3} RT) \tag{17}$$

The cross-link density obtained in this manner (ρ_x') is referenced to the swollen polymer volume at the time of compression. For comparison to theory or to similar polymers swollen to different extents, the cross-link density can be redefined in terms of moles of chains per unit volume of dry polymer (ρ_x). This conversion is simply

$$\rho_x = \rho_x'/\phi_2 \tag{18}$$

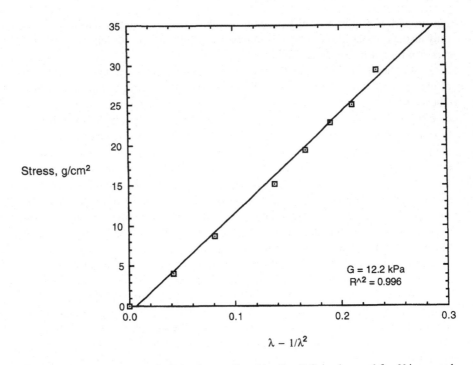

Figure 6 The stress–strain behavior predicted by Eq. (16) is observed for *N*-isopropyl acrylamide-*co*-sodium acrylate gel.

Efficiency of the cross-linking reaction can be defined as the ratio of effective cross-link density, ρ_x, to the theoretical cross-link density, ρ_t. Theoretical cross-link density assumes that all of the cross-linker added to the formulation created elastically effective cross-links; however, it cannot be defined for non-chemical methods. Thus the theoretical cross-link density, ρ_t, is given as

$$\rho_t = Cf/2 \tag{19}$$

where C = cross-linker concentration in the bulk polymer and f = cross-linker functionality [63,99]. Cross-linker functionality is defined as the number of network chains connected to the junction. For the commonly used difunctional monomers such as ethylene glycol dimethacrylate, $f = 4$. Theoretical cross-link density then reduces to $2C$. As shown earlier in Table 6, cross-linking efficiency has a complex dependence upon the synthesis conditions and is often well under 100%. However, cross-link efficiencies over 100% have been reported. This has been attributed to unintended sources of cross-links, such as preexisting order or chain transfer [110,111]. These widely varying results underscore the importance of direct measurement of cross-link density.

The ionization degree of a polyelectrolyte gel is its key swelling parameter. Even at an impurity level, ionized groups in a hydrogel network can cause a significant increase in the swelling degree, as observed in Figures 3 and 4. Since the number of ionogenic groups actually dissociated is a function of pH, the swelling degree for such gels also becomes a function of pH. Furthermore, the swelling caused by ionic groups is quite sensitive to ionic strength, in contrast to swelling caused by the inherent hydrophilicity of the polymer. Ionic groups can also bind ionic drugs, which may or may not be desirable [112,113].

Since ionic content has such an important effect on the performance of a gel in a given application, it should be accurately determined. Titration is the most direct method but is usually very tedious because of the length of time required for equilibration of the gel with the titrant. Also, many ionic hydrogels are weak acids and have an indistinct inflection in the titration curve, reducing the accuracy of the charge content determination. Thus other techniques should also be considered for the determination of charge content, although these are less generally useful [63].

These alternatives to titration for the determination of charge content include the recovery of counterions, the analysis of the sol fraction, and the use of ion exclusion. Since each charged group in the gel must have a counterion to maintain electroneutrality, if these are displaced from the gel and their number analyzed, the charge content of the gel can be determined. This can be easily done if the counterions are displaced by an excess amount of a competing counterion which does not interfere with analysis of the eluant. Or the sol fraction can be analyzed for charge content; by difference, the amount of charge incorporated into the network can be found. This works only if the sol can be quantitatively recovered from the gel after synthesis. Donnan ion exclusion can also be used to determine the charge content of the network. By this technique, the gel is immersed in a solution of an easily measured electrolyte, such as a dye, that has the same charge as the network, does not adsorb onto the gel, and is much smaller than the mesh size of the network. From the solution concentration change caused by the diffusion of the solute into the gel, the solute partition coefficient K can be determined:

$$K = c_s'/c_s \tag{20}$$

where c_s' = solute concentration in gel and c_s = solute concentration in free solution. From the well-established theory of Donnan ion exclusion, the charge concentration of the network (c_g) can be calculated as follows for a 1:1 electrolyte:

$$c_g = c_s(1/K - K) \tag{21}$$

This method is fast and has the potential for routine use, once an appropriate test solute has been identified for use with a particular class of gel, as it must be shown that the solute does not otherwise interact with the gel [63].

The presence of solutes in the aqueous medium in which the gel is immersed can also influence the swelling degree of a hydrogel. High molecular weight solutes which cannot permeate the network but do not otherwise interact with the gel may cause osmotic deswelling of the gel. The degree of deswelling can be anticipated by using Eq. (10), except that the concentration is that of the excluded solute [68,114,115]. Solutes can also have unexpected influences on the swelling degree of hydrogels. For example, poly(ethylene glycol) is known to cause deswelling of polyacrylamide gel, apparently by disrupting hydrogen bonding [116]. This can also be interpreted in terms of polymer/polymer phase separation [30,31,117]. Structure-breaking and structure-forming salts can also have pronounced effects on the swelling degrees of gels which cannot be anticipated based on the swelling theory presented earlier [85,118].

B. Swelling Kinetics

Many of the pharmaceutical applications for hydrogels depend upon the rates of solvent sorption. For example, the release rates of bioactive chemicals from delivery devices triggered by solvent contact are controlled by solvent uptake and swelling rates. A schematic of the monolithic matrix system is shown in Figure 7. In addition to swelling rates, the rates of shrinking are also key to applications proposed in recent years for environmentally responsive gels, gels whose solvent sorption capacity depends strongly upon the environmental conditions, such as temperature, pH, salt concentration, and electric field.

The kinetic swelling properties of a hydrogel which may be key to the success of a given application include the solvent sorption rate, the rate of approach to the equilibrium swelling degree, the velocity of moving solvent fronts (if present), and the transport mechanism controlling the solvent sorption (i.e., the time dependence of sorption). The solvent sorption rate is simply the mass of solvent absorbed per unit time and is directly related to the equilibrium swelling degree of the polymer. In other words, a highly swelling gel will tend to have a high sorption rate even if the rate of approach to equilibrium is slow. In contrast, the rate of approach to equilibrium is not directly tied to the swelling degree at equilibrium. For a simple diffusion-controlled system, this rate is characterized by a diffusion coefficient. When glassy polymers are immersed in a swelling solvent, moving solvent fronts quickly develop which advance toward the center of the sample. In monolithic devices, these fronts separate the swollen rubbery regions of high solute permeability from the glassy regions of immobilized solute; thus the front velocity controls the release rate [119]. The transport mechanism indicates the rate-controlling step in the solvent sorption process, which deter-

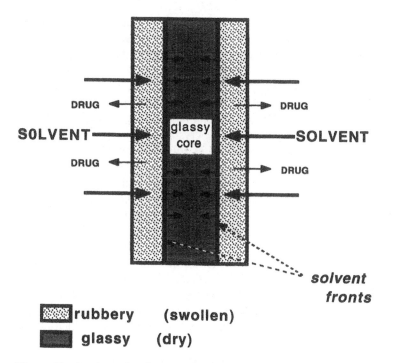

rubbery   (swollen)
glassy   (dry)

Figure 7 A schematic of a monolithic drug delivery system. (From Ref. 109.)

mines the time dependence of the sorption process. The dimensional changes which occur as the gel swells can also be very complex, displaying various maxima or minima before reaching equilibrium.

The complexity of the swelling kinetics of hydrogels means that only the simplest cases can be modeled quantitatively. Thus this section focuses on identification of rate-influencing phenomena and data analysis rather than the extensive theoretical modeling of the kinetic phenomena that has been done on this subject. Reviews of theoretical modeling include those by Peppas and Korsmeyer [119], Frisch [120], and Windle [121].

1. Rate-Limiting Steps

There are a number of phenomena which can influence the rate of swelling observed for a given hydrogel–solvent system. If a single step is rate-limiting, modeling that step alone will be sufficient to predict the rate of swelling of the material even if that step is not the sorption process. The processes which can be rate-controlling include the rate at which the stimulus causing the sorption occurs, the rate of diffusion, or the rate of polymer relaxation.

Some gels swell as the result of gradual changes in their chemical nature or physical structure. For example, cross-links may degrade over time, or hydrophobic groups may be chemically converted to hydrophilic groups; either of these processes will tend to increase the sorption capacity of the gel. If such changes occur slowly in comparison to the mutual diffusion of polymer and solvent, the observed swelling rate will be determined by the rate of cross-link degradation or the chemical reaction. One example of this phenomenon is the swelling of partially crystallized gelatin gels, where the crystallites function as cross-links, restricting the swelling of the gel. The rate of solvent uptake by these gels is dominated by the rate at which the crystallites dissolve [122].

Other gels swell due to changes in the environment which alter the equilibrium swelling degree, even in the absence of chemical reactions or structural alterations. Common examples of such gels are pH-sensitive materials. A gel with acidic groups will swell when the pH of the solution is increased from below the pK_a of the ionogenic groups to a higher pH, converting the gel from a nonionic form to the ionized form. For these gels, the rate of swelling may be controlled by the rate at which the pH changes within the gel, especially in dilute solutions. Thus the swelling and shrinking kinetics of these gels can be controlled by the ion-exchange kinetics [123]. In principle, the swelling or shrinking kinetics of any gel which responds to changes in its environment may be controlled by the rate of application of the driving stimulus. For example, the rate of heat transfer may be the rate-limiting step for a temperature-sensitive gel [124–126]. However, it has been found that for these environmentally responsive gels, the network–solvent diffusion rate is usually much slower than the stimulus rate, except for pH-sensitive or ionic strength sensitive gels or microporous gels, as mentioned earlier [25,84,123].

However, the majority of swelling-controlled systems are activated by simple immersion of the dried gel into the solution. In these cases, the rate-limiting step is either the rate of mutual diffusion of the solvent and polymer or the rate of polymer relaxation in response to the stresses caused by the invasion of the solvent. If osmotically active solutes have been incorporated into the gel, the osmotic pressure generated by dissolution can provide a significant additional driving force for swelling. Thus the rate at which a gel with embedded solute swells will be strongly influenced by the rate of solute dissolution, its concentration, and its diffusion rate out of the gel [127]. Similarly, a gel with a significant osmotically active sol fraction will usually swell faster than a gel without such a sol fraction. Furthermore, such systems will normally reach a swelling maximum prior to equilibration, followed by deswelling to the fully leached, solute-free equilibrium state.

Microporous, spongelike gels may swell and shrink by a convection-controlled process. That is, the rate-limiting step may be the rate at which the solvent is absorbed or expelled from the micropores of the sample rather than the rate of the diffusion or relaxation of the polymer network forming the pore walls [25,26,106].

2. Diffusion and Relaxation

Although many different processes can control the observed swelling kinetics, in most cases the rate at which the network expands in response to the penetration of the solvent is rate-controlling. This response can be dominated by either diffusional or relaxational processes. The random Brownian motion of solvent molecules and polymer chains down their chemical potential gradients causes diffusion of the solvent into the polymer and simultaneous migration of the polymer chains into the solvent. This is a mutual diffusion process, involving motion of both the polymer chains and solvent. Thus the observed mutual diffusion coefficient for this process is a property of both the polymer and the solvent. The relaxational processes are related to the response of the polymer to the stresses imposed upon it by the invading solvent molecules. This relaxation rate can be related to the viscoelastic properties of the dry polymer and the plasticization efficiency of the solvent [128,129].

Thus the relative rates of Fickian diffusion and polymer relaxation often determine the sorption behavior observed. Because glassy and rubbery polymers respond to mechanical stresses on much different time scales, the swelling behavior of glasses is usually quite different from that of rubbers. The rubbery state is observed when the polymer is above its glass transition temperature (T_g). In this state, the polymer is soft and elastic because the polymer chains have sufficient mobility to rearrange quickly in response to an applied stress. In other words, the chains can "relax" quickly toward new equilibrium configurations in response to stress. The characteristic time for this process is known as the relaxation time; it can be more precisely defined if desired [130–132]. Below T_g, the polymer is glassy; the polymer chains are "frozen" in position and resist deformation. Therefore a glassy polymer has a large relaxation time and thus displays a time-dependent response to stress, in contrast to the virtually instantaneous response of a rubber.

This relative importance of relaxation and diffusion has been quantified with the Deborah number, De [119,130–132]. De is defined as the ratio of a characteristic relaxation time λ to a characteristic diffusion time θ ($\theta = L^2/D$, where D is the diffusion coefficient over the characteristic length L): De $= \lambda/\theta$. Thus rubbers will have values of De less than 1 and glasses will have values of De greater than 1. If the value of De is either much greater or much less than 1, swelling kinetics can usually be correlated by Fick's law with the appropriate initial and boundary conditions. Such transport is variously referred to as "diffusion-controlled," "Fickian," or "case I" sorption. In the case of rubbery polymers well above T_g (De $\ll$ 1), substantial swelling may occur and the diffusion coefficient for the process is likely to be strongly concentration-dependent. In the case of glassy polymers well below T_g (De $\gg$ 1), the polymer will not be observed to swell macroscopically over an observable time period since the polymer chains are frozen into position on the experimental time scale. In this case, the diffusion coefficient is likely to be independent of concentration. If De $\approx$ 1, diffusion and relaxation occur on comparable time scales, leading to complex sorption behavior referred to as "non-Fickian" or "anomalous" transport. Relaxation-controlled non-Fickian transport is known as case II transport. In principle, case II transport could be observed if relaxation is much slower than diffusion but still occurs on an experimentally accessible time scale. When a glassy polymer ($T < T_g$, De > 1) is plasticized by penetrating solvent molecules and becomes rubbery ($T > T_g$, De < 1), a broad range of Deborah numbers may be observed in a single experiment, and complex non-Fickian sorption behavior may occur as a result. Although this concept is widely presented in the literature on swelling-controlled drug delivery systems, it is difficult to use in practice because of the large changes in De during swelling, typically from De $\gg$ 1 initially to De $\ll$ 1 by the end [132–135].

3.   Analysis of Sorption Data

A glassy polymer gel which becomes rubbery during solvent sorption undergoes complex dimensional changes (except for spherical samples). Soon after exposure to the solvent, well-defined penetration fronts develop and advance toward the center of the sample. These penetration fronts are usually assumed to be the point of the glass-to-rubber transition, although this might not always be true [109,136,137]. As long as a glassy core of polymer is present, the solvent-swollen, rubbery sheath swells only in the direction of diffusion since the glassy core resists stresses tending to cause isotropic expansion of the sample [55,136–139]. Thus, the volume increase of the sample caused by sorption of the solvent appears as an increase in thickness at nearly constant area. This constrained swelling results in the development of compressive stresses in the direction of diffusion and tensile stresses perpendicular to the direction of diffusion. However, once the moving fronts meet at the center of the sample, the constant-area constraint imposed by the glassy core is removed. Thus the sample dimensions quickly rearrange to relieve the anisotropic stresses, causing the area of the now rubbery sample to increase at the expense of the thickness until an isotropic state is reached. This rearrangement leads to a sudden increase in the instantaneous sorption rate because of the decrease in the diffusional path length and the increase in diffusional area. Swelling past this point occurs almost isotropically until equilibrium is reached [55,138,139]. Analogous behavior is seen for cylinders but not for spheres. In contrast to the complex sorption kinetics of glassy samples, an initially rubbery polymer swelling in solvent simply expands isotropically until the equilibrium volume is reached. Also, a sharp penetration front does not usually develop in a rubbery gel, though moving fronts might be observed, indicative of other processes such as ion exchange [84,140].

A procedure for characterizing the rates of the volume change of gels has not been uniformly adopted. Often, the kinetics are simply presented as empirical sorption/desorption curves without quantitative analysis. In other cases, only the time required for a sample of given dimensions to reach a certain percentage of equilibrium is cited. One means of reducing sorption/desorption curves to empirical parameters is to fit the first 60% of the sorption curve to the empirical expression [119,141]

$$M_t/M_\infty = kt^n \tag{22}$$

where M_t = amount (volume or mass) of solvent absorbed or desorbed at time t; M_∞ = amount (volume or mass) of solvent absorbed or desorbed at equilibrium; k = empirical rate constant; and n = transport exponent. The value of n provides an indication of the rate-controlling steps: For a flat sheet, a value of $n = 0.5$ indicates a diffusive process, while values of $n > 0.5$ usually indicate the presence of rate-influencing relaxation processes (unless other rate-limiting processes are present, as discussed in Section III.B.2). The term "anomalous transport" is usually reserved for transport exponents between 0.5 and 1.0, although it is sometimes used more broadly to cover all forms of non-Fickian transport. A value of $n = 1$ is considered evidence of relaxation-controlled case II transport. The fact that values of $n > 1$ are sometimes observed (super-case II transport) shows that the simple picture of diffusion versus relaxation rate control cannot explain transport behavior generally.

A moving front is usually observed in swelling glassy polymers. A diffusion-controlled front will advance with the square root of time, and a case II front will advance linearly with time. Deviations from this simple time dependence of the fronts may be seen in non-slab geometries due to the decrease in the area of the fronts as they advance toward the center [135,140]. Similarly, the values of the transport exponents described above for sheets will be slightly different for spherical and cylindrical geometries [141].

Modeling relaxation-influenced processes has been the subject of much theoretical work, which provides valuable insight into the physical process of solvent sorption [119]. But these models are too complex to be useful in correlating data. However, in cases where the transport exponent is 0.5, it is simple to apply a diffusion analysis to the data. Such an analysis can usually fit such data well with a single parameter and provides dimensional scaling directly, plus the rate constant—the diffusion coefficient—has more intuitive significance than an empirical parameter like k.

Two basic approaches to this problem have been developed. One is to correlate the volume change using Fick's law in a polymer-fixed reference frame, and the other is to correlate the dimensional changes using the equations of motion. While it is expected that such equations would be valid only for small displacements from equilibrium, studies of the sorption/desorption kinetics of many different systems show that these diffusion analyses are quite robust [84]. Furthermore, even when such an analysis fails, the deviations from a simple diffusion curve often help identify additional rate-influencing phenomena.

Solutions to the Fickian sorption problem for a variety of systems are given by Crank [142]. For a flat sheet with an aspect ratio greater than 10, initially free of solvent, the solvent uptake as a function of time is given as

$$\frac{M_t}{M_\infty} = 1 - \sum_{n=0}^{\infty} \left(\frac{8}{(2n+1)^2\pi^2}\right) \exp\left[-(2n+1)^2\pi^2\frac{Dt}{L^2}\right] \tag{23}$$

where L is the thickness of the sheet. A useful limiting form of this equation is

$$\frac{M_t}{M_\infty} = \left(\frac{4}{\sqrt{\pi}}\right)\left(\frac{Dt}{L^2}\right)^{0.5} \tag{24}$$

This solution is valid for the initially linear portion of the sorption (or desorption) curve when M_t/M_∞ is plotted against the square root of time. These equations also demonstrate that for Fickian processes the sorption time scales with the square of the dimension. Thus, to confirm Fickian diffusion rigorously, a plot of M_t/M_∞ vs. $\sqrt{t}/L$ should be made for samples of different thicknesses; a single master curve should be obtained. If the data for samples of different thicknesses do not overlap despite transport exponents of 0.5, the transport is designated "pseudo-Fickian."

When applied to a volume-fixed frame of reference (i.e., laboratory coordinates) with ordinary concentration units (e.g., g/cm^3), these equations are applicable only to nonswelling systems. The diffusion coefficient obtained for the swelling system is the polymer–solvent mutual diffusion coefficient in a volume-fixed reference frame, D_V. Also, the single diffusion coefficient extracted from this analysis will be some average of concentration-dependent values if the diffusion coefficient is not constant.

For an expanding or contracting sheet, a polymer-fixed reference frame should be chosen as described by Crank [142]. This converts the swelling problem to one mathematically equivalent to the constant volume case. Thus the mathematics of the swelling (or shrinking) problem are the same as in Eqs. (23) and (24), except that the length L is the initial half-thickness and the diffusion coefficient becomes that in the polymer-fixed reference frame, D_P. If the volume change occurs isotropically (as it does for rubbery gels), the relationship between D_P and the fixed reference frame diffusion coefficient D_V is

$$D_V = (V_f/V_i)^{-2/3}D_p \tag{25}$$

Billovits and Durning [143] provided a general derivation of the relationship between D_V and D_p which yields Eq. (25) as a limiting case. Equation (25) also matches the results of a less rigorous derivation presented by Westman and Lindstrom [144] and the intuitive results of Crank and Park [145].

Tanaka et al. [146] defined the diffusion coefficient of the gel network in the solvent as the ratio of the longitudinal bulk modulus of the network to the friction coefficient between the gel and fluid. This diffusion coefficient was called the "collective diffusion coefficient of the network," D_c. This definition arose from the derivation of an expression based on the equations of motion to describe the fluctuations of polymer chains around their equilibrium configurations, as measured by quasi-elastic light scattering. Following this approach, Tanaka and Fillmore [147] solved the equations of motion for a polymer gel swelling in a solvent under nonequilibrium conditions ("TF theory"). Mathematically, the Fick's law and TF approaches to the sorption problem are identical, except that the left-hand side of the TF versions of Eqs. (23) and (24) is the normalized approach to the equilibrium dimensions (L_t/L_∞) rather than the normalized approach to equilibrium volume (M_t/M_∞). In the limit of no volume change, Fick's law and TF theory become equal. Since Fick's law correlates volume changes while TF theory correlates dimensional changes according to the same function of time and diffusion coefficient, they deviate from one another by an increasingly large degree as the magnitude of the volume change increases. In terms of their ability to fit experimental data, however, both are successful, as shown in Figures 8 and 9. But both are limited theoretically; models with more solid theoretical foundations have been developed in the 1990s [148–158]. However, if one's goals are primarily correlating and interpreting experimental data, the decision must be made whether the theoretical gains of these models outweigh their price of increased complexity.

Fickian sorption in glassy polymers can also be fit using Eqs. (23) and (24), but only to the point of the disappearance of the glassy core. Because of the dimensional changes which take place when the sample becomes entirely rubbery (qualitatively illustrated in Figure 8), the sorption rate increases above the Fickian curve after the glassy core vanishes. Since this acceleration point can occur at any value of M_t/M_∞ (but typically at about 50% of equilibrium sorption), Eq. (23) should be restricted to sorption prior to this acceleration point. If it were applied to a sorption process where the sample became glassy prior to the first 60% of the uptake, the transport exponent of a Fickian process would be greater than 0.5 and erroneously indicate relaxation-influenced transport [139].

Case II transport is an interesting special case of sorption because the linear time dependence of the relaxation process means that a constant swelling rate could be observed [119,121,128,129,132–135]. Mechanistically, case II transport

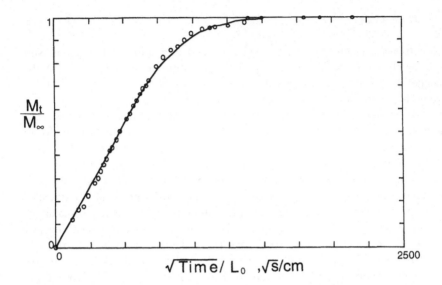

Figure 8 Fick's law correlates the swelling of poly(N-isopropyl acrylamide) gel caused by an increase in temperature from 10 to 25°C. $D_P = 2.3 \times 10^{-7}$ cm²/s; $L_0 = 1.5$ mm. (Adapted from Ref. 84.)

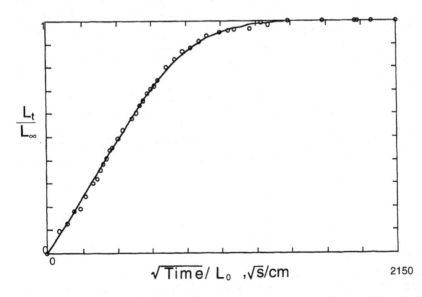

Figure 9 The equations of motion (TF theory) correlate swelling of poly(N-isopropyl acrylamide) caused by an increase in temperature from 10 to 25°C. $D = 2.5 \times 10^{-7}$ cm²/s; $L_0 = 1.5$ mm. (Adapted from Ref. 84.)

occurs when a sharp front develops (which is usually assumed to be relaxation-controlled) and moves at constant velocity. If there is a minimal concentration gradient behind the front (in the swollen portion of the gel), the constant velocity of the front results in a constant sorption rate. This has been proposed as the basis of zero-order release devices. However, true, relaxation-controlled case II sorption has not been observed in aqueous systems, although there have been reports of constant swelling rate gels ($n = 1$) where the rate is controlled by phenomena other than a relaxation-controlled moving front. For example, pseudo-case II transport may be observed if the surface concentration of the gel slowly rises toward equilibrium as the gel swells [109,159]. It must also be recognized that incorporation of drugs into a polymer alters solvent sorption kinetics. Thus, design of a constant-rate delivery system of this type can be complicated; in the next section, some of the variables which influence sorption kinetics are examined for one particular model polymer, poly(2-hydroxyethyl methacrylate) (PHEMA).

4. Swelling Kinetics of PHEMA

The swelling rates of many different polymer gels in water have been measured, but the connection between the properties of the polymers and the observed kinetics are not well established. Perhaps the most widely studied gel for swelling-controlled drug delivery devices has been poly(2-hydroxyethyl methacrylate) (PHEMA), so this will be used as an illustrative example here. Modification of the swelling properties of this material is possible by copolymerization with either more hydrophilic or hydrophobic comonomers as well as ionizable and cross-linking monomers. Because the properties of glassy polymers are history-dependent, the sample preparation conditions can also affect the sorption behavior of the gel. The incorporation of solutes into the gel also have a substantial effect on the swelling kinetics. The effects of comonomers, cross-linking, preparation, and solute incorporation on the swelling kinetics of PHEMA are briefly reviewed here.

Pure PHEMA gel is sufficiently physically cross-linked by entanglements that it swells in water without dissolving, even without covalent cross-links. Its water sorption kinetics are Fickian over a broad temperature range. As the temperature increases, the diffusion coefficient of the sorption process rises from a value of 3.2×10^{-8} cm^2/s at 4°C to 5.6×10^{-7} cm^2/s at 88°C according to an Arrhenius rate law with an activation energy of 6.1 kcal/mol. At 5°C, the sample becomes completely rubbery at 60% of the equilibrium solvent uptake ($q = 1.67$). This transition drops steadily as T_g is approached ($\approx$90°C), so that at 88°C the sample becomes entirely rubbery with less than 30% of the equilibrium uptake ($q = 1.51$) (data cited here are from Ref. 138).

For many pharmaceutical applications of gels, the gels must be preswollen, then dried. Preswelling may be required to leach low molecular weight impurities from the gel; the gel may also be swollen in a concentrated drug solution for loading purposes. Repeated cycles have little effect on subsequent swelling cycles if the gel is dried isotropically [55]. If a gel sheet is dried anisotropically, sorption kinetics may be noticeably altered after the sample becomes entirely rubbery, as shown in Figure 10. This happens because differences in stresses in the initial glassy states cause different dimensional changes once the sample becomes entirely rubbery [109,138,139]. However, the swelling rates and transport mechanisms prior to the disappearance of the glassy core are not altered by differences in stress in the glass.

Modification of the swelling behavior of PHEMA by copolymerizing HEMA with other vinyl monomers and cross-linkers has been rather extensively studied [e.g.,55,127,131,132,138,160–165]. Copolymerizing HEMA with a more hydrophilic comonomer such as N-vinylpyrrolidone will create a gel with a larger equilibrium degree of swelling than pure PHEMA and thus a faster sorption rate. The rate of approach to equilibrium also increases [131,161,162]. In contrast, copolymerization with a more hydrophobic comonomer such as methyl methacrylate reduces the swelling degree and the sorption rate. Increasing the extent of cross-linking in a PHEMA-based gel will also reduce the degree of swelling and

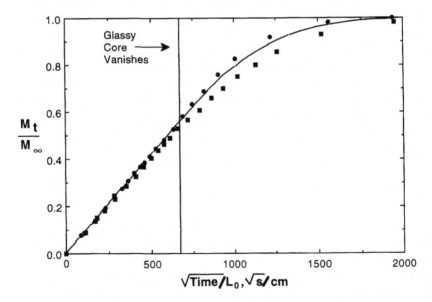

Figure 10 Swelling of PHEMA in water. (●) Isotropically dried in a vacuum oven; (■) anisotropically dried in a gel dryer; (——) Fickian curve. (Adapted from Ref. 136.)

thus the sorption rate. Increased cross-linking also slows the rate of approach to equilibrium and shifts the transport mechanism toward increasingly anomalous transport. An ionic comonomer will significantly increase the swelling degree of PHEMA even at relatively low concentrations. However, the rate of approach to equilibrium (the diffusion coefficient) and the transport mechanism are not altered by incorporation of the ionic comonomer. Thus the sorption rate simply increases in proportion to the increase in the equilibrium swelling degree [55].

Altering the degree of swelling of PHEMA by copolymerization with more hydrophobic or hydrophilic monomers or by additional cross-linking may alter most of the kinetic properties as well, but the use of an ionic comonomer to increase the swelling degree alters only the rate of sorption. This appears to be related to the relative amounts of comonomer needed to significantly alter the swelling degree. At low levels of monomer substitution, properties of the gel which influence the sorption kinetics such as the T_g, solvent diffusivity in the glass, and the relaxation rate of the chains are not significantly changed. In contrast, the use of much larger amounts of comonomers significantly changes the gel composition, and this may change other rate-influencing parameters also.

Incorporation of an osmotically active solute in the gel will also alter the swelling kinetics observed for the gel [127]. Dissolution of the solute as the gel absorbs water provides a significant driving force for swelling. Thus, increasing the concentration of solute in the dry gel increases the rate of solvent uptake. As the solute diffuses out of the gel, this driving force declines and the gel begins to deswell. Thus solute-loaded gels generally display a maximum in the swelling versus time curve, with the maximum corresponding to the greatest amount of dissolved solute in the gel.

C. Hydrogel Permeability

The widely variable and easily adjusted permeability of hydrogels is the primary attribute that makes them useful in drug delivery systems. Hydrogels can be made which are impermeable or highly permeable to any given solute, whether a low molecular weight drug, a therapeutic protein, or a gene. The most important variable which affects the permeability of a gel to a given solute is its swelling degree [166]. Thus any of the variables which can be used to adjust the swelling degree of a gel, such as cross-linking, will also directly affect the permeability. A thorough review on the subject is given in Ref. 166.

Permeability (P) is usually defined as the product of a thermodynamic property and a transport property which are, respectively, the partition or solubility coefficient, K, and the diffusion coefficient, D. This partition coefficient is defined as the ratio at equilibrium of the solute concentration inside the gel to that in solution. A value of K less than 1 indicates that the solute favors the solution phase, while a value of K greater than 1 indicates that the solute is preferentially sorbed by the gel. The diffusion coefficient is defined via Fick's law:

$$j = -D \frac{dc}{dx} \tag{26}$$

where j is the solute flux (moles per unit area per unit time) and dc/dx is the concentration gradient. Values of diffusion coefficients vary widely, from nearly the free solution value in highly swollen gels (10^{-6} to 10^{-5} cm^2/sec for most drugs) to effectively zero in a glassy matrix. Since the diffusion coefficient depends primarily upon molecular weight, and since most drugs have molecular weights within a few hundred units of one another (except for macromolecular drugs such as therapeutic proteins or DNA), this does not vary strongly for many different drugs in a given gel. However, the partition coefficient can vary widely for different solutes: K varies over a virtually unlimited range, depending upon a number of possible contributions, including electrostatics, biospecific affinity, hydrophobicity, and size.

The significance of the permeability depends upon the particular system at hand. If the hydrogel forms a barrier between a drug reservoir and a solution, the flux across the hydrogel membrane at steady state is simply

$$j = (P/L)(C_r - C_s) \tag{27}$$

where C_r is the concentration on the reservoir side, C_s is the concentration on the receiving solution side, and L is the thickness of the membrane. Thus in reservoir systems, the flux of drug is directly proportional to both the partition coefficient and the diffusion coefficient. This expression also suggests an alternative definition of permeability: $P' = P/L$. Other definitions of permeability also exist in the literature. In contrast, permeability does not appear explicitly in equations defining release from monolithic matrix systems. In matrix systems releasing to an infinite sink, the rate of approach to equilibrium is controlled by the diffusion coefficient, although the partition coefficient determines the loading conditions. Thus K and D are of more fundamental significance regarding the mobility of solutes in a gel than is the permeability P. Therefore this section examines partition and diffusion coefficients independently, rather than permeability per se. Knowing K and D allows calculation of flux from different systems, as described in texts by Baker [167], Crank [142], and Cussler [168], among others.

1. Partition Coefficients

The magnitude of the partition coefficient of a solute between a gel and a solution is determined by a number of different interactions. These interactions are usually assumed to contribute to the observed value of the partition coefficient as follows [169–171]:

$$\ln K = \ln K_0 + \ln K_{\text{size}} + \ln K_{\text{hphob}} + \ln K_{\text{el}} + \ln K_{\text{biosp}} \tag{28}$$

where the subscripts size, hphob, el, and biosp indicate the size exclusion, hydrophobic, electrical, and biospecific contributions to the partition coefficient, respectively. Any other interactions are lumped into the K_0 term. Biospecific interactions (as in antibody–antigen complexes) will not be present unless they are specifically designed into the system. Electrostatic interactions will be present only if the gel is a polyelectrolyte and the drug is ionic. If the drug and the network have the same charge, the drug will tend to be excluded from the gel due to Donnan ion exclusion [54,112,172]. If the drug and the network are oppositely charged, the system can be analyzed with ion-exchange theory [112,113,173]. Hydrophobic interactions can be very important, especially if the solute is rather hydrophobic and the gel has hydrophobic regions. Figure 11 shows some fascinating results for the partitioning of norethindrone between an aqueous solution of the drug and poly(N-isopropyl acrylamide) gel. This gel is delicately balanced

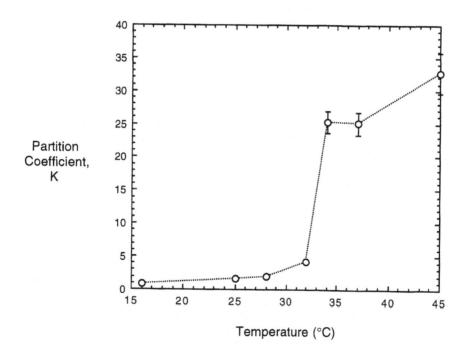

Figure 11 The partitioning behavior of norethindrone in poly(N-isopropylacrylamide) gel. This gel substantially deswells at 34°C. (Adapted from Ref. 176.)

between hydrophilic and hydrophobic character. At low temperatures, the gel is hydrophilic and highly swollen (>90% water) and has little interaction with the solute, so K is small. But above 34°C, the gel is hydrophobic, is substantially less swollen (<60% water), and strongly absorbs hydrophobic solutes such as norethindrone [166,174–176]. Similar, though less dramatic, results have been observed in other systems [177]. Partition coefficients due to hydrophobic interactions are difficult to predict from theory, but the theoretical principles have been developed [118,178–180]. The octanol/water partition coefficient $K_{o/w}$ is a commonly available parameter for different drugs used to estimate hydrophobic/hydrophilic balance. A drug with a large value of $K_{o/w}$ will be sparingly soluble in water and may be absorbed by hydrophobic gels or hydrogels with hydrophobic regions, as in a copolymer of hydrophilic and hydrophobic monomers.

If the hydrogel is inert with respect to the drug and merely obstructs solutes from freely penetrating the mesh, the partition coefficient will be due only to the size exclusion term, K_{size}. Theories have been developed which calculate the size exclusion partition coefficient as a function of solute geometry and network characteristics [181–187]. It is important to recognize that size exclusion is an entropic phenomenon which results from the fact that the orientations of a solute in a gel are restricted relative to free solution; in other words, the entropy of a solute decreases when the solute enters a gel. As a result, solutes tend not to enter a gel, even if they are not physically obstructed from doing so. Therefore, in pure size exclusion, the partition coefficient is always less than 1 and approaches 1 only in the limit of a point solute in a gel whose structure occupies no volume. Thus partial size exclusion ($0 < K < 1$) will be observed even for solutes smaller than pores of monodisperse dimension. Partition coefficients greater than 1 indicate that other mechanisms are present, although the converse is not true: Values of K less than 1 can be due to many reasons other than size exclusion, especially Donnan ion exclusion.

The degree of size exclusion observed for a given solute is directly related to the size and geometry of the solute. The theory advanced by Giddings et al. [182] predicts the proportionality

$$K \propto e^{-M\alpha} \tag{29}$$

where M = solute molecular weight and α = solute geometry factor (1/3 for spheres, 1/2 for random coils, 1 for rigid rods). A more recent theory for size exclusion is presented by Schnitzer [184]. For spherical solutes in a random network of rodlike fibers, the following equation applies:

$$K = \exp(-v_e^\circ) \exp\left[\frac{v_e^\circ(1 - r_m/r_f)^2}{1 - v_e^\circ}\right] \tag{30}$$

where v_e^o = network excluded volume (which is estimated from the product of the gel concentration, wt/vol, and the specific volume of the gel); r_p = solute radius; r_f = fiber radius; and $r_m = r_p + r_f$. The predictions of this equation are given in Figure 12 for solutes of different hydrodynamic radii (examples of hydrodynamic radii: urea, 0.25 nm; theophylline, 0.37 nm; sucrose, 0.47 nm; vitamin B-12, 0.84 nm; inulin, 0.67 nm; heparin, 1.50 nm [188]. As shown in this figure, size exclusion is sharply dependent upon the gel swelling degree only within certain ranges of swelling and only for small solutes. For large solutes, the degree of size exclusion changes more slowly with respect to changes in the swelling degree of the gel in the range shown (Sassi et al. [185] critiqued the quantitative features of this theory; it tends to underpredict K).

2. Diffusion Coefficients

Solute diffusion through hydrogels has been described in terms of two basic mechanisms, the "partition" and "pore" mechanisms. The partition mechanism

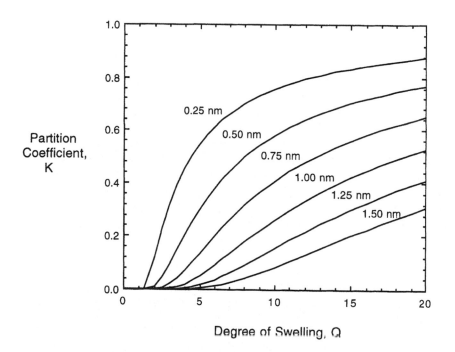

Figure 12 Theoretical predictions of solute partitioning in a gel with a fiber radius of 0.4 nm for solutes of different hydrodynamic radii, based on the size exclusion theory of Eq. (30). (From Ref. 174.)

is a diffusional process which is directly influenced by specific interactions between the gel and solute. The diffusion coefficients of solutes which interact strongly with the network are typically reduced relative to those of comparable noninteracting solutes [188–190]. If the network serves only to physically obstruct motion of the solute and the solute travels only through the water-filled regions of the gel, the pore mechanism (which is more common) will be observed. In such cases, the diffusion coefficient will depend primarily upon the relative sizes of the solute and the network pores. If the pores are very large, convection of the solute may occur in addition to diffusion [25].

Numerous models have been proposed to interpret pore diffusion through polymer networks. The most successful and most widely used model has been that of Yasuda and coworkers [191,192]. This theory has its roots in the free volume theory of Cohen and Turnbull [193] for the diffusion of hard spheres in a liquid. According to Yasuda and coworkers, the diffusion coefficient is proportional to $\exp(-V_s/V_f)$, where V_s is the characteristic volume of the solute and V_f is the free volume within the gel. Since V_f is assumed to be linearly related to the volume fraction of solvent inside the gel, the following expression is derived:

$$\frac{D_g}{D_0} = \exp\left(\frac{-Y}{Q-1}\right) \tag{31}$$

where D_g = diffusion coefficient of the solute in the hydrogel, D_0 = diffusion coefficient of the solute in the solvent, $Y = \gamma V_s/V_{f,solv}$, γ = correction factor for overlapping free volume, $V_{f,solv}$ = free volume of solvent, and Q = volume degree of swelling. The predictions of this equation for solutes of different sizes are given in Figure 13. An example of the success of this theory is illustrated in Figure 14 [174,176]. Peppas and coworkers [194–196], among others, have built upon this theory by explicitly accounting for sieving effects due to the network structure. Examples of the success of Eq. (31), without inclusion of an explicit sieving prefactor, can be found in the literature [123,191,197]. Other approaches to modeling diffusion of solutes in gels can be classified as either hydrodynamic models or obstruction models. These theories, along with the free volume theories, have been critiqued recently by Amsden, who noted a variety of practical and theoretical limitations in their use [198]. A semi-empirical approach to estimation of the diffusion coefficients of water soluble solutes in hydrogels has been developed with data for solutes between 148 and 150,000 Da in polyacrylamide and poly(vinyl pyrrolidone) hydrogels (199):

$$D_g = D_0 \exp[-(5 + 10^{-6}M)/q]$$

where M = solute molecular weight; q = mass degree of swelling.

The various diffusion equations require knowledge of the D_0, the diffusion coefficient of the solute in free solution. Measurement and estimation of these

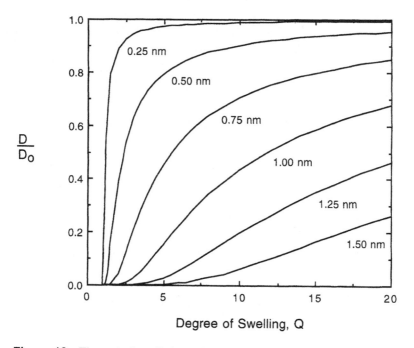

Figure 13   Theoretical predictions of Eq. (31) for diffusion in a hydrogel ($\gamma/V_{f,solv}$ = 1.8 nm^{-3}) for solutes of different hydrodynamic radii. (From Ref. 174.)

diffusion coefficients has been thoroughly discussed by Cussler [168]. In the next section, the measurement of D_g, the solute diffusion coefficient within the gel, is discussed.

3.   Measurement of Solute Diffusion Coefficients

The diffusion coefficients of solutes in swollen hydrogels can be measured from either membrane permeation or sorption/desorption experiments. The membrane permeation time-lag experiment is the most widely used technique; in fact, commercially available apparatuses are available for making these measurements [45,166,200–202]. A comprehensive review of the experimental details and data analysis is given in Ref. 166. While this is the method of choice for conventional membranes, it is not always the most effective method for use with hydrogels. Use of the membrane permeation experiment requires that the hydrogel be fashioned as a strong membrane and that steady state be established within a reasonable time period. It may be difficult or impossible to make a hydrogel membrane capable of being clamped into a commercial permeation device, especially if the

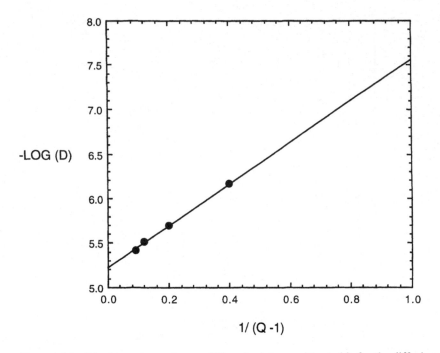

Figure 14 The free volume theory of Yasuda and coworkers holds for the diffusion of acetaminophen in swollen 10 × 4 poly(N-isopropyl acrylamide) gel. (Adapted from Ref. 176.)

hydrogel is highly swollen, since such gels typically have low mechanical strength. If the membrane must be made thick in order to overcome limited mechanical strength, the time to achieve steady state may become impractically long, especially for drugs with high molecular weights and therefore small diffusion coefficients.

Sorption or desorption experiments can be much more effective techniques for measuring the diffusion coefficients of drugs in hydrogels; they also more closely match drug delivery system conditions. These methods simply measure the rate at which solute is absorbed or desorbed by a spherical, cylindrical, or sheet gel sample as a function of time. If the solution volume is effectively infinite relative to the amount of gel, a simple expression can be used to calculate the diffusion coefficient of a drug dissolved in a hydrogel sheet of thickness L. For either sorption or desorption, D can be extracted from the slope of a plot of the fractional amount of drug absorbed or desorbed (M_t/M_∞) vs. $t^{1/2}$ using the equation [142]

$$M_t/M_\infty = (4/\sqrt{\pi})(Dt/L^2)^{1/2} \tag{33}$$

Here M_t is the mass of solute absorbed or desorbed at time t and M_∞ is the total mass of solute absorbed or desorbed at equilibrium. This equation is accurate to within 1% for $M_t/M_\infty < 0.6$.

However, if the solution volume is substantially greater than the gel volume, an extremely sensitive analytical technique is required to accurately measure the resulting small changes in bath concentration over time. In practice, the bath volume and gel volume must be comparable in magnitude. This generates a larger change in bath concentration, allowing more accurate determination of the amount of drug absorbed or desorbed by the hydrogel. Thus, the sorption/desorption experiment immerses a hydrogel sample of known geometry into a well-stirred solution of limited volume. Efficient stirring is required to minimize mass transfer resistance at the gel/solution boundary layers and to prevent development of stagnant fluid areas. For maximal accuracy, the apparatus should be sealed with minimal headspace over the solution reservoir to minimize solvent evaporation.

This method can be adapted to any type of gel, regardless of mechanical strength, and most geometries, including slabs, disks, cylinders, and spheres. Crank [142] provides the exact and limiting mathematical solutions to this problem for many different geometries and boundary conditions. Since these solutions are usually mathematically infinite series, D must be found by a nonlinear least squares fit of the appropriate equation to the data. While this is not a difficult program to develop and can be readily handled by a desktop computer of modest power using a spreadsheet program, simple, accurate approximate solutions are useful alternatives.

Lee [203] developed such approximate solutions for spheres, cylinders, and slabs. For absorption of drug by a flat sheet, the equations used are

$$F(C_t) = \left[\frac{C_0^2 - C_t^2}{2C_t^2} + \ln\left(\frac{C_t}{C_0}\right) \right]\left(\frac{3\lambda^2}{16}\right) \tag{34}$$

and

$$F(C_t) = Dt/L^2 \tag{35}$$

where C_t = concentration of drug in the external solution at time t; C_0 = initial concentration of drug in solution; C_∞ = final concentration of drug in solution; $\lambda = C_\infty/(C_0 - C_\infty)$, the effective volume ratio; and L = sheet thickness. The concentration change data are used to determine the values of $F(C_t)$ via Eq. (34); the diffusion coefficient is then found from the slope of a plot of $F(C_t)$ vs. t, as given by Eq. (35). Figure 15 illustrates the use of this method in determining D of a drug in a cylindrical gel sample [174,176]. Equation (34) is accurate for values of $\lambda < 10$. Larger values of λ correspond to nearly constant solution concentrations, given by Eq. (33). This procedure is easily adapted for desorption from gels in addition to sorption.

Finally, it should be noted that as λ decreases, the calculated value of D

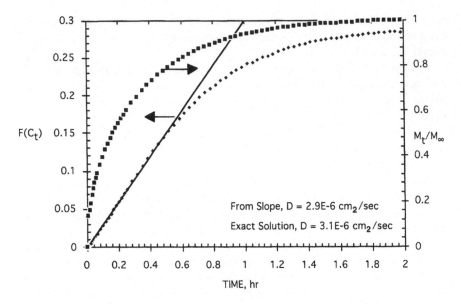

Figure 15   The sorption of acetaminophen from a solution of limited volume by 10×4 poly(*N*-isopropylacrylamide) gel, illustrating the use of Eqs. (34) and (35) to determine the diffusion coefficient of the solute. (Data from Ref. 174.)

becomes an increasingly strong function of K. This happens because under these conditions the gel absorbs most of the solute from the solution, so the concentration gradients during the experiment are strongly dependent upon K. Thus inaccuracies in the measured value of K become amplified in the calculated value of D, particularly for values of $\lambda < 1$. Therefore, it is usually desirable to design the experiments such that λ lies outside this range. In addition, more sophisticated techniques have recently been used to determine diffusion coefficients in gels; these are potentially more accurate and informative. They include pulsed gradient spin-echo nuclear magnetic resonance (PGSE-NMR), magnetic resonance imaging (MRI), and laser light scattering [204–208].

D.   In Vivo Performance Properties

1.   Biocompatibility

A device that is to be placed in contact with living tissue or biological fluids must be made of material that is biocompatible. However, ''biocompatible'' is not a precisely defined term or a measurable property. In general, biocompatibility refers to the inertness of a material under the conditions of use, especially the absence of any undesirable physiological reactions, including [209,210]

Thrombosis
Alteration of plasma proteins or enzymes
Allergic or adverse immune responses
Tissue damage
Toxicity
Cancer

Clearly, no single test exists that can quantify the compliance with all of these criteria. Nor is it likely that any given material can meet the criterion of complete biological inertness. Rather, determination of biocompatibility involves testing the material properties most relevant for a given application. For example, the material used to construct an oral dosage form would be stringently tested for carcinogens or mutagens released upon leaching impurities from the material or upon degradation of the device. This would also be of concern for a material used in an application involving blood contact such as a gel-coated catheter, but thrombus-formation properties would be even more critical [113]. In contrast, thrombogenicity would not be a concern for the oral dosage form. Since biocompatibility is application-specific, this means that no list of "biocompatible" hydrogels can be presented. Also, the FDA treats materials on a case-by-case basis; approval of a material for one application does not eliminate the need to retest the material for use in a new application.

The high water content of hydrogels makes them soft and flexible, and thus they tend not to cause tissue damage when implanted or ingested: They mimic living tissue more closely than any other synthetic material. However, high water content does not, by itself, guarantee biocompatibility. For example, some highly swollen polysaccharide gels cause serious inflammation when implanted. Since the material surface is the point of contact with the surrounding tissue, it is obvious that surface properties will also be quite important. The nature of hydrogel surfaces is complex, however [211–215]. For example, typical contact angle tests to determine the hydrophobicity and hydrophilicity of a hydrogel surface can be misleading. The high water content of hydrogels means that the polymer chains are highly mobile and can rearrange to present either hydrophilic or hydrophobic groups on the surface to the external environment. Thus, the polymer chains of poly(2-hydroxyethyl methacrylate) gel may rearrange depending upon the external solvent, exposing methyl groups to nonpolar solvents and hydroxyl groups to aqueous solutions [211].

Another important property which affects biocompatibility is permeability to low molecular weight solutes, since it is usually required that all extractable materials be leached from the device prior to use. These extractables include lower molecular weight species such as unreacted monomer, residual initiators, or the polymeric sol fraction. Highly swollen gels are typically highly permeable, especially to lower molecular weight species.

The rate of thrombus formation is the most important determinant of biocompatibility for materials in contact with blood. In vitro experiments to determine blood compatibility can measure the effect of the material on blood clotting time under well-defined conditions [209]. These tests are usually acceptable as screening tools, but because of significant differences between the test conditions and the physiological environment (e.g., exposure of the blood to air), they are not definitive. In vivo biocompatibility studies are generally performed by implantation of the device in a host animal, followed by removal and examination for any reaction after short ($\approx$ 2 hr) and long (2 weeks) periods of time. The device is then examined for adherent thrombi. Since nonadherent thrombi transported through the circulatory system cause the most adverse biological effects, the kidneys are also examined for signs of damage arising from nonadherent thrombi. Almost all implanted materials evoke some type of response; even those that are usually considered blood-compatible. For example, the material will often be rapidly covered by a thin fibrous capsule after a brief acute inflammation reaction.

The effect of the material on cellular metabolism is also an important measure of biocompatibility. To determine such effects, cultured ex vivo cells can be exposed to the polymer and the growth rates compared to controls [216,217]. The metabolic function of the cells can be tested by assay for production of a marker enzyme. An additional advantage of this type of test is that it avoids the use of live animals.

2. Biodegradability

Inertness is not always required of hydrogels; in some cases it is actually desirable that the material degrade over time. Some applications of hydrogels involve the surgical implantation of the device, which then releases the drug into the surrounding tissue. If the gel is inert, the device must be removed once the supply of drug is exhausted. In contrast, a device which would degrade and disappear passively would be distinctly preferable to one that required surgical removal [218]. For both aesthetic and practical reasons, it may also be desirable for oral dosage forms to be biodegradable.

"Biodegradable" is not a precisely defined or quantifiable property but is a context-specific term like "biocompatible." In general, biodegradable simply means degradation of a material by some biological process. Biodegradation can include both complete and incomplete metabolization of a material by either microorganisms or the host, as well as simple chemical lability at biological conditions (especially hydrolytic bond cleavage). It can be defined to include simple loss of physical identity. When applied to hydrogels, the term "biodegradable" is usually used in the broadest sense and is applied to any material which deteriorates in any sense under physiological conditions. As a result, all mechanisms of polymer degradation are covered. In an attempt to be more specific, biodegradable polymers are also described as "bioabsorbable" or "bioerodible." Implanted devices which are metabolized following release of the bioactive agent are considered "bioabsorbable." Controlled-release devices whose release rate is determined by the rate of matrix degradation by any mechanism are termed "bioerodible."

There are three mechanisms for the biodegradation of polymers, all of which may occur singly or in combination [219,220]. The first mechanism is the degradation of a gel with unstable cross-links. These cross-links split by some chemical or physical process, producing linear polymers which may then dissolve or disperse in the biological milieu or degrade further. The second mechanism is the breakdown of the linear polymer by cleavage of bonds in the macromolecular backbone, converting the polymer to low molecular weight species. Finally, chemical reactions can convert the polymer chain from an insoluble form to a soluble state. This modification typically occurs by ionization or protonation of a pendant group but may also include hydrolysis of a hydrophobic group.

In addition to the microscopic mechanism of degradation, it is also important to consider the mechanism at the macroscopic level. Degradation which occurs only at the surface of the device is termed "heterogeneous erosion." If degradation occurs throughout the entire device simultaneously, the process is considered "homogeneous" erosion. Homogeneous erosion processes cause loss of integrity of the device, as opposed to heterogeneous processes, in which the device becomes smaller as the surface erodes but otherwise retains its integrity during the degradation period. These processes are in fact idealized extremes, since degradation usually occurs by some combination of these two mechanisms. The dominant mechanism is dependent on the polymer character, especially its affinity for water. Hydrophobic devices degrade predominantly from the surface, while hydrophilic devices tend toward homogeneous degradation [22]. Devices which degrade by homogeneous processes may swell substantially prior to dissolution [122,221]. Hydrogels would generally be expected to degrade by a homogeneous process due to their inherently hydrophilic natures.

The rates of degradation can vary widely, from hours to years. The rates observed are a complex function of the permeation rate of the gel by solvent and reactive species as well as the reaction kinetics of the biodegradation reaction [222–224].

Many different types of biodegradable hydrogels can be created to decay by any combination of these degradation mechanisms, at almost any rate, as desired for the given application [225]. In any case, it is critical that all degradation products be metabolizable or excretable. Thus most biodegradable gels are made from biologically derived substances. The two basic approaches are to either cross-link stable linear polymers with unstable cross-links or to produce gels of degradable polymers. In the first case, degradation of the cross-links releases linear polymers which may be subsequently metabolized or simply excreted from the body. Examples of this type of gel include cross-linked albumin gels [221,226], starch derivatives [227], and various poly(amino acid)s [228,229]. The primary alternative is to produce gels from polymers which degrade to nontoxic, low molecular weight species in the physiological environment. Examples of such materials which hydrolyze in an aqueous environment include polyanhydrides [222–224,230,231], poly(ortho esters) [218,232,233], polyphosphazenes [234,235], polyesters [236,237], and poly(ester-amides) [238]. Use of bacterial polyesters [239,240], biosynthetic polysaccharides [241], or proteins with enzymatically cleavable bonds are also alternatives [242,243]. Note, however, that these materials have not generally been developed as hydrogels, but they could provide the basis of biodegradable hydrogels according to the synthesis principles outlined in Section II.

IV. SUMMARY

Cross-linked, hydrophilic polymeric networks—hydrogels—form a broad class of materials with diverse properties. Hydrogels can be synthesized by any of the techniques used to create cross-linked polymers. Covalently cross-linked networks can be created by copolymerization/cross-linking of monomers and by linking linear polymers with chemical reagents or through ionizing radiation. Hydrogels can also be cross-linked by noncovalent interactions, including entanglements, electrostatics, hydrogen bonds, hydrophobic interactions, and crystallites. It is also possible to produce hydrogels by the chemical conversion of one type of gel to another.

The key properties of hydrogels used for drug delivery are the equilibrium swelling degree, swelling kinetics, permeability, and in vivo performance properties. Despite the diversity in hydrogels as a class, their properties can be analyzed in terms of common properties. The swelling degree at equilibrium is the most important property of the gel, as it directly influences the other properties of the gel. Theory can make accurate qualitative predictions of the influence of experimental parameters on swelling degree but is not yet able to make quantitative predictions. For hydrogels, the swelling degree rises with an increase in the hydrophilicity of the polymer, an increase in the ionization degree of the network, or a decrease in the cross-link density. The swelling kinetics of gels can be quite complex because various combinations of physical phenomena may influence the swelling rates. The most typical rate-limiting steps are the rates of diffusion and relaxation. When diffusion is rate-limiting, a Fickian data analysis is usually successful, but deviations are observed when the sample is glassy initially but rubbery when swollen. An empirical data analysis may be required when other phenomena influence the rates. Permeability, the product of the partition and diffusion coefficients, is a particularly important property for drug delivery, since it often controls the rate at which solutes are released. Free volume theory indicates, and experiments confirm, that normally the only gel property which directly influences the diffusion coefficient is the swelling degree of the gel; diffusivity declines as swelling declines. Similarly, the partition coefficient of a solute will decline as the swelling degree of the gel declines if the partition coefficient is controlled by size exclusion. However, other effects such as electrostatic and hydrophobic interactions can strongly influence the partition coefficient, decoupling it from the swelling degree of the gel. Finally, the in vivo performance properties of hydrogels can be varied to meet the needs of a desired drug delivery application. Hydrogels display generally good biocompatibility characteristics and can be quite inert in vivo; for some types, this is true even when blood contact is required. Here the hydrogel surface properties are key. Hydrogels can also be made which biodegrade to nontoxic species, a property desirable for implantable applications. Bioerosion itself can also be used as a drug release mechanism. The mechanical properties of hydrogels are usually poor, as strength declines as the swelling degree rises. Fortunately, this is not a serious limitation in most drug delivery systems; even when high mechanical strength is required, a hydrogel can be effective when bonded to a substrate of greater mechanical integrity.

The diversity of the types of hydrogels available means that a gel can be tailored to match the requirements of almost any device requiring the control and modulation of solute transport or release, volumetric displacement, or generation of a triggering signal in response to changes in an aqueous environment. This makes hydrogels central to the development of many advanced drug delivery systems.

ACKNOWLEDGMENTS

The assistance of D. Harsh, B. Kabra, and M. Palasis in preparing the original version of this chapter, and of C. Cooper in assembling the reference list, is gratefully acknowledged.

REFERENCES

1. PJ Flory. Introductory lecture. In: Gels and Gelling Processes, Vol 57. Discussions of the Faraday Society. Aberdeen, Great Britain: University Press, 1975, pp 7–18.
2. DW Woodward, DST Hsieh. Gels for drug delivery. In: DST Hsieh, ed. Controlled Release Systems: Fabrication Technology, Vol II. Boca Raton, FL: CRC Press, 1988, pp 83–110.

3. PI Lee. Synthetic hydrogels for drug delivery: Preparation, characterization and release characteristics. In: DST Hsieh, ed. Controlled Release Systems: Fabrication Technology, Vol II. Boca Raton, FL: CRC Press, 1988, pp 83–110.

4. JA Hubbell. Hydrogel systems for barriers and local drug delivery in the control of wound healing. J Controlled Release 39:305–313, 1996.

5. SH Gehrke, PI Lee. Hydrogels for drug delivery. In: P Tyle, ed. Specialized Drug Delivery Systems, Manufacturing and Production Technology. New York: Marcel Dekker, 1990, pp 333–392.

6. DE Gregonis, CM Chen, JD Andrade. In: JD Andrade, ed. Hydrogels for Medical and Related Applications. ACS Symp Ser 31. Washington, DC: American Chemical Society, 1976, pp 88–104.

7. HG Elias. Macromolecules, Vol. 2. 2nd ed. New York: Plenum Press, 1984.

8. BG Kabra, SH Gehrke. Synthesis of fast response poly(N-isopropylacrylamide) gels. Polym Commun 32:322–323, 1991.

9. AS Hoffman, A Afrassiabi, LC Dong. Thermally reversible hydrogels: II. Delivery and selective removal of substances from aqueous solutions. J Controlled Release 4:213–222, 1986.

10. SH Gehrke, M Palasis, MK Akhtar. Effect of synthesis conditions on properties of poly(N-isopropylacrylamide) gels. Polym Int 29:29–36, 1992.

11. DR Miller, CW Macosko. A new derivation of post gel properties of network polymers. Macromolecules 9:206–211, 1976.

12. K Dusek, L Matejka, P Spacek, H Winter. Network formation in the free-radical copolymerization of a bismaleimide and styrene. Polymer 37:2233–2242, 1996.

13. K Dusek. Formation-structure relationships in polymer networks. Br Polym J 17: 185–189, 1985.

14. HH Winter, M Mours. Rheology of polymers near liquid–solid transitions. Adv Polym Sci 134:165–234, 1997.

15. T Alfrey Jr., LJ Young. The Q-e scheme. In: GE Ham, ed. Copolymerization. New York: Wiley-Interscience, 1964, pp 67–88.

16. LJ Young. Tabulation of Q-e values. In: J Brandrup, EH Immergut, W McDowell, eds. Polymer Handbook. 2nd ed. New York: Wiley-Interscience, 1975, pp 387–404.

17. NA Peppas. Hydrogels of poly(vinyl alcohol) and its copolymers. In: NA Peppas, ed. Hydrogels in Medicine and Pharmacy, Vol. II: Polymers. Boca Raton, FL: CRC Press, 1987, pp 1–48.

18. NB Graham. Poly(ethylene oxide) and related hydrogels. In: NA Peppas, ed. Hydrogels in Medicine and Pharmacy, Vol. II: Polymers. Boca Raton, FL: CRC Press, 1987, pp 95–114.

19. CA Finch. Water soluble polymers. In: CA Finch, ed. Chemistry and Technology of Water-Soluble Polymers. New York: Plenum Press, 1983, pp 81–112.

20. W Jarowenko. Starch. In: HF Mark, NG Gaylord, eds. Encyclopedia of Polymer Science and Technology, Vol. 12. New York: Wiley, 1970, pp 787–862.

21. L Chen, SH Imam, SH Gordon, RV Greene. Starch polyvinyl alcohol crosslinked film—Performance and biodegradation. J Environ Polym Degradation 5:111–117, 1997.

22. WM Kulicke, H Nottleman. Structure and swelling of some synthetic, semisyn-

thetic, and biopolymer hydrogels. In: J Glass, ed. Polymers in Aqueous Media. Adv Chem Ser. 23, Washington, DC: American Chemical Society, 1989, pp 15–44.

23. E Doelker. Water-swollen cellulose derivatives in pharmacy. In: NA Peppas, ed. Hydrogels in Medicine and Pharmacy, Vol. II: Polymers. Boca Raton, FL: CRC Press, 1987, pp 115–160.

24. DC Harsh, SH Gehrke. Controlling the swelling characteristics of temperature-sensitive cellulose ether hydrogels. J Controlled Release 17:175–186, 1991.

25. B Kabra, SH Gehrke. Rate limiting steps for solvent sorption and desorption by microporous stimuli-sensitive absorbent gels. In: FL Buchholz, NA Peppas, eds. Superabsorbent Polymers: Science and Technology. ACS Symp Ser 573. Washington, DC: American Chemical Society, 1994, pp 76–87.

26. BG Kabra, SH Gehrke, RJ Spontak. Microporous responsive HPC gels. 1. Synthesis and microstructure. Macromolecules 31:2166–2173, 1998.

27. SM O'Connor, SH Gehrke. Synthesis and characterization of temperature-sensitive HPMC gel spheres. J Appl Polym Sci 66:1279–1290, 1997.

28. O Rosén, L Picullel. Interactions between covalently crosslinked ethyl(hydroxyethyl)cellulose and SDS. Polym Gels Networks 5:185–200, 1997.

29. MV Sefton. Heparinized hydrogels. In: NA Peppas, ed. Hydrogels in Medicine and Pharmacy, Vol III: Properties and Applications. Boca Raton, FL: CRC Press, 1987, pp 17–52.

30. SH Gehrke, NR Vaid, JF McBride. Protein sorption and recovery by hydrogels using principles of two aqueous phase protein extraction. Biotech Bioeng 58:416–427, 1998.

31. SH Gehrke, LH Uhden, JF McBride. Enhanced loading and activity retention in hydrogel delivery systems. J Controlled Release. 21–33 (1998).

32. B Phillip, H Dautzenber, KJ Linow, J Kotz, W Dawydoff. Polyelectrolyte complexes—Recent developments and open problems. Prog Polym Sci 14:91–172, 1989.

33. SS Wong. Chemistry of Protein Conjugation and Crosslinking. Boca Raton, FL: CRC Press, 1993.

34. SB Ross-Murphy, H McEvoy. Fundamentals of hydrogels and gelation. Br Polym J 18:2–7, 1986.

35. NA Peppas, AG Mikos. Preparation methods and structure of hydrogels. In: NA Peppas, ed. Hydrogels in Medicine and Pharmacy, Vol. I: Fundamentals. Boca Raton, FL: CRC Press, 1986, pp 1–26.

36. AG Mikos, CG Takoudis, NA Peppas. Kinetic modeling of copolymerization/cross-linking reactions. Macromolecules 19:2174–2182, 1986.

37. A Charlesby. Gel formation and molecular weight distribution in long chain polymers. Proc Roy Soc Lond, Ser A 222:542–548, 1954.

38. O Saito. Statistical aspects of infinite network formations. Polym Eng Sci 19:234–235, 1979.

39. NA Peppas, SR Stauffer. Reinforced uncrosslinked poly(vinyl alcohol) gels produced by cyclic freezing-thawing processes—A short review. J Controlled Release 16:305–310, 1991.

40. M Suzuki, O Hirasa. An approach to artificial muscle using polymer gels formed by microphase separation. Adv Polym Sci 110:241–261, 1993.

41. WR Good, KF Mueller. Hydrogels and controlled delivery. In: SK Chandrasekaran, ed. Controlled Release Systems. AIChE Symp Ser 206. New York: American Institute of Chemical Engineers, 1981, pp 42–51.
42. CG Varelas, DG Dixon, CA Steiner. Zero-order release from biphasic polymer hydrogels. J Controlled Release 34:185–192, 1995.
43. AJ Dualeh, CA Steiner. Bulk and microscopic properties of surfactant-bridged hydrogels made from an amphiphilic graft copolymer. Macromolecules 24:112–116, 1991.
44. RF Ofstead, CI Posner. Semicrystalline poly(vinyl alcohol) hydrogels. In: J Glass, ed. Polymers in Aqueous Media. Washington, DC: American Chemical Society, 1989, pp 61–72.
45. T Okano, YH Bae, H Jacobs, SW Kim. Thermally on-off switching polymers for drug permeation and release. J Controlled Release 11:255–265, 1990.
46. PJ Flory, R Rehner Jr. Statistical mechanics of cross-linked polymer networks. J Chem Phys 11:521–525, 1943.
47. PJ Flory. Principles of Polymer Chemistry. Ithaca, NY: Cornell University Press, 1953.
48. LRG Treloar. The Physics of Rubber Elasticity. 3rd ed. Oxford, UK: Clarendon Press, 1975.
49. NA Peppas, EW Merrill. Poly(vinyl alcohol) hydrogels: Reinforcement of radiation-crosslinked networks by crystallization. J Polym Sci Polym Chem 14:441–444, 1976.
50. J Hasa, M Ilavsky, K Dusek. Deformational, swelling, and potentiometric titration of polyelectrolytes. J Polym Sci Polym Phys Ed 13:253–262, 1975.
51. J Hasa, M Ilavsky. Deformational, swelling, and potentiometric (methacrylic acid) gels. II. Experimental results. J Polym Sci Polym Phys Ed 13:263–275, 1975.
52. M Ilavsky. Effect of electrostatic interactions on phase transition in the swollen polymeric network. Polymer 22:1687–1697, 1981.
53. J Ricka, T Tanaka. Swelling of ionic gels: Quantitative performance of the Donnan theory. Macromolecules 17:2916–2921, 1984.
54. RA Siegel. Hydrophobic weak polyelectrolyte gels—Studies of swelling equilibria and kinetics. Adv Polym Sci 109:233–267, 1993.
55. BG Kabra, SH Gehrke, ST Hwang, WA Ritschel. Modification of the dynamic swelling behavior of PHEMA. J Appl Polym Sci 42:2409–2416, 1991.
56. SH Gehrke, GP Andrews, EL Cussler. Chemical aspects of gel extraction. Chem Eng Sci 41:2153–2160, 1986.
57. GP Andrews. The effect of crosslinking on extractions using polyacrylamide gel. MS Thesis, University of Minnesota, 1985.
58. L Brannon-Peppas, NA Peppas. Structural analysis of charged polymeric networks. Polym Bull 20:285–289, 1988.
59. DW Urry, SQ Peng, TM Parker, DC Gowda, RD Harris. Relative significance of electrostatic-induced and hydrophobic-induced pK_a shifts in a model protein—The aspartic acid residue. Angew Chem Int Ed 32:1440–1442, 1993.
60. DW Urry, SQ Peng, TM Parker. Delineation of electrostatic-induced and hydrophobic-induced pK_a shifts in polypentapeptides—The glutamic acid residue. J Am Chem Soc 115:7509–7510, 1993.

61. DW Urry, SQ Peng, TM Parker. Hydrophobicity-induced pK shifts in elastin protein-based polymers. Biopolymers 32:373–379, 1992.
62. DW Urry, A Pattanaik. Elastic protein-based materials in tissue reconstruction. Ann NY Acad Sci 831:32–46, 1997.
63. DC Harsh, SH Gehrke. Characterization of ionic water absorbent polymers: Determination of ionic content and effective crosslink density. In: L Brannon-Peppas, RS Harland, eds. Absorbent Polymer Technology. Amsterdam: Elsevier, 1990, pp 103–124.
64. M Ilavsky. Phase transition in swollen gels. 2. Effect of charge concentration on the collapse and mechanical behavior of polyacrylamide networks. Macromolecules 15:782–786, 1982.
65. K Dusek, D Patterson. Transition in swollen polymer networks induced by intramolecular condensation. J Polym Sci A-2 6:1209–1221, 1968.
66. T Tanaka. Collapse of gels and the critical endpoint. Phys Rev Lett 40:820–823, 1978.
67. T Tanaka. Gels. Sci Am 224:124–137, 1981.
68. S Hirotsu, Y Hirokawa, T Tanaka. Volume phase transitions of ionized N-isopropylacrylamide gels. J Chem Phys 87:1392–1395, 1987.
69. RFS Freitas, EL Cussler. Temperature sensitive gels as extraction solvents. Chem Eng Sci 41:97–105, 1986.
70. T Tanaka, I Nishio, ST Sun, S Ueno-Nishio. Collapse of gels in an electric field. Science 218:467–474, 1982.
71. T Tanaka, D Fillmore, ST Sun, I Nishio, G Swislow, A Shah. Phase transitions in ionic gels. Phys Rev Lett 45:1636–1644, 1980.
72. I Ohime, T Tanaka. Salt effects on phase transition of ionic gels. J Chem Phys 11:5725–5729, 1982.
73. KK Lee, EL Cussler, M Marchetti, MA McHugh. Pressure-dependent phase transitions in hydrogels. Chem Eng Sci 45:766–767, 1990.
74. A Suzuki, T Tanaka. Phase transition in polymer gels induced by visible light. Nature 346:345–347, 1990.
75. A Mamada, T Tanaka, D Kungwatchakun, M Irie. Photoinduced phase transition of gels. Macromolecules 23:1517–1527, 1990.
76. TA Horbett, J Kost, BD Ratner. Swelling behavior of glucose sensitive membranes. In: SW Shalaby, AS Hoffman, BD Ratner, TA Horbett, eds. Polymers as Biomaterials. New York: Plenum Press, 1984, pp 193–207.
77. K Dusek, ed. Responsive Gels: Volume Transitions I. Adv Polym Sci Ser 109. Heidelberg: Springer-Verlag, 1993.
78. K Dusek, ed. Responsive Gels: Volume Transitions II. Adv Polym Sci Ser 110. Heidelberg: Springer-Verlag, 1993.
79. SH Gehrke, DC Harsh. Environmentally responsive gels. In: JC Salamone, ed. Polymeric Materials Encyclopedia, Vol 3. Boca Raton, FL: CRC Press, 1996, pp 2093–2102.
80. YH Bae. Stimuli-sensitive drug delivery. In: K Park, ed. Controlled Drug Delivery: Challenges and Strategies. Washington, DC: American Chemical Society, 1997, pp 147–162.
81. AS Hoffman. "Intelligent" polymers. In: K Park, ed. Controlled Drug Delivery:

Challenges and Strategies. Washington, DC: American Chemical Society, 1997, pp 485–498.

82. LD Taylor, LD Cerankowski. Preparation of films exhibiting a balanced temperature dependence to permeation by aqueous solutions—A study of lower consolute behavior. J Polym Sci Polym Chem Ed 13:2551–2570, 1975.

83. PI Freeman, JS Rowlinson. Lower critical points in polymer solution. Polymer 1: 20–25, 1960.

84. SH Gehrke. Synthesis, equilibrium swelling, kinetics, permeability and applications of environmentally responsive gels. Adv Polym Sci 110:81–144, 1993.

85. X Huang, H Unno, T Akehata, O Hirasa. Effect of salt solution on swelling or shrinking behavior of poly(vinylmethylether) gel (PVMEG). J Chem Eng Jpn 21: 10–14, 1988.

86. RFS Freitas. Extraction with and phase behavior of temperature-sensitive gels. PhD Thesis, University of Minnesota, 1986.

87. F Horkay, AM Hecht, E Geissler. Effect of cross-links on the swelling equation of state: Polyacrylamide hydrogels. Macromolecules 22:2007–2009, 1989.

88. KF Freed, AI Pesci. Computation of the cross-link dependence of the effective Flory interaction parameter χ for polymer networks. Macromolecules 22:4048–4050, 1989.

89. AFM Barton. Handbook of Solubility Parameters and Other Cohesion Parameters. Boca Raton, FL: CRC Press, 1983.

90. AFM Barton, Handbook of Polymer–Liquid Interaction Parameters and Solubility Parameters. Boca Raton, FL: CRC Press, 1990.

91. MM Prange, HH Hooper, JM. Prausnitz. Thermodynamics of aqueous systems containing hydrophilic polymers or gels. AIChE J 35:803–813, 1989.

92. HH Hooper, JP Baker, HW Blanch, JM Prausnitz. Swelling equilibria for positively ionized polyacrylamide hydrogels. Macromolecules 23:1096–1104, 1990.

93. M Marchetti, S Prager, EL Cussler. Thermodynamic predictions of volume changes in temperature-sensitive gels. 1. Theory. Macromolecules 23:1760–1765, 1990.

94. M Marchetti, S Prager, EL Cussler. Thermodynamic predictions of volume changes in temperature-sensitive gels. 2. Experiments. Macromolecules 23:3445–3450, 1990.

95. DC Harsh, SH Gehrke. Modeling swelling behavior of cellulose ether hydrogels. In: M El-Nokaly, D Piatt, B Charpentier, eds. Polymeric Delivery Systems. ACS Symp Ser 520. Washington, DC: American Chemical Society, 1993, pp 105–134.

96. IC Sanchez, AC Balazs. Generalization of the lattice-fluid model for specific interactions. Macromolecules 22:2325–2331, 1989.

97. IC Sanchez, RH Lacombe. Statistical thermodynamics of polymer solutions. Macromolecules 11:1145–1156, 1978.

98. RH Lacombe, IC Sanchez. Statistical thermodynamics of fluid mixtures. J Phys Chem 80:2568–2580, 1976.

99. NA Peppas, BD Barr-Howell. Characterization of the crosslinked structure of hydrogels. In: NA Peppas, ed. Hydrogels in Medicine and Pharmacy, Vol. I: Fundamentals. Boca Raton, FL: CRC Press, 1986, pp 27–56.

100. A Galli, WH Brumage. The freely jointed chain in expanded form. J Chem Phys 79:2411–2416, 1983.

101. HM James, JE Guth. Theory of the increase in rigidity of rubber during cure. J Chem Phys 15:669–672, 1947.
102. S Candau, J Bastide, M Delsanti. Structural, elastic and dynamic properties of swollen polymer networks. Adv Polym Sci 44:27–73, 1982.
103. B Erman, JE Mark. Structures and Properties of Rubberlike Networks. New York: Oxford University Press, 1997.
104. A Silberberg. Gelled aqueous systems. In: J Glass, ed. Polymers in Aqueous Media. Washington, DC: American Chemical Society, 1989, pp 1–28.
105. J Hasa, J Janacek. Effect of diluent content during polymerization on equilibrium deformational behavior and structural parameters of polymer network. J Polym Sci Part C 16:317–328, 1967.
106. BG Kabra, MK Akhtar, SH Gehrke. Volume change kinetics of temperature-sensitive poly(vinylmethylether) gel. Polymer 33:990–995, 1992.
107. R Kishi, O Hirasa, H Ichijo. Fast responsive poly(N-isopropylacrylamide) hydrogels prepared by gamma-ray irradiation. Polym Gels Networks 5:145–151, 1997.
108. RA Orwell. The polymer–solvent interaction parameter χ. Rubber Chem Technol 50:451–456, 1977.
109. D Biren. Anomalous penetrant transport in glassy hydrogels. MS Thesis, University of Cincinnati, 1990.
110. TP Davis, MB Huglin, DCF Yip. Properties of poly(N-vinyl-2-pyrrolidone) hydrogels crosslinked with ethylene glycol dimethacrylate. Polymer 29:701–706, 1988.
111. MB Huglin, MM Rehab. Some observations on monomer reactivity ratios in aqueous and nonaqueous media. Polymer 28:2200–2206, 1987.
112. F Helfferich. Ion Exchange. New York: McGraw-Hill, 1962.
113. SH Gehrke, JF McBride, SM O'Connor, H Zhu, JP Fisher. Gel-coated catheters as drug delivery systems. Polym Mat Sci Eng 76:234–235, 1997.
114. A Peters, SJ Candau. Kinetics of swelling of polyacrylamide gels. Macromolecules 19:1952–1960, 1986.
115. T Momii, T Nose. Concentration-dependent collapse of polymer gels in solution of incompatible polymers. Macromolecules 22:1384–1389, 1989.
116. I Iliopoulos, R Audebert, C Quivoron. Reversible polymer complexes stabilized through hydrogen bonds. In: P Russo, ed. Reversible Polymeric Gels and Related Systems. ACS Symp Ser 350. Washington, DC: American Chemical Society, 1987, pp 72–86.
117. M Kurata. Thermodynamics of Polymer Solutions. New York: Harwood, 1982.
118. E Ruckenstein, V Lesins. Classification of liquid chromatographic methods based on the interaction forces: The niche of potential barrier chromatography. In: A Mizrahi, ed. Advances in Biotechnological Processes, Vol 8: Downstream Processes: Equipment and Techniques. New York: Alan R. Liss, 1988, pp 241–314.
119. NA Peppas, RW Korsmeyer. In: NA Peppas, ed. Hydrogels in Medicine and Pharmacy, Vol III: Properties and Applications. Boca Raton, FL: CRC Press, 1987, pp 109–135.
120. HL Frisch. Sorption and transport in glassy polymers. Polym Eng Sci 20:2–12, 1980.

121. AH Windle. Case II sorption. In: J Comyn, ed. Polymer Permeability. London: Elsevier Applied Science, 1985, pp 75–118.

122. CM Ofner III, H Schott. Swelling studies of gelatin. I: Gelatin without additives. J Pharm Sci 8:790–796, 1986.

123. SH Gehrke, EL Cussler. Mass transfer in pH-sensitive hydrogels. Chem Eng Sci 44:559–566, 1989.

124. X Huang, H Unno, T Akehata, O Hirasa. Analysis of kinetic behavior of temperature-sensitive water-absorbing hydrogel. J Chem Eng Jpn 20:123–128, 1987.

125. SH Gehrke, LH Lyu, MC Yang. Swelling, shrinking, and solute permeation of temperature-sensitive N-isopropylacrylamide gel. Polym Prepr 30:482–483, 1989.

126. SH Gehrke, LH Lyu, K Barnthouse. Dewatering fine coal slurries by gel extraction. Sep Sci Tech 33:1467–1485, 1998.

127. PI Lee. Dimensional changes during drug release from a glassy hydrogel matrix. Polym Commun 24:45–47, 1983.

128. NL Thomas, AH Windle. A theory of case II diffusion. Polymer 23:529–539, 1982.

129. TP Gall, RC Lasky, EJ Kramer. Case II diffusion: Effect of solvent molecule size. Polymer 31:1491–1499, 1990.

130. JS Vrentas, CM Jarzebski, JL Duda. Deborah number for diffusion in polymer–solvent systems. AIChE J 21:894–902, 1975.

131. GWR Davidson III, NA Peppas. Solute and penetrant diffusion in swellable polymers. V. Relaxation-controlled transport in P(HEMA-co-MMA) copolymers. J Controlled Release 3:243–258, 1986.

132. GWR Davidson III, NA Peppas. Solute and penetrant diffusion in swellable polymers. VI. The Deborah and swelling interface numbers as indicators of the order of biomolecule release. J Controlled Release 3:259–271, 1986.

133. JC Wu, NA Peppas. Modeling of penetrant diffusion in glassy polymers with an integral sorption Deborah number. J Polym Sci Polym Phys Ed 31:1503–1518, 1993.

134. MA Samus, G Rossi. Methanol absorption in ethylene-vinyl alcohol copolymers: Relation between solvent diffusion and changes in glass transition temperature in glassy polymeric materials. Macromolecules 29:2275–2288, 1996.

135. JX Li. Case II sorption in glassy polymers: Penetration kinetics. PhD Dissertation, University of Toronto, Toronto, Canada, 1998.

136. GF Billovits, CJ Durning. Penetrant transport in semicrystalline poly(ethylene terephthalate). Polymer 29:1468–1478, 1988.

137. N Thomas, AH Windle. Case II swelling of PMMA sheet in methanol. J Membrane Sci 3:337–346, 1978.

138. D Biren, BG Kabra, SH Gehrke. Effect of initial sample anisotropy on the solvent sorption kinetics of glassy poly(2-hydroxyethyl methacrylate). Polymer 33:554–561, 1992.

139. SH Gehrke, D Biren, JJ Hopkins. Evidence for Fickian water transport in initially glassy poly(2-hydroxyethyl methacrylate). J Biomater Sci Polym Ed 6:375–390, 1994.

140. SH Gehrke, G Agrawal, MC Yang. Moving ion exchange fronts in polyelectrolyte gels. In: RS Harland, RK Prud'homme, eds. Polyelectrolyte Gels. ACS Symp Ser 480. Washington, DC: American Chemical Society, 1992, pp 211–237.

141. PL Ritger, NA Peppas. A simple equation for description of solute release. 1. Fickian and non-Fickian release from non-swellable devices in the form of slabs, spheres, cylinders or discs. J Controlled Release 5:23–35, 1987.

142. J Crank. The Mathematics of Diffusion. 2nd ed. London: Oxford Univ Press, 1975.

143. GF Billovits, CJ Durning. Polymer material coordinates for mutual diffusion in polymer–penetrant systems. Chem Eng Commun 82:21–44, 1989.

144. L Westman, T Lindstrom. Swelling and mechanical properties of cellulose hydrogels. IV. Kinetics of swelling in liquid water. J Appl Polym Sci 26:2561–2572, 1981.

145. J Crank, GS Park, Diffusion in Polymers. New York: Academic Press, 1968.

146. T Tanaka, LO Hocker, GB Benedek. Spectrum of light scattered from a viscoelastic gel. J Chem Phys 59:5151–5159, 1973.

147. T Tanaka, DJ Fillmore. Kinetics of swelling of gels. J Chem Phys 70:1214–1218, 1979.

148. Y Li, T Tanaka. Kinetics of swelling and shrinking of gels. J Chem Phys 92:1365–1371, 1990.

149. EJ Lightfoot. Kinetic diffusion in polymer gels. Physica A 169:191–206, 1990.

150. G Rossi, KA Mazich. Kinetics of swelling for a cross-linked elastomer or gel in the presence of a good solvent. Phys Rev A 44:r4793–r4796, 1991.

151. KA Mazich, G Rossi, CA Smith. Kinetics of solvent diffusion and swelling in a model elastomeric system. Macromolecules 25:6929–6933, 1992.

152. CJ Durning, KN Morman. Nonlinear swelling of polymer gels. J Chem Phys 98:4275–4293, 1993.

153. G Rossi, KA Mazich. Macroscopic description of the kinetics of swelling for a cross-linked elastomer or a gel. Phys Rev E 48:1182–1191, 1993.

154. T Tomari, M Doi. Swelling dynamics of a gel undergoing volume transition. J Phys Soc Jpn 63:2093–2101, 1994.

155. K Yoshimura, K Sekimoto. Coupling between diffusion and deformation of gels in binary solvents—A model study. J Chem Phys 101:4407–4417, 1994.

156. T Tomari, M Doi. Hysteresis and incubation in the dynamics of volume transition of spherical gels. Macromolecules 28:8334–8343, 1995.

157. CJ Wang, Y Li, ZB Hu. Swelling kinetics of polymer gels. Macromolecules 30:4727–4732, 1997.

158. J Singh, ME Weber. Kinetics of one-dimensional gel swelling and collapse for large volume change. Chem Eng Sci 51:4499–4508, 1996.

159. FA Long, D Richman. Concentration gradients for diffusion of vapors in glassy polymers and their relation to time dependent diffusion phenomena. J Am Chem Soc 82:513–522, 1960.

160. RC Lasky, EJ Kramer, CY Hui. The initial stages of case II diffusion at low penetrant activities. Polymer 29:673–679, 1988.

161. NM Franson, NA Peppas. Influence of copolymer composition on non-Fickian water transport through glassy copolymers. J Appl Polym Sci 28:1299–1310, 1983.

162. RW Korsmeyer, EW Meerwall, NA Peppas. Solute and penetrant diffusion in swellable polymers. II. Verification of theoretical models. J Polym Sci Polym Phys Ed 24:409–419, 1986.

163. BG Kabra, SH Gehrke. Hydrogels for driving an osmotic pump. Polym Prepr 30: 490–491, 1989.

164. CR Robert, PA Buri, NA Peppas. Effect of degree of crosslinking of water transport in polymer microparticles. J Appl Polym Sci 30:301–307, 1985.

165. BD Barr-Howell, NA Peppas. Structural analysis of poly(2-hydroxyethyl methacrylate) microparticles. Eur Polym J 23:591–596, 1987.

166. SH Gehrke, M Palasis, M Lund, J Fisher. Factors determining hydrogel permeability. Ann NY Acad Sci 831:179–207, 1997.

167. R Baker. Controlled Release of Biologically Active Agents. New York: Wiley, 1987.

168. EL Cussler. Diffusion. Cambridge: Cambridge Univ Press, 1984.

169. NL Abbott, TA Hatton. Liquid–liquid extractions for protein separations. Chem Eng Prog 84:31–41, 1988.

170. PA Albertsson. Partition of Cell Particles and Macromolecules. 3rd ed. New York: Wiley-Interscience, 1986.

171. H Walter, DE Brooks, D Fisher, eds. Partitioning in Aqueous Two-Phase Systems. New York: Academic Press, 1985.

172. C Tanford. Physical Chemistry of Macromolecules. New York: Wiley, 1961.

173. EH Schacht. Hydrogel drug delivery systems physical and ionogenic drug carriers. In: JM Anderson, SW Kim, eds. Recent Advances in Drug Delivery Systems. New York: Plenum Press, 1984, pp 259–278.

174. M Palasis. The influence of interactions on the diffusion of solutes in responsive gels. PhD Dissertation, University of Cincinnati, Cincinnati, OH, 1994.

175. M Palasis, SH Gehrke. Controlling permeability in responsive hydrogels. Proc Int Symp Controlled Release Bioact Mater 17:385–386, 1990.

176. M Palasis, SH Gehrke. Permeability of temperature sensitive gels. J Controlled Release 18:1–12, 1992.

177. L Haggerty, JH Sugarman, RK Prud'homme. Diffusion of polymers through polyacrylamide gels. Polymer 29:1058–1063, 1988.

178. C Tanford. The Hydrophobic Effect. New York: Wiley, 1973.

179. PL Dubin, JM Principi. Hydrophobicity parameter of aqueous size exclusion chromatography gels. Anal Chem 61:780–781, 1989.

180. A Ben-Naim. Hydrophobic Interactions. New York: Plenum Press, 1980.

181. AG Ogston. Sedimentation in the ultracentrifuge. Trans Faraday Soc 54:1754–1757, 1958.

182. JC Giddings, E Kucera, CP Russell, MN Myers. Statistical theory for the equilibrium distribution of rigid molecules in inert porous networks. Exclusion chromatography. J Phys Chem 72:4397–4408, 1968.

183. TC Laurent, J Killander. A theory of gel filtration and its experimental verification. J Chromatogr 14:317–330, 1964.

184. JE Schnitzer. Analysis of steric partition behavior of molecules in membranes using statistical physics. Biophys J 54:1065–1076, 1988.

185. AP Sassi, HW Blanch, JM Prausnitz. Characterization of size-exclusion effects in highly swollen hydrogels: Correlation and prediction. J Appl Polym Sci 59:1337–1346, 1996.

186. S Hussain, MS Mehta, JI Kaplan, PL Dubin. Experimental evaluation of conflicting models for size exclusion chromatography. Anal Chem 63:1132–1138, 1991.

187. DA Hoagland. Unified thermodynamic model for polymer separations produced by size exclusion chromatography, hydrodynamic chromatography, and gel electrophoresis. ACS Symp Ser 635:173–188, 1996.

188. CN Satterfield, CK Colton, WH Pitcher. Restricted diffusion in liquids with fine pores. AIChE J 19:628–635, 1973.

189. W Brown, K Chitumbo. Solute diffusion in hydrated polymer networks. Chem Soc Faraday Trans I 71:1–11, 1975.

190. GM Zentner, JR Cardinal, J Feijen, SZ Song. Progestin permeation through polymer membranes. IV: Mechanism of steroid permeation and functional group contribution to diffusion through hydrogel films. J Pharm Sci 68:970–975, 1979.

191. H Yasuda, CE Lamaze, A Peterlin. Diffusive and hydraulic permeabilities of water in water-swollen polymer membranes. J Polym Sci A-2 9:1117–1131, 1971.

192. H Yasuda, CE Lamaze, LD Ikenberry. Permeability of solutes through hydrated polymer membranes. Part I. Diffusion of sodium chloride. Makromol Chem 118: 19–35, 1968.

193. MH Cohen, D Turnbull. Molecular transport in liquids and gases. J Chem Phys 31:1164–1169, 1959.

194. RW Korsmeyer, SR Lustig, NA Peppas. Solute and penetrant diffusion in swellable polymers. I. Mathematical modeling. J Polym Sci Polym Phys Ed 24:395–408, 1986.

195. NA Peppas, HJ Moynihan. Structure and physical properties of poly(2-hydroxyethyl methacrylate) hydrogels. In: NA Peppas, ed. Hydrogels in Medicine and Pharmacy, Vol II: Polymers. Boca Raton, FL: CRC Press, 1987, pp 49–64.

196. NA Peppas, CT Reinhardt. Solute diffusion in swollen membranes. Part I: A new theory. J Membrane Sci 15:275–287, 1983.

197. H Feil, YH Bae, J Feijen, SW Kim. Molecular separation by thermosensitive hydrogel membranes. J Membrane Sci 64:283–294, 1991.

198. B Amsden. Solute diffusion within hydrogels. Mechanisms and models. Macromolecules 31:8382–8395, 1998.

199. BK Davis. Diffusion in polymer gel implants. Proc Natl Acad Sci 71:3120–3123, 1974.

200. WA Ritchel, A Sabouni, SH Gehrke, ST Hwang. Permeability of [^{3}H]water across a porous polymer matrix used as a rate-limiting shell in compression-coated tablets. J Controlled Release 12:97–102, 1990.

201. A Afrassiabi, AS Hoffman, LA Cadwell. Effect of temperature on the release rate of biomolecules from thermally reversible hydrogels. J Membrane Sci 33:191–200, 1987.

202. TG Park, AS Hoffman. Immobilization of *Arthrobacter simplex* in thermally reversible hydrogels: Effect of gel hydrophobicity on steroid conversion. Biotech Prog 7:383–390, 1991.

203. PI Lee. Determination of diffusion coefficients by sorption from a constant, finite volume. In: RW Baker, ed. Controlled Release of Bioactive Materials. New York: Academic Press, 1980, pp 255–265.

204. IL Claeys, FH Arnold. Nuclear magnetic relaxation study of hindered rotational diffusion in gels. AIChE J 35:335–338, 1990.

205. IH Park, CS Johnson, DA Gabriel. Probe diffusion in polyacrylamide gels as ob-

served by means of holographic relaxation methods: Search for a universal equation. Macromolecules 23:1548–1553, 1990.

206. SJ Gibbs, EN Lightfoot, TW Root. Protein diffusion in porous gel filtration chromatography media studied by pulsed field gradient NMR spectroscopy. J Phys Chem 96:7458–7462, 1992.

207. JA Wesson, H Takezoe, H Yu, SP Chen. Dye diffusion in swollen gels by forced Rayleigh scattering. J Appl Phys 53:6513–6519, 1982.

208. DB Sellen. The diffusion of compact macromolecules within hydrogels. Br Polym J 18:28–31, 1986.

209. BD Ratner, AS Hoffman. Synthetic hydrogels for biomedical applications. In: JD Andrade, ed. Hydrogels for Medical and Related Applications. ACS Symp Ser 31. New York: American Chemical Society, 1976, pp 1–36.

210. DW Urry, CM Harris, CX Luan, CH Luan, C Gowda, TM Parker, SQ Peng, J Xu. Transductional protein-based polymers as new controlled-release vehicles. In: K Park, ed. Controlled Drug Delivery: Challenges and Strategies. Washington, DC: ACS, 1997, pp 405–437.

211. BD Ratner. Hydrogel surfaces. In: NA Peppas, ed. Hydrogels in Medicine and Pharmacy, Vol I: Fundamentals. Boca Raton, FL: CRC Press, 1986, pp 85–94.

212. WR Gombotz, AS Hoffman. Immobilization of biomolecules and cells on and within synthetic polymer hydrogels. In: NA Peppas, ed. Hydrogels in Medicine and Pharmacy, Vol I. Boca Raton, FL: CRC Press, 1986, pp 95–126.

213. TA Horbett. Protein adsorption to hydrogels. In: NA Peppas, ed. Hydrogels in Medicine and Pharmacy, Vol I. Boca Raton, FL: CRC Press, 1986, pp 127–171.

214. SM O'Connor, SH Gehrke, S Patuto, GS Retzinger. Fibrinogen-dependent adherence of macrophages to surfaces coated with poly(ethylene oxide)/poly(propylene oxide) triblock copolymers. Ann NY Acad Sci 831:138–144, 1997.

215. JD Andrade, V Hlady, SI Jeon. Poly(ethylene oxide) and protein resistance: Principles, problems, and possibilities. In: JE Glass, ed. Hydrophilic Polymers: Performance with Environmental Acceptability. Adv Chem Ser 248. Washington DC: American Chemical Society, 1996, pp 51–59.

216. R Duncan, M Bhakoo, PA Flanagan, D Sgouras. Evaluation of the biocompatibility of soluble synthetic polymers designed as drug carriers. Proc Int Symp Controlled Release Bioact Mater 15:121–122, 1988.

217. R Duncan. Soluble polymers: Use for controlled drug delivery. Proc Int Symp Controlled Release Bioact Mater 15:127–128, 1988.

218. SY Ng, T Vandamme, MS Taylor, J Heller. Controlled drug release from self-catalyzed poly(ortho esters). Ann NY Acad Sci 831:168–178, 1997.

219. J Heller. Controlled release of biologically active compounds from bioerodible polymers. Biomaterials 1:51, 1980.

220. HB Rosen, J Kohn, K Leong, R Langer. Bioerodible polymers for controlled release systems. In: DST Hsieh, ed. Controlled Release Systems: Fabrication Technology, Vol II. Boca Raton, FL: CRC Press, 1988, pp 83–110.

221. CK Sim, K Park. Examination of drug release from enzyme-digestible swelling hydrogels. Proc Int Symp Controlled Release Bioact Mater 16:219–220, 1989.

222. KW Leong, BC Brott, R Langer. Bioerodible polyanhydrides as drug-carrier ma-

trixes. I. Characterization, degradation, and release characteristics. J Biomed Mater Res 19:941–955, 1985.

223. E Mathiowitz, WM Saltzman, A Domb, P Dor, R Langer. Polyanhydride microspheres as drug carriers. II. Microencapsulation by solvent recovery. J Appl Polym Sci 35:755–774, 1988.

224. L Shieh, J Tamada, Y Tabata, A Domb, R Langer. Drug release from a new family of biodegradable polyanhydrides. J Controlled Release 29:73–82, 1994.

225. J Heller. Bioerodible hydrogels. In: NA Peppas, ed. Hydrogels in Medicine and Pharmacy, Vol III: Properties and Applications. Boca Raton, FL: CRC Press, 1987, pp 137–149.

226. H Natsume, K Sugibayashi, K Juni, Y Morimoto, T Shibata. Preparation and evaluation of biodegradable albumin microspheres containing mitomycin C. Int J Pharm 58:79–87, 1990.

227. P Stjarnkvist, I Sjoholm, T Laasko. Biodegradable microspheres. XII. Properties of the crosslinking chains in polyacryl starch microparticles. J Pharm Sci 78:52–56, 1989.

228. F Lescure, D Bichon, JM Anderson, E Doelker, ML Pelaprat, RJ Gurny. Acute histopathological response to a new biodegradable polypeptidic polymer for implantable drug delivery system. Biomed Mater Res 23:1299–1313, 1989.

229. A Nathan, J Kohn. Amino acid derived polymers. In: SW Shalaby, ed. Biomedical Polymers: Designed-to-Degrade Systems. Cincinnati, OH: Hanser/Gardner, 1994, pp 117–151.

230. AJ Domb, M Maniar. Absorbable biopolymers derived from dimer fatty acids. J Polym Sci Polym Chem Ed 31:1275–1285, 1993.

231. M Maniar, AJ Domb, A Haffer, J Shah. Controlled-release of a local anesthetic from fatty-acid dimer based polyanhydride. J Controlled Release 30:233–239, 1994.

232. J. Heller. Poly(ortho esters). Adv Polym Sci 107:41–92, 1993.

233. J. Heller, AU Daniels. Poly(orthoesters). In: SW Shalaby, ed. Biomedical Polymers: Designed-to-Degrade Systems. Cincinnati, OH: Hanser/Gardner, 1994, pp 1–34.

234. HE Allcock. Design and synthesis of new biomaterials via macromolecular substitution. Ann NY Acad Sci 831:13–31, 1997.

235. AG Scopelianos. Polyphosphazenes as new biomaterials. In: SW Shalaby, ed. Biomedical Polymers: Designed-to-Degrade Systems. Cincinnati, OH: Hanser/Gardner, 1994, pp 153–172.

236. SW Shalaby, RA Johnson. Synthetic absorbable polyesters. In: SW Shalaby, ed. Biomedical Polymers: Designed-to-Degrade Systems. Cincinnati, OH: Hanser/Gardner, 1994, pp 1–34.

237. R Bodmeier, HG Chen. Evaluation of biodegradable poly(lactide) pellets prepared by direct compression. J Pharm Sci 78:819–822, 1989.

238. TH Barrows. Bioabsorbable poly(ester-amides). In: SW Shalaby, ed. Biomedical Polymers: Designed-to-Degrade Systems. Cincinnati, OH: Hanser/Gardner, 1994, pp 97–116.

239. Y Doi, Y Kanesawa, M Kunioka, T Saito. Biodegradation of microbial copolyesters: Poly(3-hydroxybutyrate-co-3-hydroxyvalerate) and poly(3-hydroxybutyrate-co-4-hydroxybutyrate). Macromolecules 23:26–31, 1990.

240. RA Gross. Bacterial polyesters: Structural variability in microbial synthesis. In: SW Shalaby, ed. Biomedical Polymers: Designed-to-Degrade Systems. Cincinnati, OH: Hanser/Gardner, 1994, pp 173–188.

241. DL Kaplan, BJ Wiley, JM Mayer, S Arcidiacono, J Keith, SJ Lombardi, D Ball, AL Allen. Biosynthetic polysaccharides. In: SW Shalaby, ed. Biomedical Polymers: Designed-to-Degrade Systems. Cincinnati, OH: Hanser/Gardner, 1994, pp 97–116.

242. DL Hern, JA Hubbell. Incorporation of adhesion peptides into nonadhesive hydrogels useful for tissue resurfacing. J Biomed Mater Res 93:266–276, 1998.

243. JA Hubbell. Biomaterials in tissue engineering. Bio-Technology 13:565–576, 1995.

14

Stimuli-Modulated Delivery Systems

Sung Wan Kim and You Han Bae*
University of Utah, Salt Lake City, Utah

I. INTRODUCTION

The diffusional properties of some synthetic and natural polymeric membranes can be altered by swelling and conformational or morphological changes in the polymer matrices. External stimuli can generate these changes. Functional groups which are sensitive to the stimuli can be introduced; polymers bearing these groups arc defined as "stimuli-sensitive polymers." These polymers have been studied for their ability to mimic the functions of biological membranes such as active transport and permselectivity or as self-regulating systems [1]. A schematic for stimuli-modulated drug delivery is shown in Figure 1. Stimuli which are utilized for the regulation of hydrogel membranes are classified in Table 1.

II. pH-MODULATED SYSTEMS

In pH-modulated systems, typical examples include the use of polymer matrices containing pendant ionizable groups on the polymeric chains (polyelectrolytes). The pH of the external media can influence the ionization state of the pendant groups and induce swelling changes [2–4] or conformational changes, such as the helix–coil transition of polypeptides [5,6]. The pH dependence of solute permeability through polyelectrolyte gel membranes has been controlled by varying the species (pK_a, charges) and the amount of ionizable groups. Taking into account the swelling behavior of simple polyelectrolyte gels in water [7,8] with

---

* Current affiliation: Kwangju Institute of Science and Technology, Kwangju, Korea.

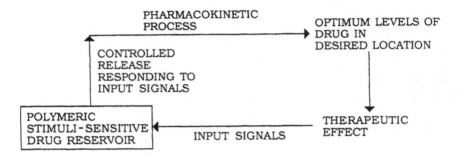

Figure 1 Schematic illustration of an ideal stimuli-modulated polymeric drug delivery system.

varying pH, ion species, and ionic strength, the permeability of neutral solutes can be understood from the role of water content in solute diffusivity in swollen gels [9,10]. In the case of charged solutes, charge–charge interactions will be an additional factor contributing to the pH effect on permeability [11].

The degree of swelling in aqueous systems is a key phenomenon for the unique behavior of pH-sensitive hydrogels. Factors influencing the degree of swelling include the pK_a of the ionizable group, cross-link density, and hydrophilicity or hydrophobicity of the polymer. Selection of adequate factors is an important aspect for the optimal design of hydrogels for specific applications. The effects of these factors on the swelling properties are summarized in Table 2.

The effect of hydrophobicity of the polymer on the permeability of poly(2-hydroxyethyl methacrylate (HEMA)-*co*-methacrylic acid (MAAc) hydrogels was studied [12]. The hydrophobicity was controlled by copolymerization with butyl methacrylate (BMA). The dependence of permeability on pH increased as the hydrophobicity increased even though the rate of diffusion decreased. Cross-link density of the hydrogel also contributed to pH-dependent permeability.

Weiss et al. [11] synthesized poly(MAAc) membranes with different degrees of cross-linking. At high pH the solute permeability was approximately the same for membranes containing different amounts of cross-linking. At low pH a membrane with a lower degree of cross-linking had a higher permeability than a highly cross-linked membrane. When solutes of different molecular weights and charge states were investigated, it was determined that the permeability of high molecular weight compounds was more dependent on pH than the permeability of low molecular weight compounds. It is well known that as pH increases and the gel becomes ionized, the degree of swelling increases. The permeability of high molecular weight solutes was shown to be more dependent on the degree of swelling, in agreement with the free volume theory [13]. pH-sensitive hydrogels consisting of *N*-isopropyl acrylamide (IPAAm), acrylic acid (AAc), and

Table 1 Stimuli-Sensitive Hydrogels and Their Mechanisms

Stimulus	Hydrogel	Mechanism
pH	Acidic or basic hydrogel	Change in pH → change in ionization → change in swelling → change in release of drug
Ionic strength	Ionic hydrogel	Change in ionic strength → change in concentration of ions inside gel → change in swelling → change in release of drug
Chemical species	Hydrogel containing electron-accepting groups	Electron-donating compounds → formation of charge transfer complex → change in swelling → change in release of drug
Enzyme substrate	Hydrogel containing immobilized enzyme	Substrate present → enzymatic conversion → product changes swelling of gel → change in release of drug
Competitive binding	Concanavalin A hydrogel containing glycosylate insulin	Increase in glucose concentration → displacement of glycosylated insulin by competitive binding → release of glycosylated insulin
Magnetic	Magnetic particles dispersed in alginate microspheres	Applied magnetic field → change in pores in gel → change in swelling → change in release of drug
Thermal	Thermosensitive hydrogel, e.g., poly(N-isopropyl acrylamide)	Change in temperature → change in polymer–polymer and water–polymer interactions → change in swelling → change in release of drug
Electrical	Polyelectrolyte hydrogel	Applied electric field → change in membrane charging and electrophoresis of charged drug → change in swelling → change in release of drug
Ultrasound irradiation	Ethylene–vinyl alcohol hydrogel	Ultrasound irradiation → temperature increase → release of drug
Photochemical	2-Hydroxyethyl methacrylate hydrogel containing azobenzene in the side chain	Photoirradiation → photoisomerization of azobenzene moiety → change in swelling → change in drug release

Source: Ref. 51.

Table 2   Factors Influencing Swelling of pH-Sensitive Hydrogels

Factor	Effect
Gel properties	
Charge of ionizable monomer	Acidic: pH ↑ → ionization ↑
	Basic: pH ↑ → ionization ↓
pK_a of ionic monomer	pK_a ↑ → pH–ionization profile shifts to ↑ pH
Degree of ionization	Ionization ↑ → swelling ↑
Concentration of ionizable monomer	Concentration ↑ → swelling in ionized state ↑
Cross-link density	Density ↑ → swelling↑
Hydrophilicity/hydrophobicity of polymer backbone	Hydrophilicity ↑ → swelling
Swelling solution	
pH	Acidic: pH ↑ → swelling ↑
	Basic: pH ↑ → swelling ↓
Ionic strength	Ionic strength ↑ → osmotic pressure inside gel ↓ → swelling ↓ (Exception: polyelectrolyte complexes)
Counterion	Effect depends on species salting-in/salting-out and effect on water structure)
Valence of counterion	Valence ↑ → swelling ↓
Coion	Usually no change

Source: Ref. 51.

vinyl-terminated polydimethylsiloxane were prepared for enteric drug delivery [14]. The designed systems exhibited temperature-sensitive swelling in addition to the pH-sensitive behavior. An indomethacin release study showed that no drug was released at pH 1.4 for 24 hr but over 90% was released at pH 7.4 during a 5-hr period.

A molecule possessing a negative charge also showed increasing permeability through acidic membranes with increasing pH [15,16]. This suggests that diffusion is more dependent on polymer swelling than on the charge density or charge repulsion.

Siegel and coworkers [17] synthesized a basic poly(methyl methacrylate (MMA)-*co*-*N*,*N*-diethylaminoethyl methacrylate (DEAEMA)) hydrogel that swells at low pH and can be used in oral delivery. At neutral pH, the gel has a low degree of swelling, and drug loaded into the gel will not be released. As pH decreases in the stomach, the swelling increases and the drug is released.

A pH-sensitive gel, poly(AAc), was used as a coating on capsules containing insulin and surfactant. The capsules protected against insulin release in the upper gastrointestinal tract. As the pH increased to 7.5, drug release increased

[18]. Insulin levels were monitored by blood glucose depression, and a 45% reduction of glucose was observed. Similar systems were designed based on capsules coated with MAAc copolymers [19]. Capsules containing insulin and a bile salt were administered in humans followed by measuring the C-peptide concentration in the blood. An observed decrease in C-peptide concentration suggests that release of endogenous insulin was suppressed by exogenous insulin.

pH-sensitive membranes have been utilized to assemble devices which release insulin in response to glucose concentration. Ishihara and coworkers [20,21] and Albin et al. [22] synthesized glucose-sensitive members composed of DEAEMA having basic polyamine groups and HEMA. Glucose oxidase enzymes were immobilized in the polymer matrix. As glucose permeated through this membrane, gluconic acid was produced, and pH subsequently decreased. This resulted in an increase in membrane swelling due to the ionization of the groups in the gel. A schematic of the concept is shown in Figure 2.

This system demonstrates regulating insulin release based on membrane reversible swelling by varying glucose concentration. However, it is difficult to mimic physiological conditions or function in vivo due to other contributing factors, such as ionic strength and proteins, in a physiological environment.

Siegel [23] proposed the use of a self-regulating mechanochemical insulin pump. The pump consists of three compartments. Compartment 3, which is in contact with body fluid, consists of a basic hydrogel membrane containing immobilized glucose oxidase. As the glucose concentration increases, the degree of swelling increases and the gel expands. The gel then pushes against a diaphragm separating compartments 2 and 3. This increases the pressure in compartments

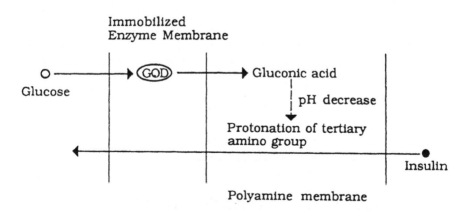

Figure 2 Schematic representation of glucose-responsive polymer membrane constituted with polyamine membrane and glucose oxidase immobilized membrane. (From Ref. 20.)

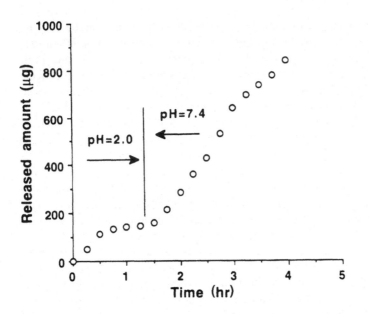

Figure 3   Acetominophen release from temperature- or pH-sensitive polymer NiPAAM/
BMA/DEAEMA at pH 2 and 7 and 37°C.

1 and 2; insulin from compartment 1 is then released through a valve. This system
avoids insulin adsorption and aggregation on the membrane surface as there is
no direct contact between the membrane and the insulin solution.

Recently we designed an oral delivery system based on the concept of
hydrogel deswelling upon a change from pH 2 to pH 7. Drug release profiles
from pH-sensitive NiPAAm/BMA/DEAEMA were affected by pH change,
which demonstrates a two-step release with an increase in the amount released
at pH 7. This system utilizes the mechanical squeezing property of pH-sensitive
hydrogel (Fig. 3) [24,25].

In the gastrointestinal tract, a mucoadhesive drug delivery system provides
advantages in prolonging the residence time of devices. The use of pH-sensitive
bioadhesive polymers has been proposed [26]. An extensive review of pH-sensitive hydrogels is given by Brøndsted and Kopecek [27].

III.   PHOTOSENSITIVE POLYMERIC MEMBRANES

Photoresponsive polymers can be prepared by coupling photochromic compounds which undergo structural changes upon irradiation with light and subse-

quent reversible changes in the physical properties of polymers. Typical photochromic compounds, such as azobenzenes, spiropyranes, and triphenylmethanes, used in the preparation of photoresponsive polymeric membranes are shown in Figure 4.

Ishihara [28] synthesized two hydrogel copolymers containing aromatic light-sensitive groups. They were poly(HEMA-*co*-*p*-phenylazoacrylanilide (PAAn)). Their structures are shown in Figure 5.

Permeation experiments were carried out using a two-compartment diffusion cell. Figure 6 represents the permeation profiles of various proteins through poly(HEMA-*co*-*p*-phenylazoethyl methacrylate (AEMA)) membranes. In the dark, the permeation values of insulin (MW 6000), lysozyme (MW 14,500), and chymotrypsin (MW 23,000) were determined. The amount of permeation of these proteins increased linearly with time. However, the permeation of albumin (MW 60,000) could not be observed. Since the permeation of solute through a polymer membrane is greatly affected by the molecular size of the solute, this finding may be attributed to the relatively large size of the albumin molecule. When

Azobenzene

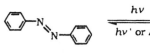

trans form		cis form

Spiropyran

closed-ring form   open-ring form

Triphenylmethane (Malachite green leuco hydroxide)

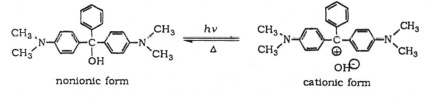

nonionic form   cationic form

Figure 4 Photoinduced structural change of photochromic compounds. (From Ref. 26.)

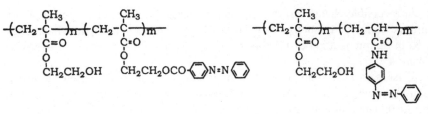

HEMA-AEMA copolymer   HEMA-PAAm copolymer

Figure 5   Azoaromatic polymers. (From Ref. 26.)

photoirradiation was applied to the membrane, the permeation of all proteins was suppressed. Lysozyme and chymotrypsin did not permeate through the polymer membrane. It is generally known that the swelling degree of polymer membranes is an important factor affecting permeability when water-soluble solutes permeate through a hydrated hydrophilic polymer membrane [29,30]. The swelling degree of the azoaromatic polymer membrane in water was decreased by UV irradiation and recovered to its original level by visible light irradiation accompanying the photoisomerization of the azobenzene moiety. The swelling degree of the membrane, defined as the ratio of the weight of solvent to the weight of dry membrane, was decreased from 0.20 to 0.12 by UV irradiation, even in buffer. Therefore, it was concluded that the decrease in the permeability of the proteins induced by UV irradiation was due to a decrease in the swelling degree of the polymer membrane.

The permeation profiles of the proteins through the polymer membrane can be approximated by two straight lines representing regions that were in the dark and under UV irradiation as shown in Figure 6. The permeation coefficients can be calculated from the slope of these straight lines by using the equation

$$Q = PACt/l$$

where Q is the amount of protein permeated, P (cm^2/sec) is the permeation coefficient, A (cm^2) is the surface area of the polymer membrane, C (g/cm^3) is initial concentration of protein, t (sec) is time, and l (cm) is the thickness of the polymer membrane.

Figure 7 shows the relationship between the permeation coefficient and the molecular weight of solute. The permeation coefficient of a low molecular weight compound, methyl orange (MW 327), took on a very large value both in the dark and under UV irradiation. However, the permeation coefficients of proteins decreased with increasing molecular weight of the proteins and became zero in the case of albumin. With UV irradiation, permeation coefficients decreased in all ranges of molecular weight tested. This change became more dramatic with increasing molecular weight; the permeation of proteins with molecular weight

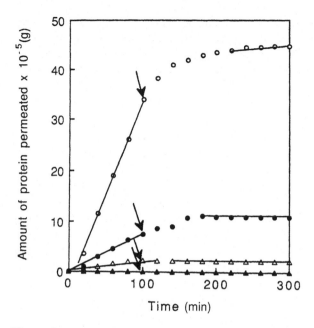

Figure 6 Permeation profiles of proteins through HEMA-AEMA copolymer membrane at 30°. The arrows represent the UV irradiation to the membrane. (○) Insulin; (●) lysozyme; (△) chymotrypsin; (▲) albumin. (From Ref. 28.)

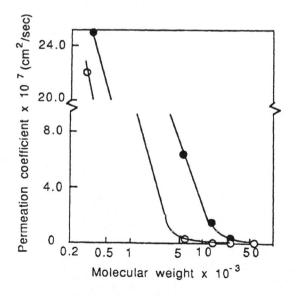

Figure 7 Permeation coefficient of solute in HEMA-AEMA copolymer membrane vs. molecular weight of solute. (●) In the dark; (○) under UV irradiation. (From Ref. 28.)

above 10^4 could not be detected. Therefore, the HEMA-AEMA copolymer membrane has the function of a permeable membrane in the dark and that of a semipermeable membrane under UV irradiation through which only low molecular weight compounds can permeate. These are clear examples of how photoinduced polymeric membranes can function.

IV. TEMPERATURE-SENSITIVE HYDROGELS

The water swelling of hydrogels depends on the strength of interactions between polymer chains and water. Hydrophilic polymers such as polyacrylamide derivatives exhibit abrupt solubility changes with temperature. When cross-linked hydrogels are used, swelling/deswelling (hydration/dehydration) occurs dramatically, with high swelling at lower temperatures and low swelling at higher temperatures, respectively. In general, in hydrogels balanced with a hydrophobic/hydrophilic nature, entropy-triggered dehydration takes place as temperature increases, which causes to precipitate polymers as shown in Figure 8. Poly(N,N-alkyl-substituted acrylamides) showing thermosensitivity in water swelling are listed in Table 3.

Temperature-sensitive polymers, depending on polymer structure and polymer–polymer interactions, generally exhibit two behaviors, lower critical solution temperature (LCST) [31] and upper critical solution temperature (UCST). Phase diagrams for these behaviors are presented in Figure 9.

The polymer–water interaction parameter (χ) of cross-linked alkyl-substituted acrylamides is a function of temperature. The degree of swelling increases as χ decreases. Thermodynamic parameters for interpreting the LCST phenomenon are described elsewhere [32]. Figure 10 is the plot of χ versus tempera-

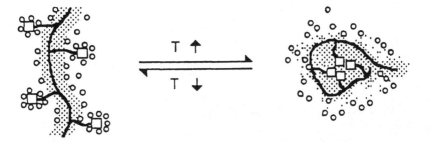

Figure 8 Schematic for the transition of lower critical solution temperature polymer. (From Ref. 29.)

Table 3 Poly(N,N-Alkyl-Substituted Acrylamides) for the Study of Alkyl Group Effects on Thermosensitivity in Aqueous Swelling

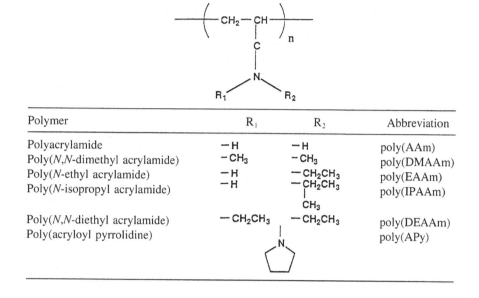

Polymer	R_1	R_2	Abbreviation
Polyacrylamide	$-H$	$-H$	poly(AAm)
Poly(N,N-dimethyl acrylamide)	$-CH_3$	$-CH_3$	poly(DMAAm)
Poly(N-ethyl acrylamide)	$-H$	$-CH_2CH_3$	poly(EAAm)
Poly(N-isopropyl acrylamide)	$-H$	$-CH_2CH_3$ $\quad\ \ CH_3$	poly(IPAAm)
Poly(N,N-diethyl acrylamide)	$-CH_2CH_3$	$-CH_2CH_3$	poly(DEAAm)
Poly(acryloyl pyrrolidine)			poly(APy)

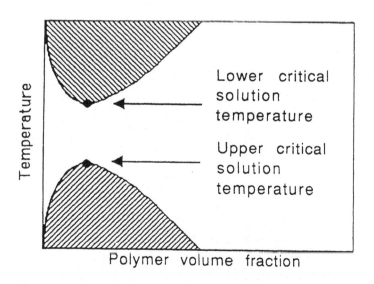

Figure 9 Qualitative phase diagram of a polymer solution showing phase separation both on heating (at the lower critical solution temperature) and on cooling (at the upper critical solution temperature). (From Ref. 31.)

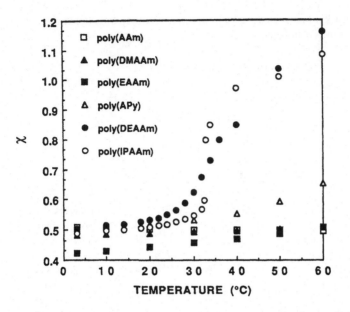

Figure 10 Polymer–water interaction parameter (χ) of cross-linked poly(N,N-alkyl-substituted acrylamide) as a function of temperature. (From Ref. 32.)

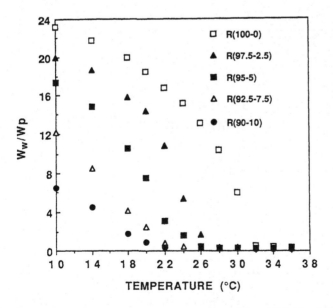

Figure 11 Aqueous equilibrium swelling of cross-linked poly(N-isopropyl acrylamide-*co*-butyl methacrylate) as a function of temperature. Ww, Wp are the weight of water and polymer of swollen polymer network respectively. (From Ref. 32.)

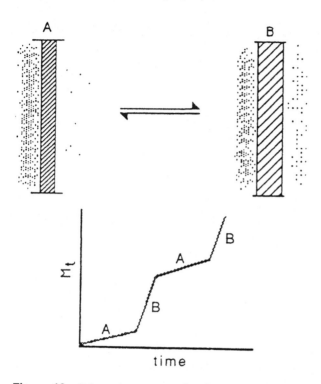

Figure 12 Schematic representation for reversibly changing drug permeation through membranes by external modulation. M_t indicates the diffused amount of drug at time t.

ture; poly(IPAAm) shows a sharp transition near 31°C, but polyacrylamide [poly(AAm)] does not have a transition temperature.

To strengthen mechanical properties and vary the transition temperature, hydrophobic monomers such as BMA were introduced into poly(IPAAm). The unique swelling properties with varying BMA composition are shown in Figure 11.

Thermosensitive polymers show large swelling changes with temperature. Figure 12 shows the mechanism for reversibly changing drug transport through membranes by external modulation. According to free volume theory, lower membrane permeability to drug exists in a deswollen state compared with higher permeability in the swollen state. Temperature affects the observed membrane permeabilities by influencing the swelling properties of the membrane, as opposed to altering the activation energies of solutes in a diffusion process.

Permeability control of solutes through thermosensitive hydrogel membranes may be thought of in terms of the degree of hydration and kinetics of

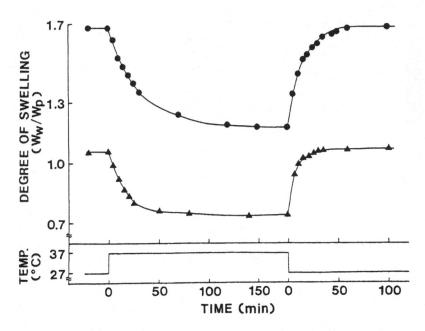

Figure 13 Typical swelling and deswelling rates of cross-linked poly(acryloyl pyrrolidine-*co*-styrene) between 27°C and 37°C. AS15 (●); AS20 (▲). The numbers indicate the content of styrene in the feed composition in moles during polymerization. Membrane thickness is 0.5 mm in the dried state. (From Ref. 34.)

membrane swelling and deswelling caused by temperature fluctuation. Two possible methods for the regulation of solute permeability control are proposed. One method is a gradually induced permeability regulation, and the other is the complete on–off regulation of solute permeability. For gradual control, acryloyl pyrrolidine (APy) copolymers with a more hydrophobic comonomer, styrene (St) were synthesized to give acceptable mechanical strength and maintain a thermosensitivity which was sufficient to induce permeability changes by swelling rather than thermal effects on solute permeability [33]. Typical examples of swelling and deswelling kinetics of these APy copolymers are shown in Figure 13. Equilibrium swelling was reached within 2 hr when the temperature was changed from 27°C to 37°C and reached again within 1 hr when the temperature was changed from 37°C to 27°C. The reswelling kinetics were faster than deswelling kinetics. These results were reflected in insulin permeation as shown in Figure 14. Gradual changes in insulin permeability through APy copolymers were observed without a noticeable lag time between 27°C and 37°C. With a decrease in temperature,

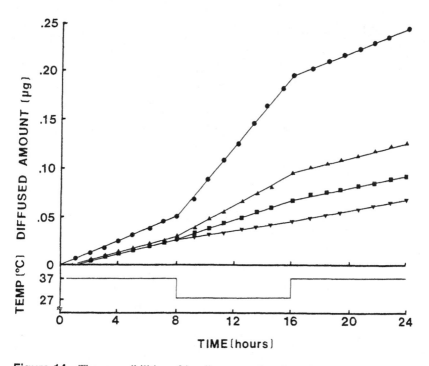

Figure 14 The reversibilities of insulin permeation through polymer membranes in a two-compartment diffusion cell: AH20 (●), AS15 (▲), AS20 (■), H (▼). Numbers indicate the content of styrene or HEMA in feed compositions in moles. H represents a cross-linked poly(2-hydroxyethyl methacrylate) (HEMA). (From Ref. 34.)

rapid hydration responses in the gel membrane eliminated lag times. With an increase in temperature, the rapidly deswollen membrane surface reduced permeability quickly, resulting in nearly the same permeability of solutes in the transition state as in the equilibrium state.

Swelling–deswelling kinetics of thermosensitive hydrogels exhibit reversible swelling changes with temperature changes. Swelling rates from a deswollen state to a swollen state are usually faster than the opposite process. The deswelling process shows an initial rapid shrinking followed by a slow deswelling. In some cases, immediate shrinking of the outer layer restricts further bulk water outflow from the interior of the gel. This response of the gel surface to temperature changes may be utilized as an on–off switch for drug release [34–36]. Figure 15 shows two schematic mechanisms of drug release control by reversible swelling–deswelling changes regulated by external modulation. In general, when a hy-

BULK MATRIX SQUEEZING

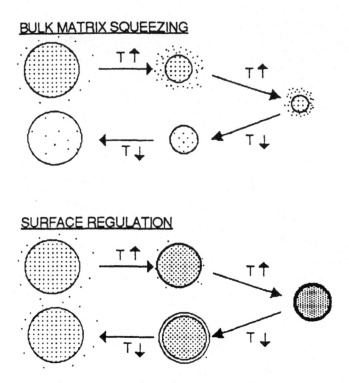

SURFACE REGULATION

Figure 15   Mechanisms of drug release from thermosensitive monolithic devices.

drogel imbibing drug solution deswells with a given stimulus, the drug solution is squeezed out of the gel matrix, a bulk squeezing mechanism. However, instant surface shrinking during a deswelling process prevents drug release.

Indomethacin was loaded into poly(IPAAM-*co*-BMA) disks containing 5 mol% BMA (dried disk size was 1.2 cm diameter, 0.12 cm thickness). A saturated solution of indomethacin 80/20 vol% ethanol/water was incorporated, and the extent of 21.4 ± 0.3 wt% indomethacin loading was obtained after drying. Figure 16 shows the results for indomethacin release for 20°C and 30°C. At 30°C a negligible release rate of indomethacin was observed, except for an initial burst effect. These results imply the effectiveness of these thermosensitive hydrogels as on–off switches for controlled drug release. The release rate for indomethacin in response to a stepwise temperature change between 20°C and 30°C (temperatures were equilibrated within 5 min) presented a pulsatile release pattern due to the reversible swelling properties of poly(IPAAm-*co*-BMA).

The permeability of poly(IPAAm-*co*-BMA) membrane for another three solutes as a function of temperature was obtained using a two-compartment diffu-

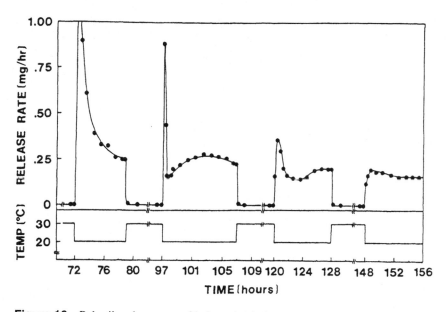

Figure 16 Pulsatile release rate of indomethacin in response to a stepwise temperature change between 20°C and 30°C in phosphate-buffered saline (pH 7.4). (From Ref. 36.)

sion cell; the result is shown in Figure 17. As expected, the permeability of uranine was the highest at all temperatures, followed by the permeability of dextran 4.4K and dextran 150K. The permeability of these solutes decreased with increasing temperature. The highest temperature at which significant permeation for uranine was possible was 27°C and decreased for solutes of increasing size (23°C for dextran 4.4K and less than 20°C for dextran 150K). No permeation of solutes was observed above 28°C. Considering swelling characteristics, the permeation of small molecules was possible at low swelling, while an increased swelling level was required for the permeation of solutes of increasing molecular size, in accordance with the free volume theory. Based on these swelling and solute size–dependent diffusion properties, membrane separation was demonstrated [37]. The temperature was maintained at 25°C during the first 100 hr and at 20°C for the remaining period. At 25°C, only permeation of uranine was significant; after 100 hr, most of the uranine (94.3%) had permeated, whereas 3.6% of dextran 4.4K and no measurable amount of dextran 150K had passed through the membrane. After 100 hr, with the temperature lowered to 20°C, permeation of dextran 4.4K became apparent, while dextran 150K still showed minimal permeation. The results are summarized in Table 4.

These thermosensitive hydrogels were utilized for the loading of heparin

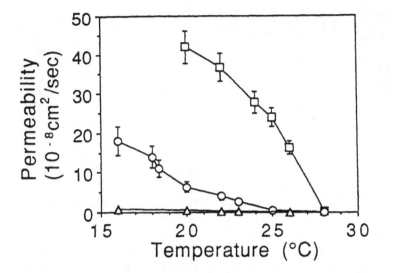

Figure 17 Permeability of uranine (□), dextran 4.4K (○), and dextran 150K (△) through cross-linked poly(N-isopropyl acrylamide-co-butyl methacrylate, 95:5 mol%) membrane. Error bars represent standard deviation in the slope of the curve of the receiver concentration of solute as a function of time at steady state. (From Ref. 37.)

Table 4 Composition of Mixtures of Solutes Permeated Through Cross-Linked P(NIPAAm-co-BMA) Membrane and Total Amounts of Each Solute Accumulated in Receiver Cell During Different Time Intervals[a]

Time interval (hr)	Composition of mixture accumulated in receiver cell (wt%)	Percentage of initial donor amount permeated
0–100	96.3% uranine 3.7% dextran 4,4K	Uranine: 94.3% Dextran 4.4K: 3.6% Dextran 150K: 0%
100–460	6.4% uranine 93.6% dextran 4,4K	Uranine: 5.7% Dextran 4.4K: 91.4% Dextran 150K: 0%

[a] Composition donor after 460 hr: 95% dextran 150K and 5% dextran 4,4K.

at lower temperatures and the release of heparin at body temperature via squeezing and slow diffusion. The method can be used in the design of nonthrombogenic polymer surfaces for blood-contacting devices [38,39].

V. ELECTRIC CURRENT–SENSITIVE POLYMERS

Electrically modulated solute permeation or release has been demonstrated by several methods.

Burgmayer and Murray [40] reported electrically controlled resistance to the transport of ions across polypyrrole membrane. The membrane was formed around a folded minigrid sheet by the anodic polymerization of pyrrole. The ionic resistance, measured by impedance, in 1.0 M aqueous KCl solution was much higher under the neutral (reduced) state of the polymers than under the positively charged (oxidized) state. The redox state of polypyrrole was electrochemically controlled; this phenomenon was termed an "ion gate," since the resistance was varied from low to high and vice versa by stepwise voltage application.

Dopamine, a neurotransmitter, was covalently coupled, via an amide bond, to a modified polystyrene having N-(2-(3,4-dihydroxyphenyl)ethyl) isonicotinamide units. The dopamine-coupled polymer was coated onto glassy carbon electrodes. In aqueous electrolyte solutions (pH 7), cathodic current caused cleavage of the amide linkage and release of dopamine at potentials more negative than 0.9 V [41]. The chemical scheme for the amide bond cleavage is presented in Figure 18.

Several researchers [11,42–44] have focused on the investigation of polyelectrolyte gels under an electric field which induces electric field–related phenomena such as electrodiffusion, pH change, electrophoresis, and electro-osmosis which result in swelling changes in the gel. Sawahata et al. [45] demonstrated the modulated release of pilocarpine hydrochloride, glucose, and insulin by using gels and microparticles of polyelectrolytes such as poly(AAc), poly(MAAc), and poly(N,N-dimethylaminopropylacrylamide). The modulated release of solute from these polyelectrolyte gels was explained by gel shrinkage under electric current application, resulting in the squeezing out of imbibed drug solution.

One effect of the electrochemical reactions in an aqueous system is a local pH change around the electrodes. By water electrolysis, hydronium ions (H_3O^+) are generated at the anode, while hydroxyl ions (OH^-) are produced at the cathode. These changes have been utilized for controlling the permeability of polyelectrolyte gel membrane or on–off solute release via ion exchange or surface erosion of interpolymer complex gels.

Weiss et al. [11] reported permeability changes for a neutral solute across a poly(MAAc) membrane. These changes were attributed to the increase in pH

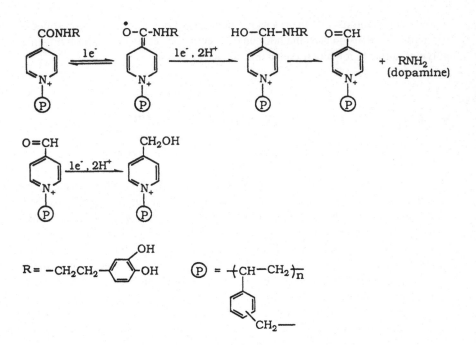

Figure 18 The mechanism of cleavage of amide bonds by cathodic current. (From Ref. 40.)

of the diffusion cell by electric current application, resulting in ionization of the membrane.

A cross-linked random copolymer of poly(2-acrylamido-2-methylpropane sulfonic acid (AMPS)-*co*-BMA) (27/73 monomer mole ratio in feed), a negatively charged network, was synthesized in a glass mold at 60°C for 3 days and then purified [46]. Edrophonium chloride, as a positively charged solute, was loaded in a sodium salt form into the gel by an ion-exchange method (18 wt% loading). To ensure that all of the positively charged solute molecules were ionically bound to the gel, physically entrapped solute was washed out with distilled and deionized water. Release experiments under an electric field were conducted in distilled deionized water. The release of positively charged solute with electrical stimulation in distilled deionized water showed a complete on–off pulsation as demonstrated in Figure 19. The release rate was regulated by the electric field. Since all physically entrapped solute was extracted by distilled deionized water and there was only ionically bound solute remaining inside the gel, we can assume that the exchange rate was generally controlled by the electric field and that all exchanged solute was released out quickly by electrostatic interactions with the

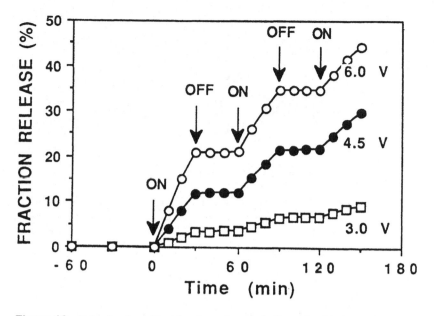

Figure 19 Edrophonium chloride release from poly(2-acrylamido-2-methylpropane sulfonic acid-*co*-butyl methacrylate) gel in distilled deionized water with various pulsatile electric currents. (From Ref. 45.)

anode, electro-osmosis, and passive diffusion. This ion-exchange mechanism is schematically illustrated in Figure 20.

Another mechanism for modulated drug release is local pH-induced surface erosion of interpolymer complex gels with entrapped solutes, as shown in Figure 21.

An insoluble polymer complex was formed between two water-soluble polymers, poly(2-ethyl-2-oxazoline) (PEOx) and poly(MAAc). PEOx and poly(MAAc) form hydrogen-bonded complexes between a carboxylic group (poly(MAAc)) and the PEOx repeating unit. There are two possible sites on the repeating unit of PEOx for hydrogen bonding to occur with carboxylic groups: the carbonyl oxygen and the nitrogen. The carbonyl oxygens are probably the dominant interactive group with a 1:1 ratio of repeating units [47]. The polymer complex was formed by adjusting the solution to pH 5; above pH 5.4 the complex dissociated. This can be attributed to the deionization and ionization of the carboxylic groups of poly(MAAc) with pH changes. Equilibrium swelling of the polymer complex in a 0.9% saline solution, pH 5.1, was 30% by weight of water uptake.

The erosion or dissolution of the polymer complex during the application

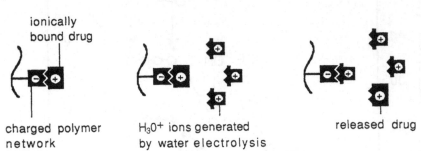

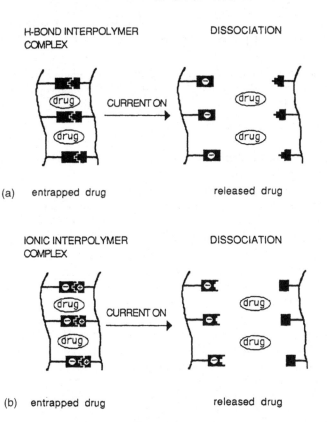

Figure 20 Ion exchange of positively charged drug bound to negatively charged polymer network with hydronium ion generated by electrolysis of water by electric current.

Figure 21 Schematic presentation of erodible polymer matrix by an electric current.

of electric current was investigated [48]. Under a constant electric current, erosion of the polymer matrix was linear until 80% of the initial disk weight was lost, after which the matrix lost its shape. This was consistent with zero-order kinetics associated with the surface erosion of slab-type polymers. The linearity of weight loss versus time implies that this process occurred through erosion of the cathode-facing surface. In addition, a stepwise weight loss was observed during the application and removal of an electric field, again until 80% total weight loss was reached.

The erosion process was a result of the hydroxyl ions produced at the cathode during electrolysis. The local pH near the cathode increased and hydrogen bonding between the two polymers was disrupted, resulting in disintegration of the polymer complex into two water-soluble polymers. To investigate modulated solute release from this erodible polymer, zinc insulin (10 mg) was suspended in 10 mL of the mixed polymer solution (pH 5.5). The drug-entrapped complex was formed by decreasing the pH below 5.0. The precipitate was filtered, swollen in acetone–water mixture, and molded into a disk shape. By applying a step function of electric current (0 and 5 mA) to the insulin-loaded matrix, insulin was released in an "on–off" manner until nearly 70% of the insulin was released. The large deviation in the release rate as shown in Figure 22 was attributed to

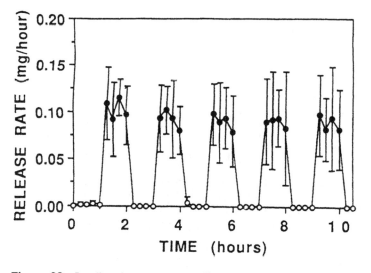

Figure 22 Insulin release rate (normalized to mg/hr per 160 mg device) from insulin-loaded matrix of polyethyloxazolin-poly(methacrylic acid) complex with the application of step-function electric current in 0.9% saline solution (mean ± from three measurements). (●) Current on (5mA); (○) current off. (From Ref. 47.)

irregular erosion of the device. The release of insulin from the matrix was minimal in the absence of electric current.

In another approach, low molecular weight heparin (MW 6000), a bioactive polyanion, was complexed with a synthetic polycation, polyallylamine (MW 50,000), for the investigation of on–off drug release from a polymer complex [49]. Each polyelectrolyte (50 mg) was dissolved separately in 10 mL of distilled water. The polymer complex formed immediately upon mixing of the two polymer solutions. The complex was filtered, washed with distilled water, and vacuum-dried overnight. The dried complex was ground into a fine powder and pressed into a disk matrix. Disks were swollen in phosphate-buffered saline (PBS) (pH 7.4), and an equilibrium swelling level of 35% water was reached; no heparin release from the device was detected during the swelling process.

To investigate heparin release in response to an electric current, the swollen heparin–polymer matrix was attached to a woven platinum cathode in a continuously stirred PBS solution (pH 7.4), and an electric current of 20 mA was applied. When the electric current was on, the polymer surface facing the cathode dissolved, thereby releasing heparin. The amount of heparin released was assayed by the Azure II method at pH 11 to prevent complexation of the two dissolved polymers. The release pattern of heparin showed a complete on–off profile in response to the applied electric field, as shown in Figure 23 [50].

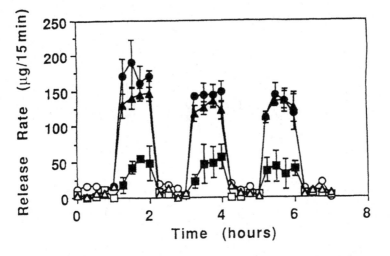

Figure 23 Heparin release rate (normalized to μg/15 min per 50 mg device) from heparin-polyallylamine complex matrix with the application of step function electric current in PBS (pH 7.4) solution (mean ± SD). (○, □, △) Off state; (●) 20 mA; (■) 15 mA; (▲) 10 mA. (From Ref. 48.)

REFERENCES

1. J Kost, ed. Pulsed and Self-Regulated Drug Delivery. Boca Raton, FL: CRC Press, 1990.
2. J Kopecek, J Vacik, D Lim. Permeability of membranes containing ionogenic groups. J Polym Sci A-1 9:2801–2815, 1971.
3. R Alhaique, M Marchetti, FM Riccieri, E Santucci. A polymeric film responding in diffusion properties to environmental pH stimuli: A model for a self-regulating drug delivery system. J Pharmacol 33:413–418, 1981.
4. Y Okahata, H Noguchi, T Seki. Functional capsule membranes. 26. Permeability control of polymer-grafted capsule membranes responding to ambient pH changes. Macromolecules 20:15–21, 1987.
5. E Pefferkorn, A Schmitt, R Varoqui. Helix-coil transition of poly(α,L-glutamic acid) at an interface: Correlation with static and dynamic membrane properties. Biopolymers 21:1451–1463, 1982.
6. D Chung, S Higuchi, M Maeda, S Inoue. pH-induced regulation of permselectivity of sugars by polymer membrane from polyvinyl-polypeptide graft copolymer. J Am Chem Soc 108:5823–5826, 1986.
7. PJ Flory. Principles of Polymer Chemistry. Ithaca, NY: Cornell University Press, 1953, p 584.
8. J Ricka, T Tanaka. Swelling of ionic gels: Quantitative performance of the Donnan theory. Macromolecules 17:2971–2921, 1984.
9. H Yasuda, CE Lamaze, LD Ikenberry. Permeability of solutes through hydrated polymer membranes. Part I. Diffusion of sodium chloride. Makromol Chem 118:19–35, 1968.
10. NA Peppas, CT Reinhart. Solute diffusion in swollen membrane. Part I. A new theory. J Membrane Sci 15:275–287, 1983.
11. AM Weiss, AJ Grodzinsky, ML Yarmush. Chemically and electrically controlled membranes: Size specific transport of fluorescent solutes through PMAA membranes. AIChE Symp Ser 82:85–98, 1986.
12. D Kirstein, H Braselmann, J Vacik, J Kopecek. Influence of medium and matrix composition on diffusivities in charged membranes. Biotech Bioeng 27:1382–1384, 1985.
13. S Sato, SW Kim. Macromolecular diffusion through polymer membranes. Int J Pharm 22:229–255, 1984.
14. L-C Dong, AS Hoffman. A novel approach for preparation of pH-sensitive hydrogels for enteric drug delivery. J Controlled Release 15:141–152, 1991.
15. PE Grimshaw, AJ Grodzinsky, ML Yarmush, DM Yarmush. Selective augmentation of macromolecular transport in gels by electrodiffusion and electrokinetics. Chem Eng Sci 45:2917–2929, 1990.
16. AJ Grodzinsky, PE Grimshaw. Electrically and chemically controlled hydrogels for drug delivery. In: J Kost, ed. Pulsed and Self-Regulated Drug Delivery. Boca Raton, FL: CRC Press, 1990, pp 47–64.
17. RA Siegel, M Falamarzian, BA Firestone, BC Moxley. pH controlled release from hydrophobic polyelectrolyte copolymer hydrogels. J Controlled Release 8:179–182, 1988.

18. JH Kou, GL Amidon, PI Lee. pH-dependent swelling and solute diffusion characteristics of poly(hydroxyethyl methacrylate-co-methacrylic acid) hydrogels. Pharm Res 5:592–597, 1988.

19. G Gwinup, AN Elias, ES Domurat. Insulin and C-peptide levels following oral administration of insulin in intestinal-enzyme protected capsules. Gen Pharmacol 22: 143–246, 1991.

20. K Ishihara, K Matsui. Glucose-responsive insulin release from polymer capsule. J Polym Sci 24:413–417, 1986.

21. K Ishihara, M Kobayashi, N Ishimaru, I Shinohara. Glucose induced permeation control of insulin through a complex membrane consisting of immobilized glucose oxidase and a poly(amine). Polym J 8:625–631, 1984.

22. G Albin, TA Horbett, BD Ratner. Glucose-sensitive membranes for controlled release of insulin. In: J Kost, ed. Pulsed and Self-Regulated Drug Delivery. Boca Raton, FL: CRC Press, 1990, pp 159–185.

23. RA Siegel. pH sensitive gels—Swelling equilibria, kinetics and applications for drug delivery. In: J Kost, ed. Pulsed and Self-Regulated Drug Delivery. Boca Raton, FL: CRC Press, 1990, pp 129–157.

24. YH Bae, SW Kim, LI Valuev. U.S. Patent 5226902, 1993.

25. A Gutowska, JS Bark, IC Kwon, YH Bae, Y Cha, SW Kim. Squeezing hydrogel for controlled oral drug delivery. J Controlled Release 48:141–148, 1997.

26. PK Gupta, S-H Leung, JR Robinson. Bioadhesive/mucoadhesives in drug delivery to the gastrointestinal tract. In: V Lenaerts, R Gurny, eds. Bioadhesive Drug Delivery Systems. Boca Raton, FL: CRC Press, 1990, pp 65–92.

27. H Brøndsted, J Kopecek. pH sensitive hydrogels: Characteristics and potential in drug delivery. ACS Symp Ser 480:285–304, 1992.

28. K Ishihara. Synthesis of stimuli responsive polymers and their biomedical applications. PhD Thesis, Waseda Univ, Japan, 1986.

29. S Wisniewski, SW Kim. Permeation of water-soluble solutes through poly(2-hydroxyethyl methacrylate) and poly(2-hydroxyethyl methacrylate) crosslinked with ethylene glycol dimethacrylate. J Membrane Sci 6:299–308, 1980.

30. SW Kim, JR Cardinal, S Wisniewski, GM Zenter. Solute permeation through hydrogel membranes, hydrophilic vs. hydrophobic solutes. ACS Symp Ser 127:347–359, 1980.

31. CA Cole, SM Schreiner, JH Priest, N Monji, AS Hoffman. Lower critical solution temperatures of aqueous copolymers of N-isopropylacrylamide and other N-substituted acrylamides. ACS Symp Ser 350:255–284, 1987.

32. YH Bae, T Okano, SW Kim. Temperature dependence of swelling of crosslinked poly(N,N-alkyl substituted acrylamides) in water. J Polym Sci B 28:923–936, 1990.

33. D Patterson. Free volume and polymer solubility. A qualitative review. Macromolecules 2:672–679, 1969.

34. YH Bae, T Okano, SW Kim. Insulin permeation through thermosensitive hydrogels. J Controlled Release 9:271–279, 1989.

35. YH Bae, T Okano, R Hsu, SW Kim. Thermo sensitive polymers as on–off switches for drug release. Makromol Chem Rapid Commun 8:581–483, 1987.

36. YH Bae, T Okano, SW Kim. "On–off" thermocontrol of solute transport. II. Solute release from thermosensitive hydrogels. Pharm Res 8:624–628, 1991.

37. H Feil, YH Bae, J Feijen, SW Kim. Molecular separation by thermosensitive hydrogel membranes. J Membrane Sci 64:283–294, 1991.
38. A Gutowska, YH Bae, SW Kim. Heparin release from thermosensitive hydrogels. J Controlled Release 22:95–99, 1992.
39. A Gutowska, YH Bae, H Jacobs, J Feijen, SW Kim. Heparin release from thermosensitive polymer coatings: In vivo studies. J Biomed Mater Research 29:811–815, 1995.
40. P Burgmayer, RW Murray. An ion gate membrane: Electrochemical control of ion permeability through a membrane with an embedded electrode. J Am Chem Soc 104:6139–6140, 1982.
41. ANK Lau, LL Miller. Electrochemical behavior of a dopamine polymer. Dopamine release as a primitive analogue of a synapse. J Am Chem Soc 105:5271–5277, 1983.
42. DE De Rossi, P Chiarelli, G Buzzigoli, C Domenici, L Lazzeri. Contractile behavior of electrically activated mechanochemical polymer actuators. Trans Am Soc Artif Intern Organ 32:157–162, 1986.
43. AJ Grodzinsky, AM Weiss. Electric field control of membrane transport and separations. Separ Purif Methods 14:1–40, 1985.
44. Y Osada. Conversion of chemical into mechanical energy by synthetic polymers (chemomechanical systems). Adv Polym Sci 81:1–46, 1987.
45. K Sawahata, M Hara, H Yasunaga, Y Osada. Electrically controlled drug delivery system using polyelectrolyte gels. J Controlled Release 14:253–262, 1990.
46. IC Kwon, YH Bae, T Okano, SW Kim. Drug release from electric current sensitive polymers. J Controlled Release 17:149–156, 1991.
47. AM Lichkus, PC Painter, MM Coleman. Hydrogen bonding in polymer blends. 5. Blends involving polymers containing methacrylic acid and oxazolime groups. Macromolecules 21:2636–2641, 1988.
48. IC Kwon, YH Bae, SW Kim. Electrically erodible polymer gel for controlled release of drugs. Nature 354:291–293, 1991.
49. YH Bae, IC Kwon, CM Pai, SW Kim. Controlled release of macromolecules from electrical and chemical stimuli-sensitive hydrogels. Makromol Chem, Macromol Symp 70:173–177, 1993.
50. IC Kwon, YH Bae, SW Kim. Heparin release from polymer complex. J Controlled Release 30:155–158, 1994.
51. H Brøndsted. PhD Dissertation, The University of Utah, Salt Lake City, UT, 1991.

15

Properties of Solids That Affect Transport

Maneesh J. Nerurkar
Merck & Company, West Point, Pennsylvania

Sarma Duddu
Inhale Therapeutics, Palo Alto, California

David J. W. Grant
University of Minnesota, Minneapolis, Minnesota

J. Howard Rytting
The University of Kansas, Lawrence, Kansas

I. INTRODUCTION

This chapter describes some of the properties of solids that affect transport across phases and membranes, with an emphasis on biological membranes. Four aspects are addressed. They include a comparison of crystalline and amorphous forms of the drug, transitions between phases, polymorphism, and hydration. With respect to transport, the major effect of each of these properties is on the apparent solubility, which then affects dissolution and consequently transport. There is often an opposite effect on the stability of the material. Generally, highly crystalline substances are more stable but have lower free energy, solubility, and dissolution characteristics than less crystalline substances. In some situations, this lower solubility and consequent dissolution rate will result in reduced bioavailability.

When going through a phase transition, a substance may also undergo a change in its thermodynamic properties, with consequent changes in its dissolution and transport characteristics. Polymorphism is common among a high percentage of solid drugs. Different polymorphs exhibit the same characteristics in solution or in the liquid state but have quite different characteristics in the solid state. These differences give rise to differences in apparent solubility, dissolution, transport, and sometimes stability.

The extent of hydration or solvation of a molecule also has a profound effect on the transport of the substance. The apparent solubility of the drug in both aqueous and nonaqueous media may be influenced by the absence or presence of moisture. Diffusion of drugs in polymeric systems may also be influenced by the hydration of the polymers and hydration of the membrane through which transport is occurring; for example, skin hydration may enhance the diffusion of drug molecules significantly.

II. CRYSTALLINITY AND AMORPHOUS MATERIALS

In the crystalline state of a substance, the molecules are arranged in a defined unit cell that is repeated in a three-dimensional lattice [1]. Since the crystal lattice can act as a diffraction grating for X-rays, the X-ray diffraction pattern of a crystal consists of a number of sharp lines or peaks, often with baseline separation. Figure 1 shows the X-ray powder diffraction pattern of the crystalline and amorphous forms of nedocromil sodium trihydrate.

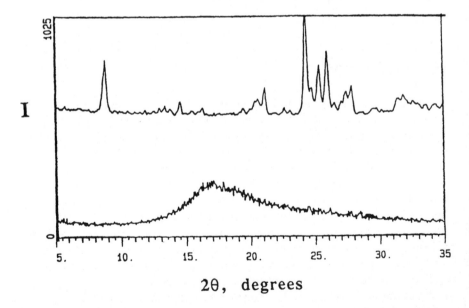

Figure 1   X-ray powder diffractograms of the crystalline and amorphous forms of nedocromil sodium trihydrate. *I* is the intensity of the diffracted beam at a diffraction angle θ. (RK Khankari and DJW Grant, unpublished observations.)

The existence of sharp peaks in the X-ray diffraction pattern is the most widely accepted test for a crystalline material. Actually, most crystals consist of tiny crystallites (grains or mosaic blocks) which are not perfectly aligned with each other. The slight misalignment of the mosaic blocks by a fraction of a degree or by a small number of degrees causes the diffracted monochromatic X-ray beams to spread out, somewhat broadening the resulting peaks. This effect, known as *asterism* in the Laue diffraction patterns, can be measured by using X-ray beams of high intensity, such as from a synchrotron source [2], and expressed as the mosaic spread. Were it not for the existence of the mosaic structure and the consequent mosaic spread, the diffracted X-ray beams would be so sharp that the well-known characteristic X-ray patterns of crystalline solids would not be discernible.

From the rate at which the crystals grow or from the stresses imposed during processing operations, imperfections or *defects* of various dimensionalities and types (Fig. 2) are introduced into the crystal lattice. These crystal lattice defects distort the crystal lattice, thereby reducing the *crystallinity* of the solid, and cause further broadening of the X-ray diffraction pattern, thereby increasing the mosaic spread [2]. Hence, a good measure of the crystallinity of the solid is the peak broadening or mosaic spread, which is an inverse function of crystallinity. Figure 3 shows the relationship between mosaic spread and crystal growth rate mentioned above.

Certain solid forms, known as *amorphous forms* or *glasses*, contain so many defects that all crystallinity is lost; the molecules are in nonuniform arrays

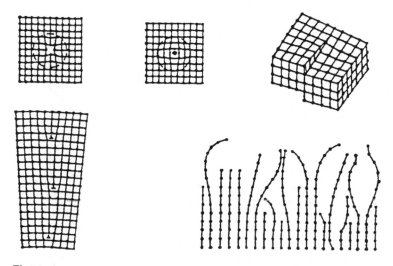

Figure 2   Crystal lattice defects of various dimensionalities. (Reprinted with permission from Ref. 5.)

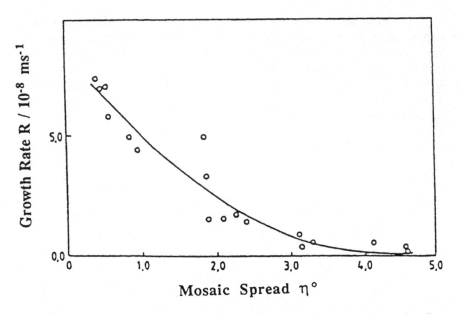

Figure 3   The variation of mosaic spread η with growth rate R for crystals of sodium chlorate, $NaClO_3$. (Reprinted with permission from Ref. 2.)

instead of in a regularly repeating crystal lattice. Figure 4 depicts two forms of silica (SiO_2): cristobalite, a crystalline form of quartz, and quartz glass, an amorphous form whose structure resembles a distorted cristobalite structure. When a substance, such as quartz or a protein, is capable of existing in both crystalline and amorphous forms, the X-ray diffraction pattern of the amorphous form [3,4] consists of a single broad, shallow peak termed an *amorphous halo*, as observed

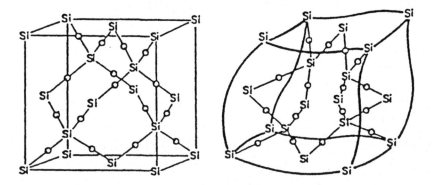

Figure 4   The structure of the two forms of silica: cristobalite (crystalline, left) and silica glass (amorphous, right). (Reprinted with permission from Ref. 4.)

in Figure 1 for the amorphous form of nedocromil sodium trihydrate. This phenomenon is also observed with liquids and indicates that the amorphous and liquid states lack the long-range order of the crystal but still retain the short-range order dictated by the predominant intermolecular interactions between the nearest neighbors.

There is actually no sharp distinction between the crystalline and amorphous states. Each sample of a pharmaceutical solid or other organic material exhibits an X-ray diffraction pattern of a certain sharpness or diffuseness corresponding to a certain mosaic spread, a certain content of crystal defects, and a certain degree of crystallinity. When comparing the X-ray diffuseness or mosaic spread of finely divided (powdered) solids, the particle size should exceed 1 μm or should be held constant. The reason is that the X-ray diffuseness increases with decreasing particle size below about 0.1 μm until the limit of molecular dimension is reached at 1–0.1 nm (10–1 Å), when the concept of the crystal with regular repetition of the unit cell ceases to be appropriate.

The nature and concentration of the crystal defects are influenced by stresses imposed on the solid by processing operations [5] such as (1) milling, grinding, or comminution, which tend to accentuate defects as well as reducing the particle size; (2) compression, which causes the defects to move during plastic flow and increase during strain and work hardening before brittle fracture; (3) crystallization and precipitation, the rate of which is controlled by the rate of desupersaturation; (4) uptake and/or loss (drying) of water or other fluid, which may also induce phase changes and particle size reduction; (5) thermal stresses, which may cause defects to move, to increase, or to decrease (anneal) or may induce phase changes or fracture; (6) electromagnetic reaction, the frequency of which determines the quantum energy and the molecular property influenced; (7) aging, which permits defects to move and decrease depending on the diffusivity in the solid; and (8) additives or impurities in solid solution [6,7], which may induce impurity defects, influence dislocations, concentrate at grain boundaries, and/or dissolve in inclusions.

Crystal defects may be directly observed using the following techniques [7]:

1. Microscopic observation of the surface of the crystals that have been etched by a suitable solvent or reactive solution [8]
2. Scanning electron microscopy (SEM)
3. Environmental scanning electron microscopy (ESEM)
4. Transmission electron microscopy (TEM)
5. X-ray topography [9]
6. Positron annihilation techniques (PAT) [7]

The influence of crystal defects on crystal properties provides indirect methods of studying defects. The following techniques may be employed:

1. Mosaic spread derived from the asterism of the diffracted X-ray beams in the Laue diffraction, mentioned above [2]
2. Atomic thermal parameters derived from single-crystal X-ray diffraction, which increase with increasing disorder and defects in the crystal [1]
3. Density or thermal expansivity [5,10]
4. Thermodynamic measurements such as heat of solution from solution calorimetry [11,12], heat of fusion from differential scanning calorimetry (DSC) [12], and entropy of processing, ΔS^P [12]
5. Fourier transform infrared spectroscopy (FTIR)
6. Solid-state nuclear magnetic resonance spectroscopy (SSNMR) [13]

The number of defects is maximal in the amorphous and liquid states. The phase diagram in Figure 5 shows the volume–temperature relationships of the liquid, the crystalline form, and the glass (vitreous state or amorphous form) [14]. The energy–temperature and enthalpy–temperature relationships are qualitatively similar.

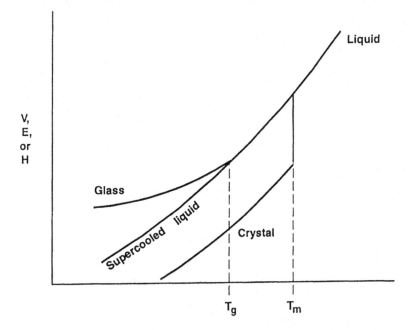

Figure 5 Changes in volume, V, energy, E, and enthalpy, H, during cooling or heating of the liquid, crystalline, and glassy (vitreous) forms of a substance. T_m is the melting point, and T_g is the glass transition temperature. (Adapted with permission from Ref. 14.)

The discontinuity between the liquid and the crystal at the melting point, T_m, is observed with many other physical properties and is a first-order transition for which the enthalpy change corresponds to the enthalpy of fusion, ΔH^f, as discussed in the next section. For certain substances under certain conditions, cooling the liquid sufficiently rapidly may allow it to bypass the melting process, becoming a supercooled liquid (Fig. 5). On further cooling, the change in slope at the glass transition temperature, T_g, is a second-order transition corresponding to a change in heat capacity as discussed in the next section. Between T_m and T_g, many polymers may exist in a rubbery state with appreciable thermal motion and mobility. Below T_g, the amorphous form exists in the glassy (vitreous) state with limited thermal motion and considerable kinetic stability.

The amorphous form (glass) can often be obtained by rapid cooling of the melt (liquid), by rapid evaporation of a liquid solution (e.g., by spray drying), or by freezing of a liquid solution followed by sublimation of the solvent by freeze drying (lyophilization). By these means, the disordered structure of the liquid is kinetically stabilized or *frozen in* before the substance has a chance to develop an ordered crystalline structure. Spray drying, by which the solvent is rapidly evaporated from the droplets of a solution, usually yields a material consisting of nearly spherical agglomerates of particles with improved flow, compatibility, and dissolution rate. For example, spray drying of lactose yields a material which is partly amorphous and partly crystalline, a combination which provides good flow and compression properties. Freeze drying (lyophilization), by which water is removed from a frozen solution via sublimation at a low temperature and pressure, produces a porous and fluffy material that rapidly dissolves when reconstituted with water. The advantages of lyophilization over spray drying are lower operating temperature, control of sterility, lower weight of the end product, and rapid reconstitution. Therefore, lyophilization is the method of choice for producing parenteral products, especially those of biological origin, despite the fact that the process itself is inherently expensive.

Depending on the processing variables, including the additives, the end product of a lyophilization process may be amorphous or crystalline or of intermediate crystallinity. As explained below the amorphous product has a higher free energy and therefore dissolves faster and tends to be less stable and more hygroscopic than the crystalline product. The choice between the crystalline and amorphous material may depend on whether an improved solubility or improved stability is required.

The ability and relative ease and difficulty with which substances form glasses vary widely [14]. Materials which readily form glasses, such as polymers, sugars, cephalosporins, and metal oxides, tend to display high viscosities at their melting points, which inhibits the ordering processes, nucleation and crystal growth. Substances with large enthalpies and entropies of fusion tend to avoid glass formation and tend to prefer the thermodynamically more favorable process, crystallization. If cooled extremely rapidly, many substances, including metals and alloys such as steel, can be cast in the vitreous state.

The distorted structure of the glass causes its energy and enthalpy to exceed those of the crystal, while the associated disorder of the glass causes its entropy to exceed that of the crystal. Since the enthalpy difference, $\Delta H = H_{(glass)} - H_{(crystal)}$, exceeds $T \Delta S$, where $\Delta S = S_{(glass)} - S_{(crystal)}$, the Gibbs free energy difference, $\Delta G = G_{(glass)} - G_{(crystal)} = \Delta H - T \Delta S$, is positive. The greater free energy of the amorphous form compared to that of the crystalline form indicates (1) that the amorphous form is unstable with respect to the crystalline form and tends to revert to the crystalline form spontaneously and (2) that the amorphous form has a higher thermodynamic activity and greater dissolution rate per unit surface area in water and in other solvents than the crystalline form. In general, similar considerations apply to partially crystalline materials, i.e., materials of reduced crystallinity. The lower the crystallinity, corresponding to a greater number and greater effectiveness of defects, the higher the enthalpy, entropy, free energy, thermodynamic activity, and dissolution rate.

The higher thermodynamic activity, a, of the amorphous form compared to that of the crystalline form explains the higher initial dissolution rate per unit surface area (intrinsic dissolution rate, J) and the higher solubility, s, of the amorphous form compared to that of the crystalline form, according to a simple form of the Noyes–Whitney equation [15],

$$J = ka \tag{1}$$

where $a \approx s$ and k is the mass transfer coefficient, which is assumed to be constant under defined conditions for a given molecule. This phenomenon is exemplified by novobiocin acid [16] (Fig. 6), for which s is represented by the plateau values while J is represented by the initial rate of increase in absorbance at 305 nm in 0.1 N hydrochloric acid at 25°C for a particle size less than 10 μm. (The salt form of novobiocin is chemically unstable in solution.) The contrast in dissolution rate and solubility is paralleled by the absorption of novobiocin from the respective solid form in the gastrointestinal tract, as measured by the plasma levels in dogs, as shown in Table 1.

When determining the solubility and dissolution rate of amorphous or partially crystalline solids, the metastability of these phases with respect to the highly crystalline solid must be considered. While the low diffusivity of the molecules in the solid state can kinetically stabilize these metastable forms, contact with the solution, for example during measurements of solubility and dissolution rate, or with the vapor, if the solid has an appreciable vapor pressure, may provide a mechanism for mass transfer and crystallization. Less crystalline material dissolves or sublimes whereas more crystalline material crystallizes out. The equilibrium solubility measured will therefore approach that of the highly crystalline solid. The initial dissolution rate of the metastable form tends to reflect its higher

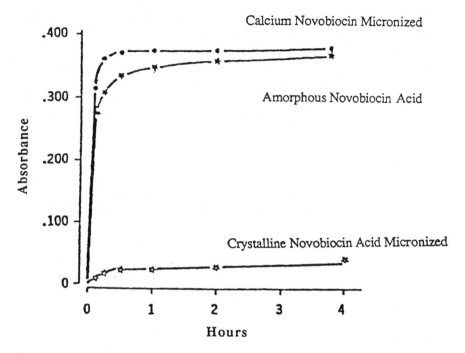

Figure 6 Increase in absorbance of various forms of novobiocin of particle size less than 10 μm dissolving in 0.1 M hydrochloric acid at 305 nm. (Reprinted with permission from Refs. 3 and 16.)

Table 1 Novobiocin Plasma Levels in Dogs Following Oral Administration of Different Solid Forms

Hours after dose[a]	Novobiocin sodium (μg/mL plasma)	Amorphous novobiocin acid (μg/mL plasma)	Crystalline novobiocin acid[b]
0.5	0.5	5	N.D.
1	0.5	40.6	N.D.
2	14.6	29.3	N.D.
3	22.2	22.3	N.D.
4	16.9	23.7	N.D.
5	10.4	20.2	N.D.
6	6.4	17.5	N.D.

[a] Dose = 12.5 mg/kg.
[b] N.D. = not detectable.
Source: Refs. 3 and 16.

thermodynamic activity and free energy. The presence of a thin film of saturated solution on the dissolving surface of the solid may facilitate the recrystallization of the stable, highly crystalline material. After a certain time, the dissolution rate will then decrease to a value characteristic of the more stable material. The slow conversion of the amorphous form in suspension or in solution to the crystalline form may be inhibited or slowed down by crystallization inhibitors, which are often polymeric, such as methylcellulose, polyvinylpyrrolidone, and alginic acid derivatives [4].

The release rate and therefore the duration of action of injectable insulin, in the form of insulin zinc, is controlled by its crystallinity coupled with its particle size. The crystallinity and particle size of insulin zinc, which is precipitated as an insoluble complex when insulin is reacted with zinc chloride, is controlled by the pH. The amorphous complex of small particle size, Prompt Insulin Zinc Suspension USP, is rapidly absorbed and has a relatively short duration of action. In contrast, the crystalline complex of large particle size, Extended Insulin Zinc Suspension USP, is slowly absorbed and has a relatively long duration of action. The intermediate form, Insulin Zinc Suspension USP, consists of seven parts of the crystalline form to three parts of the amorphous form and has an intermediate rate of absorption and duration of action.

The instability of the drug in the amorphous state becomes a critical factor for drugs, such as certain peptides, which degrade with appreciable rates in the solid state. An increase in temperature through the glass transition of the amorphous solid tends to decrease the stability of the product [17]. Furthermore, the residual water, which is retained to a greater extent in amorphous solids than in crystalline solids, reduces the glass transition temperature of the amorphous solid by acting as a plasticizer [18]. Thus, the stability of a drug in the amorphous form is a function of the storage temperature relative to the glass transition temperature and the moisture content. The stability of a material, which is assumed to be a function of its molecular mobility near its glass transition temperature, is believed to be best described by the Williams–Landel–Ferry (WLF) equation,

$$R = R_g \exp\left[\frac{C_1(T - T_g)}{C_2 + (T - T_g)}\right] \tag{2}$$

where R and R_g are the rates of degradation at the temperature of interest, T, and at the glass transition temperature, T_g, respectively, while C_1 and C_2 are universal constants. This equation shows that the rate of degradation of the material is more sensitive to temperatures near T_g than would be predicted by the classical Arrhenius equation. Roy *et al.* [17] studied the effect of moisture and temperature on the degradation of a monoclonal antibody–vinca conjugate and showed that some of the degradation processes followed WLF kinetics (Fig. 7). The role of moisture in the stability of amorphous proteins is reviewed by Hageman [19].

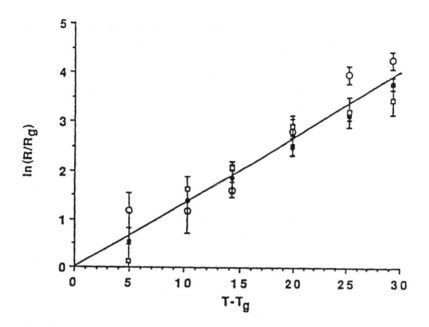

Figure 7 Analysis of the degradation of KS1/4 hydrazide conjugate according to Williams–Landel–Ferry (WLF) glass transition theory. R = rate of degradation in percent per month at a given temperature and water content; R_g = rate of degradation at T_g. R_g = 0.10 (dimer formation); 0.92 (free vinca generation); 3.9 (vinca decomposition). (O) Dimer formation; (●) free vinca generation; (□) vinca decomposition; (——) best fit of all the data to WLF theory with the values of R_g given above. (Reprinted with permission from Ref. 17.)

Many additives, such as sugars and amino acids, are used to stabilize proteins during lyophilization. Use of a certain additive may produce a crystalline lyophilized product, while use of another additive may produce an amorphous product. Although the actual mechanism is not clear, it is believed that additive either retains water molecules which are indispensable to the protein and resist overdrying or directly interacts with the protein and replaces water by hydrogen bonding with the polar groups of proteins. Izutsu et al. [20] showed that the stability of lyophilized β-galactosidase is related to the *degree of crystallinity of the additive* in the final product. The additives that were present in a crystalline form in the final product did not stabilize β-galactosidase, whereas those that existed in an amorphous state did. The authors suggest that the crystallinity of the additive affects not only the amount of water around the protein molecule but also the ordering of the water molecules.

Virtually all pharmaceutical solids contain impurities. Some impurities, such as water vapor, may be taken up by the solid from the atmosphere, while others are taken up during crystal growth in solution. The impurities or additives taken up by the crystals during their growth may be structurally related impurities or may be the solvent of crystallization itself. Crystals which contain the solvent of crystallization in their lattice, in stoichiometric or nonstoichiometric quantities, are termed *solvates*. Specifically, if the solvent is water, they are referred to as *hydrates*. The changes in the transport properties of pharmaceutical solids as a result of hydration or solvation are highlighted in Section IV. Therefore, the effect of synthetic or structurally related impurities on the physicochemical properties of crystalline solids, which affect transport processes such as the dissolution rate, are described in the next paragraph.

When present in small quantities, e.g., mole fractions less than 0.01, the impurity (guest) may be present as a solid solution, i.e., dissolved within the

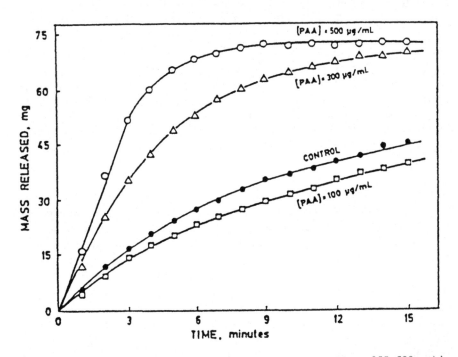

Figure 8 Dissolution–time profile of acetaminophen crystals (75 mg, 355–500 μm) in water at 4°C. The crystals were grown at 30°C and at 240 rpm from aqueous solutions containing initially 23.1 g/dm^3 acetaminophen and the following concentrations of *p*-acetoxyacetanilide: a (●) Zero; (□) 100 mg/dm^3; (△) 300 mg/dm^3; (○) 500 mg/dm^3. (Reprinted with permission from Ref. 6.)

lattice of the crystalline drug (host), and may change the properties of the crystals, both thermodynamic [21] and mechanical [22]. Although crystals preferentially incorporate molecules of the same kind, there is a large driving force for the incorporation of traces of the impurity into the growing crystals, which results from a large negative value for the partial molar free energy for the incorporation of the impurity into the crystals of the host. The impurity, because of its structural dissimilarity, introduces strain into the crystal lattice of the host, perhaps by changing the nature and concentration of crystal defects. The crystal defects, which are associated with a local increase in energy and disorder, act as high energy sites at which preferential dissolution of the crystal begins. Thus, the density of crystal defects in drug crystals may alter the dissolution rate [8,23], which is a crucial mass transport quantity for determining the bioavailability. On the other hand, impurities may also alter the dissolution rate of the host by changing the habit of the crystalline host, e.g., from prisms to faster dissolving needles. For example, a systematic habit change induced by the synthetic impurity p-acetoxyacetanilide increases the dissolution rate of acetaminophen [6] as shown in Figure 8. Duddu *et al.* [24] showed that the opposite enantiomer, present as an impurity in the crystallization medium, is taken up by crystals of (RS)-$(-)$-ephedrinium 2-naphthalenesulfonate and increases their enthalpy, entropy, and intrinsic dissolution rate, perhaps by changing the nature and concentration of crystal defects. Such changes in dissolution rate may account for the differences in the transport of the drug across the gastrointestinal tract. In principle, increases in dissolution rate brought about by excipients, as additives, may be employed to increase bioavailability.

III. PHASE TRANSITIONS

On passing through a phase transition, a substance may undergo a change, ΔY, in certain of its thermodynamic properties, e.g., from an initial value Y_1 for the initial phase, phase 1, to the final value Y_2 for the final phase, phase 2, where $\Delta Y = Y_2 - Y_1$. The most pronounced transitions in pharmaceutical solids are first-order transitions, for which the free energy, G, as a function of a given state variable (P, T, or V) is continuous, i.e., $\Delta G = 0$, whereas the first derivative of the free energy with respect to the given state variable is discontinuous [25]. Since

$$\left(\frac{\partial G}{\partial T}\right)_P = -S, \quad \left(\frac{\partial G}{\partial P}\right)_T = V, \quad \text{and} \quad \left(\frac{\partial (G/T)}{\partial (1/T)}\right)_P = H \tag{3}$$

first-order transitions exhibit discontinuities in the entropy, volume, and enthalpy of a substance at the transition temperature, so ΔS, ΔV, and ΔH are finite quantities. ΔH is positive for an endothermic transition but negative for an exothermic transition. Examples of first-order transitions include the crystallization of an amorphous form or a liquid (exotherm equivalent to the enthalpy or heat of crystallization in Fig. 9), desolvation of a solvate or dehydration of a hydrate (usually endothermic), a polymorphic transition (endothermic for an enantiotropic system below the melting point but exothermic for a monotropic system below the melting point), fusion of a crystalline solid (endotherm equivalent to the enthalpy of fusion or heat of melting in Fig. 9), vaporization of a liquid (endothermic), and sublimation of a solid (endothermic). Thermodynamic rules for enantiotropic and monotropic polymorphism have been formulated by Burger and Ramberger [26,27]. These rules comprise the various enthalpic transitions in differential scanning calorimetry (DSC) and the densities and infrared frequencies of the polymorphs of a substance. The enthalpic rules have proven to be the most applicable and reliable. Polymorphic transitions are discussed in detail later in the chapter.

Solid-state phase transitions usually take place at a slower rate than solution-mediated phase transitions because the motion of the molecules is restricted in the solid state. Processing stresses such as grinding [28], heating [28], humidity

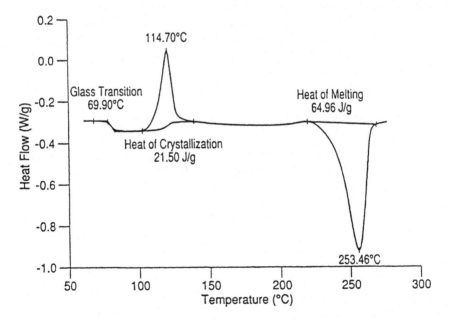

Figure 9 Differential scanning calorimetric (DSC) curve showing the phase transitions observed on heating the amorphous form (glass) of polyethylene terephthalate. (Reproduced with permission from Ref. 38.)

[29], and compaction [30] are reported to accelerate the phase transitions in pharmaceutical solids, perhaps by facilitating molecular motion in the solid state.

First-order phase transitions can be detected by various thermoanalytical techniques, such as DSC, thermogravimetric analysis (TGA), and thermomechanical analysis (TMA) [31]. Phase transitions leading to visual changes can be detected by optical methods such as microscopy [3]. Solid–solid transitions involving a change in the crystal structure can be detected by X-ray diffraction [32] or infrared spectroscopy [33]. A combination of these techniques is usually employed to study the phase transitions in organic solids such as drugs.

A second-order transition is continuous in both G and its first derivatives, S, V, and H (i.e., $\Delta G = 0$, $\Delta S = 0$, $\Delta V = 0$, and $\Delta H = 0$) but discontinuous in the second derivatives, such as heat capacity at constant pressure, C_p, compressibility, β, and thermal expansivity, α (i.e., ΔC_p, $\Delta \beta$, and $\Delta \alpha$ are finite). The change in C_p at a second-order transition is observed as a shift in a baseline on heating or cooling through the transition temperature in DSC. For an amorphous form (glass), a glass transition (Fig. 9) is a typical second-order transition. Figure 9 shows three of the typical transitions mentioned above, the substance being a polymer, polyethylene terephthalate. During the glass transition from the glassy state to the rubbery state, the molecules in the latter state have a greater mobility, causing kinetic instability, so the material crystallizes. Thus, the first-order transition, crystallization, begins before the second-order transition, glass transition, has reached completion. Examples of the superposition of first-order and second-order transitions are quite common. Besides glass transition, other second-order transitions may occur, such as relaxation in either the glassy state or the rubbery state, but these are usually less important.

Above the glass transition temperature, T_g, corresponding to the rubbery state, the higher molecular mobility permits a higher diffusivity, faster mass transport of drugs, greater reactivity, and lower stability than below T_g, corresponding to the glassy state [19,34]. For polymers, T_g is influenced by the molecular weight, copolymerization, degree of cross-linking, and the amount and nature of other substances [35,36] such as additives, plasticizers, diluents, impurities, and water. Water and other small molecules plasticize the amorphous phase, thereby lowering T_g [37]. The phenomena discussed in this paragraph are of particular importance to polymeric dosage forms, proteins, and freeze-dried products, since the drug transport in, and the stability of, the amorphous phase depend on T_g in relation to the ambient temperature.

Among crystalline solids, typical second-order transitions are associated with abrupt intermolecular conformational, rotational, and vibrational changes and/or with abrupt changes in crystalline disorder and/or defects [7]. These changes in crystalline solids are sometimes difficult to assign without the use of appropriate spectroscopic techniques such as solid-state NMR or a diffraction procedure such as single-crystal X-ray diffraction.

Several new thermal analytical techniques are potentially valuable for the study of second-order transitions in the characterization of amorphous solids and for the accurate determination of glass transition temperatures. These modern techniques can detect and characterize glass transitions and other second-order transitions that are not detectable by conventional thermal analytical techniques such as DSC, TGA, or TMA.

A. Oscillating Differential Scanning Calorimetry

While conventional DSC employs a linear temperature–time ramp, oscillating differential scanning calorimetry (ODSC) overlays the linear ramp with a sinusoidal modulation of heating and cooling [38]. Thus, the total heat flow in ODSC consists of both a thermodynamic, irreversible component which is independent of heating rate and a kinetic, reversible component which depends on the heating rate. Any overlapping reversible and irreversible transitions can be separated into their respective components, as shown in Figure 10. An example of the characterization of amorphous solids is glass transition (a reversible transition) associated with an enthalpic relaxation (an irreversible transition). The combination of these two events, when analyzed by conventional DSC, would appear as an endotherm, suggesting, incorrectly of course, a single first-order transition at that temperature.

B. Dynamic Mechanical Analysis

Dynamic mechanical analysis (DMA) measures the elastic modulus and damping properties of materials when they are deformed under a sinusoidal stress [39]. Thus, the method is useful in evaluating materials, such as amorphous solids, which are viscoelastic and exhibit mechanical properties that depend on time, frequency, or temperature. A glass transition is usually marked by significant changes in the mechanical properties of the material and can be determined accurately using DMA. Trace quantities of additives such as moisture dramatically alter the T_g of the amorphous solid and may result in a change in its physical and chemical stability. The change in elastic modulus during a glass transition is usually several orders of magnitude greater than the change in heat capacity, which is the basis for measuring T_g by DSC, or the change in volume, which is the basis for measuring T_g by TMA.

C. Dielectric Analysis

Dielectric analysis (DEA) measures the electrical properties of a material as it is subjected to a periodic electric field under various conditions. This technique provides quantitative information on the capacitance and conductance of the ma-

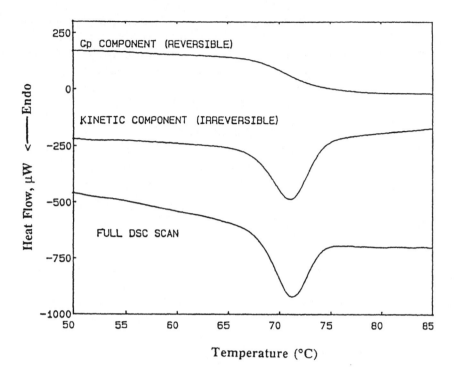

Figure 10 Oscillating differential scanning calorimetric (ODSC) curves showing the separation of the glass transition (reversible, i.e., thermodynamic component) and enthalpic relaxation (irreversible, i.e., kinetic component) which overlap in the full DSC scan. (Reprinted with permission from Ref. 38.)

terial as functions of time, temperature, or frequency [40]. Material in the glassy state is rigid and gains increased fluidity and hence increased conductance as it passes through a second-order transition. The frequency dependence of the transition temperature affords the activation energy associated with the transition.

IV. POLYMORPHISM

Regardless of the flux mechanism, it is clear upon examination of permeability expressions that flux is proportional to the concentration differential across the total barrier to mass transport. This flux is maximal in a given system for a permeant when the penetrating agent is present in the applied phase in a saturated state.

There are many situations of pharmaceutical interest where this solubility limitation plays a critical role in the transport process. For instance, the rate of dissolution of a solid in a solvent is dependent on its solubility, the barrier for dissolution being the solvent diffusion layer at the solid's surface. The absorption of slightly soluble drugs from suspensions or from solid dosage forms also involves saturation of the barrier interface in contact with the permeant. Even in uptake experiments where saturation is not attained, the solubility of the diffusant is important as a measure of the maximum driving force for transport. Furthermore, since studies concerned with determinations of diffusivity, release, etc. often involve permeability from saturated phases, it is necessary to know the solubility to calculate diffusivity or to quantitate the process [41]. The dissolution of a solid in a fluid medium is described by the Noyes–Whitney equation [15], which states

$$\frac{dM}{dt} = R_{aq}A \frac{C_s - C}{V} \tag{4}$$

where C is the bulk concentration, C_s is the solubility, M is the weight and A the surface area of solute, V is the volume of the solution, and R_{aq} is inversely proportional to the resistance of the diffusion layer. During the initial stages of dissolution, $C_s \gg C$, and the equation can be simplified to

$$\frac{dM}{dt} = R_{aq}A \frac{C_s}{V} \tag{5}$$

Thus, with all other factors held constant, during the initial stage of dissolution,

$$\frac{dM}{dt} \propto C_s \tag{6}$$

Thus, any factor that affects the solubility or apparent solubility of a drug entity has the potential of affecting the dissolution and diffusion mass transport of the molecule as shown in Eq. (6) [or Eq. (1)]. Polymorphism is one such important factor.

A polymorph is a solid crystalline phase of a compound resulting from the possibility of at least two different crystal lattice arrangements of that compound in the solid state [42]. Polymorphs of a compound are, however, identical in the liquid and vapor states. They usually melt at different temperatures but give melts of identical composition. Two polymorphs of a compound may be as different in structure and properties as crystals of two different compounds [43,44]. Apparent solubility, melting point, density, hardness, crystal shape, optical and electrical properties, vapor pressure, etc. may all vary with the polymorphic form. The polymorphs that are produced depend upon factors such as storage temperature, recrystallization solvent, and rate of cooling. Table 2 suggests the importance of polymorphism in the field of pharmaceutics [45].

Table 2 Incidence of Polymorphs in Some Common
Pharmaceuticals

Compound	Number studied	Percent having polymorphs
Barbiturates	38	63
Steroids (MP < 120°C)	48	67
Sulfonamides	40	40

Source: Ref. 45.

   The solubility difference between polymorphs reflects their free energy difference. Deliberate isolation of metastable phases is sometimes desirable because
of their advantageous processing or application properties. Many times, however,
a metastable formulation is unacceptable because of subsequent phase transformation and crystal growth which may occur on storage. If dissolution is carried
out using the metastable polymorph of the drug and the concentration of the drug
exceeds the equilibrium solubility of the less soluble form of the drug, thermodynamic instability may occur. Chance nucleation of the stable form may cause
crystallization until equilibrium is reached with respect to this form—a frequent
problem with sparingly water soluble drugs. Some examples of polymorphic transitions in aqueous suspensions have been reported. These include suspensions of
sulfathiazole, cephalexin, methylprednisolone, and tolbutamide. Knowledge of
ideal temperature and composition requirements is essential in formulation, but
it is difficult to predict the effect of variation in physicochemical parameters on
the rate at which equilibrium is approached.
   Let us now take a closer look at the theory relating polymorphism and
solubility [46]. Consider a drug that has two polymorphic forms, polymorph 1
and polymorph 2. If the subscripts 1 and 2 refer to the respective polymorphs,
we can state their free energies as

$$G_1 = H_1 - TS_1 \tag{7}$$

$$G_2 = H_2 - TS_2 \tag{8}$$

For the standard state,

$$G_0 = H_0 - TS_0 \tag{9}$$

Therefore,

$$G_1 - G_0 = (H_1 - H_0) - T(S_1 - S_0) \tag{10}$$

and

$$G_2 - G_0 = (H_2 - H_0) - T(S_2 - S_0) \tag{11}$$

So,

$$\Delta G_1 = \Delta H_1 - T \, \Delta S_1 = RT \ln(a_1 - 1) \tag{12}$$

and

$$\Delta G_2 = \Delta H_2 - T \, \Delta S_2 = RT \ln(a_2 - 1) \tag{13}$$

Subtracting,

$$G_2 - G_0 - (G_1 - G_0) = RT(\ln a_2 - \ln a_1) \tag{14}$$

or

$$G_2 - G_1 = RT(\ln a_2 - \ln a_1) \tag{15}$$

or

$$\Delta G = \Delta H - T \, \Delta S = RT \ln (a_2/a_1) \tag{16}$$

If $a_2/a_1 = g_2 m_2/g_1 m_1 \approx m_2/m_1$ (a = activity, g = molal activity coefficient, m = molality) if Henry's law is obeyed and $a = gm$, m_1, and m_2 are the molal solubilities of the polymorphs, and if the standard molar enthalpies and standard molar entropies of solution are, respectively,

$$\Delta H_1 = H_0 - H_1 \qquad \text{and} \qquad \Delta H_2 = H_0 - H_2 \tag{17}$$

$$\Delta S_1 = S_0 - S_1 \qquad \text{and} \qquad \Delta S_2 = S_0 - S_2 \tag{18}$$

then, using the Van't Hoff equation,

$$\left(\frac{\partial \ln a_1}{\partial (1/T)} \right)_P = \left(\frac{\partial \ln (g_1 m_1)}{\partial (1/T)} \right)_P = -\frac{H_1 - H_0}{R} = \frac{\Delta H_1}{R} \tag{19}$$

$$\left(\frac{\partial \ln a_2}{\partial (1/T)} \right)_P = \left(\frac{\partial \ln (g_2 m_2)}{\partial (1/T)} \right)_P = -\frac{H_2 - H_0}{R} = \frac{\Delta H_2}{R} \tag{20}$$

$$\left(\frac{\partial \ln a_2/a_1}{\partial (1/T)} \right)_P = -\frac{H_2 - H_1}{R} = \frac{\Delta H_2 - \Delta H_1}{R} \tag{21}$$

If Henry's law holds owing to the low solubility of the polymorphs in the solvent then $g_1 = g_2 = 1$. Then,

$$\left(\frac{\partial \ln(m_2/m_1)}{\partial(1/T)}\right)_p = \frac{\Delta H_2 - \Delta H_1}{R} = \frac{H_1 - H_2}{R} = \frac{-\Delta H}{R} \tag{22}$$

which on integration and assuming constant ΔH gives

$$\ln(m_2/m_1) = -\Delta H/RT + I \tag{23}$$

where I is a constant of integration.

Figures 11–13 illustrate the relationship of solubilities of polymorphs to temperature [47].

At a certain temperature of transition T_T, the two forms will be in equilibrium, ΔG will be zero, and $m_1 = m_2$. But at other temperatures the two forms will not be in equilibrium, and if $a_1 > a_2$ then because $\Delta G = RT \ln (a_2/a_1)$, ΔG will be negative and polymorph 1 will change spontaneously to polymorph 2 and will therefore be considered the less stable form and vice versa. Studies of the two polymorphic forms of methylprednisolone show that significant differences occur in the apparent solubilities of polymorphic forms and that these may be temperature-related.

Measurements of the dissolution behavior of polymorphic forms of relatively insoluble drugs are a convenient way of measuring thermodynamic parameters which, in turn, provide a rational approach to selection of the more energetic polymorphic forms of these drugs for absorption. Large differences in free energy

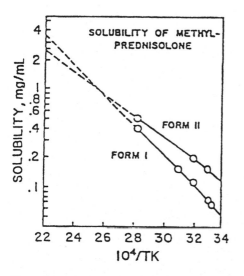

Figure 11 Van't Hoff plot of log solubility of methylprednisolone in water as a function of reciprocal water temperature. (Reprinted with permission from Ref. 47.)

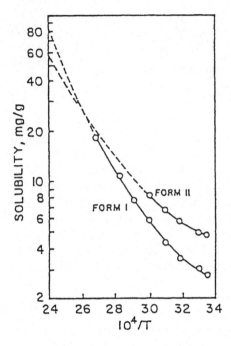

Figure 12 Van't Hoff plot of log solubility of methylprednisolone in decanol as a function of reciprocal water temperature. (Reprinted with permission from Ref. 47.)

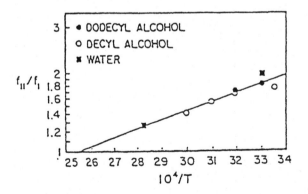

Figure 13 Plot of log of fugacity ratio or solubility ratio for methylprednisolone polymorphs as a function of solvent and reciprocal absolute temperature. (Reprinted with permission from Ref. 47.)

content can significantly affect absorption and resulting blood levels. Aguiar and Zelmer [48] studied the solubility and dissolution behavior of polymorphs of chloramphenicol palmitate and mefenemic acid. They measured the concentration achieved in solution for the polymorphic forms A and B of chloramphenicol palmitate in the presence of an excess of solid phase and under essentially constant agitation. Polymorph B dissolved faster than A and yielded solutions that were about four times more concentrated at equilibrium (Fig. 14). Although this may be partly due to geometrical factors, it was evident that the higher free energy content of this form played a significant role.

Tables 3 and 4 list thermodynamic values calculated for polymorphs of chloramphenicol palmitate and mefenamic acid, respectively. Absorption studies of chloramphenicol palmitate in humans show that suspensions containing polymorph B of chloramphenicol palmitate gave blood levels approximately 10 times higher than those produced by suspensions of polymorph A [49]. This may be due to the significant (-774 cal/mol) free energy difference between the polymorphs resulting in a substantial difference in their solubility and dissolution behavior. This theory is supported by the almost identical blood levels due to polymorphs I and II of mefenamic acid, which have a small free energy difference (-231 cal/mol) and similar solubility and dissolution behavior (Table 4).

As in the case of chloramphenicol palmitate, it may be desirable to have the more soluble (less stable) polymorph of a drug. One method, based on thermo-

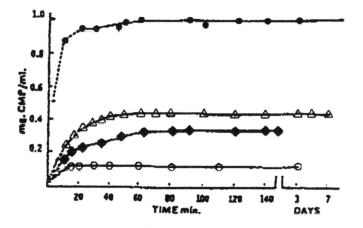

Figure 14 Dissolution curves for polymorphs A and B of chloramphenicol palmitate in 35% *tert*-butanol and water at 30°C and 38°C. (○) Polymorph A at 30°; (△) polymorph B at 30°; (◆) polymorph A at 38°; (●) polymorph B at 38°C. (Reprinted with permission from Ref. 48.)

Table 3 Thermodynamic Values for Polymorphs of Chloramphenicol Palmitate

Thermodynamic property[a]	Polymorph A	Polymorph B
T_T, °C	—	88
$\Delta H_{303(solution)}$, kcal/mol	21.8	15.4
$\Delta H_{(B\,to\,A)}$, kcal/mol	0^b	−6.4
$\Delta G_{303(B\,to\,A)}$ cal/mol	0^b	−774
$\Delta S_{303(B\,to\,A)}$, cal/(K · mol)	0^b	−19
ΔS_T, cal/(K · mol)	0^b	−18

[a] Subscript T refers to transition of B to A at 88°C = 361 K.
[b] By definition.
Source: Ref. 48.

dynamic equilibrium, involves holding crystals of one form in the temperature stability range of the desired polymorph until transformation occurs. The transformation rate is higher if the crystals are wetted with solvent and if seeds of the desired polymorph are present. A metastable form can also be prepared by achieving a high degree of supersaturation in the vapor or solution state or by supercooling in the melt state. The general objective is to crystallize the metastable form before crystals of the stable form appear. Preparation of the metastable form from the solution phase or melt is also possible, but the rate of cooling must be carefully controlled. Preparation from the vapor phase by subliming and then cooling the sublimate on a cover slip is also possible. Stabilization of the less

Table 4 Thermodynamic Values for Polymorphs of Mefenamic Acid

Thermodynamic property[a]	Polymorph I	Polymorph II
T_T, °C	—	89
$\Delta H_{303(solution)}$, kcal/mol	6.7	5.7
$\Delta H_{(II\,to\,I)}$, kcal/mol	0^b	−1.0
$\Delta G_{303(II\,to\,I)}$, cal/mol	0^b	−251
$\Delta S_{303(II\,to\,I)}$, cal/(K · mol)	0^b	−2.5
ΔS_T, cal/(K · mol)	0^b	−2.8

[a] Subscript T refers to transition of II to I at 89°C = 362 K.
[b] By definition.
Source: Ref. 48.

stable polymorph of a drug can also be achieved by using excipients and cosolvents in the drug formulation.

V. HYDRATION

Hydration can be an important factor in diffusion and mass transport phenomena in pharmaceutical systems. It may alter the apparent solubility or dissolution rate of the drug, the hydrodynamic radii of permeants, the physicochemical state of the polymeric membrane through which the permeant is moving, or the skin permeability characteristics in transdermal applications.

Hydration status of a drug can have a significant effect on its apparent solubility or dissolution rate. Shefter and Higuchi [50] made a theoretical and experimental study of the solubility of solvates. A rule applying to solubility behavior is that at room temperature solid solvates are less soluble in the solvent forming the solvate than is the original solid. Thus, hydrated crystals are less soluble in water than the corresponding anhydrous solid (Fig. 15). This is because the hydrated solid, if more soluble, cannot be prepared in a stable form, since it would spontaneously yield free water and the less soluble anhydrous form. Solvates from other water-miscible solvents are, however, more soluble in water than the corresponding anhydrous form because the negative free energy of the solvent with water makes an additional contribution to the free energy of solution.

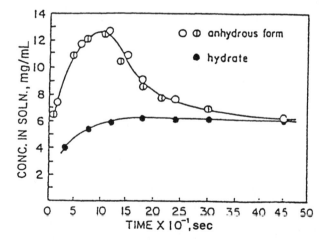

Figure 15 Dissolution of theophylline in water at 25°C. (Reproduced with permission from Ref. 50.)

Also, hydrates are more soluble in water-miscible solvents than are the corresponding anhydrous forms. For example, the solubility of caffeine hydrate is lower than that of anhydrous caffeine in water but higher in ethanol. The maximum concentration seen may be due to the solubility of the anhydrous crystalline phase or due to a temporary steady state in which the rate of dissolution of the metastable anhydrous form and the rate of crystallization of the stable hydrate are equal. The decreasing concentration represents crystallization of the stable hydrate from a solution supersaturated with respect to it. If the maximum concentration of the solute in the dissolution experiment corresponds to the solubility, then the initial increase in concentration follows the Noyes–Whitney equation [15]. Van't Hoff plots of log solubility versus the reciprocal of temperature give linear relationships (Fig. 16).

Second, hydration of drug molecules may affect their hydrodynamic radii. The diffusion coefficient D is related to the frictional resistance, f, that the diffusing particle experiences in moving through a medium by the equation [51]

$$D = RT/f \qquad (24)$$

In a homogeneous fluid the frictional resistance a particle experiences depends largely on its size and shape and on the nature of the solvent. For large molecules, where the slip factor (tendency of solvent molecules to adhere to solute) approaches infinity, the frictional resistance is

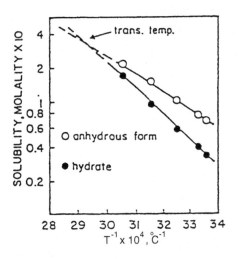

Figure 16 Van't Hoff plot of molal solubility versus reciprocal absolute temperature for theophylline in water. (Reproduced with permission from Ref. 50.)

$$f = 6\pi\eta r \tag{25}$$

while for solvent and solute molecules of similar size,

$$f = 4\pi\eta r \tag{26}$$

where η is the solvent viscosity and r the solute radius. However, the radius r is not the radius of the bare molecule but that of the hydrodynamic particle. The hydrodynamic particle consists of the diffusant molecule plus any solute or solvent molecules bound or adsorbed to the surface or within the diffusant. Therefore, hydration of the diffusant may increase the hydrodynamic radius.

The transport of many compounds takes place through interstices of polymer chains filled with aqueous medium [52]. In such cases, the rate of mass transport is directly proportional to the degree of hydration of the membranes [53]. The most widely accepted method for determining the hydration of membranes is to equilibrate the membranes in water or buffer and weigh these membranes after blotting [54]. In a newer method, the matrices to be studied are placed on a sintered glass funnel which is attached to a capillary filled with water. The absorption of water results in the movement of the capillary front [55].

Hydration of polymeric membranes may be influenced by the chemical identity of the polymers. A hydrophilic polymer has a higher potential to hydrate than a hydrophobic one. Sefton and Nishimura [56] studied the diffusive permeability of insulin in polyhydroxyethyl methacrylate (37.1% water), polyhydroxyethyl acrylate (51.8% water), polymethacrylic acid (67.5% water), and cuprophane PT-150 membranes. They found that insulin diffusivity through polyacrylate membrane was directly related to the weight fraction of water in the membrane system under investigation (Fig. 17).

Yasuda's free volume theory [57] has been proposed to explain the mechanism of permeation of solutes through hydrated homogeneous polymer membranes. The free volume theory relates the permeability coefficients in water-swollen homogeneous membranes to the degree of hydration and molecular size of the permeant by the following mathematical expression:

$$\frac{P_y}{D_{aq}} = K^*F_q^* \exp\left[-B^*\left(\frac{q}{V_f}\right)*\left(\frac{1}{H-1}\right)\right] \tag{27}$$

where P_y is the permeability of the solute in the water-swollen polymer, D_{aq} is the aqueous diffusion coefficient of the solute, K is the partition coefficient of the solute between the membrane and the solution, F_q is the probability of finding a hole within the membrane that will accommodate a solute of size q, B is a proportionality constant, V_f is the free volume of water, and H is the degree of hydration of the membrane.

Joshi and Topp [58] used Yasuda's free volume theory [57] to explain their

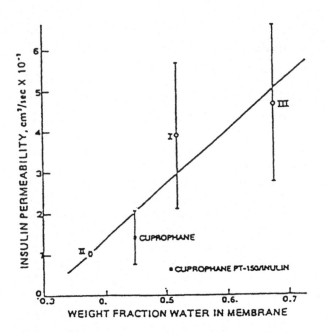

Figure 17 Effect of weight fraction of water on insulin permeability of highly swollen polyacrylate membranes. (Reprinted with permission from Ref. 56.)

observations with hyaluronic ester films. They found that films made of ethyl esters of hyaluronic acid are hydrated to a greater extent than those made of benzyl esters. Also, as the percentage of benzyl esterification increased, the percentage of hydration decreased. When they studied the diffusion of hydrocortisone, sodium hydrocortisone hemisuccinate, benzyl alcohol, mafenide acetate, and fluorescein sodium through these ethyl ester membranes, they found that the permeability coefficients of all the model drugs were significantly greater than those in the benzyl ester membranes. Furthermore, the trend for permeability coefficient values for the series of benzyl ester membranes was the same as that for the percent hydration values. Higher percent hydration was associated with higher permeability coefficient values for the model drugs studied (Fig. 18).

Similar results were obtained by Hunke and Matheson [59]. They found that copolymerization of polyurethane with PEG of different molecular weights yielded polymers with different properties than those of polyurethane. Percent hydration values increased with PEG as did the diffusion coefficients of the model drugs (Fig. 19).

Finally, in transdermal drug delivery, the hydration status of the skin can play an important role. The permeability of skin to various drug entities can be

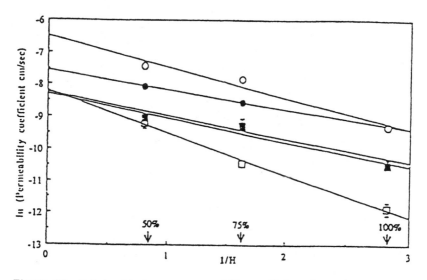

Figure 18 Relationship between permeability coefficients for various model drugs in a series of hydrated benzyl ester membranes of various degrees of esterification (100%, 75%, 50%) and reciprocal hydration. (Reprinted with permission from Ref. 58.)

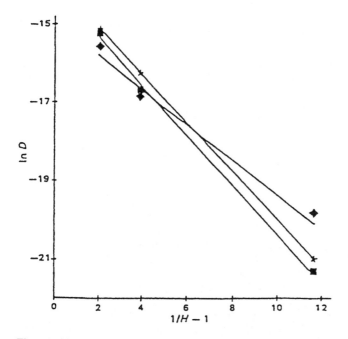

Figure 19 Free volume plot for co(polyether)polyurethane membranes at 37°C for model drugs. D is the diffusion coefficient in cm^2/sec, and H is the hydration expressed as (wet volume − dry volume)/wet volume. (Reproduced with permission from Ref. 59.)

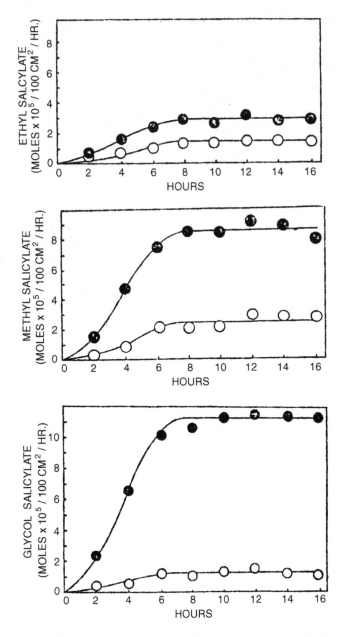

Figure 20 Urinary excretion data showing the influence of moisture on the percutaneous absorption rate of salicylates. (○) Anhydrous system rate; (●) hydrous system rate. (Reprinted with permission from Ref. 62.)

significantly influenced by the hydration of the skin. Southwell and Barry [60] studied the effect of hydration on the transport of aspirin and caffeine through human cadaver skin and found that hydration increased both the maximal penetration rate and the percent dose penetrated. In vivo investigations show that water content of the stratum corneum increases continuously with increasing depth from about 0.1 to 0.5 (w/w). In vitro studies show that the mechanism of water binding also changes over the same range [61]. This last review also describes various methods to measure skin hydration status. Wurster and Kramer [62] studied the urinary elimination rates following percutaneous absorption of three salicylates—ethyl, methyl, and glycol (Fig. 20 and Table 5).

Table 5 shows that the ratio of the excretion rate under hydrous condition to that under anhydrous condition decreases with the decreasing water solubility of the salicylate or with the oil/water distribution coefficient. It can be said that the more water soluble compound is aided to a greater degree by the presence of moisture than those of lesser solubility. This trend is, however, not a direct proportionality due to the influence of other physical phenomena such as viscosity, molecular size, and intermolecular bonding.

To summarize, the hydration status of the drug molecule and other components of a pharmaceutical formulation can affect mass transport. Solubility of drug crystals in an aqueous or nonaqueous solvent may depend on the presence or absence of moisture associated with the drug. Hydration may also determine the hydrodynamic radii of molecules. This may affect the frictional resistance and therefore the diffusion coefficient of the drug molecules. Diffusion of drugs in polymeric systems may also be influenced by the percent hydration of the polymers. This is especially true for hydrogel polymers. Finally, hydration of

Table 5 Experimental Exertion Rates and Other Physical Constants of Three Salicylates

Property	Glycol salicylate	Methyl salicylate	Ethyl salicylate
Excretion rate (hydrous system), mol/(100 cm^2 · hr)	11.7	8.6	2.9
Excretion rate (anhydrous system), mol/(100 cm^2 · hr)	1.3	2.7	1.5
Ratio (hydrous/anhydrous)	9.0	3.2	2.0
Water solubility (% w/v)	1.27	0.08	0.03
Distribution coefficient (olive oil/water)	7.7	343	1170
Relative distribution coefficient (to glycol ester)	1	45	152

Source: Ref. 62.

skin may enhance transdermal diffusion of water-soluble drugs to a significant extent.

VI. CONCLUSIONS

The mass flux (aqueous dissolution rate, or release rate, per unit surface area) of a solid drug increases with decreasing molar free energy and with increasing aqueous solubility, which is itself influenced by the crystal lattice arrangement in the solid state. For a given lattice, the solubility increases with decreasing crystallinity, reflecting increasing amorphous character arising from crystal defects introduced by the previous treatment, processing operations, additives, or impurities. For a polymorphic drug, the solubility is greater for the less stable polymorph than for the more stable polymorph. For solvates, the aqueous solubility is usually in the rank order nonaqueous solvate > non-solvate > lower hydrate > higher hydrate.

The chemical and physical stability of a solid drug decreases with decreasing crystallinity and increasing amorphous character, corresponding to an increase in molecular mobility (i.e., diffusivity) in the solid state. This phenomenon is of particular significance to proteins, peptides, and other biological materials. Certain additives other than water may stabilize proteins in the solid state, perhaps by locking in the defects.

Phase transitions, whether first-order or second-order, are potent sources of instability of solid drugs and can usually be detected and studied by thermal methods of analysis (e.g., DSC, TGA, TMA, ODSC, DMA, DEA). In crystalline solids, typical first-order transitions are polymorphic or desolvation transitions. In amorphous solids, second-order transitions, such as glass transitions, are common.

For a polymorphic drug, the polymorph obtained depends on the physical conditions, such as temperature, pressure, solvent, and the rate of desupersaturation. For a solvated drug, in addition to these conditions, the thermodynamic activity of the solvating solvent may also determine the solvate obtained. However, kinetic factors may sufficiently retard the crystallization of a stable form or the solid-state transition to the stable form that an unstable form may be rendered metastable.

The presence of a solvent, especially water, and/or other additives or impurities, often in nonstoichiometric proportions, may modify the physical properties of a solid, often through impurity defects, through changes in crystal habit (shape) or by lowering the glass transition temperature of an amorphous solid. The effects of water on the solid-state stability of proteins and peptides and the removal of water by lyophilization to produce materials of certain crystallinity are of great practical importance although still imperfectly understood.

The mass flux of many drugs through the skin or polymer membrane increases with increasing degree of hydration of the skin or membrane. The maximum extent of hydration of a polymer depends on the balance between the hydrophilic and hydrophobic groups in the constituent polymer or polymers. The mass flux of drugs of higher water solubility is aided more by the hydration of the membrane than is the mass flux of drugs of lower water solubility.

REFERENCES

1. MFC Ladd, RA Palmer. Structural Determination by X-Ray Crystallography. 2nd ed. New York: Plenum, 1985, p 502.
2. RI Ristic, JN Sherwood, K Wojciechowski. Assessment of the strain in small sodium chlorate crystals and its relation to growth rate dispersion. J Cryst Growth 91:163–168, 1988.
3. J Haleblian, W McCrone. Pharmaceutical applications of polymorphism. J Pharm Sci 58(8):911–929, 1969.
4. JK Haleblian. Characterization of habits and crystalline modification of solids and their pharmaceutical applications. J Pharm Sci 64(8):1269–1288, 1975.
5. R Huttenrauch. Molecular pharmaceutics as a basis for modern drug formulation. Acta Pharm Technol 24(6):55–127, 1978.
6. AHL Chow, PKK Chow, W Zhongshan, DJW Grant. Modification of acetaminophen crystals: Influence of growth in aqueous solutions containing p-acetoxyacetanilide on crystal properties. Int J Pharm 24:239–258, 1985.
7. JD Wright. Molecular Crystals. New York: Cambridge University Press, 1987.
8. HM Burt, AG Mitchell. Crystal defects and dissolution. Int J Pharm 9(2):137–152, 1981.
9. H Clapper. X-ray topography of organic crystals. Crystals 13:109–114, 1991.
10. WC Duncan-Hewitt, DJW Grant. True density and thermal expansivity of pharmaceutical solids—Comparison of methods of assessment of crystals. Int J Pharm 28:75–84, 1986.
11. MJ Pikal, AL Lukes, JE Lang, K. Gaines. Quantitative crystallinity determinations for beta-lactam antibiotics by solution calorimetry: Correlations with stability. J Pharm Sci 67(6):767–773, 1978.
12. P York, DJW Grant. Entropy of processing: A new quantity for comparing solid state disorder of pharmaceutical materials. Int J Pharm 30(2-3):161–180, 1986.
13. DE Bugay. Solid-state nuclear magnetic resonance spectroscopy: Theory and pharmaceutical applications. Pharm Res 10(30):317–327, 1993.
14. ED Deitz. Glassy state. Sci Technol 83:10–21, 1968
15. A Noyes, W Whitney. The rate of solution of solid substances in their own solutions. J Am Chem Soc 19:930, 1897.
16. JD Mullins, TJ Macek. Some pharmaceutical properties of novobiocin. J Am Pharm Assoc (Sci):49:245–248, 1960.
17. ML Roy, MJ Pikal, EC Rickard, A Maloney. The effects of formulation and moisture

on the stability of a freeze-dried monoclonal antibody–vinca conjugate: A test of the WLF glass transition theory. Dev Biol Stand 74:323–340, 1992.

18. G Zografi. States of water associated with solids. Drug Dev Ind Pharm 14(14):1905–1926, 1988.

19. MJ Hageman. The role of moisture in protein stability. Drug Dev Ind Pharm 14(14):2047–2070, 1988.

20. KI Izutsu, S Yoshioka, Y Takeda. Effects of additives on the stability of freeze-dried beta-galactosidase stored at elevated temperature. Int J Pharm 71:137–146, 1991.

21. MJ Pikal, DJW Grant. A theoretical treatment of changes in energy and entropy of solids caused by additives or impurities in solid solution. Int J Pharm 39:243–253, 1987.

22. P York. Crystal engineering and particle design for the powder compaction process. Drug Dev Ind Pharm 18(6-7):677–721, 1992.

23. HK Chan, DJW Grant. Influence of compaction on the intrinsic dissolution rate of modified acetaminophen and adipic acid crystals. Int J Pharm 57:117–124, 1989.

24. SP Duddu, FK Fung, DJW Grant. Effect of the opposite enantiomer on the physico-chemical properties of (−)ephedrinium 2-naphthalenesulfonate crystals. Int J Pharm 94(1-3):171–179, 1993.

25. R Swalin. Thermodynamics of Solids. 2nd ed. New York: Wiley, 1972, 416 p.

26. A Burger, R Ramberger. On the polymorphism of pharmaceuticals and other molecular crystals. I. Theory of thermodynamic rules. Mikrochim Acta (Wein) 2:259–271, 1979.

27. A Burger, R Ramberger. On the polymorphism of pharmaceuticals and other molecular crystals. II. Applicability of thermodynamic rules. Mikrochim Acta (Wein) 2:273–316, 1979.

28. MM deVilliers, JG Van der Watt, AP Lotter. Interconversion of the polymorphic forms of chloramphenicol palmitate (CaP) as a function of environmental temperature. Drug Dev Ind Pharm 17(10):1295–1303, 1991.

29. T Durig, AR Fassihi. Preformulation study of moisture effect on the physical stability of pyridoxal hydrochloride. Int J Pharm 77:315–319, 1991.

30. T Matsumoto, N Kaneniwa, S Hihuchi, M Otsuka. Effects of temperature and pressure during compression on polymorphic transformation and crushing strength of chlorpropamide tablets. J Pharm Pharmacol 43(2):74–78, 1991.

31. JL Ford, P Timmins. Pharmaceutical Thermal Analysis. Chichester, UK: Ellis Harwood, 1989, pp 136–149.

32. WW Young, R Suryanarayanan. Kinetics of transition of anhydrous carbamazepine to carbamazepine dihydrate in aqueous suspensions. J Pharm Sci 80(5):496–500, 1991.

33. SR Byrn. Solid State Chemistry of Drugs. New York: Academic Press, 1982, pp 1–186.

34. MJ Pikal. Freeze-drying of proteins, Part II: Formulation selection. BioPharm 3(9):26–30, 1990.

35. AC Tanquary, RE Lacey. Controlled release of biologically active agents. Adv Exp Med Biol 47:73–98, 1974.

36. C Shih, S Lucas, GM Zentner. Acid catalyzed poly(ortho ester) matrices for intermediate term drug delivery. J Controlled Release 15:55–63, 1991.

37. H Levine, L Slade. Water as a plasticizer: Physico-chemical aspects of low moisture polymeric systems: Water Science Reviews 3: Water Dynamics, F Franks, editor. Cambridge: Cambridge University Press 79–185, 1988.

38. Anonymous. Software for Oscillating Differential Scanning Calorimeter. Horsham, Seiko Instruments 1995.

39. Anonymous. DMA 983 Dynamic Mechanical Analyzer. Brochure. New Castle, DE: TA Instruments 1995.

40. Anonymous. DEA 2970 Dielectric Analyzer. Brochure. New Castle, DE: TA Instruments 1995.

41. GL Flynn, SH Yalkowsky, TJ Roseman. Mass transport phenomena and models: Theoretical concepts. J Pharm Sci 63(4):479–510, 1974.

42. L Ravin. Remington's Pharmaceutical Sciences. 16th ed. Easton, PA: Mack Publishing, 1980, pp 1358–1359.

43. W McCrone. Physics and Chemistry of Organic Solid State, Vol II. New York: Interscience, 1965, pp 725–767.

44. A Verma, P Krishan. Polymorphism and Polytropism in Crystals. New York: Wiley, 1966, pp 7–60.

45. MB Kuhnert. Status and future of chemical microscopy. Pure Appl Chem 10(2): 133–144, 1965.

46. DJW Grant, T Higuchi. Solubility Behavior of Organic Compounds. New York: Wiley, 1990, pp 7–60.

47. WI Higuchi, PK Lau, T Higuchi, JW Shell. Polymorphism and drug availability: Solubility relationships in the methviorednisolone system. J Pharm Sci 52:150–153, 1963.

48. A Aguiar, JE Zelmer. Dissolution behavior of polymorphs of chloramphenicol palmitate and mefenamic acid. J Pharm Sci 58(8):983–987, 1969.

49. AJ Aguiar, J Krc, AW Kinkel, JC Samyn. Effect of polymorphism on the absorption of chloramphenicol from chloramphenicol palmitate. J Pharm Sci 56:847, 1967.

50. E Shefter, T Higuchi. Dissolution behavior of crystalline solvated and nonsolvated forms of some pharmaceuticals. J Pharm Sci 52:781–791, 1963.

51. C Tanford. Physical Chemistry of Macromolecules. New York: Wiley, 1961.

52. GM Zentner, JR Cardinal, SW Kim. Progestin permeation through polymer membranes, I: Diffusion studies on plasma-soaked membranes. J Pharm Sci 67(10): 1347–1351, 1978.

53. GM Zentner, JR Cardinal, SW Kim. Progestin permeation through polymer membranes, II: Diffusion studies on hydrogel membranes. J Pharm Sci 67(10):1352–1355, 1978.

54. JA Hunt, HN Joshi, VJ Stella, EM Topp. Diffusion and drug release in polymer films prepared from ester derivatives of hyaluronic acid. J Controlled Release 12: 159–169, 1990.

55. LS Wang, PW Heng. Liquid penetration into tablets containing surfactants. Chem Pharm Bull 33(6):2569–2574, 1985.

56. MV Sefton, E Nishimura. Insulin permeability of hydrophilic polyacrylate membranes. J Pharm Sci 69(2):208–209, 1980.

57.   H Yasuda, DE Lamaze. Permselectivity of solutes in homogeneous water-swollen polymer membranes. J Macromol Sci Phys 5:111–134, 1971.
58.   HN Joshi, EM Topp. Hydration in hyaluronic acid and its esters using differential scanning calorimetry. Int J Pharm 80(2-3):213–225, 1992.
59.   WA Hunke, LE Matheson. Mass properties of Co(polyether) polyurethane membranes. Part 2. Permeability and sorption characteristics. Int J Pharm 1313–1318, 1981.
60.   D Southwell, BW Barry. Pentration enhancement in human skin. Effect of 2-pyrrolidone, dimethylformamide and increased hydration on finite dose permeation of aspirin and caffeine. Int J Pharm 22:291–298, 1984.
61.   RO Potts. Stratum corneum hydration: Experimental techniques and interpretation of results. J Soc Cosmet Chem 37:9–33, 1986.
62.   DE Wurster, SF Kramer. Investigation of some factors influencing percutaneous absorption. J Pharm Sci 50:288–293, 1961.

16

Heat and Mass Transfer in Low Pressure Gases: Applications to Freeze Drying

Michael J. Pikal
University of Connecticut, Storrs, Connecticut

I. OVERVIEW OF THE FREEZE-DRYING PROCESS

Freeze drying or "lyophilization" is a drying process in which the solvent, normally water, is first frozen and then removed by sublimation in an environment of low pressure—typically in the range 0.05–0.5 torr (6.6–66 Pa). In the pharmaceutical industry, the aqueous solution is first filled into suitable containers, usually glass vials, and the filled vials are loaded onto temperature-controlled shelves contained in a large "vacuum-tight" drying chamber. The temperature of the shelves is lowered to remove heat from the solution and ultimately transform most of the water to ice. The drug and excipients may also crystallize during this freezing stage. More often the solutes remain amorphous and are "freeze concentrated" into a rigid pseudosolid, or glass, which typically also contains roughly 20% unfrozen water.

After complete solidification, the system is evacuated by vacuum pumps, and the shelf temperature is increased to supply heat for the sublimation of ice. The stage of the process during which water is removed by sublimation of ice is termed "primary drying." Here, product temperature is maintained considerably below 0°C to allow sublimation of ice without damage to the product. As illustrated in Figure 1, the primary drying time decreases sharply as the product temperature increases—by about a factor of 2 for a 5°C increase in product temperature. Although primary drying should be carried out at the highest temperature possible to minimize process time, there is a product-dependent upper tempera-

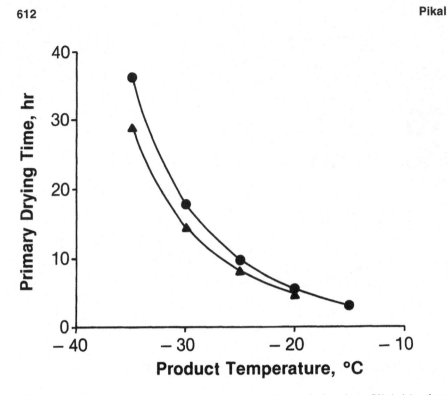

Figure 1 The effect of product temperature on primary drying time. 5% (v/v) solute; cake depth of 1 cm. (●) KCl; (▲) povidone (polyvinylpyrrolidone). (From Ref. 1.)

ture limit for primary drying referred to as the "maximum allowable temperature." This maximum temperature is the eutectic temperature for a solute system that crystallizes. For example, a frozen 1 M NaCl solution will typically exist as crystals of sodium chloride dihydrate and ice below the eutectic temperature of $-21°C$. However, above the eutectic temperature, the NaCl will exist in a concentrated solution state. Thus, below the eutectic temperature, water is removed by sublimation, leaving a solute "cake" of NaCl crystals in the same shape as the original solution. Above the eutectic temperature water is removed by evaporation (i.e., the solution "boils") and the structure of the solute cake is lost. If the solute does not crystallize but remains amorphous, the maximum allowable product temperature is the "collapse temperature." Collapse is essentially the amorphous system analogue of eutectic melt. If the product temperature rises above the collapse temperature, the amorphous solute–water phase gains sufficient fluidity to undergo viscous flow once the ice in that region of material has sublimed. That is, the amorphous solute passes through a glass-to-fluid transition which results in a loss of structure in the "dried" region adjacent to the ice/vapor boundary. This glass transition temperature is often referred to as the glass transition temperature of the "freeze concentrate," T'_g. Thus, the product temperature should be maintained close to, but slightly less than, the maximum allowable temperature to minimize process time yet ensure high product quality.

Even after all ice has been removed by sublimation, the product phase, or freeze concentrate, contains a large amount of "dissolved" water that must be removed to produce a stable product. This water is removed by desorption during secondary drying. Secondary drying is usually carried out at elevated product temperature to achieve efficient water removal.

An example of a freeze-drying process, or cycle, is given in Figure 2. A freeze-drying cycle is a record of relevant process data as a function of time during the process. Shown in Figure 2 are product temperatures, water removal rate, and residual water concentration in the product phase. The shelf temperature was maintained constant at $-8°C$ between zero and 24 hr. At the 24 hr point, the shelf temperature was increased to $50°C$ to remove the last 6% of residual moisture. (Note that because the surface above the vials was not temperature-controlled and remained close to ambient, the product temperature approaches $\approx 0°C$ at 24 hr and $\approx 40°C$ at the end of secondary drying instead of approaching the shelf temperature.) The chamber pressure was maintained constant at 0.05 torr (6.6 Pa) throughout the process by a controlled leak of air into the drying

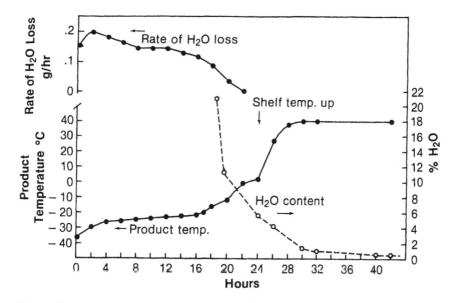

Figure 2 The rate of water loss and residual water content during freeze drying. Moxalactam disodium formulated with 12% mannitol in aqueous solution at 30% solids; 10 mL tubing vials with a solution fill depth of 1 cm. The chamber pressure was 0.05 torr (6.6 Pa). (From Ref. 2.)

chamber. The product temperature, measured at the bottom center of the vial, shows a sharp increase near 17 hr, indicating that most (or all) of the ice has been removed from that vial. Coincident with this increase in product temperature, the rate of water loss decreases sharply, but it does not drop to zero immediately. Significant water vapor transfer occurs between 17 hr and about 22 hr, reflecting rapid desorption of water during early secondary drying; it may also reflect the removal of trace quantities of ice not in actual contact with the temperature sensor. Note that the product temperature during primary drying is much lower than the shelf temperature. This is a result of the self-cooling effect of ice sublimation and the limited ability to transfer heat from the shelf to the product. The H_2O content curve is the mean moisture content of the product after primary drying at the times indicated. Note that the water content after the sharp increase in product temperature is quite high ($\approx 20\%$) and decreases very sharply with increasing time even though the product temperature is very low ($-10°C$ to $0°C$). However, after the water content decreases below about 2%, further removal of water is very slow even though the product temperature is high (40°C). Thus, the rate of water removal decreases sharply as the level of residual moisture decreases.

It should be emphasized that process control means control of the product temperature-vs-time profile during the course of the freeze-drying process. Control of shelf temperature (and chamber pressure) are important process variables that must be controlled to control product temperature. However, one freeze dries the product, not the shelf, and the common practice of specifying the shelf temperature (and perhaps the chamber pressure) is not sufficient to define the process. Factors other than shelf temperature (and chamber pressure) impact upon the product temperature. For sublimation of 1 g of ice, roughly 1000 L of water vapor must pass through the partially dried product cake, through the partially seated stopper, and into the condenser chamber. To maintain a constant product temperature, about 650 cal (2720 J) of heat per gram of ice sublimed must be supplied from the shelves to the frozen product. A given product temperature results from a particular balance between the rate of heat transfer to the product and the drying rate or mass transfer of water vapor. Thus, primary drying is a problem in coupled heat and mass transfer, and an appreciation of heat and mass transfer considerations is needed to achieve optimum process control.

II. PHENOMENOLOGICAL ASPECTS OF HEAT AND MASS TRANSFER

A. Mass Transfer and Resistance

The systematic description of heat and mass transfer applications in freeze drying is considered in this section. Mechanistic interpretations are discussed in the following sections. Mass transfer may be discussed in terms of resistance to flow of water vapor through the various mass transfer barriers, such as partially dried product, stopper openings, and chamber-to-condenser pathway. The resistance may be defined as the ratio of the driving force for sublimation (a pressure difference across the barrier) to the flow through the barrier (the sublimation rate). Figure 3 illustrates a typical pressure profile in primary drying and a corresponding bar graph showing the relative magnitudes of the individual resistances. The greatest pressure drop, corresponding to the greatest resistance, occurs across the dried product layer. This generalization is valid unless the solute concentration is extremely low (i.e., <1 wt%) or there is essentially no dried layer (i.e., at the start of primary drying). The dried product resistance, denoted R_p, depends on the cross-sectional area of the product, A_p,

$$R_p = \frac{\hat{R}_p}{A_p} \tag{1}$$

where $\hat{R}_p$ is the area-normalized product resistance, which is independent of the product area but does depend on both the thickness of the dried layer and the nature of the product. Thus, the product resistance, R_p, depends on the container

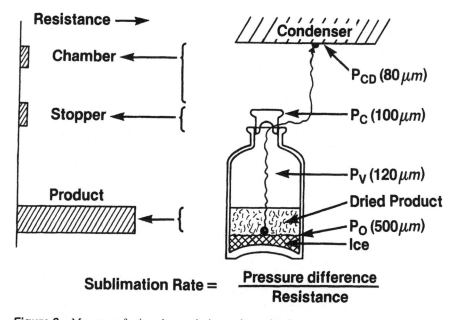

Figure 3 Mass transfer in primary drying: schematic of resistances. Pressure in μmHg. (From Ref. 3.)

used, since A_p is fixed by the container's inner diameter. Note also that for fixed fill volume, the maximum thickness of the dried product cake also depends upon the cross-sectional area. Since $\hat{R}_p$ increases as product thickness increases, a larger vial means both smaller average $\hat{R}_p$ and larger A_p, both effects decreasing the resistance, R_p. The units used for $\hat{R}_p$ are $cm^2 \cdot torr/g$ (1 torr = 133.3 Pa). With this choice of units, the numerical value of $\hat{R}_p$ represents the approximate time (in hours) to freeze dry a sample 1 cm thick at a temperature of $-20°C$ if the resistance would remain constant over all of primary drying. However, in nearly all cases, $\hat{R}_p$ increases as primary drying proceeds.

Although the product resistance does increase as the thickness of the dried layer increases, the relationship between thickness and resistance is usually not a direct proportion. Representative $\hat{R}_p$ data, determined by the microbalance procedure, illustrate the variations in $\hat{R}_p$-vs.-thickness behavior observed (Fig. 4): Type I, linear $\hat{R}_p$ vs. thickness; type II, high resistance as thickness approaches zero, followed by a decrease in $\hat{R}_p$ as drying proceeds, with finally a linear $\hat{R}_p$-vs.-thickness dependence; type III, moderate curvature showing a decrease in dependence on thickness as drying proceeds; and type IV, characterized by severe curvature. The "type" of resistance-vs.-thickness behavior shown is not intended to imply anything about the actual magnitude of resistance. For example, although the type IV curve shown happens to also have a high resistance relative to the other curves in Figure 4, many type IV curves have very low resistance. As is shown later in this chapter, simple theory based upon a uniform pore structure throughout the dried layer implies a direct proportion between thickness and resistance. The complex behavior shown by real systems in Figure 4 is likely a result of systematic variations in pore size from the top to the bottom of the product. Since the pores are created by sublimation of ice, the pores are a "template" of the size and geometry of the ice crystals formed during the freezing process. Evidently, freezing is not completely uniform throughout the sample. Type II behavior is an extreme example of heterogeneity in "cake structure." Type II behavior has been observed only with amorphous solutes. Here, the initial high resistance appears to result from a relatively impermeable surface "skin," which, after some water desorption, develops cracks and decreases in effective resistance. However, except in very rare cases of extreme Type IV behavior, where resistance behavior is dominated by a surface barrier which does not change during the drying process, the resistance increases significantly during the course of primary drying, regardless of the details of the $\hat{R}_p$-vs.-thickness behavior.

Dried product resistance normally increases with increasing solute concentration and frequently decreases as the temperature of the frozen product approaches the eutectic temperature or T_g'. Production of larger ice crystals by a lower degree of water supercooling and/or an annealing process during freezing may also decrease the resistance.

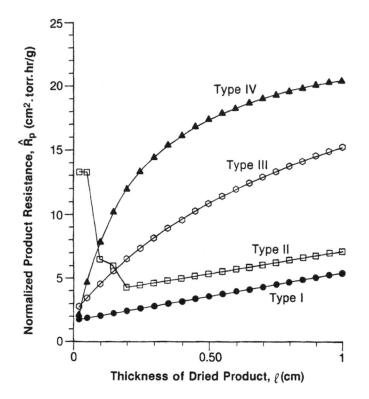

Figure 4 Dependence of dried product resistance on thickness; types of behavior. (●) 5% povidone at −26°C, type I; (□) 22.8% povidone at −26°C, type II; (◇) 5% potassium chloride at −30°C, type III; (▲) 10% potassium chloride at −20°C, type IV. (From Ref. 4.)

B. Heat Transfer Coefficients and Temperature Differences

Barriers to heat transfer produce corresponding temperature differences in a freeze-drying system, the actual temperature profile depending upon the rate of sublimation, the chamber pressure, and the container system as well as the characteristics of the freeze dryer employed. An experimental temperature profile is shown in Figure 5 for a system where vials were placed in an aluminum tray with a flat 5 mm thick bottom and a tray lid containing open channels for escape of water vapor. Here, heat transfer is determined by four barriers:

 1. The shelf itself with a temperature difference of 8°C between the shelf interior or "fluid" temperature and the shelf surface. This heat transfer

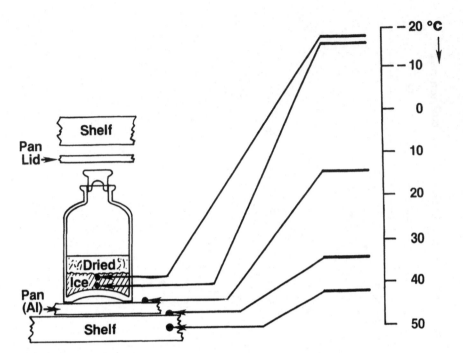

Figure 5 Temperature profile in primary drying of dobutamine HCl-mannitol (1:1), 53 mg solids/mL, 10 mL fill volume. Vials are 5304 molded glass vials (8.3 cm² cross-sectional area) which are placed in a flat aluminum tray. The heat flux is 42 cal/(cm² · hr), and the chamber pressure is 0.1 torr. (From Ref. 5.)

barrier reflects imperfect heat transfer within the shelf itself. With most modern well-maintained freeze dryers, this heat transfer barrier is about a factor of 2 or 3 smaller than found for the old dryer used to accumulate the data given in Figure 5.

2. The pan or tray, with a temperature difference of 20°C between the shelf surface and the top surface of the tray bottom.

3. The vial, with a temperature difference of 30°C between the pan surface and the product in the bottom of the vial.

4. The frozen product, with a temperature difference of about 2°C between the ice in the bottom of the vial and the ice at the sublimation interface.

The thermal resistances producing the large temperature differences between shelf surface and pan surface and between pan surface and product at the vial bottom are due almost entirely to the insulation of the vapor boundary which exists between two solid surfaces not in perfect contact. The temperature differences across the pan bottom and across the glass in the vial bottom are very small.

A heat transfer coefficient is defined as the ratio of the area-normalized heat flow to the temperature difference between the heat source (the shelf) and the heat sink (the frozen product). For the case of vials resting directly on the freeze dryer shelf, the vial heat transfer coefficient, K_v, is defined by

$$\frac{dQ}{dt} = A_v K_v (T_s - T_p) \tag{2}$$

where dQ/dt is the heat flow (cal/sec or J/sec) from the shelves to the product in a given vial; A_v is the cross-sectional area of the vial calculated from the vial outer diameter; T_s is the temperature of the shelf surface; and T_p is the temperature of the frozen product at the bottom center of the vial. Vial heat transfer data may be determined as a function of chamber pressure by steady-state experiments that measure the amount of ice sublimed in a fixed time at constant shelf temperature and constant chamber pressure. The calculated average sublimation rate, dm/dt, is then used to calculate the average heat flow as $dQ/dt = \Delta H_s (dm/dt)$, where ΔH_s is the heat of sublimation of ice. The measured average temperature difference, $T_s - T_p$, is then used with the calculated value of dQ/dt to evaluate K_v using Eq. (2). The effect of chamber pressure on vial heat transfer coefficients is shown in Figure 6 for several types of glass vials in a freeze dryer with stainless steel shelves. The differences in vial heat transfer coefficients at zero pressure arise from differences in the degree of actual contact between the vial and the shelf. An increase in chamber pressure causes an increase in vial heat transfer coefficient, although the pressure dependence varies between the different vial types.

C. Coupling Between Heat and Mass Transfer: Process Control

Heat and mass transfer are coupled by the constraint imposed by the conservation of energy, as shown by the equation

$$\frac{dQ}{dt} = \Delta H_s \frac{dm}{dt} + m_s c_v \frac{\partial T}{\partial t} \tag{3}$$

where dQ/dt is the heat flow from the shelves; $\Delta H_s(dm/dt)$ is the rate of heat removal by sublimation, where ΔH_s is the heat of sublimation; and $m_s c_v (\partial T/\partial t)$ is the rate of heat removal arising from the rate of temperature increase $(\partial T/\partial t)$ of the sample of mass m_s and specific heat c_v. Except for the first 10–20 min following a large change in shelf temperature, the specific heat term is small

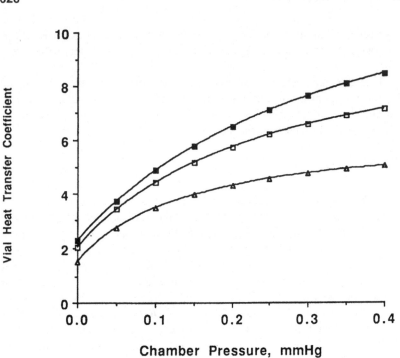

Figure 6 Pressure dependence of vial heat transfer coefficients for selected vials. (□) W5816; (■) K5816; (△) 5303. (Data from Ref. 5.)

compared to the heat of sublimation term for all of primary drying and may be ignored. The system is then in pseudo-steady state. The sublimation rate, dm/dt, may be related to the driving force for sublimation, $P_0 - P_c$, and the resistance to vapor flow from the frozen product to the drying chamber, $R_p + R_s$, by

$$\frac{dm}{dt} = \frac{P_0 - P_c}{R_p + R_s} \tag{4}$$

where P_0 is the vapor pressure of ice at the sublimation interface, P_c is the chamber pressure, and R_p and R_s are the resistances of the dried product and stopper, respectively. It should be noted that Eq. (4) assumes that the gas in the vial is essentially 100% water vapor. Both theoretical and experimental evidence indicate that during most primary drying conditions of practical interest, the gas in the vial and the gas in the drying chamber are nearly all water vapor, even when chamber pressure is being controlled by a controlled inert gas leak. As product temperature increases, the vapor pressure of ice, P_0, increases in exponential fashion, thereby increasing the sublimation rate and decreasing the time required for sublimation (i.e., see Fig. 1). Combination of Eqs. (2)–(4) and some algebraic manipulation then gives

$$\Delta H_s \frac{P_0 - P_c}{R_p + R_s} - A_v K_v (T_s - T_p) = 0 \tag{5}$$

Equation (5) is equivalent to stating that sublimation and subsequent transport of 1 g of water vapor into the chamber demands a heat input of 650 cal (2720 J) from the shelves. The vial heat transfer coefficient, K_v, depends upon the chamber pressure, P_c; and the vapor pressure of ice, P_0, depends in exponential fashion upon the product temperature, T_p. With a knowledge of the mass transfer coefficients, R_p and R_s, and the vial heat transfer coefficient, K_v, specification of the process control parameters, P_c and T_s, allows Eq. (5) to be solved for the product temperature, T_p. The product temperature, and therefore P_0, are obviously determined by a number of factors, including the nature of the product and the extent of prior drying (i.e., the cake thickness) through R_p, the nature of the container through K_v, and the process control variables P_c and T_s. With the product temperature calculated, the sublimation rate is determined by Eq. (4).

It is common practice to introduce a controlled gas leak (nitrogen) into the drying chamber to control chamber pressure. This practice not only provides process control but also decreases primary drying time. That is, due to more efficient heat transfer, the sublimation rate normally increases as the chamber pressure is increased by the gas leak. The cause of this effect involves the pressure dependence of the vial heat transfer coefficient (i.e., Fig. 6). Several examples of the effect of chamber pressure on sublimation rate are given by Figure 7. In both cases, the sublimation rate (water vapor flux) increases sharply with increasing pressure at low pressure and then begins to plateau at higher pressure. In Figure 7b, product temperature data show that as chamber pressure increases, the temperature of the frozen product increases. Of course, as the chamber pressure increases, the vial heat transfer coefficient increases (Fig. 6), thereby allowing more heat to flow from the shelves to the product at a fixed shelf temperature. Greater heat flow increases the product temperature and greatly increases the vapor pressure of ice. Thus, at least at low and moderate pressure, the net driving force for sublimation, $P_0 - P_c$, increases, which increases the sublimation rate. At higher chamber pressures, the increase in vial heat transfer coefficient for a given increase in chamber pressure is less (Fig. 6) and the net increase in driving force is smaller, thereby producing a smaller increase in sublimation rate. Eventually, at higher pressures of ~ 1 torr (133 Pa) the increase in P_0 will be less than the increase in chamber pressure, P_c, and the driving force (and sublimation rate) will actually *decrease* as the chamber pressure increases.

The objective of the primary drying stage of the freeze-drying process is to remove the ice as quickly as possible without undue risk of product loss

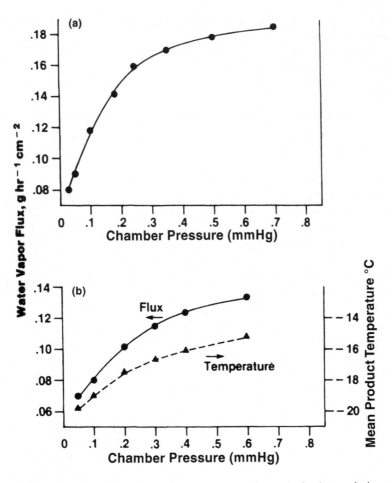

Figure 7 The effect of chamber pressure on the rate of primary drying. (a) 0.18 M methylprednisolone sodium succinate; 2 mL in molded vials (2.54 cm²), shelf temperature +45°C. (Smoothed data from Ref. 6.) (b) Dobutamine hydrochloride and mannitol (4% w/w in water), 12 mL in tubing vials (5.7 cm²) and shelf surface temperature +10°C. (MJ Pikal. Unpublished data.) (Modified from Ref. 1.)

through collapse or eutectic melt. That is, the process should be controlled such that the temperature of the product is increased to the control temperature quickly after drying is started, where the control temperature is safely below (usually 2–5°C below) the "maximum allowable product temperature." Heat input should then be adjusted such that the temperature is held constant at the control temperature for the duration of primary drying. Constant temperature throughout primary drying requires a smaller heat input near the end of primary drying than required at the beginning of this stage. Thus, the shelf temperature and/or the chamber pressure must be reduced during the course of the sublimation process. This phenomenon is a general result of the decreasing ability of the sample to "self-cool" as primary drying proceeds. Since the resistance of the dried product increases with increasing dried product thickness and therefore increases with time, the self-cooling rate from ice sublimation decreases with time throughout primary drying. Therefore, maintaining constant product temperature requires decreasing heat input. Figure 8 shows the shelf temperatures required to maintain a constant temperature of $-20°C$ in a 5% mannitol solution filled to a depth of 1 cm with the chamber pressure maintained constant at 0.15 torr. While the quantitative features of Figure 8 are valid only for the specific example given, the observation that the shelf temperature needs to decrease as drying proceeds is a general feature of nearly all freeze-drying systems if constant product temperature is a process

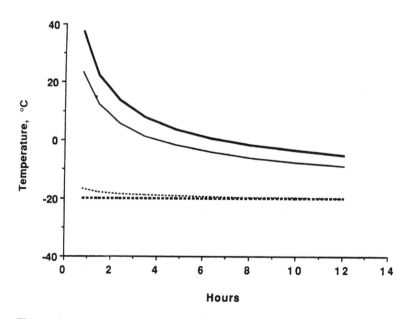

Figure 8 An example of the decreasing heat requirement during primary drying at a chamber pressure of 0.15 torr. 5% mannitol maintained at $-20°C$ during primary drying. Results obtained by computer simulation of freeze drying (see Ref. 3). Heavy curve: Shelf Fluid. Light curve: Shelf surface. Lightweight dashed curve: Product Bottom. Heavy dashed curve: Sublimation.

requirement. The decrease in shelf temperature will be most marked when the product resistance increase during drying is greatest. Mannitol shows a rather large increase in resistance as the thickness of the dried layer increases.

III. KINETIC THEORY OF GASES AND TRANSPORT PHENOMENA

A. Kinetic Theory, Collisions, and Gas Pressure

Molecules in a gas are in constant motion at speeds on the order of the speed of a rifle bullet; at equilibrium there is no net flow of gas and the motion is random. This motion produces collisions of the molecules with the walls of the vessel containing the gas, with a change in momentum of the gas molecule resulting from each collision. This change in momentum produces a force per unit area, or pressure on the wall. Consider those molecules with the component of velocity in the x direction between the value of v_x and $v_x + dv_x$. The x direction is defined as the direction normal to the wall. The fraction of molecules with the x component of velocity in this range, denoted $dN(v_x)/N$, is given by the "density function," $f(v_x)$, where

$$\frac{dN(v_x)}{N} = f(v_x)\, dv_x, \tag{6}$$

The term $f(v_x)\, dv_x$ may be thought of as the probability that the x component of the velocity will be between v_x and $v_x + dv_x$. Since the probability is unity that the velocity is between $-\infty$ and $+\infty$, we must have

$$\int_{-\infty}^{\infty} f(v_x)\, dv_x = 1 \tag{7}$$

In a time interval of dt, each gas molecule within a distance $v_x\, dt$ from the wall of area A and with a positive velocity component between v_x and $v_x + dv_x$ will strike the wall. Thus the colliding gas molecules are those contained in the volume element $Av_x\, dt$ which have an x velocity component between v_x and $v_x + dv_x$. The number of such molecules striking the wall of area A is the total number of molecules per unit volume N/V times the relevant volume ($Av_x\, dt$) times the fraction of molecules with velocity in the correct range, $f(v_x)\, dv_x$. Since the velocity before striking the wall is v_x, and the velocity after striking the wall is $-v_x$, the momentum change per collision is $2mv_x$. Thus, the momentum change for all such collisions in time dt is

$$d(mv_x) = 2mv_x \frac{N}{V} f(v_x)\, dv_x\, Av_x\, dt \tag{8}$$

Since the rate of change of momentum with time is the force of the gas molecules on the wall, the pressure, or force per unit area, from the gas molecules of velocity v_x becomes

$$dP = \frac{1}{A}\frac{d(mv_x)}{dt} = 2mv_x^2\frac{N}{V}f(v_x)\,dv_x \tag{9}$$

The total pressure coming from *all* the gas molecules directed at the wall (i.e., with positive velocity) is obtained by integration of dP in Eq. (9) over all positive velocities,

$$P = 2m\frac{N}{V}\int_0^\infty v_x^2 f(v_x)\,dv_x \tag{10}$$

The mean square velocity in the x direction, $\langle v_x^2\rangle$, defined by

$$\langle v_x^2\rangle = \int_{-\infty}^\infty v_x^2 f(v_x)\,dv_x = 2\int_0^\infty v_x^2 f(v_x)\,dv_x \tag{11}$$

where the integral from $-\infty$ to $+\infty$ has been replaced by twice the integral from zero to infinity since the distribution function, $f(v_x)$, cannot depend upon the velocity direction, i.e., $f(-v_x) = f(v_x)$. Thus, the simple kinetic theory expression for the pressure of a perfect gas is given by

$$P = Nm\frac{\langle v_x^2\rangle}{V} \tag{12}$$

The magnitude of the velocity of a gas molecule is related to the magnitudes of the component velocities by

$$v^2 = v_x^2 + v_y^2 + v_z^2 \tag{13a}$$

$$v_x^2 = v_y^2 = v_z^2 = \frac{v^2}{3} \tag{13b}$$

where the relations v_i^2 $(i = x, y, z) = v^2/3$ follow from the assumption that the gas is in equilibrium and therefore has no directional properties (i.e., is isotropic). However, since the kinetic energy of a gas molecule, E_k, is given by one-half the product of the molecule mass and the mean square of the velocity,

$$PV = \frac{2}{3}N\left(\frac{1}{2}m\langle v^2\rangle\right) = \frac{2}{3}NE_k \tag{14}$$

Since equipartition of energy requires that the kinetic energy corresponding to each transnational degree of freedom is $kT/2$, where k is Boltzmann's constant and T is the thermodynamic absolute temperature, three degrees of freedom corresponding to translation in the x, y, and z directions require that $E_k = (3/2)kT$. Thus, the ideal gas law may be written: $PV = nRT$. Here, the relationship $Nk = nN_0k = nR$, where $N_0 =$ Avogadro's number, has been used.

B. Distribution of Velocity: The Maxwell–Boltzmann Distribution

The above discussion is intended to emphasize the fact that gas pressure arises from collisions of the gas molecules with the wall of the containing vessel. The greater the velocity of the molecules, the greater the pressure. Of course, due to collisions of the molecules with the vessel and with each other, the velocity of a given molecule is constantly changing. Also, at a given instant in time, not all molecules have the same velocity, and a "snapshot" of the system would reveal a distribution of molecular velocities. For a gas at equilibrium, the net velocity vector in any given direction is zero, meaning that the gas is not flowing. In addition, the magnitudes of the individual velocities, or molecular speeds, are distributed around a mean value characteristic of the temperature and molecular mass, with most molecules having a speed close to the mean value. While the mean square speed is given by $\langle v^2 \rangle = 3kT/m$, calculation of the mean molecular speed $\langle v \rangle$ requires a knowledge of the velocity distribution function. Recall that the velocity distribution function, $f(v_x)$, is the function such that the probability of a molecule having an x component of velocity between v_x and $v_x + dv_x$ is given by the product $f(v_x)\, dv_x$. Similarly, the distribution functions for the y and z components of velocity are $f(v_y)$ and $f(v_z)$, respectively. We seek the probability of a given molecule having *simultaneously* its x component of velocity between v_x and $v_x + dv_x$, its y component between v_y and $v_y + dv_y$, and its z component between v_z and $v_z + dv_z$, denoted $F(v_x, v_y, v_z)\, dv_x\, dv_y\, dv_z$. It seems intuitive (and is also correct) to assume that the probability of having the x component of velocity in the range v_x to $v_x + dv_x$ is independent of the probability of the z component being in the range v_z to $v_z + dv_z$ and is also independent of the probability of the y component being in the interval v_y to $v_y + dv_y$. In other words, the three probabilities are independent. Thus, since the probability of the simultaneous occurrence of three independent events is given by the product of the probabilities of each of the three events, the three-dimensional distribution function $F(v_x, v_y, v_z)$ is given by the product of the three one-dimensional distribution functions. Further, since the gas is at equilibrium and is not in a state of flow, the probability of a given velocity vector (i.e., velocity components v_x, v_y, v_z) must be independent of the direction of the velocity vector and must depend only upon the magnitude of the velocity vector v. Thus, we can state that the "three-dimensional" distribution function, F, may be written as a function only of the magnitude of the velocity vector, $F(v)$, and we may write

$$F(v_x, v_y, v_z) = F(v) = f(v_x)f(v_y)f(v_z) \tag{15}$$

If we take the natural logarithm of both sides of Eq. (15) and differentiate with respect to one of the component velocities, say v_x, we have

$$\frac{\partial \ln F(v)}{\partial v_x} = \frac{\partial \ln f(v)}{\partial v_x} \tag{16}$$

where, via the chain rule for differentiation,

$$\frac{\partial \ln F(v)}{\partial v_x} = \frac{\partial \ln F(v)}{\partial v} \frac{\partial v}{\partial v_x} \tag{17}$$

From Eq. (13) we have $\partial v / \partial v_x = v_x / v$, so Eq. (17) may be rearranged to give

$$\frac{1}{v_x} \frac{\partial \ln f(v_x)}{\partial v_x} = \frac{1}{v} \frac{\partial \ln F(v)}{\partial v} \tag{18}$$

Using the same derivation as outlined above, equations equivalent to Eq. (18) may be written for the y component and the z component of velocity. Thus, we may write

$$\frac{1}{v} \frac{\partial \ln F(v)}{\partial v} = \frac{1}{v_x} \frac{\partial \ln f(v_x)}{\partial v_x} \tag{19}$$

$$= \frac{1}{v_y} \frac{\partial \ln f(v_y)}{\partial v_y} = \frac{1}{v_z} \frac{\partial \ln f(v_z)}{\partial v_z}$$

Because each of the terms in Eq. (19) involves a different independent variable but all terms are equal, each term must be equal to the same constant. Thus, we may write

$$\frac{1}{v} \frac{\partial \ln F(v)}{\partial v} = \text{constant} = -\beta \tag{20}$$

Using Eqs. (19) and (20), we may write the following differential equation for v_x:

$$\frac{\partial \ln f(v_x)}{\partial v_x} = -\beta v_x \tag{21}$$

which may be directly integrated to give

$$f(v_x) = A_x \exp(-\beta v_x^2) \tag{22}$$

where A_x is the integration constant. Since $f(v_x)\, dv_x$ is the probability that a molecule will have x-component velocity in a given interval, and since the sum of all probabilities must equal unity, the integral of $f(v_x) \, dv_x$ over all possible velocities must equal unity:

$$\int_{-\infty}^{\infty} f(v_x) \, dv_x = 1 = \int_{-\infty}^{\infty} A_x \exp(-\beta v_x^2) \, dv_x \qquad (23)$$

Solving for A_x then gives

$$A_x = \frac{1}{\displaystyle\int_{-\infty}^{\infty} \exp(-\beta v_x^2) \, dv_x} = \sqrt{\frac{\pi}{\beta}} \qquad (24)$$

Note that for the integral in Eq. (24) to be finite and real, the constant β must be positive. However, except for this condition of sign, we cannot specify β and therefore cannot completely specify the distribution function, Eq. (22), without some additional information. This additional information comes from the observation that the average kinetic energy for one degree of freedom (i.e., kinetic energy due to the x component of the velocity) is given in terms of temperature by $m\langle v_x^2 \rangle/2 = kT/2$. Thus, in principle, there is a relation between the average of the square of the molecular velocity and the absolute temperature, T. However, we must compute the average value of the square of the x-component velocity. The average value of any quantity which is a function of x, $g(x)$, is given by the mean value theorem in the form

$$\langle g \rangle = \sum_x P(x) \, g(x) = \int_{-\infty}^{\infty} P(x) \, g(x) \, dx \qquad (25)$$

where $P(x)$ is the probability that the function g has the value $g(x)$, and in the integral expression x is assumed to be a variable that varies continuously over all real values of x, positive and negative. For discrete values of x, the mean value theorem is given in terms of a sum over all allowed values of x and therefore over all allowed values of the function g. If x and $g(x)$ vary continuously, the sum may be written in terms of the definite integral over all allowed values of x. In the case of interest here, the function $g(x)$ is the x component of the kinetic energy, $(1/2)mv_x^2$. Thus, we may write

$$\left\langle \frac{1}{2} mv_x^2 \right\rangle = \int_{-\infty}^{\infty} \frac{1}{2} mv_x^2 f(v_x) \, dv_x \qquad (26)$$

$$= \int_{-\infty}^{\infty} \frac{1}{2} mv_x^2 \left(\frac{\beta}{\pi} \right)^{1/2} \exp(-\beta v_x^2) \, dv_x$$

Bringing the constant terms outside the integral and relating the mean kinetic energy to temperature then gives

$$\left\langle \frac{1}{2} m v_x^2 \right\rangle = \frac{1}{2} kT = \frac{1}{2} m \left(\frac{\beta}{\pi} \right)^{1/2} \int_{-\infty}^{\infty} v_x^2 \exp(-\beta v_x^2) \, dv_x \tag{27}$$

Since the integral in Eq. (27) is simply $\sqrt{\pi}/2\beta^{3/2}$, β is related to the absolute temperature by the expression

$$\beta = \frac{1}{2} \left(\frac{m}{kT} \right) \tag{28}$$

Thus, the distribution function $f(v_x)$ may be written

$$f(v_x) = \left(\frac{m}{2\pi kT} \right)^{1/2} \exp\left[-\frac{m v_x^2}{2kT} \right] \tag{29}$$

Similar treatments of the terms in Eq. (19) containing v_y and v_z will yield exact analogues of Eq. (29) for $f(v_y)$ and $f(v_z)$. Since the product of the three one-dimensional distribution functions gives the three-dimensional distribution function, i.e., $F(v) = f(v_x)f(v_y)f(v_z)$, we may combine Eq. (29) with similar expressions for the y and z velocity components to give

$$F(v) = \left(\frac{m}{2\pi kT} \right)^{3/2} \exp\left[-\frac{m(v_x^2 + v_y^2 + v_z^2)}{2kT} \right] = \left(\frac{m}{2\pi kT} \right)^{3/2} \exp\left[-\frac{m v^2}{2kT} \right] \tag{30}$$

Equation (30) is the Maxwell–Boltzmann distribution function in rectangular coordinates. Thus, in a system of N total molecules, the fraction of molecules, dN/N, with velocity components in the ranges x component, v_x to $v_x + dv_x$; y component, v_y to $v_y + dv_y$; and z component, v_z to $v_z + dv_z$ is given by

$$\frac{dN}{N} = F(v) \, dv_x dv_y dv_z$$

where the distribution function $F(v)$ is given by the Maxwell–Boltzmann distribution, Eq. (30).

The Maxwell–Boltzmann distribution function given by Eq. (30) may be converted to an equivalent, but generally more useful, distribution function by converting from the rectangular coordinates of Eq. (30) to spherical polar coordinates. Figure 9 illustrates the velocity vector, **v**, which forms the angle θ with the z axis and whose projection in the xy plane forms the angle ϕ with the x axis. The magnitude of the velocity vector, v, is given in terms of the x, y, and z components by Eq. (13). Thus, the velocity vector **v** may be specified by its magnitude v and the two angles θ and ϕ. The x, y, and z components of velocity may be expressed in terms of the polar coordinates (v, θ, ϕ) by the relationships

$$v_x = v \sin \theta \cos \phi, \qquad v_y = v \sin \theta \sin \phi, \qquad v_z = v \cos \theta \tag{31}$$

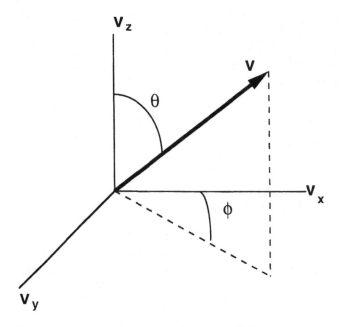

Figure 9   Spherical polar coordinates in velocity space.

where the volume element in velocity space, $dv_x \, dv_y \, dv_z$, which we denote dV, is

$$dV = dv_x dv_y dv_z = v^2 \sin\theta \, dv \, d\theta \, d\phi = v^2 \, dv \, d\Omega \tag{32}$$

where the differential solid angle $d\Omega$ is $d\Omega = \sin\theta \, d\theta \, d\phi$.

The concept of a solid angle merits some elaboration. An angle in two dimensions may be defined as the ratio of a length of circular arc to its radius. The solid angle is the corresponding concept in three dimensions. That is, a solid angle may be defined as the ratio of an area of a spherical surface to the square of its radius. The relationship $d\Omega = \sin\theta \, d\theta \, d\phi$ may be derived by the application of geometry to the above definition of solid angle.

In a system of N molecules, the fraction of molecules having speed (i.e., magnitude of velocity) between v and $v + dv$ and directional angles between θ and $\theta + d\theta$ and between ϕ and $\phi + d\phi$ is

$$\frac{dN}{N} = F_{v\theta\phi} \, dv \, d\theta \, d\phi \tag{33}$$

where $F_{v\theta\phi}$ is the distribution function in polar coordinates,

$$F_{v\theta\phi} = \left(\frac{m}{2\pi kT}\right)^{3/2} \exp\left[-\frac{mv^2}{2kT}\right] v^2 \sin\theta \tag{34}$$

We are frequently interested in the fraction of molecules, dN/N, which have speed v in a given interval, regardless of the direction of the velocity vector. The fraction of molecules having speed between v and $v + dv$ is evaluated by integration of Eq. (33) over all angles θ and ϕ, using Eq. (34) for $F_{v\theta\phi}$. Since the integral of $\sin\theta$ over all θ and ϕ is 4π, we have the result

$$\frac{dN}{N} = f_v\, dv = 4\pi \left(\frac{m}{2\pi kT}\right)^{3/2} \exp\left[-\frac{mv^2}{2kT}\right] v^2\, dv \tag{35}$$

where f_v is the distribution function for speed v, and therefore $f_v\, dv$ is the probability that a molecule will have speed in the interval between v and $v + dv$. The distribution function for speed is normalized in the sense that the integral of $f_v\, dv$ over all values of speed (i.e., zero to infinity) is unity.

Since the kinetic energy, $mv^2/2$, depends only upon the speed for a given type of molecule, we may use Eq. (35) to develop the distribution function for translational energy, E, which we denote f_E. Thus, the fraction of molecules with speed between v and $v + dv$ is also the fraction of molecules with translational energy between E and $E + dE$; i.e., $dN/N = f_E\, dE = f_v\, dv$. Making the substitutions $E = mv^2/2$ and $dv = dE/mv$ then gives

$$\frac{dN}{N} = f_E\, dE = \frac{2}{\sqrt{\pi}} \left(\frac{1}{kT}\right) \left(\frac{E}{kT}\right)^{1/2} \exp\left[-\frac{E}{kT}\right] dE \tag{36}$$

C. Calculation of Average Quantities for a Gas

As the speed v approaches zero, the distribution function for speed, f_v, will also approach zero due to the term v^2. The distribution function will also approach zero at very high values of v, since the product $v^2 \exp(-av^2)$ approaches zero as v approaches infinity (for all finite positive values of the constant a). Thus, a plot of f_v as a function of v will pass through a maximum. The condition for the maximum is $df_v/dv = 0$. The value of speed at this maximum is called the most probable speed and is denoted v_p. An expression for most probable speed may be derived by finding the value of v such that the derivative is zero, i.e., $df_v/dv = 0$ at $v = v_p$. The result is $v_p = (2kT/m)^{1/2}$.

Figure 10 shows the speed distribution function plotted as a function of "normalized" speed, v/v_p. Of course, the maximum in f_v occurs at $v/v_p = 1$. Note that the distribution function, f_v, is slightly asymmetrical in the sense that the area under the curve, or "integrated probability," is slightly greater for $v/v_p > 1$ than for $v/v_p < 1$. Thus, the average, or mean, speed will be slightly greater than the most probable speed. The average speed, v_a, may be evaluated by an application of the mean value theorem given in Eq. (25),

$$v_a = \int_0^\infty v f_v \, dv = 4\pi \left(\frac{m}{2\pi kT}\right)^{3/2} \int_0^\infty \exp\left[-\frac{mv^2}{2kT}\right] v^3 \, dv \qquad (37)$$

Note that since the mean value theorem involves an integration over all allowed values of the independent variable, the integration relevant to average speed is from zero to infinity. In terms of the mean value theorem stated in Eq. (25), the probability of a negative speed is zero, so the integral in Eq. (37) is only over positive values of v. The integral in Eq. (37) may be evaluated by making the

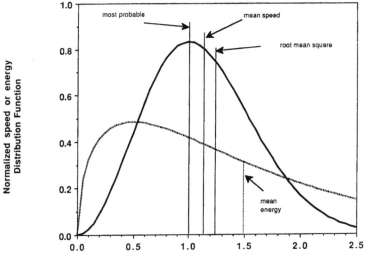

Figure 10   A graphical illustration of the Maxwell–Boltzmann distribution laws. Normalized speed——is v/v_p, and normalized energy $\cdots$ is E/kT.

substitution $z = v^2$ and then integrating by parts to give Integral $= 2(kT/m)^2$. Thus, the average velocity is found to be

$$v_a = \left(\frac{8kT}{\pi m}\right)^{1/2} \tag{38}$$

The mean value theorem may also be used to calculate the mean kinetic energy, $\langle E \rangle$, using the translational energy distribution function, f_E. Consistent with our assumptions (from equipartition of energy), the result is $\langle E \rangle = (3/2)kT$. The energy distribution function is plotted in Figure 10 as a function of the normalized energy, E/kT. Note that the mean kinetic energy $(3/2)kT$ corresponds to a normalized energy of 1.5 in Figure 10. The most probable energy, which corresponds to the energy at the maximum of the distribution function (i.e., the solution of $df_E/dE = 0$), occurs at a normalized energy of 1/2.

D. Rate of Molecules Striking a Surface: Equilibrium Gas

1. The Intensity Equation

A "Maxwellian stream" of molecules are those molecules which strike a wall of a box containing the gas sample at equilibrium or those which strike one side of a flat surface immersed in the equilibrium gas. Alternatively, the Maxwellian stream is those molecules which cross an imaginary plane in the gas sample. The intensity of the Maxwellian stream, or number of collisions with the surface per second per square centimeter, is a property which is of great use in examining transport processes in gases. The starting point for the derivation of this intensity is the slanted cylinder containing gas molecules, which is shown in Figure 11. The element of surface or wall area dA is the surface of interest and forms one end of the cylinder. The length of the cylinder is $v\,dt$, where v is the speed of the molecules and dt is a time interval. The other end of the cylinder is the projected area at solid angle Ω (i.e., direction specified by angles θ and ϕ) normal to the length. This projected area is the area of the shadow cast by area dA on a surface normal to the length vector if a flashlight is held behind area dA at direction Ω. The projected area is $dA \cos \theta$. All molecules in the slanted cylinder that are directed at dA (i.e., moving with solid angle between Ω and $\Omega + d\Omega$) will strike area dA in the time interval dt if the molecules are moving at a speed between v and $v + dv$. The "intensity" of molecules of speed v and direction Ω striking the surface is then the product of the total number of molecules in the box times the fraction moving in direction Ω times the fraction moving at speed v.

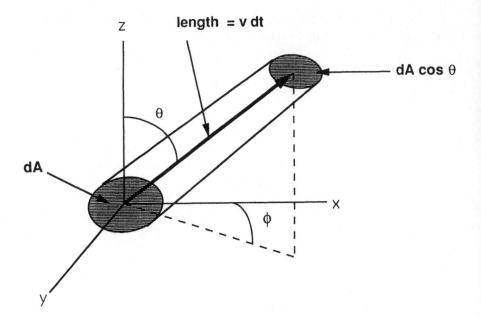

Figure 11 Slanted cylinder containing molecules that will strike a surface.

The volume of the cylinder is the product of the length and the projected area, i.e., Volume = $(v\,dt)(dA\cos\theta)$. If ρ' is the number density of molecules in the gas, N/V, then the total number of molecules in the cylinder is

Total number = $\rho'v\,dt(dA\cos\theta)$

The fraction of molecules moving in a direction between Ω and $\Omega + d\Omega$ is

$$\frac{d\Omega}{\displaystyle\int_0^{2\pi} d\phi \int_0^{\pi} \sin\theta\,d\theta} = \frac{d\Omega}{4\pi}$$

and the fraction of molecules moving at a speed between v and $v + dv$ is $f_v dv$, where f_v is the speed distribution function [i.e., Eq (35)]. Thus, the number of molecules in the cylinder that will hit the surface in dt seconds, dN, is given by

$$dN = (\rho'v\,dt\,dA\cos\theta)\left(\frac{d\Omega}{4\pi}\right)(f_v dv) \tag{39}$$

The corresponding intensity, or number of collisions per second per square centimeter, is $(1/dA)(dN/dt)$, where this intensity is really a "differential intensity," since it represents only those collisions resulting from the molecules in that particular slanted cylinder, i.e., those molecules with direction between Ω and $\Omega + d\Omega$ and speed between v and $v + dv$. We represent this differential intensity by the notation $dI(\Omega, v)$ and write

$$dI(\Omega,\ v) = (\rho' v f_v\ dv)(\cos\ \theta)\ \frac{d\Omega}{4\pi} \tag{40}$$

We are interested in the total number of collisions from *all* molecules above the surface dA. That is, we seek an expression for the intensity resulting from collisions between the surface and all molecules in the hemisphere above surface dA. This total intensity, denoted I, is given by the integral of Eq. (40) over all values of ϕ and v and those values of θ corresponding to the hemisphere above the surface, i.e., $0 \le \theta \le \pi/2$. Thus,

$$I = \int_\Omega \int_v dI(\Omega,\ v) = \frac{\rho'}{4\pi} \int_0^{2\pi} d\phi \int_0^{\pi/2} \cos\ \theta\ \sin\ \theta\ d\theta \int_0^\infty v f_v\ dv \tag{41}$$

The integral over ϕ is simply 2π, the integral over θ is $1/2$, and the integral over speed is the average speed, v_a. Equation (41) then becomes

$$I = \rho' \frac{v_a}{4} \tag{42}$$

where

$$\rho' = \frac{P}{kT} \quad (P = \text{pressure})$$

and

$$v_a = \left(\frac{8kT}{\pi m}\right)^{1/2}$$

Substituting for number density in terms of pressure and expressing mean speed in terms of absolute temperature and molecular mass m then gives the desired final result for total intensity, or number of molecules in an equilibrium gas striking a surface of unit surface area per unit time,

$$I = \text{number of collisions sec}^{-1}\text{cm}^{-2} = \frac{P}{(2\pi mkT)^{1/2}} \tag{43}$$

The proportionality between intensity and pressure is intuitively obvious, since at constant speed the number of collisions must be proportional to the number of molecules in the container and therefore proportional to the pressure. The inverse square root temperature dependence is a result of a partial cancellation between the temperature dependence of the molecular speed increasing the collision intensity in proportion to the square root of absolute temperature and the temperature dependence of the number density decreasing the intensity by an inverse proportion.

2. The Knudsen Cosine Law

In the previous derivation, the differential intensity at solid angle Ω and speed v was denoted $dI(\Omega, v)$. The *total* intensity from all molecules in the hemisphere above the surface of interest that have a given speed v is given by the integral of differential intensity, $dI(\Omega, v)$, over all solid angles in the hemisphere. Thus, the fraction of collisions with the surface coming from solid angle Ω, denoted $P(\Omega)\, d\Omega$, is the ratio of the differential intensity to the integrated intensity,

$$P(\Omega)\, d\Omega = \frac{dI(\Omega, v)}{\displaystyle\int_{\Omega} dI(\Omega, v)} = \frac{(\rho' v/4\pi)f_v\, dv \cos\theta\, d\theta\, d\Omega}{\displaystyle\int_0^{2\pi} d\phi \int_0^{\pi/2} (\rho' v/4\pi)f_v\, dv \cos\theta \sin\theta\, d\theta} \tag{44}$$

$$= \frac{\cos\theta\, d\Omega}{\displaystyle\int_0^{2\pi} d\phi \int_0^{\pi/2} \sin\theta \cos\theta\, d\theta}$$

Performing the indicated integration then leads to the probability of incident molecules striking the wall coming from angle Ω, which is $P(\Omega)\, d\Omega$, in the form of a simple cosine function,

$$P(\Omega)\, d\Omega = \frac{\cos\theta}{\pi}\, d\Omega \tag{45}$$

For a gas at equilibrium, the angular dependence of the incident molecular intensity and the angular dependence of the scattered (or rebounding) molecules must be the same. This is illustrated schematically in Figure 12, where the length of an arrow indicates the intensity at the indicated incident or scattering angle θ. If scattered and incident molecules did not have the same speed, this would imply an exchange of energy with the surface, which is the same as concluding that there is heat flow. However, in a system at equilibrium, there are no temperature gradients and no flow of heat. If scattered and incident molecules had different distributions $P(\Omega)\, d\Omega$, it would imply that the mean velocity vector of all the molecules, initially zero for the equilibrium incident gas, would be transformed into a velocity vector which does not add to zero. This result is equivalent to saying that there would be a net flow of gas molecules. However, in an equilibrium gas, there are no pressure gradients and no net flow of molecules.

For *nonequilibrium* conditions, there are temperature gradients producing heat flow in the gas and/or there exist pressure gradients which induce gas flow past the surface. Here, we cannot use the arguments given above for the equilib-

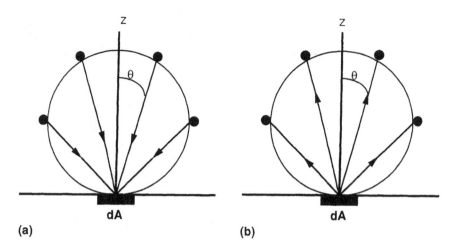

Figure 12 An illustration of the cosine law distribution for frequency of (a) incident and (b) scattered molecules. The length of an arrow is proportional to the frequency of molecules incident or scattered at the angle indicated.

rium case, and therefore we cannot conclude that the molecules scattering from a surface must follow the cosine law given by Eq. (45). However, Knudsen postulated that the scattered molecules do indeed follow the cosine law given by Eq. (45) *even if the gas is not at equilibrium*. This postulate is referred to as the Knudsen cosine law. Knudsen and others demonstrated that in many nonequilibrium systems, cosine scattering indeed does occur. However, it has also been shown that deviations from Knudsen's cosine law do occur, usually with very clean smooth surfaces. Cosine law scattering is contrasted with the opposite extreme, specular reflection, in Figure 13. While cosine law scattering does not imply random emission of molecules from a surface (i.e., scattering follows the cosine law), there is no net momentum tangential to the wall or surface. Cosine law scattering usually means that there is a complete equilibration of energy and momentum between the incident gas molecule and the surface before the molecule is emitted or scattered from the surface. With specular reflection (Fig. 13), the gas molecule does not exchange either energy or momentum with the surface. The collision is perfectly elastic, and tangential momentum of the incident molecule is preserved after emission (i.e., incoming angle θ and outbound angle θ in Fig. 13 are identical).

E. Mean Free Path

Gas molecules travel a distance far greater than their molecular diameter before encountering another gas molecule and suffering a collision. The distance trav-

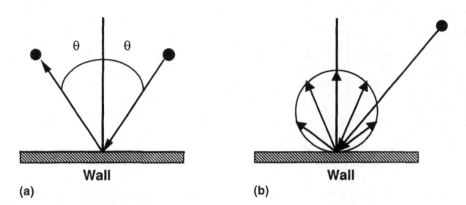

Figure 13 The contrast between (a) specular reflection and (b) cosine law scattering.

eled between collisions is termed the "free path". The mean free path is roughly the average value of the free path. More precisely, the mean free path, L, is defined by stating that the fraction of molecules traveling a distance x without a collision is given by $\exp(-x/L)$. In other words, on average, if a molecule travels a distance equal to the mean free path, i.e., $x = L$, the fraction that have *not* collided is $1/e$ ($\approx 37\%$).

The chance of a collision will obviously depend upon the number of gas molecules per unit volume or, alternatively, upon the pressure. The chance of a collision will also depend upon the size of the gas molecules. For example, the chance of two basketballs thrown toward one another undergoing a collision is much greater than the chance of having a similar collision between two golf balls. An expression for the mean free path in terms of pressure and molecular diameter may be derived from kinetic theory. We give only the result, which may be expressed as

$$L = \frac{1}{\sqrt{2}\pi\rho'd^2} \tag{46}$$

where ρ' is the number of molecules per cubic centimeter. Thus, the mean free path is in inverse proportion to the number of molecules per unit volume and is also inversely proportional to the square of the molecular diameter, or "collision diameter," d. For a given number density ρ', there is no temperature dependence in the mean free path. We may convert Eq. (46) into an expression involving pressure by expressing ρ' in terms of the pressure, $\rho' = P/kT$. Thus, Eq. (46) becomes

$$L = \frac{kT}{\sqrt{2}\pi \, Pd^2} \tag{47}$$

At 25°C, L in centimeters may be expressed as L (cm) $= 0.0696/P_{torr}d_0^2$, where the pressure is given in torr (P_{torr}) and the collision diameter d_0 is given in Angstrom units. Since collision diameters for gas molecules of interest to freeze drying are ≈ 4 Å, L in micrometers is L (μm) $\approx 44/P_{torr}$. Thus, at pressures of interest to freeze drying (i.e., in the range of 0.02–2.0 torr), the mean free path is generally larger than relevant microscopic dimensions such as "pore size" in a freeze-dried product. However, the mean free path is small compared to macroscopic dimensions, such as typical dimensions of vacuum lines in a freeze dryer, even with pressures at the lower end of interest to freeze drying (i.e., $L \approx 2$ mm at $P = 0.02$ torr). Only at extremely low pressures (i.e., $<10^{-4}$torr) does the mean free path become on the order of tens of centimeters.

F. Flow Through an Orifice: Knudsen Effusion

Mass transport of gas molecules under conditions where the pressure is low enough that collisions between molecules are infrequent and may be ignored in the context of the problem is referred to as "free molecular flow" or "Knudsen flow." Here, the mean free path must be large compared to some critical dimension of the container system. The simplest example is perhaps flow through a hole, or orifice, in a container. Such flow is illustrated by Figure 14 for two gas samples, sample 1 and sample 2, separated by a plate of thickness x containing a hole of radius a. To qualify as an orifice, the radius of the hole must be large compared to the thickness of the plate. This qualification ensures that during passage of a gas molecule through the hole, collisions between molecules and

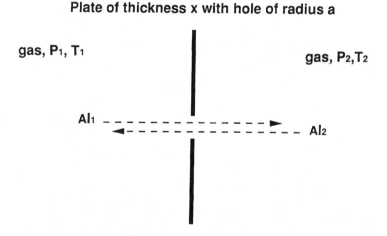

Figure 14 Free molecular flow through an orifice: Knudsen effusion.

the "wall" of the hole are infrequent enough to be ignored. At sufficiently low pressure, flow through the hole is given by the number of collisions of gas molecules with a surface of the same area as the hole, $A = \pi a^2$. Thus, the pressure must be low enough that no molecule–molecule collisions occur during passage through the hole, which implies that the mean free path must be significantly greater than the thickness of the plate. Such flow is referred to as "Knudsen effusion."

The flow of gas molecules striking the hole area is simply the product of the hole area A and the intensity of the Maxwellian stream given by Eq. (43). However, since the gas samples are characterized by different pressures, P_1 and P_2, as well as by different temperatures, T_1 and T_2, the flow of gas sample 1 is not equal to the flow of gas sample 2, and there exists a net flow through the hole,

$$\text{Net flow} = \frac{dN}{dt} = AI_{net} = A(I_1 - I_2) \tag{48}$$

If the gas samples are at the same temperature, substitution of Eq. (43) for the intensity of collisions gives

$$\frac{dN}{dt} = A\frac{P_1 - P_2}{(2\pi mkT)^{1/2}} \tag{49}$$

In many applications, the mass flow rate, G (g/sec), is desired. Using the relationship $G = m\, dN/dt$ to substitute for dN/dt in Eq. (49) then gives

$$G = A\left(\frac{m}{(2\pi kT)}\right)^{1/2} (P_1 - P_2) \tag{50}$$

Thus, in an isothermal system, the mass flow rate depends on the difference in pressures of the gas across the orifice and does not depend upon the thickness of the plate. One may define an area-normalized resistance, R, for mass transfer through the orifice using a generalization of Ohm's law, i.e., Resistance = force/flux. For Knudsen flow, the "force" is the pressure difference (analogous to voltage difference in Ohm's law) and the flux is the mass flow per unit area of the hole (analogous to the electrical current density in Ohm's law). Thus, we have

$$\text{Resistance} = \hat{R} = \frac{P_1 - P_2}{G/A} = \left(\frac{2\pi kT}{m}\right)^{1/2} \tag{51}$$

Knudsen effusion will be involved later as we discuss free molecular flow in channels and tubes. Knudsen effusion also finds application in the measurement of the vapor pressure of materials of low vapor pressure, typically in the range of 10^{-3}–10^{-8} torr. Here, a weighed sample of a solid or liquid is placed in a sealed container in which a small hole is drilled. The sample and container are then placed in a temperature-controlled vacuum chamber capable of maintaining a pressure sufficiently low to guarantee free molecular flow. Effusion is allowed to proceed for a time, and the sample and container are reweighed to measure the mass of sample effused. The mass transfer rate, G, is evaluated from the mass loss and the time of effusion, and the vapor pressure is calculated from Eq. (50), where P_1 is the equilibrium vapor pressure of the sample at the measurement temperature. Usually the pumping system is sufficient to ensure that $P_2 = 0$. Of course, it is assumed that the rate-limiting process for removal of material from the effusion apparatus is effusion through the orifice and does not involve the sublimation or evaporation process itself. If the rate of effusion is modest, this assumption is normally valid, and therefore the partial pressure of the compound inside the vessel is the equilibrium vapor pressure, as desired. An alternative and more sophisticated system employs a high vacuum recording microbalance, where mass is measured and recorded as a function of time during the effusion experiment. The slope of the mass vs. time plot gives the value of G. Usually sufficient data are evaluated in only a few hours to obtain G at a given temperature, and the temperature of the sample is changed to obtain G and vapor pressure at the new temperature setting. Thus, a complete set of vapor pressure data as a function of temperature may be generated on the same sample in a relatively short time with a minimum of effort. The technique is limited to roughly the vapor pressure range mentioned above (10^{-3}–10^{-8} torr). At much lower vapor pressures G values ultimately become too small to be measured, and at higher vapor pressures the conditions of free molecular flow are eventually violated.

G. Rates of Sublimation and Evaporation

Kinetic theory may be used as a starting point for understanding the factors which limit the rate of evaporation and the rate of sublimation. We will discuss sublimation, but nearly identical results apply for evaporation of a liquid. During sublimation a molecule must first pass from its position at the surface layer of the solid lattice into the vapor state. Next, the molecules in the vapor adjacent to the solid surface must move away from the interface and ultimately into "free space" or be deposited on the cold surface of a condenser. Kinetic theory may be used to address the second step with the aid of the diagram in Figure 15, where it is postulated that there exists a finite vapor region adjacent to the solid where the partial pressure of the component of interest is equal to the vapor pressure. In short, it is postulated that equilibrium exists between the solid and the vapor, at least in the surface region, and the molecules that leave the "saturated vapor" region in the direction of flow in Figure 15 are immediately replaced by molecules from the surface of the solid. The next stage of the analysis is essentially the same

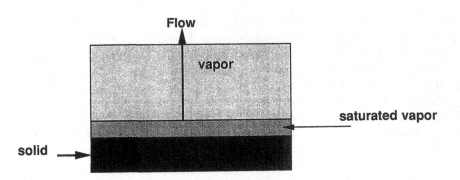

Figure 15 A schematic of the model for evaluating sublimation rate and evaporation rate from kinetic theory.

as that given when discussing flow through an orifice in the previous section. In brief, the flow of molecules out of the saturated vapor region is simply the product of the area of the surface region, A, and the intensity of the Maxwellian stream given by Eq. (43), with pressure being the equilibrium vapor pressure P_0. Flow from the region labeled "vapor" in Figure 15 into the saturated vapor region is also given by the product of area and intensity of a Maxwellian stream. In this case, the relevant pressure is the partial pressure of the component of interest in the vapor region, which we denote as P'. The net flow out of the saturated vapor region, given by the difference between flow out and flow in, is the sublimation rate. The resulting kinetic theory expression for sublimation rate is identical to the expression given by Eq. (50) for orifice flow if P_1 is replaced by the vapor pressure P_0 and P_2 is replaced by the partial pressure of the component of interest in the vapor region, P'. However, when this expression, denoted G_{theory}, is compared to the experimental sublimation rate, G_{exp}, it is found that G_{exp} is always less than G_{theory} when the sublimation rate is reasonably fast. Thus, it appears that under most experimental conditions the rate-limiting step is not the movement of molecules away from the surface through the vapor state, which is what G_{theory} calculates. Rather, the rate-limiting step is the phase change itself. It is customary to define the evaporation coefficient, α_v, by the relationship $\alpha_v = G_{\text{exp}}/G_{\text{theory}}$. Thus, the modified theoretical expression for the sublimation rate becomes

$$G = \alpha(T)A \left(\frac{M}{2\pi RT} \right)^{1/2} (P_0 - P') \tag{52}$$

where M is the molecular weight and R is the gas constant. In generating Eq. (52), substitution of $M = N_0 m$ and $R = N_0 k$ have been used, where N_0 is Avogadro's number. The notation used in Eq. (52) for the evaporation coefficient,

$\alpha_v(T)$, emphasizes the fact that the evaporation coefficient is extremely sensitive to temperature. The evaporation coefficient for ice as a function of temperature is given by Figure 16. Note that as the temperature decreases to very low values and the sublimation rate slows greatly, the evaporation coefficient approaches unity. At temperatures commonly used in freeze drying, the evaporation coefficient is significantly less than unity. For example, at an ice temperature of $-25°C$, $\alpha_v \approx 0.24$, and the sublimation rate is about 4.8 g/(hr · cm²), which is fast enough to sublime a 1 cm column of ice in only 11 min. Since even a rapidly freeze-

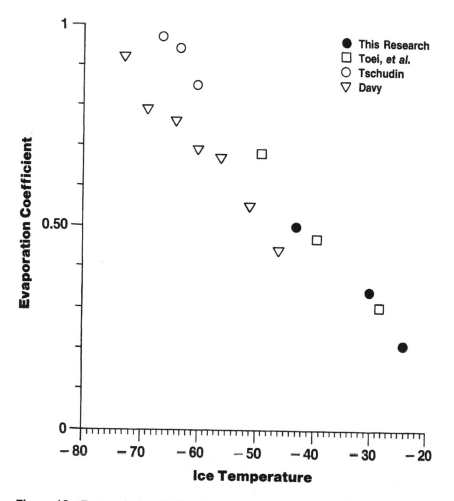

Figure 16 Evaporation coefficients for ice. (From Ref. 4. The references in the legend are identified in Ref. 4.)

drying pharmaceutical product requires on the order of hours to freeze dry, it is obvious that there is more to mass transfer than the sublimation of ice. Interaction of water molecules in the gas phase with each other and with solid surfaces (i.e., the dried product) determines mass transfer.

H.   Viscosity of Gases and Slip Flow

In a flowing gas, it is the interaction of gas molecules with each other and with surfaces of the system enclosing the gas that determines mass transfer. The interaction we speak of is not an "attractive force" between molecules, as this interaction persists even for an ideal gas. Rather, the interaction is an exchange of momentum between molecules and ultimately an exchange of momentum between the gas and the surface of the container system. Exchange of momentum between molecules moving with different velocities is responsible for viscosity and the resistance to flow of a fluid system whether the fluid is a liquid or a gas. The concept of viscosity is illustrated by Figure 17, where two planes containing gas molecules are depicted. The planes (and the gas molecules contained therein) are one mean free path, L, apart and differ in drift velocity. The term "drift velocity" is used to denote the flow velocity, or average velocity of the group of gas mole-

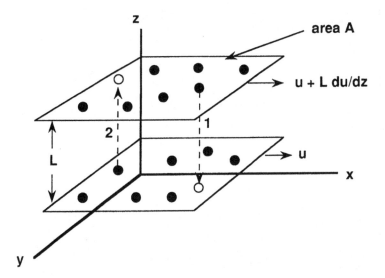

Figure 17   Kinetic theory and viscosity: A schematic illustrating momentum exchange in a flowing gas.

cules with the rapid random thermal motion averaged to zero. The bottom plane moves with average drift speed u, and the faster moving top plane moves with drift speed $u + \Delta u$, where $\Delta u = L \, du/dz$. Here, du/dz is the gradient of the drift velocity in the direction normal to the planes of molecules.

The arrows indicate molecules 1 and 2, which during random thermal motion jump the distance of one mean free path between the planes, thereby exchanging positions and producing a momentum exchange between planes. The separation between planes of one mean free path is selected to ensure that, on average, there will be no collision *during* the jump. The "fast" molecule moving from the top plane to the bottom plane brings its excess momentum to the slower moving plane, tending to speed up the slower plane. By contrast, the jump of molecule 1 from the slow plane to the fast plane brings a deficiency of momentum to the plane and tends to slow motion of the upper, faster moving plane. The net result is that molecular jumping tends to resist the relative motion between the planes. This resistance, or drag force, is what is commonly referred to as a viscous force. That is, a change in momentum per unit time is a force, and the time rate of momentum exchange between the two planes produces a viscous drag force between the two planes. Each jump produces a momentum exchange of $\Delta(mu)$ $= m\Delta u = mL \, du/dz$. The total number of exchanges per second is the sum of the Maxwellian stream flows going "up" and going "down." Thus, Total number of exchanges/sec $= A \, (I_1 + I_2)$, where A is the area of the plane, I_1 is the intensity of molecules jumping up like molecule 1, and I_2 is the intensity of molecules like molecule 2. Thus, the total number of exchanges per second is given by Total number $= (A/2) \, \rho' \, v_a$, where ρ' is the number density of the gas and v_a is the average molecular speed. Since

$$\text{Force} = (\text{no. of exchanges/per sec}) \, (\text{momentum change per exchange})$$

we have

$$\text{Force} = \frac{A}{2} \rho' v_a mL \frac{du}{dz} \tag{53}$$

However, the coefficient of viscosity, μ, is defined by

$$\text{Force} = A\mu \frac{du}{dz} \tag{54}$$

Comparison of Eqs. (53) and (54) then yields an expression for the viscosity,

$$\mu = \frac{1}{2} \rho' v_a mL \tag{55}$$

However, the mean free path, L, may be expressed in terms of the number density of gas molecules by $L = (\sqrt{2}\pi\rho' \, d^2)^{-1}$. Substituting this expression for mean free path in Eq. (55) then cancels the number density, and we may write

$$\mu = \frac{mv_a}{2\sqrt{2}\pi d^2} \tag{56}$$

Equation (56) clearly shows that the viscosity of a gas is independent of gas pressure. The physical reason for this result is that while lower pressure means fewer jumps between layers, each jump carries more momentum because at lower pressure the mean free path is longer, i.e., $\Delta mu = mL \, du/dz$.

While the derivation given above for viscosity is far from rigorous, the basic physics is correct, and the result is in surprisingly close agreement with both the results of more sophisticated theories and experimental data. The magnitude of the calculated viscosity is correct, and the predicted lack of variation with pressure is in accord with experimental observations at moderate pressure. However, at very low pressure, the viscosity begins to decrease as pressure decreases, behaving as if the gas molecules slip at the surface of the tube or channel. That is, contrary to the usual hydrodynamic concept that the molecules of the fluid stick at the surface, the tangential velocity of molecules in a low pressure gas appears *not* to be zero at the surface.

Figure 18 illustrates the difference between normal hydrodynamic flow and slip flow when a gas sample is confined between two surfaces in motion relative to each other. In each case, the top surface moves with speed u_0 relative to the bottom surface. The circles represent gas molecules, and the length of an arrow is proportional to the drift velocity for that molecule. The drift velocity variation with distance is illustrated by the plots on the right. When the ratio of the mean free path to the separation distance between surfaces is much less than unity (Fig. 18a), collisions between gas molecules are much more frequent than collisions of the gas molecules with the surfaces. Here, we have classical fluid flow or viscous flow. If the flow were flow in tubes, Poiseuille's law would be obeyed. The velocity of gas molecules at the surface is the same as the velocity of the surface, and in the case of the stationary surface the mean tangential velocity of the gas at the surface is zero.

By contrast, in the case where the mean free path is no longer small compared to the distance between the surfaces (Fig. 18b), the gas molecules no longer behave as a fluid. That is, the molecules at the surface do not move with the same velocity as the surface, and for the stationary surface the finite arrow shown for the molecule at the surface illustrates a nonzero velocity for a surface molecule. The plot of drift velocity as a function of distance likewise shows a nonzero velocity at the stationary surface and a velocity at the surface in motion which is less than the velocity of the surface.

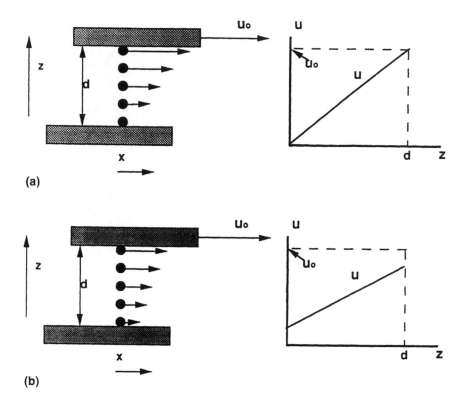

Figure 18 The contrast between (a) normal viscous flow in a fluid and (b) slip flow in a dilute gas. (a) $L/d \ll 1$; (b) $L/d > \approx 0.1$. u = mean tangential speed in the x direction. In (b), u is not zero at the stationary surface.

We now develop an expression for viscosity that reflects the slip phenomenon. First, we note that the direct proportion between u and z shown in Figure 18a is a result of collisions between molecules transferring momentum. That is, with frequent gas–gas collisions to distribute momentum, the variation of momentum with distance is smooth over the distances of the same general magnitude as the separation distance. The physical origin of slip is the failure of gas molecules to collide with each other as they approach the surface from a distance on the order of the mean free path from the surface. Thus, if a molecule carries a mean tangential velocity u_i at a position $z \approx L$, there will be no collisions to alter that velocity before the molecule hits the surface. Thus, the velocity will remain u_i right up to the surface and will, in general, be scattered from the surface at a velocity different from u_i (Fig. 19). This produces a discontinuity in drift velocity (and tangential momentum) at the surface. Of course, if the mean free path is

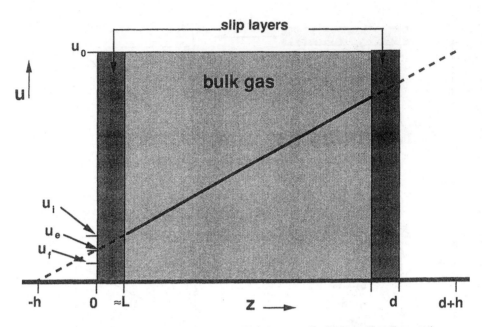

Figure 19   Velocity distribution and slip flow in a gas in the transition flow region.

tiny compared to the dimensions of the apparatus (i.e., the distance d), the resulting slip will occur over such a small distance that the resulting discontinuity in momentum will be too small to be significant. However, as the pressure decreases, the mean free path becomes a significant fraction of the separation distance, and the resulting discontinuity in momentum at the surface produces a significant effect on momentum transport, or viscosity.

In Figure 19, the incident mean tangential velocity is u_i, the mean tangential velocity of the molecules scattered from the surface is u_f, and u_e is simply the mean of the incident and scattered tangential velocities, $u_e = (u_i + u_f)/2$. If Knudsen's cosine law is obeyed, the mean tangential scattered velocity relative to the surface is zero, $u_f = 0$. The solid straight line gives the value of tangential velocity in the gas sample as a function of the distance variable z. The extrapolated velocity is indicated by the dashed line. The parameter h is the slip distance defined by Figure 19 as the z value at which the extrapolated velocity is equal to the velocity of the surface. For the surface defined by $z = 0$, the value of u reaches zero at $z = -h$, and for the surface defined by $z = d$, the value of u reaches u_0 at distance $z = d + h$. Physical intuition indicates that the slip distance h should be on the order of the mean free path, L. We now derive an expression for the slip distance, h.

The defining equation for viscous force, Eq. (54), applied to Figure 19 gives a relationship containing the slip distance,

$$\text{Force} = A\mu \left(\frac{du}{dz}\right)_{z=0} = A\mu \left(\frac{u_0}{d + 2h}\right) \tag{57}$$

where the expression in parentheses in the right-hand expression is simply the slope of the straight line in Figure 19. However, from the diagram (Fig. 19), we also may write the slope of the line as

$$\left(\frac{du}{dz}\right)_{z=0} = \frac{u_e}{h} = \frac{u_i + u_f}{2h} \tag{58}$$

The average incident tangential momentum is mu_i, while the average scattered tangential momentum is mu_f. If the gas molecule equilibrates with the surface and the scattered momentum is zero, we have Knudsen cosine scattering and complete accommodation of the incident gas molecule with the surface. On the other extreme, if specular reflection occurs, the incident momentum is retained upon scattering and $mu_i = mu_f$. The momentum accommodation coefficient, f, is introduced to describe the type of scattering that does occur, and it is defined by

$$f \equiv \frac{mu_i - mu_f}{mu_i} = 1 - \frac{u_f}{u_i} \tag{59}$$

Thus, for Knudsen cosine scattering, $f = 1$, and for specular reflection, $f = 0$. Equation (59) may be solved for the drift velocity of the scattered molecule to give $u_f = (1 - f)u_i$. The viscous force transmitted to the wall during gas collisions is the product of the number of collisions per second and the momentum change per collision,

$$\text{Force} = AIm(u_i - u_f) = A\frac{\rho' v_a}{4'} m(u_i - u_f) \tag{60}$$

where I is the intensity of the Maxwellian stream of gas molecules as given by Eq. (42). If we use the defining equation for viscous force, Force = $A\mu(du/dz)_{z=0}$, substitute Eq. (58) for the velocity gradient, and use the kinetic theory expression [Eq. (55)] for viscosity, the force may also be written in the form

$$\text{Force} = A\frac{\rho' v_a mL}{2} \left(\frac{u_i + u_f}{2h}\right) \tag{61}$$

Setting the right-hand sides of Eqs. (60) and (61) equal to each other and using the relationship $u_f = (1 - f)u_i$ then leads to the desired relationship between slip distance, h, and mean free path, L,

$$h = \frac{2 - f}{f} L \tag{62}$$

Substitution for the slip distance using Eq. (62) gives the viscous force equation

$$\text{Force} = \frac{A\{\rho' v_a mL/2\} u_0}{d + 2L\dfrac{2 - f}{f}} = \frac{A\mu^h u_0}{d\left[1 + 2\dfrac{L}{d}\left(\dfrac{2 - f}{f}\right)\right]} \tag{63}$$

where the notation μ^h is used to denote the "high pressure" kinetic theory expression for viscosity given by the collection of terms in curly brackets, { }, in Eq. (63). The force obviously decreases at low pressure; specifically, it decreases as the ratio of the mean free path to separation distance, L/d, increases from essentially zero to on the order of unity, provided the momentum accommodation coefficient is on the order of unity.

From a phenomenological or experimental viewpoint (i.e., ignoring slip), viscosity in the context of Figures 18 and 19 would be defined by an expression of the form Force $= A\mu^{\text{eff}}(u_0/d)$, where the viscosity is written as μ^{eff} to denote an "effective viscosity," which is not necessarily equal to the simple kinetic theory expression given by Eq. (55). Of course, as we have just shown, the viscous force in a gas system at low pressure (i.e., high L/d) is pressure-dependent. Comparing the definition of effective viscosity given above with the result for slip flow, Eq. (63), we see that the effective viscosity is given by

$$\mu^{\text{eff}} = \frac{\mu^h}{1 + 2[(2 - f)/f](L/d)} \tag{64}$$

At high pressure, $L/d \ll 1$, and the effective viscosity becomes independent of pressure and equal to the kinetic theory expression developed earlier. At low pressure, L/d is no longer negligible compared to unity, the slip distance becomes a significant fraction of the separation between the surfaces, and the effective viscosity decreases.

The pressure dependence of effective viscosity obviously depends upon the value of the momentum accommodation coefficient. Momentum accommodation data are relatively rare, but some representative data are given in Table 1. Note that all values are relatively close to unity. Because of this observation, momentum accommodation coefficients are normally assumed to be unity in applications

Table 1 Momentum Accommodation Coefficients

Gas	Surface	Momentum accommodation coefficient f
He	Polished oxidized Ag	1.00
O_2	Polished oxidized Ag	0.99
Air	Glass	0.89
Air	Oil	0.90
Air	Brass	1.00
CO_2	Oil	0.92
CO_2	Brass	1.00

Source: Ref. 7.

where actual data are not available. Stated differently, Knudsen cosine scattering is assumed to be valid as a good first approximation.

I. Mechanisms of Mass Flow Through Tubes and Channels

1. General Principles

The molar flux of component 1, J_1, in units of moles per square centimeter per second, is defined by $J_1 = (1/A)\, dn_1/dt$, where dn_1/dt is the number of moles of gas molecules crossing a plane in space of area A. Molar flux is also given in terms of the velocity and concentration of component 1, $J_1 = c_1 u_1$, where c_1 is the molar concentration (mol/cm^3) and u_1 is the mean velocity of component 1 relative to a stationary point on the tube or channel. It is useful to classify the flow as either "diffusion" or "bulk flow." Diffusion is the flow of components under a gradient in composition, or mole fraction, and is essentially the mixing of gas components to achieve the maximum entropy. Diffusional flow of a given component is frequently measured relative to the center of mass of the gas system. However, in freeze-drying applications, the center of mass of the gas system is usually in motion under the influence of a gradient in total pressure.

Consider the flux of component 1 in a two-component gas system. Thus, the flux relative to the apparatus, tube, or channel may be expressed as

$$J_1 = J_1^{\text{Diff}} + J_1^{\text{Bulk}}$$

with

$$J_1^{\text{Diff}} = c_1(u_1 - u_m) = -cD_{12}\,\nabla X_1 \tag{65}$$

where D_{12} is the mutual diffusion coefficient of the two components 1 and 2, and X_1 is the mole fraction of component 1 in a mixture of total concentration c, where $c = c_1 + c_2$. The bulk flux is given by

$$J_1^{\text{Bulk}} = F_c\,\Delta P_1 \tag{66}$$

where ΔP_1 is the partial pressure difference for component 1 across the tube or channel. In most freeze-drying applications where mass transport in the gas phase is important, the gas phase is largely water vapor and gradients of mole fraction are small. Consequently, true diffusion, or flow under the influence of a mole fraction gradient, is of minor importance, and we will concentrate on mechanisms of bulk flow. Depending upon the ratio of the mean free path to the radius of the tube, the mechanism of bulk flow will vary, and therefore the detailed form of the conductivity coefficient, F_c, will vary. At "moderate" vacuum conditions, the mean free path, L, will be small compared to the radius, a, of the tube or channel, and the gas will behave as a fluid and obey hydrodynamics. This is the "viscous flow region," which prevails when $L/a < 0.01$. For air at 25°C, this inequality is equivalent to stating that we have viscous flow when $aP_{\text{mtorr}} > 500$, where a is in centimeters. For flow in tubes, Poiseuille's law will be obeyed just as for liquids flowing through capillary tubes. Here, the fluid velocity is zero at the tube wall and increases in parabolic fashion as the center of the tube is approached. The result of a hydrodynamic analysis for fully developed laminar flow yields the result for the conductivity coefficient, F_c,

$$F_c = \frac{a^2}{8\mu lRT}\,P_a \tag{67}$$

where P_a is the mean pressure in the tube of radius a and length l and μ is the viscosity of the gas.

As the pressure is lowered, slip occurs, and the flow mechanism is referred to as transition flow. At pressures so low that collisions between gas molecules are rare compared to the collisions between the gas and the tube wall, the flow is said to be Knudsen flow or free molecular flow. Free molecular flow prevails when $L/a > 1$. For air at 25°C, this condition means that we have free molecular flow when $aP_{\text{mtorr}} < 5$. We now consider an intuitive derivation of the result for F_c in the free molecular flow region.

2. Free Molecular Flow

A schematic showing free molecular flow in a tube is given by Figure 20. Here, the tube radius is a and the length is l, so the tube wall area is $2\pi al$. Contrary to the observation for flow of a fluid in a tube, the drift velocity, u, in the case of free molecular flow is independent of distance from the tube wall (i.e., the arrows indicating magnitude of velocity are drawn the same length). Alterna-

Tube of length, l and radius a

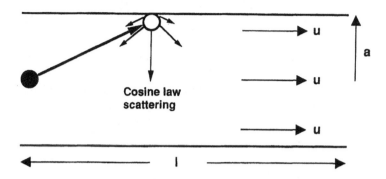

Figure 20   Mass transfer and momentum exchange during free molecular flow in a long tube.

tively, $du/dr = 0$, where r is the radial distance from the tube center. We assume cosine law scattering (i.e., $f = 0$), so while the mean incident velocity, u_i, is nonzero and equal to u, the scattered drift velocity, u_f, is zero. The momentum transferred to the wall in one collision is $\Delta mu = mu$. Thus, the force exerted by the gas on the wall (also equal to the force exerted by the wall on the gas), denoted F_{wg}, is the time rate of momentum change, $F_{wg} = d(mu)/dt$. However, this force is

$$F_{wg} = \text{(wall area) (no. collisions/(sec} \cdot \text{cm}^2))(\Delta mu \text{ per collision}) \qquad (68)$$

$$F_{wg} = 2\pi a l \frac{\rho' v_a}{4} mu$$

In steady state, the velocity of the gas is constant, and there is no net force on the gas. The force exerted on the gas by the wall, F_{wg}, is balanced by the volume force from the pressure difference. If the pressure on the high pressure end of the tube is P', and the pressure on the low pressure end is P'', the volume force is $(P' - P'')\pi a^2$. Setting the two balancing forces equal to each other then gives the equation

$$(P' - P'')\pi a^2 = 2\pi a l \frac{\rho' v_a}{4} mu \qquad (69)$$

The volume flux, denoted J_v, is the rate of volume change divided by the cross-sectional area of the tube and is identical to the drift velocity u,

$$\text{Volume flux} = J_v = \frac{1}{A_c} \frac{dV}{dt} = u \qquad (70)$$

The cross-sectional area is πa^2, and for an ideal gas,

$$\text{Ideal gas:} \quad \frac{dV}{dt} = \frac{RT}{P_a}\frac{dn}{dt} \tag{71}$$

where n is the number of moles of gas and the mean pressure, P_a, is given by $P_a = (P' + P'')/2$. Since the molar flux, J, is given by $J = (1/A_c)(dn/dt)$, the drift velocity may be written in terms of the molar flux, $u = (RT/P_a)J$. Using this relationship for drift velocity, substituting for number density with the equation $\rho' = P_a/kT$, and using the relationship developed earlier for mean velocity, $v_a = (8kT/\pi m)^{1/2}$, Eq. (69) may be rearranged to give

$$\text{Molar flux} = \frac{dn}{dt} = J = \left[\frac{\pi}{4}\left(\frac{a}{l}\right)\left(\frac{v_a}{RT}\right)\right](P' - P'') \tag{72}$$

where the grouping of terms in square brackets is the derived value of F_c in free molecular flow. This result is very close to the result obtained by Knudsen assuming a momentum accommodation coefficient of unity,

$$\text{Knudsen result:} \quad F_c = \frac{2}{3}\left(\frac{a}{l}\right)\left(\frac{v_a}{RT}\right) \tag{73}$$

differing only in the exact value of the numerical factor. Our derivation gives a numerical factor of $\pi/4$ ($= 0.78$), whereas the Knudsen result gives 0.67. Although there is some dispute over the correct numerical factor, most workers accept Knudsen's result. The main point to be made here is that it is the interaction of the gas molecules with the wall and loss of tangential momentum on collision that give rise to the resistance to flow in the free molecular flow region. Note that since the coefficient F_c is a conductivity coefficient, the corresponding resistance is given by the reciprocal of the conductivity, $R_c = 1/F_c$.

The results given above were developed for flow in long tubes. However, many applications involve tubes and channels that are relatively short, that is, where the ratio of the radius to the length is not essentially zero. The results for long tubes given by Eq. (72) and/or Eq. (73) may be modified such that the resulting expression is valid for flow in short tubes by considering a short tube as a composite of an orifice and a long tube. The total resistance for the short tube (i.e., long tube + orifice) is given by the sum of the resistance for an orifice and the resistance for a long tube. For molar flux, resistance of an orifice is given by

$$R_o = \frac{P' - P''}{G/MA} = (2\pi MRT)^{1/2}$$

where M is the molecular weight of the gas. Using the relationship $v_a = (8\,RT/\pi M)^{1/2}$, the resistance of a long tube is

$$R_{\text{long tube}} = \frac{3}{8}\left(\frac{l}{a}\right)(2\pi MRT)^{1/2}$$

The total resistance is then

$$R_{\text{total}} = (2\pi MRT)^{1/2}\left[1 + \left(\frac{3}{8}\right)\left(\frac{l}{a}\right)\right]$$

The conductance for free molecular flow in an orifice of nonzero thickness may be written

$$F_{\text{total}} = F_o\left[1 + \left(\frac{3}{8}\right)\left(\frac{l}{a}\right)\right]^{-1}, \qquad F_{\text{total}} = (2\pi MRT)^{-1/2} \qquad (74)$$

where F_o is the conductance for an orifice of zero thickness, and the term $[1 + (3/8)(l/a)]^{-1}$ is the correction for finite thickness of an orifice which is referred to as the Clausing factor. The expression in Eq. (74) is an excellent approximation for free molecular flow when l/a is either small (i.e., thin orifice) or large (i.e., long tube). When $a/l \approx 1$, the approximate expression is in error by about 10%. The conductance may also be written in a form which gives the correction for finite length of a long tube,

$$F_{\text{total}} = F_t\left[1 + \left(\frac{8}{3}\right)\left(\frac{a}{l}\right)\right]^{-1}, \qquad F_{\text{total}} = (2\pi MRT)^{-1/2} \qquad (75)$$

where F_t is the infinitely long tube expression for conductance during free molecular flow.

3. Transition Flow in a Cylindrical Tube

The velocity profile during slip flow in a cylindrical tube is shown in Figure 21. As in conventional fluid flow, the flow velocity in the z direction, $u(r)$, is parabolic, but rather than reach zero at the tube wall, slip occurs, and the velocity at the wall is greater than zero. The velocity does not reach zero until distance h from the wall surface. The derivation of the mass flux equation proceeds along the same lines as the derivation of Poiseuille's law in conventional hydrodynamics, but in slip flow, $u(r) = 0$ at $r = a + h$ instead of reaching zero at $r = a$.

Cylindrical Tube

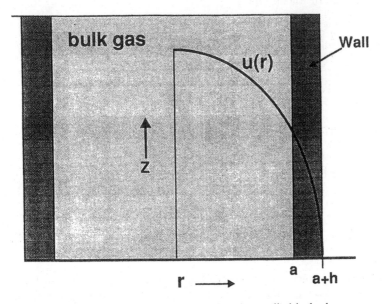

Figure 21 Velocity profile during slip flow in a cylindrical tube.

Momentum balance in a disk of thickness dr yields

$$\frac{d(r\tau_{rz})}{dr} = r\left(-\frac{dP}{dz}\right) \tag{76}$$

where

$$\tau_{rz} = \text{momentum flux} = -\mu\frac{du}{dr}$$

where μ is the high pressure value for viscosity. Note that the pressure gradient, $-dP/dz$, is a constant equal to $(P' - P'')/l$, which we symbolize as K. Integration of Eq. (76) then yields

$$r\tau_{rz} = \frac{r^2}{2}K + C_1 \tag{77}$$

where C_1 is an integration constant which must be zero since the momentum flux is finite at $r = 0$. Algebraic manipulation of Eq. (77) then gives

$$du = \frac{r}{2} \frac{K}{\mu} dr \tag{78}$$

Equation (78) may be integrated, subject to the boundary condition $u(a + h) = 0$, to give the result

$$u(r) = \frac{(a + h)^2 K}{4\mu} - \frac{r^2 K}{4\mu} \tag{79}$$

The average value of drift velocity, u_m, is given by integration of Eq. (79) over the radius of the tube according to the relevant mean value theorem,

$$u_m = \frac{\int_0^{2\pi} \int_0^a u(r)\, r\, dr\, d\theta}{\int_0^{2\pi} \int_0^a r\, dr\, d\theta} \tag{80}$$

The denominator in Eq. (80) is πa^2. The numerator is

$$\text{Numerator} = \frac{\pi}{8}\left(\frac{K}{\mu}\right) a^4 \left[1 + \frac{4h}{a} + 2\left(\frac{h}{a}\right)^2\right] \tag{81}$$

As discussed earlier, the mean drift velocity is the volume flux, J_v [see Eq. (70)]. Using the ideal gas equation to relate volume flow to molar flow [see Eq. (71)], the relationship between mean drift velocity and molar flux J may be written as $u_m = (RT/P_a)J$. With this expression for u_m, Eqs. (80) and (81) are combined to give the desired expression for molar flux,

$$J = \frac{Ka^2}{8\mu\, RT} P_a \left[1 + \frac{4h}{a} + 2\left(\frac{h}{a}\right)^2\right] \tag{82}$$

Since the conductivity coefficient, F_c, is defined by $J = F_c (P' - P'')$, the theoretical expression for F_c is obtained from this definition and Eq. (82) to yield

$$F_c = \frac{a^2}{8\mu RTl} P_a + \frac{a}{2RTl}\left(\frac{hP_a}{\mu}\right) \tag{83}$$

where the substitution, $K = (P' - P'')/l$ has been made. This derivation is self-consistent only when the slip distance h is small compared to the tube radius a. Thus, in Eq. (83) we retain only terms of first order in h/a. The first term in Eq. (83) is simply the Poiseuille's law expression for conductance in viscous flow, denoted F_v [see Eq. (67)]. As we now demonstrate, the second term is essentially the free molecular flow expression for conductance. We have shown that the slip distance is related to the mean free path, $h = (2 - f)L/f$, where L is the mean free path. Also, the viscosity is related to the mean free path by $\mu = \rho' v_a mL/2$. Finally, the number density is related to mean pressure by $\rho' = P_a/kT$, and the mean velocity is given by $v_a = (8kT/\pi m)^{1/2}$.

Making the above substitutions in Eq. (83), we find

$$F_c = \frac{a^2}{8\mu RTl}P_a + \frac{\pi}{8}\left(\frac{2-f}{f}\right)\left(\frac{a}{l}\right)\left(\frac{v_a}{RT}\right) \tag{84}$$

Knudsen's result for free molecular flow in a tube is given by Eq. (73). With $f = 1$, Knudsen's result and the second term in Eq. (84) differ only by a numerical factor. In Knudsen's result, the numerical factor is 2/3; in Eq. (84), the corresponding factor is $\pi/8$. Thus, except for a modest difference in the numerical factor, the slip term in Eq. (84) is the Knudsen free molecular flow term, and transition flow in a tube appears as a mixture of free molecular flow and viscous flow. That is, the total flow behaves approximately as a sum of two parallel flow mechanisms.

Knudsen writes the conductivity coefficient as a linear combination of the viscous flow coefficient, F_v, and the free molecular flow coefficient denoted F_m,

$$F_c = F_v + zF_m \tag{85a}$$

$$F_v = \frac{a^2}{8\mu RTl}P_a \tag{85b}$$

$$F_m = \frac{2}{3}\left(\frac{a}{l}\right)\left(\frac{v_a}{RT}\right) \tag{85c}$$

Both F_v/P_a and F_m are independent of pressure, and z is essentially a "fudge factor," which, Knudsen found experimentally, varies only slightly with pressure according to the empirical relationship

$$z = \frac{1 + 2.51\ (a/L)}{1 + 3.10\ (a/L)} \tag{86}$$

where since the mean free path, L, is inversely proportional to pressure, z depends upon pressure. As the pressure approaches zero, $F_v \to 0$, $z \to 1$, and $F_c \to F_m$. As the pressure increases, $z \to 0.83$, and for $a/L > 1$, the expression

$$F_c = K_v^c P_a + 0.83\ F_m \tag{87}$$

has approximate validity. Here, K_v^c indicates the pressure-independent part of F_v. Thus, at moderate and high pressure, the conductivity coefficient should be linear in pressure. A plot of the ratio F_c/F_m vs. a/L according to Eqs. (85) and (86) is

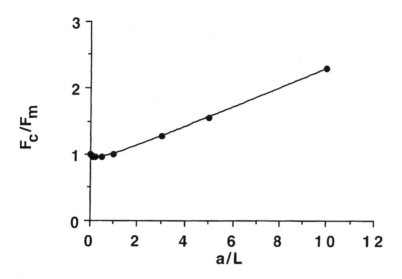

Figure 22 The pressure dependence of conductance: A plot of the ratio of the total conductance to the free molecular flow conductance as a function of the ratio of tube radius to mean free path.

shown in Figure 22. Note that at pressure low enough that $a/L < 1$, the conductivity mechanism is essentially free molecular flow, i.e., the ratio is near unity. For $a/L > 1$, the plot is nearly linear, and Eq. (87) is a good approximation.

Equations (85) and (86) and the data shown in Figure 22 represent experimental results for flow in smooth straight tubes. Experimental data for complex porous systems sometimes differ slightly from the results given by Figure 22 in the region between the low pressure limit and $a/L \approx 1$.

J. Heat Conductivity of Gases

1. Gases at High Pressure

Just as the driving force for mass transfer is a pressure difference, the driving force for heat flow is a temperature difference. The schematic given by Figure 23 shows heat flow from a hot surface at temperature T_1 to a cold surface at temperature T_2 and the resulting steady state temperature distribution. Heat flows through the gas of thermal conductivity λ which separates the two surfaces. At steady state, the temperature of the gas varies linearly with distance right up to the surfaces. The heat flow, dQ/dt, is given by

$$\frac{dQ}{dt} = A\lambda \frac{dT}{dz} = A\lambda \frac{T_1 - T_2}{d} \tag{88}$$

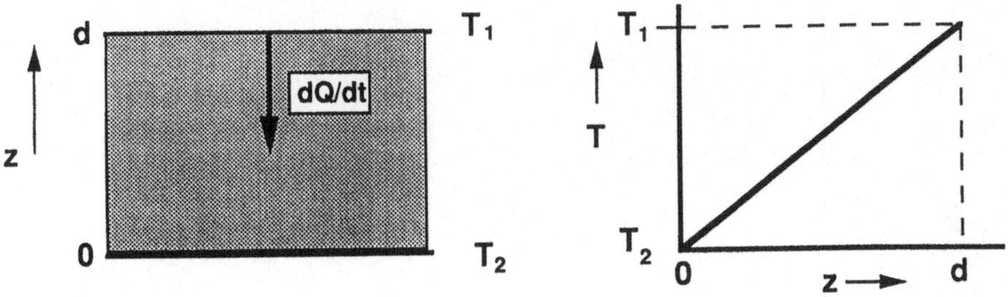

Figure 23 A schematic showing heat conduction in a fluid.

where A is the area of a surface. One can derive an expression for the thermal conductivity using the kinetic theory of gases. We will simply give the result, which may be written in the form

$$\lambda = \varepsilon\mu c_v \tag{89}$$

where

$$\varepsilon = \frac{9\gamma - 5}{4}, \qquad \gamma = \frac{c_p}{c_v}$$

where c_v is the specific heat (i.e., heat capacity per gram) at constant volume and c_p is the specific heat at constant pressure. Since neither heat capacity nor viscosity depend upon pressure in the range of moderate pressure [see Eq. (56)], the thermal conductivity also is independent of pressure at moderate pressure. Table 2 gives some viscosity and thermal conductivity data measured for selected gases at moderate pressures. Helium and hydrogen are very good thermal conductors, whereas both air and water vapor are nearly an order of magnitude less efficient in conducting heat. The pressure dependence of the effective thermal conductivity

Table 2 Experimental Viscosity and Thermal Conductivities for Gases at Moderate Pressure

Gas	$10^5 \mu$ (poise)	$10^5 \lambda(cal/(cm \cdot sec \cdot K))$
He	18.76	34.3
H_2	8.50	41.3
Air	17.22	5.6
H_2O	8.61	4.29

Source: Ref. 8.

of a gas [i.e., λ defined by Eq. (88)] is illustrated by Figure 24. The data in Figure 24 were obtained by Knudsen for hydrogen by measuring heat flow from a very fine tungsten wire (radius = 2.2 μm) to a surrounding cylinder (inner radius = 0.535 cm). Note that the thermal conductivity is essentially independent of pressure from about 100 to 1000 torr. However, below 100 torr, the thermal conductivity begins to decrease with decreasing pressure. At very low pressures, the thermal conductivity is directly proportional to pressure (see Fig. 24 insert). The basic cause for pressure dependence in thermal conductivity is the same as the cause for slip in the corresponding viscosity problem. That is, just as there will be a discontinuity in momentum within about one mean free path of a surface (i.e., slip), so also there will be a discontinuity in energy within about one mean free path of a surface (i.e., a temperature jump). At the low pressure extreme, the mean free path is much greater than the distance separating the surfaces, and since a large mean free path means few gas–gas collisions, the mean energy (and temperature) of the gas molecules is uniform throughout the gas sample, even though the two surfaces are at different temperatures. Here, energy is carried directly from the hot surface to the cold surface by gas–surface collisions, and the rate of energy transport is independent of the distance separating the surfaces (as long as $L \gg d$).

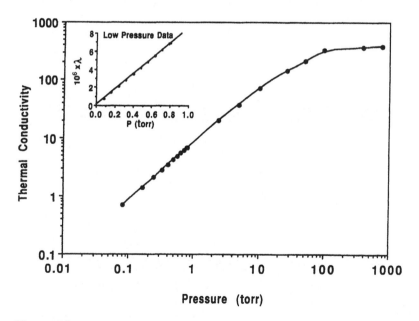

Figure 24 The thermal conductivity of a gas as a function of pressure. (Data from Ref. 9.)

2. Free Molecular Flow Heat Transfer

We now consider a simple derivation of the heat transfer through a gas at sufficiently low pressure that gas molecules collide with the hot and cold surfaces much more frequently than they collide with each other. That is, the mean free path, L, is much larger than the distance separating the two surfaces, d. A schematic of the physical model is given by Figure 25. The surface temperature of the "hot" surface is T_1, also denoted $T_s^{(1)}$ to emphasize that this is a surface temperature and not a gas temperature and refers to surface 1. The cold surface is at temperature T_2, also denoted $T_s^{(2)}$. The solid circles represent gas molecules of energy E_i or E_f. The subscript i represents a molecule incident on a surface, and the subscript f represents a molecule just scattered from the surface. The superscript in parentheses (i.e., $E^{(1)}$ or $E^{(2)}$) refers to the surface number (i.e., 1 = hot, 2 = cold). The amount of incident energy striking the surface per unit time depends on collision frequency and also depends upon the average amount of energy carried by a colliding molecule. In general, energy is carried in the form of vibrational, rotational, and translational energy. The average translational energy for a gas molecule at temperature T is $(3/2)kT$. However, since faster moving molecules (with higher energy) strike a surface more frequently than slower moving molecules, the average translational energy *of those molecules striking the surface* is somewhat greater than the average translational energy of the gas sample as a whole. It may be shown, using kinetic theory, that the mean translational energy of those gas molecules striking a surface is $2kT$ (per mole-

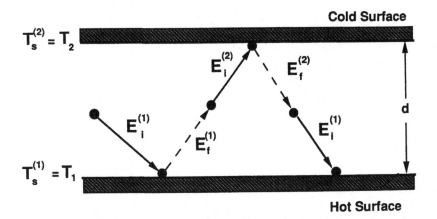

Figure 25   Energy transfer between two surfaces via a gas at low pressure ($L \gg d$) in the free molecular flow region.

cule). Thus, the ''extra'' energy amounts to $(1/2)kT$ per molecule, or an ''extra'' heat capacity of $k/2$ per molecule. Thus, if we denote the heat capacity (per molecule) at constant volume of the molecules that are striking the surface as C_v', we have $C_v' = C_v + k/2$. For an ideal gas as a whole, $C_p - C_v = k$, where C_p and C_v are the heat capacities per molecule at constant pressure and constant volume, respectively. Thus, the heat capacity of the colliding gas molecules is $C_v' = (\gamma + 1)\, C_v/2$. The corresponding energy of the colliding gas would then be the product of the heat capacity, C_v', and the temperature of the colliding gas,

$$E = \frac{\gamma + 1}{2} C_v T \tag{90}$$

The energy of an incident molecule will not normally be the same as that of the molecule when it is scattered from the surface, i.e., $E_i^{(1)} \neq E_f^{(1)}$. There will be an accommodation to the surface and an exchange of energy with the surface. Complete accommodation or equilibration with the surface would imply that the scattered molecules have the same temperature as the surface. The energy accommodation coefficient, a_c, is defined for each surface involved in the problem by the expression

$$a_c = \frac{E_i - E_f}{E_i - E_s} = \frac{T_i - T_f}{T_i - T_s} \tag{91}$$

For convenience in this derivation, we assume that the energy accommodation coefficient is the same for collisions at surfaces 1 and 2. Thus, we may write

$$a_c^{(1)} \equiv \frac{T_i^{(1)} - T_f^{(1)}}{T_i^{(1)} - T_s^{(1)}} = a_c^{(2)} \equiv \frac{T_i^{(2)} - T_f^{(2)}}{T_i^{(2)} - T_s^{(2)}} \tag{92}$$

While for a given surface the incident and scattered molecules have different energies (unless $a_c = 0$), the molecule scattered from surface 1 is the molecule that is incident for surface 2. That is, since there are no gas–gas collisions when the molecule travels between the two surfaces, there is no mechanism to change the energy that it has when it leaves surface 1. Likewise, the energy of the molecules scattered from surface 2 is the same as the energy of those molecules incident upon surface 1. Thus, the temperatures are related by

$$T_f^{(1)} = T_i^{(2)}, \qquad T_f^{(2)} = T_i^{(1)} \tag{93}$$

The heat transfer rate, dQ/dt, between two surfaces of area A is given by the product of the collision rate of gas molecules with the surface and the average change in energy per collision, which for the cold surface may be written as

$$\frac{dQ}{dt} = AI(E_i^{(2)} - E_f^{(2)})$$

$$= A\left(\frac{\rho' v_a}{4}\right)\left(\frac{\gamma + 1}{2}\right)C_v(T_i^{(2)} - T_f^{(2)}) \tag{94}$$

$$= A\frac{P}{(2\pi m\ kT)^{1/2}}\left(\frac{\gamma + 1}{2}\right)C_v(T_i^{(2)} - T_f^{(2)})$$

where I is the intensity of the Maxwellian stream [Eq. (43)], ρ' is the number density of the gas, and v_a is the mean molecular speed. Equation (94) is nearly the desired result, but we desire the heat flow in terms of the difference in temperature between the two surfaces. Combination of Eqs. (92) and (93), with a good deal of algebra, yields

$$T_i^{(2)} - T_f^{(2)} = \frac{a_c}{2 - a_c}(T_1 - T_2)$$

which may be substituted into Eq. (94) to give our desired result,

$$\frac{dQ}{dt} = A\alpha\Lambda_0 P(T_1 - T_2) \tag{95}$$

where Λ_0 is called the "free molecular flow heat conductivity" at 0°C and is defined by

$$\Lambda_0 \equiv \left(\frac{\overline{C}_v}{R} + \frac{1}{2}\right)\left(\frac{R}{2\pi M \times 273.16}\right)^{1/2} \tag{96}$$

where $\overline{C}_v$ is the molar heat capacity at constant volume, and α is a parameter which includes the energy accommodation coefficient and the absolute temperature,

$$\alpha = \left(\frac{273.16}{T}\right)^{1/2}\frac{a_c}{2 - a_c} \tag{97}$$

Thus, as observed (Fig. 24), the heat transfer rate is directly proportional to pressure at low pressure. Note also that the heat transfer rate is independent of the distance separating the two surfaces, as expected by intuition. It was assumed that the accommodation coefficient was the same on each surface. If this assumption is not valid, the expression for α must be modified to the form

$$\alpha = \left(\frac{273.16}{T}\right)^{1/2}\left(\frac{a_c^{(1)}a_c^{(2)}}{(a_c^{(1)} + a_c^{(2)}) - a_c^{(1)}a_c^{(2)}}\right) \tag{98}$$

Table 3 Experimental Data for Free
Molecular Flow Heat Conductivity

Gas	Free molecular flow heat conductivity Λ_0 (cal/(sec $\cdot$ cm^2 $\cdot$ mmHg))
H_2O	6.34×10^{-3}
N_2	3.98×10^{-3}
He	7.02×10^{-3}

Source: Ref. 8.

Free molecular flow heat conductivities and energy accommodation coefficients are given in Tables 3 and 4 for several gases. Note that free molecular flow heat conductivities of water and helium are nearly the same, but both are significantly higher than for air. Thus, while at high pressure helium is a much more efficient conductor of heat than water vapor, they are comparable at low pressure. Energy accommodation coefficients appear to be closer to unity for rough surfaces (i.e., black Pt) and for heavier gases (i.e., oxygen and carbon dioxide compared to hydrogen).

3. Heat Conductivity in Transition Flow

As the pressure decreases and the system enters the transition flow regime, the mean free path begins to be significant relative to the dimensions of the system under consideration. Just as a molecule within about one mean free path of a surface will not suffer a collision to alter the momentum, the lack of a collision will also fail to change the energy of the molecule. Thus, since average energy is proportional to temperature, the temperature of a gas will not change within a region extending a distance of about one mean free path from the surface (Fig.

Table 4 Experimental Data for Energy Accommodation
Coefficient

Surface	Values of energy accommodation coefficient a_c		
	H_2	O_2	CO_2
Platinum, polished	0.36	0.84	0.87
Platinum, black	0.71	0.96	0.98

Source: Ref. 8.

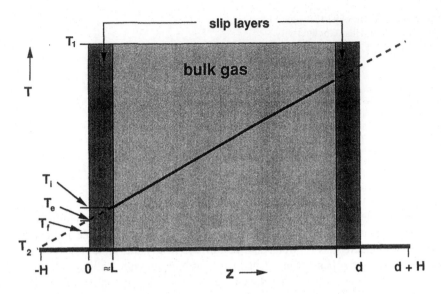

Figure 26 The temperature distribution during heat transfer between two surfaces separated by a gas in the transition flow region.

26). Thus, the temperature of the gas molecules striking the surface is T_i, the temperature of the gas molecules scattering or "emitting" from the surface is T_f, and the temperature of the surface is T_2. Because of the lack of collisions within the slip layer, $T_i > T_2$, and generally $T_f > T_2$, as illustrated by Figure 26. Thus, there is a temperature jump or discontinuity at the surface, meaning that the mean gas temperature at the cold surface, $T_e = (T_i + T_f)/2$, is greater than the temperature of the cold surface, T_2. The average temperature of the gas is given by the solid diagonal line in Figure 26, where the broken lines indicate the extrapolated gas temperature through the slip layer and beyond the actual surface. The temperature of the gas extrapolates to the temperature of the surface, T_2, at distance H beyond the surface, i.e., at position $-H$ in Figure 26. Analogous to our analysis of viscosity in slip flow, we may express the temperature gradient as the slope of the straight line in Figure 26 and write the heat flow as

$$\frac{dQ}{dt} = A\lambda \left(\frac{dT}{dz}\right)_{z=0} = A\lambda \frac{T_1 - T_2}{d + 2H} \tag{99}$$

By a derivation similar to that used to obtain a result for the slip distance in momentum transfer, h, we may relate the "thermal slip" distance, H, to mean free path by the expression

$$H = \frac{2 - a_c}{a_c}\left(\frac{2\varepsilon}{\gamma + 1}\right)L \tag{100}$$

where L is the mean free path. Substitution of Eq. (100) into Eq. (99) and using the kinetic theory expressions for thermal conductivity λ and viscosity μ, we obtain the transition regime result for heat flow,

$$\frac{dQ}{dt} = A\frac{\alpha\Lambda_0 P}{1 + (\alpha\Lambda_0/\lambda)dP}(T_1 - T_2) \tag{101}$$

At very high pressure, where the mean free path is short compared to the separation distance between surfaces, $(\alpha \Lambda_0/\lambda)dP \gg 1$, and Eq. (101) reduces to the heat flow equation for moderate pressure gases given by Eq. (88). Although the derivation given here is not self-consistent when the mean free path is roughly the same as the separation distance, it is interesting to note that as the pressure becomes very small, $(\alpha \Lambda_0/\lambda)dP \ll 1$, and Eq. (101) reduces to the free molecular flow heat flow equation at low pressure.

As the pressure increases from low values, the pressure-dependent term in the denominator of Eq. (101) becomes significant, and the heat transfer is reduced from what is predicted from the free molecular flow heat transfer equation. Physically, this reduction in heat flow is a result of gas–gas collisions interfering with direct energy transfer between the gas molecules and the surfaces. If we use the heat conductivity parameters for water vapor and assume that the energy accommodation coefficient is unity, $(\alpha\Lambda_0/\lambda)dP \approx 150\, P_{torr}d_{cm}$. Thus, at a typical pressure for freeze drying of ≈ 0.1 torr, this term is unity at $d \approx 0.7$ mm. Thus, gas–gas collisions reduce free molecular flow heat transfer by at least a factor of 2 for surfaces separated by less than 1 mm. Most heat transfer processes in freeze drying involve separation distances of at least a few tenths of a millimeter, so transition flow heat transfer is the most important mode of heat transfer through the gas.

IV. RADIATION HEAT TRANSFER

A. Blackbody Radiation

A general characteristic of matter is the ability to emit and absorb energy in the form of electromagnetic radiation. Since transfer of energy via electromagnetic radiation does not require a material medium to transport the energy, emission and absorption of radiant energy is an important mechanism for heat transfer between two bodies in the low pressure environments characteristic of freeze drying. A ''blackbody'' is a surface which emits the maximum energy via radiation for the temperature under discussion. A blackbody is also the most efficient absorber of radiant energy possible and absorbs all radiation striking it. The classic example of a blackbody is a cavity or a box with a small hole. All radiation striking the hole and entering the box is reflected around in the box until it is absorbed. Likewise, emission of radiation from the hole is the maximum amount of emission possible at that temperature. The rate at which radiant energy is emitted by a blackbody b at wavelength λ in the (θ, ϕ) direction per unit area of emitting surface normal to this direction, per unit solid angle about this direction, is denoted $I_{\lambda,b}$. The subscript λ indicates that the radiant intensity refers only to radiation at a specific wavelength, while the subscript b indicates that the object under consideration is a blackbody. The spectral hemispherical emissive power, $E_{\lambda,b}$, is the integral of $I_{\lambda,b}$ about all directions in the hemisphere above the surface. Thus, $E_{\lambda,b} = \pi I_{\lambda,b}$. The total emissive power, E_b, is the integral of $E_{\lambda,b}$ over all wavelengths. The total emissive power is the heat flux via radiation, i.e., $(1/A)\, dQ/dt = E_b$.

The dependence of spectral emissive power $E_{\lambda,b}$ on wavelength is shown in Figure 27 for several temperatures spanning the temperature range of interest to freeze drying. Note that the power reaches a maximum at a wavelength around 10 μm, with the maximum moving toward shorter wavelength at higher temperature. Since the long-wavelength limit of the visible spectrum is about 0.8 μm, the emission shown in Figure 27 is essentially all in the infrared region, and the emitted radiation is not visible to the eye. The area under a spectral emissive power curve is the total radiation emitted, or the total emissive power, E_b. The

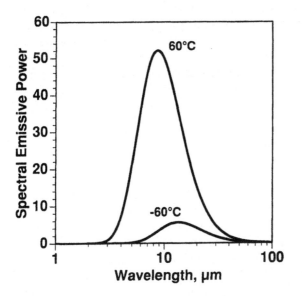

Figure 27 Spectral emissive power for a blackbody.

value of E_b obviously increases sharply with temperature. Experimentally, it was found that the total emissive power was proportional to the fourth power of the absolute temperature of the body, $E_b = A \, \sigma T^4$, where A is the area of the emitting surface and σ is the Stefan–Boltzmann constant, 1.355×10^{-12} cal/(sec $\cdot$ cm^2 $\cdot$ K^4). It is interesting to note that this result can be derived from classical thermodynamics if the radiation inside an isothermal enclosure (i.e., the cavity) is treated as a thermodynamic fluid which has energy, exerts pressure, and occupies volume similarly to a gas.

The fourth-power law, referred to as the Stefan–Boltzmann law, is developed by analysis of a Carnot engine operating with the cavity radiation as the working fluid, although the value of the Stefan–Boltzmann constant and the wavelength dependence of the spectral emissive power (Fig. 27) cannot be obtained with classical thermodynamics. However, in a work often described as the starting point for the development of quantum theory, Planck developed a statistical mechanical theory which correctly predicted both the observed wavelength dependence and the experimental value of the Stefan–Boltzmann constant. Planck treated a blackbody as a system of charges vibrating as simple harmonic oscillators and assumed that an oscillator can radiate energy only in an amount which is an integral multiple of a basic unit of energy for that oscillator, $E_n = nh\nu$, where ν is the frequency of the oscillator and the radiation, n is an integer, and h is a constant (Planck's constant). Thus, the intensity $I_{\lambda,b}$ is found to be

$$I_{\lambda,b} (\lambda, T) = \frac{2hc_0^2}{\lambda^5 [\exp(hc_0^2/\lambda kT) - 1]} \tag{102}$$

where c_0 is the speed of light in vacuum. Recall that the spectral emissive power is simply π times $I_{\lambda,b}$. The total emissive power is then given by integration over all wavelengths,

$$E_b = \pi \int_0^\infty I_{\lambda,b}(\lambda, T) \, d\lambda \tag{103}$$

Substituting for $I_{\lambda,b}$ with Eq. (102) and performing the indicated integration then yields the Stefan–Boltzmann law,

$$E_b = \sigma T^4 \tag{104}$$

with the correct value for σ.

B. Real Surfaces

The radiation emitted by a real surface is less than the radiation emitted by a blackbody, and the absorption of radiation by a real surface is incomplete. Many surfaces are excellent approximations to a blackbody, but some are not. Of the radiation incident upon a real surface, $I(\lambda, \theta)$, a portion is reflected, some is transmitted through the surface, and the remainder is absorbed by the surface (Fig. 28). Most surfaces of interest in heat transfer problems are opaque to radiation in the infrared range, which is the wavelength range important in radiation heat transfer at the moderate temperatures relevant to freeze-drying applications. Thus, we may ignore transmission, and the incident radiation is either reflected or absorbed. The fraction of incident radiation which is absorbed is termed the absorptivity, $\alpha(\lambda, \theta)$,

$$\alpha(\lambda, \theta) = \frac{I_{abs}(\lambda, \theta)}{I_i(\lambda, \theta)} \tag{105}$$

where $I_{abs}(\lambda, \theta)$ is the intensity of absorbed radiation. Since the intensities in Eq. (105) are dependent upon both wavelength λ and incident angle θ, the absorptivity is also a function of wavelength and angle, at least in principle. However, for many surfaces the angular dependence is the same for both incident and absorbed intensity and the absorptivity is independent of the angle θ. Such surfaces are "diffuse absorbers." Further, for many surfaces of practical interest, the absorptivity is also independent of the wavelength of the radiation. These surfaces are referred to as graybodies. Of course, for a blackbody, $\alpha = 1$, regardless of wavelength and angle.

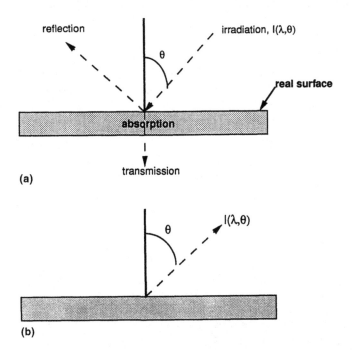

(a)

(b)

Figure 28   (a) Irradiation and absorption and (b) emission of real surfaces.

For a real surface, the scattered or emitted radiation (Fig. 28) is less than the corresponding emission of a blackbody, and the emissivity, $e(\lambda, \theta)$, is defined by the ratio of the emitted radiation, $I(\lambda, \theta)$, to the blackbody radiation, $I_{\lambda,b}$,

$$e(\lambda, \theta) = \frac{I(\lambda, \theta)}{I_{\lambda,b}} \tag{106}$$

where, in principle, the emissivity is both wavelength- and angle-dependent. Again, emissivity is often independent of angle, and such surfaces are diffuse surfaces. A diffuse surface is actually equivalent to a surface exhibiting cosine law scattering since the cosine law is built into the definition of $I_{\lambda,b}$. For a graybody the emissivity does not depend upon wavelength, and for a blackbody the emissivity is unity by definition. For a "diffuse graybody," emissivity and absorptivity are both independent of wavelength and angle and are equal to each other, $e = \alpha$. The usual assumption in applications is that the surfaces are diffuse graybodies, which is a great simplification in problems involving heat transfer between two surfaces. Fortunately, the diffuse graybody assumption is normally an excellent approximation if the two bodies do not differ greatly in temperature (i.e., by not more than a few hundred degrees). Thus, as a good approximation, most real surfaces emit and absorb radiation according to the relationship

$$\left(\frac{dQ}{dt}\right) = Ae\sigma T^4 \tag{107}$$

where A is the area and T is the absolute temperature of the body of emissivity e. The emissivity reflects the nature of the surface and does depend upon temperature, but for modest temperature ranges emissivity is a constant for a given surface.

Representative emissivity data at freeze-drying temperatures are given in Table 5 for some materials of relevance to freeze-drying problems. Emissivities

Table 5 Emissivity Data for Some Materials of Interest to Freeze Drying

Material	Emissivity e
Polished aluminum	0.04
Oxidized aluminum	0.15
Polished 316 stainless steel	0.28
Rough steel	0.66
Anodized finishes	0.95
Glass	0.94
Water	0.95
Black paint	0.95
Rubber	0.9

are usually low for highly polished metal surfaces, but as a surface oxidizes or is given a rough finish, the emissivity increases. Note that the nonmetallic materials listed all have emissivities close to unity.

V. MECHANISMS OF HEAT AND MASS TRANSFER IN FREEZE DRYING

A. Mass Transfer: Resistance

1. Resistance of Stopper and Chamber

The finite openings in the semistoppered vial and the chamber-to-condenser pathway constitute resistances to mass transfer (Fig. 3). The size of the stopper openings depend upon vial size and details of stopper design, but diameters are generally on the order of 0.3 cm, which means that the ratio of mean free path to radius of the stopper opening, L/a, is typically on the order of 0.5. Thus, mass transfer should be in the "transition flow" region, and assuming that the stopper openings behave as short tubes, one would expect the conductance (i.e., reciprocal of resistance) to be linear in pressure as shown earlier by Eq. (87) and Figure 22. Experimental data for two stoppers are presented in Figure 29. The larger stoppers (20 mm closure) have two openings with an effective diameter of 0.4 cm, and the smaller stoppers (13 mm closure) have openings of 0.2 cm diameter. As expected, the data show excellent linearity, and the conductance for the 20 mm closures is greater than for the smaller 13 mm closures. The intercept at zero pressure is roughly the free molecular flow conductance, F_m [i.e., from Eq. (87), intercept $= 0.83 F_m$]. The difference between the conductance at any nonzero pressure and the intercept is, roughly, the viscous flow contribution to conductance. For the larger stoppers, the flow mechanism is mostly viscous flow at pressures normally employed in freeze drying. Flow through the semistoppered 13 mm stoppers is mostly free molecular flow below about 0.1 torr. The experimental slopes and intercepts are relatively close to those calculated from theory using Eq. (87) with the short tube correction given by Eq. (75).

The chamber-to-condenser pathway usually involves dimensions on the order of 10 cm or greater even for laboratory freeze dryers. Here, the ratio of radius to mean free path is on the order of 0.01 or less, and flow should be essentially viscous flow. Conductance should be linear in pressure, with a zero pressure intercept near zero. Qualitative agreement with theory is observed (data not shown).

2. Resistance of Dried Product

Water vapor created by sublimation of ice must pass through the pores in the dried region that were created by prior sublimation. These pores are essentially

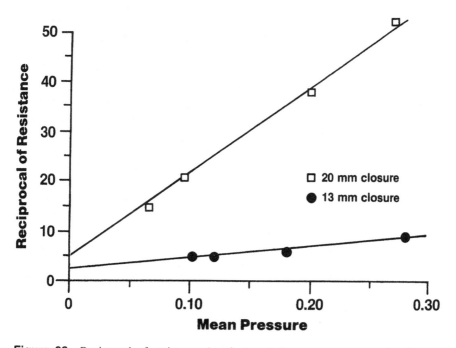

Figure 29 Reciprocal of resistance (conductance) for water vapor transfer through semistopper closures; dependence on pressure. (From Ref. 5.)

molded by the crystallization of ice and reflect the size and morphology of the ice crystals formed during the freezing process. The larger pores are normally on the order of 20 μm in diameter, frequently with a population of smaller pores originating from eutectic ice crystals, which are smaller crystals. At the product temperatures typically used in freeze drying, the pressures in the dried cake are on the order of 0.5 torr, which gives a typical L/a ratio of 10. Thus, the flow mechanism for flow through the dried product is Knudsen flow or free molecular flow. Note that significant deviations from free molecular flow do not occur until the L/a ratio is less than about 0.3 ($a/L > 3$, Fig. 22). While such a porous solid does not necessarily resemble a collection of capillary tubes, porous solids are often modeled as bundles of long tubes. Assuming that the porous cake of thickness l and porosity ε created during freeze drying is a collection of capillary tubes of diameter d, we may write the free molecular flow resistance for the dried cake as

$$\hat{R}_p = l\left(\frac{\tau^2}{\varepsilon}\right)\left(\frac{3}{2}\right)\left(\frac{f}{2-f}\right)\left(\frac{1}{d}\right)\left(\frac{\pi RT}{2M_1}\right)^{1/2} \tag{108}$$

where we have allowed for channels that are not straight by introducing the tortuosity, τ, which is the ratio of total channel length to the thickness of the porous system. Tortuosities must be greater than unity and have been measured for some freeze-dried food products to give $\tau^2 \approx 1.5$. The momentum accommodation coefficient, f, is normally assumed to be unity, since most "rough" surfaces give accommodation coefficients close to unity. Moreover, as surface impurities are introduced or the vapor is adsorbed on the surface, accommodation coefficients increase, and as the surface coverage becomes monolayer or greater, diffuse scattering is observed (i.e., $f = 1$). In freeze-drying applications, the local relative humidity is high due to the proximity of ice, and surface adsorption must be considerably in excess of a monolayer. Thus, the approximation $f = 1$ is certainly an excellent approximation for momentum transfer in the porous dried product.

For a system of 5% (v/v) solids, with a pore diameter of 20 μm, the free molecular flow resistance is calculated to be $\hat{R}_p \approx 10\, l$ (cm^2 · torr · hr/g), where l is the cake thickness in centimeters. Thus, the calculated resistance is of the same magnitude as typical dried product resistance observed (Fig. 4). The linear dependence of resistance on thickness that is predicted by Eq. (108) is frequently violated by data (Fig. 4), although experimental resistances do nearly always increase as thickness increases. The observed decrease in resistance with increasing size of ice crystals mentioned earlier in Section II is qualitatively consistent with Eq. (108). That is, resistance is predicted to be inversely proportional to pore diameter. Increasing concentration normally increases resistance. An increase in solute concentration will increase resistance by a decrease in the porosity of the porous system [Eq. (108)]. However, this is a relatively small effect, and the major effect of increasing solute concentration is to decrease the size of the ice crystals. As solute concentration increases, greater supercooling normally results, thereby increasing the nucleation rate and decreasing the size of the ice crystals ultimately formed. With a solute that forms a eutectic, a higher concentration means that a greater fraction of ice crystallizes as a salt–ice eutectic mixture. Since eutectic ice has smaller crystals than pre-eutectic ice, the "pores" formed from subliming eutectic ice are smaller, thereby giving higher resistance. Thus, these effects of thermal history appear consistent with the qualitative concepts represented by Eq. (108).

The expression for sublimation rate given by Eq. (4) for a freeze-drying system is consistent with the rate of sublimation given by Eq. (52) for sublimation from a single-phase system (i.e., ice) in that both expressions relate the mass transfer rate to the same driving force. For ice sublimation, the driving force for flow is the difference between the vapor pressure of the subliming ice and the partial pressure of water "in the background." For sublimation of pure ice, "in the background" simply means well removed from the ice/vapor interface. For freeze drying a product, "in the background" means on the low pressure side, or "other side," of the dried layer. As mentioned in the context of Eq. (4), the vapor in the vial and in the drying chamber is nearly all water vapor under most practical freeze-drying operations. The resistance to mass transfer during sublimation of pure ice arises from the finite velocity of water molecules [included in the square root term of Eq. (52)] and the barrier to the phase change that must be overcome (included in the evaporation coefficient). For a porous system as described by Eq. (108), the resistance is essentially due to momentum transfer between water molecules and the pore walls.

It should be noted that the effect of a background pressure of "inert" gas (such as air), P_{in}, on sublimation rate has not been specifically addressed. With our choice of driving force involving only the subliming component, the only effect of inert gas pressure would be to cause the resistance to depend upon P_{in}. For sublimation of pure ice, the evaporation coefficient could depend upon the pressure of inert gas; and for freeze drying a product, the dried product resistance could depend upon the pressure of inert gas. In principle, when collisions between water molecules and inert gas molecules become frequent enough to impede flow from the subliming ice surface, the evaporation coefficient will decrease as P_{in} increases. Likewise, when water–inert gas collisions become more frequent than collisions between water molecules and pore walls, the presence of the inert gas will impede the transfer of water through the pores, and the dried product resistance will increase. In fact, the evaporation coefficient for ice at $\approx -28°C$ decreases by about a factor of 2 in going from $P_{in} = 0$ to $P_{in} = 0.3$ torr. Likewise, the dried product resistance increases as P_{in} increases (Fig. 30). The data given by Figure 30 were obtained with a vacuum microbalance sublimation apparatus; sublimation rate was directly measured under conditions where the background partial pressure of water was zero. Data were accumulated such that the pressure dependence was investigated at constant dried product thickness. Selected values of P_{in} were studied by using a controlled leak of air into the apparatus. Increasing values of P_{in} cause a small but significant increase in resistance for KCl and a rather large increase in resistance for the povidone system. The resistance of KCl at $P_{in} = 0$ is much larger than the corresponding resistance for povidone, indicating that the effective pore size for KCl is smaller [see Eq. (108)]. A smaller pore size means a greater frequency of collisions between water and pore walls. Thus, a higher P_{in} will be required to achieve a frequency of water–inert gas collisions comparable to the frequency of water–pore wall collisions, and therefore the dependence of resistance on P_{in} will be less significant for KCl, as observed.

The product temperatures were $-21.2°C$ for KCl and $-31.0°C$ for povidone. The arrows mark the values of the vapor pressure of ice at the operating temperatures. The normalized dried product resistances increase smoothly with pressure as the vapor pressure of ice is approached and exceeded. Thus, the sublimation rate decreases smoothly throughout this pressure range and does not drop to zero at a total system pressure which exceeds the vapor pressure of ice. This observation is completely consistent with the theoretical concepts developed

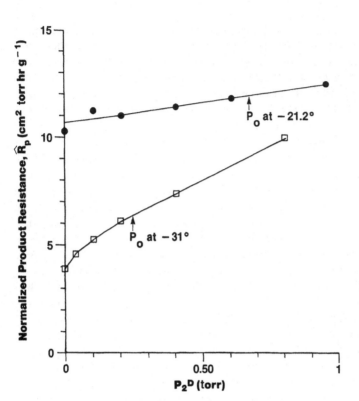

Figure 30 The effect of residual air pressure on $\hat{R}_p$ for potassium chloride 5% (v/v) at −21.2°C, $I = 0.52$ cm (●) and 5% (v/v) povidone at −31.0°C, $I = 0.70$ cm (□). The arrows indicate the vapor pressure of ice, P_0, at −21.2°C (●) and −31.0°C (□). (From Ref. 4.)

above and throughout this chapter and therefore should not be a surprise. However, this point is made to directly contradict the erroneous notion that occasionally appears in the literature which maintains that sublimation can occur only at a total pressure (water vapor plus inert gas) that is lower than the vapor pressure of ice at the sublimation temperature. This erroneous notion probably arises from the valid observation that sublimation in a conventional freeze dryer approaches zero as the chamber pressure is increased beyond the vapor pressure of the subliming ice by an inert gas leak. However, in a conventional freeze dryer, an inert gas leak increases total pressure at the pump and serves to "back up" water vapor into the drying chamber. Thus, the increase in chamber pressure is due to an increase in the partial pressure of water, even though the inert gas leak was the cause of the increased pressure. Of course, as the partial pressure of water approaches the vapor pressure of ice, the driving force for sublimation decreases sharply, resulting ultimately in a sublimation rate near zero. The sublimation rate does not actually reach zero as Eq. (4) suggests, since Eq. (4) is valid only when the gas in the chamber is mostly water vapor. As the inert gas leak increases the background partial pressure of water to more than about 90% of the vapor pressure of ice, the composition of the gas phase shifts to mostly inert gas and the rate of sublimation is governed by gas-phase diffusion rather than by Eq. (4). Diffusion is slow, but it is not zero [see Eq. (65)].

The effect of temperature on sublimation rate is sizable through the dependence of the driving force on temperature, i.e., the temperature dependence of the vapor pressure of ice. However, the effect of temperature on dried product resistance should be minimal. The only temperature effect predicted by theory is the square root of absolute temperature [Eq. (108)], which reflects the temperature dependence of gas molecule velocity. For the temperature variations relevant to freeze-drying applications, this theoretical temperature dependence is minimal. However, dried product resistance data normally show a very significant temperature dependence as the sublimation temperature approaches either the eutectic temperature or the collapse temperature. The dependence of resistance on thickness shifts from whatever low temperature behavior is characteristic of the material to a type IV behavior. In addition, the magnitude of the resistance at higher values of dried cake thickness is reduced dramatically. The data shown in Figure 31 for 5% KCl (eutectic temperature $-11°C$) were obtained with the microbalance procedure and demonstrate typical behavior. However, not all materials show significant temperature dependence in $\hat{R}_p$. For example, dried product resistances for povidone are essentially independent of temperature. The mechanism responsible for this temperature dependence is not well understood. From the perspective of Eq. (108), which reflects a free molecular flow mechanism in the pores, there are only two parameters that could possibly be temperature-dependent. The momentum accommodation coefficient could depend upon temperature. However, as discussed earlier, the momentum accommodation coefficient in porous KCl should be essentially unity even at $-30°C$, so momentum accommodation variations seem an unlikely cause. An increase in effective pore diameter as the temperature approaches a eutectic melt or the collapse temperature would decrease the resistance. Decreases in specific surface area, corresponding to increased pore size, have been measured for dried crystalline samples allowed to adsorb significant quantities of water. Since a dried crystal surface would adsorb significant quantities of water as the nearby frozen solution approaches the eutectic temperature, an increase in effective pore size is a possible interpretation. For amorphous materials, freeze drying near the glass transition temperature has been demonstrated to produce "microcollapse," resulting in increased pore size. Finally, it is also possible that the flow of gas molecules through pores is not always the major mass transfer mechanism. Under conditions of significant water

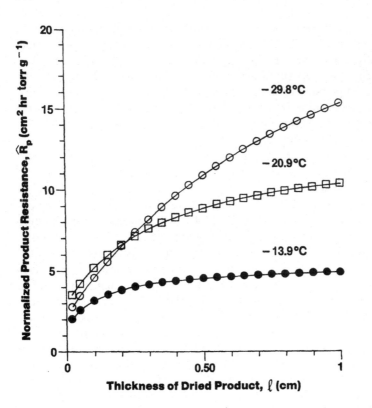

Figure 31   The effect of temperature on $\hat{R}_p$ for potassium chloride 5% (v/v). (From Ref. 4.)

adsorption on the surface of the "pore," surface diffusion and/or bulk flow of adsorbed water could provide an additional flow mechanism and thereby decrease the measured dried product resistance. Additional research is needed to resolve this question.

B.   Heat Transfer Coefficients

1.   Special Cases of Radiative Heat Transfer in Freeze Drying

Two special cases of radiative heat transfer which are applicable to heat transfer in freeze drying are illustrated by Figure 32. Heat transfer between body 1 and body 2 are illustrated for case I (Fig. 32a), where body 1 is of much greater area than body 2, $A_1 \gg A_2$, and "surrounds" body 2. The freeze-drying example is heat exchange between the top of the vial (body 2) and the freeze dryer shelf

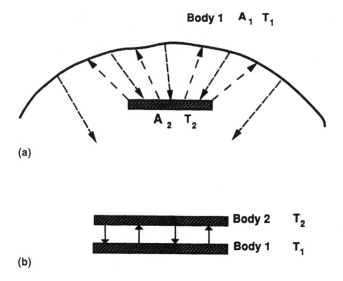

Body 1 **A₁ T₁**

A₂ T₂

(a)

Body 2 **T₂**

Body 1 **T₁**

(b)

Figure 32 Two special cases of radiation heat transfer with importance to freeze-drying applications. (a) Case I: Body 1 is the "surroundings." (b) Case II: Bodies 1 and 2 are of equal areas, and $e \approx 1$ for body 2.

(body 1) above the vial array. Case II (Fig. 32b) applies for radiation heat exchange between the bottom of the vial (body 2) and the shelf upon which the vial is resting (body 1). In general, for radiative heat exchange between bodies 1 and 2,

$$\left(\frac{dQ}{dt}\right) = \sigma A_1 \hat{F}_{12}(T_1^4 - T_2^4) \tag{109}$$

where $\hat{F}_{12}$ is a geometric factor defined in terms of areas and emissivities of the two bodies as

$$\hat{F}_{12} = \left[\frac{1}{F_{12}} + \left(\frac{1}{e_1} - 1\right) + \left(\frac{A_1}{A_2}\right)\left(\frac{1}{e_2} - 1\right)\right]^{-1} \tag{110}$$

where F_{12} is a geometric view factor and e_i ($i = 1, 2$) is the emissivity of body 1 or body 2. The geometric view factor, F_{12}, is the fraction of total radiation leaving body 1 of area A_1 that strikes body 2 of area A_2. At least one of the two geometric view factors can usually be evaluated from geometric considerations. For example, for case I, all of the radiation leaving body 2 will strike body 1, so $F_{21} = 1$. Knowledge of one of the geometric view factors often allows the other to be calculated via the reciprocal relation,

$$A_1 F_{12} = A_2 F_{21} \tag{111}$$

Since for case 1, $F_{21} = 1$ and $A_1 \gg A_2$, the radiation heat transfer rate given by Eq. (109) becomes

$$\text{Case I:} \quad \left(\frac{dQ}{dt} \right)_{\text{rad}} = A_2 e_2 \sigma (T_1^4 - T_2^4) \tag{112}$$

For case II, the two areas are equal, $F_{12} = 1$, and since the emissivity of glass is 0.94 (Table 4), $e_2 \approx 1$. Thus, Eq. (109) may be written

$$\text{Case II:} \quad \left(\frac{dQ}{dt} \right)_{\text{rad}} = A_2 e_1 \sigma (T_1^4 - T_2^4) \tag{113}$$

Note that the relevant emissivity for case I is the emissivity for body 2 (i.e., the vial top), while for case II the relevant emissivity is the emissivity for body 1 (i.e., the shelf surface).

Since the other modes of heat transfer discussed depend upon the difference in temperature, $T_1 - T_2$, it would be convenient to express radiative heat exchange in terms of $T_1 - T_2$ rather than in terms of the difference in fourth powers of temperature. For this purpose, we express the difference in fourth powers of temperature as

$$T_1^4 - T_2^4 = (T_1 - T_2)(T_1 + T_2)(T_1^2 + T_2^2)$$

As long as the absolute temperatures do not differ greatly, the arithmetic mean and geometric means are equal,

$$\frac{T_1 + T_2}{2} \cong (T_1 T_2)^{1/2}$$

$$\frac{T_1^2 + T_2^2}{2} \cong (T_1^2 T_2^2)^{1/2}$$

Combination of the above equations then gives

$$\left(\frac{dQ}{dt} \right)_{\text{rad}} = A_2 e_i \sigma 4 \overline{T}^3 (T_1 - T_2) \tag{114}$$

where $\overline{T}$ is the mean temperature. While $\overline{T}$ is not a constant in freeze-drying applications, variations are not great, and as a useful first approximation, $\overline{T} \approx$ 263 K. With this approximation and evaluation of the constants,

$$\left(\frac{dQ}{dt}\right)_{rad} = A_2 K_r (T_1 - T_2) \tag{115}$$

where

$$K_r = e_i(1.0 \times 10^{-4}), \quad cal/(s \cdot cm^2 \cdot K)$$

where K_r is the radiative heat transfer coefficient which, in units of calories per second per square centimeter per kelvin, is equal to 10^{-4} times the emissivity of either the vial top or the freeze dryer shelf, as appropriate.

2. Mechanisms of Heat Transfer in Freeze Drying

The vial heat transfer coefficient is the sum of heat transfer coefficients for three parallel heat transfer mechanisms: (1) direct conduction between glass and shelf surface at the few points of actual physical contact, K_c; (2) radiation heat exchange, K_r, which has contributions from the shelf above the vial array to the top of the vials, K_{rt}, and from the shelf upon which the vial is resting, K_{rb}; and (3) conduction via gas–surface collisions between the gas and the two surfaces, shelf and vial bottom, K_g:

$$K_v = \underbrace{K_c}_{\text{contact conduction}} + \underbrace{K_r}_{\text{radiation}} + \underbrace{K_g}_{\text{gas conduction}} \tag{116a}$$

$$K_r = \underbrace{K_{rb}}_{\text{to bottom}} + \underbrace{K_{rt}}_{\text{to top}} \tag{116b}$$

Combination of Eqs. (115) and (116) allows the radiation contribution to be written in the form

$$K_r = 1.0 \times 10^{-4}(e_v + e_s) \quad cal/(s \cdot cm^2 \cdot K) \tag{117}$$

where e_s is the emissivity of the shelf surface (0.28 for polished stainless steel) and e_v is the emissivity of the vial top (found to be 0.84 for all vials studied). Both the contact and radiation heat transfer coefficients are independent of chamber pressure. However, as noted earlier [see Eq. (101)], heat transfer via the gas phase is pressure-dependent. From Eq. (101), it is obvious that the gas heat conduction term, K_g, may be written in the form

$$K_g = \frac{\alpha \Lambda_0 P}{1 + (\alpha \Lambda_0 / \lambda_0) l P} \tag{118}$$

where the mean separation distance between vial bottom and shelf is denoted l, α is defined by Eq. (97), λ_0 is high pressure thermal conductivity of the gas, and Λ_0 is the free molecular flow heat conductivity at 0°C as given by Eq. (96). Since the gas in the freeze-drying chamber during sublimation is nearly always essentially 100% water vapor, λ_0 and Λ_0 refer to water vapor during primary drying.

Vial heat transfer coefficient data obtained as a function of pressure (Fig. 6) may be fit to Eq. (116) with K_g given by Eq. (118). Excellent fits are found, indicating agreement between theory and experiment. The extrapolated zero pressure intercept of K_v is the sum $K_c + K_r$. At any value of pressure, the value of K_g is the difference between the experimental K_v and the zero pressure intercept. From a knowledge of the emissivity of the shelf surface and the emissivity of the vial top, K_r may be calculated [Eq. (117)], and the contact heat transfer coefficient may be obtained from K_v at zero pressure and the calculated value of K_r by difference. Thus, the three heat transfer coefficients described by Eq. (116) may be evaluated from the data. The relative importance of these three heat transfer mechanisms is illustrated by Figure 33. Note that all three heat transfer mechanisms may be significant, particularly at pressures commonly used in pharmaceutical processing (i.e., ≈ 0.1 torr), although the exact distribution is vial-specific. In general, molded vials (i.e., 5303 vials) will have a greater *relative* contribution from radiation since the large separation distances and poor contact characteristic of these vials diminish the absolute contributions from contact and gas conduction. Since the radiation contribution is independent of vial "bottom geometry," the absolute contribution from radiation remains the same, and therefore the relative contribution from radiation increases. Obviously, the gas conduction term becomes more important as pressure increases.

The pressure dependence defines the energy accommodation coefficient, a_c, and the separation distance, l. It is found that the energy accommodation coefficient is $a_c \approx 0.67$, independent of the nature of the glass vial and the nature of the shelf surface. The values of l are specific to the vial. Contact heat transfer parameters, K_c, and separation distances, l, for selected vial types are compared in Figure 34. Vials 5303 and 5304 are molded vials; all others are tubing vials. Recall that a high K_c increases heat transfer while a high l decreases heat transfer. Separation distance parameters average about 2 mm, and for tubing vials there is little variation between the various vial types. Variation in contact heat transfer is substantial.

3. Variations in Heat Transfer

The heat transfer data discussed above refer only to the average behavior of a vial of a given type which is surrounded by other vials in a hexagonal packing array of vials. We now consider interval variations in heat transfer in a set of nominally equivalent vials and variations in heat transfer arising from variations in the position of the vial in the array. An experiment demonstrating such variations is described by Figure 35. Each circle represents a vial placed on a temperature-controlled shelf in a small laboratory freeze dryer. The vials contained pure

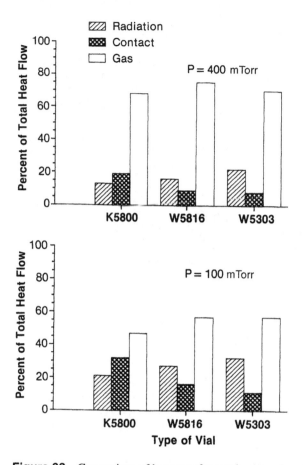

Figure 33 Comparison of heat transfer mechanisms for glass vials on a polished stainless steel shelf. Pressure is in millitorr. The vial identification numbers are manufacturers' designations. (From Ref. 1.)

water and were sealed except for a precision cut metal tube inserted into each stopper to provide a constant resistance to mass flow. Vials on the edge of the array are labeled with an E. Sublimation was carried out at a chamber pressure of 0.2 torr long enough to sublime about 25% of the ice, the freeze dryer was vented, and all vials were weighed to determine the mass loss (i.e., sublimation rate) for each vial. The only other detail that is important to this discussion is the decimal number in each circle. This number is the percent deviation in sublimation rate for that vial–position combination from the average sublimation rate for all interior vials (i.e., non-edge vials). The major source of variation is the

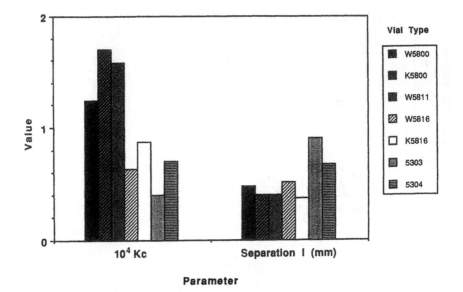

Figure 34 Experimental vial heat transfer parameters. Accommodation coefficient = 0.67; vial top emissivity = 0.84. (Data from Ref. 5.)

variation in the vial heat transfer coefficient. Variations due to vial position in the interior array are insignificant (for this freeze dryer), and pure experimental error is only a little over 1%. In general, for a given vial type, the standard deviation in K_v, denoted $\sigma(K_v)$, is in the range $\approx$ 3–6%. The standard deviation in K_v varies with vial type and chamber pressure. Due to sensitivity to variations in the contact heat transfer at low pressure and sensitivity to variations in separation distance at high pressure, $\sigma(K_v)$ passes through a broad minimum at chamber pressures in the range 0.1–0.2 torr. Values of $\sigma(K_v)$ increase very sharply at pressures below about 0.08 torr. Thus, if feasible, a freeze-drying process should be designed to utilize chamber pressures between 0.1 and 0.2 torr to minimize variability in the process.

As one examines the percent deviations from average sublimation rate for edge vials, it becomes obvious that the edge vials, as a group, are subliming faster and must be receiving more heat than a typical "interior" vial. The average edge vial sublimes about 15% faster than the average interior vial. The higher sublimation rate for edge vials arises from the atypical radiation heat transfer seen by vials at the edge of an array. Edge vials have a "view" of the shelf edge and/or the drying chamber wall on the side of the vial at the edge of the array. In this experiment, both the shelf (5°C) and the chamber wall ($\approx$15°C) were considerably warmer than a neighboring vial ($\approx$ –22°C). Consequently, the side

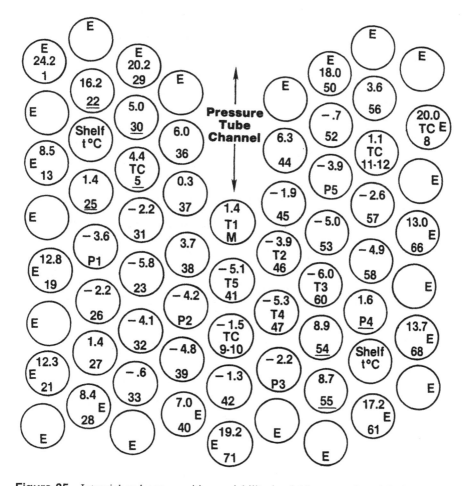

Figure 35 Interval and array position variability in vial heat transfer; vial placement on the shelf and mean percent deviation in sublimation rates. (From Ref. 5.)

of the vial with a view of a warmer surface will receive more heat than another, identical vial in the interior of the array. Edge vials adjacent to a viewport or window in the freeze dryer door will generally have a "view" of the room at ≈23°C, which will give these edge vials even a bit more heat. At the opposite extreme, edge vials with a direct view of a condenser plate (−60°C) will lose some of their energy by radiation exchange with the condenser plate and therefore will run colder and sublime slower. While most freeze dryers are constructed to minimize direct "views" of the extreme temperatures in the condenser chamber, the potential for variability in freeze drying arising from abnormal radiation effects cannot be ignored, particularly for vials on the edge of the array facing the dryer door.

REFERENCES

1. MJ Pikal. Freeze drying. In: Encyclopedia of Pharmaceutical Technology. New York: Marcel Dekker, 1992, pp 275–303.
2. MJ Pikal, S Shah, ML Roy, R Putnam. The secondary drying stage of freeze drying: Drying kinetics as a function of temperature and chamber pressure. Int J Pharm 60: 203–217, 1990.
3. MJ Pikal. Use of laboratory data in freeze drying process design: Heat and mass transfer coefficients and the computer simulation of freeze drying. J Parenter Sci Technol 39:115–138, 1985.
4. MJ Pikal. Physical chemistry of freeze-drying: Measurement of sublimation rates for frozen aqueous solutions by a microbalance technique. J Pharm Sci 72:635–650, 1983.
5. MJ Pikal, ML Roy, S Shah. Mass and heat transfer in vial freeze-drying of pharmaceutical: Role of the vial. J Pharm Sci 73:1224–1237, 1984.
6. SL Nail. J Parenter Drug Assoc Sept/Oct 1980, p 358.
7. JR Partington. An Advanced Treatise on Physical Chemistry, Vol 1. New York: Wiley, 1962.
8. S Dushman, JM Lafferty. Scientific Foundations of Vacuum Technique. 2nd ed. New York: Wiley, 1962.
9. S Dushman. Scientific Foundations of Vacuum Technique. 2nd ed. Revised by members of the research staff at General Electric Research Laboratory (JM Lafferty, ed.). New York: Wiley, 1962.

ADDITIONAL REFERENCES

Bird RB, Stewart WE, Lightfoot EN. Transport Phenomena. New York: Wiley, 1960.
Goodman F, Wachman H. Dynamics of Gas-Surface Scattering. New York: Academic Press, 1976.
Incropera FP, DeWitt DP. Fundamentals of Heat Transfer. New York: Wiley, 1981.
Moore WJ. Physical Chemistry. 4th ed. Englewood Cliffs, NJ: Prentice-Hall, 1972.

17
Heat and Mass Transport: Hygroscopicity

James Wright
Alkermes, Inc., Cambridge, Massachusetts

I. INTRODUCTION

A. Hygroscopic Materials

Water uptake causes a host of problems in drug products and the inactive and active ingredients contained in them. Moisture uptake has been shown to be an important factor in the decomposition of drug substances [1–8]. Moisture has also been shown to change surface properties of solids [9,10], alter flow characteristics of powders [11,12], and affect the compaction properties of solids [13]. This chapter discusses various mathematical models that can be used to describe moisture uptake by deliquescent materials.

A deliquescent material takes up moisture freely in an atmosphere with a relative humidity above a specific, well-defined critical point. That point for a given substance is defined as the critical relative humidity (RH_0). Relative humidity (RH) is defined as the ratio of water vapor pressure in the atmosphere divided by water vapor pressure over pure water times 100% [RH = $(P_w/P_0) \times 100\%$]. Once moisture is taken up by the material, a concentrated aqueous solution of the deliquescent solute is formed. The mathematical models used to describe the rate of moisture uptake involve both heat and mass transport.

Since heat transport is unfamiliar to many pharmaceutical scientists, this chapter begins with a discussion of vapor–liquid equilibria, heat transport in rectangular coordinates, and heat transport in spherical coordinates. Once these basic principles are established, we can build models based on heat transport. Heat transport is the dominant mechanism for moisture uptake in an atmosphere of pure water vapor. In air, however, both heat and mass transport are involved.

These models are discussed later in this chapter. The final subject is an introduction to moisture uptake in heterogeneous materials.

B. Empirical Approaches to Moisture Uptake

A rational development of models for moisture uptake begins with a description of the experimental procedure used to determine moisture uptake as a function of time. The first step in the experiment is to control the relative humidity to which a sample will be exposed. One technique to control humidity is to use saturated salt solutions. When placed in a closed system and held at a constant temperature, a saturated aqueous salt solution will provide a constant humidity (RH_0) within that system. Table 1 lists relative humidities that will be maintained over various saturated salt solutions [14].

A hygroscopic sample placed in a controlled atmosphere as described above will absorb moisture at a specific rate, and the weight gain of the sample can be determined as a function of time. The amount of moisture taken up can be determined by weight gain, by weight loss on drying, or by other analytical techniques. Empirically, the amount of moisture uptake (W) is given by Eq. (1) [15,16], where W' is the rate of moisture uptake and t is time.

$$W = W't \tag{1}$$

It has been determined experimentally that the rate of moisture uptake varies linearly with the surface area of the hygroscopic material and the relative humidity of the atmosphere. For example, if $LiCl \cdot H_2O$ were exposed to a series of saturated salt solutions, the weight of moisture uptake as a function of time

Table 1 Constant Relative Humidity of Selected Solutions at 25°C

Salt	Relative humidity (RH_0)
$LBr \cdot 2H_2O$	6%
$LiCl \cdot H_2O$	11%
$LiI \cdot 3H_2O$	18%
$MgCl_2 \cdot 6H_2O$	33%
$NaI \cdot 2H_2O$	38%
NH_4NO_3	62%
KI	69%
$NaCl$	75%
CsI	91%
K_2SO_4	97%

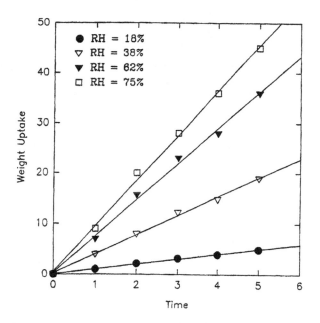

Figure 1 Moisture uptake curves for LiCl hydrate.

could be plotted as shown in Figure 1. The slope of any given line would be W', the rate of moisture uptake. If the slopes are then plotted against the relative humidities of the salt solutions used, Figure 2 is generated. The x-axis intercept in the curve shown in Figure 2 is the critical relative humidity (RH_0) of LiCl · H_2O. As the relative humidity of the system approaches the RH_0 of the hygroscopic material, the rate of moisture uptake falls to zero, because the moisture gradient falls to zero (the vapor pressure of water in the system equals the vapor pressure above the hygroscopic material). These simple plots empirically describe the behavior of deliquescing materials in high humidity environments.

II. VAPOR–LIQUID EQUILIBRIA

The discussion of moisture uptake by hygroscopic materials must include a description of the thermodynamics of vapor–liquid equilibria. For gas (g) and liquid (l) phases to be in equilibrium, the infinitesimal transfer of molecules between phases (dn_g and dn_l) must lead to a free energy change of zero.

$$dG = G_g dn_g + G_l dn_l = 0 \qquad (2)$$

where

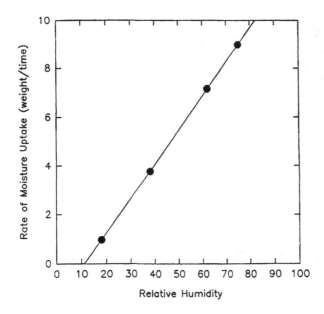

Figure 2   Rate of moisture uptake as a function of relative humidity for LiCl hydrate.

G_g is the molar free energy of the gas phase
G_l is the molar free energy of the liquid phase
dn_g is the differential change in moles in gas phase
dn_l is the differential change in moles in liquid

We know that in a closed system $dn_g = -dn_l$. If the temperature and pressure are changed by very small amounts and the system is allowed to come to equilibrium, the molar free energy changes of the two phases must be equal:

$$dG_l = dG_g \tag{3}$$

We also know that for the phase transformation at constant temperature, the change in free energy may be written as

$$dG_g = V_g dP - S_g dT \tag{4}$$
$$dG_l = V_l dP - S_l dT \tag{5}$$

where

P is pressure
T is temperature
V is the molar volume
S is molar entropy

If Eqs. (4) and (5) are set equal [according to Eq. (3)], the following equation results:

$$\frac{dP}{dT} = \frac{S_g - S_l}{V_g - V_l} \tag{6}$$

From Eq. (6) it clearly can be seen that pressure and temperature are related by the changes in entropy and volume for the phase change. This equation also shows that once temperature is chosen, pressure must be fixed accordingly. It is easy to show that the change in entropy for the phase transition is simply the heat of vaporization (ΔH_v) divided by the temperature of transition, and thus the well-known Clapeyron equation results.

$$\frac{dP}{dT} = \frac{\Delta H_v}{T(V_g - V_l)} \tag{7}$$

Since the molar volume of a gas is much larger than the molar volume of the liquid phase, the Clapeyron equation can be reduced to

$$\frac{dP}{dT} = \frac{\Delta H_v}{TV_g} \tag{8}$$

If we substitute RT/P for V_g (using the ideal gas law), we find that

$$\frac{dP}{P} = \frac{\Delta H_v}{RT^2} \tag{9}$$

Integrating Eq. (9) results in the Clausius–Clapeyron equation,

$$\ln P = -\frac{\Delta H_v}{RT} + \text{constant} \tag{10}$$

The Clausius–Clapeyron equation implies that if we plot the natural log of the pressure of the gas phase versus inverse temperature, the slope of the resulting line is the heat of vaporization divided by the gas constant (R). A plot of $\ln P$ (vapor pressure of water) versus inverse temperature is given in Figure 3. The calculated heat of vaporization (determined by multiplying the slope by R) is 10,400 cal/mol. The important aspect of Eq. (10) with regard to moisture sorption is the fact that increasing the temperature also increases the vapor pressure.

III. HEAT CONDUCTION

Building a model of moisture uptake based on heat transport requires a set of tools to describe the process of energy transport.

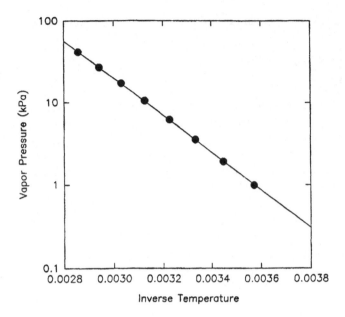

Figure 3 Vapor pressure of water as a function of temperature.

The first step in the process is to relate heat flow to a temperature gradient, just as a diffusive flux can be related to a concentration gradient. The fundamental law of heat conduction was proposed by Jean Fourier in 1807 and relates the heat flux (q) to the temperature gradient:

$$q = -\alpha\frac{dT}{dx} \tag{11}$$

$$J = -\mathbf{D}\frac{dC}{dx} \tag{12}$$

Equation (11) is known as Fourier's law of heat conduction, and α is the thermal conductivity of the material. Equation (12) is Fick's first law (which was based on the work of Fourier) and is very similar in structure to Fourier's law, with the diffusion coefficient ($\mathbf{D}$) replacing thermal conductivity (α) and the concentration gradient (dC/dx) replacing the temperature gradient (dT/dx). Since the units of heat flux are energy per unit area per unit time, the units of thermal conductivity are energy per unit length per unit time. The common units associated with thermal conduction for gases, liquids, and solids are listed below:

Table 2 Thermal Conductivity Values for Selected Materials

Material	Phase	Temperature (K)	Thermal conductivity $\alpha[cal/(cm \cdot sec \cdot K)]$
Air	Gas	273	0.00006
Oxygen	Gas	273	0.00006
Hydrogen	Gas	273	0.0004
Benzene	Liquid	293	0.0004
Ethanol	Liquid	293	0.0004
Glycerol	Liquid	293	0.0007
Water	Liquid	293	0.0014
Steel	Solid	293	0.1
Aluminum	Solid	300	0.64
Brick	Solid	293	0.0016

q	Heat flux	$cal/(cm^2 \cdot sec)$
x	Length in one dimension	cm
T	Temperature	K
α	Thermal conductivity	$cal/(cm \cdot sec \cdot K)$

To use Fourier's law of heat conduction, a thermal balance must first be constructed. The energy balance is performed over a thin element of the material, x to $x + \Delta x$ in a rectangular coordinate system. The energy balance is shown in equation 13:

$+$ Energy conducted in through the face at x

$-$ Energy conducted out through the face at $x + \Delta x$ (13)

$=$ Energy accumulated within the element from x to $x + \Delta x$

The application of Fourier's law and Eq. (13) is introduced in Sctions III.A and III.B.

A. Heat Conduction in Rectangular Coordinates

Assume that a 1 cm thick sheet of insulating material separates water maintained at 100°C from well-stirred water at 25°C (see Fig. 4). The insulating material has dimensions of 10 cm by 25 cm, and the entire surface (250 cm²) is in contact on one side with the boiling water and on the other with the water at 25°C. A set of logical questions includes

    1. What is the temperature profile in the insulating material?
    2. What is the heat flux (q)?

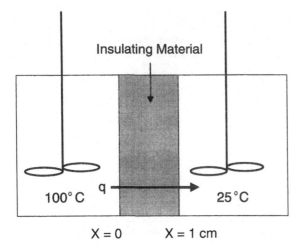

Insulating Material

100°C   q →   25°C

X = 0   X = 1 cm

Figure 4   Heat conduction in one dimension.

3.   What is the total heat flow (Q)?

The situation is illustrated in Figure 4. Assuming steady state, the energy balance over the insulating material is given by [using Eq. (13)]

$$
\begin{array}{l}
\text{Energy in} = q \times 250 \text{ cm}^2 \text{ at } x \\
\text{Energy out} = q \times 250 \text{ cm}^2 \text{ at } x + \Delta x
\end{array}
\tag{14}
$$

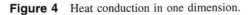

$$
\text{Total energy accumulation} = 250 \text{ cm}^2 \, (q_x - q_{(x+\Delta x)}) = 0
$$

Dividing Eq. (14) by the surface area (250 cm^2) and Δx, then taking the limit as Δx goes to zero, we find

$$
\frac{dq}{dx} = 0
\tag{15}
$$

Applying Fourier's law to Eq. (15), we see that

$$
\alpha \frac{d^2 T}{dx^2} = 0
\tag{16}
$$

Integrating Eq. (16) and applying the boundary conditions, we solve for the temperature profile and heat flux as follows:
     Boundary conditions (see Fig. 4)

$$
x = 0 \qquad T = T_h = 100°C
$$

$$x = 1 \text{ cm} \qquad T = T_{rt} = 25°C$$

Steady-state heat flux (q) of solution:

$$q = -\alpha\frac{dT}{dx} = \alpha(T_h - T_{rt})$$

The generalized form for steady-state heat conduction across a thin film (where we allow the insulating material a thickness δ) is given by

$$q = -\alpha\frac{dT}{dx} = \alpha\frac{T_h - T_{rt}}{\delta} \qquad (17)$$

where δ is the thickness of the film.

From Eq. (17) it is easy to see that the heat flux is proportional to thermal conductivity and temperature gradient and inversely related to film thickness. Not all heat conduction problems lead to such a simple solution.

B. Heat Conduction in Spherical Coordinates

The spherical coordinate system is useful for many of the moisture uptake problems encountered in this chapter. The application of the spherical coordinate system can be illustrated by the following example (see Fig. 5).

Place a hollow sphere (with an outside radius of 5 cm and a 1 cm thick insulating shell) into a large, well-stirred water bath maintained at 25°C. Assume that the inside surface of the sphere is maintained at 100°C. The same situation

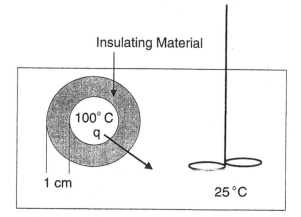

Figure 5 Heat conduction in spherical coordinates.

existed in Section III.A (see Fig. 4), except that the problem was solved for rectangular coordinates. The relevant questions now are

1. What is the temperature distribution in the insulating material covering the sphere.
2. What is the heat flux through the sphere?
3. What is the total heat flow?

An energy balance will be maintained over the sphere, and it will be assumed that there is no angular dependence on heat transport. The energy balance will be executed over a thin (Δr thickness) spherical shell and solved in essentially the same way as in Section III.A, except that we will work in spherical coordinates. The steady-state energy balance is given by

$$
\begin{aligned}
\text{Energy into the shell} &= q4\pi r^2 \text{ at } r \\
\text{Energy out of the shell} &= q4\pi(r + \Delta r)^2 \text{ at } r + \Delta r \\
\hline
\text{Total energy accumulation} &= 4\pi[q(r + \Delta r)^2 - q(r^2)]
\end{aligned}
\tag{18}
$$

Dividing by $4\pi\Delta r$ and letting Δr go to zero yields the steady-state equation

$$
\frac{d(qr^2)}{dr} = 0
\tag{19}
$$

and by integrating Eq. (19), we obtain the equation

$$
qr^2 = \text{constant}
\tag{20}
$$

Equation (20) indicates that the total heat flow (Q) is constant over each concentric sphere within the insulating layer. Application of Eq. (20) to the problem described in Figure 5 gives

Boundary conditions:

At $r = R_h$, $T = T_h = 100°C$
At $r = R_{rt}$, $T = T_{rt} = 25°C$

Steady-state heat flux (q):

$$
q = \alpha\frac{R_{rt}^{-2}(T_h - T_{rt})}{1/R_h - 1/R_{rt}}
$$

Total heat flow:

$$
Q = 4\pi\alpha\frac{T_h - T_{rt}}{1/R_h - 1/R_{rt}}
\tag{21}
$$

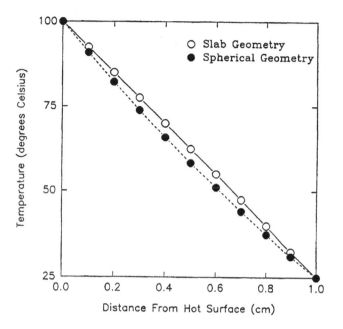

Figure 6   Temperature profiles within an insulating material. See Figures 4 and 5 for the descriptions of the two geometrics.

In Figure 6 the temperature profiles for an insulating plate and a sphere are compared. As can be seen in this figure, the temperature gradient is steeper for the sphere near the inside surface and becomes more gradual near the outside surface. Fourier's law would then predict that the heat flux is greater at the inside surface than at the outside surface, but the total heat flow is constant for each sphere. Figure 7 shows a comparison of fluxes for the sphere and plate solutions, where the ratio is the heat flux at the outside surface of a sphere compared to the heat flux for a simple slab geometry. It is seen from the profile in Figure 7 that the ratio of the heat flux at the outer shell of the sphere approaches the flux through the plate as the spherical radius gets large compared to the thickness of the shell (1 cm). Thus, spherical coordinates make a difference only when the thickness of the insulating shell is comparable in size to the radius of the sphere.

IV.   HEAT TRANSPORT CONTROLLED MOISTURE SORPTION (PURE WATER ATMOSPHERE)

A model describing moisture sorption for a deliquescent solid in a pure water vapor atmosphere is developed in this section. Under such conditions there is no

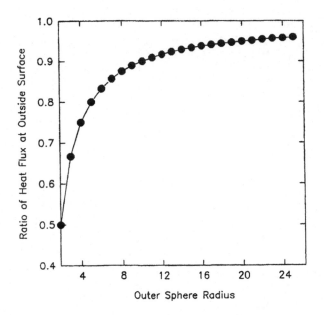

Figure 7 Heat flux ratio at the outside surface of a sphere (1 cm thick insulating material) compared to heat flux using rectangular coordinates.

mass concentration gradient and thus mass transfer resistance is not important. All resistance is assumed to come from heat transport. In the paper by VanCampen et al. [17], a heat transport model was developed in spherical coordinates, but we start in rectangular coordinates to allow the reader to focus on the physical model and not the math.

The fundamental concept of heat transport controlled moisture uptake [17] is shown in Eq. (22), where the rate of heat gained at the solid/vapor surface ($W' \Delta H$) is balanced exactly by the heat flow away from the surface (Q). The term ΔH is the heat generated by unit mass of water condensed on the surface. The two most probable sources of heat generation are the heat of water condensation and the heat of dissolution. A comparison of the heat of water condensation (0.58 cal/mg water) with the heat of dissolution for a number of salts indicates that the heat of dissolution can be neglected with little error for many materials.

$$W' \Delta H = Q \tag{22}$$

A more complete energy balance will be used that includes transport by conduction, convection, and radiation. The new energy balance equation over a small volume element takes the form

+ Energy in by conduction
+ Energy in by convection
+ Energy in by radiation
− Energy out by conduction
− Energy out by convection
− Energy out by radiation

= Total energy accumulated

The heat flux at any point is given by

$$q = -\alpha \frac{dT}{dx} - W' C_p (T - T_s) + \sigma e (T_s^4 - T_c^4) \tag{23}$$

where the first term is the conductive contribution (Fourier's law), the second is the convective component, and the last is the radiative. The convective component comes from the heat transported by mass flowing in the negative x direction (see Fig. 8). Each mole of water vapor transported into a volume element contains a given amount of heat, and thus heat is brought in by convection. The total water mass transport rate to the surface of the hygroscopic salt is W', where the heat capacity is given by C_p. The term $T - T_s$ is the temperature of the vapor (T) in reference to the surface temperature (T_s) where the vapor is headed. The product of the mass transport rate, the heat capacity, and the temperature of the vapor is the convective component of heat transport. The last term in Eq. (23) is the radiative component and is a consequence of the fact that all solid surfaces emit radiative energy. The constants σ and e are the Stefan–Boltzmann constant

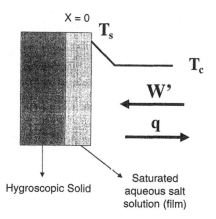

Figure 8 Moisture uptakes in rectangular coordinates.

and emissivity, respectively. Note that the radiative term is dependent only on the difference between the temperatures of the two surfaces (T_s and T_c) and is a constant contribution, independent of x.

A. Heat Flux Caused by Moisture Uptake in One Dimension

We now want to set up the problem for moisture uptake according to Figure 8. Remember from Eq. (15) that the steady-state change in heat flux for a one-dimensional problem in rectangular coordinates is given by $dq/dx = 0$, and differentiation of Eq. (23) gives

$$-\alpha\frac{d^2T}{dx^2} - W'C_p\frac{dT}{dx} = 0 \tag{24}$$

To solve for the temperature profile, heat flux, and mass transport rate, the boundary conditions must be determined as follows.

Referring to Figure 8, temperature T_c is the chamber temperature and T_s is the surface temperature at the salt solution/vapor interface. The temperature of the chamber is well defined and is an experimental variable, whereas T_s must be higher than T_c due to condensation of vapor on the saturated solution surface. We can determine T_s by applying the Clausius–Clapeyron equation to the problem. Assume that the vapor pressures of the surface and chamber are equal (no pressure gradients), indicating that the temperature must be raised at the surface (to adjust the vapor pressure lowering of the saturated solution) to P_c (at T_c) = P_s (at T_c). However, there is a difference in relative humidity between the surface and the chamber, where RH_c is the relative humidity in the chamber and RH_0 is the relative humidity of the saturated salt solution, and we obtain

$$\frac{RH_c}{RH_0} = \frac{P_c \ (\text{at } T_c)}{P_s \ (\text{at } T_c)}$$

and since we have assumed that P_s (at T_s) = P_c (at T_c) we can substitute P_s (at T_s) into the equation above to give

$$\frac{RH_c}{RH_0} = \frac{P_s \ (\text{at } T_s)}{P_s \ (\text{at } T_c)}$$

This form allows us to use the Clausius–Clapeyron equation (see Section II) so that

$$\ln\left(\frac{RH_c}{RH_0}\right) = \Delta H_v\frac{T_s - T_c}{RT_sT_c} \tag{25}$$

Solving for T_s gives

$$T_s = T_c + RT_c^2 \frac{\ln(\text{RH}_c/\text{RH}_0)}{\Delta H_v - RT_c \ln(\text{RH}_c/\text{RH}_0)} \tag{26}$$

The next step is to solve Eq. (24) by reduction in order.

$$T = T_s - (T_s - T_c)\frac{1 - \exp(-W'C_p x/\alpha)}{1 - \exp(-W'C_p A/\alpha)} \tag{27}$$

Under common experimental conditions the convective term does not contribute significantly to the temperature profile, and thus Eq. (27) can be simplified to

$$T = T_s - (T_s - T_c)x/A \tag{28}$$

It is clear that once we select rectangular coordinates and eliminate convective transport we are left with the temperature profile predicted by a simple film model (discussed in Section III.A). By solving for the heat flux in one dimension it can be seen that the temperature profile is a simple linear function of distance, chamber temperature, and surface temperature.

$$q = -\alpha\frac{dT}{dx} + \sigma e(T_s^4 - T_c^4) = \alpha\frac{T_s - T_c}{A} + \sigma e \, (T_s^4 - T_c^4) \tag{29}$$

where T_s is defined in Eq. (26).

Equation (29) is the solution for heat flux in rectangular coordinates and could be converted into a mass rate using Eq. (22). A more useful solution comes from spherical coordinates, because that coordinate system more closely matches the experimental system. In the next section we repeat the same basic steps to arrive at the heat flux and moisture uptake rate in spherical coordinates.

B. Heat Transport Limited Moisture Uptake in Spherical Coordinates

We now repeat the derivation of the steady-state heat transport limited moisture uptake model for the system described by VanCampen et al. [17]. The experimental geometry is shown in Figure 9, and the coordinate system of choice is spherical. It will be assumed that only conduction and radiation contribute significantly to heat transport (convective heat transport is negligible), and since radiative flux is assumed to be independent of position, the steady state solution for the temperature profile is derived as if it were a pure conductive heat transport problem. We have already solved this problem in Section III.B, and the derivation is summarized below. At steady state we have already shown (in spherical coordinates) that

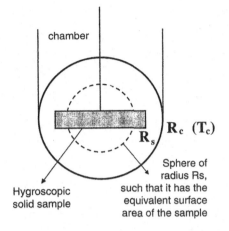

Figure 9 Moisture uptakes in spherical coordinates.

$$\frac{d(qr^2)}{dr} = 0 \tag{30}$$

This equation is solved exactly as in Section III.B, except that we now use spherical coordinates.

The boundary conditions are that at the surface of the drug ($r = R_s$) the temperature is T_s, and at the chamber radius (R_c) the temperature is T_c. By applying Fourier's law, integrating Eq. (30), and applying the boundary conditions, the temperature profile for the system is solved for:

$$T = \frac{T_s - T_c}{r/R_s - r/R_c} - \frac{T_s - T_c}{1 - R_s/R_c} + T_s \tag{31}$$

The steady-state total heat flow (Q) at $r = R_s$ is determined by using Fourier's law and *adding the radiative term*:

$$Q = 4\pi R_s^2 \left[-\alpha \frac{dT}{dr} + \sigma e(T_s^4 - T_c^4) \right] \tag{32}$$

By differentiating Eq. (31) and evaluating the solution at $r = R_s$, the temperature gradient at the surface is determined:

$$\frac{dT}{dr} \text{ at } R_s = -(T_s - T_c) \frac{R_s R_c}{R_s^2(R_s - R_c)} \tag{33}$$

The total heat flow is calculated by inserting Eq. (33) into Eq. (32):

$$Q = 4\pi \left[\alpha(T_s - T_c)\frac{R_s R_c}{R_s - R_c} + R_s^2 \sigma e(T_s^4 - T_c^4) \right] \tag{34}$$

The total heat flow (Q) at the film surface is equal to the mass flow rate (W_h') times the heat of condensation (ΔH). That is, the heat generated by condensation at the surface must be equal to the heat transported away by conduction and radiation for steady state to be achieved. In mathematical terms this results in the equation.

$$W_h' \Delta H = 4\pi \left[\alpha(T_s - T_c)\frac{R_s R_c}{R_s - R_c} + R_s^2 \sigma e(T_s^4 - T_c^4) \right] \tag{35}$$

Since T_s is not an experimentally measurable quantity, it is useful to insert the solution for T_s (from the Clausius–Clapeyron equation) and solve for W_h' as an explicit function of RH_0 and RH_c. VanCampen et al. showed (using sample algebraic approximations and conversion factors) that substituting for T_s in Eq. (35) gives the useful solution

$$W_h' = (\mathbf{C} + \mathbf{R}) \ln\left(\frac{RH_c}{RH_0}\right) \tag{36}$$

where

> $\mathbf{C}$ = conductive contribution
> $\mathbf{R}$ = radiative contribution

This solution is composed of simple radiative and conductive heat transport terms.

The solutions for moisture uptake presented in this section are based on the experimental condition of a pure water vapor atmosphere. In the next section a derivation of moisture uptake equations is based on both heat and mass transport that are characteristic of moisture uptake in air. The final section of this chapter presents the results of studies where heat transport is unimportant and mass transport dominates the process. Thus, we will have a collection of solutions covering models that are (1) heat transport limited, (2) mass transport limited, (3) heat and mass transport limited, and (4) mass transport limited with a moving boundary for the uptake of water by water-soluble substances.

V. MASS TRANSPORT CONTROLLED MOISTURE SORPTION

In Section IV it was assumed that all resistance to moisture sorption was due to heat transport. This assumption is valid only in cases where there are no other gases present. In the case where the moisture is being taken up under normal atmospheric conditions the inert gases are present at about 1 atm. The composition of air is approximately 21% oxygen, 78% nitrogen, and approximately 1% other gases. This means that for moisture uptake under normal conditions, diffusion through air will need to be accounted for. This does not imply that the models developed in the earlier sections can be forgotten, because it will be shown in Section VI that moisture uptake is limited by both mass and heat transport.

The basic assumption for a mass transport limited model is that diffusion of water vapor thorugh air provides the major resistance to moisture sorption on hygroscopic materials. The boundary conditions for the mass transport limited sorption model are that at the surface of the condensed film the partial pressure of water is given by the vapor pressure above a saturated solution of the salt (P_s) and at the edge of the diffusion boundary layer the vapor pressure is experimentally fixed to be P_c. The problem involves setting up a mass balance and solving the differential equation according to the boundary conditions (see Fig. 10).

The molar flux of water (N_w) through a unit surface area element is given [18] by

$$N_w = -cD_{w,air}\frac{dX_w}{dx} + X_wN_w \tag{37}$$

where

$$c = \text{total molar concentration of gas}$$
$$D_{w,air} = \text{diffusion coefficient of water in air}$$
$$X_w = \text{mole fraction of water vapor in air at a point } x$$

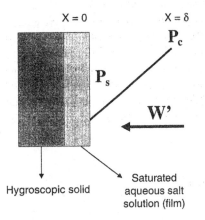

X = 0   X = δ

P_c

P_s

W'

Hygroscopic solid   Saturated aqueous salt solution (film)

Figure 10   Moisture uptakes in rectangular coordinates for the mass transport limited model.

The first term of Eq. (37) is simply Fick's first law, while the second term of Eq. (37) is the convective transport component.

By replacing the mole fraction of water with the ratio of water vapor pressure (P_w) divided by the total gas pressure (P_T), one can solve for the diffusive flux of water vapor. Also, by multiplying N_w by the molecular weight of water, the mass flux of water vapor is arrived at:

$$N_w = -\frac{cD_{w,\text{air}}}{P_T - P_w}\frac{dP_w}{dx} \tag{38}$$

$$W'_{sp} = -M_w\frac{cD_{w,\text{air}}}{P_T - P_w}\frac{dP_w}{dx} \tag{39}$$

where sp refers to specific rates of water uptake (g/cm²sec).

We now use Eq. (39) to build models for the mass transport limited uptake of water by hygroscopic materials.

A. Mass Transport Limited Moisture Uptake in One Dimension

Since it is assumed that the only limiting resistance to moisture uptake is mass transport resistance, the basis for the model is contained with Eq. (39). It is assumed that the system is at steady state and that rectangular coordinates (uptake in one dimension) are appropriate. Since the system is at steady state and we are dealing with transport in one direction, the flux into a volume element must be equal to the flux out of that element. This condition is expressed as

$$\frac{dW'_{sp}}{dx} = 0 \tag{40}$$

The boundary conditions for the system are (1) that at the surface of the hygroscopic material the partial pressure of water is determined by that of the saturated salt solution (P_s) and (2) that at a characteristic distance from the surface (δ) the partial pressure of water vapor is given by the chamber pressure (P_c).

Given the boundary conditions, the problem is solved by assuming that $D_{w,\text{air}}$ is independent of concentration and that total concentration of gas is constant.

$$\ln(P_T - P_w) = C_1 x + C_2 \tag{41}$$

where C_1 and C_2 are constants of integration. Solving for P_w gives Eq. (42), which is the vapor pressure of water as a function of distance.

$$\ln\left(\frac{P_T - P_w}{P_T - P_s}\right) = \frac{x}{\delta}\ln\left(\frac{P_T - P_c}{P_T - P_s}\right) \tag{42}$$

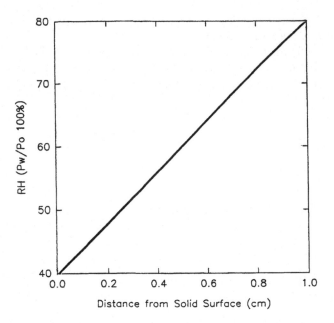

Figure 11 Vapor pressure profile for a surface relative humidity RH of 40% and chamber RH of 80%, with a diffusion film thickness of 1 cm.

An example of a vapor pressure profile is shown in Figure 11, where it is assumed that the relative humidity within the chamber is 80%, the critical relative humidity of the solid is 40%, and the thickness of the diffusion layer (δ) is 1 cm. From the figure, note that the relative humidity profile is linear and we could have made the simplifying assumption that the convective term is negligible. By ignoring the convective term, Eq. (42) simplifies to

$$P_w = P_s + \frac{x}{\delta}(P_c - P_s) \tag{43}$$

which is just the concentration profile for diffusion across a stagnant film with a thickness of δ.

The next step in developing an equation to describe moisture uptake in one dimension is to determine W'_{sp} by applying Eq. (39). Note that the partial pressure gradient (dP_w/dx) must be calculated to solve for the moisture flux. We could take the derivative of either Eq. (42) or (43) for this purpose. If we take the derivative of Eq. (42), the resulting solution is given by

$$W'_{sp} = -M_w\left(\frac{cD_{w,\text{air}}}{\delta}\right)\ln\left(\frac{P_T - P_c}{P_T - P_s}\right) \tag{44}$$

The important variables in Eq. (44) are the thickness of the diffusion layer (δ) and the partial pressures of water vapor in the chamber and above the solid surface P_c and P_s).

B. Mass Transport Limited Moisture Uptake in Spherical Coordinates

We have already solved transport problems in spherical coordinates, so the solution is only outlined here. To begin, we set the boundary conditions. At $r = a$ (distance to the surface of drug mass) the partial pressure of water is given by P_s, and at the radius of the chamber ($r = a'$) the partial pressure of water is given by the chamber vapor pressure (P_c).

The convective diffusion equation is simply cast in terms of radius (r) instead of x.

$$W'_{sp} = -\frac{M_w c D_{w,\text{air}}}{P_T - P_w}\frac{dP_w}{dr} \tag{45}$$

If a steady-state assumption is applied (total mass entering a spherical shell must equal the mass leaving the shell), the result is

$$\frac{d}{dr}\left[r^2\left(\frac{M_w c D_{w,\text{air}}}{P_T - P_w}\right)\frac{dP_w}{dr}\right] = 0 \tag{46}$$

By integrating Eq. (46) and applying the boundary conditions, the solution for the total moisture uptake limited by mass transport is found. In the solution shown in Eq. (47) the vapor pressures have been converted to relative humidities and it has been assumed that the partial pressure of water is much less than the total pressure. Under these conditions, Eq. (47) is the solution for mass transport resistance in spherical coordinates. As with transport in rectangular coordinates, the important variables are the partial pressures of the chamber and above the solid surface and the distance between the solid surface and chamber wall.

$$W'_m = -k_m(\text{RH}_c - \text{RH}_0)A \tag{47}$$

where k_m is the mass transfer coefficient and A is the surface area of drug mass.

VI. COMBINED MASS AND HEAT TRANSPORT CONTROLLED MOISTURE SORPTION

As one might expect, moisture sorption is controlled by both mass and heat transport resistance. The heat transport model discussed earlier was based on the concept that condensation of water would generate heat, which at steady state would be rate limiting to moisture uptake. As previously noted, the heat transport control model is the obvious choice for a system in which the gas phase is pure water vapor. In that system there is no mass transport resistance and all resistance results from the generation of heat. The second basic system is one in which all resistance is assumed to be from mass transport. That system would be an excellent fit if the heat of condensation of water were small, which it is not. The real world requires that both heat and mass transport resistance must be accounted for, and we will call this the combined mass–heat transport model.

The moisture uptake rate in spherical coordinates depends on both heat transport and mass transport. In a combined model there is resistance from both heat and mass transport, leading to a reduced rate of moisture sorption. Since it is predicted that the combined model will lead to a slower rate of moisture uptake, this means that there are implications concerning the boundary conditions. A lower rate of moisture uptake means a lower rate of heat generation and thus a lower temperature at the surface of the condensed film than we predicted for the simple heat transport limited model. When there is combined heat and mass transport resistance, the temperature and partial pressure gradients are reduced, as reflected by the changes in temperature and pressure at the condensed film surface. The combined model clearly calls for simultaneous solutions of the two transport equations (mass and heat) for the surface temperature (T_s) and pressure (P_s). VanCampen et al. [19] made two simplifying assumptions concerning the relationship between the temperature dependence of humidity and radiation and then solved for T_s and RH_s, where $RH_s > RH_0$. Because of these assumptions, the moisture uptake rate can be expressed by very simple expressions using heat and mass transport coefficients, as shown in Eqs. (48) and (49).

$$W' = -k_m(RH_c - RH_s)A \tag{48}$$

$$W' = -k_h(RH_s - RH_0)A \tag{49}$$

The complicating factor is that both equations include the term RH_s, which is the surface relative humidity and is greater than RH_0 and is also an unknown.

It is very easy to see that Eqs. (48) and (49) have similar form and yet they represent mass transport and heat transport resistance, respectively. What is needed is a connection between the two transport resistances, and this is accomplished by combining the two equations to eliminate RH_s. The resulting combined mass–heat transport limited rate moisture uptake (W'_{mh}) is given by

$$W'_{mh} = -k_m k_h A \frac{RH_c - RH_0}{k_m + k_h} \tag{50}$$

Equation (50) contains the two transfer coefficients, the surface area, the chamber relative humidity (RH_c), and the relative humidity above a saturated solution

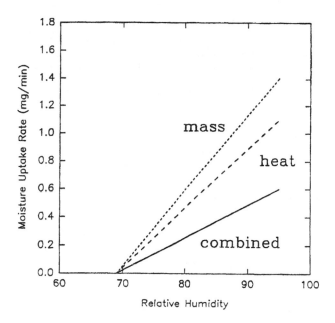

Figure 12 Moisture uptake rates predicted from the mass, heat, and combined mass—heat models.

of the hygroscopic material (RH_0). This combined heat—mass transport model developed by VanCampen et al. was clearly the best model for the prediction of the water sorption rates for choline iodide, potassium bromide, and potassium iodide disks in 1 atm air. Figure 12 shows a schematic depiction of the relative rates predicted by the heat, mass, and heat—mass transport controlled models presented in this section.

VII. DELIQUESCENT MATERIALS IN POROUS SYSTEMS

The moisture uptake models we have discussed have been concerned with pure components. The deliquescing material could be a drug substance or an excipient material. In pharmaceuticals, however, mixtures of materials are also important. One possible situation involves mixing nondeliquescing and deliquescing materials that are formed into a porous tablet or powder blend. The obvious question is, Do the models for pure components apply to porous heterogeneous materials? For pure components we have assumed that the mass and heat limiting transport processes occur in the gas phase. For heterogeneous porous materials it seems likely that diffusional resistance in liquid-filled pores will be limiting, because diffusion in a liquid system is much slower than in a gas. The model presented in this section describes the diffusional transport in the liquid phase, through aqueous solution filled pores. The model includes the description of a moving boundary between solid water-soluble material dispersed in a porous matrix and the aqueous solution phase.

A reasonable example of such a system is represented by a deliquescing drug (water-soluble salt) mixed with nonhygroscopic excipients and compressed into a porous tablet. A schematic representation of the tablet is shown in Figure 13. The schematic representation of the model shows three zones; a condensed aqueous solution of the dissolved salt at a concentration such that the relative humidity above it is equal to that maintained in the chamber, an aqueous salt solution filling pores, and the solid salt–excipient mixture. Water vapor condenses on the surface of the tablet, diffuses into the pores, and eventually causes the dissolution of the salt. As the diffusional process proceeds, the aqueous solution filled porous zone increases in thickness. The driving force for the transport of water through the matrix is the water activity gradient, and transport is assumed to be through the pores in the tablet.

The mathematical solution to moving boundary problem involves setting up a pseudo-steady-state model. The pseudo-steady-state assumption is valid as long as the boundary moves ponderously slowly compared with the time required to reach steady state. Thus, we are assuming that the boundary between the salt solution and the solid salt moves slowly in the tablet compared to the diffusion

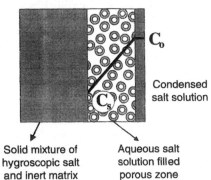

$$X = \xi \qquad X = 0$$

C_0

Condensed
salt solution

Solid mixture of
hygroscopic salt
and inert matrix

Aqueous salt
solution filled
porous zone

Figure 13 Schematic model for moisture uptakes in porous materials.

process. Using the steady-state solution allows us to use Fick's first law and perform a mass balance at the interface.

The diffusion of water through this heterogeneous system of hygroscopic salt and inert excipient is complicated by the facts that it takes place in a porous system and that there is going to be a concentration dependence associated with the salt/water binary diffusion coefficient. The concentration dependence of diffusion can be determined experimentally and included in the model [20]; however, for the sake of discussion we assume that a mean diffusion coefficient will reasonably represent the binary diffusion coefficient in the water–hygroscopic salt solution. The second issue is that the diffusion process occurs in a porous system. The porous channels will reduce the effective binary diffusion coefficient due to the twists, turns, constrictions, and complex interconnections [20]. We greatly simplify the problem by defining an effective diffusion coefficient (D_e) that takes into account both concentration dependence and the porous system. These assumptions greatly simplify the development of the model for moisture uptake in porous systems but mask the true complexity of diffusion in concentrated solutions and porous systems.

The boundary conditions are that (1) at the moving boundary (ξ) the solution is saturated with salt with a corresponding concentration of water (C_s) and (2) at the disk/atmosphere surface the concentration of water is governed by the equilibrium vapor pressure in the chamber to give a water concentration of C_0.

The steady-state flux for water is given by

$$W'_{sp} = D_e \frac{C_0 - C_s}{\xi} \tag{51}$$

where ξ is the thickness of the aqueous solution filled porous zone and is also the location of a moving boundary between the dry solid zone and the aqueous solution filled zone. The mass balance at the hygroscopic solid salt/aqueous solution boundary requires that

$$W'_{sp} = D_e \frac{C_0 - C_s}{\xi} = S_w L \frac{d\xi}{dt} \tag{52}$$

where

S_w = grams of water per grams of hygroscopic salt in a saturated solution

L = grams of salt per cubic meter of tablet (load of hygroscopic salt)

Integrating Eq. (52) and then solving for $d\xi/dt$ gives the rate at which the boundary moves:

$$\frac{d\xi}{dt} = \left[D_e \frac{C_0 - C_s}{2S_w Lt} \right]^{0.5}$$ (53)

By inserting the solution shown above into Eq. (52) and integrating over time, the amount of water taken up with time is given by the final solution,

$$W_{sp} = \left[D_e (C_0 - C_s) \frac{tS_w L}{2} \right]^{0.5}$$ (54)

The result shown in Eq. (54) is a square root of time relationship for moisture uptake. Mulski [20] demonstrated that for sodium glycinate in a hydrophobic porous matrix, moisture sorption follows Eq. (54).

VIII. SUMMARY

The chapter began with a discussion of the thermodynamics of vapor–liquid equilibria because heat transport is an important aspect of moisture uptake. In fact, in a pure water vapor atmosphere, heat transport is limiting. Heat transport controlled moisture uptake is limited by the rate of heat gained at the solid/vapor surface, where this heat generation is balanced by the heat flow away from the surface. In the case where the moisture is being taken up under normal atmospheric conditions, the inert gases are present at a pressure about 1 atm. This means that for moisture uptake under normal conditions, diffusion through air is important. Thus solutions to heat and mass transport limited moisture uptake were developed. However, moisture sorption is controlled by both mass and heat transport resistance, and solutions to the combination of transport resistances were included in the chapter. Finally, the moisture uptake in heterogeneous materials was briefly treated.

REFERENCES

1.   LJ Leeson, AM Mattocks. J Pharm Sci 74:124, 1985.
2.   JT Carstensen, M Osadca, SH Rubin. J Pharm Sci 58:549, 1969.
3.   JT Carstensen, P Pothisiri. J Pharm Sci 64:37, 1975.
4.   JT Carstensen, K Danjo, S Yoshioka, M Uchiyama. J Pharm Sci 76:548, 1987.
5.   JT Carstensen. Drug Dev Ind Pharm 14:1927, 1988.
6.   N Anderson, G Banker, G Peck. J Pharm Sci 71:3, 1982.
7.   R Teraoka, M Otsuka, Y Matsuda. J Pharm Sci 82:601, 1993.
8.   JL Wright, JT Carstensen. J Pharm Sci 75:546, 1986.
9.   M Chikazawa, W Nakajima, T Kanazawa. Funtai Kogaku Kenyu Kaishi 14:18, 1977.

10. T Kanazawa, M Chikazawa, M Kaiho, M Kikuchi. Funtai Kogaku Kenyu Kaishi 13:411, 1976.
11. BS Neumann. Adv Pharm Sci 2:181, 1967.
12. DJ Craik, BF Miller. J Pharm Pharmacol 10:136t, 1958.
13. E Shotton, JE Reese. J Pharm Pharmacol 180:160s, 1966.
14. DR Lide. Handbook of Chemistry and Physics. 75th ed. Boca Raton, FL: CRC Press, 1994–1995.
15. JT Carstensen. Pharmaceutics of Solids and Solid Dosage Forms. New York: Wiley, 1977.
16. JT Carstensen. Solid Pharmacutics, Academic Press, 1980.
17. L VanCampen, GL Amidon, G Zografi. J Pharm Sci 72:1381, 1983.
18. RB Bird, WE Stewart, EN Lightfoot. Transport Phenomena. New York: Wiley, 1960.
19. L VanCampen, GL Amidon, G Zografi. J Pharm Sci 72:1394, 1983.
20. M Mulski. Water vapor uptake in a heterogeneous matrix. PhD Thesis, University of Wisconsin, 1989.

Index

ISBN 0-8247-6610-5

EAN

9 780824 766108

90000>